HYDROGEOLOGY OF THE OCEANIC LITHOSPHERE

A comprehensive and up-to-date review of the subject of the nature, causes, and consequences of fluid flow in oceanic crust, this edited volume sets in context much recent research for the first time.

The book begins with a concise review of the relatively brief history of its subject which began shortly after the dawning of plate-tectonic theory little more than 30 years ago. It then describes the nature and important consequences of fluid flow in the sub-seafloor, ending with a summary of how the oceans are affected by the surprisingly rapid exchange of water between the crust and the water column overhead. The accompanying CD-ROM includes a full and easily navigated set of diagrams and captions, references, and photos of research vessels, submersibles, and tools used in marine hydrologic studies.

A valuable resource for graduate students and researchers of Earth Sciences and Oceanography.

EARL E. DAVIS is a senior research scientist at the Pacific Geoscience Centre, Geological Survey of Canada.

HARRY ELDERFIELD is Professor of Ocean Geochemistry and Palaeochemistry in the Department of Earth Sciences, University of Cambridge.

HYDROGEOLOGY OF THE OCEANIC LITHOSPHERE

Edited by

EARL E. DAVIS AND HARRY ELDERFIELD

CAMBRIDGE
UNIVERSITY PRESS

PUBLISHED BY THE PRESS SYNDICATE OF THE UNIVERSITY OF CAMBRIDGE
The Pitt Building, Trumpington Street, Cambridge, United Kingdom

CAMBRIDGE UNIVERSITY PRESS
The Edinburgh Building, Cambridge, CB2 2RU, UK
40 West 20th Street, New York, NY 10011–4211, USA
477 Williamstown Road, Port Melbourne, VIC 3207, Australia
Ruiz de Alarcón 13, 28014 Madrid, Spain
Dock House, The Waterfront, Cape Town 8001, South Africa

http://www.cambridge.org

First published 2004

Printed in the United Kingdom at the University Press, Cambridge

Typeface Times 10/13 pt.   *System* LATEX 2_ε [TB]

A catalog record for this book is available from the British Library

Library of Congress Cataloging in Publication data
Hydrogeology of the ocean lithosphere / edited by Earl E. Davis and Harry Elderfield.
p.   cm.
Includes bibliographical references and index.
ISBN 0 521 81929 6 (HB)
1. Hydrogeology.   2. Ocean bottom.   3. Earth – Crust.   I. Davis, Earl E., 1947–
II. Elderfield, Harry, 1943–
GB1005 H93   2004   551.46'8 – dc22   2004040410

ISBN 0 521 81929 6 hardback

Contents

Contributors

Jeffrey Alt
Department of Geological Sciences
2534 C.C. Little Building
The University of Michigan
Ann Arbor, MI 48109-1063
USA

Anne Bartetzko
Applied Geophysics
RWTH Aachen
Lochnerstr. 4-20,
52056 Aachen
Germany

Keir Becker
Rosenstiel School of Marine and
Atmospheric Science
University of Miami
4600 Rickenbacker Causeway
Miami, FL 33149-1098
USA

Mike Bickle
Department of Earth Sciences
University of Cambridge
Downing Street
Cambridge, CB2 3EQ
UK

Joe Cann
School of Earth Sciences
University of Leeds
Leeds, LS2 9JT
UK

Suzanne Carbotte
Lamont-Doherty Earth Observatory
61 Route 9W
Palisades, NY 10964
USA

Dave Chapman
Department of Geology and Geophysics
University of Utah
135 S 1460 E
Salt Lake City, UT 84112-0111
USA

Earl Davis
Pacific Geoscience Centre
Geological Survey of Canada
9860 W. Saanich Rd
Sidney, BC V8L 4B2
Canada

Harry Elderfield
Department of Earth Sciences
University of Cambridge
Downing Street
Cambridge, CB2 3EQ
UK

Andrew Fisher
Earth Sciences Department
Earth and Marine Sciences Building
University of California at Santa Cruz
Santa Cruz, CA 95064
USA

Harald Furnes
Institutt for geovitenskap
Realfagbygget, Allegt. 41
5007 Bergen
Norway

Emily Giambalvo
Sandia National Laboratories,
MS-1395
4100 National Parks Highway
Carlsbad, NM 88220
USA

Kathy Gillis
School of Earth and Ocean Sciences
University of Victoria
PO Box 3055
Victoria, BC V8W 3P6
Canada

David Goldberg
Borehole Research Group
Lamont-Doherty Earth Observatory of
Columbia University
Route 9W
Palisades, New York 10964-8000
USA

Ingo Grevemeyer
GEOMAR Research Centre
Wischhofstraße 1–3
24148 Kiel and
Department of Earth Sciences
University of Bremen
Klagenfurter Straße
28359 Bremen
Germany

Rob Harris
Department of Geology and
Geophysics
University of Utah
135 S 1460 E
Salt Lake City, UT 84112-0111
USA

Edward Hornibrook
Department of Earth Sciences
University of Bristol
Queens Rd
Bristol, BS8 IRJ
UK

Gerardo Iturrino
Borehole Research Group
Lamont-Doherty Earth Observatory of
Columbia University
Route 9W
Palisades, New York 10964-8000
USA

Miriam Kastner
Geological Research Division
Scripps Institution of Oceanography
University of California at San Diego
9500 Gilman Dr.
La Jolla, CA 92093-0212
USA

Mike Mottl
Department of Oceanography
SOEST
1000 Pope Rd
University of Hawaii
Honolulu, HI 96822
USA

R. John Parkes
School of Earth, Ocean, and Planetary
Sciences
University of Cardiff
Park Place
Cardiff, CF10 3YE
UK

Simon Peacock
Department of Geological Sciences
Box 871404
Arizona State University
Tempe, AZ 85287-1404
USA

Mark Rudnicki
Division of Earth and Ocean Sciences
103 Old Chemistry Building
Duke University
Box 90227
Durham, NC 27708-0227
USA

Dan Scheirer
Department of Geological Sciences
Box 1846
Brown University
Providence, RI 02912
USA

John Sclater
Geological Research Division
Scripps Institution of Oceanography
University of California at San Diego
9500 Gilman Dr.
La Jolla, CA 92093-0220
USA

William Seyfried, Jr.
Department of Geology and Geophysics
University of Minnesota
106 Pillsubury Hall
Pillsbury Drive, S. E.
Minneapolis, MN 55455
USA

W. C. Pat Shanks
US Geological Survey
973 Denver Federal Center
Denver, CO 80225
USA

Glenn Spinelli
Department of Geological Sciences
101 Geology Building

University of Missouri
Columbia, MO 65211
USA

Hubert Staudigel
Cecil H. and Ida M. Green
Institute of Geophysics and Planetary
Physics
Scripps Institution of Oceanography
University of California at San Diego –
0225
La Jolla, CA 92093-0225
USA

Jon Telling
Department of Earth Sciences
University of Bristol
Wills Memorial Building
Queens Rd
Bristol, BS8 IRJ
UK

Richard Von Herzen
Woods Hole Oceanographic Institution
360 Woods Hole Rd
Woods Hole, MA 02543
USA

Kelin Wang
Pacific Geoscience Centre
Geological Survey of Canada
9860 W. Saanich Rd
Sidney, BC V8L 4B2
Canada

Geoff Wheat
NURP/MLML Marine Operations
PO Box 475
Moss Landing, CA 95039
USA

Preface

Troublesome noise becomes signal and the subject of a book

This book is being published roughly 30 years after the recognition of hydrothermal circulation in the oceanic crust. A few of the contributors were involved in the early studies of this phenomenon, others followed closely on their heels, and some became engaged relatively recently. All have experienced the pleasure of designing and executing experiments, and of discovering mechanisms, controlling factors, and consequences of fluid flow beneath the ocean floor. We hope that this book conveys to readers a sense of that pleasure.

The seed for this volume was planted at the time of a workshop sponsored by the International Lithosphere Program and the Joint Oceanographic Institutions/US Science Support Program, when a group of scientists representing a broad range of disciplines gathered in December 1998 at the University of California at Santa Cruz to discuss the current state and future direction of ocean crustal hydrogeology. It was made clear by these discussions that a wealth of knowledge about fluid flow within the crust and exchange with the ocean overhead had been gained since the early 1970s, that many new challenges lay ahead, and that a summary, offering both a retrospective and prospective review of all disciplines – including theoretical and experimental physics, chemistry, and microbiology – would be timely and useful for those attending the meeting, for their peers, and for students and new interested researchers.

The importance of fluid circulation below the seafloor and exchange of water between the crust and the oceans can be easily appreciated by considering that the oceanic crust constitutes the most extensive geological formation on Earth, and that hydrologic activity within it extends from mid-ocean ridges to beneath subduction-zone accretionary prisms (Fig. 1). Its upper part is characterized by very high permeability, and it is host to huge fluxes of water. This was highlighted in a report of the second Conference on Scientific Ocean Drilling (COSOD II, 1987) which provided a succinct summary of ocean crustal hydrothermal circulation, as well as other types of sub-seafloor fluid flow as they were known in the late 1980s. The most spectacular and directly observable manifestations of flow occur at ridge axes, where heat from magmatic intrusions drives high-temperature

Hydrogeology of the Oceanic Lithosphere, eds. E. E. Davis and H. Elderfield. Published by Cambridge University Press.
© Cambridge University Press 2004.

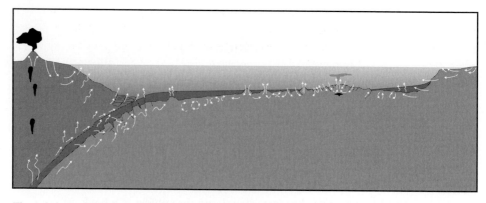

Fig. 1 Schematic cross-section depicting various types of sub-seafloor fluid flow, ranging from topographically driven flow through continental margins and consolidation-driven flow at subduction zones to thermal buoyancy driven flow at mid-ocean ridge axes and in the oceanic crust beneath broad regions of the oceans. Oceanic igneous crust and sediments are shown light gray and darker gray, respectively, and magma is shown in black; flow is depicted by white arrows. The figure was originally prepared for the Integrated Ocean Drilling Program Initial Science Plan, 2003–2013, and is reproduced courtesy of JOI, Inc.

springs at the seafloor. Although important and of fundamental interest, these hydrothermal systems are short-lived in the traveling Lagrangian reference frame of the lithospheric plate, and, when considered over the lifetime of the lithosphere, or over the full areal extent of the oceanic crust in a fixed global Eulerian reference frame, the greatest contribution to the total volumetric flow of seawater in the crust is found to occur on the flanks of mid-ocean ridges.

Hydrologic systems on ridge flanks are much more subtle than at ridge crests, and produce signals that were initially not understood, particularly since early observations tended to be too widely spaced and provided a highly incomplete picture of what are now known to be the primary signals reflecting crustal fluid flow. Early heat flux investigations were directed at comparing continental and oceanic thermal structure, and isolating thermal signatures of mantle convection and seafloor spreading. The investigators were frustrated by large "scatter" commonly present in the measurements, and by inexplicably low average values relative to levels expected on the basis of plate-tectonic theory as it was then emerging. This frustration ended in the early 1970s with the publication of a seminal paper by Lister (1972) who argued that the scatter, previously considered troublesome "noise," in fact reflected a primary "signal" from crustal fluid flow, which later work confirmed. This signal is now used frequently as a quantitative guide to the nature of local crustal fluid flow. Similarly, the shortfall in average seafloor heat flux relative to that expected from the lithosphere in a plate-tectonic environment was recognized as an indication of the amount of heat carried through the seafloor advectively, and that shortfall has become widely used as a key constraint on the rate of fluid exchange between the crust and the overlying ocean.

The COSOD II report also provided a simple and useful categorization of hydrological regimes based primarily on the dominant driving forces for flow. In simplest terms, flow at ridge crests is driven by the buoyancy generated by highly localized magmatic heat; flow in ridge flank settings is driven by the deep-seated heat from the cooling lithosphere; and flow at continental margins is driven primarily by topographically or compositionally derived "head" and by consolidation caused by gravitational or tectonic stresses. While each of these hydrologic "type settings" is interesting and important, treating them all is a greater task than could be covered in a single book. Hence we have limited ourselves in this volume to only one of these settings, the most widely distributed and volumetrically important, ridge flanks.

Structure and contents of this book

The contents are organized generally by discipline. The first chapter provides a personal perspective of the early seminal studies of hydrothermal circulation, and of the early inferences drawn largely on the basis of seafloor heat flow data collected "in the dark," i.e. with no prior knowledge of the source, meaning, or scale of signals. The second chapter discusses some of the studies that followed the initial hypothesis development and discovery phase, studies that laid the foundation for much of the quantitative aspects presented later. It also contains some personal thoughts about where future efforts should be directed.

Next are chapters devoted to physical structure, state, and processes. Chapters 3–6 provide a summary of the primary hydrologic architecture of the crust as it is created at mid-ocean ridge crests, and as it ages and is physically modified by hydrothermal alteration, mineralization, and sedimentation. Chapters 7 and 8 summarize observations of permeability, the principal property that controls the rate and direction of flow, and of pressure and temperature, the parameters of physical state that help to define buoyancy-derived driving forces and to constrain rates of flow. Chapter 9 provides a complementary view of the hydrogeology of oceanic crust gained from observations in ophiolites, where ancient crustal sections are exposed on land. Chapters 10–13 discuss inferences that can be drawn from observations and modeling concerning the routes and rates of crustal fluid flow and ocean–crust exchange, and the duration of the interval during which thermally driven flow continues within the crust after ventilation becomes insignificant.

The subsequent chapters review ocean crustal hydrologic activity and its consequences over a broad range of scales, from chemical, mineralogical, and biological perspectives. Chapters 14–16 provide reviews of current information about the hydrogeology of oceanic crust from geochemical studies of the fluid and from geochemical and mineralogical studies of the alteration products, in both the sedimentary and underlying igneous parts of the oceanic crust. Chapters 17 and 18 describe the hydrologically dependent microbial populations within and beneath the sediments. Chapter 19 provides a summary of the geochemical state of crustal water and inferred rates of flow. Among the aspects of the hydrogeology of the oceanic lithosphere that are particularly difficult to quantify is the depth and degree to which the oceanic lithosphere is altered and carries water back to subduction zones; Chapter 20 examines this, in part based on inferences that can be drawn from subduction zone characteristics, such as earthquake distributions and subduction-zone alteration and metamorphism, that imply the presence of water. Finally, Chapter 21 places the inferences drawn in many of the chapters, as well as considerations of ridge-axis hydrothermal systems, in the context of global geochemical budgets.

A few reflections

During the course of assembling what we hope is a coherent set of contributions spanning many disciplines, questions of terminology have often surfaced. In each of the fields of geophysics, geochemistry, and hydrogeology, sometimes confusing and often unnecessary jargon has been "articulated to the point of assimilation." We have tried to counter this trend. Our judgments about what is unnecessary or confusing and about what is more or less correct may be debatable, but we feel our intention in the editing process to maintain a certain level of consistency and simplicity is well justified. "Flux" is probably the most frequently used term in the book. Several other terms are commonly applied to this quantity, and the use of this term often "sounds" wrong relative to what we might have grown accustomed to. "Flux" (material or energy transfer per unit area and time) is not the same as "flow," yet flux is often used in the chemical community to mean flow, and flow is commonly used in the thermal community to mean flux. "Darcy velocity" and "specific discharge" are terms used by the hydrologic community that are synonymous with volumetric fluid flux. While not wishing to be dogmatic, we have tried to limit the number of terms, and stick to fundamental physical ones as consistently as possible. There also has been a confusing use of terms that relate to processes and behavior. For example, "convection" (with buoyancy as a driving force) and "advection" (transport with no mechanism invoked) are often used interchangeably (not here, we hope), and "free" and "forced" convection are so vague that we have avoided their use entirely. "Diffusion" is a term that describes a behavior, not a specific process. It can be applied equally well to ionic (chemical), molecular (heat), and frictional (pressure-flow) processes, as long as each is viewed at a scale large enough for the actual mechanisms to be "out of focus" – molecular diffusion has no meaning at the molecular scale, and hydrologic diffusion has no meaning at the scale of an individual water-bearing fracture. "Flow" and "permeability" denote processes and properties, but like diffusion, they must be determined or applied with proper consideration of scale. Localized flow through a fractured medium documented using chemical tracers generally will not match average flow determined from thermal perturbations, but both are correctly described as flow. In cases like these, great care must be used to discriminate the process involved, but once this is accomplished, meaning can be gained from seemingly contradictory observations.

We have attempted to express relationships with fundamental parameters, and to avoid the use of derived parameters (unless they make good physical sense). Consistent symbols have been used throughout the text; these are listed on p. **xix** following. We apologize if our editorial pen seems awkward, and more importantly if we have missed the mis-use of quantitative terms. We urge the reader to maintain a critical eye, and to cultivate a desire to reduce the propagation of loose word usage.

Reflections on the future of hydrologic studies

Although we have made some specific suggestions in Chapter 2 as to the "way forward," it is, perhaps, useful to make a few points here. The greatest challenge perhaps is to face characterizing the heterogeneity of the crust, and the distribution of flow and alteration at depth. New remote techniques are needed, and some deep drilling will be required. Regional heat deficits are defined primarily by the global heat flux data base that is far from ideally suited for this purpose. Better data are needed to improve this estimate. The next steps of improving volumetric, volume-temperature, and ultimately geochemical budgets require the combined constraints of heat flux and sediment thickness.

Huge areas of the ocean are uncharacterized. New "type areas" should be selected and studied using the multi-disciplinary surveys of North Pond, Costa Rica Rift, and Juan de Fuca as guides. Major advances are needed in the treatment of chemical elements transported in fluids. Often they are described almost as passive tracers with only a sideward glance at their reactivity. This is a reflection of the state of this field. Their description in models of fluid transport is in an immature state, except for transport in sediments (although that is treated purely as a two-dimensional process). Without improvements on all of these fronts, existing uncertainties in estimating global budgets will remain.

A new phase of characterization and hypothesis testing is about to begin with the augmented tools of the Integrated Ocean Drilling Program. As recently emphasized by the Hydrogeology Program Planning Group of the Ocean Drilling Program (Ge *et al.*, 2003), this will provide a means of penetrating deeply into the oceanic crust in a few key locations via the improved capabilities of the riser drilling vessel *Chikyu*, and offer a suite of tools that are better suited to carrying out specific and detailed hydrologic experiments on this and the non-riser vessel. This and improvements in monitoring technology will lead to much better spatial and temporal resolution of hydrologic processes that operate at scales ranging from fractures to formations, and at periods ranging from seismic to "steady state."

Finally, we would like to emphasize that hydrologic studies of the oceanic crust and upper mantle should be considered in the context of other environments, including many on land. Similar experimental strategies can be applied, and many of the processes involved in one "type setting" will have major implications for others. For example, thermal or chemical buoyancy forcing through continental margins and oceanic islands and platforms follow the same principles as forcing in the oceanic crust and overlying sediments. And effects of a highly permeable upper oceanic crust may have great consequences in the mechanical and seismogenic deformation of subduction zone accretionary and non-accretionary prisms.

References

COSOD II, 1987. *Report of the Second Conference on Scientific Ocean Drilling,* Stasbourg, 6–8 July, 1987, 142 pp.

Ge, S., Bekins, B., Bredehoeft, J., Brown, K., Davis, E. E., Gorelick, S. M., Henrey, P., Kooi, H., Moench, A. F., Ruppel, C., Sauter, M., Screaton, E., Swart, P. K., Tokunaga, T., Voss, C. I., and Whitaker, F. 2003. Fluid flow in sub-seafloor processes and future ocean drilling. *Eos, Trans. Am. Geophys. Union* **84**: 145–152.

Lister, C. R. B. 1972. On the thermal balance of a mid-ocean ridge. *Geophys. J. Roy. Astron. Soc.* **26**: 515–535.

Acknowledgments

We would like to express our appreciation to the International Lithosphere Program and the Joint Oceanographic Institutions through the US Science Support Program for funding the workshop that started our book-writing effort and for covering the costs of producing the CD-ROM that accompanies this volume. We thank especially the contributors to this volume for their willingness to take time out from their research to compile up-to-date summaries of the state of their science, as well as to provide internal reviews for chapters devoted to topics related to their own. Sally Thomas at Cambridge University Press provided us and the contributors a great deal of freedom to choose the contents, approach, and style they felt was most appropriate. Considerable assistance was provided by external reviewers (listed below) who improved the accuracy of the contents and clarity of the presentations. Sandra Last in the Department of Earth Sciences at Cambridge University provided admirably efficient assistance during all stages of the process. Ellen Kappel of Geoprose Inc. made the compilation and production of the CD-ROM seem effortless. We know her product will be admired, and we hope it will be well used by students and educators.

Reviewers of manuscripts

Roger Anderson, Lamont-Doherty Earth Observatory, Columbia University, USA
Andy Barnicoat, University of Leeds, UK
Robert Detrick, Woods Hole Oceanographic Institution, USA
Edward Irving, Geological Survey of Canada, Canada
Claude Jaupart, University of Paris, France
Joris Gieskes, Scripps Institute of Oceanography, USA
Keith Louden, Dalhousie University, Canada
Roger Morin, US Geological Survey, USA
Martin Palmer, University of Southampton, UK
Philippe Pezard, University of Montpellier, France
Elizabeth Screaton, University of Florida, USA
Leslie Smith, University of British Columbia, Canada
Tomochika Tokunaga, University of Tokyo, Japan

Damon Teagle, University of Southampton, UK
David Vanko, Towson University, USA
Karen Von Damm, University of New Hampshire, USA
William Wilcock, University of Washington, USA
Roy Wilkens, University of Hawaii, USA

Symbols and terms

Bulk modulus	$1/\beta$	Pa
Circular frequency $= 2\pi f$	ω	radians s^{-1}
Compressibility	β	Pa^{-1}
Heat flux (heat flow density)	f	W m^{-2}
Density	ρ	kg m^{-3}
Dimensionless time constant	τ	
Earthquake magnitude, body wave	m_b	
Volumetric fluid flux, Darcy flux	q	m s^{-1}
Frequency	f	s^{-1}
Gravitational acceleration	g	m s^{-2}
Heat generation rate	H	W m^{-3}
Hydraulic conductivity	K_h	m s^{-1}
Hydraulic diffusivity	η	m^2 s^{-1}
Hydraulic head	h	m
Kinematic viscosity	μ/ρ_f	m^2 s^{-1}
Length, thickness	h, L, d, b	m
Loading efficiency	γ	
Nusselt number	Nu	
Pechlet number	Pe	
Permeability	k	m^2
Poisson's ratio	ν	
Porosity	n	
Pressure	p	Pa
Radiometric decay constant	λ	s^{-1}
Rayleigh number	Ra	
Specific heat, heat capacity by mass	c	J kg^{-1} K^{-1}
Storage compressibility	ζ	Pa^{-1}
Strain	ε	
Stress	σ	Pa
Temperature	T	K, °C
Thermal conductivity	λ	W m^{-1} K^{-1}

Thermal diffusivity	κ	$m^2\,s^{-1}$
Thermal expansivity	α	K^{-1}
Time	t	s
Tortuosity	τ	
Velocity	$\boldsymbol{u}$	$m\,s^{-1}$
Viscosity, dynamic viscosity	μ	Pa s
Volumetric heat capacity	ρc	$J\,m^{-3}\,K^{-1}$
Volumetric strain	θ	
Young's modulus	E	Pa

Subscript conventions

Fluid	f
Solid	s
Solid matrix	m
Undrained formation mixture no subscript	

Chemical concentrations

mol	moles
M	molar
m	milli (10^{-3})
μ	micro (10^{-6})
n	nano (10^{-9})

Part I

Background

1

Variability of heat flux through the seafloor: discovery of hydrothermal circulation in the oceanic crust

John G. Sclater

1.1 Introduction

Never duplicate an oceanic heat flow measurement for fear it might differ
from the first by two orders of magnitude!

(Bullard's Law as stated by Maurice Hill, 1963)

The elevation of a suggestion to a hypothesis requires the prediction of
testable phenomena separate from the original discrepancy

(Lister, 1980)

The heat left over from the formation of the planet combined with that generated by radioactive decay drives the internal engine of the Earth. This heat is the cause of elevated temperatures in mines, and is the ultimate source of energy for volcanoes, hot springs, mountain building, and earthquakes. The heat flux is the product of the temperature gradient and the thermal conductivity. Scientists use temperatures measured in 100-m- to 5-km-deep boreholes coupled with conductivity measurements on the rocks penetrated to derive the heat flux through the continents. At sea, they use the temperature gradient measured by thin probes driven 3–10 m into the soft sediments of the ocean floor multiplied by the thermal conductivity of these sediments. Continental crust is both thicker and has a greater concentration of heat-producing elements than oceanic crust, so if heat is carried from the Earth's interior only by conduction, the continents should have a much greater heat flux than the oceans.

1.1.1 The early oceanic measurements

Hans Petterson, the leader of the 1947–1948 Swedish Deep Sea Expedition, made the first temperature gradient measurement on the ocean floor on November 14, 1947, near the equator between Tahiti and Hawaii (Petterson, 1949, 1957). This and another measurement near the equator on the edge of the Ontong Java Plateau yielded temperature gradients of

---

Hydrogeology of the Oceanic Lithosphere, eds. E. E. Davis and H. Elderfield. Published by Cambridge University Press.
© Cambridge University Press 2004.

approximately 40 °C km^{-1}, twice that of the average gradient on continents (Petterson, 1949). Even allowing for the differences between the conductivity of continental rocks and soft oceanic sediments, these gradients were incompatible with there being a similar deep conductive flux beneath continents and oceans. It is difficult now to overestimate the importance of the contribution of Petterson. He had recognized that stable bottom-water temperatures, coupled with the soft sediments of the ocean deeps, provided the ideal environment for the easy measurement of the heat flux through the ocean floor. In addition, his measurements stood conventional theory regarding the thermal state of continents and oceans on its head. Surprisingly, no write-up exists in English of the instrument, and Petterson did not turn the temperature gradients into heat flux. When Revelle and Maxwell (1952) and Bullard (1954) reported the first full oceanic heat flux measurements that included sediment thermal conductivity, both cited the pioneering measurements of Petterson (1949), but since then his contribution appears to have been largely forgotten.

Sir Edward Bullard had made substantial contributions to heat flux measurements on continents (Bullard, 1939) and had long wanted to make similar measurements through the ocean floor. He recognized the importance of the efforts of Petterson immediately (Bullard, 1954), encouraged the development of a heat probe device at Scripps Institution of Oceanography (SIO) in the US, and developed a similar device at the National Physical Laboratory in England. Gradients reported by Revelle and Maxwell (1952) were even higher than those of Petterson (1949). Bullard (1952) speculated that "at some not too remote time a convection current rose under the Pacific and brought hot material near the surface," and that this could account for the higher that expected oceanic flux. With additional measurements in the 1950s, an apparent equality between the mean flux through the oceans and continents emerged (Lee and Uyeda, 1965), despite the much greater contribution of radiogenic heat from continental crust. Early observations also revealed large scatter about the relatively high mean value, particularly over mid-ocean ridges. For example, Von Herzen (1959) and Von Herzen and Uyeda (1963) reported both very high and very low individual values and high average heat flux near the crest of the East Pacific Rise.

I arrived at the University of Cambridge as a graduate student in the autumn of 1962. I joined the Marine Group run by Maurice Hill and planned to work as an observational seismologist on low-frequency surface waves in the ocean crust. In addition, I volunteered, during a 1963 Cambridge expedition to the Indian Ocean on the *RRS Discovery*, to run the outrigger heat probe that Clive Lister had built as part of his Ph.D. thesis (Lister, 1963). The low-frequency seismometer that I built flooded on its first lowering, so I concentrated for the rest of the expedition and ultimately for my thesis on the measurement of heat flux through the ocean floor. I observed high heat flux within the Gulf of Aden (Sclater, 1966b), and a large scatter like that reported by Von Herzen and Uyeda (1963). The scatter troubled me and my supervisor Maurice Hill, who remained skeptical not only of the whole technique but also of the grandiloquent interpretations by Sir Edward of a measurement that could vary so much over so short a distance.

Fig. 1.1 Department of Geodesy and Geophysics, Cambridge University marine group on board the RRS Discovery in the early 50s. Back row from left: John Swallow, Maurice Hill and John Cleverly. Front row: Tony Laughton, Captain Dalgliesh and Sir Edward Bullard. Photograph by Robin Adams.

1.1.2 Bullard's law

Sir Edward Bullard (Fig. 1.1) had received his Ph.D. in the 1930s as a nuclear physicist under Rutherford at the Cavendish Laboratory in Cambridge, UK. Immediately before World War II and then entirely thereafter, Teddy Bullard had devoted himself to geophysics, ending his career as the Head of the Department of Geodesy and Geophysics at Cambridge. However, he sometimes regretted that he had had to abandon nuclear physics to find a job. He bemoaned especially the fact that unlike most of his contemporaries from the Cavendish, he had had no important physical law named after him. To rectify this oversight, Maurice Hill (Fig. 1.1) formulated Bullard's Law. Maurice, who had worked with Teddy during World War II, knew that he suffered horribly from seasickness and he eventually refused to take him to sea with the marine group for fear his illness would jeopardize the Cambridge expeditions.

In this chapter, I present a personal retrospective of the attempts to explain Bullard's Law, covering early heat flow surveys, the sedimentary observations, magnetic measurements, and physical inferences that led to the realization that fluid flowing through the oceanic crust must be responsible for the variability. I discuss the original discovery of a hydrothermal plume predicted by the physical model, and the later confirmation by Anderson *et al.* (1985) of a layer of high permeability at the top of the igneous crust. I finish by considering the advances that have come from separate considerations of advective flow and conductive thermal flow of heat through the ocean floor. As this is a personal retrospective I have restricted my comments to the papers that I believe influenced or should have influenced my own thinking on this discovery.

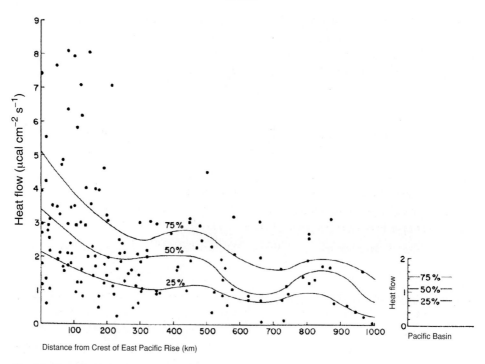

Fig. 1.2 Heat flux values versus distance from the crest of the East Pacific Rise. 75, 50, and 25 percentile bounding lines are given for values from the East Pacific Rise and Pacific Basin. (Reprinted from Lee, W. H. K. and Uyeda, S. 1965. Review of heat flow data. In *Terrestrial Heat Flow*, Geophysical Monograph Series 8, ed. W. H. K. Lee. Washington, DC: American Geophysical Union, pp. 87–190, Copyright 1965 American Geophysical Union.)

1.2 Measurement scatter and the search for its cause

Early workers such as Bullard and Day (1961) and Von Herzen and Uyeda (1963) investigated the heat flow anomaly at the crests of the mid-ocean ridges. After the accumulation of a substantial body of data, Lee and Uyeda (1965) showed that the mid-ocean ridges had an overall flux well above the global average (Fig. 1.2). However, great scatter in the measurements prevented a confident quantitative interpretation of the results.

In their compilation of heat flux measurements from the Pacific, Von Herzen and Uyeda (1963) noted that the values ranged from essentially zero to 400 mW m^{-2} and discussed in detail local factors that might affect them. Bullard *et al.* (1956) had shown already that bottom-water temperature changes had a negligible effect. Von Herzen and Uyeda (1963) examined four other factors and demonstrated that none could reasonably account for the observed scatter and individual low values. These were: (a) rapid sedimentation, (b) local sediment slumping, (c) thermal refraction through rugged basement and around sediment ponds, and (d) a downward flow of water through the sediments.

The conclusion that local environmental effects could not explain the low values set the stage for my thesis (Sclater, 1966a). In 1965, I investigated the combination of sediment ponds and sharp topographic slopes in the North Atlantic (Sclater *et al.*, 1970a), and, in 1966, I carried out a similar investigation north of the Hawaiian islands (Sclater *et al.*, 1970b). My coauthors and I concluded that only a series of recent slumps of sediment from nearby hills could explain the low values. This explanation was unrealistic, however, so the low values remained problematic.

1.3 Early thermal models of the oceanic lithosphere without good thermal constraints

Harry Hess and Tuzo Wilson spent a sabbatical at "Madingley," the English manor housing the Department of Geodesy and Geophysics at Cambridge, during my last year at Cambridge. I was so caught up in dealing with the variability of the measurements that I failed to realize the importance of either seafloor spreading (Hess, 1962) or plate tectonics (Wilson, 1965) for understanding oceanic heat flux. When the concept of seafloor spreading was proposed, heat flux measurements had played a major role. Hess suggested that huge igneous intrusions caused the high values at the crests of the mid-ocean ridges. The combination of Von Herzen's measurements showing high average values over the crest of the East Pacific Rise (Von Herzen, 1959), Bill Menard's concept of a world-encircling set of mid-ocean ridges (Menard, 1959), and the insight of Bruce Heezen (1960) regarding normal faulting at the crest of the Mid-Atlantic Ridge led Hess directly to the concept of seafloor spreading. Additional measurements by Nason and Lee (1962), Vacquier and Von Herzen (1964), and Von Herzen and Vacquier (1966) confirmed the correlation of high average heat flux (and high scatter) with the crests of all of the mid-ocean ridges.

Soon after the publication of Hess' paper, Wilson (1965) introduced the idea of transform faults from which the quantitative theory of plate tectonics developed. Lyn Sykes (1967) showed from earthquake first-motion studies that only normal faults occurred at the crest of the Mid-Atlantic Ridge and that all the strike–slip fault-plane solutions lay on the cross-cutting fracture zones, consistent with the transform fault hypothesis. At the same institution, Lamont Doherty Geological Observatory of Columbia University, Marcus Langseth (Fig. 1.3*a*), Xavier Le Pichon, and Maurice Ewing introduced the concept of a constant thickness lithosphere, created by a zone of intrusion and subsequently cooled by conduction through its upper surface (Langseth *et al.*, 1966). They pointed out that the cooling of this plate would lead to a decrease in heat flux and a subsidence due to contraction as the newly created plate moved away from the place of creation, the mid-ocean ridge. While this early model was to become a cornerstone for quantitative plate-tectonic theory, it is interesting that the authors were forced to argue against the possibility of "continuous continental drift of the spreading-floor type in the Cenozoic" in the Atlantic Ocean on the basis of the lack of a wide heat flow maximum over the Mid-Atlantic Ridge. The knowledge of unaccounted-for advective heat loss from the lithosphere was yet to come.

Stimulated by this paper, McKenzie (1967) showed that by reducing the plate thickness and the axial and lower boundary temperatures, the plate model could be made to match the heat flux data in a general way. Other attempts that followed involved more reasonable plate thicknesses and initial and lower boundary temperatures, and more sophisticated treatments

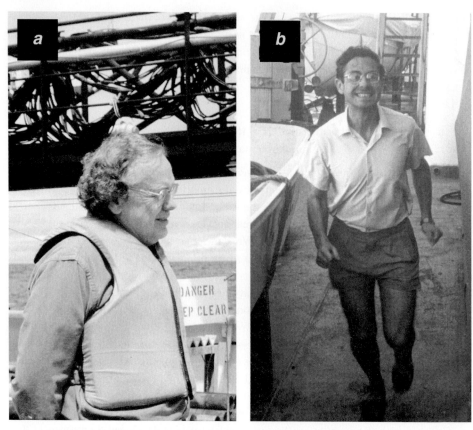

Fig. 1.3 Marcus Langseth (*a*) on board the JOIDES *Resolution* drill ship in 1991 (photograph by Andy Fisher), and Clive Lister (*b*) jogging on the *R.V. Thomas Thompson* in 1973. (photograph by Earl Davis). Reproduced with permission.

of isostatic compensation and axial boundary conditions (McKenzie and Parker, 1967; Vogt and Ostenso, 1967; Isacks *et al.*, 1968; Le Pichon, 1968; Morgan, 1968; McKenzie and Sclater, 1969; Sleep, 1969; Sclater and Francheteau, 1970). These established many quantitative aspects of plate tectonics, but all suffered from the same unrecognized problem that the heat flux data did not reflect the total flux from the lithosphere in young regions. Ironically, those that were most successful used heat flux as a constraint only semi-quantitatively or not at all (as in the case of Sclater *et al.*, 1971; Fig. 1.4).

Fig. 1.4 Average depth in the Pacific, Atlantic, and Indian Oceans plotted against age of the ocean floor. The theoretical profile (dashed line) is that for a lithosphere 100 km thick with a base temperature of 1,475 °C. The thick black line is a curve drawn by eye through the North Pacific data. (Reprinted from Sclater, J. G., Anderson, R. N., and Bell, M. L. 1971. The elevation of ridges and the evolution of the central eastern Pacific. *J. Geophys. Res.* **76**: 7,888–7,915, Copyright 1971 American Geophysical Union.)

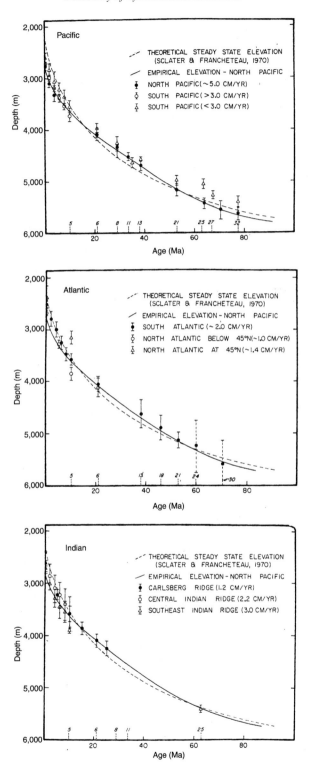

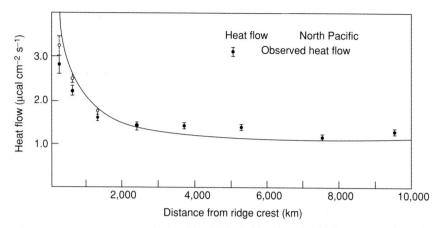

Fig. 1.5 A comparison of the observed heat flux averages in the North Pacific (filled points) with the theoretical profile for a 75-km-thick lithosphere. Open points reflect a 15% shift applied to the observed averages to account for a possible bias from environmental effects. (Reprinted from Sclater, J. G. and Francheteau, J. 1970. The implications of terrestical heat flow observations on current tectonic and geochemical models of the crust and upper mantle of the earth. *Geophys. J. Roy. Astron. Soc.* **20**: 509–537, Copyright 1970 The Royal Astronomical Society.)

1.4 Hydrothermal circulation at mid-ocean ridges

Even after omitting very low values, Sclater and Francheteau (1970) found that the mean heat flux near the ridge axes lay well below that predicted by plate theory (Fig. 1.5). An explanation that involved an arbitrary rejection of low values and even then did not match the observations was inherently unsatisfactory. To be taken seriously, the geothermal community needed an explanation of the low average values (relative to plate theory) and large scatter. The explanation came from sedimentary geochemistry and from the work of my predecessor at Cambridge, Clive Lister (Fig. 1.3*b*).

1.4.1 Geochemical evidence

The cause of variability in ocean heat flux received little attention from most of the geophysics community. However, geochemists had long been searching for an explanation for both the global and local variations in metalliferous metals in sediment samples from the ocean floor. Many (von Gumbel, 1878; Murray and Renard, 1895; Arrhenius, 1952; Petterson, 1959) had suggested hydrothermal solutions were responsible. However, none had related the phenomena to the mid-ocean ridges until Skornyakova (1964) linked the formation of iron–manganese–carbonate-rich sediments with modern volcanism and hydrothermal action at the crest of the southern East Pacific Rise. Arrhenius and Bonatti (1965) related the high barite concentration on the East Pacific Rise to active magma reservoirs, and Bonatti and Jeonsuu (1966) argued that hydrothermal activity concentrated mineral deposits right on the crest of the Rise. In their classic paper, Bostrom and Peterson (1966) related sediment enrichment in iron and manganese to the high heat flux at the crest of the Rise (Fig. 1.6). They argued for ascending solutions of deep-seated origin, and that hydrothermal activity "is the only explanation that seems reasonable."

1.4.2 Geophysical arguments

John Elder (1965) was the first geophysicist to tackle the question of hydrothermal circulation in the hard rock at mid-ocean ridges. When interpreting the Tuscan steam zone in Italy and the Taupo hydrothermal systems of New Zealand, he considered a body of freely circulating hot water and discussed the problem in terms of convection within a porous medium. He noted that the boundary conditions in an oceanic system would differ from those on land: no air–water interface, and seawater existing both above and below the ground surface. He simulated the conditions in a laboratory Hele–Shaw cell, a device that simulates flow in a porous medium with flow in a thin vertical slot bounded by parallel insulating plates. With heat applied uniformly from below, he observed that heat flux at the surface was concentrated above a narrow zone of convective up-flow. This was bordered by a broad region where the flux was zero corresponding to the area of recharge.

The Mid-Atlantic Ridge includes Iceland, where famous hot springs and geysers abound at the surface. Appropriately, the first publication linking hydrothermal circulation to the cooling of a mid-ocean ridge appeared in an Icelandic publication (Palmason, 1967). Talwani *et al.* (1971) made a similar suggestion to account for some very low values near the crest of the Reykjanes Ridge. Irving *et al.* (1970) provided further support for this hypothesis by arguing that hydrothermal alteration reduced the magnetization of the near-surface lava flows as they move away from the crest of a mid-ocean ridge.

However, only after Clive Lister (1972) carried out two detailed heat flux surveys over the variably sediment-covered Juan de Fuca and Explorer Ridges did a coherent hypothesis with testable predictions develop. As observed in other studies, the heat flux in this young area was anomalously low and variable. Carbon-14 studies of the sediments in the ponds with low heat flux showed a constant deposition rate with no evidence of major slumping. Clearly, environmental disturbances could not explain the low and variable values. Lister argued that only hydrothermal circulation in the hard rocks of the oceanic crust could explain the low values in the sediment ponds and the overall missing heat over this ridge crest. He also concluded that a thin drape of low-permeability sediments would impede water flow, and predicted in a now famous cartoon the flow pattern in three cases of differing hydrologic structure: a permeable crust where exchange with the ocean is uninhibited, a permeable crust blanketed by a uniform drape of low-permeability sediments, and a permeable crust buried by flat lying abyssal plain sediments (Fig. 1.7). He also argued that up-flow would be highly focused and create hot springs at ridge crests, but that detection of them would be difficult because of the great dilution of the springs that would take place as they mixed into the open ocean.

In a parallel development, Bodvarsson and Lowell (1972) demonstrated how easily a pipeline type of convection system could modify the heat flow of a region. Cracks as small as 3 mm wide and a kilometer deep could convect away a large amount of heat, and minute amounts of thermal contraction in the oceanic crust could generate them. This became the

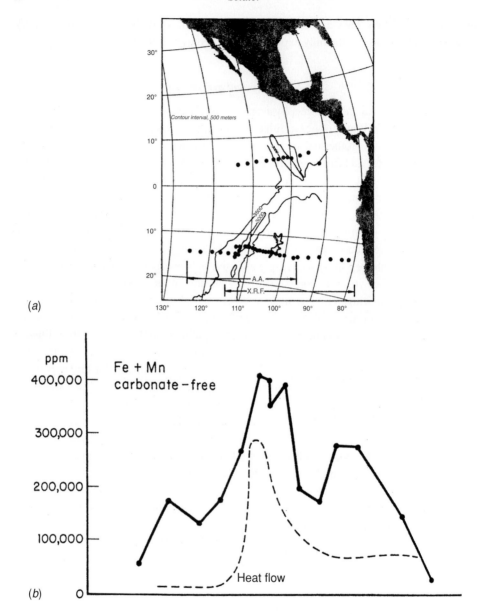

Fig. 1.6 Map of the East Pacific Rise, showing the location of sediment samples analyzed by Bostrom and Peterson (1966) (*a*), and co-variation of heat flux and combined Fe and Mn concentration (ppm) determined on a carbonate-free basis (*b*). Data are from profile A–A shown in (*a*). Heat flow data from Von Herzen and Uyeda (1963) were presented without scale in the original publication. The length of the profile is roughly 3,000 km and the maximum heat flux approximately 400 mW m^{-2} (adapted with permission from Bostrom and Peterson, 1966).

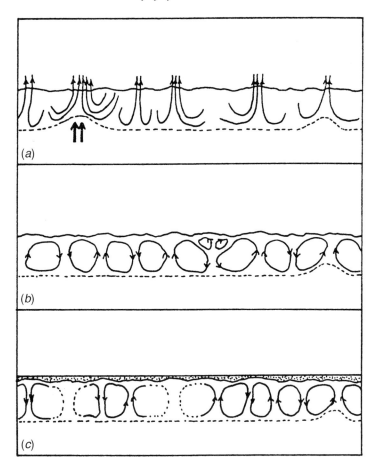

Fig. 1.7 Pattern of crustal hydrothermal convection to be expected near a ridge crest, where the permeable layer is open to the ocean (*a*). Convective pattern forced by topography in an area of permeable crust bounded by a thin impermeable blanket of sediment (*b*). The case of permeable crust with a rough boundary buried by flat lying sediments (*c*), where the varying thermal resistance of the sediment blanket overpowers the direct effect of the topography on the pattern of circulation. (Reprinted from Lister, C. R. B. 1972. On the thermal balance of a mid-ocean ridge. *Geophys. J. Roy. Astron. Soc.* **26**: 515–535, Copyright 1972 The Royal Astronomical Society.)

key to the development of the theory of water penetration into hot rock (Lister, 1974) and the horizontal flow of hydrothermal fluids over large distances (Langseth and Herman, 1981).

1.5 Tests of the concept

1.5.1 Discovery of hydrothermal plumes

In the late sixties, I remember being involved in discussions at SIO, mostly on the beach at lunch time, regarding the possible occurrence of hydrothermal circulation near the crests of

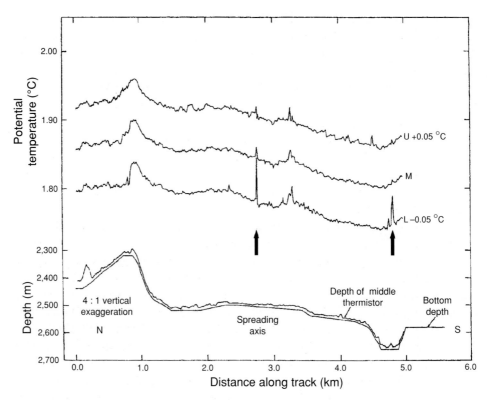

Fig. 1.8 Bottom-water potential temperatures measured by three thermistors, plotted as a function of distance on a crossing of the Galapagos Spreading Center. Thermistors of the lower pair were separated by 5.9 m and those of the upper pair by 11.2 m. The depth of the seafloor and the height of the middle thermistor above bottom are shown at the bottom of the plot. The arrows indicate the temperature anomalies (adapted from Williams *et al.*, 1974). The average horizontal speed was 1.28 km h^{-1}.

mid-ocean ridges. On one side were the geochemists who believed that such circulation must exist and pervade the entire East Pacific Rise. On the other side were the petrologists who dredged only fresh rocks from the crest of the East Pacific Rise and saw no direct evidence for such circulation. Dick Von Herzen of Woods Hole Oceanographic Institution (WHOI) persuaded me that one way out of the dilemma was for Lister (1972) to be correct and that the hot discharge areas at the ridge axis involved only narrow widely separated upwelling jets (as suggested by Elder, 1965). Von Herzen and David Williams from WHOI, and Roger Anderson and I and the Deep Tow group, all from SIO's Marine Physical Laboratory, set out to test this prediction on an expedition to the Galapagos Spreading Center in 1972 (Williams *et al.*, 1974). Because I had proposed sediment slumping to explain low heat flux values, I had a vested interest in showing that hydrothermal vents did not exist, but was open to the suggestion that I might have missed hydrothermal circulation as an explanation of the low heat flow. Von Herzen and Williams built an apparatus at WHOI that had three thermistors on a towed cable. At each, temperature was measured at high and low sensitivity over correspondingly small and large ranges. We recognized no evidence for hot water plumes during the expedition. Only on return to WHOI did Williams realize that had we failed to recognize two temperature anomalies that had gone off the highly sensitive scale used to monitor temperature in real time during the tows. Fortunately, they had been recorded on the less sensitive scale (Fig. 1.8). With these data, it became clear that we had found the isolated zones of thermal discharge predicted by Lister.

In addition to isolated thermal plumes, the Deep Tow group discovered a chain of small mounds from 5 to 20 m high in several patches 20–30 km south of the spreading axis. Klitgord and Mudie (1974) suggested that these had formed where hydrothermal fluids discharged from fissures in the basaltic crust. Their data suggested hydrothermal venting on the ocean floor but were not conclusive. On a return to the area in 1976, a conductivity–temperature–depth (CTD) instrument attached to the deep-tow vehicle sensed small plumes of warm water whose modified temperature/salinity characteristics demon- strated local geothermal heating, and a nephelometer simultaneously observed an increase in suspended matter (Weiss *et al.*, 1977). Samples from electronically triggered bottles showed the plumes to be enriched in mantle-derived He (Lupton *et al.*, 1977) and volcanogenic Mn (Klinkhammer *et al.*, 1977). The vehicle photographed yellow precipitates (Crane, 1977) and clusters of feeding animals around the discharge sites (Lonsdale, 1977). In the next year, the *DSRV ALVIN* made a close up examination of the vents, and sensed and sampled the discharging fluid before massive dilution by mixing with the ocean water had taken place. With these observations, it became clear that hydrothermal circulation is probably common beneath the ocean floor (Dive and Discover, 2002). The interdisciplinary study of seafloor hydrothermal circulation had begun in earnest.

1.5.2 Permeability measurements

Near the Galapagos Spreading Center, non-linear thermal gradients, widespread pore-water chemical anomalies, and metallogenesis showed that at this site a convection system pen- etrates the very thin sedimentary layer and therefore carries significant heat directly from the oceanic crust to the ocean (Chapter 16). However, as the seafloor spreads away from the ridge axis, the convective component of the heat transfer is gradually replaced by a totally conductive thermal regime at the seafloor. Anderson and Skillbeck (1981) modeled the crust and the overlying sediment as a two-layer porous medium as hypothesized by Lister (1972), with a low-permeability sediment layer on top of a much higher permeability basement. In this model, continued deposition increases the hydrologic resistance of the sedimentary layer, and flow of water through it stops (Chapter 6). Flow will continue beneath the sed- iments if the basement permeability is sufficiently high, but the sediments form a lid to the convection and cause a change from a convective to conductive thermal regime at the ocean floor. Alternatively, the sealing of the cracks and fractures that gives the basement its initial permeability also provided a viable mechanism for stopping circulation within the basement altogether (Chapter 5).

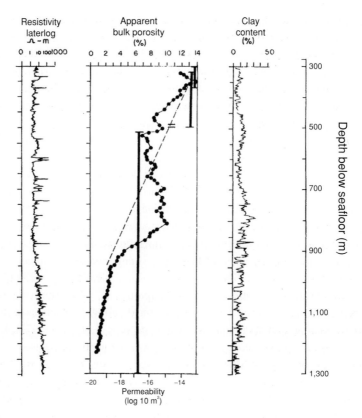

Fig. 1.9 Selected geophysical logs from DSDP Hole 504B. (Reprinted from Anderson, R. N. and Zoback, M. D. 1982. Permeability, under-pressures, and convection through the oceanic crust near the Costa Rica Rift, eastern equational Pacific. *J. Geophys. Res.* **87**: 2,860–2,868, Copyright 1982 American Geophysical Union.)

To test between these two causes for the cessation of seafloor hydrothermal heat transport, scientists made *in situ* measurements of the permeability in layer 2A of the oceanic crust by conducting pumping tests in packed intervals at DSDP Hole 504B from the *D/V Glomar Challenger* (Legs 69, 83, and 92). Anderson and Zoback (1982) reported high bulk permeabilities for their measurements in the upper pillow basalts and flows on Leg 69, and Anderson *et al.* (1985) summarized the results of the initial and subsequent two legs that showed high permeability in the upper 200 m of the basement, and a strong decrease by nearly four orders of magnitude below this depth (Fig. 1.9). These results provided another confirmation of the hydrothermal model of Lister (1972), a model that remains relevant today (Chapters 7 and 8).

1.6 Implications of hydrothermal circulation

The study of the flow of heat through the floor of the ocean separates naturally into two fields of study. One is of the advective flow of heat resulting from the circulation of seawater through the oceanic crust. The perturbed thermal state provides an excellent constraint on fluid flow, and hence gives important information about the alteration of the crust and sediments and about the effect of seawater circulation on the chemistry of oceans. The other is of the conductive flux of heat that can be used together with the depth of the ocean floor to investigate the thickness of the lithosphere and thermal convection in the mantle. The second can only be done with proper knowledge of the first. A few early attempts to understand and quantify these processes are summarized here.

1.6.1 Advective flow

In a study of the heat flux measured in the vicinity of the heavily sedimented northern Juan de Fuca Ridge, Davis and Lister (1977) found evidence for widespread hydrothermal circulation in what they referred to as "a highly heat transportive crust." As in other previously studied young regions, values of heat flux in the axial rift and rift mountains were lower than expected and highly scattered, and they attributed this to groundwater ventilation through faults and basement outcrops (as in the situation depicted in Fig. 1.7a). At the roughly 10-km spacing of the observations (considered very close at the time of that 1971 survey), no coherent pattern emerged. Within the axial rift where sediment cover was more complete, an examination of the heat flux in context of the local structure defined by seismic data did reveal a pattern: heat flux varied inversely with sediment thickness (Fig. 1.10). Errors in the heat flux and sediment-thickness determinations were large, as were navigational uncertainties, but the pattern was undeniable. They proposed that this relation was the consequence of widespread hydrothermal circulation in a highly permeable crust that maintained water temperatures at the base of the sediments within a relatively limited range. This relationship was also observed on the flank of the adjacent Explorer Ridge, but the sediment cover in that area was believed to be sufficiently thick and continuous to prevent ventilation, and they argued that the average seafloor heat flux there provided a reliable estimate of the total conductive heat loss from the underlying lithosphere.

Using a broader-scale approach, Anderson and Hobart (1976) and Anderson *et al.* (1979) noted that each ocean has a different distribution of sediment as a function of crustal age, and proposed that this controls the variation of heat flux versus age. Where the sediment cover is very thin or absent, ventilation of hydrothermal fluids through basement outcrop occurs and the mean heat flux is anomalously low (i.e. Fig. 1.7a). When the basement becomes completely covered by sediment (Fig. 1.7b, c), ventilation terminates and the observed flux rises to the predicted value (Fig. 1.11). This simple concept provided a way to account for the "missing heat" associated with ventilated circulation, and permitted calculation of both the total heat loss as a result of the creation and cooling of the lithosphere and the background heat flux from the asthenosphere (Chapter 10). Williams and Von Herzen (1974) estimated roughly 3×10^{13} W for the total heat loss of all ocean basins. For the cooling due to the advection of heat by the flow of water through the top of the lithosphere they gave a figure of 1×10^{13} W. This large value for the heat loss due to advection increased the computed total heat loss by almost 50% over that estimated by Lee and Uyeda (1965)

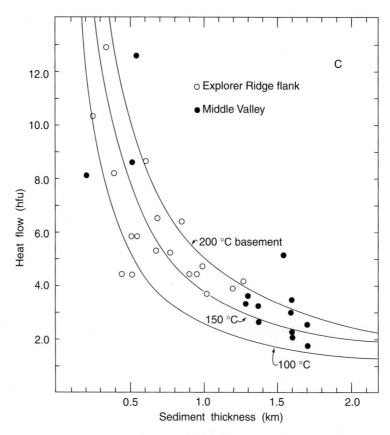

Fig. 1.10 Heat flux versus sediment thickness for fully sediment-covered regions of the Explorer Ridge flank (open circles) and the Middle Valley rift of the northern Juan de Fuca Ridge (solid circles), and relationships predicted by having a 100, 150, and 200°C basement beneath the sediments (solid lines). (Reprinted from Davis, E. E. and Lister, C. R. B. 1977. Heat flow measured over the Juan de Fuca Ridge: evidence for widespread hydrothermal circulation in a highly heat transportive crust. *J. Geophys. Res.* **82**: 4,845–4,860, Copyright 1977 American Geophysical Union.)

prior to the discoveries of hydrothermal circulation, and pointed clearly to the importance of lithospheric creation and cooling in the Earth's thermal budget.

1.6.2 Implications of conductive heat flux measurements for the lithosphere

To a large degree, the initial plate models for describing the formation and thickening of the oceanic lithosphere followed mathematical convenience; they invoked a constant-temperature basal boundary at a fixed depth. A physically more realistic model described the lithosphere as a cooling boundary layer with no specified bottom boundary (Turcotte and Oxburgh, 1967). Following Parker and Oldenburgh (1973), Davis and Lister (1974) simplified the formulation to include only vertical cooling and added a thermal-balance

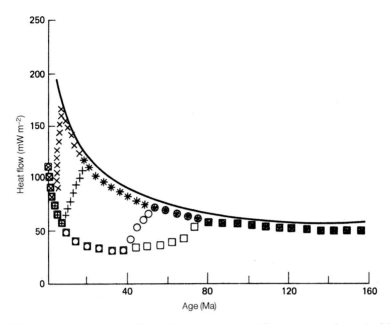

Fig. 1.11 Schematic representation of the variation in mean heat flux versus age for each of the major mid-ocean ridge segments, including those of the Galapagos ($\times$), East Pacific Rise ($+$), Indian Ocean (o), Atlantic Ocean ($\square$, and a theoretical lithospheric estimate ($-$). (Reprinted from Anderson, R. N., Hobart, M. A., and Longseth, M. G. 1979. Geothermal convection through the oceanic crust and sediments in the Indian Ocean. *Science* **204**: 3,391–3,409, Copyright 1979 American Association for the Advancement of Science.)

boundary condition at the ridge axis. Re-plotting the data of Sclater *et al.* (1971), they found a linear relation between the subsidence of the seafloor and the square-root of crustal age, just as predicted by simple cooling theory (Fig. 1.12). In the case of several individual profiles crossing the East Pacific Rise, this behavior of subsidence was observed right to the ridge axis. They considered this to be surprising, since early and later cooling, involving crust then mantle rocks of contrasting physical properties, should produce contrasting rates of subsidence. On the basis of the observed lack of such a contrast, they speculated that hydrothermal circulation may cool the crust immediately, and thus eliminate the contribution of the crust to the subsidence behavior.

Sclater *et al.* (1976) and Davis and Lister (1977) examined the average heat flux through areas covered by thick and extensive layers of sediment in the Western and Central Pacific and on the flank of the Explorer Ridge. They found a reduced scatter and a simple reduction in conductive flux with age. Sclater and Crowe (1979) tested the accuracy of this relation with a detailed heat flow survey over well-sedimented Chron-13 (35 Ma) oceanic crust. All data were consistent with boundary-layer cooling behavior, with heat flux declining linearly with the inverse square-root of crustal age (Fig. 1.13).

Parsons and Sclater (1977) compared the variations of heat flow with age, and depth with age, to the predictions of the thermal boundary layer model and the original plate

Sclater

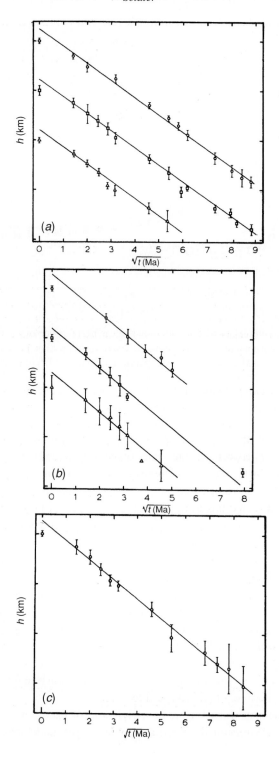

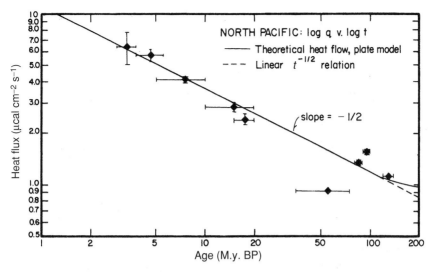

Fig. 1.13 Reliable heat flux means (calculated using measurements in well-sedimented areas) plotted on a logarithmic scale versus age also on a logarithmic scale. The theoretical curve has a slope of −1/2 until about 120 Ma. (Reprinted from Sclater, J. G. and Crowe, J. 1979. A heat flow survey at Anomaly 13 on the Reykjanes Ridge: a critical test of the relation between heat flow and age. *J. Geophys. Res.* **84**: 1,593–1,602, Copyright 1979 American Geophysical Union.)

model. They found that the plate model has the square-root of age dependence implicit in its solutions (Morgan, 1974) for sufficiently young ocean floor. For greater ages these relations break down, and the depth and heat flux exponentially approach constant values. Parsons and Sclater (1977) also extended the empirical depth versus age relation for the North Pacific and North Atlantic back to 160 Ma. They found that depths initially increase as the square-root of age, but for crust greater than 80 Ma depths approach a constant asymptotic value. They treated the heat flux in a similar manner, and showed that the heat flux should respond to the bottom boundary condition at approximately twice the age at which the depth does. This flattening in the depth and the heat flux fields requires an extra mechanism to supply heat to the base of the lithosphere.

Parsons and McKenzie (1978) suggested a simple reason for this extra heat. They proposed that the oceanic lithosphere behaves rigidly below a certain temperature, and divided the lithosphere into a rigid mechanical portion above the depth of the 1,000 °C isotherm and a viscous portion beneath. As the lithosphere moves away from the ridge crest and cools, both the mechanical and underlying viscous layers increase in thickness. Once the

Fig. 1.12 Averaged ocean depth versus the square-root of age plotted for the Pacific (*a*), Indian (*b*), and Atlantic Oceans (*c*). Depths are plotted relative to the crustal depth with the scale in 1 km divisions. (Reprinted from Davis, E. E. and Lister, C. R. B. 1974. Fundamentals of ridge crest topography. *Earth Planet. Sci. Lett.* **21**: 405–413.)

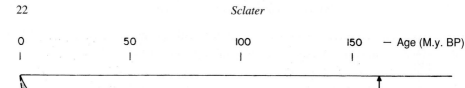

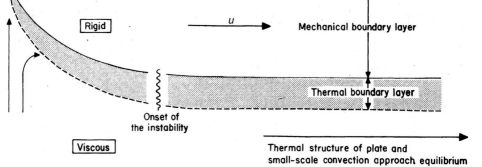

Fig. 1.14 Schematic diagram showing the division of the lithospheric plate into rigid and viscous regions, and the occurrence of an instability in the bottom viscous part of the plate once the cooled part of the viscous layer reaches a critical thickness. The boundary between the rigid and viscous region, and the bottom of the thermal boundary layer are isotherms (from Parsons and McKenzie, 1978).

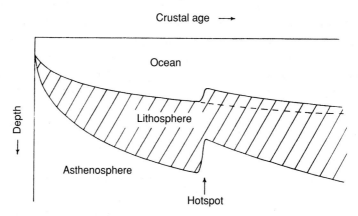

Fig. 1.15 Inferred interaction of the lithosphere with a mid-plate hot-spot. From the ridge to the hot-spot the lithosphere thickens and subsides by cooling. At the hot-spot, extra heat drives the isotherms upwards, thins the lithosphere, and causes uplift. Beyond the hot-spot, the lithosphere cools again and subsides as younger lithosphere at the same depth, rather than as normal lithosphere of the same age (dashed line). (Reprinted from Crough, S. T. 1978. Thermal origin of mid plate hot-spot swells. *Geophys. J. Roy. Astron. Soc.* **55**: 451–469, Copyright 1978 The Royal Astronomical Society.)

thickness (and thus the Rayleigh number) of the thermally defined viscous layer reaches a critical value, small-scale convection is initiated, bringing extra heat to the base of the mechanical layer under old ocean floor (Fig. 1.14).

In an alternative approach, Crough (1973) and Heestand and Crough (1981) suggested that the oceanic lithosphere cools as a thermal boundary layer, but that under old ocean floor, thermal perturbations can reset the lithosphere to a prior position on the thermal boundary layer curve (Fig. 1.15). The initial uplift is proportional to the amount of heat injected and the subsequent subsidence is proportional to the heat flux through the ocean floor.

1.7 Conclusions

Seafloor measurements of conductive heat flux provide one of the observations central to the concept of seafloor spreading. However, local variability in the observations, first seen as incoherent scatter, hindered the development of a satisfactory theory. The search for an explanation of the variability led to the discovery of hydrothermal circulation within the oceanic basement and venting of hot water to the ocean above.

Some unexpected realizations have come to me from rereading the early literature. The geochemists clearly made the first predictions of hydrothermal circulation at the mid-ocean ridges. Bostrom and Bonatti from the Arrhenius Laboratory at SIO took the lead. I am surprised that I did not take this work more seriously as Victor Vacquier's heat flux laboratory was on the same floor, and both Bostrom and Bonatti overlapped with me at SIO. It seems so obvious now to relate heat flux variability to geochemical measurements of fluid flow. However, physicists such as Bullard and Ewing who started this field did not make the correlation. Had Petterson, a radio-chemist, stayed in the field after his first measurements, the hydrothermal explanation of the variability might have appeared much sooner.

Asking the right questions and making careful measurements are important; even those that go down a wrong track can be valuable because ultimately good measurements provide clear evidence for an approaching dead end. The work by Von Herzen and Uyeda (1963), Langseth *et al.* (1966), and myself (Sclater *et al.*, 1970a,b) to seek a conductive explanation for the heat flow scatter spanned 10 years. Lister (1972) realized that this work implied the highly implausible conclusion that local slumping accounted for almost all the variability. He carefully set out to show that such slumping could not explain the low heat flux values in the ponded sediments on the flanks of the Juan de Fuca Ridge and he argued, by analogy with Iceland, that hydrothermal circulation in the oceanic crust provided the only plausible explanation. He raised the suggestions of the geochemists to the level of a hypothesis by predicting both how such circulation would occur within the igneous crust and how localized hot springs were likely to occur along the crests of the mid-ocean ridges. Testing these predictions with subsequent directed observations led to the confirmation of the hypothesis.

The flux of heat is a fundamental property of our planet. It can be measured easily in the oceans at the sediment–water interface, although physical measurements at a boundary between two media are fraught with difficulties. Bullard (1954) realized the importance of trying and continuing even when the results could not be understood. Though a light-hearted joke, Bullard's Law posed a much more profound question. "Why should we put so much time into making such a measurement when we do not understand its variability?" Members of the oceanic geothermal community have taken a very twisted path in their attempts to understand this fundamental measurement. They should have played a more significant role in the confirmation of seafloor spreading and in the development of plate tectonics, but the very active hydrologic regime on the ocean crust stood in their way as a major source of noise. However, they did discover the cause of this noise, and the oceanic hydrologic regime, the subject of this volume, has become a major field of study in its own right. Finally, in another twist, the studies presented in this volume on this regime have led to much greater confidence that, under certain sedimentary conditions, the mean value of seafloor heat flux (when properly averaged over a sufficiently large area) gives a reliable measurement of the deep conductive flux. Such measurements should enable the community to resolve better the effects of lithospheric cooling and ultimately to examine convection within the mantle, the original motivation behind the decision by Bullard (1952) to make systematic measurements of oceanic heat flux.

Acknowledgments

I would like to thank Dan McKenzie and Bullard Laboratories at Cambridge University for permission to use Fig. 1.1. Earl Davis encouraged me to write this chapter and helped keep the manuscript on focus once it was started. The original manuscript benefited greatly from a review by Ted Irving. Gustaf Arrhenius and Kurt Bostrom helped with information on the contribution of Hans Petterson.

References

Anderson, R. N. and Hobart, M. A. 1976. The relation between heat flow, sediment thickness, and age in the eastern Pacific. *J. Geophys. Res.* **81**: 2,968–2,989.

Anderson, R. N. and Skilbeck, J. N. 1981. Oceanic heat flow. In *The Sea*, Vol. 7, ed. C. Emiliani. New York: Willey-Interscience, pp. 489–524.

Anderson, R. N. and Zoback, M. D. 1982. Permeability, under-pressures, and convection through the oceanic crust near the Costa Rica Rift, eastern equatorial Pacific. *J. Geophys. Res.* **87**: 2,860–2,868.

Anderson, R. N., Hobart, M. A., and Langseth, M. G. 1979. Geothermal convection through the oceanic crust and sediments in the Indian Ocean. *Science*, **204**: 3,391–3,409.

Anderson, R. N., Zoback, M. D., Hickman, S. H., and Newmark, R. L. 1985. Permeability versus depth in the upper oceanic crust: in situ measurements, in DSDP hole 504B. *J. Geophys. Res.* **90**: 3,659–3,669.

Arrhenius, G. and Bonatti, E. 1965. Neptunism and volcanism in the ocean. In *Progress in Oceanography*, Vol. 3, ed. M. Sears. New York: Pergammon Press, pp. 7–21.

Bodvarsson, G. and Lowell, R. P. 1972. Ocean floor heat flow and the circulation of interstitial waters. *J. Geophys. Res.* **77**: 4,472–4,475.

Bonatti, G. and Joensuu, O. 1966. Deepsea iron deposits from the South Pacific. *Science* **154**: 717–730.

Bostrom, K. and Peterson, M. N. A. 1966. Precipitates from hydrothermal exhalations on the East Pacific Rise. *Econ. Geol.* **61**: 1,258–1,265.

Bullard, E. C. 1939. Heat flow in South Africa. *Proc. Roy. Soc. A* **173**: 474–502.

1952. Discussion of a paper by R. Revelle and A. E. Maxwell, heat flow through the floor of the eastern North Pacific Ocean. *Nature* **170**: 200.

1954. The flow of heat through the floor of the Atlantic Ocean. *Proc. Roy. Soc. London A* **222**: 408–429.

Bullard, E. C. and Day, A. 1961. The flow of heat through the floor of the Atlantic Ocean. *Geophys. J.* **4**: 282–292.

Bullard, E. C., Maxwell, A. E., and Revelle, R. 1956, Heat flow through the deep sea floor. *Adv. Geophys.* **3**: 153–181.

Crane, K. 1977. Hydrothermal activity and near axis structure at mid-ocean spreading centers. Ph.D. thesis, University of California, San Diego, p. 275.

Crough, S. T. 1978. Thermal origin of mid plate hot spot swells. *Geophys. J. Roy. Astron. Soc.* **55**: 451–469.

Davis, E. E. and Lister, C. R. B. 1974. Fundamentals of ridge crest topography. *Earth Planet. Sci. Lett.* **21**: 405–413.

1977. Heat flow measured over the Juan de Fuca Ridge: evidence for widespread hydrothermal circulation in a highly heat transportive crust. *J. Geophys. Res.* **82**: 4,845–4,860.

Dive and Discover 2002. The discovery of hydrothermal vents, 25th Anniversary 1977–2002. CD-ROM, Woods Hole Oceanographic Institution, Woods Hole, MA.

Elder, J. W. 1965. Physical processes in geothermal areas. *Am. Geophys. Union Monograph* **8**: 211–239.

Hess, H. H. 1962. History of ocean basins. In *Petrologic Studies: A Volume in Honor of A. F. Buddington*, eds. A. E. J. Engel, H. L. James, and B. F. Leonard. New York: Geological Society of America, pp. 599–620.

Heestand, R. L. and Crough, S. T. 1981. The effect of hot spots on the oceanic age–depth relation. *J. Geophys. Res.* **86**: 6,107–6,114.

Heezen, B. C. 1960. The rift in the ocean floor. *Sci. Am.* **203**: 98–110.

Irving, E., Park, J. K., Haggerty, S. E., Aumento, F., and Loncarovic, B. 1970. Magnetism and opaque mineralogy of basalts from the Mid-Atlantic Ridge at 45° N. *Nature* **228**: 974–976.

Isacks, B., Oliver, J., and Sykes, L. R. 1968. Seismology and the new global tectonics. *J. Geophys. Res.* **73**: 5,855–5,899.

Klinkhammer, G., Bender, M., and Weiss, R. F. 1977. Hydrothermal manganese in the Galapagos Rift. *Nature* **269**: 319–320.

Klitgord, K. D. and Mudie, J. D. 1974. The Galapagos Spreading Centre: a near bottom geophysical survey. *Geophys. J. Roy. Astron. Soc.* **39**: 563–586.

Langseth, M. G. and Herman, B. M. 1981. Heat transfer in the oceanic crust of the Brazil Basin. *J. Geophys. Res.* **86**: 10,805–10,819.

Langseth, M. G., Le Pichon, X., and Ewing, M. 1966. Crustal structure of the mid-ocean ridges, 5. Heat flow through the Atlantic Ocean floor and convection currents. *J. Geophys. Res.* **71**: 5,321–5,355.

Lee, W. H. K. and Uyeda, S. 1965. Review of heat flow data. In *Terrestial Heat Flow*, Geophysical Monograph Series, 8, ed. W. H. K. Lee. Washington, DC: American Geophysical Union, pp. 87–190.

Le Pichon, X. 1968. Sea floor spreading and continental drift. *J. Geophys. Res.* **73**: 3,661–3,697.

Lister, C. R. B. 1963. A close group of heat-flow stations. *J. Geophys. Res.* **68**: 5,569–5,573.

   1972. On the thermal balance of a mid-ocean ridge. *Geophys. J. Roy. Astron. Soc.* **26**: 515–535.

   1974. On the penetration of water into hot rock. *Geophys. J. Roy. Astron. Soc.* **39**: 465–509.

Lonsdale, P. 1977. Blustering of suspension-feeding macrobenthos near abyssal hydrothermal vents at oceanic spreading centers. *Deep-Sea Res.* **24**: 853–863.

Lupton, J. E., Weiss, R. F., and Craig, H. C. 1977. Mantle helium in hydrothermal plumes in the Galapagos Rift. *Nature* **266**: 603–604.

McKenzie, D. P. 1967. Some remarks on heat flow and gravity anomalies. *J. Geophys. Res.* **72**: 6,261–6,273.

McKenzie, D. P. and Parker, D. L. 1967. The North Pacific: an example of tectonics on a sphere. *Nature* **216**: 1,276–1,280.

McKenzie, D. P. and Sclater, J. G. 1969. Heat flow in the eastern Pacific and sea-floor spreading. *Bull. Volcanologique* **33–1**: 101–118.

Menard, H. W. 1959. Geology of the Pacific sea floor. *Experientia* **15**: 205–213.

Morgan, W. J. 1968. Rises, trenches, great faults, and crustal blocks, *J. Geophys. Res.* **73**: 1,959–1,982.

   1974. Heat flow and vertical movements of the crust. In *Petroleum and Global Tectonics*, eds. A. G. Fischer and S. Judson. Princeton: Princeton University Press, pp. 23–43.

Murray, J. and Renard, A. F. 1895. *Deep Sea Deposits. HMS Challenger Expedition Reports*. Edinburgh: Neill and Co., p. 525.

Nason, R. D. and Lee, W. H. 1962. Preliminary heat flow profile across the Atlantic. *Nature* **196**: 975.

Palmason, G. 1967. On heat flow in Iceland in relation to the Mid-Atlantic Ridge. In *Iceland and Mid-Ocean Ridges*, Vol. 38, ed. S. Bjornsom. Reykjavik: Soc. Sci. Islandica, pp. 11–27.

Parker, R. L. and Oldenburgh, D. W. 1973. Thermal model of ocean ridges. *Nature* **242**: 137–139.

Parsons, B. and McKenzie, D. P. 1978. Mantle convection and the thermal structure of the plates. *J. Geophys. Res.* **83**: 4,485–4,496.

Parsons, B. and Sclater, J. G. 1977. An analysis of the variation of the ocean floor bathymetry and heat flow with age. *J. Geophys. Res.* **82**: 803–827.

Petterson, H. 1949. Exploring the bed of the ocean. *Nature* **4168**: 468–470.

   1957. The voyage. Reports Swedish Deep Sea Exped., 1947–1948. *Goteburg* **1**: 1–123.

   1959. Manganes and nickel on the ocean floor. *Geochim. Cosmochim. Acta* **17**: 209–213.

Revelle, R. R. and Maxwell, A. E. 1952. Heat flow through the floor of the eastern North Pacific Ocean. *Nature* **170**: 199–202.

Sclater, J. G. 1966a. Heat flux through the ocean floor. Ph.D. thesis, Cambridge University, p. 124.

   1966b. Heat flow in the northwest Indian Ocean and Red Sea. *Phil. Trans. Roy. Soc. A* **259**: 271–278.

Sclater, J. G. and Crowe, J. 1979. A heat flow survey at Anomaly 13 on the Reykjanes Ridge: a critical test of the relation between heat flow and age. *J. Geophys. Res.* **84**: 1,593–1,602.

Sclater, J. G. and Francheteau, J. 1970. The implications of terrestrial heat flow observations on current tectonic and geochemical models of the crust and upper mantle of the earth. *Geophys. J. Roy. Astron. Soc.* **20**: 509–537.

Sclater, J. G., Jones, E. J. W., and Miller, S. P. 1970a. The relationship of heat flow, bottom topography and basement relief in Peake and Freen Deeps, northeast Atlantic. *Tectonophysics* **10**: 283–300.

Sclater, J. G., Mudie, J. D., and Harrison, C. G. A. 1970b. Detailed geophysical studies on the Hawaiian arch near 24° 25′ N, 157° 40′ W: a closely spaced suite of heat flow stations. *J. Geophys. Res.* **75**: 333–348.

Sclater, J. G., Anderson, R. N., and Bell, M. L. 1971. The elevation of ridges and the evolution of the central eastern Pacific. *J. Geophys. Res.* **76**: 7,888–7,915.

Sclater, J. G., Crowe, J., and Anderson, R. N. 1976. On the reliability of oceanic heat flow averages. *J. Geophys. Res.* **81**: 2,997–3,006.

Skornyakova, I. S. 1964. Dispersed iron and manganese in Pacific Ocean sediments. Lithology and mineral resources. *Internat. Geol. Rev.* **7**: 2,161–2,174.

Sleep, N. H. 1969. Heat flow, gravity and sea-floor spreading. *J. Geophys. Res.* **74**: 2,131–2,153.

Sykes, L. R. 1967. Mechanism of earthquakes and nature of faulting on the mid-oceanic ridges. *J. Geophys. Res.* **72**: 2,131–2,153.

Talwani, M., Windish, C., and Langseth, M. G. 1971. Reykjanes Ridge crest: a detailed geophysical study. *J. Geophys. Res.* **76**: 473–517.

Turcotte, D. L. and Oxburgh, E. R. 1967. Finite amplitude convective cells and continental drift. *J. Fluid Mech.* **28**: 29–42.

Vacquier, V. and Von Herzen, R. P. 1964. Evidence for connection between heatflow and the mid-Atlantic ridge magnetic anomaly. *J. Geophys. Res.* **69**: 1,093–1,101.

Vogt, P. R. and Ostenso, N. A. 1967. Steady state crustal spreading. *Nature* **215**: 810–817.

Von Gumbel, G. 1878. Ueber die im Stillen Ocean auf dem meeresgrunde vorkommended Manganknollen. Sitz Berichte d. K. Bayerisched Akademie d. *Matem-Physik Klasse*: Wissenschaften Munchen, pp. 189–209.

Von Herzen, R. 1959. Heat-flow values from the South-Eastern Pacific. *Nature* **183**: 882–883.

Von Herzen, R. P. and Uyeda, S. 1963. Heat flow through the eastern Pacific Ocean floor. *J. Geophys. Res.* **68**: 4,219–4,250.

Von Herzen, R. P. and Vacquier, V. 1966. Heat flow and magnetic profiles on the Mid-Indian Ocean Ridge. *Trans. Roy. Soc. A* **259**: 262–270.

Weiss, R. F., Lonsdale, P. F., Lupton, J. E., Bainbridge, A. E., and Craig, H. 1977. Hydrothermal plumes in the Galapagos Rift. *Nature* **267**: 600–603.

Williams, D. L. and Von Herzen, R. P. 1974. Heat loss from the earth: new estimate. *Geology* **3**: 327–328.

Williams, D. L., Von Herzen, R. P., Sclater, J. G., and Anderson, R. N. 1974. The Galapagos spreading center: lithospheric cooling and hydrothermal circulation. *Geophys. J. Roy. Astron. Soc.* **38**: 609–626.

Wilson, J. T. 1965. A new class of faults and their bearing on continental drift. *Nature* **207**: 343–347.

2

Foundations of research into heat, fluid, and chemical fluxes in oceanic crust

Harry Elderfield, Keir Becker, and Earl E. Davis

2.1 Introduction

The oceanic crust is the most extensive geologic formation on Earth; from a hydrologic perspective, it may be one of the most diverse. Young oceanic crust is extremely transmissive. A large majority of the heat lost by young lithosphere and large fluxes of many elements are transported to and from the oceans via ventilated hydrothermal circulation. Hydrothermal circulation, best known through black smoker venting within newly formed oceanic lithosphere at ridge crests, is not restricted to ridges. Rather, the inflow of cool seawater into the cracked permeable ocean floor, and the attendant cooling of the seafloor through mild heating of seawater is ubiquitous and characteristic of oceanic lithosphere of age 0–65 million years. Globally, this represents fully half of the area of the seafloor, or 35% of the total surface area of the planet. The heat budget available for hydrothermal circulation is approximately 11 terawatts, of which only 1.75 TW is due to the hot ridge-crest axial system. Given the contrast in temperature between axial and off-axis systems, this means that roughly a two-to-three-orders-of-magnitude greater amount of seawater circulates into and through the older part (1–65 Ma) of the oceanic lithosphere than through new oceanic lithosphere (0–1 Ma). The circulating seawater interacts in ways not yet well understood, and moderates the chemistry of the oceans and also alters the mineralogy and composition of the ocean crust.

Fluids are inferred to migrate through the upper crust at volumetric rates of tens of meters per year. The pattern of circulation may be changed by the accumulation of low-permeability sediments, but rates of flow may remain high; lateral flow in hydrothermal basement (the uppermost permeable part of the igneous crust) may serve to transport heat and solutes over distances of many tens of kilometers beneath the sediments. In old ocean basins, thick sections of sediments, along with diminished thermal buoyancy forces and reduced permeability in the igneous crust, may create highly isolated conditions, causing rates of fluid exchange between the ocean and lithosphere to diminish by many orders of magnitude. Near continents, fluids may be driven laterally by forces that originate from

Hydrogeology of the Oceanic Lithosphere, eds. E. E. Davis and H. Elderfield. Published by Cambridge University Press.
© Cambridge University Press 2004.

gravitational or tectonic compaction of sediments, by topographic head, or by chemically derived density contrasts. Such fluid flux may have important consequences in the way of sediment diagenesis and hydrocarbon migration and accumulation beyond the strict limits of continental margins. Conversely, the mere existence of high-permeability pathways in the oceanic crust adjacent to continental margins may influence the way that formation pressures develop, and exert a strong control over the mechanical behavior of accretionary prisms.

While a useful, semi-quantitative paradigm for oceanic hydrothermal circulation has emerged, much work is needed to understand the basic physics of the process better, to constrain the key parameters that control circulation over the full history from young to ancient lithosphere, and ultimately to determine integrated fluid fluxes. At the present time, geochemical fluxes are believed to be large, but they are poorly constrained; little is known about the depth to which thermally or chemically significant circulation penetrates and about the efficiency of water–rock interactions; and very little is known about just how isolated sedimentary and igneous sections become at great age. Discrete vents are difficult to find in off-axis environments, but they are known to occur, and they, like their counterparts at mid-ocean ridges, are known to support life based on chemosynthetic bacteria. It is also very possible that bacteria may be present ubiquitously at depth in the crust. These parts of the seafloor and sub-seafloor biosphere have only begun to be investigated.

Pilot studies have been completed in the Atlantic and Pacific Oceans that have included detailed transects and grids of heat flux measurements, sediment coring, pore-fluid geo-chemical and pore-pressure studies, and drilling observations, all placed in the context of observations of basement structure, sediment distribution, and crustal age. These have demonstrated the remarkable efficiency with which sediments hydrologically seal igneous oceanic crust locally on one hand, and the efficiency with which at least young crust can transmit water, and hence heat and chemical constituents, over great distances beneath the sediment seal on the other. They are also beginning to provide quantitative "calibrations" for reaction rates, for the "equilibrium" chemistry of basement water, and for the alteration products produced in the upper crust over a wide range of temperatures.

Applications of new knowledge will be critical for deriving accurate global heat and chemical budgets, understanding the contribution of off-axis hydrothermal circulation to seafloor and sub-seafloor biological communities, and assessing the viability or risk of using sub-seafloor formations as geologic repositories for hazardous waste. All these goals are fundamental and important from both scientific and societal points of view. The purpose of this review is to complement Chapter 1 of this volume, which describes the discovery of hydrothermal circulation in the oceanic crust, by summarizing the historical perspective of work at sites where focused hydrogeologic studies of the oceanic lithosphere have been carried out. Following this, we review the role of geochemical evidence that demonstrated lateral fluid flow in basement. Next we comment on "issues, themes, and problems" to highlight research directions that followed the increasing acceptance of the importance of wide-scale fluid flow within the oceanic lithosphere. Finally, we point to directions of current and future research, many of which are discussed in subsequent chapters, and suggest

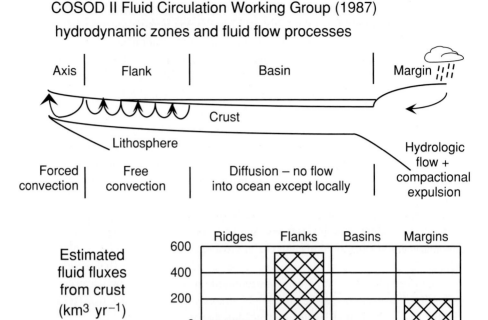

Fig. 2.1 Cartoon of fluid flow regimes in oceanic lithosphere and estimated rates of exchange between the oceans and the crust (after COSOD II, 1987).

a number of observational strategies and techniques that will be important in these efforts. In this review, we have concentrated on the ridge flanks, which are generally accepted to host by far the greatest proportion of total fluid flow through the seafloor, yet remain relatively under-studied (Fig. 2.1). Given present technology, ridge flanks also represent the best opportunities to utilize ocean drilling for hydrogeological experiments in the oceanic crust.

2.2 Historical perspective

While the significance of the hydrogeology of the oceanic crust was first made evident from reconnaissance geological and geophysical surveys through much of the ocean basins, the greatest advances in our understanding have undoubtedly arisen from detailed studies at selected "type examples." Most of these well-studied examples share one important characteristic – relatively thick sediment cover at a relatively young crustal age. This readily allows both detailed spatial mapping of the surface expressions of sub-seafloor hydrogeology using standard techniques such as heat flow and coring, as well as access to the sub-seafloor via scientific boreholes for sediment and basement cores, logging, and downhole measurements, and most recently, long-term *in situ* experiments.

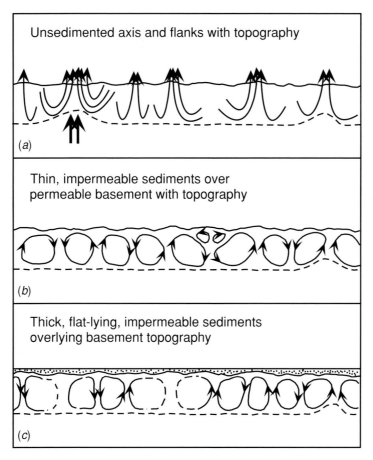

Fig. 2.2 Lister (1972) conceptual cartoon for patterns of convection for three type settings in oceanic crust: (*a*) bare-rock ridge crest and nearby flank; (*b*) thin, impermeable sediments (not shown) over-lying basement topography; and (*c*) thick, impermeable, flat-lying sediments overlying basement topgraphy.

2.2.1 The Juan de Fuca Ridge

The northernmost segment and eastern flank of the Juan de Fuca Ridge (JFR) are covered by thick turbidites shed from the nearby North American continental margin. Here, detailed heat flow surveys dating back to around 1970 (Lister, 1972; Davis and Lister, 1977) revealed the strong influence of fluid circulation and inspired some of the seminal concepts of hydrothermal circulation in oceanic crust (Fig. 2.2). The importance of co-located seismic surveying was firmly established here in understanding the influence of basement and basement structure on hydrothermal processes (e.g. Davis *et al.*, 1989, 1992; Section 2.4 below). Reconnaissance and detailed surveying has been carried out nearly continuously in the past two decades, focusing more and more on two key type examples: (a) the sedimented spreading segment at Middle Valley, and (b) the well-sedimented eastern flank of

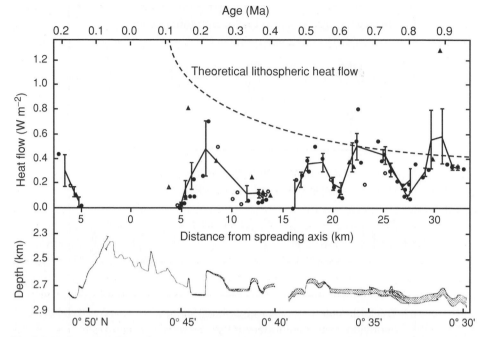

Fig. 2.3 N–S profile of heat flow, topography, and sediment thickness on the southern flank of the GSC at 86° W (after Williams *et al.*, 1974). On the heat flow profile, closed symbols are located within 2 km of the profile; open symbols are farther away. The solid line connects 2-km running averages of heat flow plotted every kilometer along the profile.

the Endeavour segment, where several classical type settings in terms of sediment-covered basement topography have been documented in detail (Davis *et al.*, 1989, 1996). Investigations at both sites have also included Ocean Drilling Program (ODP) drilling, logging, and long-term hydrological observatories (Legs 139, 168, and 169).

2.2.2 The southern flank of the Galapagos Spreading Center

On the southern flank of the Galapagos Spreading Center (GSC) at 86° W, pelagic sedimentation rates are very high, and detailed heat flow transects on young crust in the 1970s (e.g. Williams *et al.*, 1974; Green *et al.*, 1981) revealed that average seafloor heat flux reached values predicted for conductive cooling of the lithosphere at a very young age, ~1 Ma. These surveys also revealed a regular, sinusoidal variation with a similar wavelength to that of the underlying basement topography (Fig. 2.3). These observations inspired an influential set of numerical simulations of the cellular hydrothermal circulation system in young ridge flanks (Fehn *et al.*, 1983), with depths of circulation approaching the characteristic wavelength of the heat flow variation. This segment was also the site of the first discoveries of axial venting, albeit low temperature, as well as the best example of off-axis hydrothermal precipitate mounds. These were investigated with deep-tow surveys, submersible surveys, piston-coring, and the Deep Sea Drilling Project (DSDP, Legs 54 and 70). Profiles of pore-water chemistries and *in situ* temperatures determined on both piston cores and DSDP cores from the mound field show curvatures characteristic of both recharge and discharge through the sediments (Maris and Bender, 1982; Becker and Von Herzen, 1983). Since Leg 70 in 1979, there has been surprisingly little additional work done in this classic example of an off-axis hydrothermal system.

2.2.3 The southern flank of the Costa Rica Rift

On the southern flank of the Costa Rica Rift (CRR), pelagic sedimentation rates are nearly as high as at GSC, and a detailed heat flow transect in the 1970s revealed that the measured heat flux also reaches the predicted value at a young crustal age, ~6 Ma (Langseth *et al.*, 1983, 1988). Based on these hydrogeological survey results, it was surmised that the upper crust would be altered enough to be cored more easily than young crust elsewhere, and a DSDP site was established in 1979 for upper crustal hydrogeological studies – a hole that fortuitously became the deepest DSDP/ODP penetration into oceanic crust by far, Hole 504B. The multiple drilling revisits (Legs 69, 70, 83, 92, 111, 137, 140, and 148) have produced a wealth of reference information on the chemical and physical state of the upper crust, and have also inspired several return heat flow, coring, and seismic surveys, as well as complementary drill sites nearby (Fig. 2.4). Profiles of sediment pore-fluid chemistries indicate upwelling through 170 m of sediment above the basement high at Site 678 and downwelling through 310 m of sediment above the basement low at Site 677 (Mottl, 1989). Downhole experiments in Hole 504B still provide the most influential reference for the variation of porosity and permeability with depth in the oceanic crust (Fig. 2.5), and their relationship to seismic properties. For example, the results suggest a correspondence between seismic layer 2A and the most permeable section of upper basement (e.g. Becker *et al.*, 1989), a relationship that is now applied in other settings to infer sub-surface hydrologic structure from seismic surveys. Another important example is the correlation of the seismic layer 2/3 boundary not with a lithologic boundary, but instead with porosity and alteration (Detrick *et al.*, 1994; Alt *et al.*, 1996). The strong decrease in permeability and porosity with depth in Hole 504B is considerably different than the permeability structure used in early GSC-inspired numerical simulations, and has figured prominently in a revival of numerical models of off-axis circulations also inspired by JFR results (e.g. Fisher *et al.*, 1990, 1994).

2.2.4 Young crust formed at slow-spreading rates

In crust formed at slow-spreading rates, basement topography is generally very pronounced, and the typical off-axis hydrogeological setting comprises isolated sediment ponds surrounded by extensive basement exposures. The best-surveyed example is North Pond, in 7 Ma crust on the west flank of the Mid-Atlantic Ridge. This is the setting of Hole 395A,

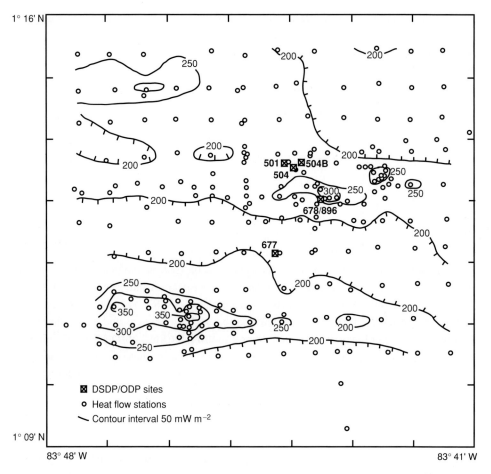

Fig. 2.4 Detailed heat flow survey (from Langseth *et al.*, 1988) and locations of DSDP/ODP sites on the southern flank of the Costa Rica Rift.

another key reference hole in upper oceanic crust, which has been revisited by DSDP and ODP several times (Legs 78B, 109, 174B) and a program of submersible-deployed wireline observations (Gable *et al.*, 1992) for a suite of logging and downhole measurements almost as complete as in Hole 504B, as well as a long-term hydrological observatory currently in operation (Becker *et al.*, 2001). Long after the hole was drilled, a detailed heat flow survey (Langseth *et al.*, 1992) was conducted over the full expanse of the sediment pond (~6 × 15 km). The results, in combination with borehole measurements, supported an earlier interpretation involving vigorous single-pass flow of ocean bottom water through the permeable upper basement that underlies the less permeable sediments (Fig. 2.6; Langseth *et al.*, 1984).

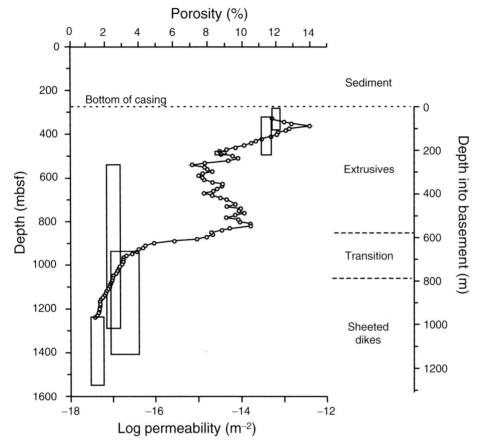

Fig. 2.5 Depth profiles of bulk permeability and apparent porosity from resistivity in Hole 504B (from Becker *et al.*, 1989).

2.2.5 *The eastern equatorial Pacific*

The regional heat flow low beneath the thick carbonate sediments in the eastern equatorial Pacific was one of the most striking and laterally extensive anomalies mapped in early reconnaissance heat flow surveys (Sclater *et al.*, 1976). Explaining such an anomaly seemed to require fluid flow in crust older than the examples above, but follow-up heat flow surveys and coring did not show coherent variations on the spatial scales suggested by the examples above. Geochemistry of pore waters from cores collected from the region showed basement pore-fluid compositions similar to bottom water, however (Fig. 2.7), and finally suggested the explanation: lateral flow in basement at regional scales comparable to the size of the anomaly (Baker *et al.*, 1991). This example is discussed in more detail in the geochemistry section below. As detailed evidence for laterally extensive basement fluid flow in the younger

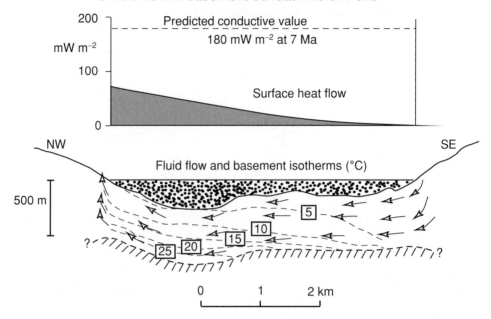

Fig. 2.6 Conceptual model of fluid flow in basement beneath North Pond and resultant seafloor heat flux (after Langseth *et al.*, 1984).

type examples has accumulated, this explanation has become more accepted despite the great scales of circulation required.

2.2.6 Middle-aged and old crust

In contrast to young seafloor environments, few detailed studies have been carried out in crust older than 10 Ma. Most inferences concerning the magnitude of hydrothermal heat and chemical fluxes in such settings have been made on the basis of relatively widely spaced heat flow observations and core samples, and with modest to poor control from seismic reflection data. Most studies in old areas have been carried out to determine regional average lithospheric heat flux, not to study crustal fluid flow. In fact, effects of fluid flow were to be avoided at all costs in many of these cases. However, instances of locally anomalous values of heat flux (i.e. values that cannot be accounted for by bottom-water temperature transients, sedimentation effects, or conductive focusing of heat through heterogeneous structure) occur with surprising frequency. Some important examples include 18–55 Ma Indian Ocean crust (Anderson *et al.*, 1977, 1979; Geller *et al.*, 1983), 20 Ma Brazil Basin crust (Langseth and Herman, 1981), and 80 Ma northwest Atlantic crust (Embley *et al.*, 1983; Fig. 2.8). While these anomalies are often subtle, their presence, even in seafloor of

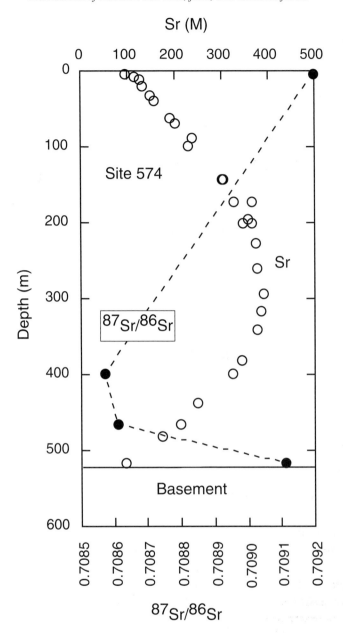

Fig. 2.7 Strontium concentrations and isotope ratios in pore-fluids from DSDP Site 574 (after Baker *et al.*, 1991).

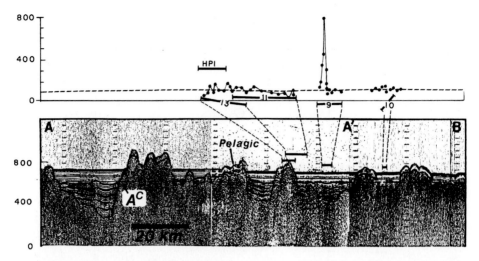

Fig. 2.8 W–E heat flux and airgun seismic profiles across the eastern Nares abyssal plain (after Embley *et al.*, 1983). The sharp heat flux high of Profile 9 occurs at a domed feature interpreted as a volcanic structure that focuses fluid flow from basement.

Cretaceous or Jurassic age, warrants further investigation. Just as in younger settings, the anomalies are often associated with seamounts or other crustal structures that are thinly buried by sediment or exposed in outcrop. The possibility of significant fluid flow in old ocean basins is also supported by high measured permeability in a borehole drilled into the upper crust of the oldest ocean basin in the Pacific (Larson *et al.*, 1993).

2.3 Geochemistry

About the time that geophysicists noted the important contribution from the ridge flanks to the global heat loss inventory, geochemists surmised that reactions between seawater and the oceanic crust were required to balance global geochemical budgets (Garrels and Mackenzie, 1971). These reactions include early diagenetic reactions that occur in the upper few meters of the sediment column (e.g. Sayles, 1979) as well as chemical exchange in basaltic basement (e.g. McDuff, 1981). In each case the geologic setting supported only diffusion as the mode of chemical transport between the crust and the oceans. The potential effects of advective transport on global geochemical budgets were not quantified until Edmond *et al.* (1979) published their calculations for hydrothermally advected chemical fluxes from mid-ocean ridge axes. These calculations spawned the idea that chemical fluxes from advective hydrothermal transport on ridge flanks may be important to global cycles; because much of the global advective heat loss occurs on the flanks at a much lower temperature than on the axis, the mass flux of seawater through the flanks must be much greater than that through the ridges. Thus, even a small chemical anomaly in hydrothermal fluids from the flanks could result in a significant global geochemical flux. At present the magnitude of these chemical fluxes is unknown, but geochemists are approaching the problem by either

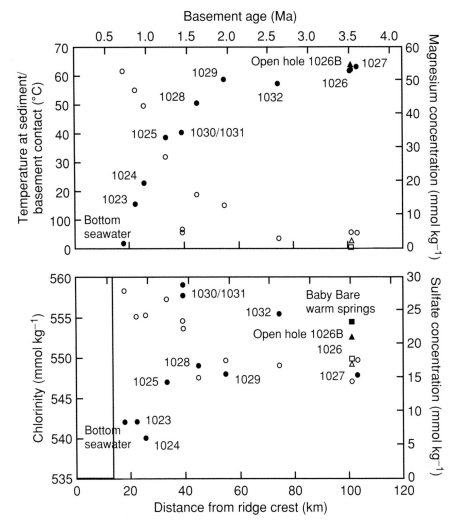

Fig. 2.9 Variations of basement temperatures and pore-water compositions at Leg 168 sites on the east flank of the Endeavour segment, Juan de Fuca Ridge. Solid symbols are temperatures in upper plot, chlorinity values in lower plot (after Davis *et al.*, 1997).

examining products of crustal alteration (e.g. Thompson, 1983) or systematic changes in the composition of crustal fluids (e.g. Mottl and Wheat, 1994; Elderfield *et al.*, 1999; Fig. 2.9 after Davis *et al.*, 1997).

Geochemistry can be used to study hydrogeologic processes using two approaches. One is through observations of altered solid phases (the aquifer) as a way to understand flow paths, hydrologic variables, and mass fluxes between the crust and seawater. The second approach is through observations of the fluid phases (liquids and gases) as a way to map flow paths and estimate rates of flow. Chemical alteration of the fluids is often a more sensitive indicator of the extent of fluid–solid reactions than alteration of the solid, because the concentrations of most elements are lower in the fluid than in the solid phases, but mass balance calculations require information on both fluid and solid compositions. Chemical and mineralogical alteration of solid phases directly influences hydrogeology, however, through its control of fundamental hydrological variables such as porosity and permeability.

Geochemistry provides a powerful tool to study the ties between microbiology, hydrogeology, and geology. These sub-disciplines are linked by inorganic and microbially mediated reactions that can alter chemical compositions of fluids and the chemical and mineralogical compositions of solids. Ocean drilling has shown that microbes can live in a wide range of environments in the ocean crust and to depths of hundreds of meters within ocean bottom sediments. Both microbial and non-microbial reactions change the chemical composition of fluids and alter rock and sediment compositions and mineralogies. Such reactions have been observed as variations in sedimentary pore-water compositions for decades (Claypool and Kaplan, 1974), and recent evidence has also been found for extensive microbial alteration of volcanic rocks (e.g. Fisk *et al.*, 1998). Conversely, fluid, rock, and sediment compositions may control the distributions of microbial species and their ecology. Deviations from seawater values of the chemical compositions of fluids can provide valuable information on the types of reactions including microbially mediated and non-microbial reactions that might have occurred. By understanding the thermodynamic conditions (temperature, pressure, and compositions) required for these reactions, it is possible to identify the origins of the fluids. Variations in fluid compositions thus can provide natural chemical tracers for fluid pathways.

2.3.1 *Fluid chemistry*

Chemical and isotopic tracers of fluid flow can be divided into several broad categories depending on the chemistry of the fluid–solid reactions, the solutes involved, and the extent of changes of the solute concentrations. Studies of the geochemistry of many of these tracers made early contributions to understanding hydrogeology of the oceanic lithosphere and show how they may be useful for future studies. Certain elements, such as Mg, Ca, and Sr are sensitive to basalt–seawater interaction, and the combination of changes in Sr concentrations and isotope ratios can be used to quantify the extent of alteration of basaltic crust and resulting mass fluxes between crust and seawater. Other elements (e.g. Cl, Br) are commonly assumed to behave conservatively in reactions between basalt and seawater. If these elements are conservative, changes in their concentrations can be used to determine mixing proportions between various water sources. Under certain circumstances, these elements may not behave conservatively. For example, measurements of Br/Cl ratios in pore water trapped in volcaniclastic sediments and deviations from seawater values of Cl isotope ratios indicate that these elements may take part in reactions involving volcanic material (Magenheim *et al.*, 1992; Martin, 1999). These reactions can provide new tools to look at the reactions and new tracers for flow.

Some dissolved components (SO_4, NO_3, H_2S) are sensitive to reactions involving organic matter and are important tracers of fluid–solid reactions and/or flow paths within the sedimentary section of the ocean crust. These components are largely controlled by microbial reactions and thus changes in their concentrations can be used to identify regions and extent of microbial activity. The alteration of organic matter within sedimentary sections, whether caused by thermal degradation or by microbially mediated reactions, can generate overpressured zones because of the volume expansion associated with the conversion from solid to liquid or gaseous hydrocarbons (Bredehoeft *et al.*, 1994; Martin *et al.*, 1997). Such overpressure has important implications for hydrogeology including providing a driving force for fluid flow in low-permeability sections, changing the rheology of sediments, varying the porosity structure, and increasing permeability through hydrofracturing.

Reactive solutes have also been used to document large-scale convection within the well-sedimented basaltic crust of the eastern equatorial Pacific (Baker *et al.*, 1991). Although diagenetic alteration changes the Ca, Mg, Sr, and SO_4 concentrations and $^{87}Sr/^{86}Sr$ ratio in pore water of the sedimentary section from seawater values, the composition of pore water at the sediment–basalt interface returns to modern seawater values (Fig. 2.7). The observed compositional profiles indicate that flow velocities are on the order of $1–10$ m yr^{-1} through the basement rocks, although the calculations require numerous assumptions including the locations of discharge and recharge.

Distribution of some radioactive elements provides chronometers as well as tracers for fluid origins. Concentrations of noble gases may be used to trace flow paths. Radiocarbon dating may be used to define fluid ages. Uranium series isotopes and tritium are also used for these purposes. For example, ^{226}Ra activities have been used to estimate groundwater fluxes to the South Atlantic Bight (Moore, 1996) and radium could be useful in other sedimented marine settings. The radium isotope quartet (^{223}Ra, ^{224}Ra, ^{226}Ra, and ^{228}Ra) provides four isotopes with half-lives ranging from 3 days to 1,600 years. Ratios of these isotopes could provide constraints on the length of time and extent of fluid vented from sedimentary sections.

Another class of fluid–solid reactions that is linked to flow involves gas hydrates, an ephemeral solid phase formed of gas molecules surrounded by a solid cage of water molecules. Dissociation of gas hydrate releases large volumes of gas, thereby generating excess pressures, similar to the microbial and thermal degradation of solid organic matter. Gas hydrates are largely confined to sedimentary sections which are rich in organic matter or where sediment pore-fluid migration is driven by tectonic forcing, and are thus unlikely to be common on ridge flanks. Sedimentary sections along continental margins, particularly those of subduction-zone accretionary prisms, are known to be important locations of gas hydrates.

2.3.2 Crustal alteration

The chemical composition of hydrothermal vent fluids provides some information about mass fluxes from the crust to the oceans, but they provide little information about the flux of mass from seawater to the crust. There is little question that diffuse low-temperature flow contributes significantly to crustal alteration, but the extent and distribution of this flow mechanism is poorly constrained. Recent evidence also suggests that much of the alteration

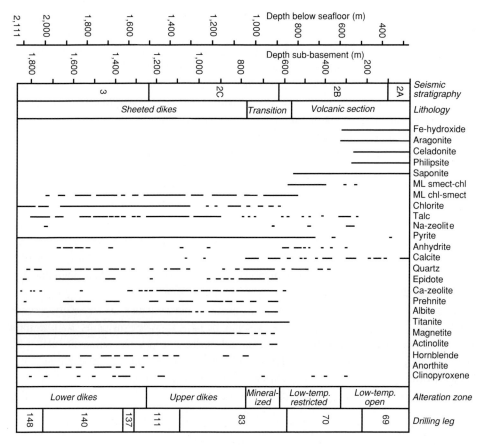

Fig. 2.10 Crustal alteration assemblage ODP Hole 504B (from Alt *et al.*, 1996).

in the shallow portions of the crust depends on microbial activities. Sampling of the altered rock matrix is critical, but because crustal alteration is heterogeneously distributed, it is difficult to observe. Even with good recovery, boreholes provide a tiny sample when the scale of heterogeneity is large. This problem is compounded by the fact that current drilling techniques recover only a small fraction of the drilled material, and that fractured units and volcanic glass, which constitute the most altered material, are poorly recovered in drill cores. Compositions of both altered and unaltered rocks are required to calculate the extent of alteration during seawater–basalt reactions. The depth of this alteration and the extent of the alteration at great depths require better recoveries of material, and an improved means to assess the representativeness of the samples.

The type of crustal alteration provides information about the distribution and timing of flow. The distribution of alteration products (e.g. celadonite and other clay minerals, micas, oxyhydroxides, pyrite, carbonate vein minerals) reflects the distribution of oxidation/reduction reactions in the crust (Fig. 2.10). The intensity of the alteration indicates at

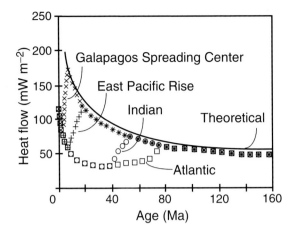

Fig. 2.11 Composite of average heat flow versus age for mid-ocean ridges (after Anderson *et al.*, 1977).

an order-of-magnitude scale the volume of water required to generate the alteration of the crust. These solid interactions provide an integrated picture of flow that when coupled with the instantaneous picture from fluid compositions can perhaps constrain the evolution of the entire hydrogeologic system.

2.4 Issues, themes, and problems

The working model of crustal hydrothermal circulation that emerged 30 years ago to explain the large scatter in marine heat flux observations, and the discrepancy between average seafloor heat flux at a given age and that expected from the cooling of oceanic lithosphere, remains generally applicable today. The model involves flow hosted by the permeable part of the igneous oceanic crust (Lister, 1972). In young areas, water is exchanged freely between the crust and the oceans. This exchange is progressively inhibited with the accumulation of low-permeability sediments, and the circulation internal to the crust is retarded by mineralization that accumulates in fractures and voids as a result of the fluid circulation. (The word "sealing" has been used in the literature in reference to both of these independent processes.) This model is consistent with a wide variety of observations, although much work is required to improve the quantification of a number of its primary aspects.

2.4.1 Distribution and nature of crustal flow

Marine hydrogeological studies in the last two decades, and particularly in the last ten years, indicate that lateral fluid flow in basement over distances of kilometers to tens of kilometers is common. That the pattern of measured (conductive) heat flux on ridge flanks is often well below expected values until the crust reaches an age of about 65 Ma (Fig. 2.11;

Anderson *et al.*, 1977; Stein and Stein, 1984) is robust evidence that large amounts of fluid must move through basement at velocities on the order of meters per year or more within much of the seafloor, and that the rate of exchange between the crust and oceans is large. As noted above, in some old crustal settings, heat flow surveys have revealed variations, often correlated with basement relief, that seem to require thermally significant fluid flow within oceanic basement continuing out to great ages. Borehole packer measurements in Jurassic basement, some of the oldest remaining *in situ* oceanic crust, indicate permeabilities similar to those measured in 3.5–7 Ma crust in other areas.

The driving forces available to move fluids laterally within oceanic basement on ridge flanks appear to be limited to pressure differences between warm and cool columns of hydrothermal fluid, and this requires that effective basement permeabilities be relatively high. Resolving the depth-scales of flow within basement of ridge flanks is difficult, but results are important; the deeper flow penetrates, the more efficient is the consequent "mining" of heat from the crust, provided there are high-permeability pathways that allow discharge and recharge. A small number of tests in boreholes that have penetrated into crust below the upper extrusive layer have yielded relatively low permeabilities, although temperature observations suggest thermally significant circulation and higher bulk permeabilities. New technologies and experiments will be required to resolve the depth distribution of permeability in oceanic basement, its variation with scale, and hence the degree of access of water to various parts of the crust.

2.4.2 *Roles of sediment cover and basement topography*

Numerous studies of oceanic sediment permeability have demonstrated that, although this property is highly variable, sediments tend to have permeabilities that are one to several orders of magnitude less than that of upper oceanic basement. Because the available driving forces are modest, even a few tens of meters of sediments are sufficient to reduce fluid discharge to levels that are below the thermal detection limit (millimeters per year). Seepage fluxes of millimeters per year are sufficient to influence the chemistry of the fluid and surrounding sediment, but certainly comprise a tiny fraction of the total fluid flow that is required to pass through the crust to result in the observed suppression of heat flow and to maintain the seawater-like composition of formation fluid often found in basement.

Basement relief commonly correlates with observed parameters (e.g. Fig. 2.12) and may play a critical role in ridge flank hydrogeology in several ways. First, basement highs tend to accumulate sediments at a lower rate than do basement lows, particularly near continental margins where turbidites are common. Thus basement highs may remain "open" to fluid exchange for a much longer time than the surrounding seafloor. Second, basement relief is often associated with faulting at the edges of abyssal hills, and these faults are commonly associated with depressed or elevated heat flow, interpreted to indicate fluid recharge or discharge, respectively. Why some faults tend to focus recharge and others focus discharge is not well understood, but is the focus of ongoing field and numerical research. Finally,

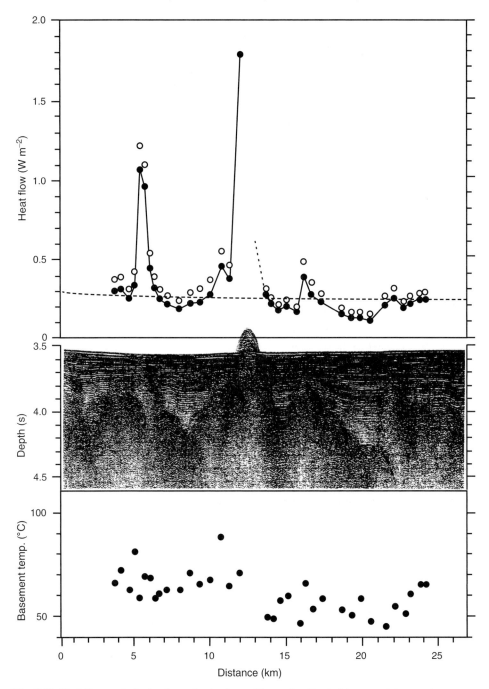

Fig. 2.12 Heat flow and single-channel seismic profiles and estimated basement temperatures on a transect over a basement "permeable penetrator" on the sedimented flank of the Juan de Fuca Ridge. In the heat flow plot, solid circles are raw values, open circles are corrected for sedimentation, and the dashed line represents the theoretical lithospheric heat flow (after Davis *et al.*, 1989).

basement relief results in the formation of tilted boundaries at the top and within the basement "aquifer," inducing instability (essentially reducing the critical Rayleigh number required to initiate hydrothermal circulation to zero). Localized convection influenced by topography can homogenize basement temperatures and fluid chemistry, and may play a role in compartmentalizing hydrothermal flows in ridge flanks.

2.4.3 Role of stratified basement permeability

High permeabilities (on the order of 10^{-13} m^2, or 0.1 darcies or more) have been measured within upper oceanic crust in several settings, and the highest permeabilities have been measured where the youngest holes (close to 1.0 Ma) have been tested. But most of these measurements have been made in the upper 100–200 m of basement, and the tests have not allowed delineation of the vertical distribution of permeability with confidence, other than on a relatively gross scale through the sheeted dikes in Hole 504B. However, other indicators of crustal layering and heterogeneity (lithologic and alteration variations, cycles in resistivity logs, differences in the density of fractures) suggest that there may be distinct permeability compartments within oceanic crust. The vertical scales of compartmentalization have not been determined, but could be as small as several tens of meters.

2.4.4 Evolution of crust and hydrogeological systems

Compilations and surveys of seismic velocities in uppermost oceanic crust (Carlson, 1998; Grevemeyer *et al.*, 1999) show a clear trend: velocity increases rapidly during the first few million years, then more slowly in older crust. Density calculations from near-bottom gravity measurements also show a rapid increase in the first 1 Ma (Holmes and Johnson, 1993), although wireline logs in older boreholes suggest only small additional increases with age. Until recently, borehole packer measurements were restricted to crustal ages <7 Ma and did not indicate any permeability evolution trend; instead, these data suggested a narrow range for uppermost oceanic basement, 10^{-13}–10^{-14} m^2. Borehole measurements are now beginning to reveal a trend in permeability evolution within uppermost basaltic basement (Fig. 2.13). Measured bulk permeability decreases rapidly as young crust ages, from $\sim 10^{-10}$ to 10^{-13} m^2 (100 to 0.1 darcies) over the first 3–4 Ma of evolution (Becker and Fisher, 2000; Fisher and Becker, 2000). Thus measurements of physical properties are consistent with initially rapid pore infilling by mechanical and magmatic processes during the earliest stages of crustal evolution, and by slower diagenetic processes as aging continues (e.g. Wilkens *et al.*, 1991; Jacobson, 1992).

As described earlier, global heat flow compilations appear to conflict with this rapid aging model because they suggest that advective heat loss continues on average out to 65 Ma despite the early permeability reduction. One possible explanation for this discrepancy is that thermally significant fluid flow in ridge flanks older than a few million years

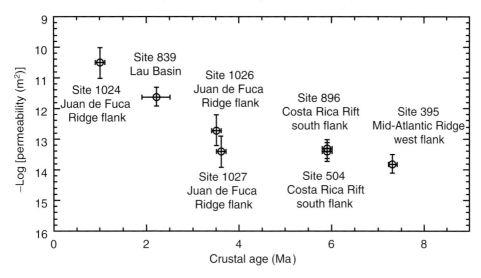

Fig. 2.13 Permeabilities measured in uppermost oceanic basement versus crustal age (Reprinted from Becker, K. and Fisher, A. 2000. Permeability of upper oceanic basement on the eastern flank of the Endeavour Ridge determined with drill-string packer experiments. *J. Geophys. Res.* **105**: 897–912, Copyright 2000 American Geophysical Union.)

is largely restricted to widely spaced channels and layers. Additional specific studies are needed, however, to resolve how permeability and flow are channelized and compartmentalized, how long flow persists in oceanic crust, and ultimately just how isolated certain sections become.

2.4.5 Global fluxes

Interaction of seawater with oceanic crust is a major mechanism for the transport of heat from the Earth's interior to the ocean. It also plays an important but poorly quantified role in geochemical budgets. High-temperature hydrothermal circulation at the axes of mid-ocean ridges is the most spectacular manifestation of this process and is thought to play a significant role in defining ocean chemistry. But the advective heat loss through the ridge axis is less than one-third of that at ridge flanks, and because this heat is lost at lower temperatures, the associated total water flux must be much greater than at ridge axes (Elderfield and Schultz, 1996; Schultz and Elderfield, 1997).

Extensive lateral fluid flow through oceanic crust off-axis on ridge flanks has been identified from heat flow and geochemical anomalies, yet there is still a very poor appreciation of the lateral scales and rates at which fluids and heat can be transported within oceanic crust and the integrated geochemical significance of these processes. Similarly, there is poor understanding of integrated fluxes in other characteristic hydrologic regimes (Figs. 1.1 and 2.1),

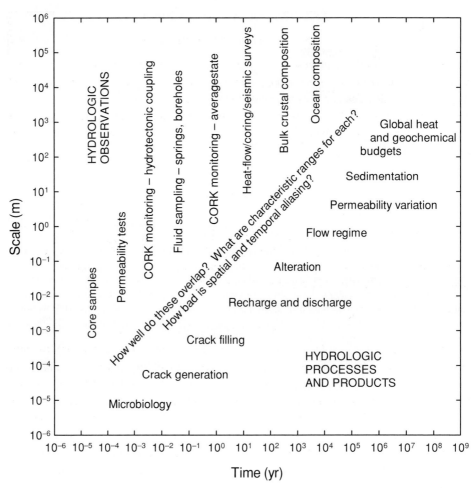

Fig. 2.14 Cartoon illustrating space–time scales of just some of the hydrologic processes and products in the oceanic lithosphere and of the observational techniques available to study them. No attempt has been made to place the categories in any "proper" characteristic position or to show the broad ranges of time and space scales that many of the processes and observational techniques span. This and the inherent limitations of the observational techniques are left to the reader to ponder.

with the net result that a great deal still needs to be done to quantify global fluid, heat, and chemical fluxes through ocean crust and sediments. Determining these integrated fluxes is one of the most important and exciting challenges in the hydrogeology of the oceanic lithosphere; but it is one of the most difficult to approach, largely because of the range of time–space scales of the processes that contribute to the integrated fluxes coupled with limitations in the time–space resolution of the techniques available (Fig. 2.14). Limitations like those represented in Fig. 2.14 apply to all the problems in this section, but they are especially significant for determining global budgets.

2.5 Directions for future research

2.5.1 Observational techniques

Understanding hydrologic processes, in which fluids move through or are confined within, and react with a host rock formation, requires determination of formation parameters, such as porosity, permeability, and constituent compressibilities, and the state of the formation, including pressure, head, temperature, stress, and composition. This must be done via a combination of techniques including remote imaging, *in situ* measurements, rock and fluid sampling, and monitoring.

Site selection

The most efficient approach to studying processes must begin with choosing sites with care, with emphasis being placed on simplicity. The greatest understanding of a particular process comes from the study of simple, and not necessarily typical, sites. Study sites should host representative processes so that results can be generalized, but excess complexity must be avoided as this can easily lead to ambiguous results.

Imaging methods

Basic hydrologic structure can be mapped in three dimensions using an appropriate combination of seismic reflection profiling and swath acoustic imagery (bathymetry and backscatter). This task is made easy in ocean crustal environments because the first-order structure is established by the upper igneous rocks and the sediments that cover them. These units differ greatly in both acoustic and hydrologic properties. Faults are also well imaged by seismic and acoustic techniques in instances where there is a history of syn-sedimentary motion. Constraints on porosity, and to a limited degree permeability, can be had from seismic velocity and gravity data. Defining deep lithologic and tectonic structure may require multi-channel seismic reflection profiling and refraction. Greater resolution and/or complementary information about hydrologic structure and parameters can be provided by near-bottom or seafloor seismic surveying, seafloor compliance, and electromagnetic profiling.

Rates of fluid flow

Constraints on thermal structure, fluid composition, and, by inference, rates of fluid flow can be gained efficiently through heat flow measurements and core samples (squeezed for their pore fluids) collected at a density appropriate for the expected scale of spatial variations. Where flow through the seafloor is suspected or identified through these or other methods (e.g. chemical and/or thermal water-column profiling), direct measurements of rates of flow and samples of fluids can be had using "benthic-barrel" devices that employ samplers driven by osmotic pumps, and flow meters that utilize mechanical, thermal, or chemical-tracer sensors.

Borehole studies

While much can learned about marine hydrogeologic systems remotely (much more efficiently than in studies of onshore systems), at some point further progress requires borehole
observations. With this step, many of the basic techniques used on land become available
for sub-seafloor studies, although a somewhat different approach must be taken since there
is a strong limit on the number of holes that can be drilled in pursuit of any specific goal.
Although improvements to a number of tools are needed, there is a broad range of tools
available. Formation testing is most commonly done with a drillstring packer that isolates
the interval from the packer to the bottom of the hole. While efficient and flexible, wireline
packers and straddle packers have been used with only limited success. In cases where
permeability is high, a combination of pumping while logging with a flow meter has been
used to constrain the distribution of permeability. Time constraints usually preclude testing
for more than a few hours, and with only minor exceptions, pumping and monitoring have
been done in the same hole. In the future, cross-hole experiments will allow permeabilities
to be determined for a much larger volume of rock, and storage properties to be inferred,
not assumed.

Just as in the case of seafloor and remote observations, it is important to integrate as
much information as possible to characterize a hydrologic system that has been penetrated by drilling. Toward this end, downhole logs are critical. Resistivity, porosity, seismic
velocity, self-potential, electrical resistivity imagery (formation microscanner) and acoustic
imagery (televiewer) logs all provide valuable constraints on the nature and heterogeneity
of formation parameters.

Ground truthing

Core samples also provide information about formation parameters, although core recovery
is usually far from complete in fractured rock, and the heterogeneity of many parameters creates a scale dependence that makes data collected from cores difficult to interpret.
Samples provide excellent information about current and past fluid flow and fluid–rock
interaction, however. Information is provided by the composition of alteration minerals,
vein-filling minerals, fluid inclusions, and interstitial pore fluids.

Long-term monitoring

While formation parameters can be measured or inferred from observations made at the
time of drilling, accurate observations of the state of the formation are nearly impossible
then because of the large perturbations generated by drilling. Temperatures can be measured
in low-permeability sediments with probes that extend below the bit, but measurements are
not possible in permeable units. Open holes create hydrologic shunts between the formation
and the water column, natural and drilling-induced pressure differentials cause flow, and as
a result, natural pressures, temperatures, and fluid compositions are usually impossible to
determine once permeable horizons are intersected. To overcome this problem, holes must

be sealed and left to re-equilibrate with long-term monitoring instrumentation left in place to observe the equilibrium state. Observations made in a number of sealed and instrumented holes demonstrate that even when holes are sealed very soon (a few days) after drilling, thermal, pressure, and compositional perturbations can last for years.

In addition to providing a means of observing the equilibrium state of a formation, long-term monitoring in sealed holes allows natural variations to be documented. A variety of natural variations have been observed that are associated with atmospheric, oceanographic (e.g. tides), and tectonic (earthquake stress change) loading. These signals allow large-scale hydrologic and elastic properties of formations to be determined, and the variations in stress related to seismic and aseismic slip to be witnessed. The most recent advances in monitoring capabilities allow multiple levels to be isolated for pressure monitoring and fluid sampling.

One of the lessons learned thus far through drilling and post-drilling observations in oceanic crust is that buried or partially buried basement edifices serve as thermal "chimneys" and as such are "over-pressured" relative to local hydrostatic conditions. Holes drilled into these edifices provide a means by which crustal fluids can be naturally produced at the seafloor for biological and chemical monitoring and sampling, and provide a situation where drilling-induced thermal and chemical perturbations are minimized.

While many of these various tools and techniques exist, there is a need to improve or expand the capabilities on a number of fronts in order to make hydrogeologic investigations more fruitful and efficient. A few examples include: (a) expanded borehole packer technology, (b) refined pumping hardware, (c) use of tracer testing, (d) execution of cross-hole hydrologic and geophysical experiments, (e) development of *in situ* biological and chemical analyzers, (f) development of autonomous vehicles for water column profiling and seafloor heat flow measurement, and (g) deep drilling.

2.5.2 Observational strategies

The type examples described in Section 2.2 share one distinguishing characteristic that is not representative of most of the oceanic lithosphere: unusually thick sediment accumulations in relatively young settings. In addition, the first three examples, which arguably include the two best-studied cases (JFR and CRR), are all in crust formed at medium spreading rate (3 cm yr^{-1} half-rate) and thus are unrepresentative of the large majority of ocean crust formed by slow and fast ridges. Therefore, while these examples are very important as process-oriented studies, they are not necessarily representative for estimating global fluxes.

Hence, a dual strategy is required for future studies: continuing detailed, process-oriented studies at a very few select sites, and at the same time conducting reconnaissance studies at other "representative" sites to quantify global fluxes. This is not a new strategy, but builds on recommendations made in the past in both COSOD (1981) and COSOD II (1987) reports and in several workshop reports (e.g. Purdy and Fryer, 1990).

Examples of "representative sites" for global fluxes would include the slow- and fast-spread, thinly sedimented crust at a range of ages that is typical of much of the ocean floor, as well as examples of important unusual settings such as large igneous provinces, carbonate platforms, old ocean basins, and regions of lithospheric flexure. Minimum requirements for reconnaissance surveys would include single-channel seismics coordinated with multi-penetration heat flux measurements and coring (both conventional coring and exploratory ocean drilling) for multi-disciplinary analyses (e.g. fluid and solid geochemistry, microbiology).

As reconnaissance sites become better studied, some might provide justification for further intensive study as type examples. In particular, a need exists for understanding processes at intensive-study sites in normally sedimented slow- and fast-spread crust and in old crust. These are critical in order to improve our currently poor understanding of the importance of low-temperature ($\sim$10 °C) circulation in crustal alteration and ocean–crust biogeochemical exchange, and to explore just how isolated are some parts of the sub-seafloor.

Finally, there is a particular need for scientific ocean drilling in both reconnaissance and intensive studies of the hydrogeology of the oceanic lithosphere. To achieve the objectives summarized in this chapter, future drilling will require development of improved core and formation-fluid recovery, expanded microbiological capabilities, refined shipboard hydrologic measurements, and sophisticated long-term *in situ* experiments.

Acknowledgments

The basis for this review is a report to the International Lithosphere Program (ILP) by H. E. and E. D. of a Workshop entitled "Hydrogeology of the Oceanic Lithosphere" convened by K. B., E. D., H. E., and Jon Martin and hosted by Andy Fisher at The University of California, Santa Cruz, USA, December 11 and 12, 1998, with sponsorship from Joint Oceanographic Institutions/US Science Support Program (JOI/USSSP) and ILP. We thank the participants and sponsors for their important contributions, especially those of our co-convenor Jon Martin.

References

Alt, J. C., Laverne, C., Vanko, D. A., Tartarotti, P., Teagle, D. A. H., Bach, W., Zuleger, E., Erzinger, J., Honnorez, J., Pezard, P., Becker, K., Salisbury, M. H., and Wilkens, R. H. 1996. Hydrothermal alteration of a section of upper oceanic crust in the eastern equatorial Pacific: a synthesis of results from Site 504 (DSDP Legs 69, 70, and 83, and ODP Legs 111, 137, 140, and 148). In *Proceedings of the Ocean Drilling Program*, Vol. 148, eds. J. C. Alt, H. Kinoshita, L. B. Stocking, *et al.* College Station, TX: Ocean Drilling Program, pp. 417–434.

Anderson, R. N., Langseth, M. G., and Sclater, J. G. 1977. The mechanisms of heat transfer through the floor of the Indian Ocean, *J. Geophys. Res.* **82**: 3,391–3,409.

Anderson, R. N., Hobart, M. A., and Lansgeth, M. G. 1979. Geothermal convection through oceanic crust and sediments in the Indian Ocean. *Science* **204**: 828–832.

Baker, P. A., Stout, P. M., Kastner, M., and Elderfield, H. 1991. Large-scale lateral advection of seawater through oceanic crust in the central equatorial Pacific. *Earth Planet. Sci. Lett.* **105**: 522–533.

Becker, K. and Fisher, A. 2000. Permeability of upper oceanic basement on the eastern flank of the Endeavor Ridge determined with drill-string packer experiments. *J. Geophys. Res.* **105**: 897–912.

Becker, K. and Von Herzen, R. P. 1983. Heat transfer through the sediments of the mounds hydrothermal area, Galapagos Spreading Center at 86° W. *J. Geophys. Res.* **88**: 995–1,008.

Becker, K., *et al.* 1989. Drilling deep into young oceanic crust, Hole 504B, Costa Rica Rift. *Rev. Geophys.* **27**: 79–102.

Becker, K., Bartetzko, A., and Davis, E. E. 2001. Leg 174B synopsis: Revisiting Hole 395A for logging and long-term monitoring of off-axis hydrothermal processes in young oceanic crust. In *Proceeding of the Ocean Drilling Program*, Vol. 174B, eds. K. Becker and M. J. Malore. College Station, TX: Ocean Drilling Program, pp. 1–12.

Bredehoeft, J. D., Wesley, J. B., and Fouch, T. D. 1994. Simulations of the origin of fluid pressure, fracture generations, and the movement of fluids in the Uinta Basin, Utah. *Am. Assoc. Petrol. Geol. Bull.* **78**: 1,729–1,747.

Carlson, R. L. 1998. Seismic velocities in the uppermost oceanic crust: age dependence and the fate of layer 2A. *J. Geophys. Res.* **103**: 7,069–7,077.

Claypool, G. and Kaplan, I. R. 1974. The origin and distribution of methane in marine sediments. In *Natural Gases in Marine Sediments*, ed. I. R. Kaplan. New York: Plenum Press, pp. 99–139.

COSOD 1981. *Report of the Conference on Scientific Ocean Drilling*, Austin, 16–18 November. Washington, DC: Joint Oceanographic Institutions, Inc.

COSOD II 1987. *Report of the Second Conference on Scientific Ocean Drilling*, Strasbourg, 6–8 July. Strasbourg: European Source Foundation.

Davis, E. E. and Lister, C. R. B. 1977. Heat flow measured over the Juan de Fuca Ridge: evidence for widespread hydrothermal circulation in a highly heat transportive crust. *J. Geophys. Res.* **82**: 4,845–4,860.

Davis, E. E., Chapman, D. S., Forster, C. B., and Villinger, H. 1989. Heat-flow variations correlated with buried basement topography on the Juan de Fuca Ridge flank. *Nature* **342**: 533–537.

Davis, E. E., Chapman, D. S., Mottl, M. J., Bentkowski, W. J., Dadey, K., Forster, C., Harris, R., Nagihara, S., Rohr, K., Wheat, G., and Whiticar, M. 1992. FlankFlux: an experiment to study the nature of hydrothermal circulation in young oceanic crust. *Can. J. Earth Sci.* **29**: 925–952.

Davis, E. E., Chapman, D. S., and Forster, C. B. 1996. Observations concerning the vigor of hydrothermal circulation in young oceanic crust. *J. Geophys. Res.* **101**: 2,927–2,942.

Davis, E. E., *et al.*, eds. 1997. *Proceedings of the Ocean Drilling Program, Initial Reports* Vol. 168. College Station, TX: Ocean Drilling Program.

Detrick, R., Collins, J., Stephen, R., and Swift, S. 1994. In situ evidence for the nature of the seismic layer 2/3 boundary in oceanic crust. *Nature* **370**: 288–290.

Edmond, J. M., Measures, C., McDuff, R. E., Chan, L. H., Collier, R., Grant, B., Gordon, L. I., and Corliss, J. B. 1979. Ridge crest hydrothermal activity and the balances of the major and minor elements in the ocean: the Galapagos data. *Earth Planet. Sci. Lett.* **46**: 1–18.

Elderfield, H. and Schultz, A. 1996. Mid-ocean ridge hydrothermal fluxes and the chemical composition of the ocean. *Ann. Rev. Earth Planet. Sci.* **24**: 191–224.

Elderfield, H., Wheat, C. G., Mottl, M. J., Monnin, C., and Spiro, B. 1999. Fluid and geochemical transport through oceanic crust: a transect across the eastern flank of the Juan de Fuca Ridge. *Earth Planet. Sci. Lett.* **172**: 151–165.

Embley, R. W., Hobart, M. A., Anderson, R. N., and Abbott, D. 1983. Anomalous heat flow in the northwest Atlantic: a case for continued hydrothermal circulation in 80-M.y. crust. *J. Geophys. Res.* **88**: 1,067–1,074.

Fehn, U., Green, K. E., Von Herzen, R. P., and Cathles, L. M. 1983. Numerical models for the hydrothermal field at the Galapagos Spreading Center. *J. Geophys. Res.* **88**: 1,033–1,048.

Fisher, A. and Becker, K. 2000. Reconciling heat flow and permeability data with a model of channelized flow in oceanic crust. *Nature* **403**: 71–74.

Fisher, A. T., Becker, K., Narasimhan, T. N., Langseth, M. G., and Mottl, M. J. 1990. Passive, off-axis convection through the southern flank of the Costa Rica Rift. *J. Geophys. Res.* **95**: 9,343–9,370.

Fisher, A. T., Becker, K., and Narasimhan, T. N. 1994. Off-axis hydrothermal circulation: parametric tests of a refined model of processes at Deep Sea Drilling Project/Ocean Drilling Program site 504. *J. Geophys. Res.* **99**: 3,097–3,121.

Fisk, M. R., Giovannoni, S. J., and Thorseth, I. H. 1998. Alteration of oceanic volcanic glass: textural evidence of microbial activity. *Science* **281**: 978–980.

Gable, R., Morin, R. H., and Becker, K. 1992. Geothermal state of DSDP Holes 333A, 395A and 534A: results from the DIANAUT program. *Geophys. Res. Lett.* **19**: 505–508.

Garrels, R. M. and Mackenzie, F. T. 1971. *Evolution of Sedimentary Rocks*, New York: W. W. Norton, 397 pp.

Geller, G. A., Weissel, J. K., and Anderson, R. N. 1983. Heat transfer and intraplate deformation in the central Indian Ocean. *J. Geophys. Res.* **88**: 1,018–1,032.

Green, K. E., Von Herzen, R. P., and Williams, D. L. 1981. The Galapagos Spreading Center at 86° W: a detailed geothermal field study. *J. Geophys. Res.* **86**: 979–986.

Grevemeyer, I., Norbert, K., Villinger, H., and Weigel, W. 1999. Hydrothermal activity and the evolution of the seismic properties of upper oceanic crust. *J. Geophys. Res.* **104**: 5,069–5,079.

Holmes, M. L. and Johnson, H. P. 1993. Upper crustal densities derived from seafloor gravity measurements: northern Juan de Fuca Ridge. *Geophys. Res. Lett.* **17**: 1,871–1,874.

Jacobson, R. S. 1992. Impact of crustal evolution on changes of the seismic properties of the uppermost oceanic crust. *Rev. Geophys.* **30**: 23–42.

Langseth, M. G. and Herman, B. M. 1981. Heat transfer in the oceanic crust of the Brazil Basin. *J. Geophys. Res.* **86**: 10,805–10,819.

Langseth, M. G., Cann, J. R., Natland, J. H., and Hobart, M. 1983. Geothermal phenomena at the Costa Rica Rift: background and objectves for drilling at Deep Sea Drilling Project Sites 501, 504, and 505. In *Initial Reports of the Deep Sea Drilling Program*, Vol. 69, eds. J. R. Cann, M. G. Langseth, J. Honnolez, *et al.* Washington, DC: US Govt. Printing Office, pp. 5–29.

Langseth, M. G., Hyndman, R. D., Becker, K., Hickman, S. H., and Salisbury, M. H. 1984. The hydrogeological regime of isolated sediment ponds in mid-oceanic ridges. In *Initial Reports of the Deep Sea Drilling Program*, Vol. 84, eds. R. Von Huene, J. Aubouin, *et al.* Washington, DC: US Govt. Printing Office, pp. 825–837.

Langseth, M. G., Mottl, M. J., Hobart, M. A., and Fisher, A. 1988. The distribution of geothermal and geochemical gradients near Site 501/504: implications for hydrothermal circulation in the oceanic crust. In *Proceedings of the Ocean Drilling Program, Initial Reports*, Vol. 111, eds. K. Becker, H. Sakai, *et al.* College Station, TX: Ocean Drilling Program, pp. 23–32.

Langseth, M. G., Becker, K., Von Herzen, R. P., and Schultheiss, P. 1992. Heat and fluid flux through sediment on the western flank of the Mid-Atlantic Ridge: a hydrogeological study of North Pond. *Geophys. Res. Lett.* **19**: 517–520.

Larson, R. L., Fisher, A. T., Jarrard, R. D., Becker, K., and Ocean Drilling Program Leg 144 Shipboard Scientific Party 1993. Highly permeable and layered Jurassic oceanic crust in the western Pacific. *Earth Planet. Sci. Lett.* **111**: 71–83.

Lister, C. R. B. 1972. On the thermal balance of a mid-ocean ridge. *Geophys. J. Roy. Astron. Soc.* **26**: 515–535.

Magenheim, A. J., Bayhurst, G., Alt, J. C., and Gieskes, J. M. 1992. ODP Leg 137, boishole fluid chemistry in Hole 504B. *Geophys. Res. Lett.* **19**: 521–524.

Maris, C. R. P. and Bender, M. L. 1982. Upwelling of hydrothermal solutions through ridge flank sediments shown by pore water profiles. *Science* **216**: 623–626.

Martin, J. B. 1999. Non-conservative behavior of Br/Cl ratios during alteration of volcaniclasitc sediments. *Geochim. Cosmochim. Acta* **63**: 383–391.

Martin, J. B., Orange, D. L., Lorensen, T. D., and Kvenvolden, K. A. 1997. Chemical and isotopic evidence of gas-influenced flow at a transform plate boundary: Monterey Bay, California. *J. Geophys. Res.* **102**: 24,903–24,915.

McDuff, R. E. 1981. Major cation gradients in DSDP interstitial waters: the role of diffusive exchange between seawater and upper oceanic crust. *Geochim. Cosmochim. Acta* **45**: 1,705–1,713.

Moore, W. S. 1996. Large groundwater inputs to coastal waters revealed by ^{226}Ra enrichments. *Nature* **380**: 612–614.

Mottl, M. J. 1989. Hydrothermal convection, reaction, and diffusion in sediments on the Costa Rica Rift flank: pore-water evidence from ODP Sites 677 and 678. In *Proceedings of the Ocean Drilling Program, Initial Reports*, Vol. III, eds. K. Becker, H. Sakai, *et al.* College Station, TX: Ocean Drilling Program, pp. 195–213.

Mottl, M. J. and Wheat, C. G. 1994. Hydrothermal circulation through mid-ocean ridge flanks: fluxes of heat and magnesium. *Geochim. Cosmochim. Acta* **58**: 2,225–2,237.

Purdy, G. M. and Fryer, G. J. 1990. *Report of a Workshop on the Physical Properties of Volcanic Seafloor*, Woods Hole, April, 24–26. Washington, DC: Joint Oceanographic Institution, Inc.

Sayles, F. L. 1979. The composition and diagenesis of interstitial solutions, I. Fluxes across the sediment–water interface in the Atlantic. *Geochim. Cosmochim. Acta* **43**: 527–545.

Schultz, A. and Elderfield, H. 1997. Controls on the physics and chemistry of seafloor hydrothermal circulation. *Phil. Trans. Roy. Soc* **355**: 387–425.

Sclater, J. G., Crowe, J., and Anderson, R. N. 1976. On the reliability of oceanic heat flow averages. *J. Geophys. Res.* **81**: 2,997–3,006.

Stein, C. A. and Stein, S. 1994. Constraints on hydrothermal heat flux through the oceanic lithosphere from global heat flow. *J. Geophys. Res.* **99**: 3,081–3,095.

Thompson, G. 1983. Basalt–seawater interaction. In *Hydrothermal Processes at Seafloor Spreading Centers*, eds. P. A. Rona, K. Bostrom, L. Laubier, and K. L. Smith. New York: Plenum Press, pp. 225–278.

Wilkens, R. H., Fryer, G. J., and Karsten, J. 1991. Evolution of porosity and seismic structure of upper oceanic crust: importance of aspect ratios. *J. Geophys. Res.* **96**: 17,981–17,995.

Williams, D. L., Von Herzen, R. P., Sclater, J. G., and Anderson, R. N. 1974. The Galapagos Spreading Center: lithospheric cooling and hydrothermal circulation. *Geophys. J. Roy. Astron. Soc.* **38**: 587–608.

Part II

Hydrologic structure, properties, and state of the oceanic crust

3

Variability of ocean crustal structure created along the global mid-ocean ridge

Suzanne M. Carbotte and Daniel S. Scheirer

3.1 Introduction

Crustal creation at mid-ocean ridges (MORs) involves a complex array of igneous and tectonic processes that vary spatially and through time, giving rise to a heterogeneous crustal section. Pervasive circulation of seawater within the oceanic crust initiates at the mid-ocean ridge and continues for millions of years as the oceanic lithosphere cools and ages. Thermal gradients associated with crustal formation and pressure gradients associated with tectonic and gravitational sources provide the driving forces for fluid circulation. Physical properties of oceanic crust, including the permeability and porosity structure, igneous stratigraphy and topography, determine the fluid pathways and mechanisms of hydrothermal flow.

The igneous structure of the oceanic crust is generated by magmatic processes operating within a zone as narrow as a few kilometers, often resulting in a layered crust comprising, from top to bottom, extrusive basalts, sheeted dikes, and layered and massive gabbroic rocks. Newly created oceanic crust is modified by brittle failure that occurs over a wider zone, extending a few tens of kilometers from the axis of spreading. These volcanic and tectonic processes at the MOR give rise to the undulating abyssal hill-and-trough terrain that characterizes the world's ocean floor. In young crust, these abyssal hills, seamounts, and the fracture zones that offset the ridge crest are important topographic elements that contribute to the pattern of fluid flow within the crust. As the crust ages and is blanketed by sediment, seamounts and elevated seafloor at fracture zones become increasingly important as isolated sites where open exchange between crustal fluids and the water column can persist.

The intrinsic permeability and porosity structure of the crust determines the distribution and depth extent of pathways that can be exploited by fluids, and it results from both magmatic accretion and tectonic processes. Within the extrusive section, volcanic brecciation, lava flow contacts, and development of void space due to lava drainback and volcanic collapse all contribute to crustal porosity. Permeability within this uppermost layer arises from

Hydrogeology of the Oceanic Lithosphere, eds. E. E. Davis and H. Elderfield. Published by Cambridge University Press.
© Cambridge University Press 2004.

interconnected porosity associated with these magmatic processes, as well as from fissures and the normal faults associated with the formation of abyssal hills. Deeper in the crust, primary crustal permeability is provided by normal faults and cracks formed by thermal contraction as crustal rocks cool.

The rate of crustal formation and the pattern of melt delivery to the ridge axis play fundamental roles in determining the physical properties and structure of oceanic crust created at the MOR. First-order variations are observed in the magnitude of seafloor relief, the extent of faulting, and the igneous architecture of oceanic crust at different spreading rates. As a result of these differences, we expect the distribution of permeable zones within the crust to be quite different for crust created at fast- and slow-spreading rates. At all spreading rates, spatial variations in magma supply from the mantle generate variations in topography and crustal structure along individual MOR spreading segments. Magma delivery to the ridge crest also varies through time, leading to magmatic cycles that impart ridge-perpendicular heterogeneity to crustal structure. These spatial and temporal variations in crustal production give rise to gradients in crustal properties, which in turn will contribute to the geometry of fluid circulation through the crust.

The focus of this chapter is on the primary structure of oceanic crust created at mid-ocean ridges and the global variability of this structure, related both to spreading rate and to temporal and spatial changes in magma delivery from the mantle. The crustal properties most important for hydrothermal flow within the crust, permeability and porosity, are expected to evolve with ongoing fluid circulation. However, the primary structure inherited from accretion at the MOR defines the initial conditions at the onset of fluid circulation and must influence fluid flow long after crustal formation. We begin with an overview of the active processes of crustal creation and modification that occur at spreading ridges (Section 3.2). Then we discuss the major structural features of the seafloor associated with these processes including the ridge crest, ridge discontinuities, abyssal hills, and seamounts (Section 3.3). While the seafloor is readily accessible to direct observation via visual, seafloor sampling, and sonar investigations, the internal structure of the crust is accessible primarily through remote sensing methods such as seismology and potential field studies, with essential ground-truth provided by isolated tectonic windows, crustal drilling, and terrestrial analogs. In Section 3.4, we summarize current understanding of the sub-surface structure of newly created crust derived primarily from seismic studies, with a focus on observations relevant to crustal composition, porosity, and permeability structure. Finally, in Section 3.5 we discuss variations in the internal crustal structure associated with accretion at the MOR and implications for the hydrologic structure of the oceanic crust. This chapter will not focus on the aging of crust with ridge flank circulation, observations of crustal properties from borehole studies, or terrestrial analogs; these topics are discussed in detail in later chapters. Because of the breadth of this chapter, we highlight references that are recent and comprehensive or ones that are pivotal to the development of particular concepts, and we commend the reader to explore those references for links to other publications and ideas.

3.2 Active processes at the mid-ocean ridge

3.2.1 A census of mid-ocean ridges

The mid-ocean ridge system is interconnected globally (Fig. 3.1*a*), with a cumulative length of spreading segments of about 53,000 km. Combined with the offsets of transform faults and other ridge discontinuities, the MOR plate boundary is about 91,000 km long. Some portions of the plate boundary are dominated by large transform faults, such as along the equatorial Mid-Atlantic Ridge (MAR), the Southwest Indian Ridge (SWIR) south of Africa, and the Chile Rise; other sections are free of large axial discontinuities, such as other portions of the SWIR, much of the southern East Pacific Rise (EPR), and the Reykjanes Ridge south of Iceland.

A histogram of present-day spreading rates (Fig. 3.1*b*) exhibits three distinct populations, leading to a classification of spreading rates as slow ($<$40 km M.y.$^{-1}$ full-rate), intermediate (40–80 km M.y.$^{-1}$), and fast ($>$80km M.y.$^{-1}$). Other authors have chosen slightly different spreading rate boundaries for these sub-divisions, and many have applied names such as "ultraslow" and "ultrafast" to extreme spreading rates. In the global picture, these end-member classes fit into a smoother spectrum, although there is significant geographic clustering. The Pacific Ocean basin does not have any slow-spreading MORs (except near the tips of propagating ridges); the Atlantic and Arctic Oceans have only slow-spreading ridges. The Indian Ocean has a bimodal distribution of slow- and intermediate-spreading ridges. Half of the present-day MOR system spreads at slow rates, $<$40 km M.y.$^{-1}$ full-rate (Fig. 3.1*c*). However, because the volume of crust produced is also a function of the spreading rate, much more than half of the ocean crust presently created is formed at intermediate and fast rates.

A compilation of half-spreading rates by Muller *et al.* (1998) allows a related histogram to be created for all seafloor that has been dated with magnetic reversal anomalies (Fig. 3.1*d*). The histogram of crustal production in the past is centered on intermediate rates, but it is skewed toward fast rates, including rates faster than along the present-day MOR (Fig. 3.1*b*). While there are greater uncertainties in the data contributing to this histogram than in the recent case, the geographic clustering of spreading rate noted for the present day also existed in the past.

Near-ridge hot-spots may influence profoundly the topography, geochemistry, and internal structure of oceanic crust formed at spreading centers. For example, the Iceland hot-spot coincides with the northern MAR and has influenced ocean crustal accretion since the opening of the north Atlantic. Other hot-spots are located near spreading centers in all of the ocean basins. The topographic shoaling of the ridge axis and thickening of the crust typically associated with hot-spots are most prominent for large hot-spots located within a few hundred kilometers of the spreading axis (e.g. Ito and Lin, 1995), but more subtle topographic and significant geochemical interactions may be present at ridge–hot-spot distances of 1,000 km and more (Schilling, 1991). Hot-spots and MORs have a close spatial association in general (Fig. 3.1*a*), and a significant percentage of the present-day ridge is located

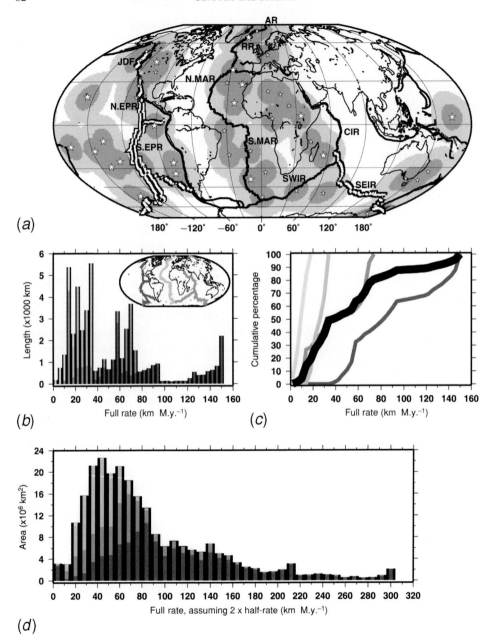

Fig. 3.1 (*a*) Global map of the tectonic plate boundaries (thick lines) and of hot-spots (stars). Black lines indicate mid-ocean ridge spreading centers; a single black line indicates spreading at full rates <40 km M.y.$^{-1}$ (slow), a double line indicates spreading at 40–80 km M.y.$^{-1}$ (intermediate), and three lines indicate spreading at >80 km M.y.$^{-1}$ (fast). Plate boundary from the PLATES project (UTIG) and spreading rates based on DeMets *et al.* (1994). Hot-spot distribution from Steinberger (2000); larger stars denote hot-spots with production >1,000 kg s^{-1}.

within 1,000 km of the nearest hot-spot. A majority of the ridge system is located within 2,000 km of the nearest hot-spot. While the source depth of hot-spots is controversial, at the Earth's surface they produce substantial volcanic anomalies and large perturbations to the mechanical and thermal structure of oceanic crust and lithosphere.

3.2.2 Igneous processes

The separation of tectonic plates at a spreading center leads to upwelling of the underlying mantle, accompanied by partial melting of the mantle mineral assemblage as it decompresses. While the mechanisms of melt delivery to the crust are active topics of research, it is well established that this melt forms the oceanic crust where it cools at shallow levels. Typical cross-sections of the oceanic crust contain both intrusive and extrusive rocks (Figs. 3.2 and 3.3). *In situ* geophysical observations and the geochemistry of erupted basalts indicate that melt generally cools in reservoirs at shallow crustal levels. These magma and mush chambers may be ephemeral in some places and steady state elsewhere, and there is evidence both for multiple replenishments of magma into chambers from below as well as for magma interaction with wall-rock and seawater (e.g. Grove *et al.*, 1992; Langmuir *et al.*, 1992).

Magma is delivered to the seafloor via dikes. These dikes form a significant portion of the oceanic upper crust, and dikes that reach the seafloor may produce recognizable eruptive fissures. The diking process at mid-ocean ridges is similar to that at terrestrial basaltic systems such as in Hawaii and Iceland; dikes propagate upward and horizontally in an orientation perpendicular to the least compressive stress (e.g. Curewitz and Karson, 1998). Dike migration may occur over lateral distances of tens of kilometers and serves to distribute magmas along the ridge. As on land, a diking episode at a spreading center is recognized by the migration of micro-earthquake swarms over short time periods (e.g. Dziak and Fox, 1999).

The style of seafloor eruption is governed by factors related to the magma composition, extrusion rate, and local topography (Gregg and Fink, 1995). Eruptions at typical mid-ocean

←——

Dark gray shades denote areas within 1,000 km of the nearest hot-spot; light gray shades denote areas from 1,000 to 2,000 km distant from the nearest hot-spot. (*b*) Histogram of the length of present-day spreading segments versus their spreading rate; their cumulative length is 53,000 km. Black bars indicate the total histogram, and gray bars sub-divide the contributions from different ocean basins (inset): darkest gray = Pacific, medium gray = Indian, light gray = Atlantic, lightest gray = Arctic. (*c*) Cumulative histograms corresponding to the data in (*b*). (*d*) Histogram of the area of global seafloor formed at different rates, where magnetic anomalies allow half-rate estimation (based on data from Muller *et al.*, 1998). Crust formed at spreading rates much higher than modern rates are from old plate where seafloor isochron picks are uncertain and where the steady symmetric spreading assumption may be violated. JDF, Juan de Fuca; EPR, East Pacific Rise; MAR, Mid-Atlantic Ridge; RR, Reykjanes Ridge; AR, Arctic Ridges; SWIR, Southwest Indian Ridge; CIR, Central Indian Ridge; SEIR, Southeast Indian Ridge. See accompanying CD-ROM for color version of figure.

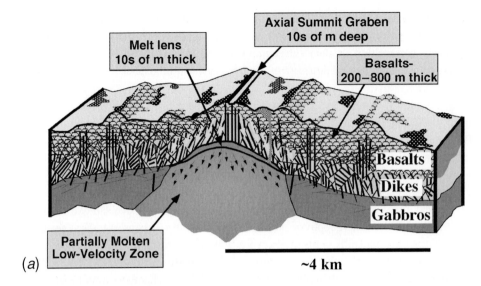

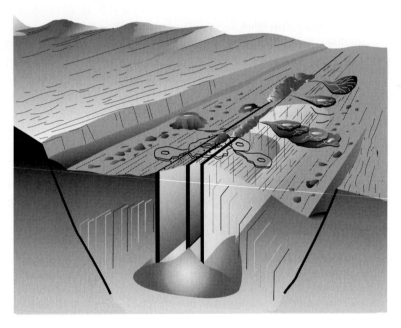

Fig. 3.2 Cartoons of mid-ocean ridges illustrating accretion processes within the plate boundary zone at different spreading rates. (*a*) At fast-spreading ridges, volcanic activity is commonly focused within a narrow axial summit trough located at the center of a 200–500-m-high axial ridge. A steady-state crustal magma body is found within the mid-crust beneath the crest, which feeds the dikes and volcanic eruptions that build the upper crust. Ridge-parallel normal faulting initiates beyond the

ridge depths are quiescent, producing characteristic submarine lava flow morphologies ranging from pillow lavas to sheet flows. Eruptions build volcanic constructions at some spreading centers (e.g. axial volcanic ridges found within the floor of axial valleys at slow-spreading centers), while sheet and lobate eruptions generate minimal relief at many fast-spreading ridges. Based on seismic evidence and suites of rocks sampled from the seafloor, some portions of the MOR system are devoid of volcanic rocks. In these cases, investigators have inferred that tectonic processes removed a volcanic carapace or that no igneous crust was formed (Dick, 1989; Cannat, 1993).

3.2.3 *Tectonic processes*

Fracturing occurs at all scales in MOR environments, and its expression varies significantly along the MOR system. At small scales, the cooling of lava produces cracks on the boundaries of flows or through entire units, generating porosity that may persist or may be reduced by in-fill with subsequent lava flows, sediment, hydrothermal deposits, or by consolidation with burial. Far-field tension, related to the forces separating the tectonic plates, produces fissures in the uppermost, brittle layer of the oceanic crust at a seafloor spreading center. Fissures can also form above dikes that do not reach the seafloor, and these fissures may have very different geometric characteristics from tectonic fissures (Wright, 1998).

Where the brittle layer is sufficiently thick and strong, extension leads to shear failure on normal faults. The dimensions, geometry, and spatial frequency of faults vary significantly as functions of both spreading rate and position within a spreading segment. With decreasing spreading rate, the magnitude of faulting increases, and the topographic fabric is dominated by fault scarps. At faster-spreading MORs, faults are smaller and may be partially buried by lava flows (Macdonald *et al.*, 1996). Most tectonic features are closely aligned with

---

axial high and results in the undulating, low-relief abyssal hill topography of the flanks. Inflation and deflation of this crustal magma reservoir may give rise to extensive brittle deformation within the upper crust, which is buried by late-stage volcanic eruptions as seafloor is rafted away from the ridge axis. (Reproduced with permission from Karson, J. A., Klein, E. M., Hurst, S. D., Lee, C. E., Rivizzigno, P. A., Curewitz, D., Morris, A. R., and Hess Deep '99 Scientific Party 2002. Structure of uppermost fast-spread oceanic crust exposed at the Hess Deep Rift: implications for subaxial processes at the East Pacific Rise. *Geochemi. Geophys. Geosys.* **3** (Paper number 2001GC000155), Copyright 2002 American Geophysical Union.) (*b*) At slow-spreading ridges, magmatism occurs within a fault-bounded axial valley. Hummocky ridges and small flat-topped seamounts are typically found along an axial volcanic ridge centered within the valley floor. The cartoon shows three dikes rising and propagating laterally; where the dikes breech the seafloor, volcanic eruptions occur. (Reproduced with permission from Smith, D. K. and Cann, J. R. 1999. Constructing the upper crust of the Mid-Atlantic Ridge: a reinterpretation based on the Puna Ridge, Kilauea Volcano. *J. Geophys. Res.* **104**: 25,379–25,399, Copyright 1999 American Geophysical Union.) See accompanying CD-ROM for color version of figure.

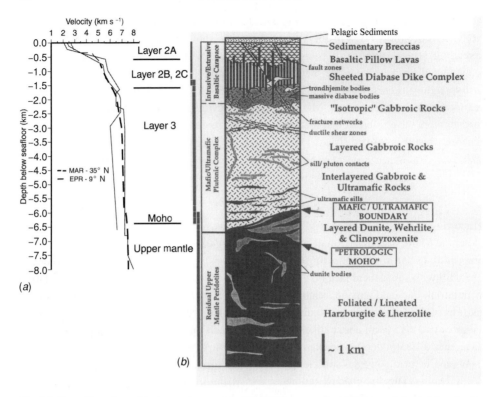

Fig. 3.3 Prevailing view of the internal structure of oceanic crust and correlations with seismic velocity profiles. (*a*) Typical seismic velocity structure of young fast- and slow-spread crust with seismic layers labeled. Data from Vera *et al.* (1990), Hosford *et al.* (2001), and Hussenoeder *et al.* (2002a). (*b*) Generalized igneous stratigraphy derived from ophiolites. Measurements of seismic velocities of rock samples from ophiolites and oceanic drill cores are commonly used to relate the velocity structure of oceanic crust to crustal lithology. (Reproduced with permission from Karson, J. A. 1998. Internal structure of oceanic lithosphere: a perspective from tectonic windows. In *Faulting and Magmatism at Mid-Ocean Ridges*, Geophysical Monograph 106, eds. W. R. Buck, *et al.*, Washington, DC: American Geophysical Union, pp. 177–218. Copyright 1998 American Geophysical Union.)

the axis of seafloor spreading, although the complicated stress patterns near ridge-axis discontinuities lead to rotated tectonic fabric in their vicinity.

3.2.4 *Interactions among volcanic and tectonic processes*

Igneous accretion and tectonic deformation both accommodate the separation of tectonic plates at a MOR. The region about the axis of spreading where the plates on either side accelerate to reach their long-term velocities is considered to be the plate boundary zone. Observations of young lava flows and active faulting at MORs suggest that the plate boundary zone is at most a few tens of kilometers wide and that faulting extends to greater distances from the axis than magmatism (e.g. Klitgord *et al.*, 1975; Bicknell *et al.*, 1987). The partitioning between magmatism and tectonism varies with spreading rate along individual spreading segments and with time at a given spreading center.

3.3 Seafloor structures

3.3.1 Morphology of the spreading axis

Detailed bathymetry surveys of MORs allow correlation of seafloor morphology with patterns of volcanism and deformation during crustal accretion in different settings (e.g. Small, 1998). Figure 3.4 illustrates the seafloor morphology of the spreading axis and young flanks of typical spreading centers that spread at fast, intermediate, and slow rates. To facilitate intercomparison of these areas, the map panels each span 350 × 275 km and have similar bathymetry color fill and shading.

Fast-spreading ridges are characterized by an axial ridge typically 200–500 m high and 4–10 km wide (Fig. 3.4*a*). Slower-spreading ridges are characterized by axial valleys up to several kilometers deep and 5–20 km wide (Fig. 3.4*c*). Intermediate-spreading ridges have variable morphology between these extremes (Fig. 3.4*b*). Important exceptions to these overall trends are typically associated with proximity to mantle hot-spot sources; for example, the slow-spreading Reykjanes Ridge (Fig. 3.4*d*) immediately south of the Iceland hot-spot does not exhibit an axial valley. Here, slow spreading occurs along an axial ridge, with the elevated axial morphology due to enhanced magma supply associated with the Iceland hot-spot. Along the intermediate-spreading Galapagos Spreading Center, the influence of the Galapagos hot-spot may account for the observed variations in axial morphology from an axial high to a shallow rift valley (Canales *et al.*, 1997).

The dimensions and shape of axial valleys and ridges vary significantly along-axis, especially in the vicinity of MOR discontinuities (Fig. 3.4). Shipboard magnetic and gravity anomalies (Fig. 3.4*e*–*g*) also vary systematically along-axis, illustrating that the variability seen in the surface morphology corresponds with changes in crustal properties at depth. These variations are thought to reflect lateral changes in the supply of magma to the ridge axis and in the tectonic disruption of the crust (e.g. Macdonald *et al.*, 1988) over distances of tens of kilometers. Ridge-axis discontinuities vary in size from small offsets of less than one kilometer to large-offset transform faults tens to hundreds of kilometers long. These offsets of the ridge axis define individual spreading segments that also exist at a range of scales from tens to hundreds of kilometers (e.g. Langmuir *et al.*, 1986; Macdonald *et al.*, 1988). MOR offsets typically leave pronounced traces on the ridge flanks in both seafloor morphology and other crustal properties, and they are an important source of spatial heterogeneity in oceanic crustal structure.

In detail, fast-spreading axial ridges are often capped by a small axial trough formed by magmatic processes at the axis (Fig. 3.5*a*; Fornari *et al.*, 1998). In contrast, the floor of slow-spreading axial valleys is often populated by mounds of pillow lava and accumulations of pillow hummocks that form axial volcanic ridges, Fig. 3.5*b* (e.g. Smith and Cann, 1999).

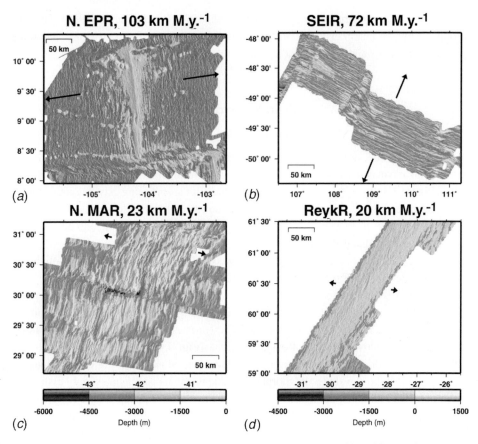

Fig. 3.4 Bathymetry and mantle Bouguer anomaly (MBA) maps and profiles of four typical areas of the MOR: (*a*) the fast-spreading northern EPR, (*b*) the intermediate-spreading SEIR, (*c*) the slow-spreading northern MAR, and (*d*) the slow-spreading Reykjanes Ridge south of Iceland. The map areas are equal-sized, 350 × 275 km, to facilitate comparison. Arrows indicate the plate separation directions, with lengths scaled to spreading rate as calculated from DeMets *et al.* (1994). The gray-scales of (*a*), (*b*), and (*c*) are identical and are shown below (*c*); the color-scale of (*d*) is shifted to depths 1,500 m shallower. False-shading of the topography is added to emphasize the morphologic fabric; topography is illuminated from the plate-spreading direction in each case. The intensity-scale of shading is constant among the panels to facilitate comparison of seafloor slopes. Bathymetry has grid spacings of 300 m (N. EPR) and 500 m (other areas), and we obtained these grids from the RIDGE Multibeam Data Synthesis (http://ocean-ridge.ldeo.columbia.edu/). The amount of spreading history illustrated in each panel varies as a function of spreading rate and off-axis coverage: for the N. EPR, coverage extends on-average to crust formed at 3.2 Ma; SEIR = 1.4 Ma; N. MAR = 11 Ma; Reykjanes Ridge = 3.9 Ma. MBA maps in (*e*), (*f*), and (*g*) correspond to the same areas shown in (*a*), (*b*), and (*c*), respectively. Contour interval is 10 mgal. MBA grids of the N.EPR (J. P. Canales and R. S Detrick, personal comm. 2003), of the SEIR (Cochran *et al.*, 1997), and the N. MAR (Pariso *et al.*, 1995) were calculated at ∼2-km horizontal grid spacing, and mean MBA values were removed to facilitate comparison. Cross-axis profiles of the topography and MBA for the N. EPR, SEIR, and N. MAR areas are shown in (*h*). The profiles are centered on the axis of spreading and are collinear with the plate spreading arrows in (*e*), (*f*), and (*g*). Vertical exaggeration of the topographic profiles in (*h*) is ∼×8. See accompanying CD-ROM for color version of figure.

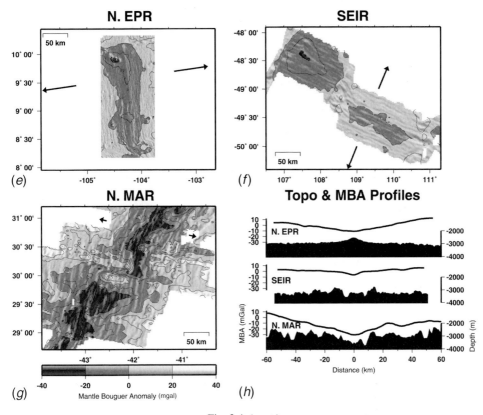

Fig. 3.4 (cont.)

Intermediate-rate spreading centers display characteristics of fine-scale morphology ranging between that typical of fast- and slow-spreading centers (e.g. Embley *et al.*, 1991; Karsten *et al.*, 1986).

The micro-relief of the neovolcanic zone largely reflects the types of lava flows issued from the seafloor, which vary systematically from pillow-dominated to sheet-flow-dominated with increasing spreading rate (Fig. 3.6). At fast-spreading centers, drain-back/drainout processes associated with effusive eruptions give rise to large-scale void space and collapse within lobate flows, which may account for much of the high porosity of the young extrusive crust. It is not known how this primary volcanic porosity evolves with time, although presumably in-fill with collapse debris, other lavas, and hydrothermal deposits reduces porosity as the crust ages.

3.3.2 Abyssal hills

The most prevalent features on the flanks of MORs are abyssal hills, ridge-parallel structures formed by tectonic and, in some cases, volcanic processes (Figs. 3.4 and 3.7). Abyssal

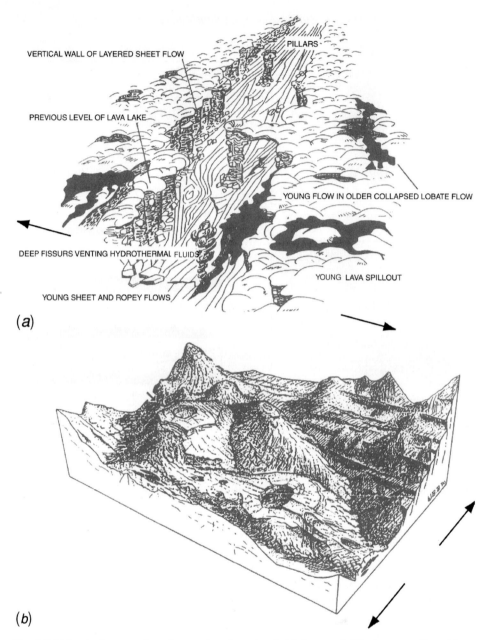

Fig. 3.5 Volcanic features that characterize the axis of (*a*) fast- and (*b*) slow-spreading ridges. (*a*) Within the inner volcanic zone along fast-spreading ridges, a collapsed volcanic terrain of lobate and sheet flows with isolated lava pillars is often found within a narrow shallow depression. (*b*) In contrast, at the slow-spreading MAR, the axial valley is floored by numerous small volcanic cones comprised of bulbous and elongate pillows. (Reprinted with permission from Perfit, M. R. and Chadwick, W. W., Jr. 1998. Magmatism at mid-ocean ridges: constraints from volcanological and geochemical investigations. In *Faulting and Magmatism at Mid-Ocean Ridges*, Geophysical Monograph 106, eds. W. R. Buck *et al.*, Washington, DC: American Geophysical Union, pp. 59–115. Copyright 1998 American Geophysical Union.)

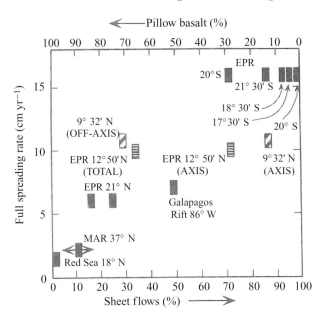

Fig. 3.6 Lava flow morphology versus spreading rate. Sheet flows are more commonly observed at fast-spreading ridges whereas pillow flows predominate at slow-spreading ridges. (Reprinted with permission from Perfit, M. R. and Chadwick, W. W., Jr. 1998. Magmatism at mid-ocean ridges: constraints from volcanological and geochemical investigations. In *Faulting and Magmatism at Mid-Ocean Ridges*, Geophysical Monograph 106, eds. W. R. Buck *et al.*, Washington, DC: American Geophysical Union, pp. 59–115, Copyright 1998 American Geophysical Union.)

hills vary in relief across the spreading-rate spectrum, with larger hills found at slower spreading rates (Fig. 3.4). Seafloor roughness, which provides a measure of abyssal hill relief, is nearly constant at fast-spreading rates, and it increases at slower rates (Fig. 3.8). The major scarps defining abyssal hills are normal faults formed by plate extension and bending within the plate boundary zone (e.g. Cowie, 1998; Bohnenstiehl and Kleinrock, 2000). Their continuation at depth and possible role in fluid circulation within the crust is described in Section 3.5. The dimensions of abyssal hills vary along-axis, especially in proximity to transform faults where abyssal hill faults increase in relief (e.g. Goff, 1991) and where changes in stress orientation yield rotated abyssal hill fabric.

The size and density of mature abyssal hill fault populations off-axis vary substantially, both with spreading rate and with position along-axis of individual spreading segments. Fault scarps at slow-spreading rates are typically hundreds of meters high, whereas at faster-spreading rates they are several tens of meters (Fig. 3.9). Estimates of tectonic strain are derived from measures of fault displacements and fault dip. At fast-spreading MORs, the fault strain reaches values of only 2–4% (Cowie *et al.*, 1993; Bohnenstiehl and Carbotte, 2001), while at slower-spreading MORs, larger strains of 10–15% are reported (Macdonald and Luyendyk, 1977; Solomon *et al.*, 1988). At slow-spreading ridges, fault parameters

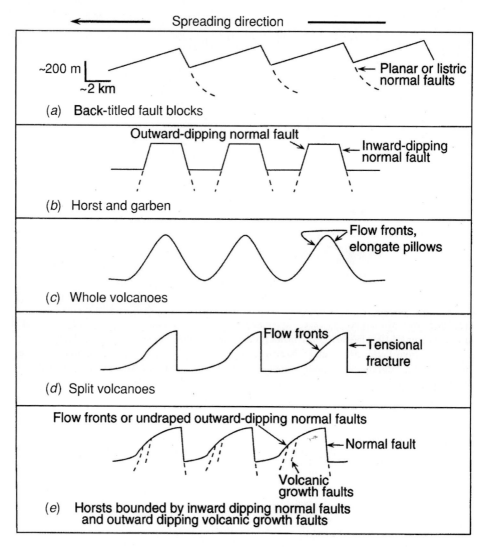

Fig. 3.7 Five models for the development of abyssal hills. (*a*) Back-tilted fault blocks formed during near-axis normal faulting along inward-dipping faults; (*b*) horsts and grabens formed during near-axis normal faulting of both inward- and outward-dipping faults; (*c*) whole volcanoes formed during episodic volcanism on-axis; (*d*) split volcanoes formed during episodic volcanic and tectonic phases on axis, and (*e*) horsts bounded by inward-dipping normal faults and outward-dipping volcanic growth faults. (Reprinted with permission from Macdonald, K. C., Fox, P. J., Alexander, R. T., Pockalny, R. A., and Gente, P. 1996. Volcanic growth faults and the origin of abyssal hills on the flanks of the East Pacific Rise. *Nature* **380**: 125–129, Copyright 1996 Nature.)

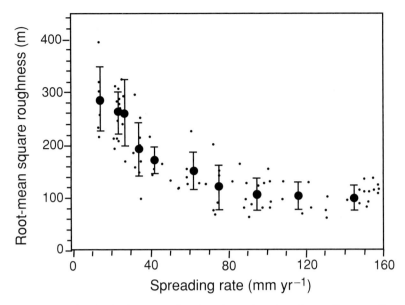

Fig. 3.8 Roughness of ridge flank topography plotted as a function of spreading rate. Topographic roughness provides a measure of average abyssal hill height and is inversely correlated with spreading rate for slow to intermediate rates of 10–70 km M.y.$^{-1}$ and is approximately constant at faster rates. Small dots correspond with calculated root-mean square roughness for individual bathymetric profiles. Large dots are averages for several profiles in spreading rate bins. (Reprinted with permission from Malinverno, A. 1991. Inverse square-root dependence of mid-ocean-ridge flank roughness on spreading rate. *Nature* **352**: 58–60, Copyright 1996 Nature.)

may vary systematically along ridge segments, even where little change in tectonic strain is observed with larger throw and more widely spaced faults near segment ends (Shaw, 1992; Escartin *et al.*, 1999). Larger throw faults may reflect a thicker brittle lithosphere toward segment ends (Shaw, 1992; Shaw and Lin, 1996), consistent with microearthquake and teleseismic studies that indicate deeper earthquakes near MOR discontinuities. Alternatively, reduced fault strength due to serpentinization near discontinuities may enhance strain localization resulting in larger throw faults and wider fault spacings at segment ends (Escartin *et al.*, 1997).

3.3.3 Seamounts

Seamounts and volcanic ridges are prominent topographic features on oceanic crust globally (Wessel, 2001). Seamounts that form near spreading centers are a common, but volumetrically small, constituent of young oceanic crust (Batiza and Vanko, 1983; Smith and Jordan, 1987). While hot-spot volcanoes such as Hawaii or Iceland perturb the thermal and mechanical structures of oceanic lithosphere of all ages, near-axis volcanoes constitute a class of mid-plate volcanism restricted to the vicinity of the MOR plate boundary.

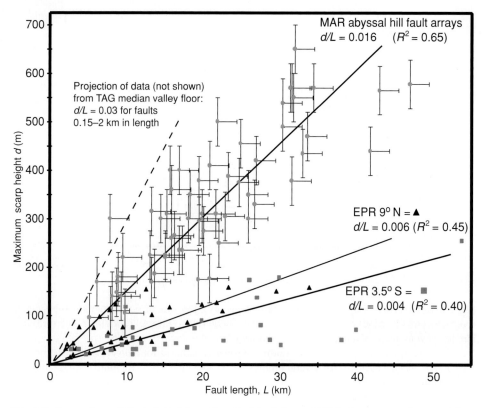

Fig. 3.9 Comparison of the height (d) and length (L) of abyssal hill fault scarps at slow- and fast-spreading ridges. Faults on the MAR (spreading rates 10–40 mm yr^{-1}) are typically several 100s of meters high whereas average fault throws are $\sim$50 m on the fast-spreading EPR (rates 90–140 mm yr^{-1}). (Modified from Bohnenstiehl, D. R. and Kleinrock, M. C. 2000. Evidence for spreading-rate dependence is seen in the displacement–length ratios of abyssal hill faults at mid-ocean ridges. *Geology* **28**: 395–398, Copyright 1996 Geological Society of America.)

Their importance for hydrothermal flow in the oceanic crust is threefold: they may host hydrothermal systems associated with their own volcanic systems, they reflect the positions of off-axis heat sources, and they may act as transmissive pathways for heat and fluid flow in areas where the surrounding crust is blanketed by sediments.

Near-axis seamounts are more prevalent on the flanks of fast-spreading centers than at slower-spreading centers (Figs. 3.4 and 3.10). They are associated more with segments having an axial ridge than with other morphologies (Fig. 3.10; Scheirer *et al.*, 1996). Very few seamounts are situated atop the axis of spreading itself; Axial Volcano, on the Juan de Fuca Ridge, is a notable exception. Rather, near-axis seamounts typically form in narrow zones on either side of the spreading axis, often producing chains of discrete volcanic cones and in some cases, volcanic ridges. Along most of the EPR, seamounts appear to form and grow between $\sim$10 and 40 km off-axis (Scheirer *et al.*, 1996). Seamount distribution

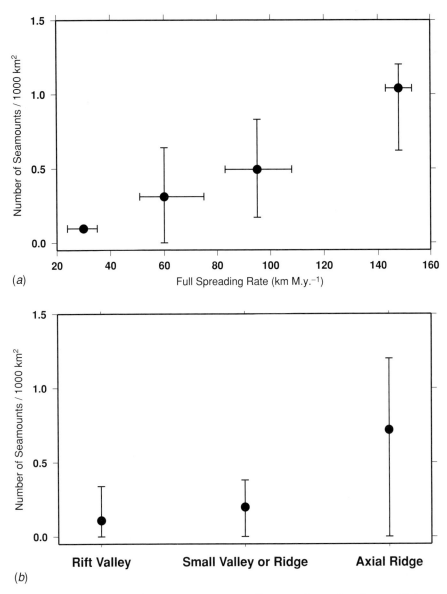

Fig. 3.10 (*a*) Areal abundance of seamounts versus spreading rate, where values from multiple spreading segments are combined into four spreading-rate ranges. Error bars indicate the full ranges of abundance and spreading rate. (*b*) Abundance of seamounts versus axial morphology. (Reprinted with permission from Scheirer, D. S. and Macdonald, K. C. 1995. Near-axis seamounts on the flanks of the East Pacific Rise, 8° N to 17° N. *J. Geophys. Res.* **100**: 2,239–2,259, Copyright 1995 American Geophysical Union.)

is sometimes asymmetric about the spreading center (e.g. Davis and Karsten, 1986). This, and the observation that the summits of seamounts are commonly shallower than the adjacent spreading centers (e.g. Fig. 3.4*a*), indicate that the magmatic plumbing systems feeding the MOR and off-axis seamounts are not directly connected, although their similar geochemistry indicates a similar mantle source (e.g. Fornari *et al.*, 1988).

The internal structure of near-axis seamounts is not known, for example, the balance of intrusive units versus extrusive flows. Caldera faults are present in many of the larger seamounts, and some edifices exhibit radial features reminiscent of terrestrial rift zones. Like basaltic island volcanoes, mass-wasting appears to be important in modifying the shape of the edifices of the larger near-axis seamounts. Because the porosity, permeability, and thermal conductivity of igneous units are greater than those of sedimentary units, near-axis seamounts will act as efficient conductors of heat and fluid flow from depth, as the ocean crust ages and is blanketed by sediments (e.g. Mottl *et al.*, 1998). Near-axis seamounts reach 1–2 km in height (Smith, 1988; Scheirer *et al.*, 1996), and thus they may facilitate mass and energy flow from the oceanic crust to the water column for many tens of millions of years.

3.4 The layered structure of oceanic crust

Since the early days of seafloor exploration, seismological methods have provided the primary source of information on the internal structure of *in situ* oceanic crust. These studies reveal a remarkably homogeneous structure compared with continental crust (e.g. Raitt, 1963; Spudich and Orcutt, 1980) comprising two igneous crustal layers: layer 2 characterized by low velocities and steep velocity gradients, and layer 3 by a high velocity and low gradient (Fig. 3.3; Chapter 5). Comparisons with ophiolites and laboratory measurements of dredged rocks and drill core samples have been used to relate the seismic velocity structure of oceanic crust to crustal lithology. A generalized view of crustal structure has emerged from these comparisons, where seismic layer 2 is commonly attributed to the upper crust comprising extrusive rocks and dikes, and where seismic layer 3 is associated with the lower crust of massive and cumulate gabbroic rocks.

Although this view is widely accepted, the geological significance of the velocity transition zones between seismic layers is a topic of ongoing debate. Seismic velocities are dependent on the physical properties of crustal rocks, including rock density and elastic moduli, which are in turn related to bulk porosity and rock composition. Within the upper crust, velocity boundaries are likely to correspond with porosity boundaries, which may or may not coincide with lithologic contacts. Furthermore, changes in crustal properties associated with seismic layer boundaries may migrate within the crust as the crust ages, and they may correspond with lithologic boundaries at some crustal ages but not at others. In this section, we summarize current understanding of oceanic crustal structure derived primarily from seismic studies, with a focus on inferences for crustal composition, porosity, and permeability. We pay particular attention to velocity transition zones because the gradients

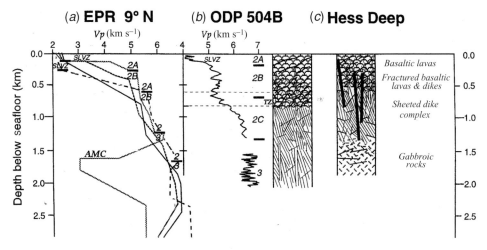

Fig. 3.11 Comparison of *in situ* seismic structure with the internal structure of the upper crust observed in drill core and tectonic exposures. (*a*) Seismic velocity with depth for newly formed crust at the East Pacific Rise. Layers 2A and 2B and the low velocities associated with the axial magma chamber (AMC) are identified. Data from Vera *et al.* (1990). (*b*) Comparison of P-wave velocities from *in situ* sonic logging within deep sea drilling Hole 504B and the lithologic units observed within the hole (Anderson *et al.*, 1982; Becker *et al.*, 1988; Schouten and Denham, 2000). (*c*) Lithologic cross-section for the upper crust at Hess Deep derived from submersible observations (Karson *et al.*, 2002). (Compiled figure from Karson, J. A. and Christeson, G. 2002. Comparison of geological and seismic structure of uppermost fast-spread oceanic crust: insights from a crustal cross-section at the Hess Deep Rift. In *Heterogeneity in the Upper Mantle: Nature, Scaling, and Seismic Properties*, eds. J. Goff and K. Holliger. New York: Kluwer Academic.)

in crustal properties associated with these boundaries are likely to play an important role in fluid flow within the crust. Gravity studies provide information on crustal densities from which porosities can be inferred and will be discussed briefly in the context of upper crustal structure.

3.4.1 Upper crust

At young ages, the shallowest crust is a thin layer (a few hundred meters) of low compressional (P-wave) velocity, <2.5 km s^{-1}, and high porosity, 10–30%, known as seismic layer 2A, Figs. 3.3 and 3.11 (e.g. Houtz, 1976; Berge *et al.*, 1992; Christeson *et al.*, 1996). Layer 2A is often correlated with the extrusive section of oceanic crust, and the high porosities are believed to reflect voids generated by volcanism and extensive fracturing of the crust. As will be discussed in detail in Chapter 5, the velocity of layer 2A increases as crust ages away from the spreading axis, reaching levels typical of deeper crustal rocks by <10 Ma (Grevemeyer and Weigel, 1996; Carlson, 1998). This increase in velocity is attributed to closure of cracks and volcanic porosity by compaction and to precipitation of hydrothermal products within voids in the extrusive section.

Seismic layer 2A is underlain by a 1.5–2-km-thick layer with seismic velocities of ~5.5–6.0 km s^{-1} known as layer 2B, which corresponds with some or all of the sheeted dike section (Fig. 3.11); in some studies this layer is subdivided into 2B and 2C. In zero-age crust, the boundary between layers 2A and 2B is defined by a steep velocity gradient within which P-wave velocities increase rapidly from <3 to 5 km s^{-1} over a distance of 100–300 m (e.g. Vera *et al.*, 1990; Christeson *et al.*, 1996). This increase in velocity corresponds to a decrease in porosity, reduced to ~10% at the top of layer 2B (Berge *et al.*, 1992). The velocity gradient zone defines the layer 2A/2B boundary and is typically sufficiently steep that it can be detected with seismic reflection methods (e.g. Harding *et al.*, 1993; Kent *et al.*, 1994; Rohr, 1994) and has been readily mapped in multi-channel seismic reflection observations of young crust created at both fast- and slow-spreading ridges (Figs. 3.12 and 3.13). In a number of recent studies, near-continuous images of this seismic boundary have been used to draw inferences on accretionary processes in different tectonic settings (e.g. Hooft *et al.*, 1996; Carbotte *et al.*, 1997). However, the geological significance of this boundary is not well understood. Two hypotheses have been advanced for the velocity gradient: that it corresponds to a mixed layer of extrusives and dikes located at the base of the extrusive section (e.g. Toomey *et al.*, 1990; Christeson *et al.*, 1996), or that it represents a porosity boundary within the extrusives associated with a fracture front or with hydrothermal alteration (e.g. McClain *et al.*, 1985; Wilcock *et al.*, 1992a). The primary evidence for a lithologic origin comes from observations along fault exposures at Hess Deep in the equatorial Pacific (Fig. 3.11). Here, the boundary between the dikes and extrusives lies at a similar depth range (200–800 m below seafloor) as the seismic layer 2A/2B boundary in young fast-spread crust (Karson *et al.*, 2002). However, the Hess Deep studies also reveal significant relief and complexity of the extrusive/dike boundary that occurs over horizontal distances of only a few hundreds of meters. This variability is well below the spatial resolution of seismic reflection studies, and it is clear that seismic observations can provide only a smoothed view of the geologic heterogeneity that may be present (Karson and Christeson, 2002).

In several recent studies, near-bottom gravity anomalies have been used to measure the density of the uppermost crust along the MOR (Fig. 3.14) and combined with rock-sample densities, to estimate the porosity of the extrusive layer. From the remarkably low crustal densities measured for several young flows on the Juan de Fuca Ridge (2280–2450 kg m^{-3}; Fig. 3.14), high porosities of 29–36% are calculated which are attributed to volcanically generated voids resulting from lava drainback and collapse (Pruis and Johnson, 1998). Similar low densities and high porosities are inferred for shallow basalts on the MAR at the TAG hydrothermal field (Evans, 1996) and on the EPR at 9° 50′ N (Cochran *et al.*, 1999). Johnson *et al.* (2000) use submersible gravity observations to infer porosities of 24% for the shallow extrusive section exposed along the Blanco Transform Fault and lower porosities of 5% for the deeper mixed extrusive and dike zone. These estimates of primary porosity are consistent with inferences from the low seismic velocities measured at the MOR

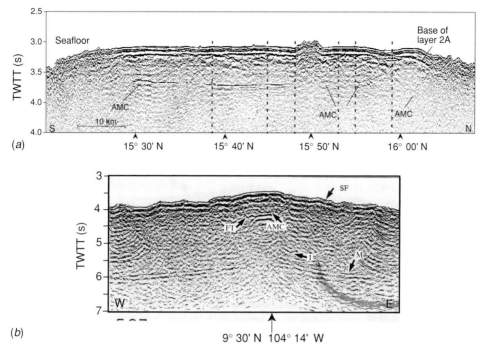

Fig. 3.12 Internal structure of young fast-spread crust imaged with multi-channel seismic reflection data. (*a*) Upper crustal profile oriented along the ridge axis showing reflections from the base of the extrusive crust (layer 2A) and the top of the axial magma chamber (AMC). The dashed, vertical lines on the seismic section mark the locations of very small offsets of the narrow depression along the axis where most volcanism is concentrated. (Reprinted with permission from Carbotte, S. M., Ponce-Correa, G., and Solomon, A. 2000. Evaluation of morphological indicators of magma supply and segmentation from a seismic reflection study of the EPR 15° 30′–17° N. *J. Geophys. Res.* **105**: 2,737–2,759, Copyright 2000 American Geophysical Union.) (*b*) Profile crossing the ridge axis at 9° 30′ N showing the seafloor (SF), axial magma chamber (AMC), other intra-crustal reflections (FT, I), and the Moho reflection (M). (Reprinted with permission from Barth, G. A. and Mutter, J. C. 1996. Variability in oceanic crustal thickness and structure: multichannel seismic reflection results from the northern East Pacific Rise. *J. Geophys. Res.* **101**: 1,7951–17,975, Copyright 1996 American Geophysical Union.)

(Fig. 3.3) and require the presence of large-scale voids at shallow levels in young extrusive crust.

3.4.2 *Lower crust*

The boundary between seismic layers 2 and 3 is defined as the change from a moderate to a low velocity gradient, which occurs at mid-crustal levels and at velocities of ~6.5–6.9 km s^{-1} (Figs. 3.3 and 3.11). Although this boundary has commonly been attributed to the dike–gabbro transition (e.g. Fox *et al.*, 1973), there is also evidence supporting an origin related to a change in metamorphic grade within the sheeted dikes (Carlson and

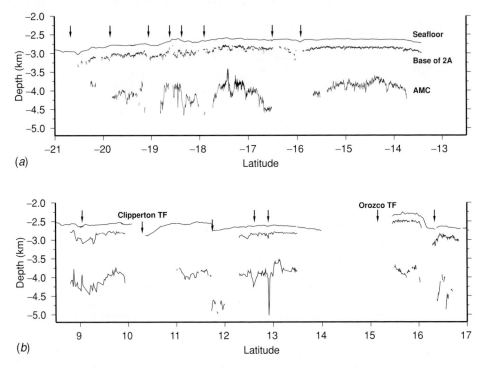

Fig. 3.13 Cross-section along the axis of the southern (*a*) and northern (*b*) East Pacific Rise showing depth to seafloor, the base of the extrusive crust (layer 2A), and the axial magma chamber (AMC) reflection. This compilation includes results from all multi-channel reflection data available along this ridge. Labeled arrows show the locations of transform faults; other arrows mark the locations of overlapping spreading centers. Data are from Detrick *et al.* (1987), Kent *et al.* (1993), Hooft *et al.* (1997), Babcock *et al.* (1998), Carbotte *et al.* (2000).

Herrick, 1990) or to the depth limit of a hydrothermal cracking front (Shaw, 1994). Evidence in support of the lithologic contact is strongest for young fast-spread crust where a bright reflection from the top of an axial magma body is imaged beneath the ridge axis (e.g. Detrick *et al.*, 1987, 1993a; Figs. 3.12 and 3.13) at approximately the same depths as the layer 2/3 boundary on the ridge flanks (1400–1800 m; Christeson *et al.*, 1997; Grevemeyer *et al.*, 1998). This magma body is believed to be the magma source of the upper crust and possibly much of the lower crust; where present, it marks the dike–gabbro contact. Visual observations at Hess Deep reveal an abrupt contact, less than 100 m thick, between intensely fractured dikes and less fractured massive gabbros (Karson *et al.*, 2002). Although no co-located seismic observations are yet available, the change in crustal porosity across this narrow lithologic contact would probably generate a change in seismic velocity structure consistent with the layer 2/3 boundary. At the Troodos ophiolite, a sub-horizontal brecciated zone is found within the dike–gabbro transition, which appears to accommodate contrasting deformation within the dike and gabbro sections (Agar and

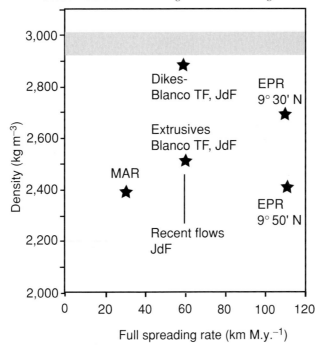

Fig. 3.14 Density of the uppermost crust at mid-ocean made from near-bottom gravity studies. Gray area denotes approximate range of rock sample densities for massive basalts. The low densities measured for *in situ* rock result from high porosities of 20–35% within the uppermost crust. Shown are data from the MAR (Evans, 1996); recent lava flows on the Juan de Fuca (JdF) Ridge (Pruis and Johnson, 1998); and two sites along the northern EPR (Cochran *et al.*, 1999). Measurements for the Blanco Transform Fault (TF) correspond with averages for the basalt and dike sections exposed along the fracture zone walls (Johnson *et al.*, 2000).

Klitgord, 1995). As at Hess Deep, the contrast in physical properties across this zone is likely to be sufficient to generate the change in crustal velocities associated with the layer 2/3 boundary.

The lithologic explanation of the seismic layer 2/3 boundary is opposed by observations of ~6 Ma crust drilled at Hole 504B, where the base of layer 2 lies well within the sheeted dike section (Fig. 3.11; Detrick *et al.*, 1994). At this site, the seismic layer 2/3 boundary corresponds with a gradual change in crustal properties, including bulk porosity and crustal alteration, over a depth interval of several hundreds of meters. Using downhole geophysical observations at 504B, Pezard (2000) infers a change in stress regime at the 2/3 boundary, with the axis of least compressive stress changing from horizontal to vertical, compatible with a change in crack closure and hence crustal porosity at this depth interval. Comparison of these studies with the near-ridge observations suggests that although the layer 2/3 boundary may coincide with a primary igneous contact in some locations (e.g. in young crust and in crust created at fast-spreading rates), this boundary is likely to evolve and migrate as the crust ages (Detrick *et al.*, 1994; Pezard, 2000).

The velocity of layer 3 (6.5–7.4 km s^{-1}) is generally similar to values measured for gabbroic rock samples (Christensen and Salisbury, 1975), and the lower crust is commonly inferred to be unfractured with low porosities (Figs. 3.3 and 3.11). Recently, however, this assumption has been called into question by studies indicating that the most commonly measured seismic velocities for the lower crust (6.9–7.0 km s^{-1}; White *et al.*, 1992) are 0.1–0.3 m s^{-1} too slow for expected lower crustal compositions (Behn, 2002); the presence of microcracks within the gabbroic crust has been proposed. Seismic reflection studies reveal bright, shallowly dipping reflections roughly 10–20 km apart within the lower section of both mature and young crust created at slow spreading rates (Fig. 3.15*a*; e.g. Banda *et al.*, 1992; Mutter and Karson, 1992). These reflections have been attributed to ductile shear zones associated with low-angle detachment faults that extend through the crust (Ranero and Reston, 1999). Similar studies of young, fast-spread crust indicate little coherent reflectivity within layer 3 (Barth and Mutter, 1996). However, prominent ridgeward-dipping lower crustal reflections spaced 5–10 km apart are observed in old fast-spread crust in the Pacific (e.g. Reston *et al.*, 1999; Fig. 3.15*b*), similar to those observed within the Atlantic. The origin of these Pacific reflectors is unclear; they cannot be connected to reflections in the upper crust nor to basement offsets, and they do not appear to be tectonic in origin. Reston *et al.* (1999) suggest that these events arise from mafic–ultramafic layering within the lower gabbros or from ductile shear zones associated with basal drag at the upper mantle/crust boundary.

The Mohorovičić Discontinuity (Moho) lies at the base of seismic layer 3 and corresponds to the transition from lower crustal to mantle velocities (7.6–8.0 km s^{-1}, Fig. 3.3). This change in P-wave velocity is often sufficiently abrupt that a sub-horizontal Moho reflection is observed in seismic reflection studies, allowing the base of the crust to be mapped. At the EPR, reflection Moho is commonly imaged to within a few kilometers of the ridge axis and occasionally beneath the magma lens reflection itself (Fig 3.12*b*; Barth and Mutter, 1996). The character of the Moho reflection in EPR crust varies from a single reflector to a diffuse or shingled event (Kent *et al.*, 1994). These variations presumably reflect changes in the structure and composition of the crust–mantle transition as observed in ophiolites, where the base of the crust can vary from a wide band of alternating lenses of mafic and ultramafic rocks to an abrupt and simple transition zone. The Moho is difficult to identify from seismic reflection data at the slow-spreading MAR. This difference may reflect poor imaging conditions associated with rough seafloor topography or a difference in the geological nature of this boundary.

Upper mantle rocks, including serpentinized ultramafics, are commonly recovered along the slow-spreading ridges near ridge discontinuities and from fracture-zone walls (e.g. Bonatti, 1976; Cannat, 1993; Tucholke and Lin, 1994). In these settings, mantle rocks may be emplaced at shallow crustal levels by faulting or serpentinite diapirism. Along some parts of the ultraslow-spreading ridges serpentinites dominate seafloor samples, and little igneous crust may be present due to very low melt production (Cannat, 1996). The presence

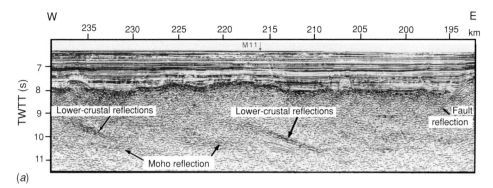

(a)

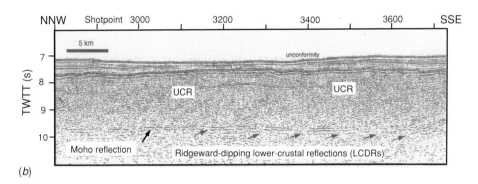

(b)

Fig. 3.15 Multi-channel seismic reflection lines showing events observed within mature crust created at (*a*) slow and (*b*) fast-spreading ridges. (*a*) Migrated sections from Mesozioc crust in the central Atlantic created at spreading rates of 20–34 km M.y.$^{-1}$. High-amplitude dipping reflections are imaged in the lower crust; these die out at the intermittent Moho reflection. Fault reflections and other events are observed in the upper crust. (Reprinted with permission from Banda, E., Ranero, C. R., Danobeitia, J. J., and Rivero, A. 1992. Seismic boundaries of the eastern central Atlantic Mesozoic crust from multichannel seismic data. *Geol. Soc. Am. Bull.* **104**: 1,340–1,349, Copyright 1992 Geological Society of America.) (*b*) Cretaceous age crust in the NW Pacific created at 100 km M.y.$^{-1}$. A bright upper crustal reflection (UCR) is observed at approximately the layer 2/3 boundary. In contrast to the Atlantic crust, a bright laterally continuous Moho (M) reflection is also observed. Dipping reflections are imaged within the lower crust, similar to those observed in the Atlantic although less prominent. (Reprinted with permission from Reston, T. J., Ranero, C. R., and Belykh, I. 1999. The structure of Cretaceous oceanic crust of the NW Pacific: constraints on processes at fast spreading centers. *J. Geophys. Res.* **104**: 629–644, Copyright 1999 American Geophysical Union.)

of serpentinites indicates that fluid pathways must exist from the seafloor to the mantle, and the distribution of these ultramafic rocks has important implications for patterns of crustal fluid flow. The densities and compressional wave velocities of serpentinized mantle rocks overlap with the range for lower crustal rocks (Christensen, 1972), and it can be difficult to distinguish between these rock types with geophysical studies. This ambiguity has lead to debate over the abundance of serpentinite within the crust. However, recent compilations of seismic velocities indicate oceanic gabbros and serpentinized peridotites can be distinguished and confirm the widely held view that, away from fracture zones, serpentinite is not a primary component of oceanic crust (Carlson and Miller, 1997).

3.5 Variations in crustal properties and implications for hydrologic structure

In this section we discuss variations in the internal structure of the crust created at different spreading rates, within ridge segments and with proximity to ridge-axis discontinuities, and due to episodic magmatic and tectonic phases. We end this section with a discussion of what is known of the geometry and distribution of faults within the crust. Although crustal properties cannot be directly inferred from gravity data without additional observations, gravity methods provide important constraints on lateral heterogeneity in the density and thickness of the crust and will be discussed briefly. In this section we also speculate on implications of crustal properties for the hydrologic structure of the crust. Although the geometry of fluid circulation within the crust is poorly understood (see Chapter 11), it must be closely linked to the gradients in crustal properties and the distributions of faults and fractures that form during crustal accretion at the MOR.

3.5.1 *Variations in primary igneous architecture of the crust at different spreading rates*

From detailed gravity and seismic experiments carried out since the mid 1980s, the influence of spreading rate on oceanic crustal structure is now well understood. Large variations in gravity anomalies are observed along the axis of slow-spreading ridges (Fig. 3.4*g*) indicating significant changes in crustal thickness and in crustal and mantle densities, which have been attributed to a three-dimensional pattern of mantle upwelling at these rates (Lin *et al.*, 1990). In comparison, only minor variations in gravity signals are observed along the fast-spreading ridges (Fig. 3.4*e*), and crustal accretion is believed to be a more two-dimensional process at these rates (Lin and Phipps Morgan, 1992). Seismic studies confirm these first-order differences and suggest other differences in the layered structure of oceanic crust. Fast- and slow-spread crust differ in the thickness of the upper crustal section (seismic layer 2) and in the internal architecture of the gabbroic and dike/lava layers; intermediate-spread crust can have properties across the spreading-rate spectrum. Nonetheless, some crustal properties are surprisingly similar throughout nearly the entire spreading-rate range, including the average thickness of the volcanic section inferred from seismic data and the average crustal thickness.

Upper crust

Similar low seismic velocities (2.3–2.5 km s^{-1}; Fig. 3.3) and low densities (2400–2700 kg m^{-3}; Fig. 3.14) are measured for the shallowest crust at fast- and slow-spreading ridges and the bulk porosity of the extrusive layer inferred from these observations appears to be insensitive to spreading rate. On the flanks of the MOR, the average thickness of the extrusive layer measured from seismic studies also appears to be insensitive to spreading rate and ranges from ~350 to 650 m (Fig. 3.16). However, the growth of this layer within the axial region differs significantly at fast- and slow-spreading ridges, with important implication for the distribution of upper crustal porosity inherited from ridge crest volcanic processes. Along the fast-spreading EPR, seismic layer 2A is remarkably uniform. It thickens from <200 m at the innermost axial zone to 400–500 m within a few kilometers of the axis (Harding *et al.*, 1993; Christeson *et al.*, 1996; Hooft *et al.*, 1997). This cross-axis thickening is attributed to the accumulation of extrusives via frequent eruptions of thin lobate and sheet flows that travel up to several kilometers from their eruption sites at the axis. Transport away from eruptive vents may occur on the seafloor or through sub-surface lava tubes. This transport is predicted to generate significant sub-horizontal layering as well as extensive zones of volcanic collapse within the extrusive section (e.g. Hooft *et al.*, 1996; Karson and Christeson, 2002). In contrast, layer 2A is slightly thicker on average and more variable (200–550 m) along the axis of the intermediate- to slow-spreading ridges (Fig. 3.16*a*), and to date the systematic cross-axis thickening found at the EPR has not been observed at these slower ridges (McDonald *et al.*, 1994; Hussenoeder *et al.*, 2002a). Full accumulation of layer 2A at the axis of these ridges may reflect the predominance of pillow flow eruptions that build steep-sided constructs with tens to hundreds of meters of relief in close proximity to eruption sites. Sub-horizontal layering within the extrusive layer may be much less prevalent within slow-spread crust.

Lower crust

The depth of the dike–gabbro transition at MORs, inferred from available seismic observations of mid-crustal magma bodies, is systematically related to spreading rate (Fig. 3.16*c*; Purdy *et al.*, 1992; Phipps Morgan and Chen, 1993). Whereas the combined thickness of the low-porosity extrusive and dike sections is only 1.2–1.5 km at fast-spreading ridges, a thicker layer of 2.5–3.5 km is formed at intermediate- to slow-spreading ridges. Ridge crest studies focused on the nature of the magma source for the crust suggest that the architecture of the underlying gabbroic section may be very different at fast- and slow-spreading centers. Two models have been proposed for construction of the lower crust at fast-spreading ridges. One view is that the lower crust forms via downward and outward transport of crystallized material from the mid-crustal magma lens (e.g. Henstock *et al.*, 1993; Phipps Morgan and Chen, 1993). This model predicts increasing crustal strain down-section that is compatible with the ridgeward-dipping layers observed within the gabbroic section of some ophiolites. The other hypothesis is that the lower layered gabbros form *in situ* through injection of sills that tap magma directly from upper mantle sources (e.g. Keleman *et al.*, 1997). At fast-spreading ridges there is evidence from both seismic and seafloor-compliance studies for melt accumulation at the base of the crust, either within melt-rich sills or with a broader partially molten zone (Dunn *et al.*, 2000; Crawford *et al.*, 1999). Magma may be

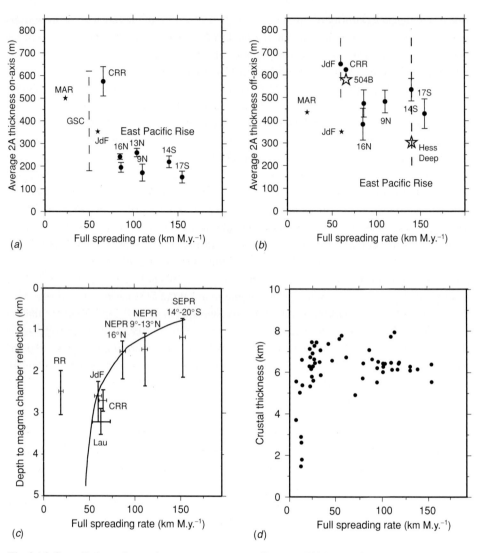

Fig. 3.16 Compilations of crustal structure versus spreading rate. Thickness of the seismically inferred extrusive crust (*a*) at the ridge axis and (*b*) off-axis. For data obtained from detailed reflection surveys, average thickness is shown with black dots and standard deviations where available (solid line) or thickness range (dotted line). East Pacific Rise data are labeled by survey location and are from: (16° N) Carbotte *et al.* (2000); (13° N) Babcock *et al.* (1998); (9° N) Harding *et al.* (1993); (14° S) Kent *et al.* (1994); and (17° S) Carbotte *et al.* (1997). Galapagos Spreading Center (GSC) from Detrick *et al.* (2002) converted from TWTT assuming a velocity of 2.5 km s⁻¹; Costa Rica Rift (CRR) from Buck *et al.* (1997) and Floyd *et al.* (2002); Juan de Fuca Ridge (JdF) from Rohr (1994). Filled stars show data derived from other seismic methods; Mid-Atlantic Ridge (MAR) from Hussenoeder *et al.* (2002b) and JdF from McDonald *et al.* (1994). Open stars show direct observations from Hole 504B (Becker *et al.*, 1988) and average thickness and total range from Hess Deep (Karson *et al.*, 2002). (*c*) Average depth of magma lens reflections beneath ridges versus spreading rate. The curved line

periodically injected from these sub-crustal bodies into the mid-crustal magma lens found at the base of layer 2 or into deeper intrusive bodies within the lower crust.

Persistent crustal magma bodies like those imaged at the EPR are not present at slow-spreading ridges (e.g. Sinton and Detrick, 1992). Instead, volcanic eruptions and emplacement of discrete plutonic bodies within the lower crust are envisioned to occur during episodic magma injection events from the mantle. Magma reservoirs may reside within the crust for short time periods and may be restricted to the shallow central portions of ridge segments, with magma delivered to the segment ends primarily through lateral dike propagation within the crust (e.g. Smith and Cann, 1999). Direct observations from seafloor fault exposures of crust formed at slow-spreading rates reveal a heterogeneous and highly variable crustal section that does not conform to the simple layered igneous structure derived from ophiolites (Karson, 1998). In places, the sheeted dike complex is absent, and lavas unconformably overlie altered and deformed gabbroic and ultramafic rocks. From these observations crustal accretion at slow-spreading ridges is inferred to be a highly discontinuous process with crustal architecture strongly influenced by tectonic disruption.

Global compilations of depth to seismic Moho reveal that the average thickness of the oceanic crust away from hot-spots is remarkably consistent, 6–7 km (Fig. 3.16*d*), over the entire spreading-rate range above 15 km M.y.$^{-1}$ (full-rate; e.g. Bown and White, 1994; Chen, 1992). Below spreading rates of 15 km M.y.$^{-1}$, the crustal thickness is typically small ($\sim$4 km) and more variable, possibly reflecting reduced melting due to enhanced conductive heat loss within the uppermost mantle (Bown and White, 1994). At all spreading rates there is no evidence for thickening of igneous crust away from the ridge crest; the crust appears to be fully formed within a few kilometers of the axis (Henstock *et al.*, 1993). The perturbation to oceanic crustal structure of a hot-spot volcanic anomaly is striking, in both proximal and distal settings. For example, the MAR-centered Iceland hot-spot produces crust up to 40 km thick beneath the island (Darbyshire *et al.*, 1998), and geochemical and morphological anomalies extend along the MAR for hundreds of kilometers from the center of the hot-spot (e.g. Schilling, 1991).

These observations of igneous structure with spreading rate have a number of implications for the hydrologic properties of the crust. Seismic and near-bottom gravity studies suggest little difference in the bulk porosity of the upper crust with spreading rate. However, there

---

←

shows the depth to the 1200 °C isotherm calculated from the ridge thermal model of Phipps Morgan and Chen (1993). Data from different ridges are labeled: Reykjanes Ridge (RR), Juan de Fuca Ridge (JdF), Costa Rica Rift (CRR), Lau Basin (Lau), northern and southern East Pacific Rise (NEPR and SEPR, respectively). (Reprinted with permission from Carbotte, S. M., Mutter, C., Mutter, J., and Ponce-Correa, G. 1998. Influence of magma supply and spreading rate on crustal magma bodies and emplacement of the extrusive layer: insights from the East Pacific Rise at lat 16° N. *Geology*, **26**: 455–458, Copyright 1998 Geological Society of America.) (*d*) Crustal thickness versus spreading rate from seismic studies away from fracture zones. (Reprinted with permission from Bown, J. W. and White, R. S. 1994. Variation with spreading rate of oceanic crustal thickness and geochemistry. *Earth Planet. Sci. Lett.* **121**: 435–449, Copyright 1994 Elsevier.)

may be systematic differences in the morphology and stratigraphy of lavas within the extrusive layer with spreading rate, and with related differences in the distribution of porosity inherited from volcanic processes. Extensive zones of sub-horizontal porosity associated with collapsed lobate flows and flat-lying flow contacts may be common within fast-spread crust, whereas the porosity of the lava layer may be more heterogeneous at slow-spread crust. Although the average thickness of the extrusive layer appears insensitive to spreading rate, a thicker combined extrusive-plus-dike section forms at slow-spreading rates and we expect the low-porosity layer of the crust to be correspondingly thicker. Finally, crustal accretion models predict differences with spreading rate in the igneous fabrics and strain patterns within the gabbroic section which could influence fluid pathways within the lower crust, with a more heterogeneous gabbro layer expected at slow rates.

3.5.2 *Variations in crustal structure associated with ridge segmentation*

Seafloor studies reveal systematic variations along ridge segments in a wide range of parameters including ridge morphology and depth, lava composition and morphology, and fault characteristics. Seismic and gravity studies indicate that changes in internal crustal structure also accompany these variations in seafloor properties. Overall, variations in crustal structure at fast-spreading ridges are modest compared with those at slow ridges. However, at all spreading rates, both short-wavelength variations associated with ridge-axis discontinuities and longer-wavelength variations within ridge segments are observed which give rise to along-axis lateral heterogeneity in crustal structure, at scales of tens to hundreds of kilometers. These along-axis variations are often more pronounced than the differences in average crustal properties observed with spreading rate, and the geometry of fluid circulation within the crust must be linked to this fundamental segmentation of the MOR (see Chapter 11).

Slow-spreading ridges

Along the slow-spreading ridges, numerous gravity and seismic studies indicate that anomalously thin crust is present within both large- and small-offset fracture zones (Fig. 3.17b). Reductions in crustal thickness of ~50% are inferred from gravity anomalies measured at a number of MAR fracture zones (Detrick *et al.*, 1995; Lin *et al.*, 1990). Seismic studies reveal crustal thicknesses as low as 1–2 km, as well as reduced P-wave velocities throughout the crust and the absence of a normal seismic layer 3 (Detrick *et al.*, 1993b). In addition to this extreme crustal thinning within the fracture zone trace, gradual thinning over several tens of kilometers toward ridge discontinuities is also observed (Fig. 3.17b). These along-axis changes primarily reflect a decrease in the thickness of seismic layer 3 (Tolstoy *et al.*, 1993). Layer 2 remains roughly constant in thickness, although seismic velocities within the uppermost crust are reduced near discontinuities.

The large-scale changes in crustal thickness found along the axis of the slow-spreading ridges are commonly attributed to focused delivery of magma from the mantle to the central portions of ridge segments (e.g. Lin *et al.*, 1990) along with enhanced tectonic stretching

East Pacific Rise (8.5°–10° N), Cocos plate

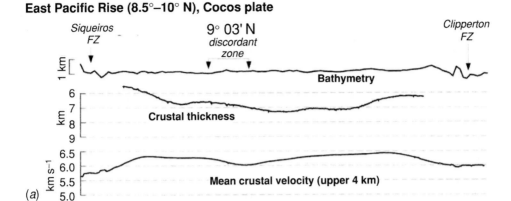

Mid-Atlantic Ridge (-35° N)

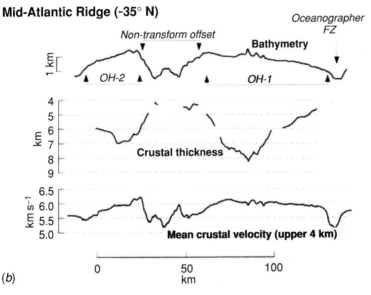

Fig. 3.17 Comparison of segmentation and crustal structure at (*a*) fast- and (*b*) slow-spread crust showing along-axis variations in crustal thickness and the mean seismic velocity of the upper crust. Markedly thinned crust is found at fracture zones and discordant zones associated with non-transform offsets at slow rates, whereas more modest changes in crustal thickness are observed at fast rates. Reduced seismic velocities near discontinuities are measured throughout the crust at all rates. (Reprinted with permission from Canales, J. P., Detrick, R. S., Toomey, D. R., and Wilcock, W. S. D. 2003. Segment-scale variations in crustal structure of 150- to 300-k.y.-old fast spreading oceanic crust (East Pacific Rise, 8° 15′ N-10° 15′ N) from wide-angle seismic refraction profiles. *Geophys. J. Int.*, Copyright 2002 The Royal Astronomical Society.)

near segment offsets. Plutonic and ultramafic rocks are frequently recovered near ridge segment ends (Karson, 1990); their presence has been attributed to low-angle detachment faulting at the "inside corners" of ridge–transform intersections interpreted from seismic images (Fig. 3.18, Mutter and Karson, 1992; Tucholke and Lin, 1994; Ranero and Reston, 1999). From these exposures of lower crustal and upper mantle rocks, extensive tectonic thinning of the crust is inferred. Reduced crustal production and tectonic extension near segment ends are likely to be closely linked with greater stretching of a thicker brittle lithosphere present where a thin crustal section is produced.

Fast-spreading ridges

Although fewer studies have been carried out at fast-spreading EPR transforms, available data provide little evidence for the extreme crustal thinning observed at the MAR. The Clipperton transform fault is the best studied of the fast-slipping transforms; here crustal thicknesses within the transform zone vary by less than 300 m, and seismic layer 3 is present (Van Avendonk *et al.*, 2001). However, as observed at MAR fracture zones, anomalously low P-wave velocities are observed throughout the crust and into the upper mantle near the Clipperton, which presumably reflects shearing and alteration within the transform zone. Low seismic velocities within layer 2 extend for over 10 km north of Clipperton transform and are associated with the increased brittle deformation observed north of the transform in seafloor fault studies (Carbotte and Macdonald, 1994). A more regional study of the EPR from the Siqueiros to Clipperton fracture zones reveals crustal thinning along the ridge segment of ~1 km and lower crustal velocities toward both transform faults (Canales *et al.*, 2003; Fig. 3.17*a*). Reduced seismic velocities within the upper and mid crust are also associated with the discordant zone trace left by the southward migrating overlapping spreading center (OSC) at 9° 03′ N (Christeson *et al.*, 1997; Canales *et al.*, 2003). These reduced velocities could reflect both a thicker extrusive section formed through ponding of lavas within the low-lying overlap depressions, as well as increased fracturing and rotation of crust within the overlap zone due to OSC migration. Similar reductions in crustal velocities are observed at other OSCs on the southern EPR (Bazin *et al.*, 1998). Variations in crustal structure associated with smaller ridge-axis offsets (of less than a few kilometers) have not been documented.

Away from transform faults and OSCs, gravity and seismic studies suggest that changes in crustal structure along the fast-spreading ridges are minor and there is general agreement that crustal accretion is more uniform at fast rates. However, the relationship between crustal production and the pattern of melt delivery from the mantle at fast-spreading ridges is a topic of ongoing debate. Systematic, although subtle, variations in crustal and seafloor properties are observed at fast ridges, and these have been attributed to three-dimensional mantle upwelling (e.g. Wang and Cochran, 1993), as inferred for slow-spreading ridges. Within the well-studied EPR 9°–10° N area, crust is thinnest within the central portion of the segment where ridge crest observations indicate active crustal accretion is focused (Barth and Mutter, 1996). This observation was originally attributed to focused mantle

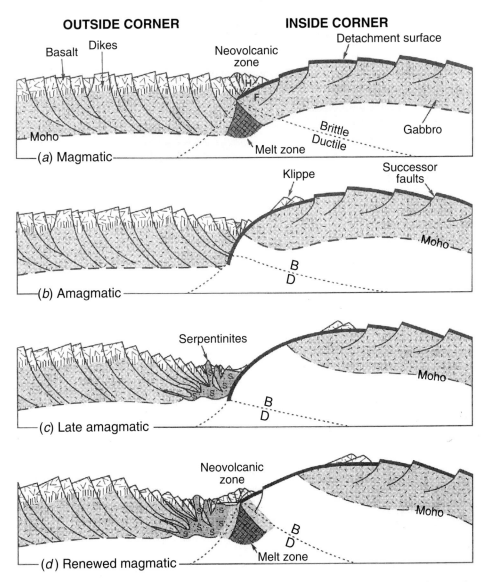

OUTSIDE CORNER   **INSIDE CORNER**

Fig. 3.18 Schematic cross-section of a slow-spreading ridge axis near a ridge–transform intersection showing the evolution of a detachment fault through a cycle of magmatic/amagmatic extension. H, hanging wall: F, foot wall: and S, serpentinites. (Reprinted with permission from Tucholke, B. E. and Lin, J. 1994. A geological model for the structure of ridge segments in slow spreading ocean crust. *J. Geophys. Res.* **99**: 11,937–11,958, Copyright 1994 American Geophysical Union.)

upwelling beneath this region, with the presence of thicker crust to the south due to either changes in the structure of the crust–mantle transition zone or crustal flow away from a mantle diapir. Recent studies indicate crustal thickness variations are less pronounced than originally inferred (Fig. 3.17*a*), and accretion processes associated with the southward migrating OSC are proposed to account for the crustal thickness variations in the region (Canales *et al.*, 2003).

These comparisons of along-axis variations in crustal structure at MOR suggest that pervasive fracturing and alteration of the crust occurs at transform and some non-transform offsets of the ridge at all spreading rates. Fracture zones provide potential fluid pathways to the mantle at all spreading rates and, on average, greater volumes of the crust may be exposed to migrating fluids at discontinuities than elsewhere along a ridge segment. Larger gradients in crustal properties are present in a ridge-parallel direction within crust formed at slow rates, and we expect the geometry of hydrothermal flow to be more three-dimensional at these rates.

3.5.3 Cyclicity in magmatic and amagmatic extension

In addition to variations in crustal structure inherited from ridge segmentation, lateral heterogeneity in crustal structure is also present in a direction perpendicular to the ridges due to fluctuations in crustal production through time. Early observations of the detailed volcanic and tectonic morphology of ridge segments revealed dramatic variations from one segment to the next. These observations gave rise to the notion that volcanism at the axis is episodic and that crustal formation involves periods of volcanism which alternate with periods of tectonic extension (e.g. Fig. 3.19; Lewis, 1979; Gente *et al.*, 1986; Kappel and Ryan, 1986). Models of episodic crustal accretion predict variability in the internal structure and composition of oceanic crust both down-section at a single location, as well as along crustal flowlines. Pezard *et al.* (1992) identify systematic changes in alteration and lava morphology within Hole 504B from which they infer that the basaltic section was formed during two major eruptive periods. Kappel and Ryan (1986) attribute the asymmetrical bow-form abyssal hills found on the flanks of the intermediate-spreading Endeavour segment to episodes of volcanic construction at the axis separated by tectonic phases. Along the slow-spreading MAR, the variable exposure of basaltic-plutonic and ultramafic rocks along the rift valley walls (Cannat, 1993) and flow-line variations in crustal thickness (1–2 km) inferred from gravity studies (Tucholke and Lin, 1994) are attributed to alternating magmatic and amagmatic periods at the ridge crest.

The duration of magmatic cycles at mid-ocean ridges is poorly constrained but appears to vary inversely with spreading rate from a few hundred thousand years to a few million years on the slow-spreading ridges (Cannat, 1993; Tucholke and Lin, 1994), to tens to hundreds of thousands of years at intermediate-spreading rates (Kappel and Ryan, 1986; Pezard *et al.*, 1992), and only a few thousand years on fast-spreading ridges (Haymon *et al.*, 1991; Reynolds *et al.*, 1992). The volumes of melt associated with these phases are also poorly

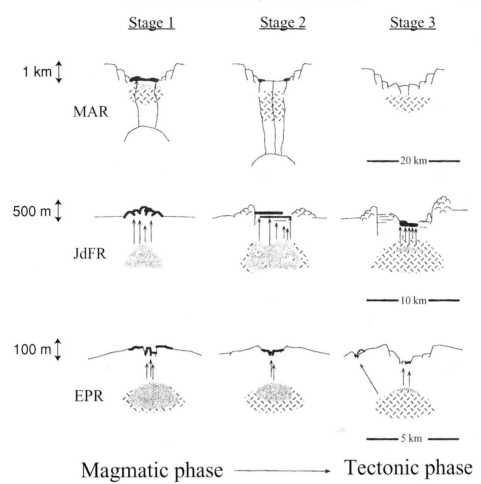

Fig. 3.19 Schematic representation of tectonomagmatic cycles at slow-, intermediate-, and fast-spreading ridges. In each case, the ridge axis evolves from a robust magmatic stage to one dominated by tectonism. Patterned region shows the sub-axial molten (gray) and partially molten (stippled) magma reservoir. (Reprinted with permission from Perfit, M. R. and Chadwick, W. W., Jr. 1998. Magmatism at mid-ocean ridges; constraints from volcanological and geochemical investigations. In *Faulting and Magmatism at Mid-Ocean Ridges, Geophysical Monograph 106*, eds. W. R. Buck, *et al.*, Washington, DC: American Geophysical Union, pp. 59–115, Copyright 1998 American Geophysical Union.)

known. However, larger volume fluctuations in melt delivery are expected at slow-spreading ridges where a steady-state crustal magma reservoir does not exist and where large episodic injections of magma from the mantle are needed to build the crust. Stronger horizontal gradients in crustal properties are predicted for slow spreading, with zones of high fault density where migrating fluids may be channelized alternating with less-disrupted crustal sections.

3.5.4 Fractures within the crust

For oceanic crust created away from ridge-axis discontinuities, the ridge-parallel faults and fractures formed within the axial zone provide primary pathways for fluid migration within the crust. Although these abyssal hill faults are smaller-scale features than transform faults, they influence more of the crust and represent the most abundant tectonic structures of the seafloor. These faults provide localized zones of crustal permeability (e.g. Becker *et al.*, 1994; see Chapter 7), which in some settings may extend through the entire crust, and where open connection to the water column may be maintained until the crust is buried by sediment. Fissures, joints, and finer-scale cracks within the upper crust form during dike injection, cooling, and stretching; these fractures also contribute to crustal porosity within the section. Microcracks due to cooling and thermal contraction may be present in the lower crust as well. The geometry, spatial density, and depth extent of fractures of all sizes formed during crustal accretion are important governing factors for ridge flank circulation. Fault geometry within the crust, whether planar or listric, will determine the igneous units that are exposed to migrating fluids along these major pathways. Fracture density and connectivity will control crustal permeability through much of the crust and the mass fraction of the crust that can be accessed by water. The depth extent of fractures is directly linked to the possible depth extent of hydrothermal circulation.

Fault plane geometry and depth extent

At slow-spreading ridges, abyssal hill faults may commonly extend through the entire crust, and there is evidence for both planar and listric fault-plane geometries at depth. Teleseismic and microseismicity studies indicate that earthquakes occur throughout the crust within the axial valley of the MAR (Huang and Solomon, 1988; Toomey *et al.*, 1988). High-resolution bathymetric observations of fault scarps (Macdonald and Luyendyk, 1977) and studies of near-ridge earthquakes (Huang and Solomon, 1988; Toomey *et al.*, 1988) suggest that planar fault surfaces dipping at 45–50° are typical both at the seafloor and within the crust. As noted above, lower angle (<30°) detachment faults are found near ridge-axis discontinuities (e.g. Tucholke and Lin, 1994; Karson, 1998). These faults expose lower crustal and upper mantle rocks and require faulting to extend through the entire crustal section. The planar and arcuate shallow-dipping seismic reflections, observed within layer 3 in slow-spread crust, flatten near the base of the crust and may be ductile shear zones associated with these detachment faults (Fig. 3.15*a*; Mutter and Karson, 1992). Reflection profiles of Early Cretaceous crust in the North Atlantic provide evidence for both steeply dipping normal faults and low-angle detachment faults that may comprise a "rolling-hinge" extension system (Reston *et al.*, 1996).

In contrast, at fast-spreading ridges, seismic imaging, earthquake studies, and near-bottom observations indicate that abyssal hill faults may be largely confined to the upper crust. Shallow focal depths of only 2 km are measured from microseismicity studies on the EPR (Riedesel *et al.*, 1982; Wilcock *et al.*, 1992b). Ridge-axis earthquakes large enough to be recorded teleseismically (earthquake magnitude, $m_b > 4$) have only been observed at slow-spreading ridges. In comparison, small earthquakes of magnitude ~ 1–3 are observed at fast-spreading ridges, reflecting the lower strength and thickness of the brittle layer in these regions. Seismic studies of Poisson's ratio at 9° N on the EPR suggest that fractures are confined to the upper crust and do not extend below depths of 1.5–1.7 km (Christeson *et al.*, 1997). The limited depth extent of fractures in fast-spread crust is also supported by observations of tectonic exposures at Hess Deep which reveal an abrupt decrease in brittle deformation across a narrow dike/gabbro contact located at a similar depth of 1.4 km (Karson *et al.*, 2002). Sub-horizontal intra-crustal reflections are observed in mature Pacific crust which indicate an abrupt change in crustal properties at close to the expected depths for the layer 2/3 boundary (Reston *et al.*, 1999; Fig. 3.15*b*). These reflections have been attributed to a structural discontinuity related to the maximum depth of hydrothermal circulation at the spreading center or the base of the sheeted dikes. The shallow-dipping reflections observed deeper within the lower crust in these same data cannot be traced into the upper crust, and Reston *et al.* (1999) argue against a tectonic origin for these structures.

Near-bottom observations reveal that fault planes are near vertical at the seafloor in fast-spread crust (e.g. Macdonald and Luyendyk, 1985). However, seafloor fault patterns indicate that significant fault-plane curvature must occur at depth at the fastest spreading rates, where clusters of small-throw closely spaced outward-facing faults are found antithetic to larger-throw, master inward-facing faults. From the pattern of hanging wall deformation, listric fault-plane geometries at shallow depths of ~ 1 km are inferred (Bohnenstiehl and Carbotte, 2001). At the somewhat slower-spreading northern EPR, inward and outward facing faults form horst and graben structures, and planar fault geometries are inferred. Bohnenstiehl and Carbotte (2001) attribute these differences in fault properties to changes in the local stress trajectories in the crust with proximity to the axial magma chamber.

The picture emerging from these studies is that for crust generated at fast-spreading ridges, ridge-parallel faults and fractures are likely to be largely confined to seismic layer 2 and faults may flatten at the layer 2/3 boundary. At slow-spreading rates, major faults can extend through the entire crustal section and possibly include ductile shear zones within the lower crust that flatten near the base of the crust. With these differences in fault characteristics, hydrothermal flow may be largely confined to the upper crust at fast rates whereas longer fluid pathways that extend through the entire crust are expected to be common at slow rates. Hence we expect that different igneous units will be exposed to migrating fluids within crust formed at different spreading rates, and the potential for interaction between fluids and mantle rocks will also vary with the rate of crustal formation.

Fracture density

Volcanic burial of tectonic relief appears to be an important process at fast-spreading ridges; hence fracture density inferred from seafloor observations must underestimate the extent of brittle deformation within the sub-surface. Volcanic draping of fault scarps is seen in submersible observations (Macdonald *et al.*, 1996) and significant fault burial is inferred from images of the base of seismic layer 2A (Carbotte *et al.*, 1997). Hess Deep observations reveal extensive brittle deformation extending through the lower lavas and dike section, including block rotations attributed to magmatic subsidence and inflation at the ridge axis (Fig. 3.11; Karson *et al.*, 2002). Evidence of caldera collapse within the zone of active crustal accretion will become largely buried by volcanic flows as seafloor is rafted onto the ridge flanks and hence will not be visible via seafloor studies (Fig. 3.2*a*). These observations suggest that brittle deformation within the upper section of fast-spread crust is inherited from both tectonic extension and magmatic processes, and it may be much greater than previously appreciated. Although larger-scale faults and higher extensional strains are found at slow-spread crust, the uppermost crust generated at fast-spreading rates may be more pervasively fractured, with perhaps a larger percentage of the upper crustal volume exposed to fluid percolation.

Seismic anisotropy studies provide a potential method to detect the abundance and alignment of open cracks within the crust, with the fast direction of seismic propagation parallel to the direction of crack alignment and the magnitude of anisotropy related to crack density. Studies of P-wave velocity reveal surprisingly similar anisotropies of 1–6% confined to the upper 2–3 km of fast-, intermediate-, and slow-spread crust (Sohn *et al.*, 1997; Barclay *et al.*, 1998; Dunn and Toomey, 2001). Larger anisotropies of 7–30%, also confined to the upper crust, have been measured in a recent study of shear wave splitting on the MAR (Barclay and Toomey, 2003). Differences in the magnitude of seismic anisotropy measured from S- and P-waves in this area may reflect the greater sensitivity of shear wave propagation to the highly fractured extrusive section. Although areal coverage is limited in these studies, the presence of abyssal hill faults does not appear to contribute to the anisotropy measured. These fault zones may be too widely spaced to be detected given the seismic wavelengths. The velocity anisotropy detected in these studies is attributed to broadly distributed, aligned fissures in the extrusives and to the presence of joints and microcracks within the sheeted dike section, formed by crustal extension and thermal contraction. These anisotropy studies highlight the potential importance of pervasive small-scale cracks, in addition to the larger-scale normal faults, as zones of fine-scale permeability, which if connected, may provide routes for fluid circulation.

3.6 Summary

In this chapter we have focused on the material properties of oceanic crust generated during crustal accretion. A complex array of magmatic and tectonic processes at mid-ocean ridges produces oceanic crust with significant heterogeneities in lithology and physical properties. Although our knowledge of the volcanic and tectonic architecture of the crust is considerable, comparatively little is known about how to relate this structure to the hydrologic properties of the crust. As described in Chapter 7, a key issue has been the difficulty of carrying out experiments in the submarine environment at the range of scales needed to characterize fluid flow. New experiments designed to characterize hydrologic properties in a range of tectonic settings are required as well as modeling studies that incorporate the complexity known to exist from geophysical characterization of the crust. Patterns and rates of fluid flow within the oceanic crust are expected to evolve through time as some flow paths seal, as new paths develop, and as the igneous crust becomes progressively isolated from direct seawater exchange by sediment burial. Understanding the evolution of fluid flow patterns through time provides a further challenge in defining the hydrologic structure of the crust. However, the physical properties of the crust resulting from crustal accretion define the initial boundary conditions for fluid flow and are expected to exert a strong influence on the patterns and rate of circulation both at the onset of fluid flow and long after crustal formation. Below, we summarize the intrinsic properties of the crust discussed in this chapter that contribute to crustal hydrologic structure.

The highest bulk porosities within oceanic crust (20–40%) are associated with the thin extrusive section that caps the crust along much of the mid-ocean ridge. Porosities inferred from seismic and gravity observations decrease to <10% at the top of the dikes and are believed to be low to negligible within the lower crust. High porosities within the extrusive layer are inherited from lava flow emplacement processes and from fissuring and fracturing associated with dike injection and brittle extension. Seafloor observations reveal differences in lava flow morphology and fracturing of young crust associated with spreading rate and position along a ridge segment, which will produce systematic differences in upper crustal porosity. Zones of sub-horizontal porosity associated with lobate and sheet flow contacts and volcanic collapse are expected within fast-spread crust, whereas the distribution of porosity within the extrusives inherited from volcanic processes may be more heterogeneous at slow rates. The available measurements of upper crustal porosity do not suggest a spreading rate dependence, although systematic studies as a function of the full spectrum of crustal accretion variables awaits further study.

Ridge-parallel normal faults bound the undulating abyssal hill topography of the ridge flanks and represent zones of primary crustal permeability within both shallow and, at slow-spreading rates, deeper crust. On the flanks of slow-spreading centers, abyssal hill faults with hundreds of meters of relief are common, and large tectonic strains of 10–15% are measured from seafloor fault populations. In these settings, active normal faults are associated with earthquakes located throughout the crust, and serpentinized lower crustal and upper mantle rocks recovered in seafloor samples indicate that significant crustal extension occurs and that some fluid pathways transect the entire crustal section. Tectonic relief is more subdued at fast-spread crust where abyssal hill faults are typically only a few tens of meters in height and where lower tectonic strains of 2–4% are measured. However, volcanic burial of tectonic relief is likely to be important at fast-spreading ridges and the extent of brittle deformation in the sub-surface may be much greater than is typically appreciated from seafloor studies. Intense deformation within the crustal lid above the axial magma chamber is expected due to magmatic processes at the axis, such as dike injection and inflation or subsidence of the magma source region; this strain will be largely masked by later-stage volcanic eruptions as seafloor is rafted out of the axial zone. Unlike slow-spread crust, fractures and abyssal hill faults may be largely confined to the upper crust at fast-spreading ridges, and the layer 2/3 boundary appears to be a major structural boundary. These observations suggest that fluid circulation within crust created at fast rates may be limited to the upper crust, at least for young ages, whereas flow paths within slow-spread crust may commonly extend through the entire crust.

At all spreading rates, ridge-axis discontinuities represent significant heterogeneities, both in seafloor morphology and in internal crustal structure. Fracture zones are regions of high-amplitude seafloor relief and reduced crustal velocities, reflecting extensive fracturing and shearing of the crust. As seafloor ages and becomes blanketed by sediment, seamounts and isolated topographic highs within fracture zones are likely to become increasingly important as sites of fluid venting to the overlying water column. Variations in crustal structure at fracture zones and other discontinuities are more pronounced at slow-spreading ridges, but they are observed at all rates and are associated with both transform and smaller non-transform offsets. From the larger gradients in crustal structure present within slow-spread crust, we expect that the geometry of fluid flow may be more strongly three-dimensional at these rates. However, gradients in crustal properties are observed with ridge segmentation at all spreading rates, and we expect the geometry of fluid circulation will be closely linked with this segmentation. Variations in crustal structure are also observed in the axis-perpendicular direction and arise from temporal changes in the supply of magma delivered to the mid-ocean ridge. Both the time scales associated with these magmatic cycles and the volumes of melt involved are poorly constrained. However, evidence suggests that longer time periods and larger volume fluctuations in magma delivery are associated with magmatic cycles at slow-spreading ridges, and greater crustal heterogeneity is expected for slow-spread crust.

The presence of neighboring hot-spot volcanic anomalies significantly perturbs the physical and thermal structure of MOR crust, as is clearly evident in young seafloor morphology. Hot-spots have modified ocean crustal structure independent of spreading rate in all ocean basins throughout the preserved record of ocean spreading. The specific influences of hot-spots on oceanic hydrothermal circulation are yet to be explored in detail, but hot-spots are likely to be important in both near-MOR and older oceanic settings.

Acknowledgments

We thank Del Bohnenstiehl, William Ryan, Rob Dunn, and Andrew Barclay for their insights and interesting discussions on crustal creation processes during the writing of this paper and Allegra Hosford-Scheirer for an early review of the manuscript. We thank Ingo Grevemeyer, Bob Detrick, and Earl Davis for their thorough reviews, which improved the manuscript. This paper benefited from research supported by US National Science Foundation grants OCE 94-02172 and OCE 98-11483 to SMC and OCE 96-15204 and OCE 99-11729 to DSS. Lamont-Doherty Earth Observatory contribution #6415.

References

Agar, S. M. and Klitgord, K. D. 1995. A mechanism for decoupling within the oceanic lithosphere revealed in the Troodos ophiolite. *Nature* **374**: 232–238.

Anderson, R. N., Honnorez, J., Becker, K., Adamson, A. C., Alt, J. C., Emmermann, R., Kempton, P. D., Kinoshita, H., Laverne, C., Motle, N. S., and Newmark, R. L. 1982. DSDP Hole 504B, the first reference section over 1 km through Layer 2 of the oceanic crust. *Nature* **300**: 589–594.

Babcock, J. M., Harding, A. J., Kent, G. M., and Orcutt, J. A. 1998. An examination of along-axis variation of magma chamber width and crustal structure on the East Pacific Rise between 13° 30′ N and 12° 20′ N. *J. Geophys. Res.* **103**: 30,451–30,467.

Banda, E., Ranero, C. R., Danobeitia, J. J., and Rivero, A. 1992. Seismic boundaries of the eastern central Atlantic Mesozoic crust from multichannel seismic data. *Geol. Soc. Am. Bull.* **104**: 1,340–1,349.

Barclay, A. H. and Toomey, D. S. R. 2003. Shear wave splitting and crustal anisotropy at the Mid-Atlantic Ridge, 35° N. *J. Geophys. Res.* **108**: 2378, doi: 10.1029/2001JB000918.

Barclay, A. H., Toomey, D. S. R., and Solomon, S. C. 1998. Seismic structure and crustal magmatism at the Mid-Atlantic Ridge, 35 degrees N. *J. Geophys. Res.* **103**: 17,827–17,844.

Barth, G. A. and Mutter, J. C. 1996. Variability in oceanic crustal thickness and structure: multichannel seismic reflection results from the northern East Pacific Rise. *J. Geophys. Res.* **101**: 17,951–17,975.

Batiza, R. and Vanko, D. 1983. Volcanic development of small oceanic central volcanoes on the flanks of the East Pacific Rise inferred from narrow-beam echo-sounder surveys. *Mar. Geol.* **54**: 53–90.

Bazin, S., van Avendonk, H., Harding, A. J., Orcutt, J. A., Canales, J. A., and Detrick, R. S. 1998. Crustal structure of the flanks of the East Pacific Rise: implications for overlapping spreading centers. *Geophys. Res. Lett.* **25**: 2,213–2,216.

Becker, K., *et al.* 1988. Site 504, Costa Rica Rift. In *Proceedings of the Ocean Drilling Program, Initial Reports, Part A*, Vol. 111. eds. K. Becker, H. Sakai, *et al.* College Station, TX: Ocean Drilling Program, pp. 35–251.

Becker, K., Morin, R. H., and Davis, E. E. 1994. Permeabilities in the Middle Valley hydrothermal system measured with packer and flowmeter experiments. In *Proceedings of the Ocean Drilling Program, Scientific Results*, Vol. 139, eds. M. J. Mottl, E. E. Davis, A. T. Fisher, and J. F. Slack. College Station, TX: Ocean Drilling Program, pp. 613–626.

Behn, M. D. 2002. The evolution of lithospheric deformation and crustal structure from continental margins to oceanic spreading centers, Ph.D. thesis MIT/WHOI Joint Program.

Berge, P. A., Fryer, G. J., and Wilkens, R. H. 1992. Velocity–porosity relationships in the upper oceanic crust: theoretical considerations. *J. Geophys. Res.* **97**: 15,239–15,254.

Bicknell, J. D., Sempere, J.-C., Macdonald, K. C., and Fox, P. J. 1987. Tectonics of a fast spreading center: a deep-tow and seabeam survey at EPR 19° 30′ S. *Mar. Geophys. Res.* **9**: 25–45.

Bohnenstiehl, D. R. and Carbotte, S. M. 2001. Faulting patterns near 19° 30′ S on the East Pacific Rise: fault formation and growth at a superfast-spreading center. *Geochem. Geophys. Geosyst.* paper no. 2001GC000156.

Bohnenstiehl, D. R. and Kleinrock, M. C. 2000. Evidence for spreading-rate dependence in the displacement–length ratios of abyssal hill faults at mid-ocean ridges. *Geology* **28**: 395–398.

Bonatti, E. 1976. Serpentinite protrusions in the oceanic crust. *Earth Planet. Sci. Lett.* **32**: 107–113.

Bown, J. W. and White, R. S. 1994. Variation with spreading rate of oceanic crustal thickness and geochemistry. *Earth Planet. Sci. Lett.* **121**: 435–449.

Buck, R. W., Carbotte, S. M., and Mutter, C. Z. 1997. Controls on extrusion at mid-ocean ridges. *Geology* **25**: 935–938.

Canales, J. P., Detrick, R. S., Danobeitia, J. J., Hooft, E. E. E., and Nar, D. 1997. Variations in axial morphology along the Galapagos Spreading Center and the influence of the Galapagos Hotspot. *J. Geophys. Res.* **102**: 27,341–27,354.

Canales, J. P., Detrick, R. S., Toomey, D. R., and Wilcock, W. S. D. 2003. Segment-scale variations in crustal structure of 150- to 300-k.y.-old fast spreading oceanic crust (East Pacific Rise, 8° 15′ N–10° 15′ N) from wide-angle seismic refraction profiles. *Geophys. J. Int.* **152**: 766–794.

Cannat, M. 1993. Emplacement of mantle rocks in the seafloor at mid-ocean ridges. *J. Geophys. Res.* **98**: 4,163–4,172.

  1996. How thick is the magmatic crust at slow spreading oceanic ridges? *J. Geophys. Res.* **101**: 2,847–2,857.

Carbotte, S. M. and Macdonald, K. C. 1994. Comparison of seafloor tectonic fabric at intermediate, fast, and super fast-spreading ridges: influence of spreading rate, plate motions, and ridge segmentation on fault patterns. *J. Geophys Res.* **99**: 13,609–13,633.

Carbotte, S. M., Mutter, J. C., and Xu, L. 1997. Contribution of volcanism and tectonism to axial and flank morphology of the southern East Pacific Rise, 17° 10′–17° 40′ S, from a study of layer 2A geometry. *J. Geophys. Res.* **102**: 10,165–10,184.

Carbotte, S. M., Mutter, J. C., Mutter, J. and Ponce-Correa, G. 1998. Influence of magma supply and spreading rode on crosta magma bodies and emplacement of the extrusive layer: insights from the East Pacific Rise at lat 16° N. *Geology* **26**: 455–458.

Carbotte, S. M., Ponce-Correa, G., and Solomon, A. 2000. Evaluation of morphological indicators of magma supply and segmentation from a seismic reflection study of the EPR 15° 30′–17° N. *J. Geophys. Res.* **105**: 2,737–2,759.

Carlson, R. L. 1998. Seismic velocities in the uppermost oceanic crust: age dependence and the fate of layer 2A. *J. Geophys. Res.* **103**: 7,069–7,078.

Carlson, R. L. and Herrick, C. N. 1990. Densities and porosities in the oceanic crust and their variations with depth and age. *J. Geophys. Res.* **95**: 9,153–9,170.

Carlson, R. L. and Miller, D. J. 1997. A new assessment of the abundance of serpentinite in the oceanic crust. *Geophys. Res. Lett.* **24**: 457–460.

Chen, Y. J. 1992. Oceanic crustal thickness verses spreading rate. *Geophys. Res. Lett.* **19**: 753–756.

Christensen, N. I. 1972. The abundance of serpentinites in the oceanic crust. *J. Geol.* **80**: 709–719.

Christensen, N. I. and Salisbury, M. H. 1975. Structure and constitution of the lower oceanic crust. *Rev. Geophys. Space Phys.* **13**: 57–86.

Christeson, G. L., Kent, G. M., Purdy, G. M., and Detrick, R. S. 1996. Extrusive thickness variability at the East Pacific Rise, 9°–10° N: constraints from seismic techniques. *J. Geophys. Res.* **101**: 2,859–2,873.

Christeson, G. L., Shaw, P. R., and Garmany, J. D. 1997. Shear and compressional wave structure of the East Pacific Rise, 9°–10° N. *J. Geophys. Res.* **102**: 7,821–7,835.

Cochran, J. R., Sempere, J.-C., and SEIR Scientific Team 1997. The Southeast Indian Ridge between 88° E and 118° E: gravity anomalies and crustal accretion at intermediate spreading rates. *J. Geophys. Res.* **102**: 15,463–15,487.

Cochran, J. R., Fornari, D. J., Coakley, B. J., Herr, R., and Tivey, M. A. 1999. Continuous near-bottom gravity measurements made with a BGM-3 gravimeter in *DSV Alvin* on the East Pacific Rise crest near 9° 31′ N and 9° 50′ N. *J. Geophys. Res.* **104**: 10,841–10,861.

Cowie, P. A. 1998. Normal fault growth in three dimensions in continental and oceanic crust. In *Faulting and Magmatism at Mid-Ocean Ridges*, Geophysical Monograph **106**, eds. W. R. Buck, *et al.* Washington, DC: American Geophysical Union, pp. 325–348.

Cowie, P. A., Scholz, C. H., Edwards, M., and Malinverno, A. 1993. Fault strain and mechanical coupling on midocean ridges. *J. Geophys. Res.* **98**: 17,911–17,920.

Crawford, W. C., Webb, S. C., and Hildebrand, J. A. 1999. Constraints on melt in the lower crustal and Moho at the East Pacific Rise, 9 degrees 48′ N, using seafloor compliance measurements. *J. Geophys. Res.* **104**: 2,923–2,939.

Curewitz, D. and Karson, J. A. 1998. Geological consequences of dike intrusion at mid-ocean ridge spreading centers. In *Faulting and Magmatism at Mid-Ocean Ridges*, Geophysical Monograph **106**, eds. W. R. Buck, *et al.* Washington, DC: American Geophysical Union, pp. 117–136.

Darbyshire, F., Bjarnason, I., White, R. S., and Flovenz, O. G. 1998. Crustal structure above the Iceland mantle plume images by the ICEMELT experiment. *Geophys. J. Int.* **135**: 1,131–1,149.

Davis, E. E. and Karsten, J. L. 1986. On the cause of the asymmetric distribution of seamounts about the Juan de Fuca Ridge: ridge-crest migration over a heterogeneous asthenosphere. *Earth Planet. Sci. Lett.* **79**: 385–396.

DeMets, C., Gordon, R. G., Argus, D. F., and Stein, S. 1994. Effect of recent revisions to the geomagnetic reversal time scale on estimates of current plate motions. *Geophys. Res. Lett.* **21**: 2,191–2,194.

Detrick, R. S., Buhl, P., Vera, E., Mutter, J., Orcutt, J., Madsen, J., and Brocher, T. 1987. Multichannel seismic imaging of a crustal magma chamber along the East Pacific Rise. *Nature* **326**: 35–41.

Detrick, R. S., Harding, A. J., Kent, G. M., Orcutt, J. A., Mutter, J. C., and Buhl, P. 1993a. Seismic structure of the southern East Pacific Rise. *Science* **259**: 499–503.

Detrick, R. S., White, R. S., and Purdy, G. M., 1993b. Crustal structure of North Atlantic fracture zones. *Rev. Geophys.* **31**: 439–458.

Detrick, R. S., Collins, J., Stephen, R., and Swift, S. 1994. In situ evidence for the nature of the seismic layer 2/3 boundary in oceanic crust. *Nature* **370**: 288–290.

Detrick, R. S., Needham, H. D., and Renard, V. 1995. Gravity anomalies and crustal thickness variations along the Mid-Atlantic Ridge between 33° N and 40° N. *J. Geophys. Res.* **100**: 3,767–3,787.

Detrick, R. S., Sinton, J. M., Ito, G., Canales, J. P., Behn, M., Blacic, T., Cushman, B., Dixon, J. E., Graham, D. W., and Mahoney, J. J. 2002. Correlated geophysical, geochemical and volcanological manifestations of plume–ridge interaction along the Galapagos Spreading Center. *Geochem. Geophys. Geosyst.* **3**(10): 8,501.

Dick, H. J. B. 1989. Abyssal peridoties, very slow-spreading ridges and ocean ridge magmatism, In *Magmatism in the Ocean Gasins*, Geological Society Special

Publication 42, eds. A. D. Sauners and M. J. Norry. London: Geological Society London, pp. 71–105.

Dunn, R. A. and Toomey, D. R. 2001. Crack-induced seismic anisotropy in the oceanic crust across the East Pacific Rise (9° 30′ N). *Earth Planet. Sci. Lett.* **189**: 9–17.

Dunn, R. A., Toomey, D. R., and Solomon, S. C. 2000. Three-dimensional seismic structure and physical properties of the crust and shallow mantle beneath the East Pacific Rise at 9° 30′ N. *J. Geophys. Res.* **105**: 23,537–23,555.

Dziak, R. P. and Fox, C. G. 1999. The January 1998 earthquake swarm at Axial Volcano, Juan de Fuca Ridge. *Geophys. Res. Lett.* **26**: 3,429–3,432.

Embley, R. W., Chadwick, W., Perfit, M. R., and Baker, E. T. 1991. Geology of the northern Cleft segment, Juan de Fuca Ridge: recent lava flows, sea-floor spreading, and the formation of megaplumes. *Geology* **19**: 771–775.

Escartin, J., Hirth, G., and Evans, B. 1997. Effects of serpentinization on the lithospheric strength and the style of normal faulting at slow-spreading ridges. *Earth Planet. Sci. Lett.* **151**: 181–190.

Escartin, J., Cowie, P. A., Searle, R. C., Allerton, S., Mitchell, N. C., Macleod, C. J., and Slootweg, A. P. 1999. Quantifying tectonic strain and magmatic accretion at a slow spreading ridge segment, Mid-Atlantic Ridge. 29° N. *J. Geophys. Res.* **104**: 10,421–10,437.

Evans, R. L. 1996. A seafloor gravity profile across the TAG hydrothermal mound. *Geophys. Res. Lett.* **23**: 3,447–3,450.

Floyd, J. S., Mutter, J. C., and Carbotte, S. M. 2002. Seismic reflection imaging of the evolution of oceanic crustal structure at the intermediate rate spreading Costa Rica Rift. *Eos, Trans. Am. Geophys. Union* **83** (47), Fall Meet. Suppl., Abstract T12B-1318.

Fornari, D. J., Perfit, M. R., Allen, J. F., Batiza, R., Haymon, R., Barone, A., Ryan, W. B. F., Smith, T., Simkin, T., and Luckman, M. A. 1988. Geochemical and structural studies of the Lamont seamounts: seamounts as indicators of mantle processes. *Earth Planet. Sci. Lett.* **89**: 63–83.

Fornari, D. J., Haymon, R. M., Perfit, M. R., Gregg, T. K. P., and Edwards, M. 1998. Axial summit trough of the East Pacific Rise 9–10°: geologic characteristics and evolution of the axial zone on fast-spreading mid-ocean ridges. *J. Geophys. Res.* **103**: 9,827–9,855.

Fox, P. J., Schreiber, E., and Peterson, J. J. 1973. The geology of the oceanic crust: compressional wave velocities of oceanic rocks. *J. Geophys. Res.* **78**: 5,155–5,172.

Gente, P., Auzende, J. M., Renard, V., Fouquet Y., and Bideau, D. 1986. Detailed geological mapping by submersible of the East Pacific Rise axial graben near 13° N. *Earth Planet. Sci. Lett.* **78**: 224–236.

Goff, J. 1991. A global and regional stochastic analysis of near-ridge abyssal hill morphology. *J. Geophys. Res.* **96**: 21,713–21,737.

Gregg, T. K. P. and Fink, J. H. 1995. Quantification of submarine lava-flow morphology through analog experiments. *Geology* **23**: 73–76.

Grevemeyer, I. and Weigel, W. 1996. Seismic velocities of the uppermost igneous crust versus age. *Geophys. J. Int.* **124**: 631–635.

Grevemeyer, I., Weigel, W., and Jennrich, C. 1998. Structure and aging of oceanic crust at 14° S on the East Pacific Rise. *Geophys. J. Int.* **135**: 573–584.

Grove, T. L., Kinzler, R. J., and Bryan, W. B. 1992. Fractionation of mid-ocean ridge basalt. In *Mantle Flow and Melt Generation at Mid-Ocean Ridges*, Geophyscial

Monograph 71, eds. J. Phipps Morgan, *et al.* Washington, DC: American Geophysical Union, pp. 281–310.

Harding, A. J., Kent, G. M., and Orcutt, J. A. 1993. A multichannel seismic investigation of upper crustal structure at 9° N on the East Pacific Rise: implications for crustal accretion. *J. Geophys. Res.* **98**: 13,925–13,944.

Haymon, R. M., Fornari, D. J., Edwards, M. H., Carbotte, S., Wright, D., and Macdonald, K. C. 1991. Hydrothermal vent distribution along the East Pacific Rise crest (9° 09′–54′ N) and its relationship to magmatic and tectonic processes on fast-spreading mid-ocean ridges. *Earth Planet. Sci. Lett.* **104**: 513–534.

Henstock, T. J., Woods, A. W., and White, R. S. 1993. The accretion of oceanic crust by episodic sill intrusion. *J. Geophys. Res.* **98**: 4,143–4,161.

Hooft, E. E. E., Schouten, H., and Detrick, R. S. 1996. Constraining crustal emplacement processes from the variation in seismic layer 2A thickness at the East Pacific Rise. *Earth Planet. Sci. Lett.* **142**: 289–309.

Hooft, E. E., Detrick, R. S., and Kent, G. M. 1997. Seismic structure and indicators of magma budget along the southern East Pacific Rise. *J. Geophys. Res.* **102**: 27,319–27,340.

Hosford, A., Lin, J., and Detrick, R. S. 2001. Crustal evolution over the last 2 m.y. at the Mid-Atlantic Ridge OH-1 segment, 35° N. *J. Geophys. Res.* **106**: 13,269–13,285.

Houtz, R. E. 1976. Seismic properties of Layer 2A in the Pacific. *J. Geophys. Res.* **81**: 6,321–6,331.

Huang, P. Y. and Solomon, S. C. 1988. Centroid depths of mid-ocean ridge earthquakes: dependence on spreading rate. *J. Geophys. Res.* **93**: 13,445–13,477.

Hussenoeder, S. A., Kent, G. M., and Detrick, R. S. 2002a. Upper crustal seismic structure of the slow spreading Mid-Atlantic Ridge, 35° N: constraints on volcanic emplacement processes. *J. Geophys. Res.* **107**: (BX), 10.1029/2001JB001691.

Hussenoeder, S. A., Detrick, R. S., Kent, G. M., Schouten, H., and Harding, A. 2002b. Fine-scale seismic structure of young upper crust at 17° 20′ S on the fast spreading East Pacific Rise. *J. Geophys. Res.* **107**: (BX), 10,1029/2001JB001688.

Ito, G. and Lin, J. 1995. Oceanic spreading center–hotspot interactions: constraints from along-isochron bathymetric and gravity anomalies. *Geology* **23**: 657–660.

Johnson, H. P., Pruis, M. J., Van-Patten, D., and Tivey, M. A. 2000. Density and porosity of the upper oceanic crust from seafloor gravity measurements. *Geophys. Res. Lett.* **27**: 1,053–1,056.

Kappel, E. S. and Ryan, W. B. F. 1986. Volcanic episodicity and a non-steady state rift valley along the northeast Pacific spreading centers: evidence from SeaMARC I. *J. Geophys. Res.* **91**: 13,925–13,940.

Karson, J. A. 1990. Seafloor spreading on the Mid-Atlantic Ridge: implications for the structure of ophiolites and oceanic lithosphere produced in slow-spreading environments. In *Proceedings of the Symposium "Troodos 1987"*, eds. J. Malpas, E. M. Moores, A. Panayiotou, and C. Xenophontos, Nicosia: Geological Survey Department, pp. 547–555.

1998. Internal structure of oceanic lithosphere: a perspective from tectonic windows. In *Faulting and Magmatism at Mid-Ocean Ridges*, Geophysical Monograph 106, eds. W. R. Buck, *et al.* Washington, DC: American Geophysical Union, pp. 177–218.

Karson, J. A. and Christeson, G. 2002. Comparison of geological and seismic structure of uppermost fast-spread oceanic crust: insights from a crustal cross-section at the Hess Deep Rift. In *Heterogeneity in the Crust and Upper Mantle: Nature, Scaling and Seismic Properties*, eds. J. Goff and K. Holliger. New York: Kluwer Academic, p. 358.

Karson, J. A., Klein, E. M., Hurst, S. D., Lee, C. E., Rivizzigno, P. A., Curewitz, D., Morris, A. R., and Hess Deep '99 Scientific Party. 2002. Structure of uppermost fast-spread oceanic crust exposed at the Hess Deep Rift: implications for subaxial processes at the East Pacific Rise. *Geochem. Geophys. Geosys.* **3**: (Paper number 2001GC000155).

Karsten, J. L., Hammond, S. R., Davis, E. E., and Currie, R. G. 1986. Detailed geomorphology and neotectonics of the Endeavor segment, Juan de Fuca Ridge: new results from Seabeam swath mapping. *Geol. Soc. Am. Bull.* **97**: 213–221.

Keleman, P. B., Koga, K., and Shimizu, N. 1997. Geochemistry of gabbro sills in the crust–mantle transition of the Oman ophiolite: implications for the origin of the lower crust. *Earth Planet. Sci. Lett.* **146**: 475–488.

Kent, G. M., Harding, A. J., and Orcutt, J. A. 1993. Distribution of magma beneath the East Pacific Rise between the Clipperton transform and the 9° 17′ N Deval from forward modeling of CDP data. *J. Geophys. Res.* **98**: 13,945–13,969.

Kent, G. M., Harding, A. J., Orcutt, J. A., Detrick, R. S., Mutter, J. C., and Buhl, P. 1994. Uniform accretion of oceanic crust south of the Garrett transform at 14° 15′ S on the East Pacific Rise. *J. Geophys. Res.* **99**: 9,097–9,116.

Klitgord, K. D., Huestis, S. P., Mudie, J. D., and Parker, R. L. 1975. An analysis of near-bottom magnetic anomalies: seafloor spreading and the magnetized layer. *Geophys. J. Roy. Astron. Soc.* **43**: 387–424.

Langmuir, C. H., Bender, J. F., and Batiza, R. 1986. Petrologic and tectonic segmentation of the East Pacific Rise, 5° 30′–14° 30′ N. *Nature* **322**: 422–429.

Langmuir, C. H., Klein, E. M., and Plank, T. 1992. Petrological systematics of mid-ocean ridge basalts: constraints on melt generation beneath ocean ridges. In *Mantle Flow and Melt Generation at Mid-Ocean Ridges*, Geophysical Monograph 71, eds. J. Phipps Morgan, *et al.* Washington, DC: Amercian Geophysical Union, pp. 183–280.

Lewis, B. T. R. 1979. Periodicities in volcanism and longitudinal magma flow on the East Pacific Rise at 23 degrees N. *Geophys. Res. Lett.* **10**: 753–756.

Lin, J. and Phipps Morgan, J. 1992. The spreading rate dependence of three-dimensional mid-ocean ridge gravity structure. *Geophys. Res. Lett.* **19**: 13–16.

Lin, J., Purdy, G. M., Schouten, H., Sempere, J. C., and Zervas, C. 1990. Evidence from gravity data for focused magmatic accretion along the Mid-Atlantic Ridge. *Nature* **344**: 627–632.

Macdonald, K. C. and Luyendyk, B. P. 1977. Deep-tow studies of the structure of the Mid-Atlantic Ridge crest near lat 37 degrees N. *Geol. Soc. Am. Bull.* **88**: 621–636.
    1985. Investigation of faulting and abyssal hill formation on the flanks of the East Pacific Rise (21° N) using Alvin. *Mar. Geophys. Res.* **7**: 515–535.

Macdonald, K. C., Fox, P. J., Perram, L. J., Eisen, M. F., Haymon, R. M., Miller, S. P., Carbotte, S. M., Cormier, M.-H., and Shor, A. N. 1988, A new view of the mid-ocean ridge from the behaviour of ridge-axis discontinuities. *Nature* **335**: 217–225.

Macdonald, K. C., Fox, P. J., Alexander, R. T., Pockalny, R. A., and Gente, P. 1996. Volcanic growth faults and the origin of abyssal hills on the flanks of the East Pacific Rise. *Nature* **380**: 125–129.

Malinverno, A. 1991. Inverse square-root dependence of mid-ocean-ridge flank roughness on spreading rate. *Nature* **352**: 58–60.

McClain, J. S., Orcutt, J. A., and Burnett, M. 1985. The East Pacific Rise in cross-section: a seismic model. *J. Geophys. Res.* **90**: 8,627–8,639.

McDonald, M. A., Webb, S. C., Hildebrand, J. A., Cornuelle, B. D., and Fox, C. G. 1994. Seismic structure and anisotropy of the Juan de Fuca Ridge at 45° N. *J. Geophys. Res.* **99**: 4,857–4,873.

Mottl, M. J. *et al.* 1998. Warm springs discovered on 3.5 Ma oceanic crust, eastern flank of the Juan de Fuca Ridge. *Geology* **26**: 51–54.

Muller, R. D., Roest, W. R., and Royer, J.-Y. 1998. Asymmetric sea-floor spreading caused by ridge-plume interactions. *Nature* **396**: 455–459.

Mutter, J. C. and Karson, J. A. 1992. Structural processes at slow-spreading ridges. *Science* **257**: 627–634.

Pariso, J. E., Sempere, J.-C., and Rommevaux, C. 1995. Temporal and spatial variations in crustal accretion along the Mid-Atlantic Ridge (29deg–31deg 30′ N) over the last 10 m.y.: implications from a three-dimensional gravity study. *J. Geophys. Res.* **100**: 17,781–17,794.

Perfit, M. R. and Chadwick, W. W., Jr. 1998. Magmatism at mid-ocean ridges: constraints from volcanological and geochemical investigations. In *Faulting and Magmatism at Mid-Ocean Ridges*, Geophysical Monograph 106, eds. W. R. Buck, *et al.* Washington, DC: Amercian Geophysical Union, pp. 59–115.

Pezard, P. A. 2000. On the boundary between seismic layers 2 and 3: a stress change? In *Ophiolites and Oceanic Crust: New Insights from Field Studies and the Ocean Drilling Program*, Geological Society of America Special Paper 349, eds. Y. Dilek, E. M. Moores, D. Etthon, and A. Nicolas. Boulder, CO: Geological Society of America, pp. 195–202.

Pezard, P. A., Anderson, R. N., Ryan, W. B. F., Becker, K., Alt, J. C., and Gente, P. 1992. Accretion, structure and hydrology of intermediate-spreading-rate oceanic crust from drillhole experiments and seafloor observations. *Mar. Geophys. Res.* **14**: 93–123.

Phipps Morgan, J. and Chen, Y. J. 1993. The genesis of oceanic crust: magma injection, hydrothermal circulation, and crustal flow. *J. Geophys. Res.* **98**: 6,283–6,297.

Pruis, M. J. and Johnson, H. P. 1998. Porosity of very young oceanic crust from sea floor gravity measurements. *Geophys. Res. Lett.* **25**: 1,959–1,962.

Purdy, G. M., Kong, L. S. L., Christeson, G. L., and Solomon, S. C. 1992. Relationship between spreading rate and the seismic structure of mid-ocean ridges, *Nature* **355**: 815–817.

Raitt, R. W. 1963. The crustal rocks. In *The Sea*, Vol. 3, ed. M. N. Hill. New York: Wiley-Interscience, pp. 85–102.

Ranero, C. R. and Reston, T. J. 1999. Detachment faulting at oceanic core complexes. *Geology* **27**: 983–986.

Reston, T. J., Ruoff, O., McBride, J. H., Ranero, C. R., and White, R. S. 1996. Detachment and steep normal faulting in Atlantic oceanic crust west of Africa. *Geology* **24**: 811–814.

Reston, T. J., Ranero, C. R., and Belykh, I. 1999. The structure of Cretaceous oceanic crust of the NW Pacific: constraints on processes at fast spreading centers. *J. Geophys. Res.* **104**: 629–644.

Reynolds, J. R., Langmuir, C. H., Bender, J. F., Kastens, K. A., and Ryan, W. B. F. 1992. Spatial and temporal variability in the geochemistry of basalts from the East Pacific Rise. *Nature* **359**: 493–499.

Riedesel, M., Orcutt, J. A., Macdonald, K. C., and McClain, J. S. 1982. Microearthquakes in the black smoker hydrothermal field, East Pacific Rise at 21° N. *J. Geophys. Res.* **87**: 10,613–10,623.

Rohr, K. M. M. 1994. Increase of seismic velocities in the upper oceanic crust and hydrothermal circulation in the Juan de Fuca Plate. *Geophys. Res. Lett.* **21**: 2,163–2,166.

Scheirer, D. S. and Macdonald, K. C. 1995. Near-axis seamounts on the flanks of the East Pacific Rise, 8° N to 17° N. *J. Geophys. Res.* **100**: 2,239–2,259.

Scheirer, D. S., Macdonald, K. C., Forsyth, D. W., and Shen, Y. 1996. Abundant
    seamounts of the Rano Rahi seamount field near the southern East Pacific Rise, 15° S
    to 19° S. *Mar. Geophys. Res.* **18**: 13–52.
Schouten, H. and Denham, C. R. 2000. Comparison of volcanic construction in the
    Troodos ophiolite and oceanic crust using paleomagnetic inclinations from Cyprus
    Crustal Study Project (CCSP) CY-1 and CY-1A and Ocean Drilling Program (ODP)
    504B drill cores. In *Ophiolites and Oceanic Crust: New Insights from Field Studies
    and the Ocean Drilling program*, Geological Society of America Special Paper 349,
    eds. Y. Dilek, E. M. Moores, D. Elthon, and A. Nicolas. Boulder, CO: Geological
    Society of America, pp. 181–194.
Schilling, J.-G. 1991. Fluxes and excess temperatures of mantle plumes inferred from
    their interaction with migrating mid-ocean ridges. *Nature* **242**: 565–571.
Shaw, P. R. 1992. Ridge segmentation, faulting and crustal thickness in the Atlantic
    ocean. *Nature* **358**: 490–493.
    1994. Age variations of oceanic crust Poisson's ratio: inversion and a porosity evolution
    model. *J. Geophys. Res.* **99**: 3,057–3,066.
Shaw, W. J. and Lin, J. 1996. Models of oceanic ridge lithospheric deformation:
    dependence on crustal thickness, spreading rate, and segmentation. *J. Geophys. Res.*
    101: 17,977–17,993.
Sinton, J. M. and Detrick, R. S. 1992. Mid-ocean ridge magma chambers. *J. Geophys.
    Res.* **97**: 197–216.
Small, C. 1998. Global systematics of mid-ocean ridge morphology. In *Faulting and
    Magmatism at Mid-Ocean Ridges*, Geophysical Monograph 106, eds. W. R. Buck,
    et al. Washington, DC: America Geophysical Union, pp. 1–26.
Smith, D. K. 1988. Shape analysis of Pacific seamounts. *Earth Planet. Sci. Lett.* **90**:
    457–466.
Smith, D. K. and Cann, J. R. 1999. Constructing the upper crust of the Mid-Atlantic
    Ridge: a reinterpretation based on the Puna Ridge, Kilauea Volcano. *J. Geophys. Res.*
    104: 25,379–25,399.
Smith, D. K. and Jordan, T. H. 1987. The size distribution of Pacific seamounts. *Geophys.
    Res. Lett.* **14**: 1,119–1,122.
Sohn, R. A., Webb, S. C., Hildebrand, J. A., and Cornuelle, B. D. 1997.
    Three-dimensional tomographic velocity structure of upper crust, CoAxial segment,
    Juan de Fuca Ridge: implications for on-axis evolution and hydrothermal circulation.
    J. Geophys. Res. **102**: 17,679–17,695.
Solomon, S. C., Huang, P. Y., and Meinke, L. 1988. The seismic moment budget of slowly
    spreading ridges. *Nature* **334**: 58–61.
Spudich, P. and Orcutt, J. A. 1980. A new look at the seismic structure of oceanic crust.
    Rev. Geophys. **18**: 627–645.
Steinberger, B. 2000. Plumes in a convecting mantle: models and observations of
    individual hotspots. *J. Geophys. Res.* **105**: 11,127–11,152.
Tolstoy, M., Harding, A. J., and Orcutt, J. A. 1993. Crustal thickness on the Mid-Atlantic
    Ridge: bull's-eye gravity anomalies and focused accretion. *Science* **262**: 726–729.
Toomey, D. R., Solomon, S. C., and Purdy, G. M. 1988. Microearthquakes beneath the
    median valley of the Mid-Atlantic Ridge near 23° N: tomography and tectonics. *J.
    Geophys. Res.* **93**: 9,093–9,112.
Toomey, D. R. Purdy, G. M., Solomon, S. C., and Wilcock, W. S. D. 1990. The
    three-dimensional seismic velocity structure of the East Pacific Rise near latitude
    9° 30′ N. *Nature* **347**: 639–645.

Tucholke, B. E. and Lin, J. 1994. A geological model for the structure of ridge segments in slow spreading ocean crust. *J. Geophys. Res.* **99**: 11,937–11,958.

Van Avendonk, H. J. A., Harding, A. J., Orcutt, J. A., and McClain, J. S. 2001. Contrast in crustal structure across the Clipperton transform fault from travel time tomography. *J. Geophys. Res.* **106**: 10,961–10,981.

Vera, E. E., Mutter, J. C., Buhl, P., Orcutt, J. A., Harding, A. J., Kappus, M. E., Detrick, R. S., and Brocher, T. M. 1990. The structure of 0–0.2-m.y.-old oceanic crust at 9° N on the East Pacific Rise from expanded spread profiles. *J. Geophys. Res.* **95**: 15,529–15,556.

Wang, X. and Cochran, J. R. 1993. Gravity anomalies, isostasy, and mantle flow at the East Pacific Rise crest. *J. Geophys. Res.* **98**: 19,505–19,531.

Wessel, P. 2001. Global distribution of seamounts inferred from gridded Geosat/ERS-1 altimetry. *J. Geophys. Res.* **106**: 19,431–19,441.

White, R. S., McKenzie, D. P., and O'Nions, R. K. 1992. Oceanic crustal thickness from seismic measurements and rare earth element inversions. *J. Geophys. Res.* **97**: 19,683–19,715.

Wilcock, W. S. D., Solomon, S. C., Purdy, G. M., and Toomey, D. R. 1992a. The seismic attenuation structure of a fast-spreading mid-ocean ridge. *Science* **258**: 1,470–1,474.

Wilcock, W. S. D., Purdy, G. M., Solomon, S. C., DuBois, D. L., and Toomey, D. R. 1992b. Microearthquakes on and near the East Pacific Rise, 9°–10° N. *Geophys. Res. Lett.* **19**: 2,131–2,134.

Wright, D. 1998. Formation and development of fissures at the East Pacific Rise: implications for faulting and magmatism at mid-ocean ridges. In *Faulting and Magmatism at Mid-Ocean Ridges*, Geophysical Monograph 106, eds. W. R. Buck, *et al.* Washington, DC: American Geophysical Union, pp. 137–152.

4

Fracturing and fluid flow in the oceanic crust: insights from borehole imaging and other downhole measurements

David Goldberg, Gerardo J. Iturrino, and Keir Becker

4.1 Introduction

During the past two decades, ocean drilling has provided new insight into the spatial and age-dependent variations in the hydrological properties of the crust. Studies of recovered core samples and remote seismic experiments provide information about the composition and structure of the ocean crust, but the *in situ* measurements made downhole have become viewed as essential and complementary to these studies. Boreholes are necessary for certain measurements, such as temperature, that must be made *in situ*. The process of drilling a hole in the ocean crust can perturb the hydrological state for periods of days to months, and because a route for the exchange of formation fluid and ocean bottom water is created, flow through the hole into or out of the formation may be induced, potentially masking signals from natural fluid flow within the crust. Thermal signatures of natural and induced flow are often distinguishable, however, and may be investigated by repeating temperature measurements over time. Drilling multiple holes in the ocean crust can yield insights into hydrological flow systems on a local scale, or into variations of hydrological properties of the crust on a regional scale, for example as they may vary with age.

Relatively new instruments that image the borehole wall with high resolution can reveal layers and faults that could be seen previously only in core sections, and only if that layer was recovered. Visualization of sub-surface structure can identify naturally open fractures, and can characterize sub-surface aquifers in a way that is not possible with only cores. Together with measurements of fluid temperature, natural gamma-ray, and other downhole logs, such images provide critical information about both active and relic fluid flow in the crust.

The scope of this review focuses on specific logging measurements that provide insight into the hydrogeology of the ocean crust, including: (1) measurements made in a borehole by instruments lowered on a wireline, (2) logging measurements made while drilling (LWD), and (3) wireline measurements repeated over time. Other methods of measuring *in situ* properties below the seafloor, such as those deployed for long-term observations (usually one year or more), and how hydrologic properties of the crust may be interpreted from those data, are discussed in detail in Chapters 7 and 8.

Hydrogeology of the Oceanic Lithosphere, eds. E. E. Davis and H. Elderfield. Published by Cambridge University Press.
© Cambridge University Press 2004.

Table 4.1 *Recent drilling and logging at crustal sites*

	North Atlantic	Mid-Atlantic	East Pacific	Mid-Pacific	West Pacific	West Indian
Holes drilled	20	16	2	16	26	3
Holes logged	0	1	2	3	7	2
Depth logged (m)	0.0	490.0	2078.8	146.5	367.0	572.0
Penetration (mbsf)[a]	125.7	490.0	2111.0	154.5	386.7	1508.0

[a]Penetration measured in meters below seafloor.

In over 30 years of ocean drilling expeditions, more than 1,250 sites have been occupied, but relatively few have included drilling and logging in the ocean crust. The deepest hole has been drilled to a depth greater than 2 km below the seafloor, yet it penetrates through only 30% of the oceanic crust. Table 4.1 summarizes the drilling and logging expeditions to crustal site locations since 1993. These encompass a variety of oceanic crustal environments from ridge axes to old mid-plate locations, flood basalt provinces, and highly deformed gabbro. Wireline logging has been conducted in fewer than 20% of these holes, and LWD data have been acquired in only two. With typical core recovery well-below 50%, downhole imaging and logging in these formations have been essential to evaluate the hydrogeological regime in the crust. Borehole resistivity imaging and temperature and gamma-ray logs provide information about fracturing and *in situ* fluid flow as a function of depth. In this review, four examples are presented where downhole imaging and logging measurements have been recorded. The locations have been selected to be representative of contrasting hydrothermal environments – two low-temperature off-axis sites in the northeastern Pacific and northern Atlantic, and two high-temperature axial sites in the northeastern and southwestern Pacific. A short background discussion is also given to define the range of measurements that can currently be made by logging.

4.2 Background

Most downhole measurements use technology developed by the oil and gas exploration industry. Parallel advances in the data transmission and computer technologies have led to a remarkable increase in the quantity and speed of downhole data acquisition and processing. Development of high-bandwidth and memory-intensive downhole imaging devices has expanded capabilities further. Several good reviews of downhole technologies have been published in a variety of industry journals (e.g. Snyder and Fleming, 1985; Prensky, 1994), and Prensky (1999) provides an excellent summary of current downhole imaging technologies. Doveton (1986), Serra (1987), Ellis (1987), Goldberg (1997), and others have summarized the applications. The interested reader is referred to these publications for more detailed discussions.

The following summary of downhole logging measurements is included to provide an overview of the fundamental methods that are used in fluid flow studies – temperature, natural gamma-ray, and electrical logging measurements – and the range of data and their measurement resolutions. Electrical methods include downhole imaging. All can be done by wireline and LWD techniques.

4.2.1 Temperature tools

Temperature logging, either from a drillship or by wireline re-entry, provides profiles of the borehole fluid temperature. This method correctly reflects *in situ* formation temperature if the fluid temperature has fully recovered from drilling induced disturbances. Alternatively, the equilibrium temperature can be extrapolated from measurements at several intermediate times. Persistent temperature perturbations are associated with zones in which fluid moves either into or out of the borehole through fractures (e.g. Goldberg *et al.*, 1991).

4.2.2 Natural gamma-ray activity tools

These instruments measure naturally occurring radioisotopes and mineral constituents of the formation. Perhaps the most common nuclear logging measurements are done with gamma-ray activity tools; they rely on simple statistical counting of sub-atomic particles emitted from scattered gamma rays generated by the decay of natural isotopes of potassium, uranium, and thorium in the Earth. They detect the radioactive decay using a scintillation counter and a crystal detector. The response of the detector is a simple function of the concentration by weight of the radioisotopes and the formation density. The average depth of penetration of the measurement into the borehole wall is about 0.5 m and vertical resolution is approximately 0.3 m (Allen *et al.*, 1988). New tool designs may improve this resolution by a factor of three–four (Goldberg *et al.*, 2001). The natural gamma-ray log responds primarily to clay content in a sedimentary formation, where naturally radioactive elements concentrate, or to alteration products in crustal and basement rocks that include clay minerals, oxides, and other diagenetic compounds. The concentration of the natural radioisotopes in the ocean crust is therefore largely controlled by alteration and diagenesis, which often results from chemical water–rock interactions due to *in situ* fluid flow.

Natural gamma-ray logs may be recorded with conventional wireline tools or using new LWD technology. LWD tools use sensors placed just above the drill bit to detect natural gamma rays, among other logging sensors, as drilling progresses (e.g. Allen *et al.*, 1989; Bonner *et al.*, 1992; Murphy, 1993). LWD logs are therefore made minutes after the bit cuts through formation so that the ephemeral *in situ* properties can be measured. Depending on the drilling rate, LWD, and conventional wireline gamma-ray logs can have nearly equal vertical resolution, although physical differences in the measurements often must be taken into account (Evans, 1991). An important advantage of LWD tools is that data usually can be acquired without gaps below the seafloor, at the bottom of the drill-hole, and through unstable intervals that do not stay open between drilling and wireline logging operations. Because core recovery in fractured basalt is often very poor, LWD technology promises to be effective for the acquisition of downhole logs in the upper igneous crust.

4.2.3 Electrical resistivity and imaging tools

Formation electrical properties are determined with tools that measure currents propagating through the borehole and pore fluids in the surrounding rock and sediment layers. Water is a ubiquitous and conductive fluid underground; its electrical conductivity is orders of magnitude greater than the host rock, and it increases with the concentration of Na^+ and Cl^- ions and with temperature. Electrical resistivity measurements in a formation thus allow estimates of porosity, fluid content, and the degree of fracturing. Clays also contribute to the measured electrical conductivity because of the negative ions that are commonly associated with the molecular structure of various Al-bearing minerals found in many types of clay.

Self-potential (SP) devices measure the electrical potential generated by ions flowing between the borehole and pore fluids. The SP measurement is sensitive to active fluid flow in the borehole and to the clay content of the formation (Revil and Glover, 1998). SP values are typically high where resistivities are low, such as in clay-rich formations, and the measurement is poor when seawater fills both the borehole and formation pore space. Induction devices for resistivity determination are used to measure lower (<100 Ωm) resistivities and have vertical resolutions ranging between 0.5 and 2.0 m; current-generating devices (laterologs) are used to measure higher resistivities (>100 Ωm) that may occur in the low-porosity calcareous and igneous rocks encountered in the ocean crust. The depth of sensitivity of these measurements into the borehole wall is related to the spacing between electrodes on the tools.

Electrical imaging tools deliver high-resolution pictures of the wall of a borehole using fine-scale measurements of electrical conductivity. Electrical images are oriented with respect to the geomagnetic field measured downhole. The wireline imaging tool used by the ODP, the Formation Mircroscanner™ (FMS), measures the borehole's surface conductivity on four pads pressed against the borehole wall with vertical resolution of approximately 5 mm, nearly 100 times finer than most other downhole measurements. The Azimuthal Array Imager™ (ARI) generates a full 360°, but coarser, image of the borehole wall using laterolog electrodes. LWD imaging with the Resistivity-at-bit™ (RAB) tool, a relatively new technology developed over the last five–ten years, uses sensors placed just above the drill bit which measure electrical conductivity as the drilling bit rotates (Lovell *et al.*, 1995). These sensors produce a full 360° scan of the borehole wall with vertical image resolution of approximately 5 cm, significantly coarser resolution than the FMS, although full wall coverage is usually achieved over the entire drilled interval.

4.2.4 Porosity and porosity logging tools

Measurements of the *in situ* porosity have been used extensively to evaluate the layered stratigraphy of the oceanic crust and to link other rock properties, such as velocity and permeability, to seismic data. Unlike resistivity, gamma-ray, and temperature tools, porosity logging tools provide indirect measurements of the formation properties. Several types of porosity tools exist that utilize neutron, sonic, and electrical methods, but it is not clear which type of log yields the best porosity estimate. We refer the reader to the summaries by Ellis (1987) and Goldberg (1997) for further reading on the estimation of porosity using logging tools.

Porosity variations in the ocean crust have been extracted from sonic logs (e.g. Moos *et al.*, 1990), but electrical methods have been used most often to estimate porosity in the ocean crust. An empirical formulation is typically employed to determine the apparent formation porosity which reflects both pore spaces filled with seawater and any conductive alteration minerals (Archie, 1942; Pezard, 1990). Neutron logging tools measure the scattering of gamma radiation from hydrogen nuclei and are commonly used to estimate porosity in sedimentary environments. However, in ocean crust it is difficult to distinguish hydrogen bound in clay and alteration minerals from free fluid porosity with neutron logging tools (Broglia and Ellis, 1990).

The estimation of porosity using any logging tool is complex and involves several assumptions. Porosity logs are often used jointly to determine a composite estimate, and such composite estimates have been related to variations in seismic velocities with depth (e.g. Detrick *et al.*, 1994). We do not further address these complexities in this paper, but instead focus on direct measurements of resistivity, conductivity imaging, and temperature using logs to infer the hydrological structure of the ocean crust.

4.3 Low-temperature off-axis hydrothermal environments

4.3.1 Costa Rica Rift flank

Hole 504B was drilled into 5.9 Ma oceanic crust in the eastern equatorial Pacific Ocean about 200 km south of the spreading axis of the Costa Rica Rift (Anderson *et al.*, 1982; Becker *et al.*, 1989; Alt *et al.*, 1993). Visited eight times by DSDP and ODP drilling expeditions (e.g. CRRUST, 1982), it provides the deepest penetration into oceanic crust and is an important reference site for the hydrological properties of oceanic crust. Alt *et al.* (1993) estimate the equilibrium bottom-hole temperature to be about 195 °C at a depth of 2,111 m below the seafloor. During each revisit, initial operations included acquiring temperature logs long after dissipation of any disturbances due to drilling circulation during the previous visit. An example of temperature logs collected during four of the revisits spanning 6.5 years is shown in Fig. 4.1. These logs illustrate both a stable gradient deep in the hole and, in the

-->

Fig. 4.1 Temperature, resistivity, gamma-ray, derived fracture porosity, and FMS logs in the upper 600 m of Hole 504B on the flank of the Costa Rica Rift. The successive temperature logs indicate a decrease, then an increase with time in downflow of ocean bottom water to a depth of 300–400 m into the crust. Resistivity logs are used to compute vertical and horizontal fracturing profiles, and indicate that horizontal fracturing occurs only at shallow depths in Hole 504B. Gamma-ray logs were recorded through the drill pipe above 288 mbsf and in open hole below. Increases in gamma-ray values in the porous pillow and flow units above ~300–400 mbsf indicate the presence of low-temperature, K-rich alteration minerals. FMS images confirm both distinct steep and shallow fracture occurrences that serve as conduits for fluid circulation in the shallow ocean crust.

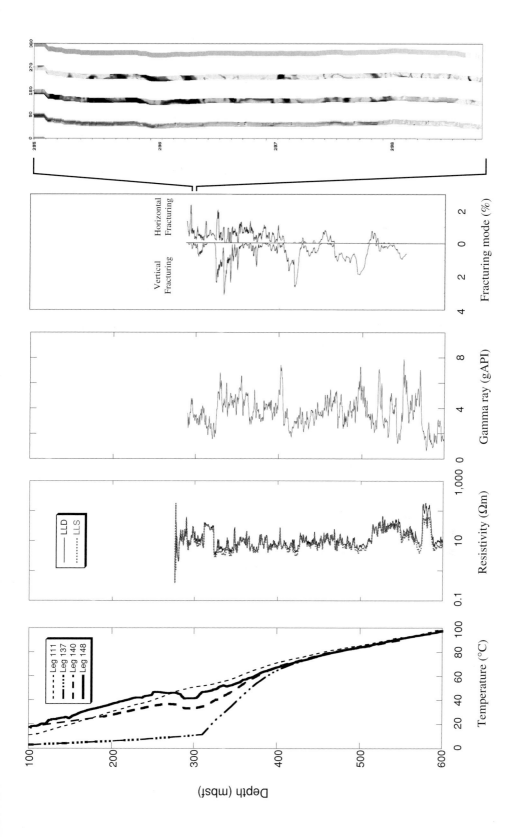

shallow section, the effects of a time-varying downhole flow of ocean bottom water into permeable uppermost basement (e.g. Becker *et al.*, 1983; Gable *et al.*, 1989). This flow was probably initially induced by drilling, but it persisted long after drilling operations ended. The high intrinsic permeability of the porous and fractured uppermost pillow lavas and the natural local sub-hydrostatic pressure promote such downhole flow (e.g. Becker *et al.*, 1983; Chapter 7), and also cause vigorous hydrothermal circulation in the upper basement (e.g. Fisher *et al.*, 1997; Chapter 8). This circulation may be vertically compartmentalized; Pezard *et al.* (1992) suggest that low permeability massive flow units in the uppermost basement section may act as hydrological barriers between higher permeability pillow units.

Resistivity measurements have been used most often to estimate the porosity and fracturing in Hole 504B. Becker (1985) used a large-scale electrode array to measure average resistivities over 10–50-m intervals, and then estimated bulk porosities over these intervals. Subsequent lateral resistivity logs generally agree with the large-scale resistivities but provide more detail. Pezard (1990) estimated the amount of fracturing using the lateral resistivity log and FMS imaging (Fig. 4.1). Brewer *et al.* (1999) analyzed the character of the FMS logs recorded in Hole 504B and in nearby Hole 896A, and identified the most fractured and brecciated zones in the upper 300–400 m of pillow lavas. Boldreel *et al.* (1995) derived a relationship between temperature and fracture porosity from electrical resistivity. Their analysis shows a correlation between fracturing and temperature gradient in the shallow interval above 300 mbsf that is attributed to fluid flow through open pore space and permeable pathways. Underlying the pillow lava section are sheeted dikes with resistivity-determined porosities less than 1% and permeabilities of the order of 10^{-17} m^2, suggesting that only the uppermost few hundred meters of basement at the site is significantly permeable (Becker *et al.*, 1989; Alt *et al.*, 1993). Other studies, such as those presented in Chapters 7 and 8, suggest greater permeabilities, although all results from logging and downhole measurements indicate that the highest permeability values exist in the shallowest crustal layers.

Fluid flow through permeable layers in the oceanic crust is important in controlling the chemistry of the rock. Anderson *et al.* (1985) measured permeability in Hole 504B using a packer and concluded that fracturing contributes most to the open porosity and permeability and extends only to a shallow depth. The interaction of seawater flowing through permeable cracks and faults results in chemical exchanges with the rock (Alt *et al.*, 1986). In fresh basalt subjected to low-temperature alteration, K and Mg from seawater typically replace Ca, Fe, and Si. Anderson *et al.* (1989) estimated the bulk elemental distribution from geochemical logging data and computed the relative chemical change that the rocks had undergone relative to fresh basalt from newly formed oceanic ridges. K was found to be enriched by 20% in the upper crust and depleted by 10% in the lower crust. Figure 4.1 illustrates the natural gamma-ray log response in Hole 504B, which is dominated by increased K concentrations in altered basalts. These findings suggest that hydrothermal fluids have transported K from deeper to shallower levels, and deposited K-rich alteration products in high-porosity and permeable zones that later become sealed.

When drilling in Hole 504B was abandoned in 1993, the companion re-entry Hole 896A was drilled about 1 km away on a sediment-covered basement knoll with elevated seafloor heat flow (Alt *et al.*, 1993). This hole penetrates 469 mbsf, 290 m into igneous basement, and was logged with temperature and electrical resistivity and imaging tools, as was Hole 504B. In general, the physical state of the crust in both holes is similar, with high permeability through the fractured pillows and flow in the upper basement (Becker, 1996). Eight years after both holes were last drilled, however, wireline re-entry temperature logging in Hole 896A indicated that it is producing warm fluids from at least three zones in the upper basement, whereas Hole 504B continues to draw ocean bottom water downhole at a slow rate (Becker *et al.*, in press). Hole 896A is one of several documented examples of producing holes in the upper ocean crust, where warm formation fluids or superhydrostatic pressures are present; all occur in settings much like that of Hole 896A – sediment-covered basement highs (Davis and Becker, 1994; Fisher *et al.*, 1997; Becker and Davis, 2003). The so-called "chimney" effect in such settings is discussed further in Chapter 8.

4.3.2 North Pond, Mid-Atlantic Ridge flank

Hole 395A was drilled in "North Pond," a small (150 km^2) sedimentary basin on approximately 7 Ma crust, 110 km from the Mid-Atlantic Ridge at 23° N. Analysis of the recovered drill core (19% average recovery) shows that it consists of alternating layers of aphyric basalts, phyric basalts, breccias (mainly basaltic), and intrusive dolerites, with formation porosity values typically ranging from 10 to 35% (Melson *et al.*, 1978). Subsequent visits to the site for downhole logging have taken place during three ODP and DSDP legs, as well as during the DIANAUT diving series (Gable *et al.*, 1992). Figure 4.2 shows the temperature, resistivity, gamma-ray, SP, and ARI and FMS image logs recorded in Hole 395A. Temperature logs show strongly depressed borehole temperatures that are nearly isothermal to a depth of 300–400 m into basement (Becker *et al.*, 1984; Kopietz *et al.*, 1990; Gable *et al.*, 1992). The near-isothermal temperatures, as well as packer and flowmeter experiments, indicate that this section of basement is much more permeable that the underlying formation and that there is a strong downhole flow of ocean bottom water into the permeable upper basement (Hickman *et al.*, 1984; Becker, 1990; Morin *et al.*, 1992). Recent monitoring experiments have confirmed that cool seawater had flowed downhole and then laterally in the shallow brecciated and fractured intervals in Hole 395A, and have showed that the upper part of the igneous section continues to be kept relatively isothermal by natural circulation even now that downhole flow has been stopped (Becker *et al.*, 1998, 2001).

In Fig. 4.2, the resistivity and image logs define pillow lava and massive flow layer thicknesses. The layered structure of pillows and flows is apparent, as well as the electrically conductive (dark colored) fluid-filled contacts and fractures within them. The flow sequence identified by gradual upward increases in the gamma-ray and resistivity logs confirms the presence of cyclic alteration trends that correspond to the historical phases of volcanic quiescence and eruption (Hyndman and Salisbury, 1984; Moos, 1989; Bartetzko *et al.*, 2001). These cycles arise from the combined processes of volcanism, tectonism, and diagenesis that replace basalt and fills void spaces with low-temperature, K-rich alteration minerals. All of these processes contribute to the physical structure of the shallow crust, although higher porosity and fracturing above 400 mbsf allows for more extensive fluid exchange

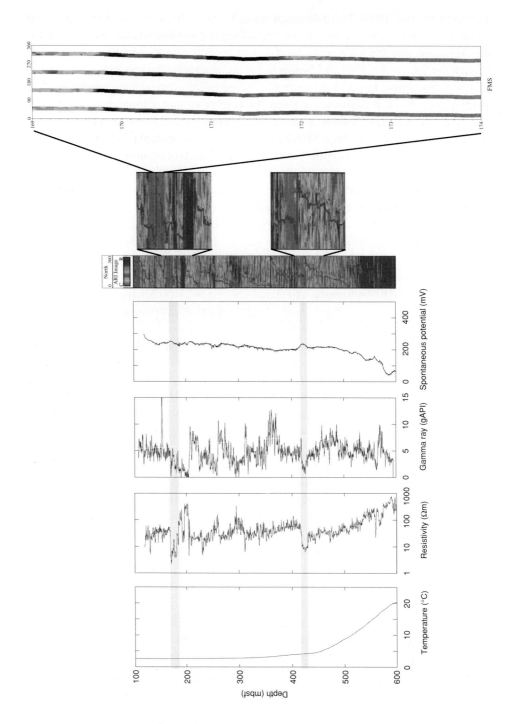

between the basalt and seawater. This supports the model for lateral fluid flow through a 400-m-thick crustal layer with high permeability and porosity, sandwiched between the North Pond sediments above and a deeper layer with low permeability and porosity (Langseth *et al.*, 1992).

In low-temperature off-axis environments, such as these sites near the Costa Rica Rift and Mid-Atlantic Ridge, the ocean crust remains permeable to several hundred meters depth for at least several million years. The interaction of seawater flowing through permeable layers in the shallow crust continues to alter the basalt chemistry and fill open fractures and pore space until the permeable pathways are closed. The process of drilling boreholes in this environment opens the exchange of cool bottom water with warmer formation fluids in the crust. Downflow may persist over long periods of time in certain cases such as at 395A, or subside as at 504B, but the temperature logs initially recorded at these sites all indicate borehole temperatures that are strongly cooled by the downflow of ocean bottom water. Hole sealing is required for conditions to return to an unperturbed state.

4.4 High-temperature hydrothermal environments

4.4.1 Middle Valley, Juan de Fuca Ridge

Two active hydrothermal vent systems located within the sediment-filled Middle Valley rift on the northern Juan de Fuca Ridge were drilled during two ODP cruises (Davis *et al.*, 1992a,b; Fouquet *et al.*, 1998). Site 858 was drilled in the sediment-hosed Dead Dog vent field, situated above a local basement edifice, and Site 856 targeted the Bent Hill massive sulfide deposit located about 3 km away. A third site, 857, drilled through the sediment fill of the valley into basement rocks that were inferred to be the source of hot water for both vent sites (Davis and Fisher, 1994).

Hole 858G in the Dead Dog vent field was sealed and instrumented with a CORK hydrologic borehole observatory (Davis and Becker, 1994). A temperature log was collected a few hours after the instrumentation had been removed for refurbishment; it recorded a rapid increase in temperature from bottom-water values at the seafloor to approximately 228 °C at 9 mbsf (Fig. 4.3*a*; Fouquet *et al.*, 1998), demonstrating that the hole was producing hot fluid, in contrast to the initial downhole flow of cool ocean water when the hole was first drilled.

Fig. 4.2 Temperature, resistivity, gamma-ray, SP, ARI, and FMS images in the upper 600 m of Hole 395A near the Mid-Atlantic Ridge. Temperature is nearly isothermal down to 400 mbsf due to the drawdown of cold ocean bottom water into the permeable formation. The resistivity and SP logs indicate high porosity and active fluid flow in several isolated zones above this depth. Upward increases in the gamma-ray and resistivity logs in several cycles indicate relic low-temperature alteration within porous and permeable rocks. Both ARI and FMS images show conductive zones (dark banding) near 180 and 420 mbsf, indicating open fractures and brecciated intervals that likely serve as major conduits for natural fluid flow as well as drawdown into the crust via the borehole.

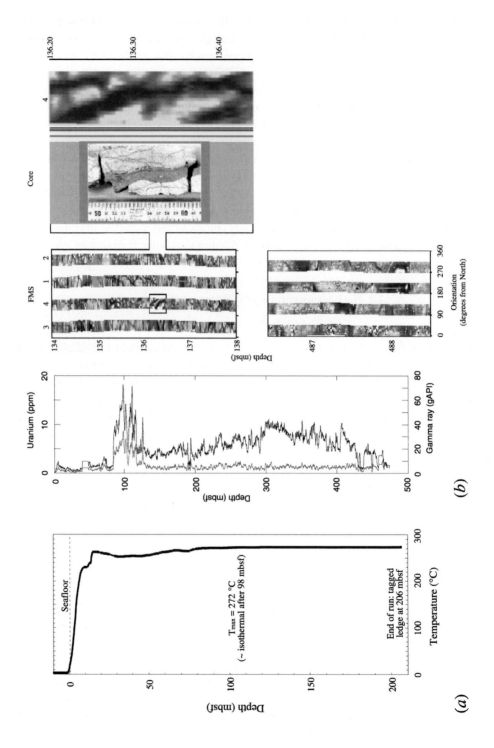

Core

136.20

136.30

136.40

4

FMS

2

1

4

3

Depth (mbsf)

134

135

136

137

138

487

488

0   90   180   270   360
Orientation
(degrees from North)

Uranium (ppm)

0   10   20

Gamma ray (gAPI)

0   20   40   60   80

Depth (mbsf)

0   100   200   300   400   500

(b)

Temperature (°C)

0   100   200   300

Seafloor

T_{max} = 272 °C
(~ isothermal after 98 mbsf)

End of run: tagged
ledge at 206 mbsf

Depth (mbsf)

0   50   100   150   200

(a)

Drilling in the Bent Hill area included the 500-m-deep Hole 856H, which penetrated a 4-m layer of unconsolidated clastic sulfides overlying a 94-m massive sulfide unit, a sulfide feeder zone, and a sequence of basaltic sill and flow units (Davis *et al.*, 1992a,b; Fouquet *et al.*, 1998). Core recovery was poor (less than 13%) in most intervals, but much about the structure, fracturing, and flow regimes in this high-temperature environment was revealed by downhole measurements. High temperatures limited wireline logging operations at this and other sites in Middle Valley, but hole cooling techniques allowed logging for short periods of time.

The recorded downhole logs and FMS provided oriented views of the structure in Hole 856H (Fig. 4.3*b*). The FMS images showed contrasts between high- and low-conductivity features, such as water-filled fractures or fine-scale variations in alteration, and enabled the sparsely recovered pieces of core to be located at the correct depth and orientation (Iturrino *et al.*, 2000). The dip, strike, width, and depth of such geological features in the core could thus be correlated with the sinusoidal features in the images (e.g. Luthi, 1990; Paillet *et al.*, 1990). An example of this is shown in Fig. 4.3*b*, where a sub-vertical massive sulfide vein cross-cutting the hydrothermally altered formation is seen in both a core photograph and a corresponding FMS image. Elevated gamma-ray values and U concentrations toward the top of the sulfide feeder zone correspond to fractured intervals where high-U minerals were likely deposited by hydrothermal fluids. A series of steeply dipping fractures are also highlighted in the FMS images. These persist through the sulfide feeder zone and into the underlying pillow and flow basalts imaged at the bottom of the logged interval.

Site 857 was drilled about 2 km south and 2 km west of the Dead Dog and Bent Hill Sites 858 and 856, respectively. The age of the igneous crust is estimated to be approximately 0.25 Ma (Davis *et al.*, 1992a). Hole 857D penetrated 470 m of sediment and into a permeable sill–sediment complex to a depth of 936 mbsf. Downhole temperature measurements show a depressed gradient to ~ 700 mbsf, indicating a strong downflow of cold ocean bottom water into the hole (Fig. 4.4). Flowmeter tests measured a drawdown of over 10,000 l min^{-1}, or a downward flow rate of nearly 2.5 m s^{-1}, mostly in a zone between 610 and 620 mbsf (Davis *et al.*, 1992a). Resistivity and SP logs decrease in thin high-conductivity (high porosity) zones at these same depths (Becker *et al.*, 1994), consistent with the influx of seawater into the formation. The mineralogical, geochemical, thermal, and pressure similarities between

---

Fig. 4.3 (*a*) Temperature profile from Hole 858G in the Dead Dog vent field in the Middle Valley sedimented rift, northern Juan de Fuca Ridge. The log shows that a very steep thermal gradient exists in the upper 10 m of the hole. A maximum temperature of 272 °C is reached at 98 mbsf, and the hole is nearly isothermal below this depth. (*b*) Gamma-ray log, Formation MicroScanner (FMS), and core data from Hole 856H, in the Bent Hill area of the Juan de Fuca Ridge. The FMS image from the sulfide feeder zone correlates with a cored sub-vertical massive sulfide vein (photograph). High gamma-ray values and U concentrations toward the top of the sulfide feeder zone correspond to fractured intervals where high-U deposits were likely mobilized by high-temperature hydrothermal fluids. A single-quadrant FMS image (within the upper box) enables approximate orientation of the dip and strike of the sulfide vein. Pillow basalts penetrated at the bottom of the hole (lower box) may also be oriented using the FMS images.

the venting areas and the sill–sediment complex are consistent with efficient hydrothermal fluid flow along permeable conduits between the sites (Becker *et al.*, 1994; Davis and Becker, 1994; Fouquet *et al.*, 1998).

4.4.2 Manus Basin, a western Pacific back-arc spreading center

The Manus Basin in the SW Pacific has been formed by back-arc spreading north of the present-day subduction zone along the New Britain Trench (Taylor *et al.*, 1995; Tregoning *et al.*, 1998). Site 1188 was drilled into altered basaltic rocks in a hydrothermal area characterized by diffuse venting of fluids at the seafloor and the presence of extensive white bacterial mats (Binns *et al.*, 2002). The average core recovery at this site was 13.9%, and accurate depths of samples were difficult to determine. Mineral assemblages arising from hydrothermal alteration vary considerably with depth, and the presence of anhydrite veins and other assemblages suggest that several episodes of veining and fluid flow occurred. In particular, the appearance of corrensite clay in the recovered cores denotes relatively high-temperature alteration (Binns *et al.*, 2002).

Temperature, resistivity, and image logs were obtained from Hole 1188B and in the lower section of Hole 1188F (Fig. 4.5*a*). A sequence of temperature measurements recorded a maximum temperature of 313 °C at the bottom of the borehole, the highest in any ODP hole. Bottom-hole temperature increased by approximately 213 °C in a period of seven days after initial wireline logging. In the upper 200 m of the borehole, cool borehole temperatures persisted over this same period, suggesting that cold ocean bottom water was being drawn downhole to this depth, much like the situation elsewhere in the ocean crust where *in situ* pressure is less than the hydrostat (e.g. as in Hole 857D). Downhole image logs recorded using the FMS and RAB (Fig. 4.5*b*) revealed a large number of fractures, complicated fracture patterns, and several highly brecciated intervals. The images characterize both resistive and conductive features over this interval, the latter corresponding to porous formations such as fractures and breccias that remain open. These porous structures may act as permeable conduits for the drawdown of cold seawater (Iturrino and Bartetzko, 2002). Zones of conspicuous bleaching and high-gamma-ray altered rocks persist over many meters depth. Low resistivity and high gamma-ray and U concentrations correspond to fractured intervals evident as dark bands in the RAB and FMS images. Alteration materials were likely deposited by high-temperature hydrothermal fluid migration to several hundred meters depth into the ocean crust. Elevated temperatures and the rapid rate of thermal rebound at the bottom of this hole suggest that downhole flow does not persist to that depth.

Fig. 4.4 Downhole temperature log from Hole 857D obtained several hours after drilling operations had been completed, indicating a strong downflow of cold ocean bottom water. An expanded temperature profile is compared with resistivity, spontaneous potential, and flowmeter logs over the interval from 550 to 700 mbsf. Circles denote average flowmeter readings with packers inflated and pump injection at approximately 2800 l min^{-1}. Squares represent the much higher flow recorded when the packer was deflated and in-flow of ocean bottom water was allowed. Electrical log response decreases in high porosity and high permeability in-flow zones (adapted from Becker *et al.*, 1994).

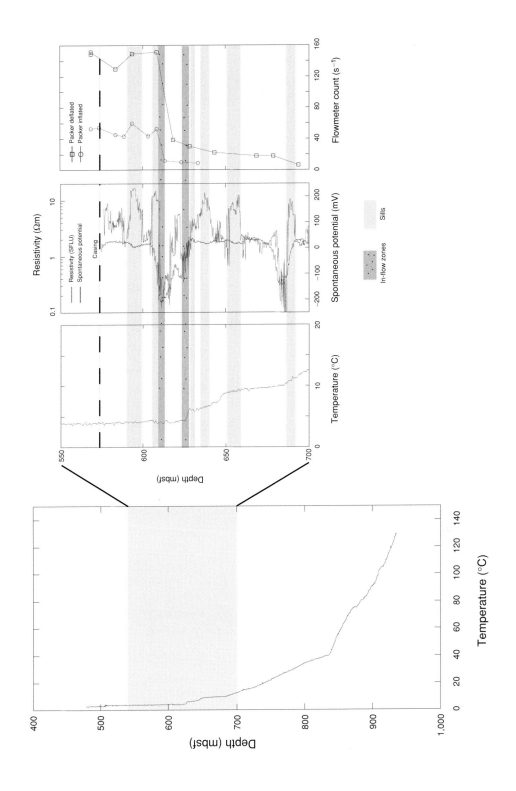

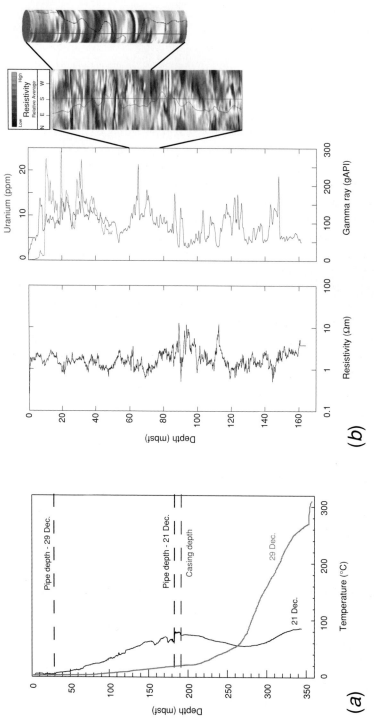

Fig. 4.5 (*a*) Downhole temperature logs obtained in the Snowcap area of the Manus Basin. Bottom-hole temperatures in Hole 1188F increased by approximately 213 °C over a period of seven days to a maximum of 313 °C. The drawdown of cold ocean bottom water into the permeable formation accounts for the cooling above 270 mbsf. (*b*) Resistivity and gamma-ray logs and RAB images from Hole 1189C in the Roman Ruins active venting area in the Manus Basin. The conventional wireline U log is shown with the LWD gamma-ray log in the shallow interval below seafloor. Both two- and three-dimensional image perspectives of a series of fractures are shown with the LWD gamma-ray log superimposed. Low resistivity, high gamma-ray values, and U concentrations correspond to fractured intervals where high-U alteration deposits were mobilized by high-temperature hydrothermal fluids.

4.5 Conclusions

In recent years, borehole resistivity, resistivity imaging, geochemical, and temperature logs have contributed significantly to studies of the hydrologic structure of upper oceanic crust. Downhole temperature profiles are rarely representative of natural thermal conditions because temperature perturbations can persist for weeks after drilling, and because downflow of ocean water may persist in open boreholes for years until the natural thermal state recovers. Open-hole temperature logs provide excellent sensitivity to where fluids enter or exit the formation, however, and may be used to provide quantitative constraints on the rates of open-hole flow and hence on permeability (e.g. Chapter 7). Other logging observations allow for the imaging of fractures or other permeable structures that may act as conduits for hydrothermal fluids, indicate which of these host active flow, and provide information about the alteration products that may be concentrated in these structures. Hydrothermal flow alters the chemistry of the crust through seawater–rock interaction, until over time alteration minerals fill open fractures and permeable pathways. Examples in low-temperature environments on the flanks of the Mid-Atlantic Ridge and Costa Rica Rift and in high-temperature environments on the Juan de Fuca Ridge and in the Manus Basin show a downflow of cool bottom water into the formation that indicates that the ocean crust remains permeable to natural flow up to several hundred meters depth. Fluid flow at low-temperature sites can remain active for many millions of years. Steep thermal gradients cause significant alteration of the shallow crust in high-temperature environments. The collection and integration of wireline logging and logging-while-drilling parameters provides valuable constraints on all these processes.

During the next decade, downhole measurements will play an increasing role in research into the hydrogeology of the ocean crust. Existing downhole technologies will be deployed by wireline re-entry and by drill ships to produce images of the *in situ* properties of the ocean crust with improved resolution. As drilling technologies continue to advance, new instruments will be developed for logging in even higher-temperature environments. Using LWD downhole imaging in lateral and deviated holes in conjunction with a riser – something that has yet to be attempted in the deep ocean – and in shallow crustal environments where holes have been unstable and difficult to drill in the past, will yield new insight into the fractured ocean crust and its complex hydrothermal systems.

Acknowledgments

We acknowledge the shipboard scientists and crews of the *JOIDES Resolution* who have contributed to the collection and use of downhole data in oceanic drill holes. We appreciate the invitation to contribute to this volume and editorial assistance from E. Davis and H. Elderfield. The Ocean Drilling Program funded through the US National Science Foundation and its international partners provided the data presented in this review.

References

Allen, D., Barber, T., Flaum, C., Hemingway, J., Anderson, B., des Ligneris, S., Everett, B., and Morriss, C. 1988. Advances in high-resolution logging. *Tech. Rev.* **36**: 4–15.

Allen, D., Bergt, D., Best, D., Clark, B., Falconer, I., Hache, J. M., Kienitz, C., Lesage, M., Rasmus, J., Roulet, C., and Wraight, P. 1989. Logging while drilling. *Oilfield Rev.* (April): 4–17.

Alt, J. C., Honnorez, J., Laverne, C., and Emmermann, R. 1986. Hydrothermal alteration of a 1 km section through the upper oceanic crust, Deep Sea Drilling Project Hole 504B: mineralogy, chemistry, and evolution of seawater–basalt interactions. *J. Geophys. Res.* **91**: 10,309–10,335.

Alt, J. C., Kinoshita, H., Stokking, L. B., *et al.*, eds. 1993. *Proceedings of the Ocean Drilling Program, Initial Reports*, Vol. 148. College Station, TX: Ocean Drilling Program.

Anderson, R. N., Honnorez, J., Becker, K., Adamson, A. C., Alt, J. C., Emmermann, R., Kempton, P. D., Kinoshita, H., Laverne, C., Mottl, M. J., and Newmark, R. L., 1982. DSDP Hole 504B, the first reference section over 1 km through Layer 2 of the oceanic crust. *Nature* **300**: 589–594.

Anderson, R. N., Alt, J. C., and Malpas, J. 1989. Geochemical well logs and the determination integrated geochemical fluxes in Hole 504B, eastern equatorial Pacific. In *Proceedings of the Ocean Drilling Program, Scientific Results*, Vol. 111, eds. K. Becker, H. Sakai, *et al.* College Station, TX: Ocean Drilling Program, pp. 119–132.

Archie, G. E. 1942. The electrical resistivity log as an aid in determining some reservoir characteristics. *J. Pet. Technol.* **6**: 1–8.

Bartetzko, A., Pezard, P., Goldberg, D., Sun, Y. F., and Becker, K. 2001. Volcanic stratigraphy DSDP/ODP Hole 395A: an interpretation using well-log data. *Mar. Geophys. Res.* **22**(–2): 111–127.

Becker, K. 1985. Large-scale electrical resistivity and bulk porosity of the oceanic crust, Deep Sea Drilling Project Hole 504B, Costa Rica Rift. In *Initial Reports of the Deep Sea Drilling Project*, Vol. 83, eds. R. N. Anderson, J. Honnorez, K. Becker, *et al.* Washington: US Govt. Printing Office, pp. 419–427.

Becker, K. 1990. Measurements of permeability of the upper oceanic crust at Hole 395A, ODP Leg 109. In *Proceedings of the Ocean Drilling Program, Initial Reports*, Vol. 106/109, eds. R. Detrick, J. Honnorez, W. B. Bryan, T. Juteau, *et al.* College Station, TX: Ocean Drilling Program, pp. 213–222.

Becker, K. 1996. Permeability measurements in Hole 896A, Leg 148, and implications for the lateral variability of upper crustal permeability at Sites 504 and 896. In *Proceedings of the Ocean Drilling Program, Scientific Results*, Vol. 148, eds. J. C. Alt, H. Kinoshita, L. B. Stokking, and P. J. Michael. College Station, TX: Ocean Drilling Program, pp. 353–363.

Becker, K. and Davis, E. 2003. New evidence for age variation and scale effects of permeabilities of young oceanic crust from borehole thermal and pressure measurements. *Earth Planet. Sci. Lett.* **210**: 499–508.

Becker, K. and Malone, M. J., eds. 1998. *Proceedings of the Ocean Drilling Program, Initial Reports*, Vol. 174B. College Station, TX: Ocean Drilling Program.

Becker, K., Langseth, M. G., Von Herzen, R. P., and Anderson, R. N. 1983. Deep crustal geothermal measurements, Hole 504B, Costa Rica Rift. *J. Geophys. Res.* **88**: 3,447–3,457.

Becker, K., Langseth, M. G., and Hickman, S. H. 1984. Temperature measurements in Hole 395A, Leg78B. In *Initial Reports of the Deep Sea Drilling Project*, Vol. 78B, eds. R. D. Hyndman, M. H. Salisbury, *et al.* Washington, DC: US Govt. Printing Office, pp. 689–698.

Becker, K., *et al*. 1989. Drilling deep into young oceanic crust at Hole 504B, Costa Rica Rift. *Rev. Geophys.* **24**: 79–102.

Becker, K., Morin, R. H., and Davis, E. 1994. Permeabilities in the Middle Valley hydrothermal system measured with packer and flowmeter experiments. In *Proceedings of the Ocean Drilling Program, Scientific Results*, Vol. 139, eds. M. J. Mottl, E. E. Davis, A. T. Fisher, and J. F. Slack. College Station, TX: Ocean Drilling Program, pp. 613–626.

Becker, K., Bartetzko, A., and Davis, E. E. 2001. Leg 174B synopsis. Revisiting Hole 395A for logging and long-term monitoring of off-axis hydrothermal processes in young oceanic crust. In *Proceeding of the Ocean Drilling Program, Scientific Results*, Vol. 174B, eds. K. Becker and M. J. Malone. College Station, TX: Ocean Drilling Program, pp. 1–13.

Becker, K., Davis, E. E., Spiess, F. N. and de Moustier, C. P. in press. Temperature and video logs from the upper oceanic crust, Holes 504B and 896A, Costa Rica Rift Flank: implications for the permeability of upper oceanic crust. *Earth Planet. Sci. Lett.*

Binns, R. A., Barriga, F. J. A. S., Miller, D. J., eds. 2002. *Proceedings of the Ocean Drilling Program, Initial Reports*, Vol. 193. College Station, TX: Ocean Drilling Program.

Bonner, S., Clark, B., Holenka, J., Voisin, B., Dusang, J., Hansen, R., White, J., and Walsgrove, T. 1992. Logging while drilling: a three-year perspective. *Schlumberger Oilfield Rev.* **4**: 4–21.

Boldreel, L. O., Harvey, P. K. H., Pezard, P., and Iturrino, G. J. 1995. Lithostratigraphy, fracturing and fluid flow in the upper oceanic crust. In *Proceedings of the Ocean Drilling Program, Scientific Results*, Vol. 137/140, eds. J. Erzinger, K. Becker, H. J. B. Dick, and L. B. Stokking. College Station, TX: Ocean Drilling Program, pp. 313–319.

Brewer, T. S., Harvey, P. K., Haggas, S., Pezard P. A., and Goldberg, D. 1999. The role of borehole images in constraining the structure of the ocean crust: case histories from the Ocean Drilling Program. In *Borehole Imaging: Applications and Case Histories*, Special Publication No. 159, eds. M. A. Lovell, G. Williamson, and P. Harvey. London: Geological Society of London, pp. 283–294.

Broglia, C. and Ellis, D. 1990. Effect of alteration, formation absorption, and standoff on the response of the thermal neutron porosity log in gabbros and basalts: examples from Deep Sea Drilling Project–Ocean Drilling Program sites. *J. Geophys. Res.* **95**: 9,171–9,188.

CRRUST (Costa Rica Rift United Scientific Team) 1982. Geothermal regimes of the Costa Rica Rift, East Pacific, investigated by drilling, DSDP–IPOD legs 68, 69, and 70. *Geol. Soc. Am. Bull.* **93**: 862–875.

Davis, E. E. and Fisher, A. T. 1994. On the nature and consequences of hydrothermal circulation in the Middle Valley sedimented rift: inferences from geophysical and geochemical observations, Leg 139. In *Proceedings of the Ocean Drilling Program, Scientific Results*, Vol. 139, eds. M. J. Mottl, E. E. Davis, A. T. Fisher, and J. F. Slack. College Station, TX: Ocean Drilling Program, pp. 695–717.

Davis, E. E., Becker, K., Pettigrew, T., Carson, B., and Macdonald, R. 1992a. CORK: a hydrologic seal and downhole observatory for deep-ocean boreholes. In *Proceedings of the Ocean Drilling Program, Initial Reports*, Vol. 139, eds. E. E. Davis, M. J. Mottl, A. T. Fisher, *et al*. College Station, TX: Ocean Drilling Program, pp. 43–53.

Davis, E. E., Mottl, M. J., Fisher A. T., *et al.*, eds. 1992b. *Proceedings of the Ocean Drilling Program, Initial Reports*, Vol. 139. College Station, TX: Ocean Drilling Program.

Detrick, R., Collins, J., Stephen, R., and Swift, S. 1994. In situ evidence for the nature of the seismic layer 2/3 boundary in oceanic crust. *Nature* **370**: 288–290.

Doveton, J. H. 1986. *Log Analysis of Subsurface Geology: Concepts and Computer Methods*. New York: Wiley.

Ellis, D. V. 1987. *Well Logging for Earth Scientists*. New York: Elsevier.

Evans, H. B. 1991. Evaluating differences between wireline and MWD systems. *World Oil* **212**: 51–61.

Fisher, A. T., Becker, K., and Davis, E. E. 1997. The permeability of young oceanic crust east of Juan de Fuca Ridge determined using borehole thermal measurements. *Geophys. Res. Lett.* **24**: 1,311–1,314.

Fouquet, Y., Zierenberg, R. A., Miller, D. J., *et al.*, eds. 1998. *Proceedings of the Ocean Drilling Program, Initial Reports*, Vol. 169. College Station, TX: Ocean Drilling Program.

Gable, R., Morin, R. H., and Becker, K. 1989. Geothermal state of Hole 504B: ODP Leg 111 overview. In *Proceedings of the Ocean Drilling Program, Scientific Results*, Vol. 111, eds. K. Becker, H. Sakai, *et al.* College Station, TX: Ocean Drilling Program, pp. 87–96.

Gable, R., Morin, R. H., and Becker, K. 1992. Geothermal state of Holes 333A, 395A, and 534A: results from the DIANAUT program. *Geophys. Res. Lett.* **19**: 505–508.

Goldberg, D. 1997. The role of downhole measurements in marine geology and geophysics. *Rev. Geophys.* **35**(3): 315–342.

Goldberg, D., Broglia, C., and Becker, K. 1991. Fracturing, alteration, and permeability: in situ properties in Hole 735B. In *Proceedings of the Ocean Drilling Program, Scientific Results*, Vol. 118, eds. R. P. Von Herzen, P. T. Robinson, *et al.* College Station, TX: Ocean Drilling Program, pp. 261–269.

Goldberg, D. Meltser, A., and ODP Leg 191 Shipboard Scientific Party 2001. High vertical resolution spectral gamma ray logging: a new tool development and field test results. *Soc. Prof. Well Log Analysts 42nd Ann. Sympos.*, Houston, TX.

Hickman, S. H., Langseth, M. G., and Svitek, J. F. 1984. In situ permeability and pore-pressure measurements near the Mid-Atlantic Ridge, Deep Sea Drilling Project, Hole 13 Mid-Atlantic Ridge, Deep Sea Drilling Project Hole 395A. In *Initial Reports of the Deep Sea Drilling Project*, Vol. 78B, eds. R. D. Hyndman, M. H. Salisbury, *et al.* Washington, DC: US Govt. Printing Office, pp. 839–848.

Hyndman, R. D. and Salisbury, M. H. 1984. The physical nature of young oceanic crust on the Mid-Atlantic Ridge, Deep Sea Drilling Project Hole 395A. In *Initial Reports of the Deep Sea Drilling Project*, Vol. 78B, eds. R. D. Hyndman and M. H. Salisbury. Washington, DC: US Govt. Printing Office, pp. 839–848.

Iturrino, G. J. and Bartetzko, A. 2002. subsurface fracture patterns in the PACMANUS hydrothermal system identified from downhole measurements and their potential implications for fluid circulation. Transactions of the American Geophysical Union EOS, Fall AGU poster presentation.

Iturrino, G. J., Davis, E. E., Johnson, J., Gröschel-Becker H., Lewis T., Chapman, D., and Cermak, V. 2000. Permeability, electrical, and thermal properties of sulfide, sedimentary and basaltic units from the Bent Hill area in Middle Valley, Juan de Fuca Ridge. In *Proceedings of the Ocean Drilling Program, Scientific Results*, Vol. 169, eds. R. A. Zierenberg, Y. Fouchet, D. J. Miller, and W. R. Normark. College Station, TX: Ocean Drilling Program.

Kopietz, J., Becker, K., and Hamono, Y. 1990. Temperature measurements at Site 395, ODP Leg 109. In *Proceedings of the Ocean Drilling Program, Initial Reports*, Vol. 106/109, eds. R. Detrick, J. Honnorez, W. B. Bryan, T. Juteau, *et al.* College Station, TX: Ocean Drilling Program, pp. 197–203.

Langseth, M. G., Becker, K., Von Herzen, R. P., and Schultheiss, P. 1992. Heat and fluid flux through the sediment on the western flank of the Mid-Atlantic Ridge: a hydrological study of North Pond. *Geophys. Res. Lett.* **19**: 517–520.

Lovell, J. R., Young, R. A., Rosthal, R. A., Buffington, L., and Arceneaux, Jr. C. L. 1995. *Structural Interpretation of Resistivity-at-the-Bit Images, Transactions of the SPWLA 36th Annual Logging Symposium*, June 26–29, Paris, paper TT.

Luthi, S. 1990. Sedimentary structures of clastic rocks identified from electrical borehole images. In *Geological Applications of Wireline Logs*, Special Publication No. 48. London: Geological Society of London, pp. 3–10.

Melson, W. G., Rabinowitz, P. D., *et al.*, eds. 1978. *Initial Reports of Deep Sea Drilling Project*, Vol. 45. Washington, DC: US Govt. Printing Office.

Morin, R. H., Hess, A. E., and Becker, K. 1992. In situ measurements of fluid flow in DSDP Holes 395A and 534A: results from the DIANAUT program. *Geophys. Res. Lett.* **19**: 509–512.

Moos, D. 1990. Petrophysical results from logging in DSDP Hole 395A, ODP Leg 109. In *Proceedings of the Ocean Drilling Program, Scientific Results*, Vol. 106/109, eds. R. Detrick, J. Honnorez, J. B. Bryan, T. Juteau, *et al.* College Station, TX: Ocean Drilling Program, pp. 49–62.

Murphy, D. P. 1993. What's new in MWD and information evaluation. *World Oil* **214**: 47–52.

Paillet, F. L., Barton, C., Luthi, S., Rambow, F., and Zemanek, J. 1990. Borehole imaging and its application in well logging: a review. In *Borehole Imaging*, eds. F. Paillet, C. Barton S. Luthi, F. Rambow, and J. Zemanek. Houston, TX: Society of Professional Well Log Analysts, pp. 1–23.

Pezard, P. A. 1990. Electrical properties of mid-ocean ridge basalt and implications for the structure of the upper oceanic crust in Hole 504B. *J. Geophys. Res.* **95**: 9,237–9,264.

Pezard, P. A., Anderson, R. N., Ryan, W. B., Becker, K., Alt, J., and Pascal, G. 1992. Accretion, structure and hydrology of intermediate spreading-rate oceanic crust from drillhole experiments and seafloor observations. *Mar. Geophys. Res.* **14**(2): 93–123.

Prensky, S. E. 1994. A survey of recent developments and emerging technology in well logging and rock characterization. *The Log Analyst* March–April: 15–45.

  1999. Advances in borehole imaging technology and applications. In *Borehole Imaging: Applications and Case Histories*, Special Publication No. 159, eds. M. A. Lovell, G. Williamson, and P. Harvey. London: Geological Society of London, pp. 1–43.

Revil, A. and Glover, P. W. 1998. Nature of surface electrical conductivity in natural sands, sandstones, and clays. *Geophys. Res. Lett.* **25**: 691–694.

Serra, O. 1987. *Fundamentals of Well-Log Interpretation, 2. The Interpretation of Logging Data*. Amsterdam: Elsevier.

Snyder, D. and Fleming, D. 1985. Well logging: a 25 year perspective. *Geophysics* **50**: 2,504–2,529.

Taylor, B., Goodliffe, A., Martinez, F., and Hey, R. 1995. Continental rifting and initial sea-floor spreading in the Woodlark basin. *Nature* **374**: 534–537.

Tregoning, P., Lambeck, K., Stolz, A., Morgan, P., McClusky, S., van der Beek, P., McQueen, H., Jackson, R., Little, R., Laing, A., and Murphy, B. 1998. Estimation of current plate motions in Papua New Guinea from Global Positioning System observations. *J. Geophys. Res.* **103**: 12,181–12,203.

5

Hydrothermal aging of oceanic crust: inferences from seismic refraction and borehole studies

Ingo Grevemeyer and Anne Bartetzko

5.1 Introduction

The oceanic crust is created continuously by seafloor spreading at mid-ocean ridges (Chapter 3) and covers nearly 60% of the Earth's surface. The processes responsible for shaping the structure and properties of that crust are of fundamental importance in marine Earth sciences. The primary source of our knowledge of the structure of oceanic crust is the interpretation of seismic refraction experiments. During the past several decades our knowledge of the seismic velocity structure of the oceanic crust has improved considerably over the classical three-layer model proposed by Raitt (1963). Advances in shooting and recording technology, in particular the development of large and highly repetitive airgun sources and ocean bottom seismometers, have led to a new picture of oceanic crust, suggesting that the upper crust (layer 2) is a region of strong velocity gradients. The lower crust (layer 3) is relatively homogeneous, although it does show an increase in velocity with depth (e.g. Kennett and Orcutt, 1976; Whitmarsh, 1978). More recent studies have subdivided the upper crust into layer 2A, composed of extruded basalts, and layer 2B, formed by basaltic sheeted dikes (e.g. Christeson *et al.*, 1994). The lower crust, or layer 3, often called the "oceanic layer," is inferred to be composed of gabbros (e.g. White *et al.*, 1992).

One of the most striking features of the oceanic crust emerging from all studies is the variability of layer 2. Early interpretations by Raitt (1963) demonstrated that layer 2 compressional-wave velocities range widely, from less than 3.0 km s^{-1} to more than 6.0 km s^{-1}. Subsequent work on layer 2 by Houtz and Ewing (1976) show an uppermost crust with velocities ranging from 3.0 to 4.0 km s^{-1} underlain by rocks with velocities between 4.8 km s^{-1} and 6.4 km s^{-1}. Within this context, the variations in layer 2 velocity observed by Raitt can be interpreted as being due to the presence or absence of one or both of the sub-layers and regional variations in their thickness and velocity. Since laboratory measurements on fresh basalts yield velocities of about 6 km s^{-1}, low seafloor velocities can only be explained by abundant porosity at a scale larger than that sampled by drilling or dredging (e.g. Hyndman and Drury, 1976). The increase in velocity with depth is generally believed to result from a reduction of this large-scale porosity (Whitmarsh, 1978;

Hydrogeology of the Oceanic Lithosphere, eds. E. E. Davis and H. Elderfield. Published by Cambridge University Press.
© Cambridge University Press 2004.

Spudich and Orcutt, 1980). In addition, Houtz and Ewing (1976) and Houtz (1976) found that refraction velocities in layer 2A not only increase with depth but increase also with lithospheric age, from about 3.3 km s^{-1} at the ridge crests to that of layer 2B in crust about 40 M.y. old. Houtz and Ewing (1976) suggested that secondary minerals, associated with hydrothermal circulation, deposited in cracks and fissures within the uppermost basalts, are the most likely explanation for the velocity increase. Evidence for hydrothermal circulation in the upper oceanic crust on the ridge flanks is derived from patterns of local heat flow variability (Chapters 8, 11, and 13; Anderson *et al.*, 1977; Stein and Stein, 1994) and from the occurrence of low-temperature alteration products in numerous Deep Sea Drilling Project (DSDP) and Ocean Drilling Program (ODP) sites (Chapter 15; e.g. Alt *et al.*, 1986).

For roughly two decades little progress was made in refining and understanding the interaction between hydrothermal activity and the evolution of the seismic properties of oceanic crust. The Houtz and Ewing data set remained the only clear evidence for a systematic, age-dependent variation in crustal structure (Purdy and Ewing, 1986). Re-examinations of the data set exacerbated the problem by showing that some of the early interpretations were compromised either by basement topography (Diebold and Carlson, 1993) or by assumptions in their interpretation methods (Carlson and Jacobson, 1994). However, more recent studies (Purdy, 1987; Grevemeyer and Weigel, 1996; Carlson, 1998) and particularly systematic studies (Rohr, 1994; Grevemeyer and Weigel, 1997; Grevemeyer *et al.*, 1999) have found that upper crustal velocity structure is indeed a function of plate age; seismic velocities increase by a factor of 2 from approximately 2.2 km s^{-1} near mid-ocean ridges within less than 10 M.y. The intent of this chapter is to review the impact of these recent discoveries on our ability to understand and describe processes controlling crustal evolution and hence aging. In addition to work done in recent years, we will include evidence from downhole measurements and logging from ODP drill sites and link hydrogeological constraints with changes in physical properties of the crust. Purdy and Ewing (1986) pointed out that the layer 2A velocity increase with age remains the only known change in the structure of the oceanic crust. However, recent evidence from the Mid-Atlantic Ridge suggested that serpentinite is a common feature under slow-spreading ridges. Seawater is required to form serpentinite, and hence pervasive fracturing of the entire crust is suggested. Indeed, fracturing can establish a hydrogeological regime even in aged crust. We will therefore also re-visit the lower oceanic crust to search for evidence for age-dependent features.

5.2 A review of uppermost crustal structure as a function of plate age

Although crustal structure is similar everywhere in the deep oceans, seismic experiments near spreading ridges indicate that seismic velocities in the top of the igneous crust are typically much lower than those in mature oceanic crust, and thus indicate an evolutionary process that affects the physical properties of oceanic crust. The first attempt to define systematic age-dependent trends of the seismic velocity structure was the work of Houtz and Ewing (1976) and Houtz (1976), based on the compilation of archival sonobuoy seismic refraction profiles. The most important results are that velocities of the uppermost crust, layer 2A, increase while the layer thickness decreases with crustal age until about 40 M.y. In addition, Anderson et al. (1977) recognized a close relationship between the continuation of hydrothermal flow through the crust and the apparent disappearance of layer 2A and therefore suggested a strong link between hydrothermal circulation and the evolution of the seismic properties. Purdy and Ewing (1986) pointed out, however, that the Houtz and Ewing data set was the only clear evidence for a systematic, age-dependent variation in the structure of crust.

Carlson and Jacobson (1994) took the data set of Houtz (1976) and re-examined the claim that: (1) layer 2A velocities increase with age over millions of years until they cannot be distinguished from velocities characteristic of layer 2B; and (2) layer 2A apparently thins with age, disappearing when velocities approximate those in layer 2B. They found that: (i) if they consider only the data identified by Houtz as layer 2A velocities, they could find no significant dependence of either velocity or thickness on basement age. Assuming that: (ii) the variation of layer 2A velocities with age is defined by the first refraction branch from the acoustic basement for ages up to 50 M.y. they found that there is a strong correlation between velocity and basement age (Fig. 5.1a). However, the apparent thinning of layer 2A was an artificial consequence.

Independent evidence for an age-dependence of upper crustal velocities was found at abandoned spreading ridges, where seafloor spreading ceased millions of years ago. Extinct ridges have been identified, for example, in the Labrador Sea and in the Norwegian Basin. Seismic profiles shot along the abandoned ridges provided velocities of 3.8–5.4 km s^{-1} at the top of the basement, which are more typical for mature than juvenile crust (Osler and Louden, 1995; Grevemeyer et al., 1997). The most likely explanation is that, after extinction, crustal evolution continued over millions of years just as in the case of crust moved off-axis by continued spreading. Hydrothermal precipitation of secondary minerals has likely decreased porosity in extruded basalts, causing layer 2A velocity to increase from its initial value to values more characteristic for mature crust.

To explore further the relationship between crustal age and seismic velocities of the uppermost igneous crust, Grevemeyer and Weigel (1996) compiled a new data set of seismic refraction velocities which generally does not include values used by Houtz and Ewing (1976) or Houtz (1976). Using seismic refraction data from post-1970 studies, they were able to show that seismic velocities in young crust depend strongly on basement age and increase rapidly, doubling in less than 10 M.y. but do not increase significantly in older crust. Carlson (1998) increased that data set and carried out a rigorous statistical test of the hypothesis that most of the increase in seismic velocity occurs in <10 M.y. That study indicated that in crust <1 Ma, seismic velocities are commonly <3 km s^{-1}, while in crust between 1 and 5 Ma, velocities range from 2 to ~5 km s^{-1}. Velocities <3 km s^{-1} are not observed in crust older than 5 M.y. and there was no detectable systematic change in the seismic properties of the uppermost oceanic crust after 10 M.y. (Fig. 5.1b).

The first systematic observations related to crustal aging resulted from an unusual seismic refraction experiment where both the source and the receiver were placed on the seafloor (Purdy, 1987). The experiment provided evidence that crustal aging increases the seismic

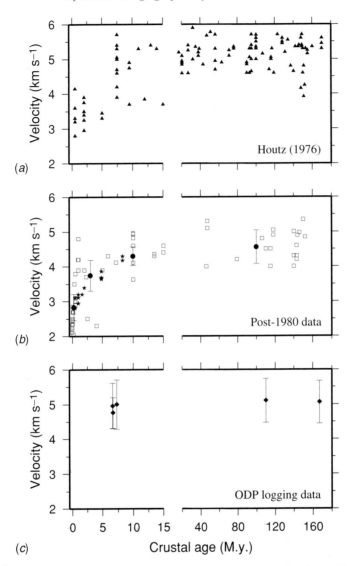

Fig. 5.1 Upper crustal or layer 2A compressional-wave velocity as a function of basement age. Note that the scale of the age axis changes at 15 M.y. (*a*) velocities derived from archival sonobuoy seismic refraction profiles (Houtz, 1976). (*b*) Compilation of velocities from post-1980 studies (open squares) and mean velocities (solid circles) from Carlson (1998) are shown along with velocities of an experiment (stars) designed to study crustal aging (Grevemeyer *et al.*, 1999). (*c*) Mean velocities and error bars calculated from the drilled portion of DSDP and ODP Holes 395A, 418A, 504B, 801C, and 896A.

velocity in layer 2A. Within the median valley of the Mid-Atlantic Ridge velocities of 2.1 km s^{-1} were determined at the top of the basement, while about 120 km away from the ridge axis in 7 Ma crust velocities of 4.0 km s^{-1} were detected. In another experiment Rohr (1994) used constraints from stacking velocities of a multi-channel seismic reflection flow line profile on the sedimented eastern flank of the Juan de Fuca Ridge to show that seismic velocities increase rapidly close to the ridge axis and reach a constant value thereafter. The next systematic survey designed to study age-dependent trends of the seismic properties of upper oceanic crust was carried out in 1995 on the eastern flank of the East Pacific Rise near 14° S (Grevemeyer and Weigel, 1997). This work clearly supports the global trends revealed by Grevemeyer and Weigel (1996) and Carlson (1998). Seismic refraction data show a two-stage evolution (Fig. 5.1*b*): values increase rapidly by about 45–50% within only 0.5–1 M.y. ($\sim$0.8–1 km s^{-1} 1 M.y.$^{-1}$), and increase slowly thereafter (0.1–0.2 km s^{-1} 1 M.y.$^{-1}$) to double within 8 M.y. or less (Grevemeyer and Weigel, 1997; Grevemeyer *et al.*, 1999).

While these recent efforts have revealed that hydrothermal activity clearly affects the evolution of the seismic properties in young oceanic crust, little is known about crustal aging in crust older than approximately 10 M.y. Hydrothermal circulation, however, is known to control heat loss (and hence possibly crustal alteration) over millions of years and may persist in very old crust (Chapter 13). Compilations of seismic velocity data (Fig. 5.1*a,b*) show a considerable degree of variability in older crust, but show no systematic changes (Carlson, 1998).

Additional constraints on changes of the seismic properties of mature crust can be drawn from direct measurements in ODP wells. Five holes have been drilled and logged several hundred meters into the basaltic layer 2A. ODP holes 504B, 896A, 395A, 418A, and 801C were drilled into 6.6, 6.7, 7.3, 110, and 167 Ma oceanic crust, respectively. For all holes except 896A a depth interval of 20–300 m below the top of the igneous crust was chosen for the compilation here. To avoid bias from breccias, logging data in 896A was analysed between a 20 and 200 m sub-basement depth. In Hole 801C, the rocks of the uppermost part consists of alkali basalts and a strongly hydrothermally altered crust. This zone, which is believed to be related to off-axis volcanism, was excluded from the logging data analysis. Data quality and the logging program of the five holes are described by Salisbury *et al.* (1988), Alt *et al.* (1993), Plank *et al.* (2000), and Bartetzko *et al.* (2001). Mean velocities calculated from the logging data are shown in Fig. 5.1*c*. Like the global seismic refraction data set, little evidence is provided that crustal aging affects the seismic properties far away from the ridge crests.

5.3 Scale-dependence of hydrothermal aging

In contrast to seismic field studies (Fig. 5.1), laboratory measurements on basaltic samples from DSDP and ODP sites (Fig. 5.2) document a decrease of velocity with age (Christensen and Salisbury, 1972, 1973). This decrease is associated with a change of mineralogy, and of bulk-rock geochemistry (Johnson and Christensen, 1997), with the content of CaO decreasing, and K$_2$O increasing as the rocks age (Fig. 5.2). The chemical reactions destroy

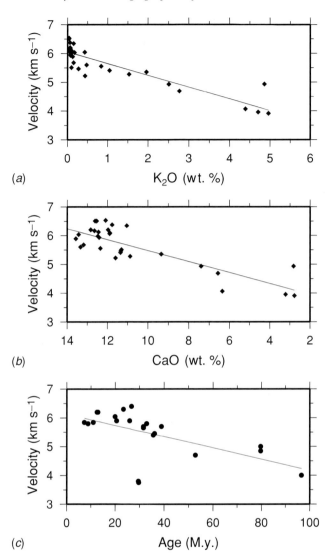

Fig. 5.2 Laboratory measurements of compressional velocities (at elevated confining pressure) from DSDP and ODP basement core samples as a function of increasing K_2O (*a*) and decreasing CaO (*b*) content (Johnson and Christensen, 1997), indicating the effect of alteration on the basalts that results in decreased velocities (*c*) with increasing basement age (Christensen and Salisbury, 1972, 1973).

the fabric of solid basalts and decrease the velocity of the rock samples. At a larger scale, the generation of secondary alteration products has a significant impact on the evolution of the large-scale porosity structure (Gillis and Sapp, 1997). The two effects work against one another.

Alteration is a complex topic (Chapters 9 and 15; Alt *et al.*, 1986); it is treated here in a very simple bi-modal manner. During alteration, the rocks take up H_2O, forming phillipsite, smectite, and Fe–Mn oxides. Geochemical alteration proceeds inward, particularly along cracks and veins. Secondary minerals formed during weathering of the rocks are smectites, celadonite, analcite, and carbonates. All are characterized by low intrinsic velocities, so replacement minerals progressively lower the velocity on a small scale. Over millions of years, hydrothermal mineralization also fills cracks, fissures, and veins. Thereby, initial porosity of 14–17% estimated to occur in juvenile uppermost crust at ridge crests is reduced to 4–8% in mature uppermost crust (Gillis and Sapp, 1997). The reduction of the bulk porosity structure is inherently related to changes in the elastic properties of crustal rocks, and seismic velocity increases as porosity is decreased (Kuster and Toksöz, 1974; Wilkens *et al.*, 1991). Therefore, alteration of basalts can explain both the decrease of velocities measured at a specimen scale in the laboratory (via replacement mineralization; Fig. 5.2) and the increase of velocities measured by seismic refraction (via large-scale porosity and compressibility reduction). The seismic refraction velocity increase suggests that hydrothermal filling of pore spaces overwhelms any effect of replacement mineralization in changing seismic properties over time.

Since the seismic response of a porous medium is affected by porosity at all scales, it is important to consider the pore space distribution of the lava pile. Laboratory determinations of basalt velocities from DSDP and ODP cores are much higher than either seismic or downhole sonic logs. Figure 5.3 shows the range of velocities measured by different tools and indicates the effect of crustal evolution at different scales. Holes 504B and 418A sampled 6.6 and 110 Ma crust, respectively. While the velocities measured on the drill cores are generally lower in the older crust (primary in the uppermost 200 m) the velocities measured by downhole logging and by seismic techniques are higher in the older hole. This discrepancy is related to the different frequencies used to determine velocities and hence is related to the scale of pore spaces that affects the measurements. Laboratory techniques use sample sizes of a few centimeters and frequencies in the megahertz range, and thus sample microfractures on the scale of millimeters or less. Sonic logs used in downhole measurements apply frequencies in the kilohertz region and sample approximately 0.5 m into the formation and are presumably sensitive to void spaces up to that size. Refraction seismic techniques use sources with frequencies ranging from 5 to 50 Hz which average over tens to hundreds of meters, and hence are sensitive to almost all pore sizes. Because the sonic log measurements are largely comparable to seismic determinations (Fig. 5.3), it appears that the distribution of pore spaces responsible for the early increase of seismic velocities with age spans a spatial scale larger than the core samples, but not so large as to be missed by logging data (e.g. Jacobson, 1992).

5.4 Crustal aging as a function of depth

5.4.1 Uppermost crust, layer 2A

The greatest change of the seismic properties with age is the compressional-wave velocity at the top of the igneous lava pile, layer 2A (Fig. 5.1), although comparing velocity–depth

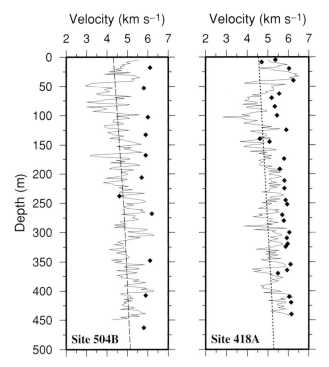

Fig. 5.3 Upper crustal velocity–depth profiles from seismic refraction techniques (broken lines), sonic downhole logging (thin solid lines), and laboratory measurements at elevated confining pressure on samples (diamonds) from 6 Ma (Hole 504B) and 110 Ma (Hole 418A) crust (Data sources: Hole 504B – Little and Stephen, 1985; Salisbury *et al.*, 1985; Alt *et al.*, 1993. Hole 418A – Salisbury *et al.*, 1988; Swift and Stephen, 1989.)

solutions revealed by studies on young and mature oceanic crust indicates structural changes deeper in layer 2A as well (Fig. 5.4). The best resolution in juvenile crust is perhaps provided by unusual seismic experiments which placed both the seismic source and the receiver close to the seabed (Purdy, 1987; Christeson *et al.*, 1994). Studies with similar resolution in older crust are provided by oblique seismic experiments (OSE) in ODP drill sites (Little and Stephen, 1985; Swift and Stephen, 1989). The OSE approximates the structure revealed by logging reasonably well (Fig. 5.3). Profiles on juvenile and young crust yield an 80–150 m thick surficial low-velocity layer (layer 2A) and a 200–300-m-thick transition zone reaching 5 km s^{-1} at its bottom (layer 2B; Fig. 5.4). Studies with less resolution provide a roughly 400-m-thick layer 2A with a steep velocity gradient, underlain by a ∼1-km-thick layer 2B with velocities of 4.5–5 km s^{-1} at the top and ∼6.6–6.8 km s^{-1} at the bottom (Grevemeyer *et al.*, 1998, 1999). In older crust, however, the well-defined boundary between layers 2A and 2B is not evident (Figs. 5.3 and 5.4). This change in velocity contrast at the layer 2A/2B boundary is even apparent in multi-channel seismic reflection data. On axis and in young oceanic crust created at fast spreading rates, the base of layer 2A can be imaged

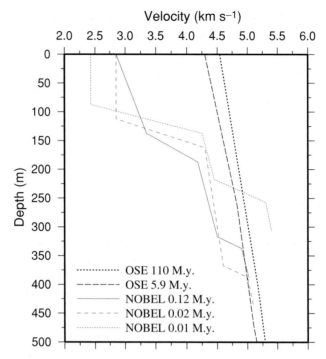

Fig. 5.4 Upper crustal velocity–depth profiles from seismic refraction techniques obtained in crust of different basement ages, showing a homogenization of seismic velocities with increasing crustal age and a disappearance of the velocity contrast at about 400 m depth, i.e. at the layer 2A/2B boundary. (Data sources: NOBEL – Christeson *et al.*, 1994. OSE – Little and Stephen, 1985; Swift and Stephen, 1989.)

(e.g. Kent *et al.*, 1994; Rohr, 1994), while in old oceanic crust no such layer is detected (Reston *et al.*, 1999). Thus, a strong velocity contrast within layer 2 is confined to young crust.

Hussenroeder *et al.* (2002) show for the Mid-Atlantic Ridge that the layer 2A/2B transition zone thins close to the ridge axis, over approximately 2 M.y. They propose that hydrothermal processes contribute to the thinning through the preferential deposition of alteration products at the base of layer 2A. They suggest that the interfingering of high-porosity extrusives with low-porosity sheeted dikes may affect hydrothermal mineralization. Alteration studies of DSDP/ODP Hole 504B and of ophiolites have led researchers to postulate that this interval was once the locus of intense alteration, where upwelling of fluids along dike fissures and fractures mix with cooler seawater circulating through the permeable lava pile (Alt *et al.*, 1986). Thus, the most likely cause for the disappearance of the layer 2A/2B transition is a progressive upward sealing of oceanic crust by secondary minerals. This evolution is consistent with the alteration history of basalts from several DSDP sites (Peterson *et al.*, 1986), although constraints from the Troodos ophiolite seem to contradict the model;

dating of alteration minerals did not provide any evidence for a progressive upward sealing (Gallahan and Duncan, 1994). Nevertheless, porosity and permeability decrease with depth (Chapter 7), suggesting that the uppermost crust is more accessible for fluids and hence alteration than the lower portion of crust. The higher water/rock ratio may therefore promote precipitation of secondary minerals in the uppermost crust and may explain why velocities of rocks near the seafloor increase by a much greater factor, and why seismic velocities in layer 2 become more uniform as crust ages.

5.4.2 Layer 2B and the oceanic layer 3

DSDP Hole 504B is the deepest borehole in oceanic crust. Although it did not reach layer 3, results suggest that the layer 2/3 transition zone is caused by a gradual downward change in crustal porosity and alteration (Detrick *et al.*, 1994), implying that hydrothermal circulation can affect physical properties well below the extrusive lava pile. Purdy and Detrick (1986) investigated zero-age crust on the Mid-Atlantic Ridge north of the Kane Fracture Zone. They found a crustal structure with some characteristics of simple mature oceanic crust, like a low-gradient lower crust and a well-defined Moho transition zone, but with a slightly lower layer 3 velocity. Crust a few thousand or tens of thousands of years older already provided typical layer 3 properties. Purdy and Detrick therefore suggested that hydrothermal circulation in juvenile crust can penetrate down to the base of the crust and that precipitation of secondary minerals into crustal cracks and fissures may increase the velocity by several percent. Thus, hydrothermal aging may be an important evolutionary process in governing even lower crustal seismic properties. Hydrothermal activity should be strongest during the first few hundred thousand years after crustal emplacement, and indeed, little evidence has been found for changes of the seismic properties in deep crust older than a few hundreds of thousands of years. Grevemeyer *et al.* (1998) investigated age-dependent features in 0.5 to ~10 Ma oceanic crust at the East Pacific Rise. While they found the expected large increase in layer 2A velocities, they found no evidence for any systematic change in seismic structure in layers 2B or 3 (Fig. 5.5). Therefore, mid and lower crustal permeability seem to be too low in aged fast-spreading crust to allow significant continuous hydrothermal alteration.

Because hydrothermalism is strongly dependent on permeability (Chapter 7), faults and fissures (Chapter 3) will affect the evolution of crust. Mid-ocean ridges dominated by tectonism may experience more extensive hydrothermal alteration, because faults and fractures provide pathways for fluids to enter the crust. Huang and Solomon (1988) show that a systematic relationship exists between spreading rate and the maximum centroid depth of ridge-crest earthquakes; depth increases as the spreading rate decreases. The maximum depth of ridge-crest earthquakes is sensitive to the thickness of the mechanically strong layer, which is prone to brittle failure in response to the extensional stresses associated with the diverging plates. The penetration of faults to greater depths may increase the amount of open cracks and fissures in the lower crust, which in turn increases the bulk porosity and

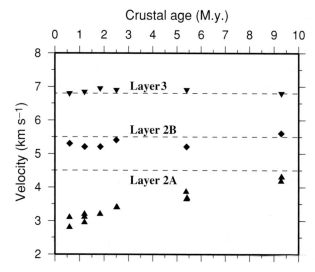

Fig. 5.5 Crustal velocities as a function of depth and age. Data are from fast-spreading East Pacific Rise near 14° S (Grevemeyer *et al.*, 1998). We characterize each layer by its velocity at the top of the layer. Except for upper crustal velocities, little evidence for age-dependent features is revealed over the 0.5–10 M.y. interval. Velocities of 4.5, 5.5, and 6.8 km s^{-1} are shown by broken lines for reference.

decreases the seismic velocity. These cracks allow hydrothermal fluids to reach the lower crust and upper mantle, which has the two known effects of increasing the seismic velocities by filling open pore spaces or decreasing seismic velocities by weathering the rock itself. Studies from extinct spreading ridges reveal that lower crustal velocities are well below those of typical mature oceanic crust (Osler and Louden, 1995; Grevemeyer *et al.*, 1997). This has been attributed to an increasing degree of tectonism preceding extinction, although alteration of lower crustal rocks may contribute. Little is known about deep crustal aging from so-called normal oceanic crust. Sleep and Barth (1997) provided an initial assessment by re-visiting velocity–depth solutions compiled by White *et al.* (1992). They divided the data set into Pacific and Atlantic crust, and into young and old crust, and re-sampled the solutions into 1-km bins (Fig. 5.6). For the Pacific Ocean, these average profiles show the well-documented increase in upper crustal velocities and constant velocities in the lower crust. They provide no indications for off-axis hydrothermal circulation affecting the lower crust in fast-spreading lithosphere. For the slow-spreading Atlantic crust, however, upper crustal velocities increase, while velocities in the lower crust decrease significantly with age.

The observations from the Atlantic Ocean suggest that faulting at the ridge axis (and before extinction) creates pathways for hydrothermal circulation and alteration through-out the entire crust, which results in alteration of lower crustal gabbroic rocks and hence decreasing seismic velocities. In addition, fluids may even penetrate into the upper mantle to alter mantle peridotite. This may explain reduced upper mantle velocities at extinct ridges

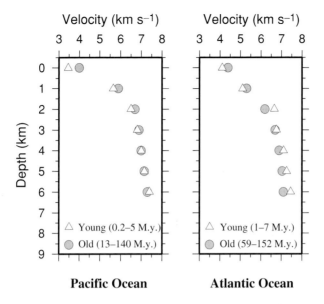

Pacific Ocean   **Atlantic Ocean**

Fig. 5.6 Averaged compressional-wave velocity versus depth in the oceanic crust. Using velocity–depth solutions of White *et al.* (1992) the aging of oceanic crust was re-visited by obtaining velocities picked at 1-km-depth intervals (Sleep and Barth, 1997). There is evidence for lower crustal aging only in the Atlantic data, formed at slow-spreading rates. This may indicate that if crust suffers from pervasive fracturing of the entire crust during emplacement, hydrothermal alteration can act over millions of years on the lower crustal rocks to reduce seismic velocities.

(Osler and Louden, 1995; Grevemeyer *et al.*, 1997) and at the ends of spreading segments (e.g. Reston *et al.*, 2002).

5.5 Factors governing crustal aging

Application of rock physics, like the Kuster and Toksöz (1974) model for cracks with varying pore geometry, can be used to investigate the age-dependent trends in seismic velocities. Wilkens *et al.* (1991) have investigated basalt velocity–porosity relationships for a wide range of pore sizes. Their efforts concentrated on how seismic velocities can change by crack filling. Their model suggested a pore space modification with crustal age, where small-aspect ratio pore spaces (small and thin cracks) are bridged first and then large-aspect ratio voids (wider cracks). In this way seismic velocities can easily be doubled with a small 5% reduction of effective porosity. Although the porosity in the Troodos ophiolite (Gillis and Sapp, 1997) did not indicate that the small-aspect ratio pore spaces are prone to seal first, these calculations indicate that clogging of pore spaces by hydrothermal precipitates is an important process in changing seismic properties with age.

The mechanisms that affect the vigour of hydrothermal circulation, and hence changes in upper crustal seismic structure, are of great interest. Anderson and Hobart (1976) suggested that fluid exchange between the ocean and crust is restricted when the basement is buried with 150–200 m of sediment. With a high degree of continuity, an even thinner section may be required (Chapter 6). The volume of water passing through the crustal rocks may also decrease with age by decreased porosity and hence permeability of the crust (Chapter 7) due to hydrothermal deposition of secondary minerals (Anderson *et al.*, 1977). In this case the reduction in hydrothermal exchange should correlate with increases in crustal seismic velocity. This idea was consistent with the Houtz and Ewing (1976) study, but it is contradicted by recent discoveries (Fig. 5.1), which indicate that changes in crustal structure occur primarily in the first few millions of years after crustal generation (<10 M.y.), while hydrothermal circulation continues tens of millions of years longer (Chapters 10 and 13). However, permeability shows a similar trend: at the scale sampled by borehole testing, permeability decreases rapidly as young crust ages (Chapter 7), suggesting a close relationship between the evolution of seismic velocity and bulk permeability.

Grevemeyer and Weigel (1997) suggest that the two-stage evolution of seismic properties of layer 2A (rapid change at young ages, gradual thereafter) is caused by the ridge crest and ridge flank hydrothermal circulation system. Changes in physical properties are believed to be caused by seawater–rock interactions, which are controlled by the water/rock ratio (Chapter 14) and the temperature of reaction. In general, the higher the temperature, the faster and greater the extent of the reaction. This implies that the high-temperature axial system, with its rapidly circulating hydrothermal fluids, is accompanied by a compara-tively large amount of hydrothermal alteration and sealing of open void spaces. Alteration studies support this model, by indicating that most of the hydrothermal alteration occurs within the first 3 M.y. (e.g. Thompson, 1983). In the ridge flank system, chemical reactivity is reduced at the much lower temperatures. Mineralization and alteration continue away from the ridge crest, but at a slower rate, particularly when the flow of seawater into the crust becomes restricted by several hundreds of meters of sediment (Anderson and Hobart, 1976; Jacobson, 1992) or when the permeability of crust is reduced by crack clogging (Chapter 7).

Seafloor gravity measurements provide valuable constraints on the porosity structure of young volcanic rocks (Holmes and Johnson, 1993; Pruis and Johnson, 1998). They show an abrupt increase in density (and hence velocity) of the upper 200 m of crust within ~150,000 years after crustal construction. This density change is interpreted as a systematic reduction in bulk porosity of the upper crustal section, from 23–36% for the axial ridge to 10% for the flanking abyssal hills. Holmes and Johnson (1993) suggest that the most likely cause of this dramatic decrease in porosity is elimination of large-scale voids known to be present at volcanically active ridges (Francheteau *et al.*, 1979). The largest-scale porosity, like hollow pillows and lava tubes, may collapse as the lithosphere moves away from the ridge axis. The idea that tectonically controlled large-scale porosity reduction occurs in young crust is supported by porosity estimates from the Troodos ophiolite, indicating an initial mean macroscopic porosity of 14–17% (Gillis and Sapp, 1997), i.e. lower than the *in situ* estimates from seafloor gravity measurements at axial locations by about a factor of two. The ophiolite values are higher than gravity-estimated values at flank locations; this discrepancy may provide an indication of the fraction of pore space reduction driven by the physical collapse of large-scale void spaces. Thus, in addition to hydrothermal alteration, tectonically and gravitationally controlled large-scale porosity reduction may contribute to the rapid change in seismic velocity at young ages. The gradual change of seismic velocities in aged crust, however, is likely to be solely related to hydrothermal alteration, decreasing porosity to 4–8% in mature uppermost crust (Gillis and Sapp, 1997). We therefore suggest that in juvenile crust both the collapse of large-scale porosity and secondary mineralization related to the vigorous on-axis circulation system may control the rapid change in seismic velocities (Grevemeyer and Weigel, 1997; Grevemeyer *et al.*, 1999). In the age range of ~1 to ~10 M.y. hydrothermal precipitation of minerals alone is suggested to govern the long-term increase of seismic velocities.

The Grevemeyer and Weigel (1997) study and global compilations (Grevemeyer and Weigel, 1996; Carlson, 1998) investigated regions where basement is generally not isolated from the ocean by a thick sediment layer. In regions with significant sediment cover, the evolution of crustal porosity is quite different. Burial of basement rocks reduces the degree of ventilation of hydrothermal flow through the crust and, in turn, temperatures in the crust increase. Alteration of basalts is intensified with increasing temperature, and hence the formation of metamorphic minerals is accelerated. Under such conditions, sealing of open void spaces by secondary minerals, and hence reduction of porosity, are enhanced. Stephen and Harding (1983) proposed this mechanism to explain high velocities in young upper oceanic crust in the Gulf of California (crustal age ~1 Ma). Temperature was also suggested by Rohr (1994) to be the controlling factor of crustal evolution at the sedimented eastern flank of the Juan de Fuca Ridge. At the Juan de Fuca Ridge, some recent seismic refraction measurements sampled velocities in the along-axis direction (Rosenberger and Fechner, 1997). Velocities of about 4.3 km s^{-1} occur in 1.5–2.0 Ma basement rocks. Crust of the same age at the sparsely sedimented East Pacific Rise yields velocities of <3.4 km s^{-1} (Fig. 5.7) and a value of 4.3 km s^{-1} is reached much later, after ~8 M.y. of crustal evolution. Thus, crustal evolution took three times longer. The major difference between both areas is indeed the temperature within the igneous basement. While crustal temperatures at the sedimented eastern flank of the Juan de Fuca Ridge rapidly reach values above 60 °C in ~1.5 Ma crust, the temperature in the sparsely sedimented East Pacific Rise crust remain well below 10 °C (Fig. 5.7). As a consequence, chemical reactivity is reduced. Therefore, in addition to the water/rock ratio (Chapter 14), basement temperature, itself a function of basal heat flow, sediment thickness, and sediment continuity, may control the evolution of the seismic properties of upper oceanic crust (Stephen and Harding, 1983; Rohr, 1994). Figure 5.7 serves well to illustrate that the temperature is probably among the key properties that define the time scale of crustal evolution.

5.6 Time scale of crustal aging

Seismic velocities reveal little or no crustal aging beyond 10 Ma (Fig. 5.1), and this is consistent with limited information about the time scale of hydrothermal alteration provided

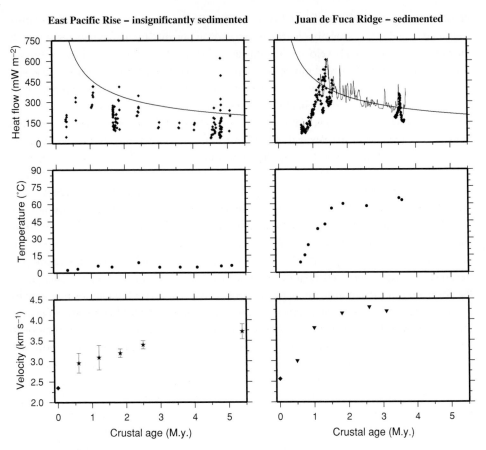

Fig. 5.7 Hydrogeological and seismic data from the sparsely sedimented East Pacific Rise (left) and well-sedimented Juan de Fuca Ridge (right). Heat flow data from the East Pacific Rise (Villinger *et al.*, 2002) have been used along with seismic constraints on the sediment thickness (Hauschild *et al.*, 2003) to calculate the temperature at the top of layer 2A. Temperatures are only a few degrees above the bottom-water temperature and well below 10 °C. Seismic velocities along this transect reach ~3.4 km s^{-1} within 2.5 M.y. (Tolstoy *et al.*, 1997; Grevemeyer *et al.*, 1999). Heat flow data and basement temperatures from the Juan de Fuca Ridge (Davis *et al.*, 1999) indicate rapidly increasing basement temperature along this sediment-sealed crustal transect. Temperatures at the top of the igneous basement reach 60 °C in crust only a little over 1.5 Ma. Seismic velocities increase rapidly within 2 M.y. to values >4.0 km s^{-1} (Cudrak and Clowes, 1993; Rohr, 1994; Rosenberger and Fechner, 1997). It is therefore reasonable to suggest that basement temperature controls the evolution of oceanic crust.

by rock samples. Although precise dating of alteration history is generally unavailable, a great deal of information exists regarding the composition of the altered upper crust. K/Ar and ^{87}Rb/^{86}Sr dating of celadonites, a near-final stage low-temperature alteration mineral, provided an initial assessment of the time interval within which alteration, and hence the evolution of seismic properties, can occur. Results from several DSDP and ODP holes as well as from the Troodos ophiolite suggest that hydrothermal alteration in the upper igneous crust continued for at least 7 M.y. and possibly up to 15 M.y. (Staudigel and Hart, 1985; Peterson *et al.*, 1986; Staudigel *et al.*, 1986; Hart *et al.*, 1994). In contrast, global heat flow estimates suggest that ventilated hydrothermal circulation may affect crust and lithosphere as old as 65 M.y. (Chapter 10; Stein and Stein, 1994) and closed circulation may continue even longer (Chapter 13). Hydrothermal heat transfer, and in particular exchange with the ocean, is the principal cause of hydrothermal alteration of crustal rocks and the formation of alteration products. Thus there is an apparent discrepancy between the time scale for hydrothermal circulation from global heat flow and alteration studies. More recent studies in the Troodos ophiolite indicate that the low-temperature water/rock chemical exchange continued for at least 40 M.y. after crustal formation (Gallahan and Duncan, 1994). This represents a 100% increase over previous estimates and makes the age range of crustal alteration more consistent with global heat flow data. Nevertheless, the younger age suggested for most alteration and for the end of the resolvable increase in seismic velocity suggests that alteration in crust older than 10 M.y. is minor.

Although not as easily determined, bulk compositional indicators and physical properties other than the seismic velocity might be more appropriate for detecting and quantifying the impact of crustal evolution in old crust. Among the variety of parameters provided by downhole logging (Chapter 4), observations of the natural radioactivity (gamma ray) and electrical resistivity are perhaps most diagnostic. K usually gives the largest contribution to the total gamma-ray spectrum, but the amount of K in pristine oceanic crust is very low. The crust is a sink for seawater K (Fig. 5.2) through interaction between seawater and basalts (e.g. Alt *et al.*, 1986). Consequently, on-going low-temperature weathering of the crust will increase the amount of K in the formation and hence increase the total gamma-ray radiation for highly altered crust. Alteration also affects the electrical properties of crustal rocks. Clay minerals are among the first minerals to be precipitated; they are stable and abundant in young crust. With the passage of the time and decreasing temperatures, carbonate minerals such as calcite commonly fill veins and fractures. Other alteration minerals such as zeolites and K-feldspar also occur during later stages of alteration, but are generally less important than calcite. In contrast to clay minerals, carbonates, zeolites, and K-feldspar are electrical insulators. The growth of carbonates in the fracture network may interrupt interconnected fluid pathways and cause the electrical resistivity to increase with age. Consequently, both gamma-ray and resistivity logs can provide useful proxies for crustal alteration and aging.

Downhole logging data from Holes 395A, 418A, 504B, 801C, and 896A have sampled crust ranging from 6 to 167 Ma (Salisbury *et al.*, 1988; Alt *et al.*, 1993; Plank *et al.*, 2000; Bartetzko *et al.*, 2001). In crust younger than 7 M.y., resistivities are generally in the range of 10–50 Ωm, gamma-ray radiation values are less than 5 API, and P-wave velocities are between 4–5.5 km s^{-1} (Fig. 5.8). As crust ages, the scatter of values increases significantly. In the oldest crust (i.e. Hole 801C) resistivities are between 10 and 3,000 Ωm, the gamma ray indicates radioactivity of 5–20 API, and seismic velocities are between 4–6 km s^{-1}. To facilitate interpretation of these compositional and physical indicators, we use cross plots (Fig. 5.8). Using the hole drilled into the youngest crust as the reference hole (Hole 504B drilled into 6 Ma crust), and depth as common factor, the ratio revealed by the cross plots

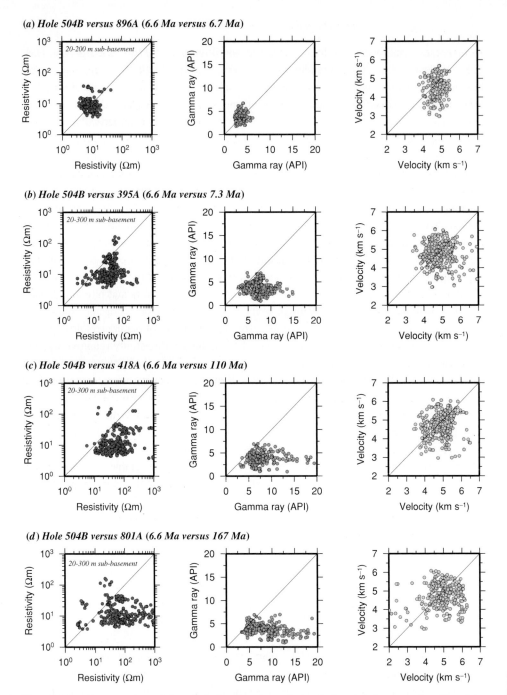

(a) Hole 504B versus 896A (6.6 Ma versus 6.7 Ma)

(b) Hole 504B versus 395A (6.6 Ma versus 7.3 Ma)

(c) Hole 504B versus 418A (6.6 Ma versus 110 Ma)

(d) Hole 504B versus 801A (6.6 Ma versus 167 Ma)

Fig. 5.8 Pseudo-cross plots of downhole logging data of electrical resistivity, natural radioactivity (gamma ray), and seismic velocity. Sub-basement depth is used as common factor. Data have been resampled at 1-m intervals. From top to bottom, data from Hole 504B are plotted against data from older crust (896A, 395A, 418A, and 801C). As crust becomes older, the variability increases and the electrical resistivity and gamma ray shift to higher values, indicating cross-exchange of elements between seawater and basalts and the change in alteration mineralogy with ongoing hydrothermal circulation as crust ages (see text for discussion).

should shift systematically to higher values if the differences between the logs are solely due to crustal aging. In this way, changes of the physical properties as a function of plate age can be detected. Hole 896A sampled crust of virtually the same age as Hole 504B, and those cross plots show a roughly one to one relationship, as expected if both sites had experienced similar alteration histories. As sites become progressively older, the electrical resistivity and gamma-ray ratios decrease, while the velocity plots show little change except that the scatter increases.

The most important conclusion for the time scale of crustal aging is that seismic velocities show little evidence for change in crust older than that at Hole 504B ($\sim$6 Ma), while other physical properties indicate significant changes in crust older than this, even between 110 and 167 Ma crust. Although we cannot rule out regional effects (e.g. changes in the primary concentration of K), it is most likely that hydrothermal activity, and hence crustal alteration, are acting on the oceanic crust over many tens of millions of years. Thus, the fate of crust is perhaps to age over its whole lifetime until it is subducted and recycled at a convergent plate boundary.

5.7 Summary

In young oceanic crust, values of heat flow are highly variable and fall below those expected for cooling solely by thermal conduction. This heat flow discrepancy reflects significant heat transport from the crust through the seafloor by circulating fluids. Hydrogeological flow is most vigorous in juvenile crust where venting of fluids at high temperatures occurs. As the lithosphere is moved off-axis the vigour of hydrothermal circulation is reduced, but fluid exchange with the ocean can continue over tens of millions of years. Hydrothermal activity inherently affects the physical properties of the oceanic crust by closing pore space with secondary hydrothermal alteration products. This is inferred to contribute to an observed rapid increase of seismic velocities in the uppermost basaltic crust, layer 2A, by 45–50% within only 0.5–1 M.y. after crustal emplacement. Values increase more slowly thereafter, doubling within 8–10 M.y. then remain relatively unchanged in older crust. Porosity estimates from seafloor gravity measurements suggest that the large-scale bulk porosity structure of the crust is reduced rapidly near ridge-axes, from 23–36% for the axial ridge to 10% beneath ridge flank abyssal hills. Ophiolites reveal porosities of 4–8%. Basement temperature and the water/rock ratio are important factors governing the rate at which initially water-filled void spaces are filled with authigenic minerals. The most likely scenario for the evolution of the porosity and hence seismic velocity structure is that tectonic and gravitational collapse of large-scale porosity and a vigorous hydrothermal system control the rapid change of seismic velocity close to ridge crests, while low-temperature hydrothermal precipitation controls the long-term increase of seismic velocities in older crust.

Although the off-axis hydrothermal system is active for tens of millions of years, seismic velocities change little after 10 M.y. of crustal evolution. Other properties, like the natural radioactivity and the electrical resistivity of the crust, are more sensitive to age-dependent changes than the seismic velocity; they are directly affected by the alteration mineralogy and cross-exchange of elements between the seawater and basalts. Constraints on the natural radioactivity and resistivity from five drill sites of the Ocean Drilling Program suggest that crustal aging may continue over the whole life of the crust until it is carried into a subduction zone and recycled. However, aging is strongest and fastest within a few million years after crustal emplacement.

Changes of crustal properties by hydrothermal circulation are generally restricted to the permeable uppermost crust where circulating fluids can easily enter the crust before a continuous sediment layer isolates the igneous basement from seawater. In zero-age slow-spreading crust, seismic measurements indicate velocities in the lower crust well below those expected for gabbroic crust, although they reach values typical for layer 3 within a very short time. Therefore, existing cracks and fissures in the lower crust must be rapidly closed after crustal emplacement, and remain unchanged until the crust is subducted. However, if the crust is pervasively fractured due to tectonic stresses, alteration of lower crustal rocks may affect velocities in layer 3. Evidence for aging in the way of decreasing lower crustal velocities has only been found within tectonically dominated slow-spreading crust. In such settings, fluids may even penetrate into the upper mantle to alter peridotite and form serpentinite. This may explain reduced velocities associated with extinct ridges and at the ends of spreading segments. The long-term decrease of layer 3 velocities in slow-spreading crust suggests that fluids can circulate at deep levels in the crust for millions of years.

Seismic data constraining the nature of crustal alteration is difficult to obtain and inherently sparse. Thus future work must be carefully targeted to refine some of the major conclusions that are summarized here. For example, the off-axis increase in the extrusive layer 2A velocity has been documented systematically in only two locations. It is believed to be linked to the degree to which the crust is isolated from fluid exchange with the ocean above, but whether temperature or chemistry (e.g. the transition from oxic to anoxic conditions) is the key factor is not yet well resolved. Other well-located systematic studies can address this question, and whether the increase may be postponed in other settings. Data from old crustal areas are even more sparse. New high-quality information is needed to improve constraints on the manner and degree of deep crustal aging.

Acknowledgments

The paper benefited from the research done by many marine refraction seismologists over the last four to five decades. Major results shown in Figs. 5.1, 5.5, and 5.7 have been obtained during research sponsored by the German Ministry of Education, Science, Research, and Technology (BMBF grants 03G0105A, 03G0111A, 03G0145A). In addition, this study used data provided by the Ocean Drilling Program (ODP). The ODP is sponsored by the US National Science Foundation (NSF) and participating countries under management of the Joint Oceanographic Institution (JOI). We are grateful to the German Science Foundation (DFG) for supporting the ODP. Constructive reviews by Suzanne Carbotte, Earl Davis, and Roy Wilkens were much appreciated.

References

Alt, J. C., Honnorez, J., Laverne, C., and Emmermann, R. 1986. Hydrothermal alteration of a 1 km section through the upper oceanic crust, Deep Sea Drilling Project Hole 504B: mineralogy, chemistry, and evolution of seawater–basalt interactions. *J. Geophys. Res.* **91**: 10,309–10,335.

Alt, J. C., Kinoshita, H., Stokking, L. B., eds. 1993. *Proceedings of the Ocean Drilling Program, Initial Results*, Vol. 148. College Station, TX: Ocean Drilling Program.

Anderson, R. N. and Hobart, M. A. 1976. The relation between heat flow, sediment thickness, and age in the eastern Pacific. *J. Geophys. Res.* **81**: 2,968–2,989.

Anderson, R. N., Langseth, M. G., and Sclater, J. G. 1977. The mechanism of heat transfer through the floor of the Indian Ocean. *J. Geophys. Res.* **82**: 3,391–3,409.

Bartetzko, A., Pezard, P., Goldberg, D., Sun, Y.-F., and Becker, K. 2001. Volcanic stratigraphy of DSDP/ODP hole 395A: an interpretation using well-logging data. *Mar. Geophys. Res.* **22**: 111–127.

Carlson, R. L. 1998. Seismic velocities in the uppermost oceanic crust: age dependence and the fate of layer 2A. *J. Geophys. Res.* **103**: 7,069–7,077.

Carlson, R. L. and Jacobson, R. S. 1994. Comment on "Upper crustal structure as a function of plate age" by R. Houtz and J. Ewing. *J. Geophys. Res.* **99**: 3,135–3,138.

Christensen, N. I. and Salisbury, M. H. 1972. Sea floor spreading, progressive alteration of layer 2 basalts, and associated changes of in seismic velocities. *Earth. Planet. Sci. Lett.* **15**: 367–375.

　1973. Velocities, elastic moduli and weathering-age relations for Pacific layer 2 basalts. *Earth. Planet. Sci. Lett.* **19**: 461–470.

Christeson, G. L., Purdy, G. M., and Fryer, G. J. 1994. Seismic constraints on shallow crustal emplacement processes at the fast spreading East Pacific Rise. *J. Geophys. Res.* **99**: 17,957–17,973.

Cudrak, C. F. and Clowes, R. M. 1993. Crustal structure of Endeavour ridge segment, Juan de Fuca Ridge, from a detailed seismic refraction survey. *J. Geophys. Res.* **98**: 6,329–6,349.

Davis, E. E., Chapman, D. S., Wang, K., Villinger, H., Fisher, A. T., Robinson, S. W., Griegel, J., Pribnow, D., Stein, J., and Becker, K. 1999. Regional heat flow variation across the sedimented Juan de Fuca Ridge eastern flank: constraints on lithospheric cooling and lateral hydrothermal heat transport. *J. Geophys. Res.* **104**: 17,675–17,688.

Detrick, R. S., Collins, J. A., Stephen, R. A., and Swift, S. A. 1994. In situ evidence for the nature of the seismic layer 2/3 boundary in oceanic crust. *Nature* **370**: 288–290.

Diebold, J. and Carlson, R. 1993. Layer 2A revisited (abstract). *Eos, Trans. Am. Geophys. Union* **74**(43): Fall Meet. Suppl., 603.

Francheteau, J., Juteau, T., and Rangan, C. 1979. Basaltic pillars in collapsed lava pools on the deep ocean floor. *Nature* **281**: 209–211.

Gallahan, W. E. and Duncan, R. A. 1994. Spatial and temporal variability in crystallization of celadonites within the Troodos ophiolite, Cyprus: implications for low-temperature alteration of the oceanic crust. *J. Geophys. Res.* **99**: 3,147–3,161.

Gillis, K. M. and Sapp, K. 1997. Distribution of porosity in a section of upper oceanic crust exposed in the Troodos ophiolite. *J. Geophys. Res.* **102**: 10,133–10,149.

Grevemeyer, I. and Weigel, W. 1996. Seismic velocities of the uppermost igneous crust versus age. *Geophys. J. Int.* **124**: 631–635.

1997. Increase of seismic velocities in upper oceanic crust: the "superfast" spreading East Pacific Rise at 14°14' S. *Geophys. Res. Lett.* **24**: 217–220.

Grevemeyer, I., Weigel, W., Whitmarsh, R. B., Avedik, F., and Dehghani, G. A. 1997. The Aegir Rift: crustal structure of an extinct spreading axis. *Mar. Geophys. Res.* **19**: 1–23.

Grevemeyer, I., Weigel, W., and Jennrich, C. 1998. Structure and ageing of oceanic crust at 14° S on the East Pacific Rise. *Geophys. J. Int.* **135**: 573–584.

Grevemeyer, I., Kaul, N., Villinger, H., and Weigel, W. 1999. Hydrothermal activity and the evolution of the seismic properties of upper oceanic crust. *J. Geophys. Res.* **104**: 5,069–5,079.

Hart, S. R., Blusztajn, J., Dick, H. J. B., and Lawrence, J. R. 1994. Fluid circulation in the oceanic crust, contrast between volcanic and plutonic regimes. *J. Geophys. Res.* **99**: 3,163–3,173.

Hauschild, J., Grevemeyer, I., Kaul, N., and Villinger, H. 2003. Asymmetric sedimentation on young ocean floor at the East Pacific Rise. *Mar. Geol.* **193**: 49–59.

Holmes, M. L. and Johnson, H. P. 1993. Upper crustal densities derived from seafloor gravity measurements: northern Juan de Fuca Ridge. *Geophys. Res. Lett.* **20**: 1,871–1,874.

Houtz, R. 1976. Seismic properties of layer 2A in the Pacific. *J. Geophys. Res.* **81**: 6,321–6,331.

Houtz, R. and Ewing, J. 1976. Upper crustal structure as a function of plate age. *J. Geophys. Res.* **81**: 2,490–2,498.

Huang, P. Y. and Solomom, J. C. 1988. Centroid depth of the mid-ocean ridge earthquakes: dependence on spreading rate. *J. Geophys. Res.* **93**: 13,445–13,477.

Hussenroeder, S. A., Kent, G. M., and Detrick, R. S. 2002. Upper crustal seismic structure of the slow spreading Mid-Atlantic Ridge, 35° N: constraints on volcanic emplacement processes. *J. Geophys. Res.* **107**: 10.1029/2001JB001691.

Hyndman, R. D. and Drury, M. J. 1976. The physical properties of oceanic basement rocks from deep drilling on the Mid-Atlantic Ridge. *J. Geophys. Res.* **81**: 4,042–4,060.

Jacobson, R. S. 1992. Impact of crustal evolution on changes of the seismic properties of the uppermost ocean crust. *Rev. Geophys.* **30**: 23–42.

Johnson, J. E. and Christensen, N. I. 1997. Seismic properties of layer 2 basalts. *Geophys. J. Int.* **128**: 285–300.

Kuster, G. T. and Toksöz, M. N. 1974. Velocity and attenuation of seismic waves in two-phase media, Part I, Theoretical formulations. *Geophysics* **39**: 587–606.

Kennett, B. L. N. and Orcutt, J. A. 1976. A comparison of travel time inversions for marine refraction profiles. *J. Geophys. Res.* **81**: 4,061–4,070.

Kent, G. M., Harding, A. J., Orcutt, J. A., Detrick, R. S., Mutter, J. C., and Buhl, P. 1994. Uniform accretion of oceanic crust south of the Garrett transform at 14°15' S on the East Pacific Rise. *J. Geophys. Res.* **99**: 9,097–9,116.

Little, S. A. and Stephen, R. A. 1985. Costa Rica Rift borehole seismic experiment, Deep Sea Drilling Project Hole 504B, leg 92. In *Initial Reports of the Deep Sea Drilling Project*, Vol. 83, eds. R. N. Anderson, J. Honnorez, K. Becker, *et al.* Washington, DC: US Govt. Printing Office, pp. 517–528.

Osler, J. C. and Louden, K. E. 1995. The extinct spreading centre in the Labrador Sea: I - crustal structure from a 2-D seismic refraction velocity model. *J. Geophys. Res.* **100**: 2,261–2,278.

Peterson, C., Duncan, R., and Scheidegger, K. F. 1986. Sequence and longevity of basalt alteration at Deep Sea Drilling Project site 597. In *Initial Reports of the Deep Sea*

Drilling Project, Vol. 92, eds. M. Leinen, D. K. Rea, *et al.* Washington, DC: US Govt. Printing Office, pp. 505–515.

Plank, T., Ludden, J. N., Escutia, C., *et al.*, eds. 2000. *Proceedings of the Ocean Drilling Program, Initial Results*, Vol. 185. College Station, TX: Ocean Drilling Program.

Purdy, G. M. 1987. New observations of the shallow seismic structure of young oceanic crust. *J. Geophys. Res.* **92**: 9,351–9,362.

Purdy, G. M. and Detrick, R. S. 1986. Crustal structure of the Mid-Atlantic Ridge at 23° N from seismic refraction studies. *J. Geophys. Res.* **91**: 3,739–3,762.

Purdy, G. M. and Ewing, J. 1986. Seismic structure of oceanic crust. In *The Geology of North America, Vol. M, The Western North Atlantic Region*, eds. P. R. Vogt and B. E. Tucholke. Boulder, CO: Geological Society of America, pp. 313–330.

Puris, M. J. and Johnson, H. P. 1998. Porosity of very young oceanic crust from sea floor gravity measurements. *Geophys. Res. Lett.* **25**: 1,959–1,962.

Raitt, R. W. 1963. The crustal rocks, in *The Sea*, Vol. 3, ed. M. N. Hill. New York: Wiley-Interscience, pp. 85–102.

Reston, T. J., Ranera, C. R., and Belykh, I. 1999. The structure of Cretaceous oceanic crust of the NW Pacific: constraints on processes at fast spreading centers. *J. Geophys. Res.* **104**: 629–644.

Reston, T., Weinrebe, W., Grevemeyer, I., Flueh, E. R., Mitchell, N. C., Kristein, L., Kopp, C., Kopp, H., and Gershwin Scientific Party 2002. A rifted inside corner massif on the Mid-Atlantic Ridge at 5° S. *Earth Planet. Sci. Lett.* **200**: 255–269.

Rohr, K. M. M. 1994. Increase of seismic velocities in upper oceanic crust and hydrothermal circulation in the Juan de Fuca plate. *Geophys. Res. Lett.* **21**: 2,163–2,166.

Rosenberger, A. and Fechner, N. 1997. Increase of seismic velocities with crustal age: Juan de Fuca Ridge (abstract). *Eos, Trans. Am. Geophys. Union* **78**: Fall meeting suppl., 469.

Salisbury, M. H., Christensen, N. I., and Becker, K. 1985. The velocity structure of layer 2 at Deep Sea Drilling Project Site 504 from logging and laboratory experiments. In *Initial Reports of the Deep Sea Drilling Project*, Vol. 83, eds. R. N. Anderson, J. Honnorez, K. Becker, *et al.* Washington, DC: US Govt. Printing Office, pp. 529–539.

Salisbury, M. H., Scott, J. H., Auroux, C., Becker, K., Bosum, W., Broglia, C., Carlson, R., Christensen, N. I., Fisher, A., Gieskes, J., Holmes, M. A., Hoskins, H., Moos, D., Stephen, R., and Wilkens, R. 1988. Old oceanic crust: synthesis of logging, laboratory, and seismic data from leg 102. In *Proceedings of The Ocean Drilling Program, Scientific Results*, Vol. 102, eds. M. H. Salisbury, J. H. Scott, *et al.* College Station, TX: Ocean Drilling Program, pp. 155–180.

Sleep, N. H. and Barth, G. A. 1997. The nature of lower crust and shallow mantle emplaced at low spreading rates. *Tectonophysics* **279**: 181–191.

Staudigel, H. and Hart, S. R. 1985. Dating of ocean crust hydrothermal alteration: strontium isotope ratios from hole 504B carbonates and a reinterpretation of Sr isotope data from Deep Sea Drilling Project sites 105, 332, 417, and 418. In *Initial Reports of the Deep Sea Drilling Project*, Vol. 83, eds. R. N. Anderson, J. Honnorez, K. Becker, *et al.* Washington, DC: US Govt. Printing Office, pp. 297–303.

Staudigel, H., Gillis, K., and Duncan, R. 1986. K/Ar and Rb/Sr ages of celadonites from Troodos ophiolite, Cyprus. *Geology* **14**: 72–75.

Stein, C. A. and Stein, S. 1994. Constraints on hydrothermal heat flux through the oceanic lithosphere from global heat flow. *J. Geophys. Res.* **99**: 3,081–3,095.

Stephen, R. A. and Harding, A. J. 1983. Travel time analysis of borehole seismic data. *J. Geophys. Res.* **88**: 8,289–8,298.

Spudich, P. and Orcutt, J. R. 1980. A new look at the seismic structure of the oceanic crust. *Rev. Geophys.* **18**: 627–645.

Swift, S. A. and Stephen, R. A. 1989. Lateral heterogenity in the seismic structure of upper oceanic crust, western North Atlantic. *J. Geophys. Res.* **94**: 9,303–9,322.

Thompson, G. 1983. Hydrothermal fluxes in the ocean. *Chem. Oceanog.* **8**: 271–337.

Tolstoy, M., Harding, A. J., Orcutt, J. A., and the TERA Group 1997. Deepening of axial magma chamber on the southern East Pacific Rise toward the Garrett fracture zone. *J. Geophys. Res.* **102**: 3,097–3,108.

Villinger, H., Grevemeyer, I., Kaul, N., Hauschild, J., and Pfender, M. 2002. Hydrothermal heat flux through aged oceanic crust: where does the heat escape? *Earth Planet. Sci. Lett.* **202**: 159–170.

White, R. S., McKenzie, D. and O'Nions, R. K. 1992. Oceanic crustal thickness from seismic measurements and rare earth element inversion. *J. Geophys. Res.* **97**: 19,683–19,715.

Whitmarsh, R. B. 1978. Seismic refraction studies of the upper igneous crust in the North Atlantic and porosity estimates for layer 2. *Earth Plant. Sci. Lett.* **37**: 451–464.

Wilkens, R. H., Fryer, G. J., and Karsten, J. 1991. Evolution of porosity and seismic structure of upper oceanic crust: importance of aspect ratios. *J. Geophys. Res.* **96**: 17,981–17,995.

6

Sediment permeability, distribution, and influence on fluxes in oceanic basement

Glenn A. Spinelli, Emily R. Giambalvo, and Andrew T. Fisher

6.1 Introduction

Sediments blanketing oceanic igneous basement rocks control the communication between fluid within the crust and the oceans. Seafloor sediments have low permeability (e.g. Pearson and Lister, 1973; Bryant *et al.*, 1975; Marine Geotechnical Consortium, 1985; Saffer *et al.*, 2000) relative to the extrusive basalt that comprises the upper permeable portion of the basaltic basement (Fisher, 1998; Chapter 7). Therefore, sediments can be viewed as a leaky confining layer that hydrologically isolates the basement from the overlying ocean. As sediments gradually accumulate on oceanic crust, they change the nature of fluid circulation. Where sediment cover on basaltic basement is not continuous, fluid often flows laterally in the upper portion of the basement between basalt outcrops (e.g. Williams *et al.*, 1979; Langseth *et al.*, 1984). Where sediment cover is continuous, the sediment hydraulic impedance (resistance to fluid flow) exerts an important control on the degree to which the basement aquifer and the ocean remain coupled (Karato and Becker, 1983; Fisher *et al.*, 1994; Snelgrove and Forster, 1996; Giambalvo *et al.*, 2000).

In changing the nature of hydrothermal circulation, sediment accumulation potentially affects the transition from advective to conductive heat flow through the seafloor (Anderson and Hobart, 1976; Sclater *et al.*, 1976; Stein and Stein, 1994a; Chapter 10), the nature of hydrothermal alteration of the crust (Alt and Teagle, 1999; Chapter 15), the evolution of crustal properties such as seismic velocity (Jacobson, 1992; Chapter 5), and the chemistry of the ocean via the exchange of solutes (e.g. Mottl and Wheat, 1994; Chapters 19 and 21). The degree to which sediments isolate the basement influences basement fluid temperature, circulation patterns, and residence time (Chapters 8 and 11) – each of which affect the exchange of reactive solutes (e.g. Ca, Mg, Na, and K) between the crust and ocean (e.g. Maris and Bender, 1982; Mottl, 1989; Wheat and Mottl, 1994; Chapters 19 and 21). Fluids that react with sediment while entering or exiting the crust, modify the sediment and affect the exchange of additional reactive solutes (e.g. SiO_2, SO_4, PO_4, and NH_4; Williams *et al.*, 1979; Mottl, 1989; Wheat and McDuff, 1994, 1995; Wheat and Mottl, 1994; Davis *et al.*, 1997a,b; Giambalvo *et al.*, 2002; Chapter 16).

Hydrogeology of the Oceanic Lithosphere, eds. E. E. Davis and H. Elderfield. Published by Cambridge University Press.
© Cambridge University Press 2004.

The permeability of typical seafloor sediments ranges over at least seven orders of magnitude, from $\sim 10^{-19}$ to $\sim 10^{-12}$ m^2 (e.g. Bryant and Rack, 1990), depending on sediment type and porosity. Sediment permeability (the capacity of the sediment to transmit fluid) is controlled by size and interconnectedness of interstitial pores, and the tortuosity of fluid flow paths; such factors are in turn controlled by the grain-size distribution, grain shape, and porosity. Sediment hydraulic impedance, which controls the rate of fluid flow through a sediment column, is a function of both sediment permeability and thickness. For sediment columns ranging from 10 to 100 m thick, the hydraulic impedance can vary by five orders of magnitude for typical sediment types (Section 6.5.2). As a result, the nature of fluid circulation in partially or continuously sedimented regions can depend strongly on the distribution of sediment thickness, which varies laterally, and on sediment type, which can vary both laterally and with depth in the section (e.g. Snelgrove and Forster, 1996; Giambalvo *et al.*, 2000). The general influence of sediments on fluid and heat exchange between the crust and the ocean has been recognized for many years (e.g. Lister, 1972). Corresponding trends of increasing sediment thickness with increasing lithospheric age, and increasing observed heat flux relative to a reference conductive cooling model with increasing lithospheric age, suggest that as their thickness increases, sediments hydraulically seal the crust and limit thermally significant fluid flow (Anderson and Hobart, 1976). Some controversy exists over the nature of sealing, for while average heat flux at some young well-sedimented sites are in agreement with conductive cooling models (e.g. Sclater *et al.*, 1976; Davis and Lister, 1977; Langseth *et al.*, 1983, 1988; Davis *et al.*, 1999), in a global average sense there is no clear difference between heat flux in well-sedimented and poorly sedimented areas (Stein and Stein, 1994b). Nevertheless, it is clear that sediments affect crustal fluid flow in two ways: (1) by impacting the rate of diffuse fluid flow (and therefore heat transport) through the seafloor, and (2) by controlling the pattern of fluid flow from the basaltic basement, including the focusing of fluid discharge to outcrops.

In this chapter, we will discuss the distribution of marine sediments, explain how sediment permeability measurements are made, and define sediment characteristics that influence hydraulic properties. We place sediment distribution and property variations in perspective of global and local flow by estimating limits on fluid, heat, and chemical fluxes through sediments of various types and thicknesses. Finally, we discuss the potential influence of sediments on the nature of fluid circulation in the basement aquifer.

6.2 Distribution of sediment in ocean basins

6.2.1 Global distribution by type

The pattern of ocean sediment deposition and accumulation is a function of the location of sediment sources, the nature of sediment transport, and the effects of weathering and diagenesis. We discuss the distribution of six general sediment types: three types of pelagic sediment that accumulate in most ocean areas (calcareous sediments, biogenic siliceous sediments, and pelagic red clays), two sediment types that are locally significant (terrigenous and volcanogenic sediments), and one sediment type restricted to high latitudes (ice-rafted sediments).

Calcareous sediments represent 74% of the mass of pelagic sediments, and approximately 41% of the mass of all ocean sediments (Hay *et al.*, 1988). Calcareous sediments are composed primarily of calcium carbonate tests (invertebrate shells or skeletal remains). The regional distribution of calcareous sediments is controlled by surface productivity and the preservation of material at depth (Barron and Whitman, 1981). Surface production of carbonate is controlled by the supply of limiting nutrients, and therefore by ocean circulation patterns. High productivity is found in areas of intense upwelling (e.g. along western continental margins) and divergent ocean currents (e.g. the equatorial Pacific; Millero, 1996). Because calcium carbonate solubility increases with depth (i.e. with increasing pressure, decreasing temperature, and increasing partial pressure of CO_2) the oceans become undersaturated with respect to calcium carbonate at great depth (Millero, 1996). Below the carbonate compensation depth (the CCD, typically lying at about 4,000 m) the rate of carbonate dissolution exceeds the supply of carbonate into the system. Well-preserved calcium carbonate tests in surface sediments above the CCD indicate that most carbonate dissolution occurs at the seafloor; there is little dissolution of particles as they settle through the water column (Parker and Berger, 1971; Edmond, 1974).

Calcareous sediments are the dominant pelagic sediment type where the seafloor is above the CCD, generally close to mid-ocean ridge axes (Fig. 6.1; Barron and Whitman, 1981). They also accumulate along the Hawaii–Emperor seamount chain, which rises above the surrounding north Pacific basin. As young crust moves away from the mid-ocean ridge, it cools and subsides; the seafloor passes below the CCD and the accumulation of non-carbonate sediment dominates. The combination of subsidence and the depth-dependence of carbonate stability results in the base of many sediment columns throughout the ocean being dominated by calcareous sediment.

Siliceous sediment, composed of biogenic silica, is the dominant sediment type where surface plankton productivity is high and the seafloor lies below the CCD. Siliceous sediments are composed mostly of opaline silica tests, primarily from diatoms and radiolarians. High surface productivity areas for plankton with siliceous tests are generally the same as those for carbonates. Radiolarians dominate equatorial belts in the Pacific and Indian Oceans, and both radiolarians and diatoms are common in coastal upwelling zones (Kastner, 1981). High diatom productivity extends further into high latitudes than the zones of high carbonate productivity, resulting in a continuous belt of siliceous sediments at $\sim 60°$ S latitude (Fig. 6.1; Kastner, 1981). The shape of the Atlantic Ocean prohibits the development of equatorial current divergence. As a result, biogenic silica production is relatively low, and there is no Atlantic equatorial belt of siliceous sediments (Barron and Whitman, 1981). Siliceous sediments make up approximately 6% of pelagic sediments by mass, and 3% of the mass of all seafloor sediments (Hay *et al.*, 1988).

Pelagic red clays make up the remainder ($\sim 19\%$ by mass) of the pelagic sediments, or $\sim 10\%$ of the mass of all seafloor sediments (Hay *et al.*, 1988). Pelagic red clay is derived from the settling of airborne dust into the oceans. Large dust sources are located in

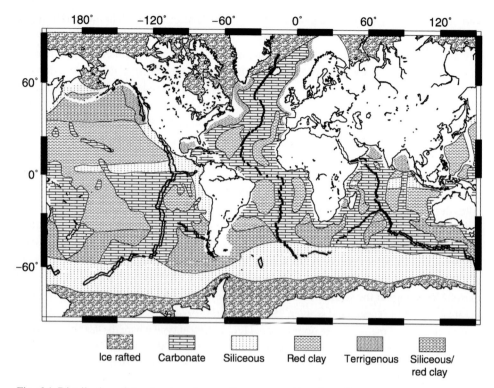

Fig. 6.1 Distribution of dominant seafloor sediment types (data from Barron and Whitman, 1981). Sediment hydrologic properties vary with sediment type, therefore the distribution of sediment types influences the fluid communication between the ocean and lithosphere. Carbonates primarily accumulate in relatively shallow waters near mid-ocean ridges, where the seafloor is above the carbonate compensation depth. Belts of siliceous sediments accumulate in the equatorial Pacific and Indian Oceans and in the southern ocean, where biological productivity is high. Pelagic red clays accumulate in relatively deep water near 30° N and 30° S.

sub-tropical dry zones (centered on the descending branches of Hadley cells, ∼30° N and 30° S latitudes) and leeward of large mountain ranges. The composition of pelagic red clays can vary depending on the composition of the source area, but they frequently contain large amounts of illite and quartz (products of arid mechanical weathering; Barron and Whitman, 1981). Their distribution on the seafloor is controlled by wind patterns, wind strength, particle size, and precipitation patterns. Pelagic red clay is the dominant sediment where the seafloor is below the CCD and biogenic silica productivity in the surface water is relatively low, generally in deep water near 30° N and 30° S (Fig. 6.1; Barron and Whitman, 1981).

Terrigenous sediments constitute the largest mass fraction (47%) of seafloor sediments (Hay *et al.*, 1988), but they cover a limited area relative to pelagic sediments. Terrigenous sediments are the products of mechanical and chemical weathering input to the oceans by rivers. Their composition varies with the lithology and climate of the source area. Thick accumulations of terrigenous sediments are concentrated on and adjacent to continental margins, near the mouths of major rivers (Fig. 6.1). The amount of sediment input from rivers depends on topography, climate, and the efficiency of erosion (Vail *et al.*, 1991). For example, as sea level fell periodically throughout much of the Pleistocene (Shackleton, 1987), rivers incised their valleys in response to lowered base levels (Vail *et al.*, 1991). As a result, average Pleistocene sediment discharge from rivers was large relative to the present (Barron and Whitman, 1981). Sediment discharge varied dramatically between high and low sea level stands, and times of transgression and regression; modern rates are relatively low. In the discussion of sediment physical properties that follows (Section 6.4), terrigenous sediments are sub-divided into turbidites and hemipelagic sediment. The turbidites that we discuss are generally fine-grained (silty) sediments deposited from turbidity currents or surges, but can include coarse, highly permeable intervals. The fine-grained turbidites tend to have a much lower permeability than the sandy turbidites (see Section 6.4.2). For vertical fluid flow through layered sediments, the effective permeability of the sediment column is controlled by the low-permeability layers; high-permeability sand layers have very little impact on vertical fluid flow (e.g. Giambalvo *et al.*, 2000). Hemipelagic sediment is fine grained with a significant terrigenous component that is shed off the continental margin and is transported by means other than turbid gravity flow.

Ice-rafted sediments are mechanically weathered sediments transported by icebergs. Their composition varies with the source rocks; they often contain lithic fragments, quartz, chlorite, and illite (products of mechanical weathering; Barron and Whitman, 1981). Because their distribution is restricted to areas of the ocean commonly containing icebergs, the accumulation of ice-rafted sediment is limited in both time and space. Significant deposition of ice-rafted sediment occurs only when continental glaciers reach the ocean; the oldest ice-rafted ocean sediments are Miocene (Barron and Whitman, 1981). Ice-rafted sediments are the dominant sediment type poleward of ~60° N and 60° S (Fig. 6.1).

Volcanogenic sediments, primarily ash and glass, are derived from both subaerial and submarine volcanoes. They are concentrated near subduction zones and mid-ocean ridges. Volcanogenic sediments are predominantly glass, which alters to smectite and zeolites (Kastner, 1981). Because of the large number of volcanoes surrounding the Pacific and Indian Oceans, their sediments have higher smectite content than Atlantic sediments. Subaerial volcanoes can distribute ash regionally or globally, but volcanogenic sediments are generally a minor component of marine sediments.

6.2.2 Global distribution by thickness

Sediment thickness complements sediment type as a major control on the hydraulic behavior and function of seafloor sediments. Oceanic sediment thickness varies widely, from effectively zero on most mid-ocean ridges to accumulations occasionally > 10,000 m in sedimentary basins on continental margins (Fig. 6.2). In general, sediment thickness increases systematically with increasing lithospheric age, and thus with distance from mid-ocean ridges. The thickest sediment accumulations are on or adjacent to continental margins that

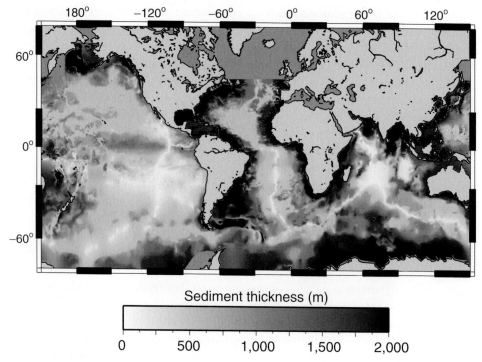

Fig. 6.2 Global distribution of sediment thickness (data from Divins, 2002). Sediment thickness data are compiled from ocean drilling results and reflection seismic profiles, and gridded with a spacing of $5 \times 5'$. Sediment thickness varies as a function of lithospheric age (distance from mid-ocean ridge), proximity to zones of high biological productivity, and proximity to terrigenous sediment sources. Sediment greater than 2,000 m thick, primarily on continental margins, is shown in black.

are proximal to large sediment sources, where high-volume supply combined with tectonic deformation and/or isostatic compensation allow for the development of deep sedimentary basins.

While sediment thickness generally increases with lithospheric age, there is significant variability in this trend resulting from variations in the proximity of the seafloor to large terrigenous sediment sources, the distribution of high-biological-productivity zones, and the nature of sediment transport. We compare the global distribution of sediment thickness (Divins, 2002) and lithospheric age (Muller *et al.*, 1997) in Fig. 6.3*a*. The data (grouped into 5 Ma bins) tend to be skewed, with most of the values concentrated at smaller sediment thickness and very few relatively large sediment thickness values. In the Atlantic, lithospheric age increases monotonically from the center of the ocean basin to the continents. The resulting trend of sediment thickness versus age reflects both the greater time for sediment accumulation on older lithosphere and the large terrigenous sediment input to old lithosphere on passive margins (which has been adjacent to the continents throughout the opening of the Atlantic Ocean basin). The trend in the Pacific is more complex. On

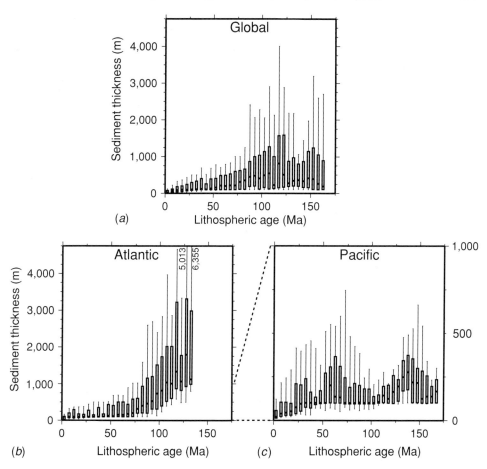

Fig. 6.3 Sediment thickness (data from Divins, 2002) versus lithospheric age (data from Muller *et al.*, 1997). Pairs of sediment thickness and lithospheric age values are determined on a grid with a spacing of $1.67 \times 1.67°$, then grouped by age into 5 Ma bins. The box plot for each 5-Ma age bin shows the median sediment thickness (bold horizontal line), the 25th and 75th percentile (bottom and top of the box, respectively), and the 10th and 90th percentile (end of the vertical bars extending from the bottom and top of the boxes). For the global sediment thickness plot (*a*), the 5 Ma bins between 0 and 35 Ma have ~500 data points, with the number of points per bin gradually decreasing to ~100 for bins >120 Ma. The global sediment thickness (*a*) reflects both the relatively simple trend of increasing sediment thickness with age in the Atlantic (*b*) and more complex relationships in the Pacific (*c*), Indian, and Southern Oceans (not shown).

young lithosphere (<50 Ma), generally in the eastern Pacific and on the Philippine Plate, sediment thickness is comparable to the Atlantic (note different vertical scale). Sediment thickness remains relatively uniform with age in the central Pacific (~65–105 Ma), where the plate is far from terrigenous sediment sources or coastal zones of high biological productivity. Sediment thickness increases with age on very old (>125 Ma) lithosphere, as a result of proximity to western Pacific continental margins, although not nearly as much as the increase on older lithosphere of the Atlantic.

6.2.3 Local distribution by thickness

Like regional variations, local sediment thickness variations can also be large, especially in areas of rough basement topography. Local variations are important hydrologically because of the strong tendency for seafloor fluid flow to be focused through sections where sediments are thin or absent. The exchange allowed by focused flow can affect large volumes of surrounding igneous crust that is well sedimented. Local sediment thickness is influenced by original depositional processes and by sediment redistribution, and both are controlled by local seafloor topography. Seafloor topography in ocean basins is commonly expressed as seamounts produced by localized volcanic activity, or linear basaltic ridges generated by normal faulting and temporally variable volcanic supply at seafloor spreading centers (e.g. Kappel and Ryan, 1986; Chapter 3). Seafloor roughness is greater in slow-spreading ocean basins than in fast-spreading basins (200–300 versus ~120 m, respectively; Malinverno, 1991; Chapter 3). Therefore, on lithosphere <65 Ma, where average sediment thicknesses in the Atlantic and Pacific might be comparable, seafloor topography may create greater sediment thickness variability and less continuity of coverage in the Atlantic than in the Pacific.

Both terrigenous and pelagic sediment accumulation are affected by seafloor topography. Turbidity currents tend to flow into topographic lows and deposit relatively flat turbidites that onlap local topographic highs (e.g. Pickering and Hiscott, 1985; Kneller *et al.*, 1991; Pickering *et al.*, 1992; Alexander and Morris, 1994; Haughton, 1994). In pelagic depositional environments, seafloor topography may influence sediment depositional patterns by causing variations in fluid shear velocity in the bottom boundary layer (Lister, 1976; Mitchell *et al.*, 1998) or by providing conditions for the redistribution of sediments by local gravity flows (Mitchell *et al.*, 1998).

Terrigenous deposition environment

In terrigenous depositional environments, the interaction of turbidity currents with topography can cause local variations in both sediment thickness and type. Basaltic basement highs that rise above surrounding basins (i.e. seamounts or ridges) are often sediment free or thinly draped with sediment (Chapter 8). Dense turbidity currents preferentially fill low areas between topographic highs, concentrating turbidite deposition in basins. While turbidity currents are typically blocked or diverted by topography, hemipelagic sediment accumulation can be relatively uniform. The resulting pattern of thin drapes of hemipelagic sediment on basement highs and thicker blankets of mixed terrigenous sediment (hemipelagic and turbidite) in local basins is found in many locations, including the Alarcon Basin within the Gulf of California (Fig. 6.4*a–c*). Thus, local variations in sediment thickness and type in terrigenous environments can be dramatic.

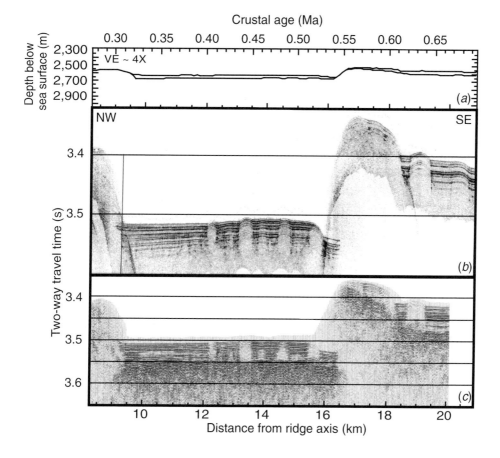

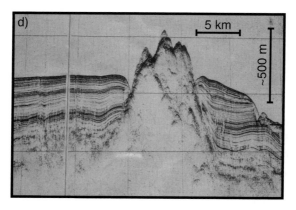

Fig. 6.4 Local variability in sediment thickness in the Alarcon Basin, Gulf of California (*a–c*), and in the Panama Basin (*d*). In the Alarcon Basin (*a* – line drawing; *b* – 3.5-kHz seismic profile; *c* – single-channel seismic profile), a terrigenous depositional environment, basement relief affects sediment accumulation patterns. Turbidites accumulate between and onlap basement highs. Hemipelagic sediments drape basement topography uniformly. In the Panama Basin (*d* – high-resolution reflection seismic profile from Davis, personal communication), a pelagic depositional environment, sedimentation is inhibited on large topographic features, possibly by the interaction of bottom currents with the topography, or by remobilization and mass wasting.

Pelagic deposition environment

In areas dominated by pelagic sedimentation, topography can also impact sediment accumulation patterns, despite the relatively uniform rain of particles from above. A high-resolution reflection seismic line across a seamount in the Panama Basin shows sediment cover on the seamount to be extremely thin (Fig. 6.4d). Local variability in pelagic sediment thickness like this is common, and may be due to tidal current interactions with topography or redistribution by gravity driven flow. On the East Pacific Rise, where topography is generally subdued, sediment thickness increases uniformly with distance from the ridge crest, but where topographic features are present that are greater than 300 m high (and slopes on the sides of the topographic highs are ~5–7°), they are bare or thinly covered with sediment. These large topographic features may magnify the peak boundary velocity (and therefore shear stress) associated with internal waves due to diurnal and semi-diurnal tides, and effectively inhibit the deposition of settling particles on topographic highs (Lister, 1976).

Another mechanism for redistributing pelagic sediment is mass wasting. Young oceanic crust (<2 Ma) on the Mid-Atlantic Ridge at 29° N is uniformly draped with sediment (Mitchell *et al.*, 1998), but as sediment continues to accumulate, ponded turbidites become more common. Pelagic material on local highs and steeper slopes may become unstable, resulting in sediment failures which redistribute sediments locally. Local depressions collect and pond sediments transported from surrounding steep slopes by mass movements (Mitchell *et al.*, 1998). The resulting discontinuous burial of basement becomes similar to that produced in terrigenous settings, with high-standing basement structures remaining relatively thinly sedimented. In both cases, caution must be used when applying the average sediment thickness distributions presented in Figs. 6.2 and 6.3 to models for crustal hydrogeology. Locally thicknesses are likely to be highly variable, and cover is likely to be discontinuous.

6.3 Methods for determining the permeability of sediments

6.3.1 Definitions

The permeability of marine sediments, a function of sediment type and burial depth, controls the nature of fluid flow through sediments and the degree to which sediments can hydraulically isolate the basement aquifer. The volumetric flux of fluid through a porous medium is related to the gradient in non-hydrostatic pressure (often expressed as hydraulic head) through a constant of proportionality known as hydraulic conductivity:

$$q = -K_h \frac{dh}{dL} \tag{6.1}$$

where q (L T^{-1}) is volumetric fluid flux (often referred to as Darcy velocity, or specific discharge), K_h (L T^{-1}) is hydraulic conductivity, and dh/dL is head gradient (for a review of basic hydrogeology terminology see Chapter 12 of this volume and references therein). Hydraulic conductivity is a function of both fluid properties and properties of the porous medium:

$$K_h = k\frac{\rho g}{\mu} \tag{6.2}$$

where k (L^2) is permeability, ρg is the specific weight of the fluid, and μ is fluid dynamic viscosity. Permeability describes the intrinsic capacity of sediment (or other porous media) to transmit fluid. In general, sediment permeability varies with direction and thus must be expressed as a tensor. Permeability is usually determined as a single value, but this is strictly correct only in the isotropic case, where $k_x = k_y = k_z$ (Bear, 1972). Both consolidation and permeability testing are usually done in one dimension, with the direction of strain and fluid flow oriented perpendicular to bedding. Multi-directional laboratory tests have found no permeability anisotropy in consolidated clays, despite the presence of anisotropic grain fabric (Dewhurst *et al.*, 1996). Sheared silty clays have large permeability anisotropy (k_x up to 16 times larger than k_z; Dewhurst *et al.*, 1996). When considering fluid flow from the oceanic crustal basement aquifer through sediments in hydrothermal systems, vertical permeability is likely the most relevant parameter.

The vertical hydraulic impedance (Karato and Becker, 1983) describes the resistance to fluid flow through an entire sediment column:

$$I = \int_0^{z'} \frac{dz}{k(z)} \tag{6.3}$$

where z is depth. Low-permeability layers dominate the vertical hydraulic impedance. In a section of uniform lithology, sediment permeability decreases with depth, and the bulk permeability (effective permeability) of a sediment column,

$$k_{\text{bulk}} = \frac{z}{I} \tag{6.4}$$

decreases with increasing sediment thickness. When comparing resistance to fluid flow between sediment sections of different lithologies, hydraulic impedance and bulk permeability are more useful than permeability at a point (or points) in the sediment column.

6.3.2 Laboratory tests

The permeability of samples of seafloor sediments can be determined in the laboratory by direct or indirect methods. Direct methods entail forcing water through a sediment sample and monitoring the volumetric flux and pressure gradient in order to solve Darcy's Law (6.1) for permeability (via hydraulic conductivity). Three types of direct laboratory permeability tests have been commonly employed: constant head, falling head, and constant flux tests. Tests can be performed in conjunction with consolidation testing to estimate the effect of porosity (and equivalent burial depth) on permeability as an alternative to testing samples from a given lithology that have been collected over a range of depths. Sediment permeability can be derived from consolidation tests by monitoring the rate at which water is expelled during consolidation. Consolidation tests can be performed either by incrementally increasing the load on a sediment sample and monitoring its thickness or by deforming the sediment at a constant rate and monitoring the effective stress (σ_e):

$$\sigma_e = \sigma_1 - P \tag{6.5}$$

where σ_1 is the axial load and P is the pore pressure.

In a constant-head permeability test, a confined sediment sample is subject to a constant head that is maintained by a replenished column of water. The volume flux of water through the sample is measured, allowing the hydraulic conductivity (and permeability, given the fluid properties) to be determined. Simple constant-head permeability tests are not back-pressured (i.e. fluid flows out of the sample to atmospheric pressure). Non-back-pressured constant-head permeability tests are best suited for sediments with high permeability ($>1 \times 10^{-13}$ m^2, i.e. within the typical range of permeabilities for silty sand; Freeze and Cherry, 1979). Simple constant-head permeability tests require long times and/or high head gradients to determine the permeability of fine-grained sediments. This causes complications associated with possible evaporation (affecting q), variations in temperature (affecting ρ, μ), and sediment particle creep induced by high-pressure gradients (changing k). Falling-head permeability tests (in which the column of water driving flow is not replenished; Remy, 1973; Freeze and Cherry, 1979) suffer the same problems as non-sealed constant-head tests, and provide reliable results only for high-permeability samples. Sealed, back-pressured constant-head permeability systems have been developed to prevent effects of evaporation and of trapped air bubbles within sediment pores. Lower heads can be used to force water slowly through a fine-grained sample. Back pressured constant-head tests can be used to test samples with lower permeabilities (1×10^{-15} to 1×10^{-18} m^2; Silva *et al.*, 1981).

In constant-flow permeability tests, water is forced through a confined sample at a known rate and the head induced across the sediment sample is measured (Olsen, 1966; Olsen *et al.*, 1985). The constant flow is maintained by a smoothly moving piston or syringe. Two flow pumps, one on either side of the sample pushing water in opposite directions at equal speeds, can be used to ensure conservation of fluid storage in the system (e.g. Taylor and Fisher, 1993; Fisher *et al.*, 1994), and if desired the system can be back-pressured.

In a typical one-dimensional consolidation test, a laterally confined sediment sample is subjected to an incrementally increasing effective stress (via one-dimensional axial loading) and the sample thickness is monitored (ASTM, 1989). Porosity is calculated for each increment of effective stress. Standard consolidation testing involves doubling the effective stress every 24 hours. During each 24-hour period the sediment undergoes both primary and secondary consolidation. Primary consolidation occurs during the initial transient drainage of pore fluid in response to the increased load. It is usually completed in seconds or minutes, with the time constant depending on permeability (Lowe *et al.*, 1964; Lambe and Whitman, 1969; Mitchell, 1993). In most standard laboratory tests, primary consolidation accounts for the bulk of the change in porosity. Secondary consolidation occurs under drained conditions as soil particles slowly creep under constant load. The relative importance of primary and secondary consolidation depends on the sediment type, the thickness of the sample, and the magnitude of the applied load (Lambe and Whitman, 1969). After primary consolidation is complete, a permeability test can be performed to determine permeability at a new porosity. Consolidation tests should be back-pressured to avoid trapped interstitial air which can impede fluid flow and alter the sediment consolidation behavior through its extremely high compressibility (Lowe *et al.*, 1964).

Sediment permeability can be estimated from the results of one-dimensional consolidation tests by using the change in sample thickness with time during primary consolidation as a proxy for the expulsion of water from the sediment (Lambe and Whitman, 1969; Mitchell, 1993). Although this is efficient, direct permeability measurements are generally preferable for two reasons. First, the porosity and permeability of sediment samples change during primary consolidation. The calculated permeability is an effective permeability over the duration of primary consolidation, rather than the permeability at a fixed porosity. Second, the time dependence of consolidation depends on both elastic properties (consolidation coefficient) and permeability, and the partitioning cannot be determined uniquely. This leads to inherent errors in the determination of permeability from consolidation behavior (Lambe and Whitman, 1969).

Continuous deformation tests can also be used to estimate the dependence of permeability on porosity. In this test a sample is consolidated at a constant rate; the effective stress is monitored and both consolidation parameters and permeability are determined (The Geotechnical Consortium, 1984). Constant deformation rate tests can be completed rapidly (hours per test) relative to incremental consolidation tests (~15 days per test; The Geotechnical Consortium, 1984; Marine Geotechnical Consortium, 1985), but transient-state behavior is difficult to detect and take into account.

6.3.3 *Estimates of permeability from inversion of field data*

A variety of techniques have been used to estimate sediment permeability *in situ* at a broad range of scales greater than that of laboratory core samples. One uses the transient decay of excess pore fluid pressure generated by penetrating sediments with a probe (e.g. Schultheiss and McPhail, 1986; Davis *et al.*, 1991; Langseth *et al.*, 1992; Fang *et al.*, 1993). In this method the consolidation coefficient of the sediment is estimated from the magnitude of the insertion pressure pulse, and the permeability is estimated from the time constant of the subsequent decay. This provides an estimate of permeability in the radial direction over a scale of roughly tens of centimeters, depending on elastic properties and permeability. Probes that are left in sediments over a number of days record pressure variations in response to tidal fluctuations. The tidally induced pore pressure variations can also be used to estimate sediment compressibility and permeability in the vertical direction over a scale that also depends on permeability, typically a few meters (Fang *et al.*, 1993; Wang and Davis, 1996; Chapters 7 and 12).

Sediment-column-scale (tens to hundreds of meters) permeability estimates can be made where both the fluid flow rate and driving force are known or can be estimated (e.g. Langseth *et al.*, 1988; Davis and Becker, 2002; Chapter 8; Section 6.5.1). This provides an excellent means of permeability determination at a very appropriate scale for seafloor hydrologic processes. Unfortunately very few locations on the seafloor have both flow rate and driving force data available. On an even larger scale of kilometers, inverse modeling has been used to estimate permeability, for example of the décollement zones of the Barbados (Bekins *et al.*, 1995) and Nankai accretionary prisms (Saffer and Bekins, 1998). In these cases, the permeability (including transient permeability variations) of modeled décollement zones are determined by matching modeled and observed fluid chemistry anomalies within the décollement, fluid pressures within the décollement, or fluid flow rates at seafloor cold seeps (Bekins *et al.*, 1995; Saffer and Bekins, 1998). Inverse modeling of sediment permeability has also been applied to large shale units in terrestrial basins (e.g. Neuzil, 1994). Permeability estimates from regional models like these apply to lateral scales of a few to hundreds of kilometers.

6.3.4 Sediment permeability estimates from borehole testing

Estimates of the *in situ* permeability of sediments have been made in Ocean Drilling Program (ODP) boreholes on the active margins of Barbados (Fisher and Zwart, 1996; Screaton *et al.*, 1997, 2000) and Oregon (Screaton *et al.*, 1995). These *in situ* sediment permeability tests have been performed in the décollement zone (Barbados) and along a thrust fault within the accretionary prism (Oregon), where sediments are subjected to numerous processes (e.g. faulting, deformation, rapid vertical loading) not typical of most seafloor sediments on mid-ocean ridge flanks or in abyssal settings. However, the techniques used to determine sediment permeability in these instances may be applied in more quiescent settings.

Borehole determinations employ both slug and continuous pumping tests. During a slug test, a section of the borehole is isolated (usually an interval from an inflated packer seal to the bottom of the hole), a discrete pulse of fluid is injected into that section, and the permeability of the surrounding sediment is estimated from the decay of the pressure pulse (e.g. Screaton *et al.*, 1995; Fisher and Zwart, 1996). Slug tests characterize the hydraulic properties of the sediment within a relatively small radius around the borehole. The hydraulic properties of a larger volume around the borehole can be determined by sustained pumping of fluid into a borehole (e.g. Screaton *et al.*, 1995; Fisher and Zwart, 1996) or releasing overpressured fluids from the borehole to the overlying ocean (e.g. Screaton *et al.*, 1997). In one case, the drilling of an ODP site into the Barbados décollement zone perturbed the fluid pressure within a nearby monitored ODP borehole ($\sim$45 m away). The recorded pressure perturbation was used to estimate the sediment permeability along the décollement between the two sites (Screaton *et al.*, 2000). Two-hole experiments such as this can characterize the permeability of larger volumes of sediment than either slug or pumping tests in single holes. The results of a variety of these tests indicate that the permeability of sediment within décollement or thrust-fault zones increases with increasing fluid pressure (decreasing effective stress; Screaton *et al.*, 1995; Fisher and Zwart, 1996; Screaton *et al.*, 1997, 2000), and that the values determined *in situ* at near-hydrostatic fluid pressures are comparable to laboratory permeability measurements of samples from the same ODP sites (Taylor and Leonard, 1990; Moran *et al.*, 1995).

6.3.5 Applicability of laboratory-derived consolidation and permeability data to natural settings

Sediment consolidation tests are important for predicting the response of sediment to burial, although there are natural limitations in applying laboratory results to *in situ* sediment consolidation. At the slow increases in stress that occur in nature, secondary consolidation plays a larger role than it does in the laboratory (Mitchell, 1993). For thick sediment columns deposited at high rates, undrained conditions may be present, and primary consolidation takes longer and probably occurs concurrently with secondary consolidation (Mitchell, 1993). In natural settings, the development of interparticle bonds and cementation may also increase the strength of sediment and impede consolidation.

Despite these uncertainties, permeability as a function of depth can be predicted by applying permeability versus porosity relationships to either measured porosity versus depth profiles (e.g. Giambalvo *et al.*, 2000), or to porosity versus depth profiles predicted from consolidation behavior (e.g. Spinelli *et al.*, 2003). Differences between measured porosity–depth profiles and predicted porosity profiles from consolidation behavior can be less than 5% (e.g. Davis *et al.*, 1997; Giambalvo *et al.*, 2000; Spinelli *et al.*, in press), although they can be larger for very shallow (< 1 m) extremely high-porosity sediments (which are difficult to load into a consolidation cell) or for sediments that undergo significant strengthening due to cementation. Estimated permeability versus depth profiles can then be integrated to determine hydraulic impedance and bulk permeability, which can be used to compare the resistance to vertical fluid flow at different sites.

Estimates of the basin-scale (up to a few kilometers) permeability of shales are similar to laboratory measurements of shale and clay permeability (Neuzil, 1994). This suggests that fracture permeability is not important at the scale of a few kilometers (Neuzil, 1994). The presence of fractures appears to affect estimates of shale permeability only at larger scales (Bredehoeft *et al.*, 1983; Neuzil, 1994). Because shale permeability measured in the laboratory can be applied to the regional scale (up to a few kilometers), it is reasonable to apply laboratory determinations of permeability for soft sediment to regions where sediment is fairly homogeneous. Soft sediments are even less likely to be affected by fractures than shale, although within overpressured environments or extensional settings, fracturing of soft material is possible, potentially increasing the effective permeability greatly (e.g. Brown and Behrmann, 1990; Moore *et al.*, 1995; Fisher and Zwart, 1996). Applying core-sample measurements of sediment permeability to larger areas should be done cautiously in any case, but particularly in layered depositional systems or regions where depositional processes create lateral heterogeneity.

6.4 Sediment permeability data

6.4.1 Data summary

The permeability of typical marine sediments varies from $\sim 10^{-19}$ to 10^{-12} m^2 generally according to porosity and sediment type. More than 600 measurements of sediment permeability are summarized in Fig. 6.5. The compiled data include both direct measurements

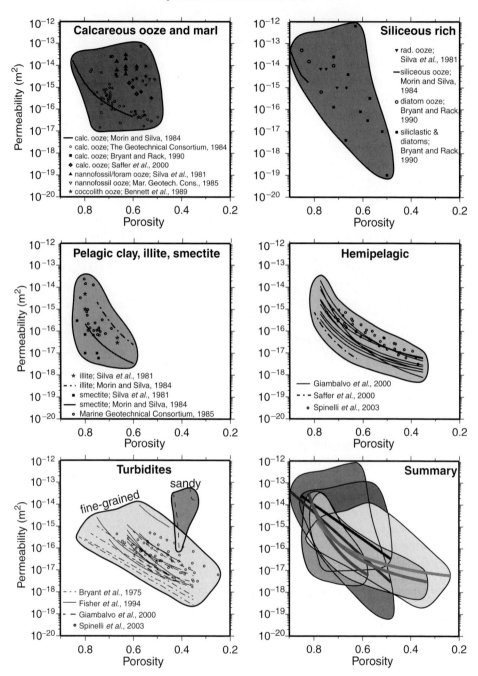

Fig. 6.5 Sediment permeability versus porosity (see Table 6.1 for references and information about measurement methods). Permeability tends to decrease with decreasing porosity; the trend is weaker in calcareous sediments than in other types. Bold lines on the summary plot are best fit functions to permeability versus porosity relationships; they are used to estimate the hydraulic impedance for sediment columns of each sediment type.

of permeability (e.g. constant flow rate and constant head tests) and estimates of permeability determined from consolidation tests (see Table 6.1 for a summary of references and information about the data). The permeability measurements are grouped by sediment type: calcareous, siliceous, pelagic red clay, and terrigenous (sub-divided into hemipelagic and turbidite). In each group permeability decreases with decreasing porosity. This trend is least prominent for calcareous sediments. The tendency for permeability to increase with grain size can be seen by comparing pelagic red clays (finest) to hemipelagic sediments and turbidites (coarsest). There is considerable overlap between the sediment types, but permeability at a given porosity (see summary plot on Fig. 6.5) typically increases from pelagic red clay (red) to hemipelagic sediments (green) to turbidites (yellow).

Relationships between permeability and porosity have been proposed for sub-sets of the data plotted in Fig. 6.5 (e.g. Bryant *et al.*, 1975; Morin and Silva, 1984; Giambalvo *et al.*, 2000). For each sediment type, we fit globally representative trends of permeability as a function of porosity to the compiled data (bold lines on summary plot on Fig. 6.5; equations in Table 6.2). The forms of the equations that provide the best fits to the data have previously been used to summarize permeability–porosity data (e.g. exponential of porosity, Bryant *et al.*, 1974; exponential of void ratio, Giambalvo *et al.*, 2000; void ratio power law, Morin and Silva, 1984). For most sediment types there are strong trends of both decreasing sediment permeability with decreasing porosity, and decreasing porosity with depth (Fig. 6.6; equations and references in Table 6.2). Differences in permeability as a function of depth between sediment columns of different types can be greater than three orders of magnitude (Fig. 6.7).

6.4.2 Factors that determine sediment hydraulic properties

A number of physical attributes, including sediment grain size, grain shape, fabric, and cementation, can affect sediment consolidation and permeability. Attempts have been made to define relationships between relatively easy-to-determine parameters (e.g. grain-size distribution, or formation factor) and permeability, which is much more time-consuming and labor-intensive to measure (Eggleston and Rojstaczer, 2001). No simple set of such relations can provide accurate results, although approximate estimates can be made through the two general relationships between porosity and permeability observed in Fig. 6.5. First, for a given sediment type, permeability decreases with decreasing porosity. This, coupled with increasing consolidation with burial, results in a general decrease in permeability with depth in sections of uniform lithology (Fig. 6.7). Second, permeability varies with lithology. For example, nannofossil ooze sediments from the Lau Basin have porosities of 72–78% and permeabilities of 4.2×10^{-17} to 5.6×10^{-16} m^2, while vitric sandy silt from the same area has a lower porosity (66%), but permeability more than two orders of magnitude higher (1.5×10^{-13} m^2; Fischer and Lavoie, 1994). This dependence is due primarily to the size and shape of interstitial pore space.

Table 6.1 *References and information for permeability data plotted in Fig. 6.5*

Reference	Location	Lithology	Consolidation method	Permeability method	# of Samples/ measurements	Region on plot represents
Bennett et al. (1989)	Exuma Sound, offshore Bahama	calcareous (coccolith matrix supporting foraminifera)	back-pressured(?)	from Cv (?)	7/NA	k versus *in situ* n (all samples)
Bryant et al. (1975)	Gulf of Mexico	siliciclastic	back-pressured	from Cv	>100/NA	k versus *in situ* n (all samples)
Bryant and Rack (1990)	Weddell Sea	siliceous ooze (84–88% diatoms)	back-pressured	from Cv	4/NA	k versus *in situ* n (all samples)
		calc. ooze(10–20% diatoms)			3/NA	
		hemipel. (0–45% diatoms)			10/NA	
Fisher et al. (1994)	Middle Valley, Juan de Fuca Ridge	turbidites (silty clay to sand/silt 50:50)	back-pressured	constant flow	10/~75	k versus n all measurements
Geotechnical Consortium (1984)	Walvis Ridge	calcareous and siliceous ooze (11–76% CaCO3)	back-pressured and non-back-pressured standard consol., constant rate of deformation (CRD)	falling head, from Cv, from CRD	51/NA	k versus *in situ* n (all samples)
Giambalvo et al. (2000)	Juan de Fuca Ridge Flank	turbidites (sand to clay)	back-pressured	constant flow	9/~65	k versus n all measurements
		hemipelagic (silty clay to clay, 2–40% CaCO3)			9/~70	
Marine Geotechnical Consortium (1985)	Northwest Pacific	pelagic clay (fine-grained, IB)	back-pressured and non-back-pressured standard consol., constant rate of deformation (CRD)	falling head, const. head, const. flow, from Cv, and from CRD	32/NA	k versus n at preconsolidation stress (all samples)
		pelagic clay (silty, IA)			27/NA	
		calc. ooze (nannofossil, II)			8/NA	

Reference	Location	Sediment	Method	Flow type	k vs n	Note
Morin and Silva (1984)	North Central Pacific	pelagic red clay (illitic, 64% clay)	back-pressured	constant gradient	4/~20	k versus n all measurements
	North Central Pacific	pelagic red clay (smectite-rich, 72% clay, reconst.)			4/~20	
	Gulf of California	siliceous ooze (54% SiO2, 8% CaCO3)			4/~30	
	Galapagos Spreading Center	calcareous ooze (65% CaCO3, 21% SiO2)			4/~30	
Saffer et al. (2000)	Costa Rica sub. zone (seaward of wedge)	siliceous hemipelagic (interbedded diatom. sediment and ash)	back-pressured	constant flow	3/13	k versus n all measurements
		calcareous ooze			1/6	
Silva et al. (1981)	Location not given	illite (reconstituted)	back-pressured	constant gradient	?/4	k versus n all measurements
		smectite (reconstituted)			?/5	
		calc. ooze (nanno., reconst.)			?/3	
		calc. ooze (foram., reconst.)			?/4	
		sil. ooze (rad., reconst.)			?/4	
Spinelli et al. (2003)	Juan de Fuca Ridge Flank	silty turbidites	back-pressured	constant flow	8/55	k versus n all measurements
		hemipelagic			6/40	

Table 6.2 *Estimated porosity–depth and permeability–porosity relationships for various sediment types*

Sediment type	Location	Porosity	Permeability (m²); fit to compiled data in Fig. 6.5	Porosity reference
Siliceous ooze	central Pacific; Gulf of California	$n = 0.9 - 0.016 \frac{z}{10^3} - 3.854 \left[\frac{z}{10^3} \right]^2$	$k = 4.6 \times 10^{-23}\, e^{23n}$	Hamilton (1976)
Pelagic clay	central Pacific; north Pacific	$n = 0.814 - 0.813 \frac{z}{10^3} - 0.164 \left[\frac{z}{10^3} \right]^2$	$k = \dfrac{10^{1.15(n/(1-n))-13.8}}{10^5}$	Hamilton (1976)
Hemipelagic	Juan de Fuca Ridge flank	$n = 0.909 z^{-0.073}$	$k = 1.1 \times 10^{-18}\, e^{2.2(n/(1-n))}$	Giambalvo et al. (2000)
Calcareous ooze	central Pacific; Galapagos	$n = 0.72 - 0.987 \frac{z}{10^3} - 0.83 \left[\frac{z}{10^3} \right]^2$	$k = 5.6 \times 10^{-21}\, e^{18n}$	Hamilton (1976)
Fine-grained turbidite	Juan de Fuca Ridge flank; Gulf of Mexico	$n = 0.84 z^{-0.125}$	$k = 3.7 \times 10^{-18}\, e^{1.7(n/(1-n))}$	Giambalvo et al. (2000); Bryant et al. (1975); Bouma et al. (1986)

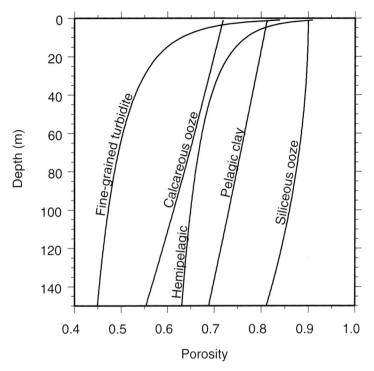

Fig. 6.6 Typical sediment porosity profiles for common seafloor sediment types (see Table 6.2 for equations and references). Porosity decreases with depth; the trend is strongest for shallow, fine-grained, terrigenous sediments.

Factors affecting consolidation behavior

Laboratory consolidation test results can be used to estimate sediment properties in locations where mechanical compaction is the primary contributor to sediment consolidation with burial (i.e. where sediments are still "soft," and cementation has not occurred). The degree of cementation in sediments varies with time, temperature, pressure, sediment composition, fluid chemistry, and fluid flux; the depth at which significant diagenetic reactions occur varies accordingly. For example, in siliceous sediments, the transition from opal A (amorphous silica) to opal CT (porcelanite) results in a dramatic change in physical properties, including a drop in porosity from 70–80% to 35–45% (Kastner, 1981). This transition occurs at temperatures of 15–56 °C (Kastner, 1981), and over a wide range of depths (100–1100 mbsf) depending on geothermal gradient and fluid flow rate (Wheat and Tribble, 1994). Similarly, the depth at which carbonate sediment transforms to chalk can vary widely with location (Kastner *et al.*, 1986; Mottl, 1989). As a result, the accuracy with which a mechanical model of sediment consolidation can be applied will vary widely.

Where a mechanical model of sediment compaction is valid, sediment fabric, grain size, and mineralogy are the primary factors that affect sediment consolidation. Sediment fabric

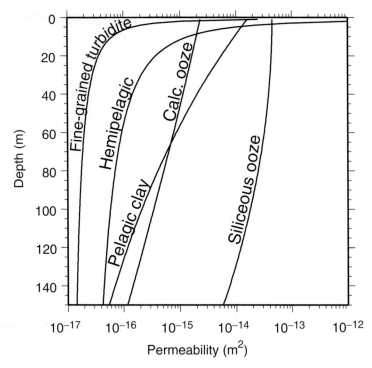

Fig. 6.7 Sediment permeability profiles derived from porosity versus depth and permeability versus porosity relationships (see Table 6.2 for equations and references). Permeability decreases with depth, weakly within siliceous and calcareous oozes, and most dramatically in shallow, fine-grained, terrigenous sediments.

impacts consolidation behavior by influencing the interaction between particles as grains are rearranged and realigned. In the central equatorial Pacific, carbonate sediments have relatively low porosity, as platy nannofossils are easily aligned during consolidation, whereas siliceous sediments at the same sites have relatively high porosity due to the open structure of the constituent radiolarians (Wilkens and Handyside, 1985). Throughout consolidation, platy clay particles progress from random orientations, providing numerous open spaces and high porosity, to tightly packed, well-aligned arrangements with low porosity (e.g. Bennett *et al.*, 1981; Mitchell, 1993). Grain size can also affect sediment consolidation, with finer sediments generally consolidating more easily (Lambe and Whitman, 1969; Mitchell, 1993). Less consolidation in silts than in clays may result from a greater strength of the framework of silty sediments, due to the way the larger (and more rounded) silt grains interact (Dewhurst and Aplin, 1998). Within terrigenous (turbidite and hemipelagic) sediments on the Juan de Fuca Ridge flank, the ease of consolidation increases with increasing proportion of hemipelagic sediment, which is characteristically finer grained (Spinelli *et al.*, 2003).

Factors affecting permeability at a given porosity

Sediment permeability can vary by orders of magnitude, independent of variations in porosity (Fig. 6.5). These permeability variations, which occur both between and within sediment types, can result from differences in sediment fabric or grain size. Carbonate tests often contain large void spaces within them (i.e. large intraparticle porosity), and their packing arrangement can control sediment permeability (Bennett *et al.*, 1989). Vitric sandy silt and nannofossil oozes in the Lau Basin have similar grain size, but due to variations in sediment fabric the sandy silt has larger interconnected pore spaces and larger connections between pores. As a result the permeability of the sandy silt is more than two orders of magnitude larger (Fischer and Lavoie, 1994). Sandy and silty sediments have larger pores and connections between pores than clays at the same porosity, which leads to higher permeability for the sandy and silty sediments (Dewhurst and Aplin, 1998). In Juan de Fuca Ridge flank sediments, permeability (at 50% porosity) decreases approximately two orders of magnitude with a decrease in grain size from primarily silt-sized particles to clay-sized particles (Spinelli *et al.*, 2003).

6.5 Fluid and advective heat fluxes through sedimented seafloor

6.5.1 Typical observed flow rates through sediment

Fluid flow rates through seafloor sediments have been determined at several locations on the basis of curvature in pore-water chemistry profiles (e.g. Maris and Bender, 1982), curvature in sediment temperature profiles (Bredehoeft and Papadopulos, 1965), and results from seepage meters (e.g. Tryon *et al.*, 2001). Fluid flow rates vary with sediment type, sediment thickness, deep heat flux, and hydrologic structure. Lithology and thickness control the bulk permeability of the sediment column. The deep-seated heat flux (a function of lithospheric age; Parsons and Sclater, 1977; Chapter 10), and the local hydrologic structure control the thermal-buoyancy-derived driving force for sediment pore-fluid flux (Chapter 8).

Relatively rapid fluid flow rates have been determined from shallow sediment temperature and pore-water chemistry profiles collected from $\sim$30-m-thick calcareous sediments at the 0.7 Ma Galapagos Mounds (Williams *et al.*, 1979; Becker and Von Herzen, 1983; Maris and Bender, 1982; Maris *et al.*, 1984), a site that is notable for yielding highly complementary chemical and thermal constraints. Seepage rates through the gravel-bearing mounds themselves (200–300 mm yr^{-1}) are 20–30 times larger than rates through the surrounding finer calcareous sediments (10 mm yr^{-1}). Fluid flux through terrigenous sediments overlying 3.5 Ma crust on the Juan de Fuca Ridge flank vary inversely with sediment thickness, from <1 mm yr^{-1} where sediments are greater than about 160 m thick up to 80 mm yr^{-1} over buried basement edifices where sediments are $\sim$10 m thick (Davis *et al.*, 1992; Wheat and Mottl, 1994, 2000). This systematic variation is probably a result of the combination of decreasing pressure differential (Chapter 8) and increasing hydraulic impedance with increasing sediment thickness and basement depth. On older (5.9 Ma) crust of the Costa Rica Rift flank, flux through calcareous and siliceous sediments also varies inversely with sediment thickness, ranging from <1 to about 7 mm yr^{-1} through sections ranging from over 200 m to about 120 m thick (Langseth *et al.*, 1988; Mottl, 1989). For a given sediment thickness, fluxes through the sediments on the Costa Rica Rift flank are higher than through those on the Juan de Fuca Ridge flank; if the nature and magnitude of driving forces are similar, this suggests a contrast in permeability between the pelagic and terrigenous sections that is consistent with the results summarized in Fig. 6.7. Very low values of fluid flux (up to 0.2 mm yr^{-1}) through carbonate sediments blanketing 40 Ma crust of the flank of the East Pacific Rise have been estimated on the basis of pore-water strontium (Richter, 1993). Whether these low rates are associated with the relatively great age of the lithosphere, the thickness of the section ($\sim$500 m), or peculiarities of local hydrologic structure, is not possible to determine. Extremely large fluxes (up to 2 m yr^{-1}) have been estimated from corer–outrigger temperature profiles in sediments of the Indian Ocean (Anderson *et al.*, 1979; Abbott *et al.*, 1981; Geller *et al.*, 1983). Unfortunately, these have not been confirmed with geochemical observations, and local hydrologic structure is poorly defined.

Except for the highest fluxes through the thinnest sediment sections, the rates of flow estimated in these examples are insufficient to carry a significant quantity of heat. In order to account for the difference between heat expected from cooling oceanic lithosphere and the average conducted through the seafloor (Stein and Stein, 1994; Chapter 10), rates of fluid flow are required that are much greater than those observed. For example, Wheat and Mottl (1994) have pointed out that in order to account for the conductive heat flow deficit on ridge flanks with seepage of <20 °C fluid distributed over as much as 30% of seafloor <65 Ma in age, rates must average $\sim$100 mm yr^{-1}. Fluxes this large are extremely rare, and it is unlikely that sediment-hosted flow contributes significantly to the global heat deficits observed. Well-documented examples of flow at any rate are sparse, however, so to examine the role of sediments in hosting flow further, we turn next to implications that can be drawn from sediment permeability measurements.

6.5.2 Sediment thickness required to prevent thermally significant vertical fluid flow

For a given driving force, the fluid seepage rate through a sediment column and the associated advective chemical and heat fluxes are functions of sediment thickness and type or, more explicitly, of sediment hydraulic impedance. The impedance of sediment columns of different types can be calculated from permeability–depth profiles by numerically integrating Eq. 6.3. For a given thickness, the hydraulic impedance of different sediment types can range over roughly three orders of magnitude (Fig. 6.8).

For terrigenous sediments, the hydraulic impedance has been estimated for hemipelagic and fine-grained turbidite sediment columns, but not layered sand-bearing turbidites. For vertical fluid flow through layered turbidites, the relatively low-permeability fine-grained constituent is the dominant contributor to the total hydraulic impedance. This is illustrated in Fig. 6.9, where the impedance is shown for a 100-m-thick sediment column comprising interbedded fine-grained and sandy layers of varying proportion (porosity and permeability

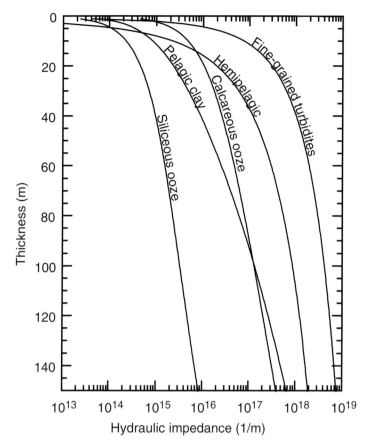

Fig. 6.8 Sediment hydraulic impedance as a function of sediment thickness for common seafloor sediment types, derived from permeability versus depth profiles shown in Fig. 6.7. At a given sediment thickness, hydraulic impedance can vary over about three orders of magnitude due to differences in sediment type.

given in Table 6.2 and Fig. 6.5). As the cumulative thickness of sandy layers increases, the total hydraulic impedance of the sediment column decreases, but the effect is relatively small for a wide range of sand-layer fraction. High-permeability sandy layers can significantly enhance the horizontal permeability of a layered sediment section, but not the vertical (Davis and Fisher, 1994). The same can be said for a layered section of other contrasting constituents such as siliceous and carbonate oozes; the collective vertical impedance will be strongly controlled by the most resistive, lowest permeability component.

Hydraulic impedance can be combined with estimates of the force driving fluid flow through sediments to obtain estimates of fluid seepage rate as a function of sediment thickness for each sediment type (pure end members are considered here). In ridge flank settings the driving forces for fluid flow through sediments (i.e. basement overpressure) result from

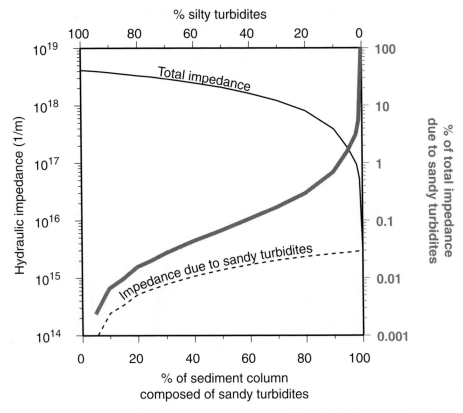

Fig. 6.9 Vertical hydraulic impedance (left axis) through a 100-m-thick column of turbidites comprising sandy ($k = 1.5 \times 10^{-13}$ m^2) and fine-grained layers (permeability given in Fig. 6.7 and Table 6.2) as a function of the proportion of sand. As the sand-layer fraction increases, the impedance due solely to the sandy turbidite layers (dashed line) increases, but the total impedance (solid black line) decreases. The contribution of the sandy turbidites to the total hydraulic impedance (bold gray line; right axis) is less than 1% until the sand fraction exceeds 90%. Some 10 m of fine-grained turbidites account for 99% of hydraulic impedance through the 100-m-thick sediment column.

differences in buoyancy of water columns at different temperatures (and therefore different densities) arising to a large degree from basement topography. Non-hydrostatic basement fluid pressures (relative to local sediment geotherm hydrostats) measured in four boreholes on the Juan de Fuca Ridge flank range from a few to roughly 25 kPa (Davis and Becker, 2002; Chapter 8). Basement overpressure at another site, inferred from fluid seepage rate and sediment hydraulic impedance, is ~13 kPa (Spinelli *et al.*, 2003). We use a differential pressure of 20 kPa to estimate rates of flow, and compare these to the limiting thickness of sediment sections that can support chemically detectable and thermally detectable fluid flow (Fig. 6.10). For a given sediment thickness, estimated fluid flow rates through different sediment types vary by more than two orders of magnitude.

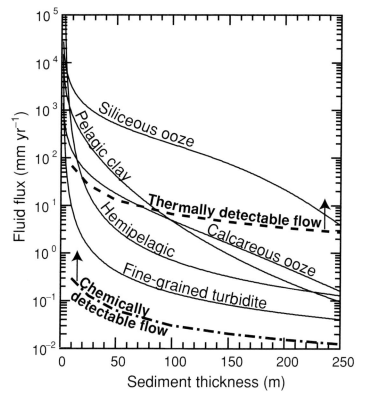

Fig. 6.10 Fluid seepage rate versus sediment thickness for the major sediment types of Fig. 6.8 calculated with a differential pressure of 20 kPa. With this driving force, fluid flux is chemically detectable (above dot–dash line) through all sediment types up to 250 m sediment thickness. Flux at thermally detectable rates (above dashed line) is restricted once a few tens of meters of sediment have accumulated except in the case of highly permeable siliceous oozes. Chemical and thermal detection thresholds are defined by Peclet numbers (the ratio of advective flux relative to diffusive flux in the absence of advection; see Chapter 12) of 0.2 and 0.1, respectively, assuming chemical and thermal diffusivities of 6.3×10^{-6} cm^2 s^{-1} and 2×10^{-3} cm^2 s^{-1}, respectively (see text).

For these calculations, the thresholds for thermally and chemically detectable flow (dashed lines) are defined by thermal Peclet numbers of 0.1 and 0.2, respectively. These fluxes cause Mg concentrations to deviate by up to 1.2 mM from a diffusive profile between constant concentration boundaries at 4 and 52 mM (typical Mg concentrations in warm ridge flank basement water and ocean bottom water, respectively; e.g. Davis *et al.*, 1997; Chapters 19 and 21), and temperatures to deviate by up to 0.6 °C from a diffusive profile between constant temperature boundaries at 2 and 50 °C . The threshold for thermally detectable flow is almost three orders of magnitude greater than that for chemical detectability because values of thermal diffusivity are ∼1000 times larger than values of ion diffusion coefficients (e.g. thermal diffusivity $\kappa \sim 2 \times 10^{-3}$ cm^2 s^{-1}, and for pore-water Mg,

$D_{sed} \sim 6 \times 10^{-6}$ cm^2 s^{-1}; Li and Gregory, 1974). These calculated chemically and thermally significant fluid flow thresholds are somewhat arbitrary, but reasonable. Thresholds shown in Fig. 6.10 can be scaled by the Peclet number for particular situations (e.g. higher values for detectability where observational errors are large or the depth of observations is small, and much higher values where the threshold for advective dominance is to be considered) but the position of the curves will change little.

These calculations demonstrate quantitatively what has been noted previously from studies in several specific settings – a relatively thin sediment layer reduces fluid seepage sufficiently so as to limit advective heat loss greatly. With the assumed basement overpressure of 20 kPa, only siliceous ooze is sufficiently permeable to allow fluid flow at rates that are thermally significant for the range of sediment thickness shown (up to 250 m). The seepage rate estimate of 100 mm yr^{-1} (through 30% of the seafloor <65 Ma) of Wheat and Mottl (1994) cited above, could be supported by flow through 135 m of siliceous ooze, 35 m of pelagic clay, or <15 m of calcareous ooze or terrigenous sediment. While the median sediment thickness for most (~80%) ocean lithosphere <65 Ma is less than 135 m (Fig. 6.3), only a small fraction of those sediments are siliceous ooze. Thus, the potentially high seepage rates through siliceous ooze will be realized rarely. Very little of the seafloor has less than 35 m or 15 m of sediment cover (~5 and ~2%, respectively). Therefore, it is clear that only a small fraction of advective heat loss on ridge flanks can occur through sediments; the vast majority of advective heat loss must occur by fluid flow directly from basement rocks to the ocean.

In contrast, all sediment types can support chemically detectable flow over the range of thickness shown, which means that a "calibration" of the impedance–thickness–lithology relationships shown in Fig. 6.10 should be possible. Unfortunately, estimates of fluid flow rates through sediment sections where driving forces are constrained are rare, so such a calibration is not currently possible.

6.5.3 Influence of sediment type and thickness on the nature of circulation in basement

In addition to impeding flow through the seafloor and limiting the exchange of water between the crust and ocean, sediment cover – especially discontinuous sediment cover – exerts a major influence on fluid circulation patterns within the basement aquifer. Heat flow observations indicate that fluid can enter and escape the basement aquifer where sediments are absent or very thin, and can flow laterally for long distances where sediments have accumulated in ponds and small basins (Langseth and Herman, 1981; Langseth *et al.*, 1992; Davis *et al.*, 1999; Fisher *et al.*, 2003). Furthermore, model results suggest that in the presence of strong lateral flow, local cellular convection may be suppressed (Wang *et al.*, 1997; Giambalvo, 2001). This is illustrated in Fig. 6.11, where the effects of accumulation in a 3.5-km-wide sediment pond are simulated. Where the hydraulic impedance of sediment cover is small, cellular convection dominates flow in basement, and exchange between the crust and ocean is unrestricted (Fig. 6.11a). As sediment accumulates and its hydraulic

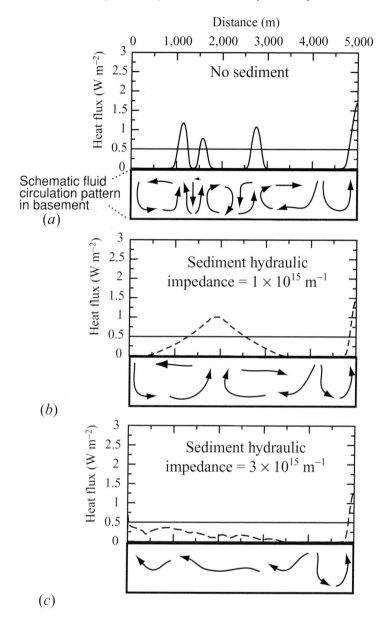

Fig. 6.11 Model results showing the possible effect of sediment cover on circulation in basement. The upper portion of each panel shows the conductive seafloor heat flux (after Giambalvo, 2001). The lower portion of each panel shows a cartoon of the basement circulation pattern. Sediment covers the seafloor from near the left side of the model (where a permeable window permits exchange with the ocean) to 3,500 m. From 3,500 to 5,000 m the seafloor has no sediment cover. At the left edge of the model, the upper basement is connected to the seafloor by a column of high permeability nodes, which facilitates recharge or discharge. The heat flux into the base of the model is 500 mW m^{-2} (horizontal line on heat flux plots). As the hydraulic impedance of the sediments increases (*b* and *c*), fluid circulation within the basement changes from primarily cellular to lateral flow.

impedance increases, both the dimension of convection cells within basement and the degree to which fluid flows through the sediments and around the convection cells change (Fig. 6.11*b*; Snelgrove and Forster, 1996). When the impedance of the sediment cover becomes large, flow in basement beneath the sediment becomes predominantly lateral, with water flowing from one basement outcrop to the other (Fig. 6.11*c*; Giambalvo, 2001), or between areas of contrasting basement topography and sediment burial in regions of completely buried basement (Fisher and Becker, 1995; Wang *et al.*, 1997). Ultimately, as sediments become thicker and more continuous, exchange of crustal and ocean water will be diminished, the conductive insulation provided by the sediments will increase, and average temperatures in basement will increase. Thus by controlling the nature of basement fluid flow (cellular convection versus lateral flow) and the degree of communication between the basement aquifer and the overlying ocean, sediments will affect the transport of both heat and solutes within and from the oceanic lithosphere, and change the conditions under which water–rock reactions occur.

6.6 Summary and recommendations

It is well known that marine sediments, generally characterized by low permeabilities relative to the upper igneous crust on which they accumulate, play an important role in ocean crustal hydrothermal circulation. In this chapter we have provided a review of measured sediment permeabilities and their variation with lithology and consolidation state, as well as a summary of the global distribution of sediments by average thickness and type, in order to understand more quantitatively the role of sediments in ocean crustal hydrogeology. Sediment type varies regionally in a coherent way, but variations with depth are less well characterized. Sediment thickness generally increases with lithospheric age, although commonly large local sediment thickness variations make the relationship between average thickness and lithospheric age difficult to apply hydrologically in a general way because of the extreme importance of sediment cover continuity.

Sediment permeabilities vary over many orders of magnitude with lithology and burial depth, and as a result, the rate of fluid flow through sediments is strongly dependent on sediment type and thickness. For most sediment types, a few tens of meters of sediment cover is sufficient to inhibit thermally significant fluid flow. Although there is no clear difference within the global heat flow data set between well and poorly sedimented areas (Stein and Stein, 1994b), specific examples where sediments have accumulated rapidly on young crust (e.g. the flanks of the Costa Rica Rift, Juan de Fuca Ridge, and southern East Pacific Rise) show that sediment thickness, type, and continuity do exert a dominant control on the exchange of water between the crust and the oceans, and on the nature of circulation within the crust. Where sediment cover is not continuous, sediment thickness and properties are probably not as important as the availability and distribution of permeable basement outcrops in determining seafloor heat and fluid fluxes.

While the general influence of sedimentation on hydrothermal circulation is well established, a number of uncertainties remain. For example:

(1) Typical permeabilities measured for siliceous ooze suggest that thermally significant flow may be permitted through up to 200 m or more of these sediments under differential pressures of a few tens of kilopascals. However, it is rare for sections to be mono-lithologic; silicate sections commonly include significant carbonates, particularly at depth (where sediments were deposited before the seafloor subsided below the CCD), and thus the extreme rates suggested in Fig. 6.10 for siliceous oozes are probably never realized. Conversely, carbonate sections commonly contain silicates and thus may permit flow at rates higher than estimated in Fig. 6.10. For rates to be estimated in the manner described here more accurately, the lithologic details of the sections must be considered.

(2) Rapid rates of flow suggested by thermal data at sites in the Indian Ocean fall at the extreme end of the range suggested in Fig. 6.10, and would require large driving forces and very high permeabilities. Confirmation of these rates with geochemical data, and determination of the cause of such rapid flow with good site characterization are called for.

(3) If the results of Fig. 6.10 are taken at face value, most of the field of predicted fluid flux falls more than an order of magnitude above the geochemically detectable threshold. Geochemical constraints are relatively rare, but existing data suggest that such rates of flow are not common. It is likely that pressures of the magnitude assumed (20 kPa, based on observations in the high-relief, warm basement of the Juan de Fuca Ridge flank) are atypical. It is also possible that the bulk permeabilities of actual mixed lithologies are lower than the mono-lithologic sections assumed.

To address these and other questions better, studies are needed that allow direct bulk permeability determinations for full sediment sections based on observed pressure differentials and geochemically constrained flow, along with complementary laboratory permeability determinations using samples collected from the same sections.

Further study is also required to determine whether there is significant feedback between fluid flow and sediment properties, i.e. to determine whether a significant history of flow can enhance or decrease the permeability of a seepage pathway. Taken to an extreme, for example, sediment alteration caused by seepage may strengthen the sediments to the point where brittle failure and consequent channelized fluid flow can occur. This is known to happen in high-temperature environments in sedimented ridge-axis settings (e.g. Davis and Fisher, 1994), and it may be possible in cooler sections on ridge flanks and accretionary prisms.

References

Abbott, D., Menke, W., Hobart, M., and Anderson, R. 1981. Evidence for excess pore pressures in Southwest Indian Ocean sediments. *J. Geophys. Res.* **86**: 1,813–1,827.

Alexander, J. and Morris, S. 1994. Observations on experimental, nonchannelized, high concentration turbidity currents and variations in deposits around obstacles. *J. Sed. Res.* **A64**(4): 899–909.

Alt, J. C. and Teagle, D. A. H. 1999. The uptake of carbon during alteration of ocean crust. *Geochim. Cosmochim. Acta* **63**(10): 1,527–1,535.

Anderson, R. N. and Hobart, M. A. 1976. The relation between heat flow sediment thickness, and age in the Eastern Pacific. *J. Geophys. Res.* **81**: 2,968–2,989.

Anderson, R. N., Hobart, M. A., and Langseth, M. G. 1979. Geothermal convection through oceanic crust and sediments in the Indian Ocean. *Science* **204**: 828–832.

ASTM, 1989. Standard test method for one-dimensional consolidation properties of soils. In *Annual Book of ASTM Standards, Section 4, Construction*. Philadelphia: ASTM, pp. 283–287.

Barron, E. J. and Whitman, J. M. 1981. Ocean sediments in space and time. In *The Oceanic Lithosphere*, ed. C. Emiliani. New York: Wiley, pp. 689–731.

Bear, J. 1972. *Dynamics of Fluids in Porous Media*, New York: Elsevier.

Becker, K. and Von Herzen, R. P. 1983. Heat transfer through sediments of the Mounds hydrothermal area, Galapagos Spreading Center at 86° W. *J. Geophys. Res.* **88**(B2): 995–1,008.

Bekins, B. A., McCaffrey, A. M., and Dreiss, S. J. 1995. Episodic and constant flow models for the origin of low-chloride waters in a modern accretionary complex. *Water Resources Res.* **31**(12): 3,205–3,215.

Bennett R. H., Bryant W. R., and Keller G. H. 1981. Clay fabric of selected submarine sediments: fundamental properties and models. *J. Sed. Pet.* **51**(1): 217–232.

Bennett, R. H., Fischer, K. M., Lavoie, D. L., Bryant, W. R., and Rezak, R. 1989. Porometry and fabric of marine clay and carbonate sediments: determinants of permeability. *Mar. Geol.* **89**: 127–152.

Bouma, A. H., Coleman, J. M., Meyer, A. W., *et al.* 1986. *Initial Reports of the Deep Sean Drilling Project*, Vol. 96. Washington, DC: US Govt. Printing Office, 824 pp.

Bredehoeft, J. D. and Papadopulos, I. S. 1965. Rates of vertical groundwater movement estimated from the earth's thermal profile, *Water Resources Res.* **1**(2): 325–328.

Bredehoeft, J. D., Neuzil, C. E., and Milly, P. C. D. 1983. *Regional Flow in the Dakota Aquifer: a Study of the Role of Confining Layers*. US Geol. Surv. Water Supply Paper 2237. Reston, VA: US Geological Survey.

Brown, K. M. and Behrmann, J. H. 1990. Genesis and evolution of small-scale structures in the toe of the Barbados Ridge accretionary wedge. In *Proceedings of the Ocean Drilling Program, Scientific Results*, Vol. 110, eds. J. C. Moore, A. Mascle, and C. Auroux. College Station, TX: Ocean Drilling Program, pp. 229–244.

Bryant, W. R. and Rack, F. R. 1990. Consolidation characteristics of Weddell Sea sediments. In *Proceedings of the Ocean Drilling Program, Scientific Results*, Vol. 113, eds. P. F. Barker and J. P. Kennet. College Station, TX: Ocean Drilling Program, pp. 211–223.

Bryant, W. R., Deflanche, A. P., and Trabant, P. H. 1974. Consolidation of marine clays and carbonates. In *Deep-sea Sediments, Physical and Mechanical Properties*, ed. A. L. Inderbitzen. New York: Plenum Press, pp. 209–244.

Bryant, W. R., Hottman, W., and Trabant, P. 1975. Permeability of unconsolidated and consolidated marine sediments, Gulf of Mexico. *Mar. Geotech.* **1**(1): 1–14.

  2002. Formation pressures and temperatures associated with fluid flow in young oceanic crust: results of long-term borehole monitoring on the Juan de Fuca Ridge flank. *Earth. Planet. Sci. Lett.* **204**: 231–248.

Davis, E. E. and Fisher, A. T. 1994. On the nature and consequences of hydrothermal circulation in the Middle Valley sedimented rift: Inferences from geophysical and geochemical observations, Leg 139. In *Proceedings of the Ocean Drilling Program, Scientific Results*, Vol. 139, eds. M. J. Motte, E. E. Davis, A. T. Fisher, and J. F. Slack. College Station, TX: Ocean Drilling Program, pp. 695–717.

Davis, E. E. and Lister, C. R. B. 1977. Heat flow measured over the Juan de Fuca Ridge: evidence for widespread hydrothermal circulation in a highly heat transportive crust. *J. Geophys. Res.* **82**: 4,845–4,860.

Davis, E. E., Horel, G. C., MacDonald, R. D., Villinger, H., Bennett, R. H., and Li, H. 1991. Pore pressures and permeabilities measured in marine sediments with a tethered probe. *J. Geophys. Res.* **96**(B4): 5,975–5,984.

Davis, E. E., Chapman, D. S., Mottl, M. J., Bentkowski, W. J., Dadey, K., Forster, C., Harris, R., Nagihara, S., Rohr, K., Wheat, G., and Whiticar, M. 1992. FlankFlux: an experiment to study the nature of hydrothermal circulation in young oceanic crust. *Can. J. Earth Sci.* **29**: 925–952.

Davis, E. E., Chapman, D. S., Villinger, H., Robinson, W., Grigel, J., Rosenberger, A., and Pribnow, D. 1997a. Seafloor heat flow on the eastern flank of the Juan de Fuca Ridge: data from "FlankFlux" studies through 1995. In *Proceedings of the Ocean Drilling Program, Initial Reports*, Vol. 168, eds. E. E. Davis, A. T. Fisher, J. V. Firth, *et al.* College Station, TX: Ocean Drilling Program, pp. 1–11.

Davis, E. E., Fisher, A. T., Firth, J. V., and the Shipboard Scientific Party 1997b. Buried basement transect (Sites 1028, 1029, 1030, 1031, and 1032). *In Proceedings of the Ocean Drilling Program, Initial Reports*, Vol. 168, eds. E. E. Davis, A. T. Fisher, J. V. Firth, *et al.* College Station, TX: Ocean Drilling Program, pp. 161–212.

Davis, E. E., Chapman, D. S., Wang, K., Villinger, H., Fisher, A. T., Robinson, S. W., Grigel, J., Pribnow, D., Stein, J. S., and Becker, K. 1999. Regional heat flow variations on the sedimented Juan de Fuca Ridge eastern flank: constraints on lithospheric cooling and lateral hydrothermal heat transport. *J. Geophys. Res.* **104**: 17,675–17,688.

Dewhurst, D. N. and Aplin, A. C. 1998. Compaction-driven evolution of porosity and permeability in natural mudstones: an experimental study. *J. Geophys. Res.* **103**(B1): 651–661.

Dewhurst, D. N., Brown, K. M., Clennel, M. B., and Westbrook, G. K. 1996. A comparison of the fabric and permeability anisotropy of consolidated and sheared silty clay. *Engng Geol.* **42**: 253–267.

Divins, D. L. 2002. Total sediment thickness of the world's oceans and marginal seas, NOAA National Geophysical Data Center. See http://www.ngdc.noaa.gov/mgg/sedthick/sedthick.html and references therein.

Edmond, J. M. 1974. On the dissolution of carbonate and silicate in the deep ocean. *Deep-Sea Res.* **21**: 455–480.

Eggleston, J. and Rojstaczer, S. 2001. The value of grain-size hydraulic conductivity estimates: comparison with high resolution in situ field hydraulic conductivity. *Geophys. Res. Lett.* **28**(22): 4,255–4,258.

Fang, W. W., Langseth, M. G., and Schultheiss, P. J. 1993. Analysis and application of in situ pore pressure measurements in marine sediments, *J. Geophys. Res.* **98**(B5): 7,921–7,938.

Fisher, A. T. 1998. Permeability within basaltic oceanic crust. *Rev. Geophys.* **36**: 143–182.

Fisher, A. T. and Becker, K. 1995. Correlation between seafloor heat flow and basement relief: observational and numerical examples and implications for upper crustal permeability. *J. Geophys. Res.* **100**: 12,641–12,657.

Fischer, K. M. and Lavoie, D. L. 1994. Geotechnical properties of Lau Basin sediments from a microfabric perspective. In *Proceedings of the Ocean Drilling Program, Sceintific Results*, Vol. 135, eds. J. Hawkins, L. Parsons, J. Allan, *et al.* College Station, TX: Ocean Drilling Program, pp. 797–804.

Fisher, A. and Zwart, G. 1996. The relation between permeability and effective stress along a plate-boundary fault, Barbados accretionary complex. *Geology* **24**: 307–311.

Fisher, A. T., Fischer, K., Lavoie, D., Langseth, M., and Xu, J. 1994. Geotechnical and hydrogeological properties of sediments from Middle Valley, northern Juan de Fuca ridge. In *Proceedings of the Ocean Drilling Program, Scientific Results*, Vol. 139, eds. M. J. Motte, E. E. Davis, A. T. Fisher, and J. F. Slack. College Station, TX: Ocean Drilling Program, pp. 627–647.

Fisher, A. T. Stein, C. A., Harris, R. N., Wang, K., Silver, E. A., Pfender, M., Hutnak, M. Cherkaovi, A., Bodzin, R., and Villinger, H. 2003. Abrupt thermal transition reveals hydrothermal boundary and role of seamounts within the Cacos Plate. *Geophys. Res. Lett.* **30**(11): 1550, doi: 10.1029/2002GL016766.

Freeze, R. A. and Cherry, J. A. 1979. *Groundwater*. Englewood Cliffs, NJ: Prentice Hall.

Geller, G. A., Weissel, J. K., and Anderson, R. N. 1983. Heat transfer and intraplate deformation in the central Indian Ocean. *J. Geophys. Res.* **88**: 1,018–1,032.

Giambalvo, E. R. 2001. Factors controlling fluxes of fluid, heat, and solutes from sedimented ridge-flank hydrothermal systems. Ph.D. thesis, University of California, Santa Cruz.

Giambalvo, E. R., Fisher, A. T., Martin, J. T., Darty, L., and Lowell, R. 2000. Origin of elevated sediment permeability in a hydrothermal seepage zone, eastern flank of the Juan de Fuca Ridge, and implications for transport of fluid and heat. *J. Geophys. Res.* **105**: 913–928.

Giambalvo, E. R., Steefel, C. I., Fisher, A. T., Rosenberg, N. D., and Wheat, C. G. 2002. Effect of fluid–sediment reaction on hydrothermal fluxes of major elements, eastern flank of the Juan de Fuca Ridge. *Geochim. Cosmochim. Acta* **66**(10): 1,739–1,757.

 1976. Variations of density and porosity with depth in deep-sea sediments. *J. Sed. Pet.* **46**: 280–300.

Hamilton, E. L. 1976. Variations of density and porosity with depth in deep-sea sediments. *J. Sed. Pet.* **46**: 280–300.

Haughton, P. D. W. 1994. Deposits of deflected and ponded turbidity currents, Sorbas Basin, southeast Spain. *J. Sed. Res.* **A64**(2): 233–246.

Hay, W. W., Sloan, J. L., and Wold, C. N. 1988. Mass/age distribution and composition of sediments on the ocean floor and the global rate of sediment subduction. *J. Geophys. Res.* **93**: 14,933–14,940.

Jacobson, R. S. 1992. Impact of crustal evolution of changes of the seismic properties of the uppermost ocean crust. *Rev. Geophys.* **30**: 23–42.

Kappel, E. S. and Ryan, W. B. F., 1986. Volcanic episodicity and a non-steady state rift valley along Northeast Pacific spreading centers: evidence from Sea MARC I. *J. Geophys. Res.* **91**: 13,925–13,940.

Karato, S.-I. and Becker, K. 1983. Porosity and hydraulic properties of sediments from the Galapagos spreading center and their relationship to hydrothermal circulation in the oceanic crust. *J. Geophys. Res.* **88**: 1,009–1,017.

Kastner, M. 1981. Authigenic silicates in deep-sea sediments: formation and diagenesis. In *The Oceanic Lithosphere*, eds. C. Emiliani. New York: Wiley, pp. 915–980.

Kastner, M., Gieskes, J. M., and Hu, J.-Y. 1986. Carbonate recrystallization in basal sediments: evidence for convective fluid flow on a ridge flank. *Nature* **321**: 158–160.

Kneller, B., Edwards, D., McCaffrey, W., and Moore, R. 1991. Oblique reflection of turbidity currents. *Geology* **14**: 250–252.

Lambe, T. W. and Whitman, R. V. 1969. *Soil Mechanics*. New York: Wiley.

Langseth, M. G. and Herman, B. M. 1981. Heat transfer in the oceanic crust of the Brazil Basin. *J. Geophys. Res.* **86**: 10,805–10,819.

Langseth, M. G., Cann, J. R., Natland, J. H., and Hobart, M. 1983. Geothermal
phenomena at the Costa Rica Rift: background and objectives for drilling at Deep
Sea Drilling Project Sites 501, 504, and 505. In *Initial Reports of the Deep Sea
Drilling Project*, Vol. 69, eds. J. R. Cann, M. G. Langseth, J. Honnorez, *et al.*
Washington, DC: US Govt. Printing Office, pp. 5–29.

Langseth, M. G., Hyndman, R. D., Becker, K., Hickman, S. H., and Salisbury, M. H.
1984. The hydrogeological regime of isolated sediment ponds in mid-oceanic ridges.
In *Initial Reports of the Deep Sea Drilling Project*, Vol. 78B, eds. R. H. Hyndman,
and M. H. Salisbury. Washington, DC: US Govt. Printing Office, pp. 825–837.

Langseth, M. G., Mottl, M. J., Hobart, M. A., and Fisher, A. T. 1988. The distribution of
geothermal and geochemical gradients near Site 501/504, implications for
hydrothermal circulation in the oceanic crust. In *Proceedings of the Ocean Drilling
Program, Initial Reports*, Vol. 111, eds. Becker, K., Sakai, H., *et al.* College Station,
TX: Ocean Drilling Program, pp. 23–32.

Langseth, M. G., Becker, K., Von Herzen, R. P., and Schultheiss, P. 1992. Heat and fluid
flux through sediment on the western flank of the Mid-Atlantic Ridge: a
hydrogeological study of North Pond. *Geophys. Res. Lett.* **19**: 517–520.

Li, Y.-H. and Gregroy, S. 1974. Diffusion of ions in seawater and in deep-sea sediments.
Geochim. Cosmochim. Acta **38**: 703–714.

Lister, C. R. B. 1972. On the thermal balance of a mid-ocean ridge. *Geophys. J. Roy.
Astron. Soc.* **26**: 515–535.

1976. Control of pelagic sediment distribution by internal waves of tidal period:
possible interpretation of data from the southern East Pacific Rise. *Mar. Geol.* **20**:
297–313.

Lowe, J., Zaccheo, P. F., and Feldman, H. S. 1964. Consolidation testing with back
pressure. *J. Soil Mech. and Found. Div., Proc. Am. Soc. Civil Eng.* **90**: 69–86.

Malinverno, A. 1991. Inverse square-root dependence of mid-ocean-ridge flank roughness
on spreading rate. *Nature* **352**: 58–60.

Marine Geotechnical Consortium 1985. Geotechnical properties of northwest Pacific
pelagic clays: Deep Sea Drilling Project Leg 86, Hole 576A. In *Initial Reports of the
Deep Sea Drilling Project*, Vol. 86, eds. Heath, G. R., Burckle, L. H., *et al.*
Washington, DC: US Govt. Printing Office, pp. 723–758.

Maris, C. R. P. and Bender, M. L. 1982. Upwelling of hydrothermal solutions through
ridge flank sediments shown by pore water profiles. *Science* **216**: 623–626.

Maris, C. R. P., Bender, M. L., Froelich, P. N., Barnes, R., and Luedtke, N. A. 1984.
Chemical evidence for advection of hydrothermal solutions in the sediments of the
Galapagos Mounds Hydrothermal Field. *Geochim. Cosmochim. Acta* **48**:
2,331–2,346.

Millero, F. J. 1996. *Chemical Oceanography*. Boca Raton, FL: CRC Press.

Mitchell, J. K. 1993. *Fundamentals of Soil Behavior*. New York: Wiley.

Mitchell, N. C., Allerton, S., and Escartín, J. 1998. Sedimentation on young ocean floor at
the Mid-Atlantic Ridge, 29° N. *Mar. Geol.* **148**: 1–8.

Moore, J. C., Shipley, T. H., Goldberg, D., Ogawa, Y., Filice, F., Fisher, A. T., Jurado,
M.-J., Moore, G. F., Rabute, A., Yin, H., Zwart, G., Brueckmann, W., *et al.* 1995.
Abnormal fluid pressures and fault zone dilation in the Barbados accretionary prism:
evidence from logging while drilling. *Geology* **23**: 605–608.

Moran, K., Gray, B., and Jarrett, K. 1995. Compressibility, permeability and stress history
of sediment from the Cascadia Margin. In *Proceedings of the Ocean Drilling
Program, Scientific Results*, Vol. 146B, eds. B. Carson, G. K. Westbrook, R. J.
Musgrave, and E. Suess. College Station, TX: Ocean Drilling Program.

Morin, R. and Silva, A. J. 1984. The effects of high pressure and high temperature on some physical properties of ocean sediments. *J. Geophys. Res.* **89**: 511–526.

Mottl, M. J. 1989. Hydrothermal convection, reaction and diffusion in sediments on the Costa Rica Rift flank, pore water evidence from ODP Sites 677 and 678. In *Proceedings of the Ocean Drilling Program, Scientific Results*, Vol. 111, eds. K. Becker, H. Sakai, *et al.* College Station, TX: Ocean Drilling Program, pp. 195–214.

Mottl, M. J. and Wheat, C. G. 1994. Hydrothermal circulation through mid-ocean ridge flanks: fluxes of heat and magnesium. *Geochim. Cosmochim. Acta* **58**: 2,225–2,237.

Muller, R. D., Roest, W. R., Royer, J. Y., Gahagan, L. M., and Sclater, J. G. 1997. Digital isochrons of the world's ocean floor. *J. Geophys. Res.* **102**: 3,211–3,214.

Neuzil, C. E. 1994. How permeable are clays and shales? *Water Resources Res.* **30**(2): 145–150.

Olsen, H. W. 1966. Darcy's law in saturated kaolinite. *Water Resources Res.* **2**: 287–295.

Olsen, H. W., Nichols, R. W., and Rice, T. L. 1985. Low gradient permeability measurements in a triaxial system. *Geotechnique* **35**(2): 145–157.

Parker, F. L. and Berger, W. H. 1971. Faunal and solution patterns of planktonic foraminifera in surface sediments of the South Pacific. *Deep-Sea Res.* **13**: 73–107.

Parsons, B. and Sclater, J. G. 1977. An analysis of the variation of ocean floor bathymetry and heat flow with age. *J. Geophys. Res.* **82**: 803–827.

Pearson, W. C. and Lister, C. R. B. 1973. Permeability measurements on a deep-sea core. *J. Geophys. Res.* **78**: 7,786–7,787.

Pickering, K. T. and Hiscott, R. N. 1985. Contained (reflected) turbidity currents from the Middle Ordovician Cloridorme Formation, Quebec Canada: an alternative to the antidune hypothesis. *Sedimentology* **32**: 373–394.

Pickering, K. T., Underwood, M. B., and Taira, A. 1992. Open-ocean to trench turbidity-current flow in the Nankai Trough: flow collapse and reflection. *Geology* **20**: 1,099–1,102.

Remy, J. P. 1973. Measurement of small permeabilities in the laboratory. *Geotechnique* **23**(3): 454–458.

Richter, F. M. 1993. Fluid flow in deep-sea carbonates: estimates based on porewater Sr. *Earth Planet. Sci. Lett.* **119**: 133–141.

Saffer, D. M. and Bekins, B. A. 1998. Episodic fluid flow in the Nankai accretionary complex: timescale, geochemistry, flow rates, and fluid budgets. *J. Geophys. Res.* **103**(B12): 30,351–30,370.

Saffer, D. M., Silver, E. A., Fisher, A. T., Tobin, H., and Moran, K. 2000. Inferred pore pressures at the Costa Rica subduction zone: implications for dewatering processes. *Earth Planet. Sci. Lett.* **177**: 193–207.

Schultheiss, P. and McPhail, B. 1986. Direct indication of pore-water advection from pore pressure measurements in Madeira Abyssal Plain. *Nature* **320**: 348–350.

Sclater, J. G., Crowe, J., and Anderson, R. N. 1976. On the reliability of oceanic heat flow averages. *J. Geophys. Res.* **81**: 2,997–3,006.

Screaton, E. J., Carson, B., and Lennon, G. P. 1995. Hydrogeologic properties of a thrust fault within the Oregon accretionary prism. *J. Geophys. Res.* **100**(B10): 20,025–20,035.

Screaton, E. J., Fisher, A. T., Carson, B., and Becker, K. 1997. Barbados Ridge hydrogeologic tests: implications for fluid migration along an active décollement. *Geology* **25**(3): 239–242.

Screaton, E. J., Carson, B., Davis, E., and Becker, K. 2000. Permeability of a décollement zone: results from a two-well experiment in the Barbados accretionary complex. *J. Geophys. Res.* **105**(B9): 21,403–21,410.

Shackleton, N. J. 1987. Oxygen isotopes, ice volume, and sea level. *Quat. Sci. Rev.* **6**: 183–190.

Silva, A. J., Hetherman, J. R., and Calnan, D. I. 1981. Low-gradient permeability testing of fine-grained marine sediments. In *Permeability and Groundwater Contaminant Transport*, eds. T. F. Zimmie and C. O. Riggs. Philadelphia: American Society for Testing Materials, pp. 121–136.

Snelgrove, S. H. and Forster, C. B. 1996. Impact of seafloor sediment permeability and thickness on off-axis hydrothermal circulation: Juan de Fuca Ridge eastern flank. *J. Geophys. Res.* **101**: 2,915–2,925.

Spinelli, G. A., Zühlsdorff, L., Fisher, A. T., Wheat, C. G., Mottl, M. J., Spiess V. and Giambalvo, E. R. in press. Hydrothermal seepage patterns above a buried basement ridge, eastern flank of the Juan de Fuca Ridge. *J. Geophys. Res.*

Stein, C. A. and Stein, S. 1994a. Comparison of plate and asthenospheric flow models for the thermal evolution of oceanic lithosphere. *Geophys. Res. Lett.* **21**: 709–712.

1994b. Constraints on hydrothermal heat flux through the oceanic lithosphere from global heat flow. *J. Geophys. Res.* **99**: 3,081–3,095.

Taylor, E. and Fisher, A. 1993. Sediment permeability at the Nankai accretionary prism, Site 808. In *Proceedings of the Ocean Drilling Program, Scientific Results*, Vol. 131, eds. I. A. Hill, A. Taira, and J. V. Firth. College Station, TX: Ocean Drilling Program, pp. 235–245.

Taylor, E. and Leonard, J. 1990. Sediment consolidation and permeability at the Barbados forearc. In *Proceedings of the Ocean Drilling Program, Sceintific Results*, Vol. 110B, College Station, TX: Ocean Drilling Program, pp. 129–140.

The Geotechnical Consortium 1984. Geotechnical properties of sediments from Walvis Ridge, Deep Sea Drilling Project, Leg 75, Hole 532A. *Mar. Georesources Geotech.* **12**: 297–339.

Tryon, M. D., Brown, K. M., Dorman, L., and Sauter, A. 2001. A new benthic aqueous flux meter for very low to moderate discharge rates. *Deep-Sea Res.* **48**(9): 2,121–2,146.

Vail, P. R., Audemard, F., Bowman, S. A., Eisner, P. N., and Perez-Cruz, C. 1991. The stratigraphic signatures of tectonics, eustasy and sedimentology: an overview. In *Cycles and Events in Stratigraphy*, eds. G. Einsele, W. Ricken, and A. Seilacher. Berlin: Springer-Verlag, pp. 617–659.

Wang, K. and Davis, E. E. 1996. Theory for the propagation of tidally induced pore pressure variations in layered sub-seafloor formations. *J. Geophys. Res.* **101**(B5): 11,483–11,495.

Wang, K., He, J., and Davis, E. E. 1997. Influence of basement topography on hydrothermal circulation in sediment-buried igneous oceanic crust. *Earth Planet. Sci. Lett.* **146**: 151–164.

Wheat, C. G. and McDuff, R. E. 1994. Hydrothermal flow through the Mariana Mounds: dissolution of amorphous silica and degradation of organic matter on a mid-ocean ridge flank. *Geochim. Cosmochim. Acta* **58**: 2,461–2,475.

1995. Mapping the fluid flow of the Mariana Mounds ridge flank hydrothermal system: pore water chemical tracers. *J. Geophys. Res.* **100**: 8,115–8,131.

Wheat, C. G. and Mottl, M. J. 1994. Hydrothermal circulation, Juan de Fuca Ridge eastern flank: factors controlling basement water composition. *J. Geophys. Res.* **99**: 3,067–3,080.

2000. Composition of pore and spring waters from Baby Bare: global implications of geochemical fluxes from a ridge flank hydrothermal system. *Geochim. Cosmochim. Acta* **64**(4): 629–642.

Wheat, C. G. and Tribble, J. S. 1994. Diagenesis of amorphous silica in Middle Valley, Juan de Fuca Ridge. In *Proceedings of the Ocean Drilling Program, Scientific Results*, Vol. 139, eds. M. J. Mottl, E. E. Davis, A. T. Fisher, and J. F. Slack. College Station, TX: Ocean Drilling Program, pp. 341–349.

Wilkens, R. H. and Handyside, T. 1985. Physical properties of equatorial Pacific sediments. In *Initial Reports of the Deep Sea Drilling Project*, Vol. 85, eds. L. Mayer, F. Theyer *et al*. Washington, DC: US Govt. Printing Office, pp. 839–847.

Williams, D. L., Green, K., van Andel, T. H., Von Herzen, R. P., Dymond, J. R., and Crane, K. 1979. The hydrothermal mounds of the Galapagos Rift: observations with *DSRV Alvin* and detailed heat flow studies. *J. Geophys. Res.* **84**: 7,467–7,484.

7

In situ determinations of the permeability of the igneous oceanic crust

Keir Becker and Earl E. Davis

7.1 Introduction

The permeability of the igneous oceanic crust is the property that most directly controls the rates and patterns of the fluid circulation that is so widely observed through the crust and is the overall subject of this entire volume. In basic, intuitive terms, permeability is a measure of the capability of the formation to allow the passage of fluids along a non-hydrostatic pressure gradient. For fractured crystalline formations like the oceanic crust, however, it is surprisingly difficult to advance beyond this intuitive concept to a rigorous definition and meaningfully representative measurement of permeability. The reasons are many, but largely center on the highly heterogeneous nature of the fluid-occupied pore, void, and fracture space in the oceanic crust, on scales ranging from millimeters to kilometers. The effects are also many, and include a huge range of reported permeabilities in oceanic crust – more than ten orders of magnitude – as well as recent indications of significant scale dependence of the results.

Recently, Fisher (1998) comprehensively reviewed the state of understanding of permeability of the basaltic upper oceanic crust. The intent of this chapter is not to repeat that review, but to build on it, focusing on interpretations of permeability from direct measurements of various types within oceanic crust and highlighting some important recent advances. The reader is referred to Fisher (1998) for more detailed exposition of many of the matters discussed herein. In Chapter 12, Wang describes complementary numerical modeling techniques to constrain permeability and fluid fluxes in oceanic crust. In Chapter 8, Davis and Becker consider the thermal perturbations that are the consequences of fluid flow as well as the pressure differences that drive fluid flow through the permeable igneous oceanic crust – differences that in some settings arise from gravitational and tectonic factors, but are more often hydrothermal in the oceanic crust. In Chapter 6, Spinelli, Giambalvo, and Fisher consider the permeability of the sediments that overlie the igneous oceanic crust everywhere except near unsedimented spreading centers.

To set the framework for this chapter, we adopt Fisher's (1998) conceptual permeocentric view of state and coupled processes in the igneous oceanic crust (Fig. 7.1). This vision

Hydrogeology of the Oceanic Lithosphere, eds. E. E. Davis and H. Elderfield. Published by Cambridge University Press.

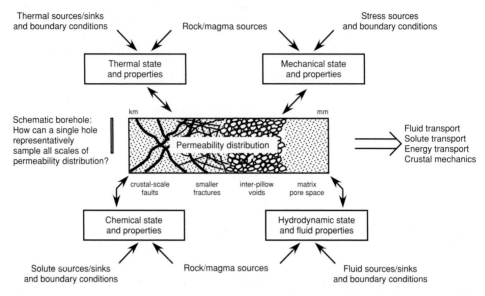

Fig. 7.1 Permeocentric vision of state and processes in the igneous oceanic crust, after Fisher (1998). The conceptualized borehole intended to represent the limitations on sampling multiple scales of the permeability distribution from isolated holes, which may explain much of the apparent scale dependence of reported permeabilities from various measurements in the same isolated holes.

reflects the underlying bias of this chapter in at least two ways. First, almost nowhere in the igneous oceanic crust can permeability be fully represented as a single-valued isotropic parameter; instead, it is the heterogeneous permeability distribution that must be understood in a representative fashion. Second, this heterogeneous permeability distribution is of utmost importance as a transform among processes and state of the igneous oceanic crust – which represents the overall subject of this volume. Figure 7.1 also conceptually illustrates the basic limitations in sampling from isolated boreholes the full range of scales of the permeability distribution of the igneous oceanic crust.

   The main focus of this chapter is on direct, in situ measurements, mainly of permeability of the igneous oceanic crust but also of closely related effects that are diagnostic of in situ permeability. This is a slightly more restrictive view than taken by Fisher (1998), who also reviews the considerable range of more indirect inferences about oceanic crustal permeability from observations in ophiolites and interpretations from resistivity- or velocity-derived porosities coupled with assumed porosity–permeability relationships. We begin with some basic definitions and assumptions and then proceed with a description of methods and their results, generally following the chronological order in which the methods were first applied over the past 20+ years. It so happens that this historical order also generally corresponds to an ordering in terms of greater and greater scales of investigation. Hence this historical exposition in terms of methods also serves to illustrate the development of the main scientific issues – which ranged from initially simply documenting typical values for upper-crustal permeability in any oceanic setting to now trying to understand the dependence of the permeability of the igneous oceanic crust on crustal age, depth, and scale of measurement.

7.2 Definitions, assumptions, and practical limitations

The original concept of permeability – what is sometimes called Darcy permeability – was defined in the laboratory for a column of uniform porous medium saturated with water, by monitoring steady-state laminar flow in the direction of an applied head differential. (Head is the potential energy per unit weight of the fluid and is therefore related to pressure by the scale factor of fluid specific weight $\rho_f g$.) The original Darcy (1856) equation relates the fluid flux to the head differential with a proportionality constant K_h called hydraulic conductivity. K_h was later shown to depend both on properties of the fluid (specific weight $\rho_f g$ and dynamic viscosity μ) and the intrinsic permeability k of the porous medium, $K_h = k\rho_f g/\mu$. Units for hydraulic conductivity are L T^{-1} and units of permeability are L^2.

The definition of Darcy permeability is strictly applicable only to an isotropic, uniformly porous medium, under conditions of laminar fluid flux in the direction of an applied head differential. These conditions are sometimes approximated reasonably in marine sediments (Chapter 6), but are rarely approached in the fractured and faulted igneous oceanic crust with its many irregular scales of interconnected voids that allow flow of fluids. Strictly speaking, the intrinsic permeability of the igneous oceanic crust should be described as a tensor that varies throughout the crust, but most methods cannot resolve separate components. Similarly, the head gradient that drives flow is a vector. There is no inherent reason that the principal components of permeability in the oceanic crust should be aligned with the components of the head gradient (i.e. the non-hydrostatic pressure gradient). As fluids tend to follow the most permeable pathway, the distribution and form of permeability will generally exert more control over the fluid flux vector than will the non-hydrostatic pressure gradient (Norton and Knapp, 1977). This is probably true to a great degree in the layered and heterogeneous oceanic crust.

Nevertheless, many of the direct measurement methods of assessing ocean crustal permeability rest on applications of the Darcy equation or the pressure diffusion equation from which it can be derived – and hence depend on an assumption that the tested volume of crust can be represented by an equivalent porous medium with a single-valued effective permeability. For this assumption to have any validity requires that the tested volumes representatively sample the heterogeneities in both porosity and rock matrix within the oceanic crust. The last two sentences constitute an intuitive statement of the concept of "representative elemental volume" (REV) formalized in hydrogeology, e.g. as described by Bear (1993) and Fisher (1998). Some of the recent advances described later in this chapter are in the realm of assessing permeability at increasingly large spatial scales in the oceanic crust, and the surprising differences in effective permeabilities at different scales are being explained essentially by differences in the nature of the interconnected porosity existing within the tested volumes. In other words, the representative properties of the oceanic crust may vary considerably over a broad range of scales of fluid flow, such that the concept of an REV may not be strictly applicable for the oceanic crust.

As is described in Chapter 6, oceanic sediments are commonly much less permeable than the igneous oceanic crust on any scale larger than hand samples of the rock matrix. Thus, in off-axis settings the upper oceanic crust is often conceptualized as a saturated aquifer of high average permeability capped by sediments of much lower average permeability. A confined aquifer of thickness b and uniform permeability is characterized by a transmissivity, T (units $L^2 T^{-1}$) $= bK_h = bk\rho g/\mu$, which is typically the quantity most directly measured in some of the test methods described below. In those cases, effective average permeability is calculated from the measured transmissivity assuming knowledge of the thickness of the discrete permeable zone(s) and the fluid properties, of which viscosity is particularly dependent on temperature (see Fig. 12.4). When more than one aquifer is encountered in a hole, their transmissivities are cumulative even if their permeabilities are quite different. In the extreme case sometimes encountered in the oceanic crust, in which an isolated fracture transmits significant fluxes of fluids through generally less-permeable host rock, it is more proper to describe the transmissivity of the fracture rather than its "permeability." If the average permeability of the formation including the isolated transmissive fracture is calculated, it will vary inversely with the assumed thickness of the aquifer – a "scale effect" of opposite sense as the effects recently observed and described below.

While it is intuitive that permeability should be related to interconnected porosity, developing a valid transform between porosity (which is more amenable to estimation by remote means) and permeability is an elusive goal, not just in ocean crustal investigations but also in hydrocarbon reservoir and waste disposal studies. No matter what the relationship, the parameter porosity does not enter directly into the determination of transmissivity and permeability. Porosity does enter into the other important property of an aquifer, its storativity, S, which is a measure of the ability of the aquifer to store or release fluid in response to a change in head. For a confined aquifer of thickness b, storativity is defined as the product of the aquifer thickness and average dynamic storage coefficient, $S = b\Delta V/V/\Delta h$. Within saturated sub-seafloor systems like the oceanic crust, rock grain compressibility is generally negligible, and the storage coefficent is often assumed to be controlled by aquifer (matrix framework) compressibility and fluid compressibility β_f. With the important exception of young, poorly consolidated extrusives, the effects of fluid compressibility usually predominate in the igneous oceanic crust, i.e. pore-space changes are generally small. In that case, the storativity reduces to the product $bn\beta_f\rho_f g$ of aquifer thickness b and a simplified dynamic storage coefficient, and clearly includes the effects of porosity n – but again, this is valid only for the isotropic case with a single-valued porosity that is representative of the entire aquifer.

Nearly all of the experiments described below share the common feature of utilizing boreholes that penetrate the oceanic crust for *in situ* testing of various sorts. There are only a limited number of DSDP/ODP crustal holes appropriate for these techniques (Fig. 7.2, Table 7.1), so results from these few holes figure prominently in our understanding of the hydrogeology of the igneous oceanic crust. The number of appropriate holes is relatively

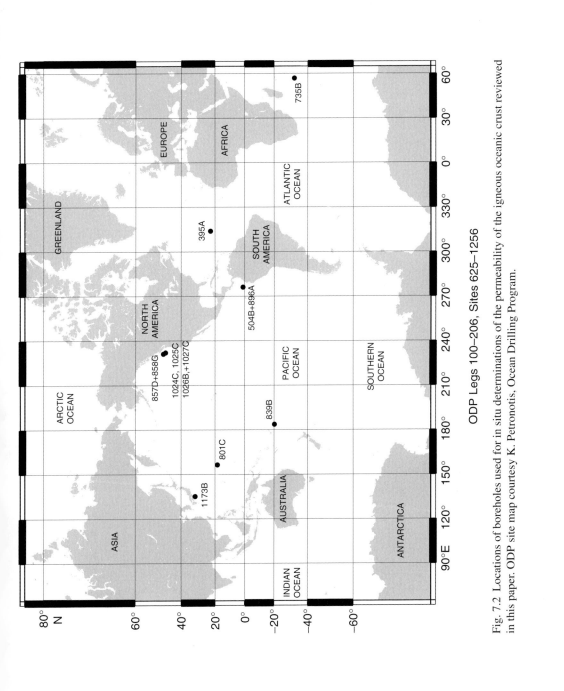

ODP Legs 100–206, Sites 625–1256

Fig. 7.2 Locations of boreholes used for in situ determinations of the permeability of the igneous oceanic crust reviewed in this paper. ODP site map courtesy K. Petronotis, Ocean Drilling Program.

Table 7.1 *Summary of DSDP/ODP site characteristics and measurements reviewed in this chapter.*

Hole	Location/setting/lithology	Depth/thickness (m)[a]			Measurement type and date[b]				Citation
		Sed.	Cas.	Basement		P	F	C	
504B	S flank Costa Rica Rift, 5.9 Ma, medium-rate crust, thickly sedimented, extrusive basalts underlain by sheeted dikes	274	276	214	1979 – Leg 69	✓			Anderson and Zoback (1982)
				1013	1981 – Leg 83	✓			Anderson et al. (1985)
				1132/1273	1986 – Leg 111	✓			Becker (1989)
				214	1979 – Leg 69 + 70		✓		Becker et al. (1983a)
				562	1981 – Leg 83		✓		Becker et al. (1983b)
				1076	1983 – Leg 92		✓		Becker et al. (1985)
				1076	1986 – Leg 111		✓		Gable et al. (1989)
				1288/1347	1991 – Legs 137+140		✓		Gable et al. (1995)
				1726	1993 – Leg 148		✓		Guerin et al. (1996)
				1836	2001 by wireline			✓	Becker et al. (2004)
395A	W flank mid-Atlantic Ridge, 7.3 Ma, slow-rate crust, 10-km-wide sediment pond, extrusive basalts	93	112	571	1981 – Leg 78B	✓			Hickman et al. (1984)
				513	1986 – Leg 109	✓			Becker (1990)
				513	1981 – Leg 78B		✓		Becker et al. (1984)
				513	1986 – Leg 109		✓		Kopietz et al. (1990)
				513	1990 – DIANAUT		✓		Morin et al. (1992)
				513	1997 – Leg 174B		✓		Becker et al. (2001)
				513	1997 – Leg 174B			✓	Davis et al. (2000)
735B	SW Indian Ridge, 11 Ma, tectonically exposed gabbros	0	0	500	1987 – Leg 118	✓			Becker (1991)
839B	Lau Basin, 2 Ma, back-arc crust, extrusive basalts	214	0	283	1991 – Leg 135	✓			Bruns and Lavoie (1994)
857D	Juan de Fuca Ridge, Middle Valley sedimented rift, intercalated sill/sediments	480	574	(456)	1991 – Leg 139	✓			Becker et al. (1994)
					1991 – Leg 139			✓	Davis and Becker, (1994)
					1996 – Leg 169			✓	Davis et al. (2001)

Hole	Description	Sed.	Cas.	Basement			Date – Leg	References
858G	Juan de Fuca Ridge, Middle Valley sedimented rift, extrusive basalts	258	270	175	√		1991 – Leg 139	Becker et al. (1994)
							1991 – Leg 139	Fisher (1998)
						√	1991 – Leg 139	Davis and Becker (1994)
							1996 – Leg 169	
801C	Western Pacific, Jurassic crust, extrusive basalts	462	481	132	√		1992 – Leg 143	Larson et al. (1993)
896A	S flank Costa Rica Rift, 5.9 Ma, medium-rate crust, extrusive basalts, sedimented basement high	179	195	390	√	√	1993 – Leg 148	Becker (1996)
						√	2001 by wireline	Becker et al. (in review)
1024C	E flank Juan de Fuca Ridge, 0.9 Ma, medium-rate crust, extrusive basalts	152	167	24	√		1996 – Leg 168	Becker and Fisher (2000)
					√		1996 – Leg 168	Becker and Davis (2003)
						√	1996 – Leg 168	Davis et al. (2000, 2001)
1025C	E flank Juan de Fuca Ridge, 1.3 Ma, medium-rate crust, extrusive basalts	101	102	47			1996 – Leg 168	Becker and Davis (2003)
						√	1996 – Leg 168	Davis et al. (2000, 2001)
1026B	E flank Juan de Fuca Ridge, 3.5 Ma, medium-rate crust, thickly sedimented, extrusive basalts, basement high	247	249	48	√		1996 – Leg 168	Becker and Fisher (2000)
					√		1996 – Leg 168	Fisher et al. (1997)
						√	1996 – Leg 168	Davis et al. (2000)
1027C	E flank Juan de Fuca Ridge, 3.6 Ma, medium-rate crust, thickly sedimented, sill + extrusive basalts, basement valley	576	579	57	√		1996 – Leg 168	Becker and Fisher (2000)
					√		1996 – Leg 168	Becker and Davis (2003)
						√	1996 – Leg 168	Davis et al. (2000, 2001)
1173B	Shikoku Basin offshore Nankai Trough, SW Japan, 15 Ma crust, extrusive basalts,	731	728	20		√	2001 – Leg 196	

[a] Sed., sediment thickness; Cas., casing depth; Basement, basement depth cored as of time of measurement. (For Hole 857D, "basement" comprises an intercalated sill/sediment sequence.)

[b] P, packer; F, borehole flow; C, CORK.

limited because a significant investment is required to establish the necessary seafloor re-entry cone and steel casing through the sediments to allow basement coring. (Although ODP has developed the technology to drill in unsedimented environments, the great majority of re-enterable crustal holes have been established in sedimented environments utilizing more traditional operational methods that depend on the sediments and/or casing through sediments to support the drillstring as crustal drilling begins.)

Given the limitations inherent in single-hole testing within a highly heterogeneous medium, the results of the methods described below are often accurate to no better than an order of magnitude. Nevertheless, given the large range of reported permeabilities in igneous oceanic crust, results good to within an order of magnitude are still very useful. In subaerial testing, considerably more sophisticated methods can be utilized to resolve permeabilities to much better precision. Thankfully, sub-seafloor testing offers a few simplifying advantages over subaerial conditions: except for geologically brief "moments" during accretion at the spreading axis, the igneous oceanic crust can be taken to be fully saturated with seawater, and subject to a hydrostatic boundary condition at the seafloor. And on ridge flanks, where sediment accumulation becomes significant, the permeability of the sediments is so much lower than that of the igneous crust that the confined aquifer concept can generally be applied.

7.3 Packer measurements

7.3.1 Packer methods

Drillstring packers have been used for permeability measurements in DSDP/ODP re-entry holes that penetrate basement since 1979. The packers used in DSDP and ODP have all incorporated tough rubber inflatable elements that seal the hole when inflated with seawater pumped down from the rig floor (Fig. 7.3a). They can be set either within casing or in open hole in zones where the formation provides a good mechanical and hydraulic packer "seat." They can theoretically be used in tandem, to isolate a section of formation straddled by two seals, but more typically they have been used by ODP in single-seal mode. In that mode, the hydraulically isolated and tested section of formation comprises the zone between the packer and the bottom of the hole. When the packer is inflated in casing, the tested zone includes the entire open-hole section below casing, and the results depend on the quality of the cement bond between casing and formation. The packer used during ODP can be reset at different depths as long as the element maintains its integrity, so in some cases it has been used for multiple inflations during a single testing sequence, assessing different intervals of the formation bounded by the packer and the bottom of the hole. Further details on the tools and techniques employed in ODP packer testing can be found in Becker (1986, 1988), Fisher (1998), and the appendix of Becker and Fisher (2000).

Once the formation is isolated with the inflated packer, the hydrological properties of the formation can be tested by assessing the response to controlled injection of seawater from the rig floor into the formation. The primary data are pressures, recorded by pressure gauges both at the rig floor and *in situ* within the isolated section of borehole; the latter

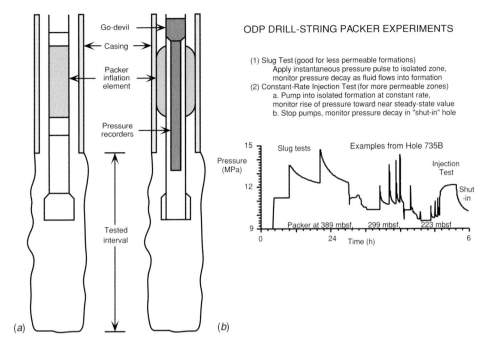

Fig. 7.3 (*a*) Schematic of ODP drillstring packer operations and (*b*) an actual pressure record recorded with a downhole gauge during packer operations in gabbros in Hole 735B (Becker, 1991). The pressure record illustrates: (1) the capability of the packer for multiple sets at different depths in the hole, and (2) the need to shift from slowly decaying slug tests in the relatively impermeable deeper section of the hole to constant-rate injection test when the packer was set in the more permeable shallower section of the hole.

are carried in the so-called "go-devil" that first allows activation of the packer inflation process, then allows communication from rig floor to the isolated formation (Fig. 7.3*a*). Note that DSDP/ODP packer testing to date has not allowed for producing fluids from the formation (i.e. flowing or pumping from the formation) and assessing the pressure response, an approach that is often preferable in subaerial testing because it minimizes the risk of affecting permeability by "opening" the formation through excess fluid pressure. Note also that temperature logs in holes that produce naturally allow permeability estimates without utilizing a packer, as is described in Section 7.4.

Two types of pumping tests have been applied in DSDP and ODP to assess formation hydrological response: pressure pulse or "slug" tests and constant-rate injection tests (Fig. 7.3*b*). They are both interpreted with mathematical methods that arise from a radial pressure diffusion equation under the assumption of azimuthally and radially uniform formation response, and both require knowledge of the fluid properties (compressibility and viscosity).

For the slug test, the "modified" procedure of Bredehoeft and Papadopulos (1980) is used: a brief pulse of fluids is pumped into the sealed hole, with the aim of producing a sharp but controlled pressure pulse, and then the decay of this pulse is monitored as pressurized fluids flow from the borehole into the isolated zone. The decay of an instantaneous pressure pulse in a borehole that penetrates a uniform aquifer is described mathematically by the complicated integral function F("α," "β"), and permeability is estimated by fitting measured pressures to type curves of this function. (Quotes are used here to denote the original historical usage; note that "α" and "β" are not the same as the parameters α and β as used elsewhere in this volume.) In the slug test formulation, "α" is a dimensionless parameter that depends on storativity and fluid parameters and the dimensionless "β" depends on transmissivity, time, and fluid parameters. Transmissivity and then permeability are determined from a slug test by the method of Cooper *et al.* (1967) and Papadopulos *et al.* (1973), fitting the measured data to type curves spanning a range of values of "α." In general, the slug test method is relatively insensitive to the value of "α" but much more sensitive to changes in "β" and hence it resolves transmissivity and permeability much better than storage coefficient and porosity.

The permeability calculated from a slug test depends directly on the effective compressibility of the fluids in the pressurized testing system, which Neuzil (1982) showed may often be up to an order of magnitude greater than that of the pure fluid in the laboratory. Care is required in comparing slug test results from different studies to date in oceanic crust (summarized in Section 7.3.2), because earlier investigators assumed the laboratory value for the compressibility of water, whereas later investigators followed Neuzil's (1982) methods and calculated effective compressibilities from the amplitude of the pressure rise produced from pumping a known volume of fluids during the slug test.

In a relatively transmissive zone, a pressure pulse will decay too quickly to resolve the transmissivity, and an injection test must be conducted. In this experiment, fluids are pumped into the isolated zone at a constant rate, and the slow build-up of pressure is monitored; after pumping is stopped, the return of pressure to original conditions also allows an assessment of permeability. The rise of pressure during injection and the return to original conditions after pumping are described by the integral "well function" W("u"), in which the parameter "u" is a function of transmissivity, storativity, time, and radius (Theis, 1935; Cooper and Jacob, 1946). For ODP injection tests, the small-radius, long-time approximation is applied, in which the pressure change becomes linear with log time (e.g. Horner, 1951, Matthews and Russell, 1967); transmissivity and permeability are determined from the slope of the straight-line segment of the pressure versus log time plot. The results are less sensitive than slug test results to the values of the fluid parameters. At long pumping times, the pressure increase may effectively stabilize, in which case the steady-state approximation "Glover" formula (Snow, 1968) can be used to calculate transmissivity and permeability.

In subaerial testing, slug and injection tests are also relatively standard hydrogeological techniques, although two important limitations must be noted for applications in oceanic crust. (1) In subaerial testing, these techniques are commonly applied in a source well and signals recorded in one or more separate observation wells. To date, technology has limited packer testing in the oceanic crust to single-hole situations, in which the source well is also the observation well. This geometry requires a key assumption of azimuthal and radial uniformity of the formation properties and response, and also generally requires interpretation using small-radius approximations to the full equations describing the pressure behavior.

(2) As noted above, controlled production from the formation is not generally possible in ODP packer work, but it is utilized more commonly in subaerial testing, as it involves less risk of altering effective permeability through excessive fluid pressure.

7.3.2 Permeabilities of igneous oceanic crust determined from packer tests

When Fisher (1998) compiled his review of permeability of basaltic oceanic crust, results from packer experiments were available from about two-thirds of the sites listed in Table 7.1 – those up to but not including the Leg 168 sites in young crust on the flank of the Juan de Fuca Ridge. Like the authors whose work he summarized, Fisher (1998) graphically presented the direct packer measurements on a plot of log of permeability versus depth, with the depth axis oriented vertically. In such a plot, an individual permeability determination appears as a rectangle, with the vertical extent of the rectangle representing the interval over which a so-called "bulk" permeability is averaged (e.g. the section isolated by the packer), and the horizontal extent representing the estimated uncertainty. For the packer methods described above, uncertainties arising from measurement limitations and assumptions of ideal geometries and parameter values are typically up to an order of magnitude in permeability.

Despite these uncertainties and the relatively wide range of lithologic and geographic settings, the results through 1996 (Fig. 7.4) seemed remarkably coherent. Nearly all of those results indicated typical measured permeabilities of 10^{-14}–10^{-12} m^2 in the upper 200–400 m of igneous crust, including the tectonically exposed gabbros in Hole 735B. The similarity of permeabilities in all these upper sections of basement led to the suggestion that this can be taken as a representative average value for much of the upper igneous crust (e.g. Becker, 1991, 1996). Furthermore, the similarity of values across different lithologies also suggested that fracture porosity is more important in controlling upper crustal permeability than lithology.

Deeper than the seemingly characteristic permeable interval in the uppermost crust, results from three holes (504B, 395A, and 735B) suggested relatively sharp transitions to much lower (two–three orders of magnitude) permeability values. Especially influential were very low permeabilities reported by Hickman *et al.* (1984) in the deepest extrusive section in Hole 395A and Anderson *et al.* (1985) in a long zone spanning the lower extrusives and upper dikes in Hole 504B. These reports suggested a permeability reduction with depth that is much sharper than had been utilized in early models of hydrothermal systems, with the transition seeming to correspond to that between seismic layers 2A and 2B (e.g. Becker *et al.*, 1982). This has led to the approximation utilized in Chapter 8: that the thickness of seismic layer 2A can serve as a proxy for the thickness of the most permeable and hydrologically active "hydrothermal basement" (e.g. Rohr, 1994). Furthermore, the combination of low-permeability sediments above and low-permeability crust below lends some validity to the confined, horizontal aquifer approach employed in interpreting permeability measurements from within this hydrothermal basement.

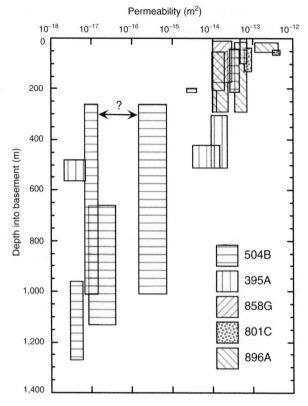

Fig. 7.4 Bulk permeabilities determined with packer experiments in the igneous oceanic crust through 1996. The question mark denotes the reinterpretation toward a higher bulk permeability by Becker (1996) of the slug test results and a previously discounted injection test of Anderson *et al.* (1985) in the lower extrusives and upper dikes of Hole 504B.

In subsequent studies, Becker (1989) reported similarly low permeabilities in deeper sections of the sheeted dikes in Hole 504B, but Becker (1996) suggested that the original results of Anderson *et al.* (1985) for the deep extrusive and shallow sheeted dike sections may have been biased toward too low a permeability because of the assumed low value of fluid compressibility. If the latter is the case, then a smoother reduction of permeability with depth in Hole 504B would be indicated, with an intermediate value in the deeper extrusives (Fig. 7.4). Nevertheless, the basic result – permeability reducing by orders of magnitude from the extrusives into the sheeted dikes – seems to hold up. Clearly, further work is needed in documenting the variation of permeability with depth in the igneous oceanic crust, e.g. refining the results in 504B and making new measurements in other deep holes, but no such opportunities have arisen since the late 1980s.

Since Fisher's (1998) review, an important new set of packer results has been reported: the determinations of Becker and Fisher (2000) and Fisher and Becker (2000) in relatively young crust of the Juan de Fuca Ridge (Table 7.1). These are important because the sites

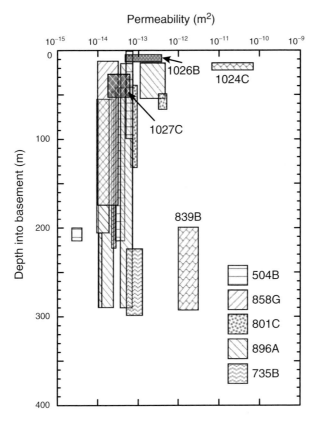

Fig. 7.5 Bulk permeabilities measured with packer experiments in the upper extrusives of the igneous oceanic crust through 2000.

are in "normal" oceanic crust younger than sites of prior permeability measurements and the results provide additional understanding of the variation of upper crustal permeability versus crustal age. Becker and Fisher (2000) reported results of slug and injection tests in uppermost basement at three sites, all of which were cased through sediments, drilled a few meters to tens of meters into basement, and ultimately instrumented with "CORK" sealed-hole observatories (Fig. 8.4) after the packer testing. Two of the sites were in crust of ~3.5–3.6 Ma, and they yielded upper crustal permeabilities fully consistent with prior indications of "typical" upper crustal values of 10^{-14}–10^{-12} m^2. The third site was in younger crust, 0.9 Ma, and it was notable for yielding a permeability greater by at least an order of magnitude. In combination with the measurement by Bruns and Lavoie (1994) in young back-arc crust, the new results indicated an expansion of the previously smaller range of upper crustal permeabilities (Fig. 7.5). More important, they provided the first indications of a coherent variation of upper crustal permeability with age (Fig. 7.6) that Fisher and Becker (2000) noted is consistent with the known variation of upper crustal seismic velocity with age. We return to these results below, because these holes also yielded

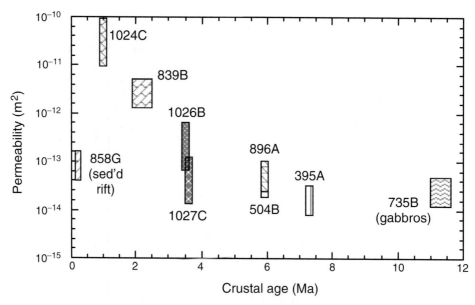

Fig. 7.6 Permeabilities determined from packer experiments in uppermost oceanic crust versus crustal age.

permeability estimates utilizing three other methods (Sections 7.4 and 7.5), and the sites therefore provide a prime dataset for a consideration of scale effects of permeabilities in the upper oceanic crust (Section 7.7).

7.4 Borehole flow revealed by temperature logs: implications for permeability

In the mid to late 1970s, it was noted that temperature logs in a series of re-entry holes drilled through sediments into oceanic crust showed nearly isothermal profiles through the cased sections, and these observations were interpreted to be due to the effects of flow of cold ocean bottom water down the holes and into the formation (e.g. Hyndman *et al.*, 1976). However, the circulation of seawater to flush cuttings during drilling has a similar effect on borehole temperatures, so only this qualitative interpretation was possible from temperature logs taken during the drilling leg itself; it was not until DSDP began revisiting some of the holes for logging and deepening at significant times after previous drilling episodes that any quantitative interpretation of flow in holes could be derived from the temperature profiles. Becker *et al.* (1983a,b, 1984), adapting a gas industry model from Lesem *et al.* (1957), showed how borehole temperatures measured during re-visits to the classic reference Holes 504B and 395A could be used first to estimate the downhole flow rate and then to infer the average permeability of the formation accepting the downhole flow.

Figure 7.7 shows sample observational data and the basic model used to interpret the data. This figure and the wording of this section are for the case of downhole flow, which

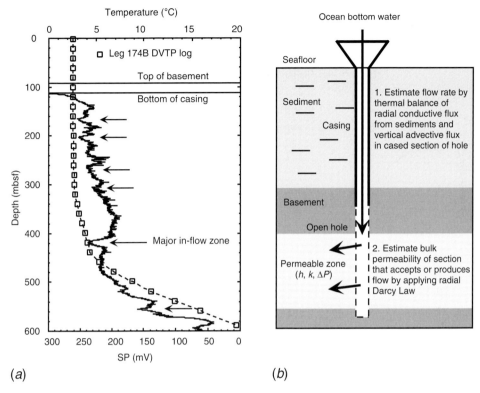

(a)   (b)

Fig. 7.7 (*a*) A typical temperature profile indicative of downhole flow of ocean bottom water into upper levels of basement. This example is from Hole 395A, about 21 years after the hole penetrated through sediments into igneous oceanic crust. Also shown is a log of spontaneous potential, which is sensitive to fluid flow and suggests several zones that accept the downhole flow. (*b*) A schematic of the physical basis for the mathematical models used to: (1) interpret the temperature log to estimate flow rate, and then (2) to estimate average permeability of the zone(s) accepting the flow.

has been observed fairly often in sites drilled into sedimented basement topographic lows. However, the calculation is also valid for the case of up-hole flow (see Fisher *et al.*, 1997), which has now been observed in at least four sites that penetrate basement topographic highs covered with sediments. (See Chapter 8 for further discussion of this "chimney" effect.) If the undisturbed temperature profile in the sediments behind the casing is known, it provides both the far-field boundary condition and also the initial condition that governs borehole temperatures after the emplacement of the hole and the beginning of flow down into the basement formation beneath the sediments. By balancing the radial heat flow from sediments into the hole with that advected in the cased section of hole, type curves for borehole fluid temperatures within the casing can be calculated, and the actual flow rate can be estimated by matching observed temperatures to these type curves (Becker *et al.*, 1983a). In the case of Hole 395A, the thermally based flow rate estimates were later verified with a flowmeter log by Morin *et al.* (1992).

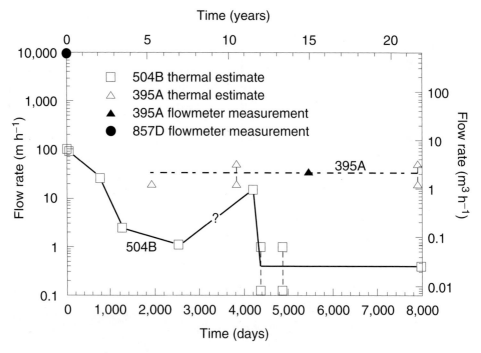

Fig. 7.8 Twenty-year histories of downhole flow rates determined on various re-entries of Holes 504B and 395A.

In the igneous crust below the casing, the downhole flow may exit into the formation at various transmissive intervals, and inflection points in the temperature log or responses of other logs can be interpreted to mark the boundaries of the transmissive zones (e.g. Fig. 7.7*a*). If the thickness of the aquifer can be estimated from the temperature profile, flowmeter log, and/or other logs, then the average permeability of the aquifer can be calculated as a function of the head differential driving the flow. This calculation utilizes a radial form of the Darcy equation, so the result is again a bulk permeability averaging transmissivity through the entire interval. The head differential may be due to either (or a combination) of two effects: actual non-hydrostatic formation pressures or a differential induced by drilling with fluids of different temperature and density than formation fluids. If the head differential can be estimated, e.g. from the difference in borehole pressures measured when setting a packer in the hole or when sealing the hole with long-term instrumentation (Section 7.5), then the average permeability of the formation can be calculated.

The two longest-studied examples of downhole flow are Holes 504B and 395A (see references in Table 7.1), and a comparison is important in at least two respects. First, the holes differ in their long-term (>20 year) flow behavior, as summarized in Fig. 7.8. While periodic revisits to Hole 395A indicated that the rate of its downhole flow remained relatively constant, the revisits to Hole 504B sampled much greater long-term variability in downhole flow rates. During the first seven years of observations in Hole 504B (1979–1986), the rate of downhole flow decreased smoothly from an initially high rate to a thermally nearly negligible rate; also the temperature–depth inflection point marking the limit of thermal effects was about 100 m into basement, below a thick massive flow unit that has been interpreted as a permeability barrier within the extrusive section (Pezard *et al.*, 1992). After a five-year hiatus, the temperature log on the next re-visit in 1991 was again more isothermal in the upper section, indicating a renewed downhole flow, and the inflection point appeared to have moved to a level above the massive flow unit. Since then, temperatures have rebounded again toward the conductive profile, indicating a decay of the downhole flow rate to the very slow rate estimated from temperatures logged in 2001 (Becker *et al.*, 2004). Second, although crustal ages are similar, crustal temperatures are considerably cooler at Hole 395A than at Hole 504B, owing to the more continuous sediment cover in the region of Hole 504B (Chapter 8); as a result, the long-lived pressure differential at Hole 395A was much more clearly due to formation underpressure than drilling-induced effects. It is unclear whether the changes in downhole flow in Hole 504B reflect changes in pressure differentials and/or formation permeability; examples of temporal changes in both are described in Chapter 8.

Figure 7.9 shows all reported permeabilities estimated from borehole flow determinations, plotted at the same scale as packer-determined upper crustal measurements shown in Fig. 7.5. Calculations of permeabilities based on borehole flow have been reported for Holes 504B, 395A, 1026B, and 858G (see references in Table 7.1). For Hole 504B, two values are reported for two interpretations of the thickness and position of the upper crustal aquifer based on inflection points in the temperature logs (Becker *et al.*, 1983a,b). For Holes 1026B and 858G, Fisher *et al.* (1997) and Fisher (1998) reported likely permeability ranges for reasonable ranges of estimates of the driving pressure differentials. The installation of CORKs in four holes on the flank of the Juan de Fuca Ridge (1024C–1027C in Table 7.1) provided the temperature and pressure data at the moment of sealing to allow both flow estimates from temperatures and coupled permeability calculations (Becker and Davis, 2003).

For Holes 504B and 395A, the driving pressure differentials were originally estimated from relatively large (150–1,000 kPa) deviations of sealed-hole pressures from hydrostatic values estimated from packer experiments conducted by Anderson and Zoback (1982) and Hickman *et al.* (1984) months to years prior to the temperature logs used to obtain flow-rate estimates. However, much smaller pressure differentials were measured years later at the time the holes were sealed with CORKs (Becker *et al.*, in press), suggesting that smaller pressure differentials were actually effective at the time the flow rates in Holes 395A and 504B were estimated. This would imply proportionally larger permeability estimates. For example, in estimating permeability from a flowmeter log conducted in Hole 395A in 1990, Morin *et al.* (1992) used the pressure differential estimated by Hickman *et al.* (1984) during packer experiments conducted in 2001. However, a differential pressure two orders of magnitude lower was measured on installation of a CORK in Hole 395A in 1997; hence the actual value at the time of the flowmeter log was probably much less than the value reported by Hickman *et al.* (1984). Figure 7.9 shows recalculated values for Hole 395A.

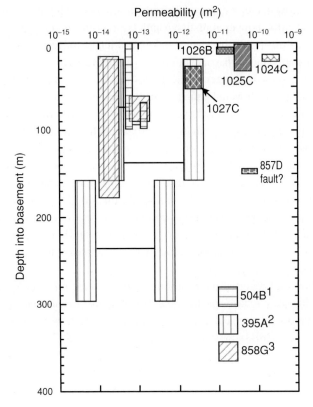

Fig. 7.9 All upper crustal permeabilities estimated from borehole flow calculations, plotted at the same scale as Fig. 7.6. [1]For Hole 504B, two possible interpretations are shown, depending on whether the aquifer is assumed to be a 30-m zone or the upper 100 m of basement (Becker *et al.*, 1983a,b). [2]For Hole 395A, two possible interpretations are shown based on the flowmeter logs of Morin *et al.* (1992), depending on the assumed value of head differential driving the flow. [3]For Hole 858G, two possible interpretations are shown, depending on the thickness assumed for the aquifer producing uphole flow (Fisher, 1998).

Finally, in 2001 CORKs were installed by wireline re-entry in Hole 504B and the nearby Hole 896A. Temperatures were logged in both holes, showing that Hole 504B was still drawing bottom water at a very slow rate whereas Hole 896A was producing formation fluids at 58–65 °C from at least three zones in the uppermost igneous crust. Pressure data recovered in late 2002 have allowed calculation of flow-based permeabilities in both holes (Becker *et al.*, in press).

7.5 Inferences from long-term pressure observations in sealed holes

Around 1990, the "CORK" sealed-hole hydrogeological observatory was developed by Davis *et al.* (1992; see also Davis and Becker, 1998, 2002), and several important crustal holes have been instrumented with variants of this system (Table 7.1; Fig. 8.4). In most crustal CORK installations, the hole is sealed near the top of the casing, which extends through the sediment column into open hole in igneous basement, and a thermistor string, pressure gauge, and in some cases fluid samplers are suspended in the sealed hole from a long-term data logger accessible from the seafloor. *In situ* data collected with these installations have important implications for permeability in a number of ways. First, as is summarized in the previous section, down- or up-hole flow has been indicated in several cases by borehole temperatures logged during deployment of the CORK instrumentation in the moments before actual sealing. Second, as is described in Chapter 8, the long-term pressure and temperature data have been very important in deducing *in situ* conditions and the head differentials that can drive flow – differentials which in some important cases are so small that very large average permeabilities are required to produce the flow rates necessary to generate the nearly isothermal conditions observed in uppermost basement. Deductions about the required permeability values are summarized in Chapter 8 and Section 7.6 below.

In this section, we focus on the transient response of formation pressures to external forcing, as recorded in CORK pressures – a method that has proven very illuminating in constraining permeability over much larger scales than can be tested with packer methods. Two excellent examples have been reported in the literature: (a) the effects of seafloor tidal loading on sub-seafloor pressures, and (b) the *in situ* pressure response to plate strains and the subsequent hydraulic relaxation of the pressure change.

These are both cases of the response of the formation pressures to a time-varying load, a response that reflects a combination of elastic and diffusive effects. The situation is schematically illustrated in Fig. 7.10, in which the sediment or the permeable igneous basement (at formation scale large enough to represent all elements of porosity) can be considered to consist of a solid matrix hosting fluid in its pore spaces. If there is a change in external forcing, both the matrix and fluid deform elastically, but they respond by different amounts according to their elastic properties. In the usual case, the matrix, being less compressible, develops more internal stress to take up the load, while the more-compressible fluid takes up less load. How the total load is instantaneously divided between matrix and fluid also depends on porosity, with a higher porosity enhancing the effect of fluid compressibility. The fraction of the total incremental load instantaneously taken up by the change in fluid pressure is called the loading efficiency γ, which is defined rigorously in terms of the elastic constants of the matrix and fluid constituents (see van der Kamp and Gale, 1983; Wang and Davis, 1996, 2003; a definition for the simple one-dimensional case is given in the next section). A more compressible matrix (e.g. sediments as compared to igneous oceanic crust) leads to a larger loading efficiency because the fluid takes up a greater share of the incremental load.

Wherever there are different elastic responses in adjacent sub-seafloor formations, i.e. differences in loading efficiency, loading will give rise to non-hydrostatic pressure differences across the interface. At low loading frequencies, such pressure gradients may cause fluid flow in the direction of the pressure differentials, flow that further modifies the pressure field. This flow is governed by Darcy's Law, and the fluid pressure change associated

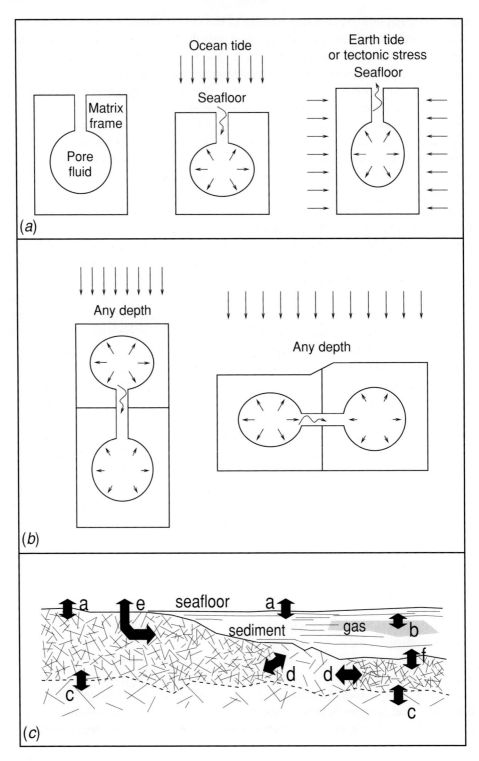

with it is called the diffusive component of the pressure variation due to the loading. For a high-frequency periodic loading function, as is usually the case for the passage of a seismic wave, the elastic pressure gradient is reversed too quickly and the loading effectively causes no fluid flow; under such "undrained" conditions, the fluid pressure is simply the total pressure times loading efficiency. However, when the loading frequency is low, as in the case of tidal loading or a tectonically induced step change, the diffusive part of the response to loading may be significant. The propagation of this diffusive part is governed by the hydraulic diffusivity, η, which depends on permeability, fluid viscosity, the elastic properties of the constituents of the medium, and whether matrix deformation is one- or three-dimensional. The hydraulic diffusivity is commonly defined as the ratio of transmissivity to storativity, which can be rewritten to bring out dependence on permeability as $\eta = k/\mu\zeta$, where ζ is storage compressibility or kinematic storage coefficient (see Wang and Davis, 2003; Chapter 12).

Hence, permeability at formation scales can be constrained by resolution of the diffusive part of the sub-seafloor pressure response to a known loading function, as long as the propagation pathway for the diffusion wave can be determined.

7.5.1 Seafloor tidal loading method

Wang and Davis (1996) provide the basic theory for estimating permeability from the formation response to seafloor tidal loading (see also Chapter 12). Given a sinusoidal tidal loading of period F and amplitude A at the seafloor, the fluid pressure that would be observed within the sub-surface is a combination of elastic and diffusive responses described above, as is illustrated in Fig. 7.11. In a uniform half-space, the elastic response is instantaneous, i.e. always in phase with the tidal loading, and its amplitude is the product of the surface amplitude and the loading efficiency of the medium. For this uniform, one-dimensional case, the one-dimensional loading efficiency γ' is defined as $\gamma' = (1 + 3n\beta_f(1 - \nu)/\beta(1 + \nu))^{-1}$, where n is the porosity of the medium, ν is Poisson's ratio of the medium, and β and β_f are compressibilities of the frame and fluid, respectively. A more complex expression

---

Fig. 7.10 Cartoon illustrating the poroelastic response of the oceanic crust to loads, redrawn from Wang and Davis (2003). (*a*) A sub-sea formation can be represented as a poroelastic medium consisting of a solid matrix frame and pore fluid. When the medium is compressed, the fluid and frame share the load, and the fraction taken up by the fluid is given by the loading efficiency γ. The loading efficiency is by definition unity in the water column above the seafloor but is less below the seafloor. The ensuing differential pressure across the seafloor induces Darcy flow (wiggly arrows), adding a diffusive component to the overall pore fluid pressure response. (*b*) Wherever there are contrasts in loading efficiencies (vertical, horizontal, or any other orientation), the differential fluid pressures tend to induce fluid flow in response to loads. (*c*) In sub-sea formations, formation boundaries, variations in fracture density and porosity, and changes in fluid elastic properties all promote loading-induced fluid flow. Arrows represent various examples of such flow: a and e, flow across the seafloor; b, flow at boundaries of gas-rich sediments; c, d, and f, flow between regions of different frame compressibilities.

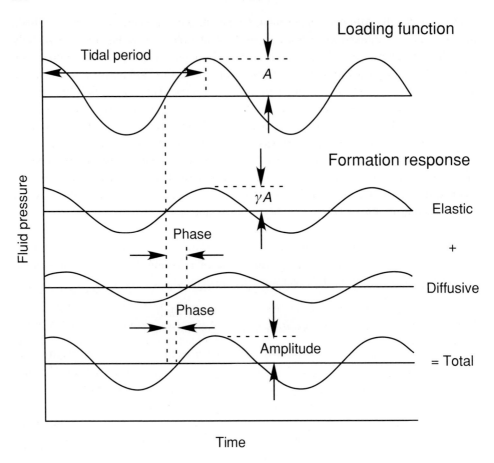

Fig. 7.11 Schematic illustration of the elastic and diffusive components of the overall pressure response to periodic tidal loading.

arises if the compressibility of the material constituting the matrix is significant relative to that of the fluid and framework.

The diffusive component propagates into the half-space as a diffusion wave, with amplitude that decreases exponentially with depth and phase lag that varies regularly with depth. The depth q equivalent to half a wavelength into the medium is defined as the penetration length of the pressure wave and depends on diffusivity and period, $q = (\pi \eta F)^{1/2}$. For a uniform half-space, there is no phase lag at depths that are multiples of the penetration depth and at each additional penetration depth the amplitude of the diffusion wave is attenuated by $e^{-\pi}$; the overall response becomes essentially elastic at depths greater than q. Figure 7.12a illustrates an example calculation of pressure amplitude and phase as functions of depth for a uniform half-space under 12-hour tidal loading at the surface. The zero-phase crossing defines q, which yields a solution for hydraulic diffusivity and then allows permeability to be calculated if other parameters are known.

Davis *et al.* (2000) applied this method to upper crustal pressures measured in CORKs in sedimented settings on ridge flanks. The sediments were thick enough to attenuate fully the diffusive response that propagates vertically from the seafloor, and the propagation pathway for the diffusive component clearly defined in the CORK pressures measured in upper basement was inferred to be through permeable upper basement, from the nearest basement outcrop to each CORK. Hence, the one-dimensional half-space model could be applied in a horizontal sense with propagation distance defined as distance to nearest unsedimented outcrop. Amplitude and phase of pressures measured in three CORKs (at two different sedimented ridge flanks) at distances of 1, 6, and 14 km from exposed basement fit this simple model surprisingly well (Fig. 7.12*b*). Given reasonable assumed values for other parameters, the analysis indicates horizontal permeabilities at the 1–14-km scale on the order of 10^{-10} m^2.

7.5.2 Pressure response to tectonic strain events

If a discrete strain event occurs in the lithosphere, it will induce a stress change within the surrounding poroelastic medium, and this stress change will induce a change in pore pressure. Once the tidal effects and other periodic signals were filtered from pressures recorded in three CORKs on the Juan de Fuca Ridge flank, Davis *et al.* (2001) showed that these data defined the pressure response to a known earthquake swarm on the axis of the ridge arising from an extensional event that caused compression of the crust along the transect of CORKs normal to the spreading axis. The pressures from the three installations show the elastic response – an instantaneous pressure rise with magnitude decreasing with distance from the ridge. They also show continuing pressure rise followed by eventual decay, which are consequences of Darcy flow that follows the diffusion equation and hence they depend on hydraulic diffusivity and permeability in a similar manner as the diffusive response to tidal loading described above. The analysis is described in detail by Davis *et al.* (2001) and in Chapter 8. It revealed that the strain event was largely aseismic and requires a formation-scale permeability over tens of kilometers on the order of 10^{-9}–10^{-10} m^2 to match the observed diffusive pressure response along the transect of CORKs.

7.6 Inferences from nearly isothermal upper basement temperatures

As is described in Chapter 8, geothermal observations in a number of sedimented ridge flank hydrothermal systems indicate a significant degree of isothermality of temperatures in uppermost, permeable basement. This was originally indicated by strong correlations between surface heat flow values and sediment thickness and has been confirmed by *in situ* temperatures measured at several ridge flank CORK sites. CORK pairs in two basement ridge/trough settings (1026B/1027C on the Juan de Fuca Ridge and 896A/504B on the Costa Rica Rift) indicate in situ temperatures even more strongly isothermal than estimated from extrapolations of surface heat values (Davis and Becker, 2002; Chapter 8). This requires vigorous fluid flow over large lateral scales within the uppermost basement to effectively homogenize temperatures (e.g. Davis *et al.*, 1997). Pressures at one of these CORK pairs indicate that the driving pressure differentials are also surprisingly small (about 2 kPa over 2 km lateral scale), leading to the intuitive conclusion that very high upper crustal permeabilities (10^{-10}–10^{-9} m^2) on regional scales are required to allow such vigorous convection (Davis and Becker, 2002).

Constraints on the actual values of the regional-scale upper crustal permeabilities can be obtained by reconciling the observations with theoretical considerations and systematic modeling results (Davis *et al.*, 1997; Wang *et al.*, 1997; Wang and Davis, 2003; Chapter 12). The vigor of buoyancy-driven thermal convection in a porous medium is characterized by the Rayleigh number, *Ra*, a measure of thermal buoyancy versus viscous resistance to flow. For a permeable layer, the Rayleigh number depends directly on permeability, layer thickness, and temperature difference across the layer, as well as on various fluid properties. More specifically, for a flat-lying isotropic layer with vertical temperature difference ΔT across thickness b, the Rayleigh number can be expressed as $Ra = \alpha gkb\rho_f\Delta T/\mu\kappa$, where α is fluid thermal expansivity, μ/ρ_f is fluid kinematic viscosity, k is permeability, and κ is thermal diffusivity of the fluid-saturated porous medium. In the igneous oceanic crust, none of these parameters is uniform, but permeability displays by far the widest range of values. Therefore the vigor of thermal convection is most strongly affected by the permeability of the formation, and hence the degree of isothermality of upper basement temperatures is to first order diagnostic of the permeability of that section (the "hydrothermal basement" as defined in Chapter 8).

For a flat layer with uniform properties, the system becomes unstable, and thermally significant convection takes place, when *Ra* exceeds a critical value of $4\pi^2$ or about 40. Where topography is present, as in the oceanic crust, the potential for convection is enhanced and flow can be stimulated below the critical value, with very little thermal effect but possibly significant geochemical effects when integrated over geological time. For *Ra* near and somewhat greater than the critical value, steady-state convection generally develops. The effects of such steady-state convection can be simulated with numerical models

--→

Fig. 7.12 (*a*) Pressure amplitude (normalized to tidal signal at seafloor) and phase as functions of depth for a homogeneous poroelastic half-space under surface tidal loading at 12-hour period, calculated following the methods of Wang and Davis (1996) for permeability of 1.7×10^{-10} m^2 and reasonable values of other parameters. (*b*) Upper panel: simplified geometry for interpreting tidally induced formation fluid pressure variations observed at CORKs in sedimented sites that vary in distance to nearest basement outcrops. At such sites, the basement formation responds instantaneously to loading directly from above, but the diffusive component propagates horizontally into the permeable basement layer from the basement outcrop. Lower panel: amplitudes (normalized to seafloor tidal load) and phases of formation fluid pressure variations observed in the CORKs at Holes 1024C and 1025C, presented as 12-hour period equivalents at appropriate lateral scales from outcrop. Note similarities between observed response and the half-space calculation, suggesting basement permeabilities on the order of 10^{-10} m^2. After Davis *et al.* (2000) and Wang and Davis (2003).

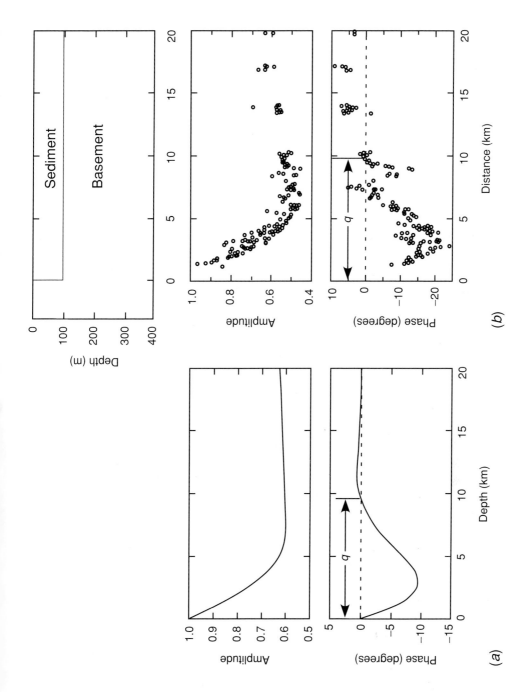

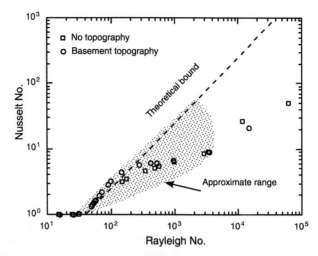

Fig. 7.13 Plot summarizing the relationship between Nusselt number and Rayleigh number. Squares and circles illustrate results of simulations described in Section 8.1.2, calculated for material properties and vertical temperature difference of 125 °C. Speckled area illustrates the range of results from various studies, as summarized by Lister (1990) and Bejan (1984). At high values, it appears that any given *Nu* value could be consistent with convection at *Ra* values ranging over at least an order of magnitude; there would be similar uncertainties in estimated permeabilities.

(see Fisher and Becker, 1995; Wang *et al.*, 1997; and Chapter 12 for further discussion of actual models and simulations).

For *Ra* much greater than the critical value, convection will become unsteady, with periodic or even chaotic temporal variability that cannot be characterized accurately with repeatable numerical simulations. Instead, for a vigorous, possibly unsteady system, it is useful to characterize the overall thermal budget in terms of the Nusselt number *Nu*, which is defined as the ratio of total heat *Q* actually transferred to the heat that would be transferred only by conductive processes, $Nu = Q/(\lambda dT/dz)$, where λ is thermal conductivity. As convection becomes more vigorous, more heat is transferred by convection, so *Nu* increases as *Ra* increases. The relationship between *Nu* and *Ra* at high values is poorly understood and possibly not universal. It has been constrained by laboratory and numerical experiments at moderate *Nu* and *Ra* values; Lister (1990) summarized these and developed a "semi-empirical" functional relationship (see Chapter 12). However, the actual relationship between *Nu* and *Ra* probably depends on various matrix and fluid properties, such that only limits can be relied upon (Fig. 7.13).

In the presence of vigorous convection, the thermal effect can be equivalently described using a hypothetical purely conductive regime with thermal conductivity that is *Nu* times the actual conductivity. Hence the overall thermal effect of vigorous convection can be assessed with conductive models with very high effective conductivities, and such models can be used to constrain the *Nu* required to produce the observed nearly isothermal basement temperatures under the observed conditions of sediment-covered basement relief

Table 7.2 *Approximate lateral scales of investigation for methods used to determine permeabilities of igneous oceanic crust*

Measurement method	Approximate scale of investigation (km)
Slug tests	0.1–1
Injection tests	0.1–1 (tests up to 30 min. duration)
Borehole flow thermal methods	1.0–10 (depending on duration of flow)
CORK methods and inferences from basement isothermality	5.0–100 (requires multiple observation points or knowledge of drainage path)

(Davis *et al.*, 1997). For the 1026B/1027C CORK pair and a 600-m thick permeable "hydrothermal basement" corresponding to seismic layer 2A (see Chapter 8), such models indicate that a *Nu* of about 100 is required to produce the observed temperatures isothermal to within 2 K. Lister's relationship would suggest that *Nu* of 100 would require $Ra > 4,000$; the limiting relationship derived by Bejan (1984) would suggest $Ra \approx 4,000$ (Fig. 7.13). Given reasonable values for other parameters, such a high value for Ra in turn suggests an overall permeability of 10^{-10}–10^{-9} m^2, a value that would be effective for the large lateral scales over which the convection effectively homogenizes upper basement temperatures. Similar conclusions have been reached from observations on the somewhat older Costa Rica Rift flank near Hole 504B (Davis *et al.*, in press).

7.7 Measurement scales and effects of scale on reported permeabilities

It seems intuitive that the effective scales of investigation of the methods described above progressively increase from slug tests to injection tests to results of longer-term natural in-hole flow to the formation response recorded in CORK pressures to broadly distributed signals. However, except for the last, where the signal source is broadly distributed through the formation, it is difficult to quantify the scale of investigation to better than an order of magnitude. This is true for a number of reasons. As is noted by Mathews and Russell (1967), "a precise answer cannot be given to this question since any pressure disturbance is felt to a small extent throughout the reservoir." Hence, defining a scale of investigation requires specifying some characteristic fraction of the source signal and calculating a characteristic radial length scale to which that fractional pressure is transmitted away from the borehole. In the absence of observation wells distinct from the source hole, such calculations depend on the same kind of assumptions as many of the other calculations reviewed in this chapter: radial uniformity of an isotropic permeability which cannot adequately represent the effects of any large-scale fracturing.

A number of authors discuss the means to estimate effective scales of investigation, and Fisher (1998) reviews some of the calculations, which are difficult to compare across test types. For the permeabilities typically reported in the igneous oceanic crust, effective radii of investigation for the methods discussed above are given in Table 7.2, good at best only to order of magnitude. To reiterate, for the single-borehole methods, these are based on the assumption of radial uniformity of an isotropic permeability; much larger (but azimuthally varying) radii of investigation are possible if pressure signals are transmitted via isolated, highly transmissive structures.

Fisher (1998) also summarized results of laboratory measurements of samples from the igneous oceanic crust. Such samples are typically small, solid rock matrix that severely underrepresents any large-scale porosity that may dominate *in situ* permeability at scales larger than centimeters. Nevertheless, the typical values – on the order of 10^{-18}–10^{-20} m^2 – are very useful as end-member values for *in situ* permeability of the rock matrix of the crust. This low-matrix permeability may control the local exchange of heat and chemical species over a large fraction of the crust, i.e. in integral volumes of rock away from any fractures or interconnected porosity. Results of *in situ* measurements described here all yield higher permeability values because they tend to capture the effect of irregular interconnected porosity to greater extent.

It is commonly observed that permeabilities measured on crystalline rocks and rock formations range over many orders of magnitude (Brace, 1984; Clauser, 1992). In particular, laboratory-scale measurements on competent rock specimens that do not sample natural fractures are often several orders of magnitude less than values obtained with *in situ* borehole testing (Brace, 1984; Renshaw, 1998). A direct relationship between measured permeability and test scale is suggested, but it is unclear whether this extends to scales greater than hundreds of meters, at which it is generally more difficult to test permeability. For example, in a compilation of all available data from crystalline rocks at all test scales, Clauser (1992) noted a scale effect up to the scales investigated with borehole testing (meters to hundreds of meters), but not at larger scales. Similar effects are noted in continental hydrology in sedimentary formations as well (e.g. Garven, 1986, for fractured carbonates), although there remains debate as to what extent scale effects at large scales are real or artifacts of the methods (e.g. Rovey and Cherkauer, 1995; Butler and Healey, 1998a,b; Rovey and Niemann, 1998). Finally, theoretical considerations for percolation theory in fracture networks even suggest that there should be an inverse relationship between scale and permeability at scales of borehole testing and larger (meters and above; Renshaw, 1998).

For the three Juan de Fuca flank holes for which Becker and Fisher (2000) report packer-derived permeability results, the longer-term flow-based permeability determinations (Becker and Davis, 2003) are larger by roughly an order of magnitude (Fig. 7.14). The scale of investigation of these flow-based methods is also greater by approximately an order of magnitude. It should be noted that the flow-determined permeability estimates are quite sensitive to the actual values of the *in situ* formation pressures, but the larger non-hydrostatic formation pressures required to bring the flow-based permeabilities at the Juan de Fuca sites into agreement with the packer permeabilities would be completely inconsistent with the CORK pressure data. Previous reports of flow-determined permeabilities in Holes 504B and 395A seemed more consistent with the packer-determined values (Anderson and Zoback, 1982; Becker *et al.*, 1983a,b, 1984; Morin *et al.*, 1992), but as noted in Section 7.4, the large

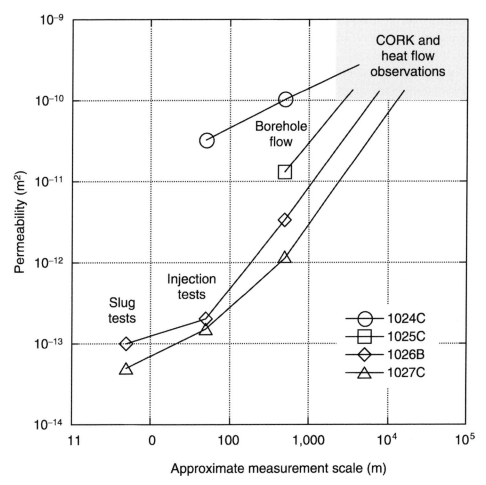

Fig. 7.14 Permeabilities determined with multiple techniques at four sites on the flank of the Juan de Fuca Ridge versus approximate scale of penetration of the methods.

pressure differentials used for these calculations may have been unrealistic. Recent results from sealed-hole observatories later installed in these holes suggest that the actual pressure differentials were much smaller, which would require that the flow-based permeabilities be proportionally greater (Fig. 7.9).

As described in Sections 7.5 and 7.6, even larger permeability values at even larger scales are suggested by the tidal loading and strain relaxation analyses of long-term CORK pressure data, and by model requirements to simulate *in situ* pressure and temperature values in uppermost basement recorded with the CORKs. The results are generally consistent, and suggest permeabilities in the range of 10^{-10}–10^{-9} m^2. Uncertainties (e.g. arising from using two-dimensional models to simulate convective heat transport in the crust, and lack of constraints on the depth–extent of permeability) are difficult to assess, but like the borehole determinations described above, the estimates are probably good to within an order of magnitude. Thus, there is little doubt that they lend further support to an extension of the scale effect of ocean crustal permeability over a very large range.

An apparent permeability of 10^{-10} m^2 has been interpreted in only one instance of ODP borehole measurements to date (Fig. 7.9), in Hole 857D which penetrates an intercalated sediment/sill sequence in the Middle Valley sedimented rift (Becker *et al.*, 1994). This came from assigning the entire transmissivity determined with a packer experiment over a 362-m interval to a 5-m zone that a flowmeter log indicated was accepting nearly all of a huge downhole flow (Fig. 7.8) and appeared to be a major fault zone from all other circumstantial evidence. We therefore infer that regional-scale crustal permeabilities of the same order would require open networks of such faults/fractures of great lateral extent. If this is the rule in young oceanic crust, then the issue arises: why do not more boreholes penetrate such open faults and yield extraordinarily high permeabilities (see Fig. 7.1)? For the Juan de Fuca sites that form our prime dataset for assessing scale effects, this dilemma can be reconciled by invoking the very shallow penetration depths of the holes into upper oceanic basement as well as the low probability of intersecting steeply dipping fractures or faults with a vertical borehole (Terzaghi, 1965). A full test will be possible if these sites are deepened through the upper basement, under which conditions it should be expected that packer permeability measurements should become more representative of crustal-scale conditions.

The range of methods applied in the Juan de Fuca flank sites also suggests a variable permeability–age relationship depending on the scale of investigation (Fig. 7.14); permeabilities at smaller measurement scales apparently show a stronger reduction in permeability over the age range of those sites (Becker and Davis, 2003). This supports an interpretation in which the smaller-scale fractures and voids are the first to be sealed during the aging process, while the larger-scale fault systems presumably remain open to greater crustal ages. This interpretation is consistent with the model of Wilkens *et al.* (1991; see also Chapter 4) for evolution of the porosity and seismic structure of the upper oceanic crust by initial sealing of only small-scale thin cracks and voids, with little effect on interconnected fracture porosity and large-scale permeability.

7.8 Conclusions and recommendations for future work

From an intuitive perspective, permeability is both straightforward to conceptualize and also central to nearly all understanding of the hydrogeology of the oceanic crust and lithosphere. However, understanding quantitatively the permeability distribution within the igneous oceanic crust has proven quite difficult. One important reason is that all of the methods for directly assessing permeability discussed in this chapter depend on accessing the oceanic crust via boreholes that individually cannot sample representatively all the scales of the permeability distribution. Like many investigations of nature, our advances in understanding the permeability of the igneous oceanic crust are often accompanied by many new questions and uncertainties.

Nevertheless, more than 30 years of study in DSDP and ODP boreholes has established a conceptual model for the permeability structure of the oceanic crust, comprising: (1) low-permeability sediments, overlying (2) a few hundred meters of highly permeable (10^{-14}–10^{-12} m^2 at borehole-testing scales) extrusives that seem to correspond to seismic layer 2A, grading into (3) much less permeable (order 10^{-16} m^2 or less) sealed extrusives and sheeted dikes. This basic model has been very useful in constraining simulations of fluid flow processes in the igneous oceanic crust (see Chapter 12). Recent results have provided important new advances, including:

(a) indications of an age variation in the permeability of the uppermost extrusives that is consistent with other data relating the effects of aging to other physical properties (Becker and Fisher, 2000; Fisher and Becker, 2000; Becker and Davis, 2003), and
(b) indications of hydrological responses extending over regional scales that seem to require even higher permeabilities (order 10^{-10} m^2) on those scales (Davis *et al.*, 2000, 2001).

However, there remains much work to be done in understanding the permeability distribution within the igneous oceanic crust. Among the key outstanding issues are the following:

1. Understanding the depth variation of permeability in the igneous oceanic crust, both in more locations than the two key early examples (Holes 395A and 504B) and at greater depth than either of these holes extends (e.g. see Section 8.2.2. about debate regarding whether higher permeabilities to provide super-critical *Ra* conditions are required at Site 504 deeper than Hole 504B now extends).
2. Understanding the spatial variability of the permeability distribution in a local sense, i.e. breaking beyond the assumption of local lateral radial/azimuthal uniformity that applies to all determinations to date.
3. Understanding the spatial variations in the permeability distribution in a regional sense, i.e. the variation with crustal age and the still untested prospects for lateral variations in crust of the same age.
4. Understanding the scale effects inherent in assessing permeability from measurements in boreholes, not only for the basic relevance to the permeability distribution within the oceanic crust but also for key implications for overall fluid–rock interactions within the oceanic crust.
5. Evaluating the distribution of storativity and compressibility within the igneous oceanic crust and its relationship with transmissivity and permeability.

Addressing these issues will almost certainly require the use of boreholes, both old holes and new holes to be drilled by the Integrated Ocean Drilling Program. It will also require new techniques that should be feasible with dedicated effort. Some of the approaches we would recommend for the future include:

1. Drilling new holes in a range of characteristic environments, e.g. sites in the age range completely unsampled to date (12–150 Ma).
2. Deepening some of the well-studied existing holes (e.g. Juan de Fuca sites that penetrate only a few tens of meters into igneous crust) to assess the depth variations in the permeability distributions at those sites.
3. Drilling arrays of holes at characteristic sites and developing hole-to-hole testing capabilities.

4. Developing better packer systems, e.g. routine straddle packer systems to isolate narrower zones and packer systems that can be utilized in single-bit holes without re-entry structures.

5. Developing the capability for longer-term testing using packers that goes beyond the typical three–six hours generally provided from expensive drill ships.

6. Applying multiple techniques in each of the relatively rare holes utilized primarily for hydrogeological studies.

7. Improving the integration of borehole and remote geophysical techniques, to allow spatial extension of borehole permeability results to the formation away from boreholes using remote geophysics.

Acknowledgments

The packer and CORK results reviewed in this chapter would not have been possible without considerable support on many fronts. We gratefully acknowledge: (1) the pioneering packer work of Roger Anderson, Mark Zoback, and Steve Hickman during DSDP; (2) at-sea support provided by the drilling crews, technicians, and engineers of *DV Glomar Challenger* and *DV JOIDES Resolution*, and the pilots and ship's crews of the manned and unmanned submersibles used to recover CORK data; (3) exceptional CORK engineering support from Tom Pettigrew, Bob Macdonald, and Bob Meldrum; and (4) strong and consistent financial support from the US National Science Foundation throughout ODP and from the Geological Survey of Canada since 1989. For their help specifically as we prepared this review, we thank Andy Fisher for careful discussions and Roger Anderson and Roger Morin for their thorough reviews.

References

Anderson, R. N. and Zoback, M. D. 1982. Permeability, underpressures, and convection in the oceanic crust near the Costa Rica Rift, eastern equatorial Pacific. *J. Geophys. Res.* **87**: 2,860–2,868.

Anderson, R. N., Zoback, M. D., Hickman, S. H., and Newmark, R. L. 1985. Permeability versus depth in the upper oceanic crust: in-situ measurements in DSDP Hole 504B, eastern equatorial Pacific. *J. Geophys. Res.* **90**: 3,659–3,669.

Bear, J. 1993. Modeling flow and contaminant transport in fractured rocks. In *Flow and Contaminant Transport in Fractured Rocks*, eds. J. Bear, C.-F. Tsang, and G. de Marsily. San Diego, CA: Academic Press, pp. 1–37.

Becker, K. 1986. Special report: development and use of packers in ODP. *JOIDES J.* **12**(2): 51–57.

    1988. *A Guide to ODP Tools for Downhole Measurements*, ODP Technical Note, No. 10. College Station, TX: Ocean Drilling Program.

    1989. Measurements of the permeability of the sheeted dikes in Hole 504B, ODP Leg 111. In *Proceedings of the Ocean Drilling Program, Scientific Results*, Vol. 111, eds. K. Becker, H. Sakai, *et al.* College Station, TX: Ocean Drilling Program, pp. 317–325.

    1990. Measurements of the permeability of the upper oceanic crust at Hole 395A, ODP Leg 109. In *Proceedings of the Ocean Drilling Program, Scientific Results*, Vol. 106/109, eds. R. Detrick, J. Honnovez, J. B. Bryan, T. Juteau, *et al.* College Station, TX: Ocean Drilling Program, pp. 213–222.

1991. In-situ bulk permeability of oceanic gabbros in Hole 735B, ODP Leg 118. In *Proceedings of the Ocean Drilling Program, Scientific Results*, Vol. 118, eds. R. P. Von Herzen, P. T. Robinson, *et al.* College Station, TX: Ocean Drilling Program, pp. 333–347.

1996. Permeability measurements in Hole 896A and implications for the lateral variability of upper crustal permeability at Sites 504 and 896. In *Proceedings of the Ocean Drilling Program, Scientific Results*, Vol. 147, eds. C. Mevel, K. M. Gillis, J. F. Allan, and P. S. Meyer. College Station, TX: Ocean Drilling Program, pp. 353–363.

Becker, K. and Davis, E. E. 2003. New evidence for age variation and scale effects of permeabilities of young oceanic crust from borehole thermal and pressure measurements. *Earth Planet. Sci. Lett.* **210**: 499–508.

Becker, K. and Fisher, A. T. 2000. Permeability of upper oceanic basement on the eastern flank of the Juan de Fuca Ridge determined with drill-string packer experiments. *J. Geophys. Res.* **105**: 897–912.

Becker, K., Von Herzen, R. P., Francis, T. J. G., *et al.* 1982. In situ electrical resistivity and bulk porosity of the oceanic crust, Costa Rica Rift. *Nature* **300**: 594–598.

Becker, K., Langseth, M. G., and Von Herzen, R. P. 1983a. Deep crustal geothermal measurements, Hole 504B, Deep Sea Drilling Project Legs 69 and 70. In *Initial Reports of the Deep Sea Drilling Project*, Vol. 69, eds. J. R. Carr, M. G. Langseth, J. Honnorez, *et al.* Washington, DC: US Govt. Printing Office, pp. 223–236.

Becker, K., Langseth, M. G., Von Herzen, R. P., and Anderson, R. N. 1983b. Deep crustal geothermal measurements, Hole 504B, Costa Rica Rift. *J. Geophys. Res.* **88**: 3,447–3,457.

Becker, K., Langseth, M. G., and Hyndman, R. D. 1984. Temperature measurements in Hole 395A, Leg 78B. In *Initial Reports of the Deep Sea Drilling Project*, Vol. 78B, eds. R. D. Hyndman, M. H. Salisbury, *et al.* Washington, DC: US Govt. Printing Office, pp. 689–698.

Becker, K., Langseth, M. G., Von Herzen, R. P., Anderson, R. N., and Hobart, M. A. 1985. Deep crustal geothermal measurements, Hole 504B, Deep Sea Drilling Project Legs 69, 70, 83, and 92. In *Initial Reports of the Deep Sea Drilling Project*, Vol. 83, eds. R. N. Anderson, J. Honnorez, K. Becker *et al.* Washington, DC: US Govt. Printing Office, pp. 405–418.

Becker, K., Morin, R. H., and Davis, E. E. 1994. Permeabilities in the Middle Valley hydrothermal system measured with packer and flowmeter experiments. In *Proceedings of the Ocean Drilling Program, Scientific Results*, Vol. 139, eds. M. J. Mottl, E. E. Davis, A. T. Fisher, and J. F. Slack. College Station, TX: Ocean Drilling Program, pp. 613–626.

Becker, K., Bartetzko, A., and Davis, E. E. 2001. Leg 174B synopsis: revisiting Hole 395A for logging and long-term monitoring of off-axis hydrothermal processes in young oceanic crust. In *Proceedings of the Ocean Drilling Program, Scientific Results*, Vol. 174B, eds. K. Becker and M. J. Malone. College Station, TX: Ocean Drilling Program, pp. 1–12.

Becker, K., Davis, E. E., Spiess, F. N., and de Moustier, C. P. 2004. Temperature and video logs from the upper oceanic crust, flank Holes 504B and 896A, Costa Rica Rift: implications for the permeability of upper oceanic crust. *Earth Planet. Sci. Lett.* **222**: 881–896.

Bejan, A. 1984. *Convection Heat Transfer.* New York: Wiley.

Brace, W. F., 1984. Permeability of crystalline rocks: new in situ measurements. *J. Geophys. Res.* **89**: 4,327–4,330.

Bredehoeft, J. D. and Papadopulos, S. S. 1980. A method for determining the hydraulic properties of tight formations. *Water Resources Res.* **16**: 223–238.

Bruns, T. D. and Lavoie, D. L. 1994. Bulk permeability of young backarc basalt in the Lau Basin from a downhole packer experiment (Hole 839B). In *Proceedings of the Ocean Drilling Program, Scientific Results*, Vol. 135, eds. J. Hawkins, L. Povsons, J. Allan, *et al.* College Station, TX: Ocean Drilling Program, pp. 805–816.

Butler, J. J., Jr. and Healey, J. M. 1998a. Relationship between pumping-test and slug-test parameters: scale effect or artifact. *Ground Water* **36**: 305–313.

1998b. Authors' reply. *Ground Water* **36**: 867–868.

Clauser, C. 1992. Permeability of crystalline rocks. *Eos, Trans. Am. Geophys. Union* **73**: 233, 237–238.

Cooper, H. H., Jr. and Jacob, C. E. 1946. A generalized graphical method for evaluating formation constants and summarizing well-field history. *Trans. Am. Geophys. Union* **27**: 526–534.

Cooper, H. H., Jr., Bredehoeft, J. D., and Papadopulos, I. S. 1967. Response of a finite diameter well to an instantaneous charge of water. *Water Resources Res.* **3**: 267–269.

Darcy, H. 1856. *Les Fontaines Publiques de la Ville de Dijon*. Paris: Victor Dalmont.

Davis, E. E. and Becker, K. 1994. Formation temperatures and pressures in a sedimented rift hydrothermal system: 10 months of CORK observations, Holes 857D and 858G. In *Proceedings of the Ocean Drilling Program, Scientific Results*, Vol. 139, eds. M. J. Mottl, E. E. Davis, A. T. Fisher, and J. F. Slack. College Station, TX: Ocean Drilling Program, pp. 649–666.

1998. Borehole observatories record driving forces for hydrothermal circulation in young oceanic crust. *Eos, Trans. Am. Geophys. Union* **79**: 369, 377–378.

2002. Observations of natural-state fluid pressures and temperatures in young oceanic crust and inferences regarding hydrothermal circulation. *Earth Planet. Sci. Lett.* **204**: 231–248.

Davis, E. E., Becker, K., Pettigrew, T., Carson, B., and Macdonald, R. 1992. CORK: a hydrologic seal and downhole observatory for deep-sea boreholes. In *Proceedings of the Ocean Drilling Program, Initial Reports*, Vol. 139, eds. E. E. Davis, M. J. Mottl, A. T. Fisher, *et al.* College Station, TX: Ocean Drilling Program, pp. 45–53.

Davis, E. E., Wang, K., He, J., Chapman, D. S., Villinger, H., and Rosenberger, A. 1997. An unequivocal case for high Nusselt number hydrothermal convection in sediment-buried igneous oceanic crust. *Earth Planet. Sci. Lett.* **146**: 137–150.

Davis, E. E., Wang, K., Becker, K., and Thomson, R. E. 2000. Formation-scale hydraulic and mechanical properties of oceanic crust inferred from pore pressure response to periodic seafloor loading. *J. Geophys. Res.* **105**: 13, 423–13, 435.

Davis, E. E., Wang, K., Thomson, R. E., Becker, K., and Cassidy, J. F. 2001. An episode of seafloor spreading and associated plate deformation inferred from crustal fluid pressure transients. *J. Geophys. Res.* **106**: 21, 953–21, 963.

Davis, E. E., Becker, K., and He, J. 2004. Costa Rica Rift revisited: constraints on shallow and deep hydrothermal circulation in oceanic crust. *Earth Planet. Sci. Lett.* **222**: 863–879.

Fisher, A. T. 1998. Permeability within basaltic oceanic crust. *Rev. Geophys.* **36**: 143–182.

Fisher, A. T. and Becker, K. 1995. Correlation between seafloor heat flow and basement relief: observational and numerical examples and implications for upper crustal permeability. *J. Geophys. Res.* **100**: 12,641–12,657.

2000. Channelized fluid flow in oceanic crust reconciles heat-flow and permeability data. *Nature* **403**: 71–74.

Fisher, A. T., Becker, K., and Davis, E. E. 1997. The permeability of young oceanic crust east of Juan de Fuca Ridge determined using borehole thermal measurements. *Geophys. Res. Lett.* **24**: 1,311–1,314.

Fouchet, Y., Zierenberg, R. A., Miller, D. J., *et al.*, eds. 1998. *Proceedings of the Ocean Drilling Program, Initial Reports*, Vol. 169. College Station, TX: Ocean Drilling Program.

Gable, R., Morin, R., and Becker, K. 1989. Geothermal state of Hole 504B: Leg 111 overview. In *Proceedings of the Ocean Drilling Program, Scientific Results*, Vol. 111, eds. K. Becker, H. Sakai, *et al.* College Station, TX: Ocean Drilling Program, pp. 87–96.

Gable, R., Morin, R., Becker, K., and Pezard, P. 1995. Heat flow in the upper part of the oceanic crust: synthesis of in situ temperature measurements in Hole 504B. In *Proceedings of the Ocean Drilling Program, Scientific Results*, Vol. 137, eds. J. Eizinger, K. Becker, H. J. B. Dick, and L. B. Stokking. College Station, TX: Ocean Drilling Program, pp. 140, 321–324.

Garven, G. 1986. The role of regional fluid flow in the genesis of the Pine Point deposit, western Canada sedimentary basin: a reply. *Econ. Geol.* **81**: 1,015–1,020.

Guerin, G., Becker, K., Gable, R., and Pezard, P. A., 1996. Temperature measurements and heat flow analysis in Hole 504B. In *Proceedings of the Ocean Drilling Program, Scientific Results*, Vol. 148, eds. J. C. Alt, H. Kinoshita, L. B. Stokking, and P. J. Michael. College Station, TX: Ocean Drilling Program, pp. 291–296.

Hickman, S. H., Langseth, M. G., and Svitek, J. F., 1984. In situ permeability and pore-pressure measurements near the Mid-Atlantic Ridge, Deep Sea Drilling Project Hole 395A. In Initial Reports of the Deep Sea Drilling Project, Vol. 78B, eds. R. D. Hyndman, M. H. Salisbury, *et al.* Washington, DC: US Govt. Printing Office, pp. 699–708.

Horner, D. R. 1951. Pressure build-up in wells. In *Proceedings of the Third World Pet. Congress*, Vol. II. Leiden: E. J. Brill, p. 503–521.

Hyndman, R. D., Von Herzen, R. P., Erickson, A. J., and Jolivet, J. 1976. Heat flow measurements in deep crustal holes on the Mid-Atlantic Ridge. *J. Geophys Res.* **81**: 4,053–4,060.

Kopietz, J., Becker, K., and Hamano, Y. 1990. Temperature measurements at Site 395, ODP Leg 109. In *Proceedings of the Ocean Drilling Program, Scientific Results*, Vol. 106/109, eds. R. Detrick, J. Honnorez, J. B. Bryan, T. Juteau, *et al.* College Station, TX: Ocean Drilling Program, pp. 197–203.

Larson, R. L, Fisher, A. T., Jarrard, R., and Becker, K. 1993. Highly layered and permeable Jurassic oceanic crust in the western Pacific. *Earth Planet. Sci. Lett.* **119**: 71–83.

Lesem, L. B., Greytok, F., Marotta, F., and McKetta, Jr., J. 1957. A method of calculating the distribution of temperature in flowing gas wells. *Trans. Am. Inst. Min. Metall. Pet. Eng.* **210**: 169–176.

Lister, C. R. B. 1990. An explanation for the multivalued heat transport found experimentally for convection in a porous medium. *J. Fluid Mech.* **214**: 287–320.

Matthews, C. S. and Russell, D. G. 1967. *Pressure Buildup and Flow Tests in Wells*, Monograph No. 1. Dallas: SPE-AIME.

Mikada, H., Becker, K., Moore, J. C., Klaus, A., *et al.* (eds.) 2002. *Proceedings of the Ocean Drilling Program, Initial Reports*, Vol. 196. College Station, TX: Ocean Drilling Program.

Morin, R. H., Hess, A. E., and Becker, K. 1992. In situ measurements of fluid flow in DSDP Holes 395A and 534A: results from the DIANAUT Program. *Geophys. Res. Lett.* **19**: 509–512.

Neuzil, C. E. 1982. On conducting the modified 'slug' test in tight formations. *Water Resources Res.* **18**: 439–441.

Norton, D. and Knapp, R. 1977. Transport phenomena in hydrothermal systems: the nature of porosity. *Am. J. Sci.* **277**: 913–936.

Papadopulos, S. S., Bredehoeft, J. D., and Cooper, H. H. 1973. On the analysis of 'slug test' data. *Water Resources Res.* **9**: 1,087–1,089.

Pezard, P. A., Anderson, R. N., Ryan, W. B. F., Becker, K., Alt, J. C., and Gente, P. 1992. Accretion, structure, and hydrology of intermediate spreading-rate oceanic crust from drillhole experiments and seafloor observations. *Mar. Geophys. Res.* **14**: 93–123.

Renshaw, C. E. 1998. Sample bias and the scaling of hydraulic conductivity in fractured rock. *Geophys. Res. Lett.* **25**: 121–124.

Rohr, K. 1994. Increase of seismic velocities in upper oceanic crust and hydrothermal circulation in the Juan de Fuca plate. *Geophys. Res. Lett.* **21**: 2,163–2,166.

Rovey, C. W. and Cherkauer, D. S. 1995. Scale dependency of hydraulic conductivity measurements. *Ground Water* **33**: 769–780.

Rovey, C. W. and Niemann, W. I. 1998. Discussion. *Ground Water* **36**: 866–867.

Snow, D. T. 1968. Rock fracture spacing, openings and porosities. *J. Soil Mech. Foundation Div. ASCE Proc*, **94**: 73–91.

Terzaghi, R. 1965. Sources of error in joint surveys. *Geotechnique* **15**: 287–304.

Theis, C. V. 1935. The relationship between the lowering of the piezometric surface and the rate and duration of discharge of a well using ground-water storage. *Trans. Am. Geophys. Union* **16**: 519–524.

Van der Kamp, G. and Gale, J. E. 1983. Theory of Earth tide and barometric effects in porous formations with compressible grains. *Water Resources Res.* **19**: 538–544.

Wang, K. and Davis, E. E. 1996. Theory for the propagation of tidally induced pore pressure variations in layered subseafloor formations. *J. Geophys. Res.* **101**: 11,483–11,495.

   2003. High permeability of young oceanic crust constrained by thermal and pressure observations. In *Land and Marine Hydrogeology*, eds. M. Taniguchi, K. Wang, and T. Gamo. Amsterdam: Elsevier, pp. 165–188.

Wang, K., He, J., and Davis, E. E. 1997. Influence of basement topography on hydrothermal circulation in sediment-buried igneous oceanic crust. *Earth Planet. Sci Lett.* **146**: 151–164.

Wilkens, R. H., Fryer, G. J., and Karsten, J. 1991. Evolution of porosity and seismic structure of upper oceanic crust: importance of aspect rations. *J. Geophys. Res.* **96**: 17,981–17,995.

8

Observations of temperature and pressure: constraints on ocean crustal hydrologic state, properties, and flow

Earl E. Davis and Keir Becker

8.1 Introduction

8.1.1 Background

Temperature and pressure are key parameters for understanding hydrologic processes. Observations of pressure provide direct constraints on driving forces for flow, and temperature variations provide a useful proxy for fluid flow when rates are sufficiently high to cause advective heat transport to become significant relative to conduction. When combined, pressure and temperature observations provide strong constraints on formation hydrologic properties, and knowledge of temperature is necessary for understanding chemical reactions between water and its host rock matrix and for assessing the potential for microbiological activity.

In this chapter we summarize a variety of observations and estimates of temperature and pressure in the oceanic crust. Thermal data are most efficiently acquired with wireline-deployed, gravity-driven probes that penetrate a few meters into sediments and measure thermal gradient and conductivity. In most deep-ocean settings, water temperature fluctuations are small (Davis *et al.*, 2003) and deep thermal structure can be determined through extrapolation of short-probe data wherever conductive heat transport dominates. Where advective thermal transport within the formation becomes significant, *in situ* conditions can be determined with observations in boreholes, although these must be made with the possibility for flow within the boreholes themselves always in mind.

Estimates of formation pressure can also be had with shallow gravity-driven sediment probes, although extrapolations of shallow gradients are unreliable because of the large uncertainty associated with estimating hydraulic resistance. Thus, reliable determination of pressure at depth also requires borehole observations. Fluid pressures in sub-seafloor formations can originate from a variety of sources including physical consolidation, chemical dewatering, elastic deformation, and thermal buoyancy. The last is most important in the oceanic crust, although strain associated with seafloor (e.g. tidal) and tectonic (e.g. fault

Hydrogeology of the Oceanic Lithosphere, eds. E. E. Davis and H. Elderfield. Published by Cambridge University Press.
© Cambridge University Press 2004.

dislocation) loading has been found to create significant driving forces for fluid flow as well.

8.1.2 A working model for ocean crustal hydrothermal circulation

Ever since the recognition three decades ago that hydrothermal circulation in the oceanic crust was responsible for large local variations and low average values of seafloor heat flux (Chapter 2), the nature of the variations and the degree to which the average level of heat flux is depressed relative to that expected from underlying lithosphere have been used to infer the spatial pattern and depth of cellular convection and to estimate rates of fluid and heat exchange between the ocean and crust (Lister, 1972; Williams *et al.*, 1974; Anderson and Hobart, 1976; Anderson *et al.*, 1977; Chapter 2). Great credit must be given to the early models for being conceptually correct, although attempts to use heat flux variations in the absence of good structural controls, and an inability to determine the nature of variations in an unaliased manner with sparse measurements, led to incomplete and inaccurate conclusions. A strongly layered hydrologic structure was inferred from the fundamental seismic structure (Chapters 3 and 5), with some portion of the crust possessing sufficiently high permeability to host convection of pore fluid (e.g. Lowell, 1980; Anderson and Skilbeck, 1981; Hartline and Lister, 1981; Langseth and Herman, 1981). This permeable igneous section, referred to here as "hydrothermal basement," was interpreted to be sandwiched between deeper rocks of low permeability and an accumulation of similarly low-permeability sediments that thicken and become progressively more continuous with time (Chapters 6, 7, and 11). The amount of heat lost by ventilated convection diminishes as the crust ages, cools, is hydrothermally altered, and is buried by sediments (Lister, 1972; Williams *et al.*, 1974; Anderson and Hobart, 1976; Sclater *et al.*, 1976; Anderson *et al.*, 1977; Stein and Stein, 1994; Chapters 2 and 10). Crustal circulation may persist to great crustal age, causing local heat flux variations to be larger than can be accounted for by conductive effects (Chapter 13), but with a continuous sediment seal overhead, the circulation becomes isolated from the overlying ocean (Snelgrove and Forster, 1996; Chapter 6). The average conductive heat flux through the seafloor then correctly reflects the deep-seated lithospheric heat flux (Chapter 10), and geochemical exchange between the crust and ocean is greatly reduced or eliminated (Chapter 21). This basic conceptual model has been fully verified in the three decades since its inception, and observational and modeling efforts have been focused primarily on understanding better the hydrologic structure of the crust, and quantifying material properties, rates and distributions of flow, and chemical exchange between the crust and the oceans.

   Figures 8.1 and 8.2 demonstrate how two important parameters, permeability and topography of a confined permeable layer, influence the hydrologic regime, and illustrate how observations of temperature and pressure like those described in this chapter can be used to constrain material properties. Other important factors include depth-dependence or fracture-control of permeability (Williams *et al.*, 1986; Yang *et al.*, 1996), and routes for recharge and discharge (Davis *et al.*, 1999; Becker *et al.*, 2000; Rosenberg *et al.*, 2000a). Discussions

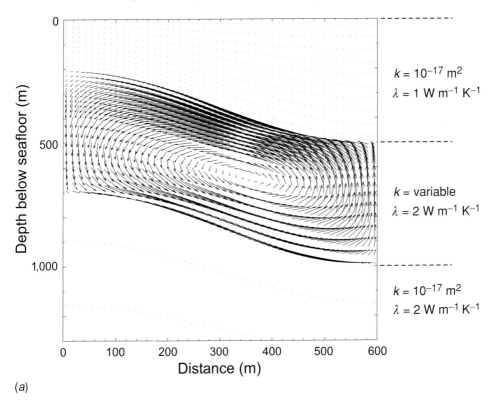

(*a*)

Fig. 8.1 Convection simulation results showing general nature of circulation in a 500-m-thick permeable basement layer bounded above and below by low-permeability sediments (averaging 350 m thick) and rock, respectively, and supplied with a constant heat flux from below (250 mW m^{-2}). Values of permeability and thermal conductivity are shown in (*a*). Cases are considered in domains with no basement relief, and with 300-m-amplitude basement topography as shown in (*a*). The 600-m horizontal domain dimension was chosen to limit flow to a single convection cell (see flow vectors), so that the effects of only topography and permeability could be examined. Vertical profiles of temperature (*b* and *d*) and pressure (*c* and *e*) are shown for the topographic domain at low and high permeabilities (see Fig. 8.2, solid points) at the right (up-flow) and left (down-flow) ends of the domain. Pressures are shown relative to the local hydrostatic pressure computed from the local geotherm, and thus represent pressure gradients available to drive flow in the vertical direction. Simulations were computed by Jiangheng He (pers. comm., 2002) following the scheme presented in Wang *et al.* (1997).

of modeling approaches and reviews of efforts to investigate these parameters are provided in Chapters 7, 11, and 12.

At "low" permeabilities (i.e. at near-critical Rayleigh number conditions as defined in Chapter 12), thermal buoyancy forces are relatively large, but flow is slow, and the thermal state differs only modestly from conductive conditions (Figs. 8.1*b* and 8.2). Cellular convection, signaled by temperature and pressure variations along the upper basement surface, begins when the Rayleigh number (the ratio of the buoyancy force to flow resistance, and

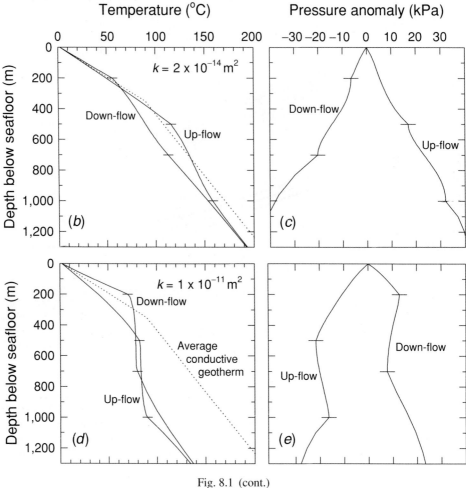

Fig. 8.1 (cont.)

thus proportional to permeability) reaches a critical value of 40 ($4\pi^2$ Fig. 8.2; Chapter 12). Topography or other heterogeneities (probably always present in nature) stimulate sub-critical flow (Lowell, 1980), and reduce the abruptness of the transition from sub-critical to critical conditions. The geometry of flow is influenced by the thermal structure that arises from the combination of basement topography (with a tendency for up-flow to be stimulated beneath ridges) and insulative sediment cover (which causes warmer temperatures, and hence up-flow, to be located beneath the thick sections in sediment-filled valleys).

With increasing basement permeabilities, increasing rates of flow cause basement temperatures to become lower on average and more homogenized. Pressure variations also diminish. At very high permeabilities ($\sim$100 times critical), lateral variations in both temperature and pressure along the upper boundary of a flat permeable horizon become very small (Davis *et al.*, 1996; Fig. 8.2*b*), making sub-critical and high Rayleigh number situations difficult to discriminate without deep observations.

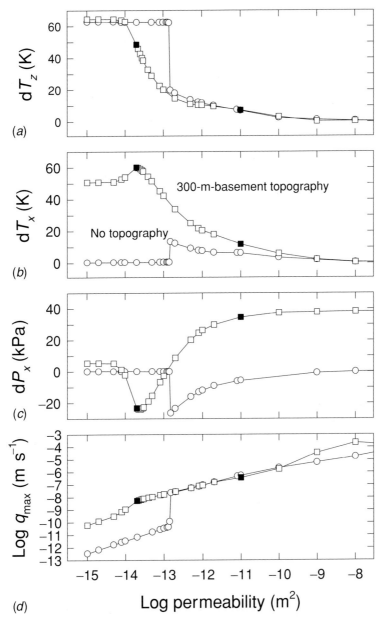

Fig. 8.2 Summary of simulation results in flat and topographic domains described in Fig. 8.1, showing effect of basement-layer permeability on hydrothermal structure and fluid flow, including: (*a*) the difference between the average of temperatures along the upper and lower boundaries of the permeable layers (inversely proportional to the convective Nusselt number); (*b*) lateral temperature variation along the top of the permeable basement layer determined as the difference between temperatures at the right (valley) and left (ridge) ends of the domains; (*c*) lateral pressure variation along the top of the permeable basement layer determined as the difference between the non-hydrostatic pressures at the right (valley) and left (ridge) ends of the domains; (*d*) maximum fluid flux, q. Solid points correspond with results plotted in Fig. 8.1*b*–*e*. The onset of convection occurs at a Rayleigh number ≈ 40 as defined by fluid properties at the base of the permeable layer.

The presence of basement topography and associated sediment thickness variations allows low and high Rayleigh number conditions to be more easily distinguished: As in the flat-layer case, vigorous convection at high permeabilities causes temperatures to be homogenized, but this leads to large heat flux variations at the seafloor that are correlated inversely with sediment thickness. At low permeabilities, heat flux variations generated by refraction effects are small, and uppermost basement temperature variations are large (Fig. 8.2*a,b*). Pressure variations are also diagnostic. At low permeability, the effects of cellular circulation dominate, creating high pressure at the top of the crust/base of the sediment section above rising convective limbs (Fig. 8.1*c*). At high permeabilities, pressures available to drive flow within basement (i.e. deviations from local hydrostatic pressure gradients defined by the thermally controlled density along the path of flow) are small, and the "chimney effect" created in warmed basement edifices becomes large. This elevates pressures relative to the hydrostats defined through the overlying sediment (and thus causes fluids to be driven upwards through the sediment section) above high-standing basement regardless of the direction of circulation beneath (Davis and Becker, 1994; 2002; Davis *et al.*, 1997a; Figs. 8.1*c,e* and 8.2*c*). Under such conditions, observations that describe conditions at the top of a highly permeable upper igneous crust become insensitive to the vertical distribution of permeability and extent of convection in basement. In the case of the layer itself, a reduction in thickness can be balanced by an increase in permeability to create an equivalent degree of thermal homogeneity and uniformity of upper crustal non-hydrostatic pressure (Davis *et al.*, 1997a). If sufficiently permeable, a layer of any thickness can mask signals from below, and effectively decouple flow through sediments and flow deep in the crust. Thus, a broad range of observations both at the top and within the igneous section is required to characterize fully the magnitude and distribution of permeability in the crust.

8.2 Observations and estimates of steady-state temperature

8.2.1 Inferences from seafloor heat flux data

Some of the best constraints on fluid circulation in context of this basic model have been derived from seafloor heat flux transects that were completed with measurements spaced sufficiently close together to characterize the nature of heat flux variability, and positioned in context of local and regional hydrologic structure defined by seismic reflection data. Several such transects are provided in Fig. 8.3 to illustrate the nature of heat flux variability, and, more importantly, the extrapolated uppermost basement thermal regime, in a variety of ocean crustal hydrothermal environments. In these examples, temperature profiles were extrapolated using observed seafloor heat flux and thermal conductivity and seismic velocity measurements on nearby drill cores collected through the full sediment sections (e.g. Wilkens and Langseth, 1983; Davis *et al.*, 1999; Pribnow *et al.*, 2000), although reasonable estimates can be had without such constraints since these key physical properties of semi-to unconsolidated sediments fall within a fairly narrow range.

The first transect (Fig. 8.3*a*) crosses a sediment pond bounded by thinly sedimented basement ridges. Estimated basement temperatures are seen to be highest, over 30 °C, near the center of the pond, 10–20 K warmer than near the edges. Heat flux is highest at the rim of the pond but estimated basement temperatures remain low; the rapid increase in heat flux appears to be only the consequence of the thinning of the sediment seal over a relatively uniform-temperature basement. The average heat flux is only about 20% of that expected from the young lithosphere in this setting. In most respects, the thermal regime of this pond is analogous to that of the well-studied North Pond on the Mid-Atlantic Ridge flank (Langseth *et al.*, 1992), where ventilated hydrothermal circulation cools igneous crust exposed at the seafloor, and lateral advective heat transport in the upper basement section efficiently "mines" the majority of heat supplied conductively deep beneath the pond. It is also probably representative of countless other locations where selective sediment deposition covers only part of the seafloor and leaves abyssal hills and seamounts free of sediment (Lister, 1972; Chapter 6).

Thermal "mining" by lateral advective transport in the crust has been recognized as an important process on the basis of early seafloor heat flow observations (Lister, 1972; Langseth and Herman, 1981; Langseth *et al.*, 1984; Chapter 2), and confirmed by observations of basement fluid compositions which often show strong seawater signatures (Baker *et al.*, 1991; Wheat and Mottl, 1994; Oyun *et al.*, 1995; Chapter 20). An excellent indication of the lateral scale over which this process can operate is revealed by the transect shown in Figure 8.3*b*, which extends from an outcropping basement exposure over an extensive region of sediment-sealed crust. This example effectively constitutes a "semi-infinite" sediment pond. Heat flux is only a fraction (less than 15%) of that expected from the lithosphere near the basement outcrop, and increases toward the full lithospheric value over a distance of roughly 20 km. The thermal effect of ventilation through the outcrop is inferred to extend this far. Estimated temperatures of uppermost basement also increase systematically with distance from the area of outcrop. Observations in boreholes here (15 °C at Ocean Drilling Program (ODP) Site 1023, 23 °C at Site 1024, and 37 °C at Site 1025; Fig. 8.3*b*; Davis *et al.*, 1999; Becker and Davis, 2003) and further along this same ridge-perpendicular transect confirm this trend and document a strong geochemical signature of fresh seawater ventilation over an even greater scale (Davis *et al.*, 1997b; Elderfield *et al.*, 1999; Chapter 20), consistent with the relative importance of thermal diffusion, chemical diffusion, and advection.

The exact nature of the fluid flow responsible for long-distance lateral heat and chemical transport is poorly understood. Models used to account for similar but less precise observations in the Atlantic Ocean were based on simple or dispersive lateral flow (the latter represented by the "well-mixed aquifer" of Langseth and Herman, 1981). This model is meaningful in the case of a sediment pond bounded by outcrops (Langseth *et al.*, 1984, 1992), but not in the case of the semi-infinite pond illustrated in Fig. 8.3*b*, where unidirectional "d.c." flow through basement has nowhere to go in the plane of the transect. Highly dispersive and time-dependent convection must play a strong role in the lateral transport process, although the 20-km scale of the thermal effect becomes enigmatic when the limited

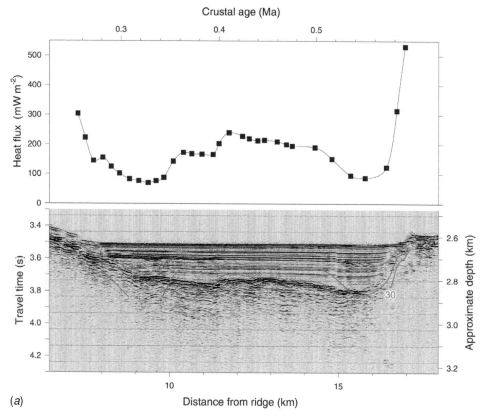

(a)

Fig. 8.3 Seafloor heat flux and estimates of temperatures in the vicinity of uppermost basement along transects *a–g* represent a range of hydrologic "type sections." Temperatures, shown at 10-K intervals, are estimated from seafloor heat flux using sediment physical properties (seismic velocity and thermal conductivity) constrained by drilling data (e.g. Wilkens and Langseth, 1983; Davis *et al.*, 1999; Pribnow *et al.*, 2000), and are meaningful only within the sediment section where advective heat transport is insignificant. Basement relief is dominantly two-dimensional and crossed at right angles by the heat flow/seismic transects, with the exception of seamounts crossed by the transects in (*f*) and (*g*) which are circular. Expected lithospheric heat flux (see Chapter 10) is shown for each transect by dashed lines (off-scale in (*a*)), with no account for the effects of sedimentation. Locations are shown of co-located boreholes (projected in the cases of Sites 504 and 896) where deep temperature and pressure observations have been made. Examples comprise: (*a*) a sediment pond close to the Juan de Fuca Ridge axis; (*b*) a "semi-infinite" sediment pond on the Juan de Fuca Ridge flank, where the igneous crust is blanketed continuously beneath turbidite sediments of Cascadia Basin east of the point of onlap seen at the west end of the transect; well-sealed crust on the flanks of the Juan de Fuca Ridge (*c*) and Costa Rica Rift (*d, e*); Grizzly Bare (*f*) and Baby Bare (*g*) seamounts on the Juan de Fuca Ridge flank. Heat flux data sources: Villinger and Fahrtteilnehmer (1996), (*a*); Davis *et al.* (1997c), (*b, c, g*); Fisher *et al.* (2003), (*f*); and Davis *et al.* (2004) (*d, e*; reproduced with permission of Elsevier Science). Seismic profiles: Rosenberger *et al.* (2000), (*a, b, c, f, g*); Swift *et al.* (1998) and Stephen Swift (pers. comm., 2001), (*d, e*).

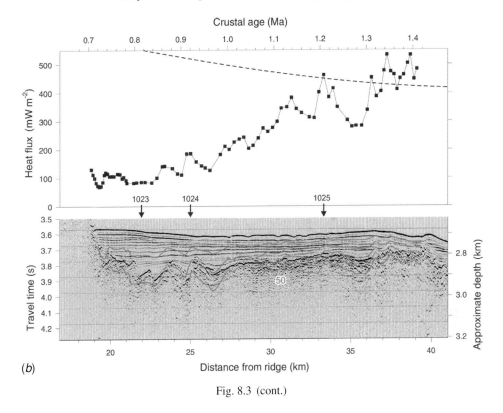

Fig. 8.3 (cont.)

thickness of the crust that is likely to host most of the flow is considered (a few hundred meters or less; see p. 455). Some form of unidirectional transport seems to be required (Davis *et al.*, 1999; Rosenberg *et al.*, 2000; Chapter 11), possibly in a direction orthogonal to this transect.

In addition to the large-scale gradient in heat flux caused by dispersive lateral flow, this transect exhibits another interesting form of heat flow variability: Locally, heat flux can be seen to be correlated with basement topography, or more strictly, inversely with sediment thickness variations. It has been known ever since early seafloor heat flux measurements were made that local variability commonly far exceeds that which would be produced by conductive refraction and sedimentation effects. The relationship between sediment thickness and heat flux began to be recognized as soon as detailed heat flux transects were completed in proper context of seismic structure (e.g. Davis *et al.*, 1989; Davis *et al.*, 1992a; Langseth *et al.*, 1992; Swift *et al.*, 1998), and most local variability could be attributed to the tendency of local convective circulation to maintain the top of igneous basement at a much more uniform temperature than would be the case under conductive conditions. Much better examples of this are seen in Fig. 8.3*c–e*, at locations well away from sites of significant basement outcrop (and thus where the local variability is not

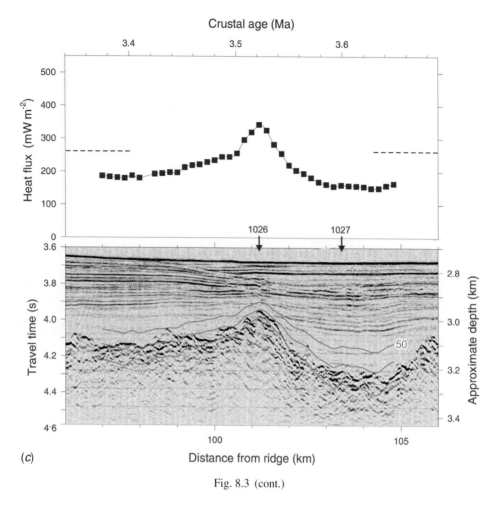

Fig. 8.3 (cont.)

superimposed on a ventilation-induced trend). Isothermal upper basement conditions have been confirmed by temperature measurements in pairs of boreholes that intersect the permeable crust beneath contrasting amounts of sediment along or near these same transects (see p. 455). The degree of isothermality is striking in light of the large (up to 3 : 1) variations in sediment thickness. The estimated thermal structure can be used with a well-mixed aquifer formulation to estimate the effective fluid flux along the undulating basement layer (e.g. Davis and Becker, 2002), and with convective models to estimate the product of the layer thickness and permeability (Davis *et al.*, 1997a; Fig. 8.2*b*; Chapters 7 and 12).

It is important to note that because sediment thickness variations so dominate local heat flux variations as a consequence of the hydrothermal "short circuit" created by the permeable uppermost igneous crust, nothing can be confidently inferred from seafloor heat flux observations about any possible cellular structure of circulation (either in plan or section;

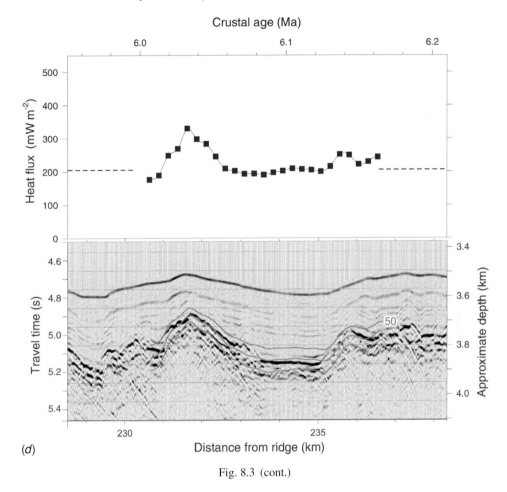

Fig. 8.3 (cont.)

Davis and Chapman, 1996) as was the hope in many studies (e.g. Williams *et al.*, 1974). This problem is compounded by the likelihood that convection in a layer as permeable as estimated from the thermal structure is likely to be unsteady (Wang *et al.*, 1997), and thus difficult to image. Vigorous upper crustal circulation is known also to homogenize fluid compositions (Chapter 20), so detecting signals associated with deep, higher temperature water–rock interactions via geochemical observations is equivalently difficult.

The two final transects of Fig. 8.3 illustrate cases where isolated basement outcrops (seamounts, circular in planform) are surrounded by crust covered by regionally continuous sediments. The seamount in Fig. 8.3*f* appears to be sufficiently large to support some recharge; this is evident in the decrease in heat flux above the buried flank of the edifice. Only at the measurements most proximal to the outcrop (less than 100 m from the point of sediment onlap) does the decrease in sediment thickness cause the heat flux to increase

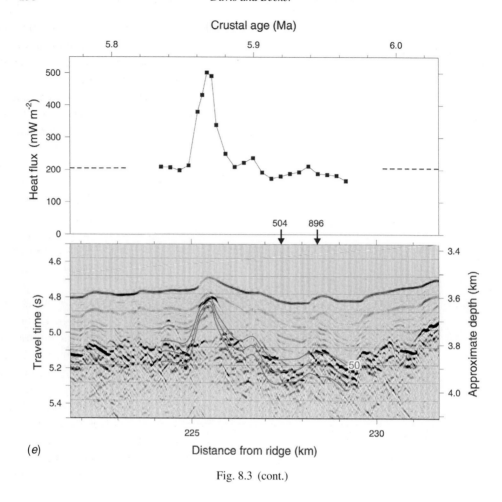

Fig. 8.3 (cont.)

(an effect similar to but smaller than that seen at the points of onlap along the transects of Fig. 8.3*a,b*). The magnitude of recharge must be modest at this site, however, for nothing like the depression of seafloor heat flux seen in Fig. 8.3*b* occurs. After thermal effects of sedimentation are accounted for, the background heat flux only a few kilometers away from the edifice is found to be very close to the expected lithospheric value. Geochemical effects of recharge may be important, however, over distances of many tens of kilometers (Fisher *et al.*, 2003).

Once outcrops become much smaller than this, they appear to serve only as chimneys for discharge. An extreme case is provided in Fig. 8.3*g*, where all but the upper 80-m summit relief of a small seamount is buried by sediment. The sub-seafloor thermal structure appears to be dominated by some combination of vigorous intracrustal convective circulation (the same circulation that creates the structure defined in Fig. 8.3*c*–*e*) and discharge via the exposed edifice (see example simulation in Fig. 12.2). Discharge has been identified at this

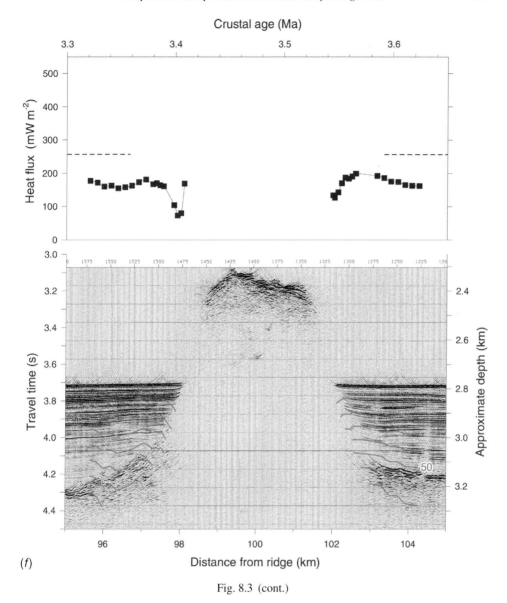

(*f*)

Fig. 8.3 (cont.)

location on the basis of thermal and particulate plumes in the water column (Thomson *et al.*, 1995; Wheat *et al.*, 1997), geochemically anomalous pore-fluid compositions in onlapping sediments, and springs at the seafloor (Mottl *et al.*, 1998; Becker *et al.*, 2000).

While few other sites are as well characterized as these, low average heat flux and large variability are commonly observed (Von Herzen and Uyeda, 1963; Anderson and Hobart, 1976; Anderson *et al.*, 1977; Sclater *et al.*, 1974, 1976; Stein and Stein, 1994; Chapter 10), and thus it is likely that the examples provided in Fig. 8.3 are globally representative.

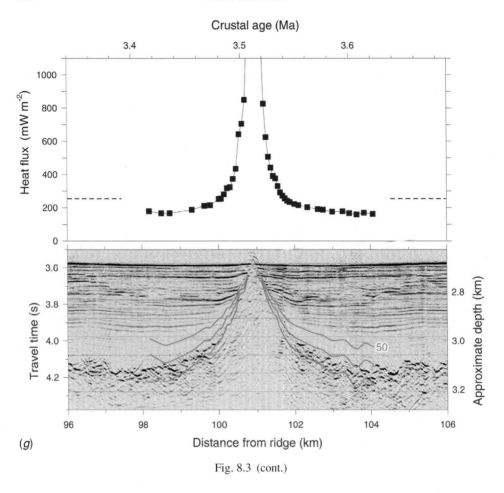

Fig. 8.3 (cont.)

Similar studies involving detailed heat flux observations made in context of simple and well-constrained structure would allow the generality to be confirmed, and constraints to be placed on permeability and volumetric flux in a wider range of lithospheric settings.

8.2.2 Observations in boreholes

Shallow observations

Using "remote" data to estimate sub-seafloor thermal and hydrologic structure is efficient, but some uncertainties are unavoidable, and there are strict limits to which extrapolations apply. Borehole data can eliminate errors associated with extrapolating observed seafloor heat flux, demonstrate whether heat flux is uniform with depth (and provide constraints on pore-water advection if it is not), provide an unequivocal determination of the depth to the top of hydrothermal basement (which may be deeper than seismic basement if massive low-permeability flows or sills occur in the section), and allow observations within permeable units where conductive conditions may not prevail. Within unconsolidated or semi-consolidated sediments, temperatures are measured with sensors mounted in coring shoes or in probes that are pushed in with the drillstring sufficiently far below the last-drilled depth to avoid drilling-induced perturbations that diffuse into the formation below the holes. Observations in lithified sediment or igneous rock require a different approach. Where permeable horizons are not encountered, wireline temperature logs provide good information once drilling perturbations have dissipated, although this can take weeks to months (Chapter 4). Holes intersecting permeable horizons must be sealed to break the connection and stop flow between the formation and the ocean. Instrumented CORK (Circulation Obviation Retrofit Kit) borehole seals have been developed to stop open-hole flow and allow observations of temperature and pressure to be made long after flow-induced transients dissipate. CORK seals can be installed within casing at the seafloor (Fig. 8.4*a*), or at multiple levels within the formation to isolate several zones (Fig. 8.4*b*). Sensors and samplers can be left to monitor temperature, pressure, fluid composition, strain, and seismic motion over periods of many years (Davis *et al.*, 1992b; Mikada *et al.*, 2002).

Data from two CORKed sites (Fig. 8.5) illustrate the magnitude of the thermal perturbation commonly created by open-hole flow. These holes penetrated underpressured and overpressured extrusive sections, and rapid down- and up-hole flow produced nearly isothermal conditions prior to installation of the CORK seals. The magnitude of perturbation to the natural temperature profiles can be used to estimate the rate of flow and, with determinations of pressure differentials, the permeability of the section that produces or accepts water (Becker *et al.*, 1983a,b; Gable *et al.*, 1989; Fisher *et al.*, 1997; Becker and Davis, 2003; Chapter 7).

After being sealed for several weeks, these holes equilibrated to a state that is common in young (< 10 Ma), sediment-covered oceanic crust. Heat flux is uniform through the sediment section, verifying that any seepage through the sediment sections is thermally insignificant (≤ 1 cm yr^{-1}), and it agrees well with co-located seafloor probe values, lending confidence to extrapolations from shallow data to the base of the sediment section (Davis *et al.*, 2003). In uppermost basement, gradients are observed to be lower than those in the overlying sediments by more that can be accounted for by the 2 : 1 contrast in thermal conductivity between basalt (~ 1.8 W m^{-1} K^{-1}) and sediment (~ 0.9 W m^{-1} K^{-1}). The reduced basement gradients are probably a consequence of convection in the uppermost extrusive section.

Borehole observations like these have improved the accuracy with which uppermost basement temperatures can be determined, and this has been proven to be particularly useful in the case of pairs of holes located along the transects shown in Figs. 8.3*c* and *d*, where extrapolations from seafloor heat flux measurements suggest that upper basement temperatures are locally homogenized by vigorous flow (see previous section). Deep sediment probe measurements and CORK observations yield upper basement temperatures that differ by less than 2 K (59 versus 57.5 °C in paired Holes 504B and 896A on the Costa Rica Rift flank, and 64.5 and 63 °C in Holes 1026B and 1027C on the Juan de Fuca Ridge flank)

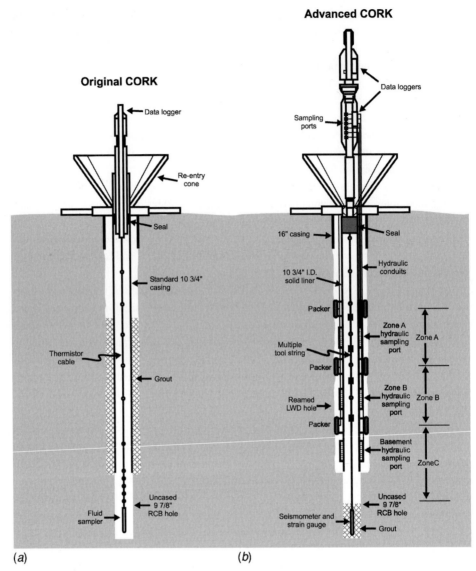

Fig. 8.4 Schematic diagrams of CORK borehole hydrologic observatories. The original configuration (*a*) allows formation pressure to be observed in a single interval spanned by open hole below casing or by a perforated section of casing (Davis *et al.*, 1992b). More recent developments allow multiple zones to be isolated and monitored; three zones are shown in (*b*). The multi-interval system shown here is drilled into the formation (Mikada *et al.*, 2002); another configuration allows installation in previously drilled holes by wireline with standard oceanographic vessels (Spiess *et al.*, 1992).

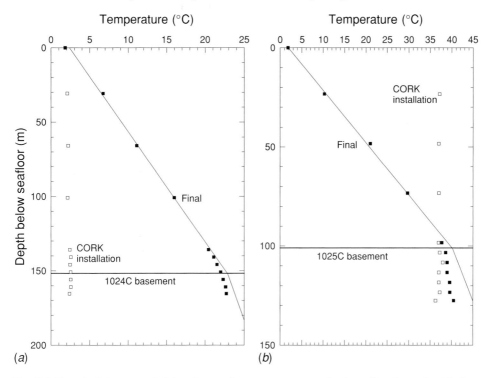

Fig. 8.5 Perturbed (open symbols) and natural-state temperature–depth profiles determined before and long after holes at ODP Sites 1024 (*a*) and 1025 (*b*) were sealed with CORK observatory instrumentation (see Fig. 8.3*b* and Table 7.1 for locations). Perturbed profiles like these constrain rates of open-hole flow (draw-down in Hole 1024 and production in Hole 1025) which, when combined with determinations of pressure differentials, allow formation permeability to be estimated (Becker and Davis, 2003; Chapter 7). Equilibrium temperatures are consistent with conductive heat transfer through the sediment section, and dominantly advective transfer in uppermost basement (data from Davis and Becker, 2002). The sediment/basement thermal interface is probably "smeared" in both examples by internal borehole convection. Reproduced with permission of Elsevier Science.

despite substantial separation (1.2 and 2.2 km, respectively) and large contrasts in sediment burial (1.5 : 1 and 2.5 : 1, respectively) between the neighboring sites of each pair. In the case of the Juan de Fuca pair, temperatures would differ by 40 K under conductive conditions (Davis *et al.*, 1997a). These observed temperature differences are considerably smaller than those that can be resolved by extrapolation of seafloor heat flux, and they provide improved constraints on both rates of flow and permeability in uppermost basement (Fig. 8.2*b,d*; Chapters 7 and 12).

It should be noted that while the determination of the temperature of the sediment–basement boundary is greatly improved with CORK data, uncertainties remain. Within the uncased sections of the holes, flow may occur between permeable igneous units that were isolated from one another prior to drilling (Davis and Becker, 2002). Within cased or naturally impermeable sections, convection within the borehole can blur otherwise sharp thermal boundaries. Given the typical diameter of Deep Sea Drilling Project (DSDP) and Ocean Drilling Program (ODP) boreholes (20–30 cm) and normal geothermal gradients, vigorous convection is inevitable (Gable *et al.*, 1989; Fisher and Becker, 1991), and heat can be carried many hole diameters toward and/or away from a boundary. This effect is probably the cause of the smooth transitions observed in the equilibrated profiles across the sediment–basement interface in Fig. 8.5, where gradients are reduced above the boundary and increased below. Thus the *in situ* CORK observations provide only a lower limit for the temperature at the sediment–basement interface and an upper limit for the natural thermal gradient in uppermost basement. A better estimate of the upper basement temperature is probably provided by the extrapolation of temperatures observed more than about 10 m from the boundary.

Deep observations

Observations deeper in oceanic crust are rare but very important for the reason discussed previously, i.e. that the existence of a highly permeable shallow horizon does not allow inferences to be made about conditions at depth without *in situ* observations. A reasonable number of holes penetrate oceanic crust (those where hydrologic observations have been made are listed in Table 7.1), but only two provide reliable thermal data below the uppermost part of the igneous crust: DSDP Hole 395A, which penetrates 93 m of sediment and 571 m of the extrusive igneous section (pillow basalts and flows), and Hole 504B, which penetrates 275 m of sediment and 1,836 m of extrusive and intrusive (sheeted dikes) igneous rock. They were begun over two decades ago (395A in December 1975 and 504B in October 1979), and have been revisited several times since for deepening in the case of 504B, for open-hole measurements, and most recently for CORK installations (at 395A in August 1997 and at 504B in August 2001).

Hole 395A is situated near the edge of an isolated sediment pond on the western flank of the Mid-Atlantic Ridge. As in the example shown in Fig. 8.3a, the average heat flux through the sediments of this pond ($\sim$40 mW m^{-2}) is only a small fraction of that expected from the underlying 7.2 Ma lithosphere (roughly 185 mW m^{-2}). The low heat flux and cool upper basement temperatures have been attributed to the effects of ventilated hydrothermal circulation (Langseth *et al.*, 1984, 1992). The hole has been re-entered four times for logging since it was drilled (1981, 1986, 1990, and 1997); these and the CORK temperature data (collected most recently in 2001) provide valuable evidence for the hydrologic state of the crust at this site (Fig. 8.6a). At all times prior to when the hole was sealed, the thermal state of the uppermost 350 m of the crust was dominated by downhole flow (Becker *et al.*, 1984; Kopietz *et al.*, 1990; Gable *et al.*, 1992), which was measured at a rate of 2300 l h^{-1} (Morin *et al.*, 1992). Below this depth, the flow became undetectable, consistent with the low permeability measured deep in the hole (Hickman *et al.*, 1984; Becker, 1990), and the thermal gradient increased to a level that appeared to be appropriate for the lithosphere (Gable *et al.*, 1992). Downhole flow was finally stopped with a CORK installation in 1997,

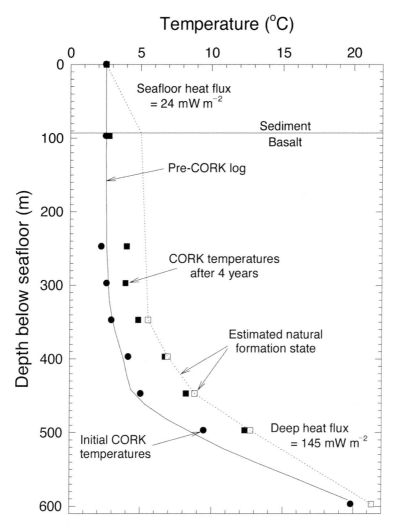

Fig. 8.6 (*a*) Temperature–depth profile at DSDP Site 395 on the western flank of the Mid-Atlantic Ridge (see Table 7.1 for location), showing open-hole logging and initial, late-stage, and estimated equilibrium CORK observations (data from Becker *et al.*, 2001; Becker and Davis, unpublished data). The unperturbed profile through the sediment section is estimated by extrapolating the closest seafloor heat flux measurement reported in Langseth *et al.* (1992). Heat flux in the deepest 150 m of the hole is calculated from the estimated natural-state temperature gradient and the formation thermal conductivity estimated by Hyndman and Salisbury (1984).

but because flow into the formation via the open hole had persisted for more than 20 years, thermal recovery proceeded very slowly. This is particularly true at shallow levels where flow of cold seawater down the hole or into the formation was rapid, but even below 350 m in the crust, significant thermal recovery is evident (Fig. 8.6*a*). This indicates that some pre-sealing downward flow occurred all the way to the bottom of the hole, which is

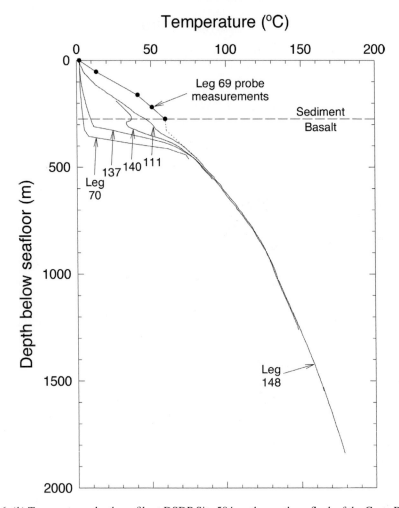

Fig. 8.6 (*b*) Temperature–depth profile at DSDP Site 504 on the southern flank of the Costa Rica Rift (see Fig. 8.3*e* and Table 7.1 for location), showing sediment probe observations, and several episodes of open-hole logging observations, each completed immediately prior to new drilling operations during the legs indicated. (Data from Becker *et al.*, 1983a; Gable *et al.*, 1989, 1995; Guerin *et al.*, 1996. Reproduced with permission of Elsevier Science.)

consistent with geochemical observations made by Gieskes and Magenheim (1992) who found bottom-seawater in the entire borehole, and no evidence of formation water or of seawater–rock interaction.

Recovery in the upper part of basement is so slow that extrapolations of the four-year CORK records to equilibrium values are not reliable. Natural-state temperature at the top of the crust is best constrained by extrapolation through the sediment section from a nearby seafloor probe measurement (Langseth *et al.*, 1992). Deeper in the hole where only small quantities of seawater have invaded the formation, time-extrapolations of the CORK records

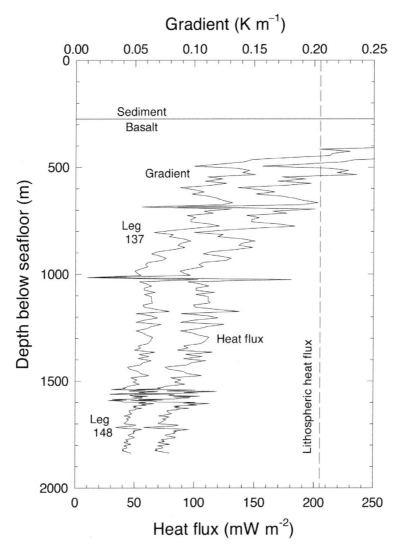

Fig. 8.6 (*c*) Temperature gradient and heat flux in Hole 504B calculated from temperature logs (*b*) and shipboard determinations of thermal conductivity (Alt *et al.*, 1993) corrected for effects of temperature using sensitivity summarized by Zoth and Haenel (1988). Reproduced with permission of Elsevier Science.

are probably accurate to within about 0.5 °C. The combination of these results (Fig. 8.6*a*, open symbols) shows the natural thermal gradient through the upper 300–350 m of the igneous section to be very low, and defines the degree and extent to which mixing (presumably by a combination of flow dispersion and convection) in this "well-mixed aquifer" operates. Although to a far lesser extent, the heat flux deep in the hole is also depressed relative to that expected from the local lithosphere (by roughly 20%). Thus, in addition to evidence for some pre-CORK downhole flow, an influence of thermally significant natural fluid circulation is suggested there as well.

A much deeper view of the thermal state of oceanic crust is provided by observations in Hole 504B, drilled into continuously sedimented 6 Ma crust on the southern flank of the Costa Rica Rift. As in the case of Hole 395A, temperature logs have been completed on several occasions, in this case immediately prior to drilling operations to deepen the hole (DSDP and ODP Legs 70 in 1979, 85 in 1982, 111 in 1986, 137 and 140 in 1991, and 148 in 1993). Also, like Hole 395A, this hole is situated where the crust is underpressured (see Section 8.3), and flow down the hole and into the uppermost permeable part of the crust persisted to varying degrees throughout the 22 years prior to CORK sealing (Becker *et al.*, 1983a,b; Gable *et al.*, 1995; Chapters 4 and 7).

The perturbing thermal effects of this downhole flow are most apparent in the uppermost 100 m of the igneous section (Fig. 8.6*b*) where high permeabilities have been measured (Anderson and Zoback, 1982; Becker, 1996; Chapter 7). Long-term CORK data will eventually help to constrain the natural thermal structure; the combination of existing sediment probe and logging observations suggests that only a few tens of meters of the uppermost crust at this site is characterized by a low vertical thermal gradient as observed at the Juan de Fuca and Mid-Atlantic sites (Fig. 8.6*b*). The permeability of this part of the section must be sufficiently high to allow rapid lateral flow and creation of the small lateral temperature variations observed at the top of basement (compare Fig. 8.2*b* with Fig. 8.3*d* and *e*). This flow may be promoted by large permeability anisotropy in the sub-horizontally layered extrusive igneous sequence (Fisher and Becker, 1995, 2000).

Deeper than about 250 m in the crust, permeability measured in the borehole is too low to permit significant downhole flow (Becker, 1989). Stagnant conditions have been confirmed by geochemical observations which show significant borehole-water–rock reaction during the long intervals between drilling and subsequent re-entry and fluid sampling operations, and no fluid interchange between the formation and the borehole (Magenheim *et al.*, 1992). Given the hole diameter and the geothermal gradient, convection within the borehole is likely (Sammel, 1968; Beck and Balling, 1988; Gable *et al.*, 1989; Fisher and Becker, 1991), but this will not cause the thermal gradient within the hole (when averaged over multiple cells that are characteristically about three hole diameters in vertical dimension) to differ from the gradient in the surrounding formation (Magenheim and Gieskes, 1996). Thus, at depths greater than about 250 m in the crust, the thermal state of the hole determined by the logging data probably represents the natural formation state accurately.

As at Hole 395A, the thermal state of the deeper crustal section at Hole 504B appears to be influenced by thermally significant natural flow in the formation. This is apparent in the temperature profile, which shows a generally continuous decrease in gradient with depth, and is quantitatively evident in plots of thermal gradient and heat flux versus depth (Fig. 8.6*c*). The latter shows a systematic reduction from a value near that expected for the local lithosphere just below the upper permeable zone (equal at this continuously sedimented site to the local area average heat flux conducted through the sediments overhead as determined by Langseth *et al.* (1988); see also Fig. 8.3*d* and *e*) to about 40% of the expected value in the deepest part of the hole. This decrease with depth has been noted previously (Becker *et al.*, 1983a,b; Gable *et al.*, 1989; Fisher and Becker, 1991; Guerin *et al.*, 1996), but not satisfactorily explained. The conclusion that the profile is a consequence of deep convective circulation in the formation seems unavoidable (Davis *et al.*, 2004), and if this is true, two additional conclusions follow. First, convection must extend to a depth much greater than that penetrated by the hole, since there is no sign of the heat flux increasing at the bottom of the hole back toward the expected lithospheric value (compare data in Fig. 8.8 with profiles in Fig. 8.1*b*, with the depth scale of the latter expanded by an order of magnitude). Second, the average permeability of the section below about 250 m in basement must be higher than that measured in Hole 504B itself. For conditions to be convectively super-critical (Chapter 12) and permit a thermal perturbation as great as observed (Fig. 8.2) the permeability would need to exceed about 10^{-15} m^2, whereas measured values average about 10^{-17} m^2 (Becker, 1989; Chapter 7). It is likely that formation permeability is created primarily by interconnected fractures; these must be sufficiently close together for the formation thermally to "look" like it is influenced by wholesale fluid flow, yet far enough apart for them to be under-represented in the borehole itself. In effect, the permeability must be scale dependent, much in the way that it has been inferred to be in the uppermost part of the crust (Davis *et al.*, 1997a; Becker and Fisher, 2000; Fisher and Becker, 2000; Becker and Davis, 2003; Chapter 7).

8.3 Observations of steady-state pressure

As in the case of determining sub-seafloor temperature, the simplest way to estimate formation pressure at depth is via vertical extrapolation of observations made with gravity-driven probes. Unfortunately, it is not practical to use wireline-tethered instruments because of the relatively long time constant that characterizes the dissipation of the penetration transient (Davis *et al.*, 1991). Typical hydraulic diffusivity of fine-grained seafloor sediments is much smaller than thermal diffusivity, and the time required for determining unperturbed pressures requires measurements spanning many hours or even days, not minutes as in the case with heat flow probes. Added to this is the requirement to account for the effect of tidal loading which generates pressure variations that depend on depth and on elastic and hydrologic properties (see Section 8.4.1 and Chapter 7). This can be done only by acquiring data that span many tidal cycles.

Both of these problems are overcome with free-fall instruments that can be left undisturbed for many days (e.g. Fig. 8.7*a*; Schultheiss and McPhail, 1986; Schultheiss, 1990). Data (e.g. Fig. 8.7*b* and *c*) are typically characterized by penetration transients that are long lived (several-hour time constant), and very large (several tens of kilopasacals) relative to the resolution required to determine local pressure gradients (in this instance non-hydrostatic pressures are only -0.3 and -1.2 kPa at the ports positioned approximately 1.5 and 3.0 m below the seafloor). Tidal loading effects are also present. This example makes it clear that extrapolations of early time data cannot be accomplished with adequate accuracy to resolve natural gradients; several days of data are essential.

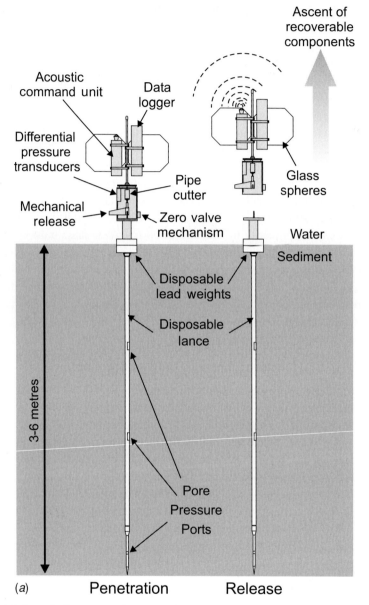

Fig. 8.7 (*a*) Schematic illustration of a gravity-driven pore pressure probe. The probe is deployed in free-fall mode, and the data logger and sensors are recovered by acoustic command (after Schultheiss, 1990).

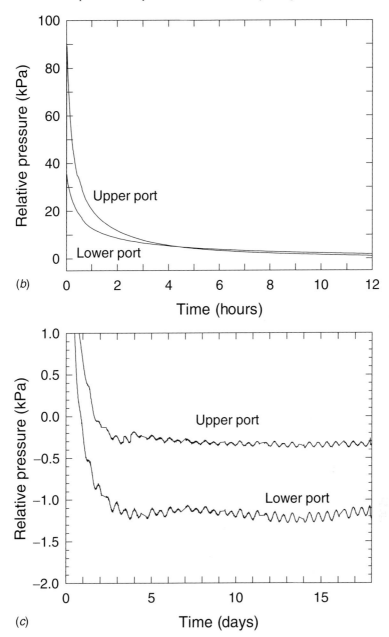

Fig. 8.7 (*b, c*) Differential pressure data (example shown is from deployment "A" near a sediment-hosted hydrothermal vent; Fouquet *et al.*, 1998) from the probe illustrated in Fig. 8.7*a*, showing early (*b*) and late (*c*) parts of the records. Data are characterized by large insertion transients and slow decays; the rates of decay depend on sediment permeability. Equilibrium conditions are reached only after several days. The attenuation and phase of tidal signals relative to absolute pressure variations at the seafloor (not recorded in this instance) depend on both the elastic and hydrologic properties of the sediment.

Even with highly accurate seafloor determinations, however, extrapolation to conditions at depth is difficult. As with thermal extrapolations, pressure must be extrapolated against hydraulic resistance, not depth, but unlike thermal conductivity, which varies over a small and normally well-constrained range, sediment permeability can vary with lithology and consolidation state by orders of magnitude (Chapter 6). Thus, the only way deep formation pressure can be determined with confidence is with *in situ* observations.

Pressures in sediment-covered igneous sections were first estimated from borehole pressures recorded when using drillstring packer seals for permeability testing (e.g. Anderson and Zoback, 1982; Hickman *et al.*, 1984). The accuracy of such estimates is severely limited, however, by thermally generated drilling perturbations that arise from the density contrast between the water filling the hole at the time of the measurement and the natural-state formation water. Often the thermal state is poorly known, particularly if a significant volume of cold seawater has invaded the formation penetrated by the hole, so it is difficult to account for this effect accurately.

The CORK sealed-hole instrumentation described in Section 8.2.2 provides a way to avoid these problems by allowing pressures to be observed for a length of time sufficient for the borehole and formation to return to natural-state conditions. In an extreme example shown in Fig. 8.8, the combination of anomalously high (cold) borehole pressure during drilling and low natural formation pressure caused a large volume of cold water to be both driven and drawn into the formation before the hole was sealed. Interconnection of originally isolated horizons may also have contributed to this particularly large and long-lived perturbation (Davis and Becker, 2002). Some recovery is evident more than two years after sealing (Fig. 8.8), despite the relatively short time (five days) the hole was open prior to sealing. Most holes sealed immediately after drilling take only a few weeks to equilibrate (the case for Holes 1024C and 1025C discussed in Section 8.2.1); holes sealed long after drilling (e.g. Holes 395B, 504B, and 896A) require an expectably much longer time.

Guided by the conceptual model discussed in Section 8.1, a number of conclusions have been reached on the basis of the observations of stabilized formation pressures in these and other holes that penetrate sediment-covered ocean crust. Because there is little hydraulic resistance to lateral flow within the high-permeability section of uppermost basement (discussed below), and because circulation is sufficiently rapid to keep high-standing basement edifices warm, the chimney effect created by buried basement topography is strong. Relative to local geotherm hydrostats, basement is consistently found to be overpressured at sediment-buried ridges or edifices, and underpressured in valleys, stimulating upward and downward pore-fluid seepage through the respective overlying sections (Figs. 8.1*e* and 8.9; Mottl, 1988; Davis *et al.*, 1992a; Wheat and Mottl, 1994). Where the sediment cover is breached, rapid flow can occur at warm springs (Fig. 8.3*g*; Thomson *et al.*, 1995; Mottl *et al.*, 1998; Becker *et al.*, 2000). Lateral pressure gradients may also be generated within the sediment section, and when integrated over long periods of time, both the vertical and lateral seepage may promote alteration. Normally, flow will be driven away from edifices into the surrounding sediment section, but in cases where there is a low-loss hydrologic pathway to the seafloor (e.g. at an isolated outcrop), the opposite sense of flow may occur.

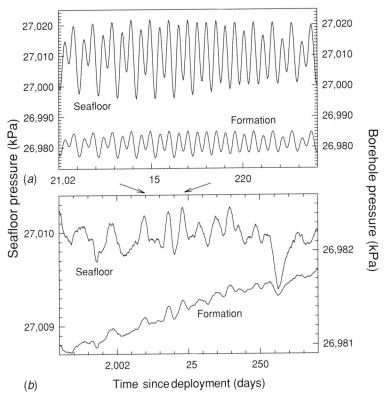

Fig. 8.8 CORK pressure and temperature data from Hole 1027C (see Fig. 8.3*c* and Table 7.1 for location), including: (*a*) a two-week raw record dominated by tides; (*b*) a three-month record with dominant diurnal and semi-diurnal tidal constituents removed, showing barometric and oceanographic signals and the recovery trend of the formation pressure following drilling perturbations; and (*c, d*) multi-year records of pressure (with seafloor loading effects removed) and temperature (at 177 m below seafloor within the sediment section, at 587 m in an intrasedimentary sill, and at the top of basement at 612 m). Long-term recovery from open-hole perturbations and response to tectonic events can be seen in both pressure and basement temperature records. Reproduced with permission of the American Geophysical Union (*a, b*) and Elsevier Science (*c, d*).

A good example of this has been observed on a small scale in the Middle Valley sedimented rift (Schultheiss, 1997; Fouquet *et al.*, 1998; Stein and Fisher, 2001), where a locally uncon-fined hot hydrostat beneath a hydrothermal vent creates a local low-pressure anomaly that draws seepage into the cooler surrounding section and toward the vent. In fact, the negative pressure gradient shown in Fig. 8.7 was observed in this setting.

Within the upper igneous crust, pressure differences available to drive flow laterally are very small, nearly too small to be resolved. This is shown for the case of Holes 1026B and 1027D in Fig. 8.9, where equilibrated formation pressures (shown relative to hydrostats calculated for local geotherms) are compared. Assuming the holes are connected along an isothermal upper basement pathway, as justified by both seafloor heat flux and borehole data

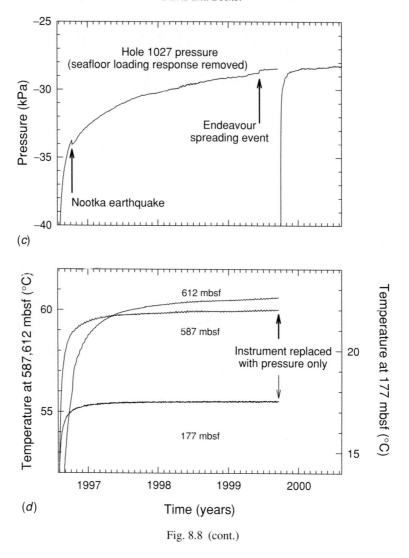

Fig. 8.8 (cont.)

(Fig. 8.3c; Section 8.2), the pressures at the two locations are seen to differ by less than 2 kPa despite their separation of over 2 km. In light of the simple model results summarized in Figs. 8.1 and 8.2, both temperature and pressure observations are seen to be consistent in suggesting that the permeability of hydrothermal basement between the sites must be very high. A similar conclusion has been reached with temperature and pressure data from Holes 857D and 858G in a much younger crustal setting (Davis and Becker, 1994; Stein and Fisher, 2001). Data from recently instrumented Hole 504B provide complementary information in a somewhat older crustal setting, where borehole data as well as seafloor heat flux observations show laterally homogeneous upper crustal temperatures (Becker and Davis, 2004; Davis et al., 2004, Fig. 8.3d,e).

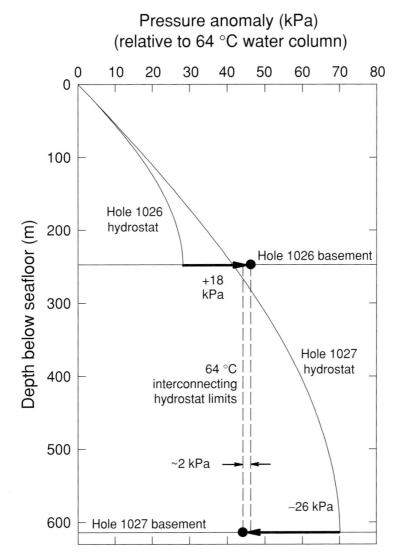

Fig. 8.9 Synthesis of natural-state basement pressure observations in Holes 1026B and 1027C (locations given in Fig. 8.3c and Table 7.1). Observed pressures (filled circles) are shown relative to the local geotherm hydrostats, which are in turn shown relative to a hydrostat defined by 64 °C water (the temperature along the upper basement pathway connecting the two holes). The difference in pressures defined in this manner provides an estimate of the pressure available to drive flow laterally between the sites, and is only of the order of 1 kPa, no larger than the uncertainties in the relative pressure determinations. Reproduced with permission of Elsevier Science.

8.4 Time-dependent signals

8.4.1 Tidal pressure variations and tidally driven flow

In addition to the steady buoyancy-derived pressures that are available to drive steady circulation within the oceanic crust and exchange of water between the crust and ocean above, significant temporal variations of pressure are revealed by long-term observations. The most obvious pressure variations arise from the poroelastic response to variable loads at the seafloor imposed by ocean tides and oceanographic and atmospheric sources (Fig. 8.8). The response has both elastic (in-phase) and diffusive (phase-shifted) components that depend on the compressibilities of the rock or sediment matrix and interstitial fluid, and on the hydraulic diffusivity (Van der Kamp and Gale, 1983; Fang *et al.*, 1993; Wang and Davis, 1996; Chapters 7 and 12).

In sediments the matrix is compliant, and the majority of any load is born by the pore fluid. This is particularly true for unconsolidated sediments near the seafloor. The small differential pressures seen in the shallow probe data in Fig. 8.7*c* show the seafloor loading signal to be attenuated by only a few percent. Diffusion may also contribute to making differential pressures small, for while the characteristic scale of diffusion in fine-grained sediment is small at tidal frequencies, diffusion may be significant over the few-meter scale of the shallow probe observations (Wang and Davis, 1996). At greater depths below the seafloor (or distances from lithologic boundaries), the response depends only on the instantaneous "loading efficiency" (the degree to which loads are delivered elastically to the interstitial water). This progressively decreases with increasing consolidation state and temperature.

Within the oceanic crust, the loading efficiency is generally less than that in sediment, and permeability is much greater. As a result, a smaller fraction of the variable seafloor load is born by interstitial water, and diffusion occurs over a much greater distance. Relative amplitudes of tidal pressure variations observed to date range from ~0.6 to 0.15 in a way that depends on crustal age (loading efficiency and inferred compliance decrease with increasing age and alteration) and temperature (higher temperatures result in higher fluid compliance, causing the rock matrix to bear an even greater majority of the load; Fig. 8.10). Where it has been defined, lateral diffusion is significant over a scale of many kilometers at tidal frequencies. The frequency-dependent spatial variations of phase and amplitude can be used to estimate formation-scale permeability and tidally driven flow velocities (Chapter 7; Wang and Davis, 1996; Davis *et al.*, 2000).

Examples of pressure differences that are generated in this manner are shown in Fig. 8.11. In this example, diffusion occurs laterally as a consequence of leakage through igneous rock outcrop 6 and 14 km away from the two observation sites (see Fig. 8.3*b*). Vertical diffusion is prevented by a locally continuous blanket of low-permeability sediment. Lateral flow velocities inferred from the amplitude and phase distribution exceed 10^{-6} m s^{-1}, and permeability is estimated to be $\approx 10^{-10}$ m^2 (Davis *et al.*, 2000; Chapter 7).

Tidal "pumping" will occur everywhere. Water can be driven not only across the seafloor, but also within the crust wherever neighboring volumes of rock possess contrasting elastic

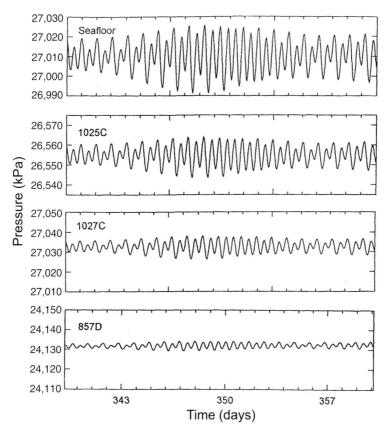

Fig. 8.10 Seafloor and formation tidal pressure variations showing contrasting responses to seafloor loading in young, compliant basement filled with cool (39 °C) water (Hole 1025C), in altered, less-compliant basement filled with warmer (63 °C) water (Hole 1027C), and in highly altered, rigid basement filled with hot (280 °C), compliant water (Hole 857D). See Fig. 8.3c,d and Table 7.1 for hole locations. Reproduced with permission of Elsevier Science.

properties (Wang and Davis, 1996; Davis and Becker, 1999; Wang *et al.*, 1999). An example of localized flow is provided by a record of temperature in Hole 1027C (Fig. 8.12), where flow is driven along a thermal gradient between two igneous units separated by a sediment layer. In this instance, flow occurs via the borehole, although any natural permeable connection between horizons of contrasting elastic properties will host flow in the same manner.

While it is only just beginning to be documented, tidally driven flow is likely to be common and important. The magnitude of the pressure gradients and consequent flow velocities are equivalent to those associated with buoyancy-driven hydrothermal circulation, and the direction of the tidal and buoyancy components will be generally in different directions. Thus, tidal flow will serve as a strong and active dispersive agent, helping to

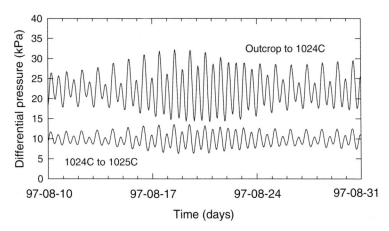

Fig. 8.11 Lateral pressure differential between Holes 1025C and 1024C (separated by 8 km), and between Hole 1024C and the seafloor of an area of extensive basement outcrop (separated by 6 km). The combination of relative phases and amplitudes allow the formation permeability (Chapter 7) and rates of flow driven by the pressure gradients to be estimated (see Fig. 8.3*b* and Table 7.1 for locations). Reproduced with permission of Elsevier Science.

homogenize crustal temperatures, to provide access for water to volumes of rock not reached by buoyancy-driven flow alone, and to stimulate more pervasive crustal alteration than would occur in its absence.

8.4.2 *Hydrologic response to tectonic strain*

Part of the load induced by tectonic strain is also taken up by interstitial water, and several events have been observed in ODP boreholes that demonstrate this. Hydrologic response to co-seismic strain has been observed on land in numerous instances (e.g. Wakita, 1975; Bower and Heaton, 1978; Muir-Wood and King, 1993; Roeloffs, 1996, 1998; Brodski *et al.*, 2003), although quantitative inferences are made difficult by the limitations of the observational techniques (e.g. well-level and stream-flow monitoring), formation heterogeneity, and the difficulty to account for environmental contributions (e.g. precipitation). Sealed ocean crustal boreholes provide much simpler observations. Environmental "noise" (generated by tidal, oceanographic, and atmospheric loading) can be removed effectively because the source is well defined by seafloor pressure data, and the response function is well constrained (see previous section). In addition, boreholes that are cased and sealed through sediments and drilled into underlying permeable igneous rocks represent relatively incompliant "transducers" (particularly when compared to water-level monitoring in open wells) and the signals that they tap represent a large volume of formation. These factors make the observations accurate and insensitive to local heterogeneities (an advantage over standard strain gauges which provide greater absolute resolution and directional information, but are highly sensitive to heterogeneity).

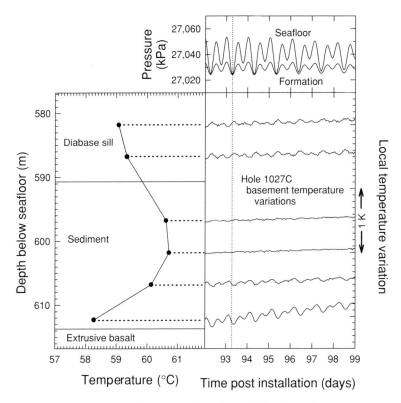

Fig. 8.12 Basement temperature variations resulting from tidally driven flow between two igneous horizons intersected by Hole 1027C (see Fig. 8.3*c* and Table 7.1 for locations). Temperature variations lag pressure by roughly 90°, suggesting that maximum flow occurs at times of maximum rate of change of pressure. The amplitude and sense of the temperature variations depend on the magnitude and sign of the local thermal gradient. Reproduced with permission of Elsevier Science.

Examples are shown in Fig. 8.13 of the transient response to a seismogenic seafloor spreading event on the Juan de Fuca Ridge. The transients are characterized by an initial rapid change in pressure followed by a slow decay back toward the unperturbed state. Pressures increase (or decrease) in response to local compression (or dilatation) at the time of the events. The transients reflect a mixture of the instantaneous (undrained elastic) response to deformation, and a diffusional hydrologic contribution (Ge and Stover, 2000; Davis *et al.*, 2001). The latter begins immediately as a further increase or decrease in pressure that is the result of lateral diffusion from larger pressure anomalies generated closer to the slip dislocations. This is then followed by a decay as the effects of lateral drainage from beneath the sediment section to areas of basement outcrop begin to dominate.

This behavior is illustrated with a simulation of the spreading event that produced pressure transients at four sites on the ridge and ridge axis in Fig. 8.14. The elastic strain generated by a simple dislocation (Fig. 8.14*a,b*) is computed using the method of Okada (1992). The theory used to estimate pressure from volumetric strain is essentially the same as that which

Davis and Becker

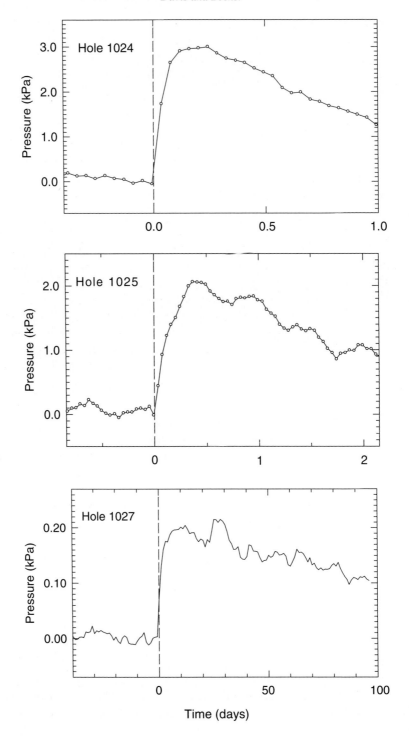

Time (days)

describes the response of pore pressure to seafloor loading (Van der Kamp and Gale, 1983; Wang and Davis, 1996; Section 8.4.1; Chapters 7 and 12). Under undrained conditions (i.e. at early times), the pore fluid pressure increase or decrease is expressed as $P = \beta \sigma_t$, where $\sigma_t = \sigma_{ii}/3$ is the total pressure change, and σ_{ii} are principal stresses. Parameter β is the three-dimensional loading efficiency, that is, the fraction of the total elastic pressure increase supported by the pore fluid (Davis *et al.*, 2001). In a confined situation in which the deformation of the matrix is purely vertical, as is commonly assumed for tidal loading, β reduces to the one-dimensional tidal loading efficiency (Wang and Davis, 1996). Parameters involved in β and in the relationship between stress and strain, namely the bulk modulus and porosity of the matrix frame (Davis *et al.*, 2000), and the bulk moduli of the fluid and the solid constituents (Hamilton, 1971; Keenan *et al.*, 1978), are well constrained, in part owing to the "calibration" provided by the observed tidal loading efficiency (which involves the same parameters plus Poisson's ratio). The resulting coefficient in the linear relationship between pressure and strain, roughly $10\,\mathrm{kPa}\,\mu\mathrm{strain}^{-1}$, results in pressure being a relatively sensitive indicator of strain and allows the elastic dislocation model to be matched quantitatively to the observed pressure transients (Fig. 8.14*b*). In this instance, it was found that the computed elastic strain is much greater than that which would be commensurate with any of the earthquakes in the associated episode of seismicity (the first earthquake of the sequence was of a moment magnitude $M_W = 4.6$, whereas the estimated dislocation would have generated an $M_W = 5.7$ event under conditions of full seismic efficiency), and this led to the conclusion that all earthquakes were effectively "aftershocks" of a discrete but dominantly aseismic spreading event (Davis *et al.*, 2001).

The change of pressure following the elastic response (Fig. 8.14*c,d*) is modeled using the elastic strain response as the initial state, and assuming that only lateral flow takes place in the sedimented crust and that drainage occurs in the region of outcrop near the ridge axis, just as in the case of tidal pressure diffusion discussed in the previous section. The time constant for the decay depends on the distance of individual sites to the sparsely sedimented area near the ridge axis where efficient drainage can occur, and on the formation-scale permeability of basement. In these instances, the drainage time constants (defined arbitrarily as the time taken for pressures to decay to 50% of the maximum amplitude) vary from one to two days at the sites near the axial region to roughly 100 days at the site furthest away (consistent in the latter case with the decay following the Nootka transform fault event visible in Fig. 8.8*c*). These time constants suggest a basement permeability between 10^{-10} and $10^{-9}\,\mathrm{m}^2$, generally consistent with other formation-scale estimates (e.g. from constraints of

---

Fig. 8.13 Pressure transients observed in Holes 1024C, 1025C, and 1027C induced by a spreading event on the adjacent Juan de Fuca Ridge axis (see Fig. 8.14*b* for relative locations). This seismogenic strain event, and another associated with a strike–slip earthquake roughly 120-km distant from Hole 1027C, can be seen in the long-term record shown in Figure 8.8a. Pressure and time axes have been adjusted for each plot to "normalize" the scaling and illustrate the fundamental similarity of the transients. Hourly observations are indicted by open symbols. Reproduced with permission of the American Geophysical Union.

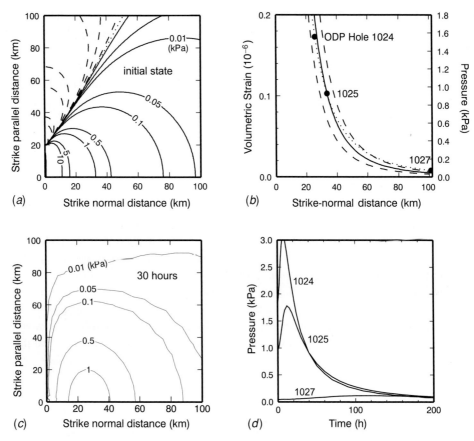

Fig. 8.14 Results of a poroelastic model that simulates the event responsible for the observations shown in Fig. 8.13: (*a*) Distribution of the instantaneous change in pressure (one quadrant shown) resulting from the elastic deformation produced by an extensional dislocation at the "ridge axis" (from −20 to +20 km along the left-hand axis). Pressure is computed from strain in the manner described in the text. Dimensions of the dislocation are constrained by the distribution of epicenters along axis (40 km total) and with depth (roughly 3 km) during a seismic swarm on the Juan de Fuca Ridge axis. The magnitude of extension is adjusted to fit the observed pressures. (*b*) Profile of strain (and estimated pressure; see text) along the horizontal axis of (*a*), computed for a 0.12-m-magnitude extensional dislocation (dashed curves are computed for ±0.4 m extension to show sensitivity), along with observed instantaneous pressures at ODP CORK Sites 1024, 1025, and 1027. (*c*) Pressure field computed for partially drained state (30 hours after strain event), assuming water is constrained to flow laterally and can flow freely through the $x = 0$ boundary (representing the unsedimented ridge axis). (*d*) Pressure vs. time (computed as in (*c*)) for the locations of Sites 1024, 1025, and 1025. The character and time constants of the curves are consistent with the CORK observations (Fig. 8.13) when computed with a permeability of 10^{-9} m^2 (also used for (*c*)). Reproduced with permission of the American Geophysical Union.

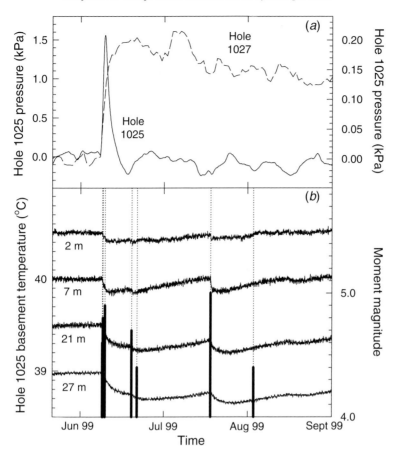

Fig. 8.15 Three-month record of basement pressure (with effects of seafloor loading removed) from the CORKs in Holes 1025C and 1027C (*a*) and temperatures in Hole 1025C (*b*), with times (dashed lines) and magnitudes (solid lines) of seismic events at the ridge axis having $M_W \geq 4.0$. The axis lies roughly 30 and 100 km away from the two sites, respectively (see Fig. 8.14 for relative locations). Temperature-sensor positions are shown as depths below the top of basement. Absolute temperature of the sensor at 21 m is correct; other records are progressively offset by 0.5 K for plotting convenience.

uppermost basement isothermality and of lateral diffusion of pressures at tidal frequencies; Chapter 7).

Just as in the case of tidally driven flow, the flow driven by pressure gradients generated by strain (either between different volumes of basement having contrasting elastic properties, or between semi-confined basement volumes and areas of basement outcrop that allow ventilation) will create a thermal signature wherever flow is sufficiently fast and in the direction of a thermal gradient. Examples where this condition is met are shown in Fig. 8.15. As in the case of the tidally driven flow (Fig. 8.12), it is not possible to ascertain to what degree flow is focused through the low-resistance path of the open sections of the

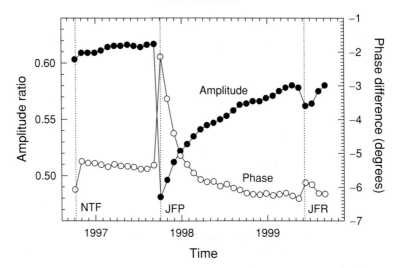

Fig. 8.16 Phase and amplitude of formation tidal signals relative to seafloor loading, calculated for 28-day windows over a three-year observation interval at ODP Hole 1025C, Juan de Fuca Ridge flank (see Fig. 8.3*b* and Table 7.1 for location). Changes in both phase and amplitude of the formation response are seen at the times of three seismogenic slip events in the region. The NTF (Nootka transform fault) and JFR (Juan de Fuca Ridge) events can be seen as pressure transients at Site 1027 further to the east (Fig. 8.8*a*). The JFR and JFP (Juan de Fuca Plate) events, but not the NTF event, are observed as pressure transients at this site. Response at the strongest M2 tidal constituent is shown; similar behavior is resolved at other dominant frequencies (e.g. K1, O1) and at other sites.

boreholes, but it is clear from these observations that flow in the formation is driven for long periods of time after the strain events. It is also noteworthy that thermally resolvable flow often occurs in response to strain too small to be resolved by pressure.

Beyond the edge of our understanding lies a final hydrologic observation that is subtle but potentially instructive: in several instances, both the amplitude and phase of the primary tidal constituents of the pressure variations in basement relative to seafloor loading are seen to change at the time of large strain events, then slowly return to former values over a period of many months (Fig. 8.16). At Site 1025, small changes in response are seen at the times of the Nootka transform fault and Juan de Fuca Ridge events discussed above, and a large change is seen at the time of an earthquake only a few kilometers from the site. Similar changes are observed at Site 1024. Changes of both amplitude and phase of the loading response suggest sensitivity of both elastic and hydrologic properties to strain. In several instances, changes in tidal loading response are observed when the more direct response of pressure to strain is not resolved. For example, the Nootka transform slip event is not resolved as a pressure transient at either Sites 1024 or 1025, whereas it can be seen by way of tidal loading response at both. The cause for this sensitivity is not understood; it may be associated with localized strain amplification by fractures or faults and related changes in hydraulic properties.

8.5 Summary and recommendations

Observations of temperature and pressure generally fall into two categories, those made "remotely" and those made in boreholes. One of the most efficient methods for constraining the hydrologic state of the crust remotely involves detailed transects of seafloor heat flux measurements, made in context of seismic reflection profiles, and a number of illustrative examples are presented in this chapter. Inferences drawn from these observations include: (1) Spatial variations of heat flux that cannot be accounted for by conductive mechanisms (sedimentation, erosion, and refraction) are evident in settings spanning a wide range of crustal ages (Chapter 13). Such variations are a basic indicator for fluid flow. (2) Heat flux near areas of extensive igneous outcrop is commonly less than 20% of the total expected. The degree to which average heat flux through the seafloor under-represents that expected from the cooling lithosphere beneath (Chapter 10) yields a quantitative estimate of the efficiency of hydrothermal ventilation (Chapter 11). (3) Systematic spatial variations in this deficiency reveal the scale over which lateral transport of heat via semi-confined flow beneath sediment cover can occur (Chapters 11 and 12). Effects of lateral flow are observed over distances of 20 km, and fluid fluxes required to carry heat over this distance are ≥ 1 m yr^{-1}. (4) Temperatures estimated at the top of sediment-sealed igneous sections by extrapolation of seafloor data are often found to be locally homogeneous, despite large variations in the thickness of overlying thermally resistive sediments. The estimated degree of thermal homogeneity of uppermost basement can be used to estimate rates of flow (also ≥ 1 m yr^{-1}) and permeability within the upper crustal section ($\geq 10^{-10}$ m^2; Chapter 7). (5) In contrast to extensive crustal outcrops which serve to cool the crust by efficient hydrothermal ventilation, small outcrops appear to serve as chimneys for discharge, drawing water from surrounding sediment-covered crust.

Observations made in boreholes are by nature more difficult, but they provide important information that cannot be inferred from seafloor data. They permit greater resolution of upper basement temperatures, allow pressures to be determined, and provide a means of determining the thermal regime within sections where advective heat transport is important. Several key observations have been made about conditions in the uppermost crust at a limited number of sites penetrated by relatively shallow holes: (1) With their reduced uncertainties, *in situ* observations have shown that the homogeneity of upper basement temperature is even greater than that estimated from seafloor heat flux extrapolation. Lateral variations as small as 2 K are observed where variations of up to 40 K would be present under conductive conditions. This refines the constraints on rates of fluid flow and on transport properties of the uppermost crust. (2) Pressures observed in the uppermost crust are consistently sub-hydrostatic (as defined by the local geotherm through the sediment section) in sediment-filled valleys, and super-hydrostatic at the tops of buried basement ridges and seamounts. This is consistent with observations that upward seepage of basement water occurs above buried ridges (Chapter 16), and that discharge occurs at isolated outcropping edifices (Chapter 20). (3) When these same pressures in neighboring pairs of holes are compared relative to the interconnecting isothermal basement hydrostats, lateral pressure differentials are found to be very low, in one instance ≈1 kPa over a distance of 2.2 km. With such small pressure gradients, very high permeabilities are required to permit lateral flow at rates sufficient to maintain uniform basement temperatures.

Deeper observations are provided by only two holes, and while it is not safe to generalize from them, the implications are intriguing: (1) Vertical gradients through the uppermost crust indicate that vigorous circulation extends over a very limited depth interval, <100 m in the eastern Pacific Hole 504B, and roughly 300 m in the Mid-Atlantic Hole 395A. This is consistent with permeability measurements that yield high values over the same limited intervals. The rapid flow responsible for basement isothermality at Site 504 and lateral thermal "mining" at Site 395 must be limited to these relatively thin intervals. (2) The intervals of reduced gradient and high permeability fall within, but do not extend to the base of the extrusive section. Mineralization, mechanical consolidation, or original volcanic processes must be responsible for restricting the highest permeability and most vigorous flow to the upper parts of the extrusive crustal sections. (3) While permeability does decrease with depth, thermal gradients at depth in both holes fall significantly below those expected for the underlying lithosphere, suggesting that significant convective flow extends well below the top of the crust, at least 2–3 km in the case of Hole 504B. The permeability measured at depth is not sufficient to permit thermally significant circulation; it appears that the value characteristic for the formation is larger than that determined in the immediate vicinity of the borehole. This scale effect is discussed in Chapter 7.

Besides providing these insights into buoyancy-driven convection, borehole observations have also allowed two other modes of flow to be recognized. Variable loads at the seafloor, including tides and barometric pressure, are imparted to pore fluids in a way that depends on the elastic and hydrologic properties of the formation. Hydraulic diffusion between volumes of rock with contrasting elastic response takes place over scales that depend on the frequency of the loading signal and permeability. Diurnal signals diffusing laterally through sediment-sealed igneous crust are seen to propagate with a characteristic scale length of about 10 km, which suggests a permeability $\geq 10^{-10}$ m^2, a value similar to that derived from thermal constraints. Tectonic events also impart loads to interstitial fluids and create transient signals. The amplitude of these transients allows local strain to be determined, and with knowledge of the hydrologic structure, the rate of decay (drainage in the case of compressional strain) provides an estimate of permeability over a potentially very large scale. Where observed, this permeability (which characterizes the least resistive path through the upper igneous crust) is again $\geq 10^{-10}$ m^2. Flow driven by these modes of loading can be observed wherever it occurs along a thermal gradient. Rates of flow are comparable to those driven by buoyancy, and in general, the directions will differ. Thus, these modes may augment buoyancy-driven flow to grant water more ubiquitous access to the crust.

It is important to note here that virtually all detailed seafloor and borehole studies completed to date have been conducted in young (less than 10 Ma) settings. Heat flow variability and low average values, the standard litmus tests for flow, are common in crust many tens of millions of years in age, but observations are insufficient in detail to allow quantitative assessment of how hydrologic properties, rates of intra-crustal flow, and rates of exchange with the ocean change as the crust alters and sediments accumulate. Additional seafloor and seismic surveys like those described in this chapter need to be carried out at older sites. Additional deep observations are also needed to assess whether the thermal signature suggestive of significant convection at great depth in the crust at Hole 504B is common. In all studies, the integration of physical observations, geochemical observations, and modeling is essential, and great care must be used in selecting sites at locations where signals will be easily interpreted.

Acknowledgments

The heat flux and CORK results reviewed in this chapter would not have been possible without the collaboration and exceptional engineering support of Tom Pettigrew, Bob Macdonald, and Bob Meldrum, the at-sea support provided by the drilling crews, technicians, and engineers of the drilling vessel *JOIDES Resolution*, the captains, officers, and deck crews of numerous research vessels including *CHS Tully*, *R. V. Thompson*, *R. V. Atlantis*, and *R. V. Revelle*, the pilots and support personnel of the manned and unmanned submersibles *Alvin*, *Nautile*, *Kaiko*, *Jason*, and *Ropos*, and the strong and consistent financial support from the US National Science Foundation through the Ocean Drilling Program and from the Geological Survey of Canada. Figs. 8.5, 8.8, 8.10, 8.12, 8.13, and 8.14 have been reproduced with the kind permission of American Geophysical Union and Elsevier Science.

References

Alt, J. C., Kinoshita, H., Stokking, L. B., *et al.*, eds. 1993. *Proceedings of the Ocean Drilling Program, Initial Reports*, Vol. 148. College Station, TX: Ocean Drilling Program.

Anderson, R. N. and Hobart, M. A. 1976. The relation between heat flow, sediment thickness, and age in the eastern Pacific. *J. Geophys. Res.* **81**: 2,968–2,989.

Anderson, R. N. and Skilbeck, J. N. 1981. Oceanic heat flow. In *The Sea*, Vol. 7, ed. C. Emiliani, New York: Wiley-Interscience, pp. 489–524.

Anderson, R. N. and Zoback, M. D. 1982. Permeability, underpressures, and convection in the oceanic crust near the Costa Rica Rift, eastern equatorial Pacific. *J. Geophys. Res.* **87**: 2,860–2,868.

Anderson, R. N., Langseth, M. G., and Sclater, J. G. 1977. The mechanisms of heat transfer through the floor of the Indian Ocean. *J. Geophys. Res.* **82**: 3,391–3,409.

Baker, P. A., Stout, M. P., Kastner, M., and Elderfield, H. 1991. Large-scale lateral advection of seawater through oceanic crust in the central equatorial Pacific. *Earth Planet. Sci. Lett.* **105**: 522–533.

Beck, A. E. and Balling, N. 1988. Determination of virgin rock temperatures. In *Handbook of Terrestrial Heat-Flow Density Determination*, eds. R. Haenel, L. Rybach, and L. Stegena. Dordrecht: Kluwer Academic Publishers, pp. 59–86.

Becker, K. 1989. Measurements of the permeability of the sheeted dikes in Hole 504B, ODP Leg 111. In *Proceedings of the Ocean Drilling Program, Scientific Results*,

Vol. 111, eds. K. Becker, H. Sakai, *et al.* College Station, TX: Ocean Drilling Program, pp. 317–325.

1990. Measurements of the permeability of the upper oceanic crust at Hole 395A, ODP Leg 109. In *Proceedings of the Ocean Drilling Program, Scientific Results*, Vol. 106/109, eds. R. Detrick, J. Honnolez, J. B. Bryan, T. Jutean *et al.* College Station, TX: Ocean Drilling Program, pp. 213–222.

1996. Permeability measurements in Hole 896A and implications for the lateral variability of upper crustal permeability at Sites 504 and 896. In *Proceedings of the Ocean Drilling Program, Scientific Results*, Vol. 148, eds. J. C. Alt, H. Kinoshita, L. B. Stokking, and P. J. Michael. College Station, TX: Ocean Drilling Program, pp. 353–363.

Becker, K. and Davis, E. E. 2003. New evidence for age variation and scale effects of permeabilities of young oceanic crust from borehole thermal and pressure measurements. *Earth Planet. Sci. Lett.* **210**: 499–508.

Becker, K. and Fisher, A. T. 2000. Permeability of upper oceanic basement on the eastern flank of the Juan de Fuca Ridge determined with drill-string packer experiments. *J. Geophys. Res.* **105**: 897–912.

Becker, K., Langseth, M. G., and Von Herzen, R. P. 1983a. Deep crustal geothermal measurements, Hole 504B, Deep Sea Drilling Project Legs 69 and 70. In *Initial Reports of the Deep Sea Drilling Project*, Vol. 69, eds. J. R. Cann, M. G. Langseth, J. Honnorez, *et al.* Washington, DC: US Govt. Printing Office, pp. 223–235.

Becker, K., Langseth, M. G., Von Herzen, R. P., and Anderson, R. N. 1983b. Deep crustal geothermal measurements, Hole 504B, Costa Rica Rift. *J. Geophys. Res.* **88**: 3,447–3,457.

Becker, K., Langseth, M. G., and Hyndman, R. D. 1984. Temperature measurements in Hole 395A, Leg 78B. In *Initial Reports of the Deep Sea Drilling Project*, Vol. 78B, eds. R. D. Hyndman, M. H. Salisbury, *et al.* Washington, DC: US Govt. Printing Office, pp. 689–698.

Becker, K., A. Bartezko, and E. E. Davis, 2001. Leg 174B synopsis: Revisiting Hole 395A for logging and long-term monitoring of off-axis hydrothermal processes in young oceanic crust. In *Proceedings of the Ocean Drilling Program, Scientific Results*, Vol. 174B, eds. K. Becker and M. J. Malone. College Station, TX: Ocean Drilling Program, pp. 1–12.

Becker, N. C., Wheat, C. G., Mottl, M. J., Karsten, J. L., and Davis, E. E. 2000. A geological and geophysical investigation of Baby Bare, locus of a ridge flank hydrothermal system in the Cascadia Basin. *J. Geophys. Res.* **105**: 23, 557–523, 568.

Bower, D. R. and Heaton, K. C. 1978. Response of an aquifer near Ottawa to tidal forcing and the Alaskan earthquake of 1964. *Can. J. Earth Sci.* **15**: 331–340.

Brodoki, E., Roeloffs, E., Wookcock, D., Gall, I., and Manga, M. 2003. A mechanism for sustained groundwater pressure changes induced by distant earthquakes. *J. Geophys. Res.* **102**, B8, 2390, doi: 10.1029/2002JB002321.

Davis, E. E. and Becker, K. 1994. Formation temperatures and pressures in a sedimented rift hydrothermal system: ten months of CORK observations, Holes 857D and 858G. In *Proceedings of the Ocean Drilling Program, Scientific Results*, Vol. 139, eds. M. J. Mottl, E. E. Davis, A. T. Fisher, and J. F. Slack. College Station, TX: Ocean Drilling Program, pp. 649–666.

1999. Tidal pumping of fluids within and from the oceanic crust: new observations and opportunities for sampling the crustal hydrosphere. *Earth Planet. Sci. Lett.* **172**: 141–149.

2002. Formation pressures and temperatures associated with fluid flow in young oceanic crust: results of long-term borehole monitoring on the Juan de Fuca Ridge flank. *Earth Planet. Sci. Lett.* **204**: 231–248.

Davis, E. E. and Chapman, D. S. 1996. Problems with imaging cellular hydrothermal convection in oceanic crust. *Geophys. Res. Lett.* **23**: 3,551–3,554.

Davis, E. E., Chapman, D. S., Forster, C. B., and Villinger, H. 1989. Heat-flow variations correlated with buried basement topography on the Juan de Fuca Ridge flank. *Nature* **342**: 533–537.

Davis, E. E., Horel, G. C., Macdonald, R. D., Villinger, H., Bennett, R. H., and Li, H. 1991. Pore pressures and permeabilities measured in marine sediments with a tethered probe. *J. Geophys. Res.* **96**: 5,975–5,984.

Davis, E. E., Chapman, D. S., Mottl, M. J., Bentkowski, W., Dadey, K., Forster, C., Harris, R., Nagihara, S., Rohr, K., Wheat, G., and Whiticar, M. 1992a. FlankFlux: an experiment to study the nature of hydrothermal circulation in young oceanic crust. *Can. J. Earth Sci.* **29**: 925–952.

Davis, E. E., Becker, K., Pettigrew, T., Carson, B., and Macdonald, R. 1992b. CORK: a hydrologic seal and downhole observatory for deep-ocean boreholes. In *Proceedings of the Ocean Drilling Program, Initial Reports*, Vol. 139, eds. E. E. Davis, M. J. Mottl, A. T. Fisher, *et al.* College Station, TX: Ocean Drilling Program, pp. 43–53.

Davis, E. E., Chapman, D. S., and Forster, C. B. 1996. Observations concerning the vigor of hydrothermal circulation in young oceanic crust. *J. Geophys. Res.* **101**: 2,927–2,942.

Davis, E. E., Wang, K., He, J., Chapman, D. S., Villinger, H., and Rosenberger, A. 1997a. An unequivocal case for high Nusselt-number hydrothermal convection in sediment-buried igneous oceanic crust. *Earth Planet. Sci. Lett.* **146**: 137–150.

Davis, E. E., Fisher, A. T., Firth, J. V., and Shipboard Scientific Party, 1997b. *Proceedings of the Ocean Drilling Program, Initial Reports*, Vol. 168, College Station, TX: Ocean Drilling Program.

Davis, E. E., Chapman, D. S., Villinger, H., Robinson, S., Grigel, J., Rosenberger, A., and Pribnow, D. 1997c. Seafloor heat flow on the eastern flank of the Juan de Fuca Ridge: data from "FlankFlux" studies through 1995. In *Proceedings of the Ocean Drilling Program, Initial Reports*, Vol. 168, eds. E. E. Davis, A. T. Fisher, J. V. Firth, *et al.* College Station, TX: Ocean Drilling Program, pp. 23–33.

Davis, E. E., Chapman, D. S., Wang, K., Villinger, H., Fisher, A. T., Robinson, S. W., Grigel, J., Pribnow, D., Stein, J., and Becker, K. 1999. Regional heat flow variations across the sedimented Juan de Fuca Ridge eastern flank: constraints on lithospheric cooling and lateral hydrothermal heat transport. *J. Geophys. Res.* **104**: 17,675–17,688.

Davis, E. E., Wang, K., Becker, K., and Thomson, R. E. 2000. Formation-scale hydraulic and mechanical properties of oceanic crust inferred from pore pressure response to periodic seafloor loading. *J. Geophys. Res.* **105**: 13,423–13,435.

Davis, E. E., Wang, K., Thomson, R. E., Becker, K., and Cassidy, J. F. 2001. An episode of seafloor spreading and associated plate deformation inferred from crustal fluid pressure transients. *J. Geophys. Res.* **106**: 21,953–21,963.

Davis, E. E., Wang, K., Becker, K., Thomson, R. E., and Yashayaev, I. 2003. Deep-ocean temperature variations and implications for errors in seafloor heat-flow determinations. *J. Geophys. Res.* **108**: 2,034, doi:10.1029/2001JB001695.

Davis, E. E., Becker, K., and He, J. 2004. Costa Rica Rift revisited: constraints on shallow and deep hydrothermal circulation in young oceanic crust. *Earth Planet. Sci. Lett.* **222**: 863–879.

Elderfield, H., Wheat, C. G., Mottl, M. J., Monnin, C., and Spiro, B. 1999. Fluid and geochemical transport through oceanic crust: a transect across the eastern flank of the Juan de Fuca Ridge. *Earth Planet Sci. Lett.* **172**: 151–165.

Fang, W., Langseth, M. G., and Schultheiss, P. J. 1993. Analysis and application of in situ pore pressure measurements in marine sediments. *J. Geophys. Res.* **98**: 7,921–7,938.

Fisher, A. T. and Becker, K. 1991. The reduction of measured heat flow with depth in DSDP 504B: evidence for convection in borehole fluids? *Scientific Drilling* **2**: 34–40.

   1995. Correlation between seafloor heat flow and basement relief: observational and numerical examples and implications for upper crustal permeability. *J. Geophys. Res.* **100**: 12,641–12,657.

   2000. Channelized fluid flow in oceanic crust reconciles heat-flow and permeability data. *Nature* **403**: 71–74.

Fisher, A. T., Becker, K., and Davis, E. E. 1997. The permeability of young oceanic crust east of Juan de Fuca Ridge, as determined using borehole thermal measurements. *Geophys. Res. Lett.* **24**: 1,311–1,314.

Fisher, A. T., Davis, E. E., Hutnak, M., Speiss, V., Zuehlsdorff, L., Cherkaoui, A., Christiansen, L., Edwards, K. M., Macdonald, R., Villinger, H., Mottl, M. J., Wheat, C. G., and Becker, K. 2003. Hydrothermal recharge and discharge across 50 m guided by seamounts on a young ridge flank. *Nature* **421**: 618–621.

Fouquet, Y., Zierenberg, R. A., Miller, D. J., *et al.* 1998. In *Proceedings of the Ocean Drilling Program, Initial Reports*, Vol. 169, eds. R. A. Zierenberg, Y. Fouquet, D. J. Miller, and W. R. Normark. College Station, TX: Ocean Drilling Program.

Gable, R., Morin, R. H., and Becker, K. 1989. Geothermal state of Hole 504B: ODP Leg 111 overview. In *Proceedings of the Ocean Drilling Program, Scientific Results*, Vol. 111, eds. K. Becker, H. Sakai, *et al.* College Station, TX: Ocean Drilling Program, pp. 87–96.

   1992. Geothermal state of DSDP Holes 333A, 395A, and 534A: results from the DIANAUT program. *Geophys. Res. Lett.* **19**: 505–508.

Gable, R., Morin, R., Becker, K., and Pezard, P. 1995. Heat flow in the upper part of the oceanic crust: synthesis of in-situ measurements in Hole 504B. In *Proceedings of the Ocean Drilling Program, Scientific Results*, Vol. 137/140, eds. J. Erzinger, K. Becker, H. J. B. Dick, and L. B. Stokking. College Station, TX: Ocean Drilling Program, pp. 321–324.

Ge, S. and Stover, S. C. 2000. Hydrodynamic response to strike- and dip-slip faulting in a half-space. *J. Geophys. Res.* **105**: 25,513–25,524.

Gieskes, J. M. and Magenheim, A. J. 1992. Borehole fluid chemistry of DSDP Holes 395A and 534A, results from Operation Dianaut. *Geophys. Res. Lett.* **19**: 513–516.

Guerin, G., Becker, K., Gable, R., and Pezard, P. A. 1996. Temperature measurements and heat-flow analysis in Hole 504B. In *Proceedings of the Ocean Drilling Program, Scientific Results*, Vol. 148, eds. J. C. Alt, H. Kinoshita, L. B. Stokking, and P. J. Michael. College Station, TX: Ocean Drilling Program, pp. 291–296.

Hamilton, E. L. 1971. Elastic properties of marine sediments. *J. Geophys. Res.* **76**: 579–604.

Hartline, B. K. and Lister, C. R. B. 1981. Topographic forcing of super-critical convection in a porous medium such as the oceanic crust. *Earth Planet. Sci. Lett.* **55**: 75–86.

Hickman, S. H., Langseth, M. G., and Sviteck, J. F. 1984. In situ permeability and pore pressure measurements near the Mid-Atlantic Ridge, DSDP Site 395. In *Initial Reports of the Deep Sea Drilling Project*, Vol. 78B, eds. R. D. Hyndman, M. H. Salisbury, *et al.* Washington, DC: US Govt. Printing Office, pp. 699–708.

Hyndman, R. D. and Salisbury, M. H. 1984. The physical nature of young upper oceanic crust on the Mid-Atlantic Ridge, Deep Sea Drilling Project Hole 395A. In *Initial Reports of the Deep Sea Drilling Project*, Vol. 78B, eds. R. D. Hyndman, M. H. Salisbury, *et al*. Washington, DC: US Govt. Printing Office, pp. 839–848.

Keenan, J. H., Keyes, F. G., Hill, P. G., and Moore, J. G. 1978. *Steam Tables*. New York: Wiley.

Kopietz, J., Becker, K., and Hamano, Y. 1990. Temperature measurements at Site 395, ODP Leg 109. In *Proceedings of the Ocean Drilling Program, Scientific Results*, Vol. 106/109, eds. R. Detrick, J. Honnorez, J. B. Bryan, T. Jutean *et al*. College Station, TX: Ocean Drilling Program, pp. 197–203.

Langseth, M. G. and Herman, B. M. 1981. Heat transfer in the oceanic crust of the Brazil Basin. *J. Geophys. Res.* **86**: 10,805–10,819.

Langseth, M. G., Hyndman, R. D., Becker, K., Hickman, S. H., and Salisbury, M. H. 1984. The hydrogeological regime of isolated sediment ponds in mid-ocean ridges. In *Initial Reports of the Deep Sea Drilling Project*, Vol. 78B, eds. R. D. Hyndman, M. H. Salisbury, *et al*. Washington, DC: US Govt. Printing Office, pp. 825–837.

Langseth, M. G. Mottl, M. J., Hobart, M. A., and Fisher, A. T. 1988. The distribution of geothermal and geochemical gradients near Site 501/504: implications for hydrothermal circulation in the oceanic crust. In *Proceedings of the Ocean Drilling Program, Scientific Results*, Vol. 111, eds. K. Becker, H. Sakai, *et al*. College Station, TX: Ocean Drilling Program, pp. 23–32.

Langseth, M. G., Becker, K., Von Herzen, R. P., and Schultheiss, P. 1992. Heat and fluid flux through sediment on the western flank of the Mid-Atlantic Ridge: a hydrogeological study of north pond. *Geophys. Res. Lett.* **19**: 517–520.

Lister, C. R. B. 1972. On the thermal balance of a mid-ocean ridge. *Geophys. J. Roy. Astron. Soc.* **26**: 515–535.

Lowell, R. P. 1980. Topographically driven sub-critical hydrothermal convection in the oceanic crust. *Earth Planet. Sci. Lett.* **49**: 21–28.

Magenheim, A. J. and Gieskes, J. M. 1996. Simulation of borehole fluid mixing on the basis of geochemical observations, Hole 504B. In *Proceedings of the Ocean Drilling Program, Scientific Results*, Vol. 148, eds. J. C. Alt, H. Kinoshita, L. B. Stokking, and P. J. Michael. College Station, TX: Ocean Drilling Program, pp. 111–118.

Magenheim, A. J., Bayhurst, G., Alt, J. C., and Gieskes, J. M. 1992. ODP Leg 137, borehole fluid chemistry in Hole 504B. *Geophys. Res. Lett.* **19**: 521–524.

Mikada, H., Becker, K., Moore, J. C., Klaus, A., and Shipboard Scientific Party (eds.) 2002. *Proceedings of the Ocean Drilling Program, Initial Reports*, Vol. 196. College Station, TX: Ocean Drilling Program.

Morin, R. H., Hess, A. E., and Becker, K. 1992. In-situ measurements of fluid flow in DSDP Holes 395A and 534A: results from the Dianaut program. *Geophys. Res. Lett.* **19**: 509–512.

Mottl, M. J. 1988. Hydrothermal convection, reaction, and diffusion in sediments on the Costa Rica Rift flank: pore-water evidence from ODP Sites 677 and 678. In *Proceedings of the Ocean Drilling Program, Scientific Result*, Vol. 111, eds. K. Becker, H. Sakai, *et al*. College Station, TX: Ocean Drilling Program, pp. 195–213.

Mottl, M. J., Wheat, C. G., Baker, E., Becker, N., Davis, E., Feely, R., Grehan, A., Kadko, D., Lilley, M., Massoth, G., Moyer, C., and Sansone, F. 1998. Warm springs discovered on 3.5 Ma-old crust, eastern flank of the Juan de Fuca Ridge. *Geology* **26**: 51–54.

Muir-Wood, R. and King, G. C. P. 1993. Hydrological signatures of earthquake strain. *J. Geophys. Res.* **98**: 22,035–22,068.

Okada, Y. 1992. Internal deformation due to shear and tensile faults in a half space. *Bull. Seismol. Soc. Am.* **82**: 1,018–1,040.

Oyun, S., Elderfield, H., and Klinkhammer, G. P. 1995. Strontium isotopes in pore waters of eastern equatorial Pacific sediments: indicators of seawater advection through oceanic crust and sediments. In *Proceedings of the Ocean Drilling Program, Scientific Results*, Vol. 138, eds. N. G. Pisias, L. A. Mayer, T. R. Janecek, A. Palmer-Julson, and T. H. van Andel. College Station, TX: Ocean Drilling Program, pp. 813–820.

Pribnow, D. F. C., Davis, E. E., and Fisher, A. T. 2000. Borehole heat flow along the eastern flank of the Juan de Fuca Ridge, including effects of anisotropy and temperature dependence of sediment thermal conductivity. *J. Geophys. Res.* **105**: 13,449–13,456.

Roeloffs, E. A. 1996. Poroelastic methods in the study of earthquake-related hydrologic phenomena. In *Advances in Geophysics*, ed. R. Dmowska. San Diego: Academic Press, pp. 135–203.

1998. Persistent water level changes in a well near Parkfield, California, due to local and distant earthquakes. *J. Geophys. Res.* **103**: 869–889.

Rosenberg, N. D., Fisher, A. T., and Stein, J. S. 2000. Large-scale lateral heat and fluid transport in the seafloor: revisiting the well-mixed aquifer model. *Earth Planet. Sci. Lett.* **182**: 93–101.

Rosenberger, A., Davis, E. E., and Villinger, H. 2000. Data report: Hydrocell-95 and -96 single-channel seismic data on the eastern Juan de Fuca Ridge flank. In *Proceedings of the Ocean Drilling Program, Scientific Result*, Vol. 168, eds. A. T. Fisher, E. E. Davis, and C. Escutia. College Station, TX: Ocean Drilling Program, pp. 9–20.

Sammel, E. A. 1968. Convective flow and its effect on temperature logging in small-diameter wells. *Geophysics* **33**: 1,004–1,012.

Schultheiss, P. J. 1990. Pore pressures in marine sediments: an overview of measurement techniques and some geological applications. *Mar. Geophys. Res.* **12**: 153–168.

1997. Advances in fluid flow determinations for marine hydrogeology using pre pressure measurements. *Eos, Trans. Am. Geophys. Union, Fall Meet. Suppl.* **78**: F672.

Schultheiss, P. J. and McPhail, S. D. 1986. Direct indication of pore water advection from pore pressure measurements in Madiera abyssal plain sediments. *Nature* **320**: 348–350.

Sclater, J. G., Von Herzen, R. P., Williams, D. L., Anderson, R. N., and Klitgord, K. 1974. The Galapagos Spreading Center: heat flow on the north flank. *Geophys. J. Roy. Astron. Soc.* **38**: 609–626.

Sclater, J. G., Crowe, J., and Anderson, R. N. 1976. On the reliability of oceanic heat flow averages. *J. Geophys. Res.* **81**: 2,997–3,006.

Snelgrove, S. and Forster, C. B. 1996. Impact of seafloor sediment permeability and thickness on off-axis hydrothermal circulation: Juan de Fuca Ridge eastern flank. *J. Geophys. Res.* **101**: 2,915–2,925.

Spiess, F. N., Boegeman, D. E., and Lowenstein, C. E. 1992. First ocean-research-ship-supported fly-in re-entry to a deep ocean drill hole. *J. Mar. Tech. Soc.* **26**: 3–10.

Stein, J. S. and Fisher, A. T. 2001. Multiple scales of hydrothermal circulation in Middle Valley, northern Juan de Fuca Ridge: physical constraints and geologic models. *J. Geophys. Res.* **106**: 8,563–8,580.

Stein, C. A. and Stein, S. 1994. Constraints on hydrothermal heat flux through the oceanic lithosphere from global heat flow. *J. Geophys. Res.* **99**: 3,081–3,095.

Swift, S. A., Kent, G. M., Detrick, R. S., Collins, J. A., and Stephen, R. A. 1998. Oceanic basement structure, sediment thickness, and heat flow near Hole 504B. *J. Geophys. Res.* **103**: 15,377–15,391.

Thomson, R. E., Davis, E. E., and Burd, B. J. 1995. Hydrothermal venting and geothermal heating in Cascadia Basin. *J. Geophys. Res.* **100**: 6,121–6,141.

Van der Kamp, G. and Gale, J. E. 1983. Theory of Earth tide and barometric effects in porous formations with compressible grains. *Water Resources Res.* **19**: 538–544.

Villinger, H. und Fahrtteilnehmer 1996. Fahrtbericht SO111, 20.08. – 16.09.1996, *Berichte Fachbereich Geowissenschaften*, Universitaat Bremen, Nr. 97, 115 S., Bremen.

Von Herzen, R. P. and Uyeda, S. 1963. Heat flow through the eastern Pacific floor. *J. Geophys. Res.* **68**: 4,219–4,250.

Wakita, H. 1975. Water wells as possible indicators of tectonic strain. *Science* **189**: 553–555.

Wang, K. and Davis, E. E. 1996. Theory for the propagation of tidally induced pore pressure variations in layered sub-seafloor formations. *J. Geophys. Res.* **101**: 11,483–11,495.

Wang, K., He, J., and Davis, E. E. 1997. Influence of basement topography on hydrothermal circulation in sediment-buried igneous oceanic crust. *Earth Planet. Sci. Lett.* **146**: 151–164.

Wang, K., van der Kamp, G., and Davis, E. E. 1999. Limits of tidal energy dissipation by fluid flow in subsea formations. *Geophys. J. Int.* **139**: 763–768.

Wheat, C. G. and Mottl, M. J. 1994. Hydrothermal circulation, Juan de Fuca eastern flank: factors controlling basement water composition. *J. Geophys. Res.* **99**: 3,067–3,080.

Wheat, C. G., Mottl, M. J., Baker, E. T., Feely, R., Lupton, J., Sansone, F., Resing, J., Lebon, G., and Becker, N. 1997. Chemical plumes from low-temperature hydrothermal venting on the eastern flank of the Juan de Fuca Ridge. *J. Geophys. Res.* **102**: 15,433–15,446.

Wilkens, R. H. and Langseth, M. G. 1983. Physical properties of sediments of the Costa Rica Rift, Deep Sea Drilling Project Sites 504–505. In *Initial Reports of the Deep Sea Drilling Project*, Vol. 69, eds. J. R. Cann, M. G. Langseth, J. Honnorez, *et al.* Washington, DC: US Govt. Printing Office, pp. 659–673.

Williams, C. F., Narasimhan, T. N., Anderson, R. N., Zoback, M. D., and Becker, K. 1986. Convection in the oceanic crust: simulation of observations from Deep Sea Drilling Project Hole 504B, Costa Rica Rift. *J. Geophys. Res.* **91**: 4,877–4,889.

Williams, D. L., Von Herzen, R. P., Sclater, J. G., and Anderson, R. N. 1974. The Galapagos Spreading Center: lithospheric cooling and hydrothermal circulation. *Geophys. J. Roy. Astron. Soc.* **38**: 587–608.

Yang J., Edwards, R. N., Molson, J. W., and Sudicky, E. A. 1996. Fracture-induced hydrothermal convection in the oceanic crust and the interpretation of heat flow data. *Geophys. Res. Lett.* **23**: 929–932.

Zoth, G. and Haenel, R. 1988. Appendix. In *Handbook of Terrestrial Heat-Flow Density Determination*, eds. R. Haenel, L. Rybach, and L. Stegena. Dordrecht: Kluwer Academic Publishers, pp. 449–466.

9

Hydrothermal insights from the
Troodos ophiolite, Cyprus

Joe Cann and Kathryn Gillis

9.1 Introduction

Ophiolites provide a crucial dimension to understanding the generation of new oceanic lithosphere at mid-ocean ridges. There are difficulties: ophiolites are not active any more, so the evidence from them must be disentangled from the superimposition of all of the processes that modify new lithosphere, and many ophiolites formed in supra-subduction zone environments rather than in open ocean basins. But there are advantages that offset these problems: they can be shown to have formed by seafloor spreading, and they preserve three-dimensional sections of crust and mantle formed by spreading which can be examined by direct observation and sampling.

The hydrothermal mineralization and alteration of ophiolites preserves records of both axial and off-axis hydrothermal activity, superimposed on each other. In some places these are difficult to disentangle, but in other places the succession of the different stages of activity are clearly distinguished. Most of this evidence is in the upper crustal section, from the upper part of the plutonics to the ancient seafloor. Within this section of a well-preserved ophiolite it is possible to determine the structure of the hydrothermal systems, and their relationships to heat sources and to structural and sedimentary features that potentially affect permeability.

Evidence from ophiolites must be interpreted in conjunction with evidence from active fast-spreading mid-ocean ridges (see Chapter 3). From these we know that: (1) The upper crust is constructed by eruption of lava flows from a narrow fissure zone no more than a few hundred meters wide. (2) The fissure zone probably represents the zone of dike injection at which a sheeted dike unit is being built. (3) The lava flows reach typically a few kilometers from the fissure zone. (4) Faults cut the lavas very close to the spreading axis, and are active within the reach of lava flows. (5) Black smoker activity is most common close to the axis, well within the area reached by lava flows. (6) Off-axis hydrothermal activity continues to crustal ages of tens of millions of years.

This evidence results in the simplified cross-section of upper crust generated at a fast-spreading center such as the East Pacific Rise which is shown in Fig. 9.1. Note that: (1) At

Hydrogeology of the Oceanic Lithosphere, eds. E. E. Davis and H. Elderfield. Published by Cambridge University Press.
© Cambridge University Press 2004.

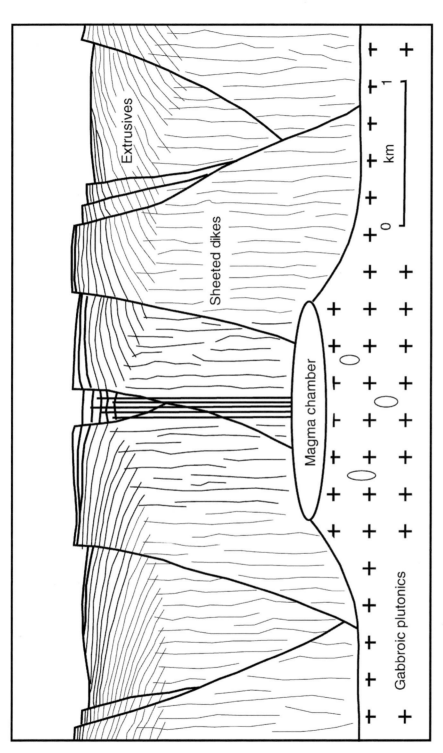

Fig. 9.1 Schematic cross-section through an active fast-spreading center, showing the relation between dikes, lavas, and faulting. The most recent features are shown in thicker lines. Note especially that lava flows, and hence exhalative sulfide deposits enclosed between lava flows, formed close to the spreading axis end up at the base of the extrusive section.

any one place, the top of the lava flow pile was generated farthest from the spreading center, while the base formed at the spreading axis. (2) The construction of the sheeted dike unit is complete much closer to the axis than that of the lava pile. (3) Active faults cut the sheeted dike unit only when its construction is nearly complete, but cut the lava unit while it is piling up, and continue to be active after the construction of the lava pile is complete. The structure of slow-spreading centers such as the Mid-Atlantic Ridge is very different, with, for example, considerable topographic relief, and the exposure of plutonic rocks at the seafloor. Neither the Troodos ophiolite nor the Oman ophiolite show these or other features typical of slow-spreading mid-ocean ridges.

In many ophiolites there is good evidence for high-temperature hydrothermal activity, in the form of massive sulfide deposits and their associated alteration zones, but most have been metamorphosed to the extent that the products of lower temperature circulation are no longer preserved. In Cyprus, the Troodos ophiolite has, for unknown reasons, avoided these later changes. Not only is the upper part of the ophiolite preserved almost undeformed, but in addition low-temperature alteration minerals such as clays and zeolites still survive from the time of their formation on the ocean floor. For this reason we base our account here principally on the evidence from Troodos, supplemented with evidence from Oman for alteration of the lower plutonic crust.

9.2 The Troodos ophiolite

The Troodos ophiolite was generated in deep water about 90 M.y. (Blome and Irwin, 1985; Mukasa and Ludden, 1987), and was first recognized as a slice of oceanic lithosphere by Gass (1968), a recognition reinforced by the work of Moores and Vine (1971). The ophiolite is exposed in the form of a broadly anticlinal dome, elongated east–west (Fig. 9.2). Within it, the trend of dikes in the sheeted dike unit, as well as those cutting the lavas, is generally north–south, indicating that the spreading axis lay in that orientation. Close to the southern margin is a major east–west fault zone, the Arakapas Fault, considered by Simonian and Gass (1978) as a transform fault, consistent with its orientation relative to that of the dikes. North of the Arakapas Fault, the ophiolite has a simple structure, typical of many ophiolites, in which seafloor sediments overlie lavas, which in turn overlie sheeted dikes and eventually plutonics. In Troodos the plutonics have a restricted outcrop at the heart of the ophiolite. South of the Arakapas Fault, the ophiolite has been affected by a major extensional episode shortly after crustal construction, perhaps associated with tectonic activity associated with the transform fault zone (Gass *et al.*, 1994; Cann *et al.*, 2001).

Lava geochemistry and plutonic petrography show that the ophiolite belongs to the supra-subduction zone class (Pearce *et al.*, 1984), and must have formed in close association with a subduction zone. But there is no sign of an associated arc, and the sediments above the ophiolite are pelagic, with no hint of any nearby land. An arc volcano did form on top of the ophiolite 20 M.y. after the ophiolite was generated, but that was probably in a different tectonic situation (Robertson and Woodcock, 1979).

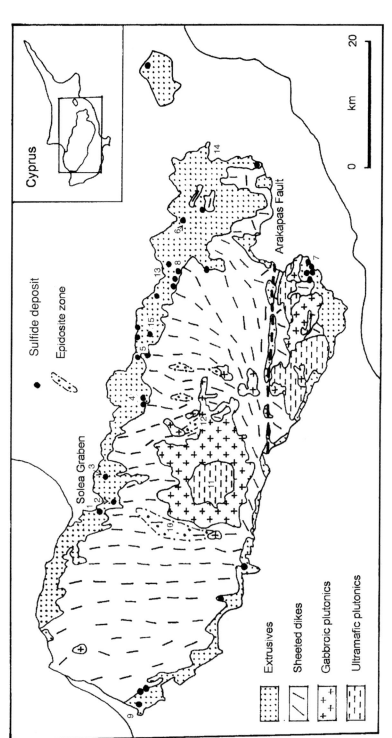

Fig. 9.2 Outline geological map of the Troodos ophiolite, showing lithological units, sulfide deposits, and epidosites zones (deposits from the *Mineral Resources Map of Cyprus* 1982, published by the Cyprus Geological Survey; see also Bear, 1960). Localities marked by numbers are mentioned in the text. Localities 1–9 are sulfide deposits or orefields: 1, Mavrovouni; 2, Apliki; 3, Skouriotissa; 4, Memi; 5, Kokkinopezoula; 6, Mathiati; 7, Kalavasos Orefield (including Mousoulos); 8, Kokkinovounaros; 9, Limni. The other localities are: 10, the best-developed epidosites on the Yerakies–Kykkos road; 11, Mount Olympus; 12, the Khandria–Platanistasa area; 13, Kambia; 14, Ayia Anna; 15, the Akaki River section. Of the 32 deposits on the Mineral Resources Map of Cyprus (1982), 22 were massive and ten were stockwork deposits.

Lava compositions in Troodos differ from mid-ocean ridge basalts (MORBs) in that they have a higher volatile content (2–3%; Muenow *et al.*, 1990) and range up to more siliceous compositions (basaltic andesite to dacite; Robinson *et al.*, 1983; Pearce *et al.*, 1984). Because of these differences, the Troodos volcanics have a broad range in vesicularity (<5 to >15%) and calculated viscosity (50–50,000 Pa s at magmatic temperatures); by contrast, MORBs have lower vesicularity (<5%) and a narrow range of viscosity (50–250 Pa s; Schmincke and Bednarz, 1990).

Pelagic sediments did not cover the ophiolite uniformly. Immediately after its generation, the ophiolite lay below the carbonate compensation depth, and the sediment preserved from that time is siliceous, and local in extent. Carbonate sediment began to accumulate toward the end of the Cretaceous. At that time the eastern end of the ophiolite was apparently deeper than its west end, since the oldest carbonate sediments lie in the east, ponded in a seafloor depression tens of kilometers across, and onlapping onto the basalts that lie to the west. In the west, carbonate sediment did not cover the ophiolite until as late as the Oligocene.

The volcanic section of Troodos varies in vertical thickness from ∽600 to 1200 m, and is constructed from a broad range of extrusive lithologies. Pillow lava flows are a large part of the lava sequence. Pillow flows form part of larger eruptive units, in which very large pillows or tubes (up to several meters in diameter) lie close to the base, grading up into normal-sized pillows or tubes (about 1 m in diameter; Schmincke and Bednarz, 1990). Pillow flows are highly porous on eruption (Gillis and Sapp, 1997), with radiating cooling cracks in individual pillows, and interpillow spaces incompletely filled with shards of broken glassy pillow margin. In places, pillow flows break up into units of pillow breccia, in which the pillows become more and more disaggregated. In extreme cases the breccia is made of entirely disjointed pillow fragments. Throughout pillow flows, the primary porosity is highly (and isotropically) connected, so that the primary permeability must have been isotropic and high.

Sheet flows are not as abundant as pillow flows, but still make up a major part of the volcanic sequence. In places, sheet flows are interbedded with pillows, but often a series of sheet flows forms a distinct unit, considered by Schmincke and Bednarz (1990) as a sheet flow volcano. Individual sheet flows are 3–15 m thick, sometimes ponding to more than 20 m. They typically have a chilled margin at the base, a central portion comprising massive columnar-jointed basalt and making up about 90% of the flow, and a rubbly flow top. Sheet flows are commonly separated by zones of hyaloclastite up to several meters thick with a high primary porosity. Overall porosity in sheet flow units is much lower than in pillow flows, but high porosity zones run parallel to the flow margins. Primary permeability within sheet flows would have been highly anisotropic. While the near-horizontal flow margins and interflow material would have been highly permeable, the permeability of the flow centers would have been much lower, with flow restricted to the narrow joints that border the columns.

Thick (5–20 m) units of hyaloclastite occur locally within the volcanic unit, composed of millimeter to centimeter–sized fragments of basaltic or andesitic glass, intruded by irregular pods of crystalline basalt. These would have had a very high primary porosity.

The volcanic sequence is cut by faults. The fault rocks and the volcanics on either side of them are typically much more oxidized than rocks farther away, indicating that the faults acted to localize fluid flow during low-temperature alteration, and were thus significantly more permeable than the surrounding volcanics at later times. The primary permeability of all of the volcanic units has been much modified by later precipitation of alteration minerals (see below).

The sheeted dike complex is made up of sub-parallel, interlacing dikes. Its upper boundary is marked by a mixed zone of dikes and lava screens, historically called the Basal Zone, which is generally tens to a few hundred meters thick. The base of the complex is commonly intruded by high-level gabbroic plutons, which may be later intruded by dike swarms (Malpas, 1990). In some locations, a narrow crush zone is present that partitions deformation between the sheeted dikes and plutonics (Agar and Klitgord, 1995). Primary porosity in the sheeted dikes is very low. As one dike intrudes another, its chilled margin is tightly welded to the host rock. Jointing is common within the dikes, both parallel to the chilled margins and as columnar joints perpendicular to the margins. Numerous small-offset faults that cut the sheeted dikes are either localized along dike margins or cut orthogonally across one or more dikes (Dietrich and Spencer, 1993). Larger-scale detachment faults, such as those that demarcate the Solea graben, may have acted to focus fluid flow but probably not during the axial high-temperature phase of hydrothermal circulation (Bettison-Varga *et al.*, 1992; Varga *et al.*, 1999). Fault rocks are composed of angular clasts, with pore space between, which has commonly been sealed with hydrothermal minerals. Except within fault zones, the primary permeability of the sheeted dike unit must have been very low.

9.3 Axial high-temperature hydrothermal systems

It was one of the triumphs of ophiolite studies that the existence of black smoker hot springs in the oceans, the temperature of the hot water, and the nature of the sulfide deposits surrounding them had been predicted from studies of the sulfide deposits of Troodos. The work of a United Nations team in Cyprus in the late 1960s demonstrated conclusively that cupriferous pyrite ores could be formed by exhalation of hydrothermal fluids onto the seafloor (Constantinou and Govett, 1972, 1973). Deposits of massive pyrite ("massive" meaning made up almost entirely of pyrite), up to a few million tonnes in total size, and grading a few percent of copper and zinc, were being mined within the volcanic rocks of the Troodos ophiolite (Bear, 1963). These deposits were enclosed above and below by seafloor volcanics. The lavas above had not been altered by hydrothermal solutions, while those beneath had been altered. Particularly critical was the recognition that the bedded yellow ochres associated with several deposits had been formed by seafloor transport and deposition of iron oxides resulting from the submarine weathering of sulfide ore. This demonstrated that the ores had been exposed, and partly weathered, at the ocean floor before the succeeding lavas had been deposited. Stockworks, networks of sulfide veins cutting the altered lavas, could be found beneath the deposits, representing the channelways through which the hydrothermal fluids had risen. The temperature of the fluids was determined from the mineral assemblages in the stockworks, and also from fluid inclusions (Spooner and Bray, 1977). The alteration petrology of the lavas and dikes was first described by Smewing *et al.* (1975), and this alteration was linked through isotopic evidence to the hydrothermal circulation that gave rise to the ores (Chapman and Spooner, 1977; Heaton and Sheppard, 1977; Spooner *et al.*, 1977). All of this work was complete before the first black smokers were discovered in the oceans.

Now that black smoker systems have been widely studied, evidence from the massive sulfide deposits and the metalliferous sediments strongly supports the view that the fossil deposits in Cyprus were formed in the same way as modern deposits of mid-ocean ridges, as will be discussed below. This strengthens the case for using the parts of these fossil systems that formed well below the seafloor as analogs for those hidden parts of active systems of the oceans today.

9.3.1 Massive sulfide deposits

More than 30 sulfide deposits have been mined, or have been identified as possible targets for mining, in the Troodos ophiolite (Bear, 1963; Cyprus Geological Survey, 1982). These range in size from a few hundred thousand tonnes up to the 14 million tonnes (Mt) of the Mavrovouni deposit; the total tonnage of sulfide in all deposits taken together was over 50 Mt. Many other smaller deposits and prospects are known, but were not regarded as economic. The deposits have been described by Constantinou and Govett (1972, 1973), Constantinou (1976, 1980), and Adamides (1980, 1990).

The sulfide deposits can be grouped into a number of distinct orefields. Eight large deposits contained 75% of the total mass of sulfide in Troodos, with about 40% being concentrated in the three large deposits of the Solea orefield (Fig. 9.2). About 1 Mt of copper has been mined, of which 75% came from the three deposits of the Solea orefield. There is no systematic relationship between size and copper grade of the ore deposits. Zinc was only present in small amounts in the Troodos deposits. Overall, Cu/Zn was about 7, and there is no clear relationship between copper and zinc grades. One of the major orefields is less than 10 km from the Arakapas Fault, showing that proximity to a transform did not inhibit ore formation.

The sulfide deposits can be found at all depths within the lava pile. Only one major deposit, Skouriotissa, lay at the top of the lava pile, and was not covered by any later lava flows after its formation. Following the cross-section of Fig. 9.1, deposits at such shallow depths must have formed at some distance from the spreading axis, possibly as far as 1–2 km away. Other deposits were found deep within the lava pile, and must have formed very close to the spreading axis, perhaps within an axial summit graben. Most deposits were found at intermediate depths, indicating formation perhaps on the edge of a summit graben, or at a distance of up to a few hundred meters from the spreading axis.

The deposits can be divided into massive deposits, in which a large proportion of the sulfide was present as massive ore (ore largely made up of sulfide), and stockwork deposits (see Section 9.3.2), in which the ore took the form of a matrix of highly altered basalt, partly replaced by sulfide and cut by veins of sulfide and quartz. Geological mapping of the pits from which sulfide has been extracted (Adamides, 1980, 1990; Richards *et al.*, 1989) shows that many deposits are bordered by a large fault, often with associated talus at the horizon of the ore, and therefore present as a scarp at the time the ore formed. The centers of the ore deposits themselves are usually not on the line of the fault, but displaced 100–200 m laterally from the fault, on the hanging-wall block.

The massive sulfide deposits show a range of mineralogy and textures closely parallel to those of modern black smoker deposits such as TAG (Humphris *et al.*, 1995). Much of the sulfide is in the form of fine-grained black or silvery pyrite with some marcasite, often with banded/colloform textures. This type of ore is highly porous. The other principal ore type is dense, coarsely crystalline yellow pyrite. The chief difference between the Troodos deposits and modern seafloor sulfide deposits such as TAG is the almost complete lack of anhydrite, which presumably dissolved in cold seawater after hydrothermal activity ceased. We interpret the assemblage of minerals and textures as the result of crystallization in an exhalative hydrothermal environment, in a number of stages: (a) primary precipitation of fine-grained sulfides and black smoker chimneys as at modern ventfields; (b) recrystallization of this primary sulfide within the sulfide mound in the presence of black smoker fluid percolating through the mound; (c) episodes of brecciation, perhaps tectonic, or perhaps hydrothermal in origin; (d) cementing of brecciated ore by further primary sulfide precipitation; (e) dissolution of anhydrite and collapse of the sulfide mass. This exhalative model is confirmed by the relationships of the ores to the surrounding metalliferous sediments, ochres (Constantinou and Govett, 1972, 1973), and umbers (Robertson, 1975; Boyle and Robertson, 1984; Boyle, 1990; see Section 9.3.3), and by the presence of fossil worm tubes and snails within the ore of several deposits (Little *et al.*, 1999).

9.3.2 Stockworks and alteration pipes

The massive ore deposits are underlain by stockworks of hydrothermally altered lavas cut by, and often partly replaced by, a network of veins containing sulfides and, often, quartz. Close to the base of the units of massive ore, the stockworks are as wide as the massive ore, and contain large percentages of sulfides. Such high pyrite percentages probably indicate sub-surface mixing of hydrothermal fluids with cold seawater percolating below the seafloor and entrained into the high-temperature flow. The stockworks narrow downwards, and become poorer in pyrite, eventually reaching a diameter of 100–200 m, and containing only a few percent of pyrite. At this level the term "stockwork" becomes inappropriate, and "alteration pipe" is used instead. The stockwork deposits that were mined in Troodos must represent either feeders of hydrothermal fluids to exhalative deposits now eroded away, or, perhaps, deposits in which mixing of hydrothermal fluids and seawater happened below the seafloor, as at the modern Galapagos deposits (Edmond *et al.*, 1979).

Alteration pipes are concentrically zoned. Two distinct types of alteration pipes were recognized by Richards *et al.* (1989). In the most common type, pipe centers are dominated by an illite mineral (usually rectorite, which is probably a replacement of original muscovite or paragonite), pyrite, and quartz, with minor anatase (Fig. 9.3*b*). More rarely, as at Mathiati, the mineral assemblage at the center of the pipe is dominantly chlorite + quartz (see also Lydon and Galley, 1986; Fig. 9.3*c*). Outward from these central zones, other phases may enter the assemblage. In the marginal zone, lavas are altered to a typical greenschist facies metabasaltic assemblage (albite + chlorite + quartz + actinolite + titanite ± epidote ± pyrite ± relict primary phases). Some pipes display late silicification, though most pipes are relatively unsilicified. The drilling at the TAG vent field in the Atlantic has shown that the stockwork there is of the most common Troodos type, dominated by quartz and paragonite (Humphris *et al.*, 1995).

Richards *et al.* (1989) interpret the concentric zonation as caused by variation in water/rock ratio across the diameter of the pipe, with the highest ratios at the center, and low ratios in the marginal zone. Since the mineral assemblages are all of greenschist facies, temperature cannot have varied much across the zones, and little cold water could have been entrained into the pipe from the surroundings. Overall the water/rock ratio must have been high, since about 10 km^3 of hydrothermal fluid must have passed through the alteration pipes or stockworks beneath the larger deposits (Humphris and Cann, 2000). Richards *et al.* conclude that the difference between the two different types of alteration pipe in Cyprus is caused by different ratios of Fe to S in the hydrothermal fluid, but this has yet to be tested by study of fluid inclusions or of modern active vents. Any lower-temperature alteration around the margins of the alteration pipes has been obscured by later widespread alteration of the volcanics.

One curious feature of the hydrothermal systems in Troodos is that alteration pipes of this kind are confined to the lava unit. None has been described from within the sheeted dikes. This is discussed further in Section 9.3.4.

9.3.3 *Metalliferous sediments*

There are two distinct types of metalliferous sediment associated with the sulfide deposits in Cyprus, umber and ochre. Least abundant is ochre (Constantinou and Govett, 1972). This is a fine-grained, yellow, iron-rich sediment, found closely associated with some of the ore deposits, and the result of seafloor oxidation of the sulfides. The best example comes from Skouriotissa, which is the only deposit that was not covered by a lava flow after it formed. Above the sulfides there, several meters of ochre are overlain by a similar thickness of umber. The sulfide at Skouriotissa lay at or very close to the seafloor for a significant period, and was subject to much greater weathering than the other sulfide deposits, which were protected by a cover of up to several hundred meters of later lava flows. Similar ochre sediments have been described from the TAG vent field on the Mid-Atlantic Ridge (Metz *et al.*, 1988; Mills and Elderfield, 1995).

The other type of metalliferous sediment, umber, is much more widespread. It is a dark brown, very fine-grained, microscopically highly porous material dominantly composed of iron and manganese oxides (Boyle and Robertson, 1984; Boyle, 1990; Robertson, 1975).

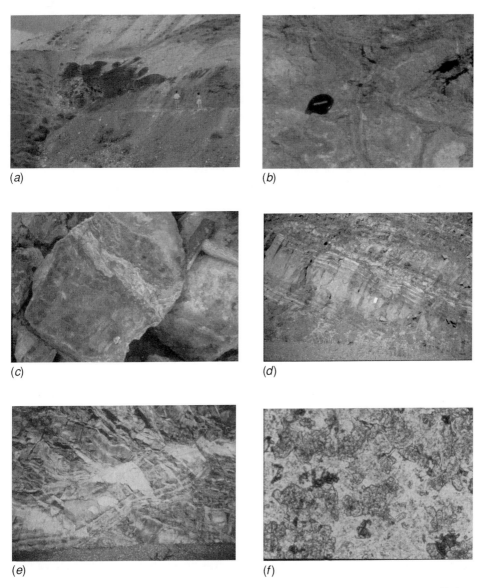

Fig. 9.3 Photographs from high-temperature axial systems. (*a*) Outcrop of umber above lava and below carbonate sediment, SE of Kambia. (*b*) Stockwork of hydrothermally altered pillow lavas at core of the Pitharokhoma alteration pipe. Junctions between pillows are defined by triangular areas of pyrite and quartz replacing interpillow sediment, and are cut by pyrite–quartz veins. Mineral assemblage is quartz–pyrite–chlorite–illite–anatase. (*c*) Silicified stockwork from beneath the Mathiati massive sulfide. Silicified altered basalt is cut by quartz–pyrite veins. Mineral assemblage is quartz–pyrite–chlorite–anatase. (*d*) Highly altered dikes at the core of an epidosite zone, Yerakies to Kykkos road. Pale stripes are epidosite (epidote–quartz–titanite) and darker stripes contain chlorite in addition. Narrow stripes are close to dike margins and broader stripes at the centers of dikes. (*e*) Alteration of cross-cutting dikes close to previous site. Note that the stripes follow the margins of the cross-cutting dike, and the stripes in earlier dikes are truncated by the margin of the later dike. (*f*) Thin section of epidosite from the same locality, showing lack of igneous texture and lack of specific pseudomorphing of one primary phase by an alteration phase.

It varies from blocky and massive, to finely laminated. Some umber occurs within the lava pile, and forms beds up to a meter thick, which lie between lava flows and can sometimes be traced to a sulfide deposit. At the top of the lava unit, umber is locally present as beds up to 20 m thick accumulated in hollows on the seafloor a few tens to a few hundred meters across (Fig. 9.3*a*). The presence of narrow mineral veins in umber which have been crumpled by compaction shows that originally umber was much thicker and more porous than it is today, and has been compacted by the weight of the overlying carbonate sediments. Since the carbonates did not begin to accumulate for millions of years after the umbers were formed, for much of the early history of the ophiolite the umbers would have been several times (perhaps even ten times) thicker than they are today. Though umbers are highly porous, the pore size is very small, so that permeability would have been very low. Umber is mineralogically and chemically very similar to the metalliferous sediments of the mid-ocean ridges (Bonatti and Joensuu, 1966; Boström and Peterson, 1966), and probably formed by the sedimentation of oxide particles from the hydrothermal plumes above black smoker ventfields (Boyle, 1990).

9.3.4 Hydrothermal reaction zones

The term "axial hydrothermal reaction zone" is used to denote a region within the crust where seawater is heated and chemically transformed into hydrothermal fluid prior to its ascent to the seafloor. Because key chemical changes may occur under different conditions during fluid flow, hydrothermal fluids may acquire aspects of their chemistry at different times and in different places, as they are heated and react with the enclosing rocks. Here we describe candidates for reaction zones associated with the axial hydrothermal circulation lying within the sheeted dike complex. These are distinguished by the pervasive replacement of previously altered dikes to epidote–quartz-rich rocks (Wilson, 1959; Schiffman *et al.*, 1987; Richardson *et al.*, 1987). These so-called epidosite zones are metal-depleted, usually lie close to the base of the sheeted dike complex, and are up to a few hundred meters thick, a few hundred meters to 2–3 km wide across axis, and extend at least a few kilometers along axis (Richardson *et al.*, 1987; Schiffman and Smith, 1988). Away from these zones, widely scattered, isolated dikes may display minor replacement by an epidosite assemblage.

Within an epidosite zone, the degree of replacement is variable from outcrop to outcrop and, within outcrops, from dike to dike. The most intensely altered rocks are true epidosites (epidote + quartz), whereas less intensely altered rocks also contain chlorite, and the least intensely altered rocks contain albite, actinolite, and some relict primary phases. Single dikes commonly display stripes of more intensely altered rock that run parallel to dike margins (Fig. 9.3*d*). In some cases these are truncated by cross-cutting dikes that are independently altered (Fig. 9.3*e*). True epidosites are composed of an equigranular mosaic of quartz and epidote, with minor chlorite and other phases (Fig. 9.3*f*). Some epidosites display faint ghosts of the original igneous texture, but there is no preferential replacement of one primary phase by one metamorphic phase.

Fluid inclusion data for epidosites indicate trapping temperatures of around 350 °C, and a wide range of salinities, with most close to seawater salinity, and others of low or very high salinity (Schiffman *et al.*, 1987; Kelley *et al.*, 1992; Cowan and Cann, 1988).

The transformation of basalt to a true epidosite involves considerable chemical change (Cowan, 1989; Schiffman *et al.*, 1990). On the scale of an epidosite zone, the sheeted dikes are consistently depleted in Na and K and are enriched in Si. The extent of Mg and Ca mobility is more variable, such that chlorite-rich epidosites are enriched in Mg and depleted in Ca, whereas chlorite-poor rocks show the opposite trends. Copper is severely depleted (by up to 90%) in epidosite zones in general, and both Zn and Mn are depleted by about 50%. Mass balance calculations (Humphris and Cann, 2000) show that these depletions are sufficient to supply the metals for the sulfide deposits and the umbers. Oxygen isotope analyses of rocks from the largest epidosite zone (Schiffman and Smith, 1988) show a decrease in $\delta^{18}O$ in the altered rocks compared to the dikes that lie outside the epidosite zones. Interestingly, the Sr-isotopic values for epidosites fall within the range of altered dikes (Fig. 9.4), and are indicative of similar time-integrated fluid fluxes (Bickle and Teagle, 1992).

A number of indicators demonstrate that the dikes in epidosite zones were altered very early in their history. Paleomagnetic evidence shows that the dikes were altered very close to the spreading axis before being tectonically rotated (Varga *et al.*, 1999). Most of the tectonic rotation of the dikes took place during the accumulation of the lavas, since the lavas are considerably less rotated than the underlying dikes. In several instances, the cross-cutting of alteration stripes in early dikes by later dikes and their alteration shows that individual dikes were hydrogeologically distinct. Almost all of the alteration apparently predates the development of jointing within the dikes, since the cross-cutting columnar joints do not localize penetrative alteration, though they may be coated with a thin skin of epidote and quartz.

Taken all together, the combination of field evidence, fluid inclusion data, and geochemistry leaves no doubt that epidosite zones have seen the passage of large volumes of hot hydrothermal fluids very early in the history of the sheeted dike complex. These characteristics, and their location close to the dike–gabbro boundary suggest that epidosite zones were the principal sites where the axial hydrothermal systems acquired their heat and chemistry (Fig. 9.5).

Two interesting problems remain. Epidosites have been very rarely recovered from mid-ocean ridge settings, although they have been discovered in a fore-arc environment (Banerjee *et al.*, 2000). It is possible that mid-ocean ridge reaction zones are characterized by other assemblages, perhaps because MORB lava chemistry is different from that of ophiolites and fore-arcs. Chlorite–quartz rocks are much more abundant from mid-ocean ridge settings (Mottl, 1983), and may play an equivalent role there to that of epidote–quartz rocks in ophiolites.

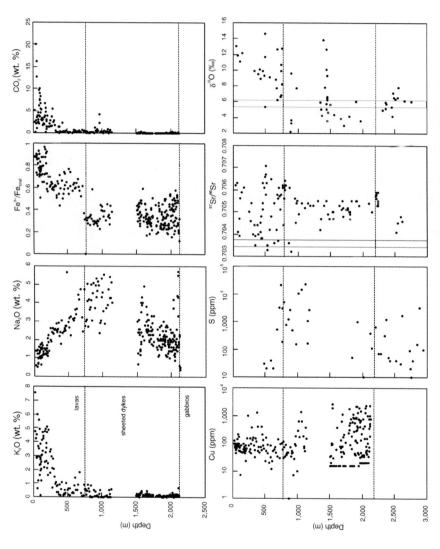

Fig. 9.4 Whole rock compositions versus depth. K_2O, Na_2O, Fe^{3+}/Fe^{T}, CO_2, and S data for drill holes CY1, CY1A, and CY4. Data sources: Gibson *et al.* (1989, 1991); Alt (1994). $\delta^{18}O$ and $^{87}Sr/^{86}Sr$ data from summary presented in Bickle and Teagle (1992 – see paper for data sources). Depth distribution of CY drill holes from Malpas *et al.* (1989). Stippled bands are values for fresh Troodos lavas.

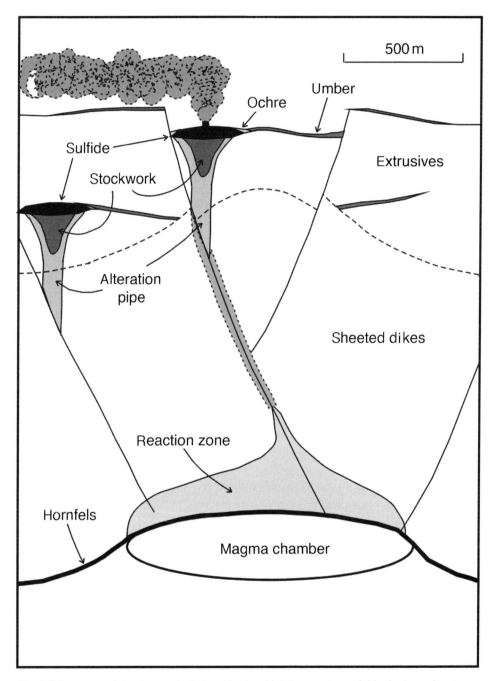

Fig. 9.5 Summary of structure and relationships in a high-temperature axial hydrothermal system.

The other problem is the nature of the link between the epidosite zones deep within the sheeted dikes and the alteration pipes in the lavas. Close to the best-developed epidosites in Troodos (SW of Yerakies) is a major fault zone within which the fault rocks are highly altered, and are transformed in places to an epidote rock. It is possible that faults of this type are the channel-ways through which hydrothermal fluids pass up through the relatively impermeable sheeted dikes until they reach the more permeable lavas, where flow continues through alteration pipes of the kind discussed in Section 9.3.2.

9.3.5 Dike–plutonic relationships

In some places the base of the sheeted dikes (containing no screens of plutonics) is truncated by gabbroic intrusions. Here, a narrow (tens of meters) contact aureole developed where hydrothermally altered dikes were recrystallized to very fine-grained hornfels at high temperatures (750–950 °C; Fig. 9.6). A complex network of tonalite veins and patches cross cut the hornfels outcrops and are indicative of localized partial melting (Gillis and Coogan, 2002). Temperatures rose abruptly (from ~400 to ≥750 °C) at the top of the contact aureole and rapidly (5–10 °C m^{-1}) within the aureole. Cross-cutting relations between the tonalite and amphibole veins, in conjunction with pressure constraints derived from fluid inclusion data, suggest that the contact aureoles were subjected to oscillations in temperature and pressure (Gillis and Roberts, 1999).

These contact aureoles developed prior to the cessation of hydrothermal flow. They are cut by broadly distributed mineralized faults (<0.5-m-wide) that record fluid flow at conditions similar to those of the epidosites (Gillis, 2002). Wider (30–40-m-wide) mineralized fault zones also occur at the same structural level as the contact aureoles. One of these appears to be linked to a small epidosite zone at a higher structural level in the dikes. A similar mineralized fault zone occurs structurally below the epidosite at Adelphi (now Madari) Ridge described by Richardson *et al.* (1987). These associations suggest that the broad mineralized zones play a role in the evolution of epidosites, though it is possible that epidosites and mineralized zones formed at different times in the complex high-temperature hydrothermal history close to the spreading axis.

In other places there is a gradual increase in the percentage of plutonic screens in the dike complex with depth, until the plutonics contain very few dikes. This is consistent with the late injection of a swarm of dikes after the plutonics had crystallized, probably injected laterally from another magma chamber along axis. Here, the thermal gradient was more gradual, peak temperatures were likely lower, and alteration is more pervasive than where contact aureoles are preserved (Baragar *et al.*, 1989; Vibetti *et al.*, 1989a).

Beneath the sheeted dikes, the upper few hundred meters of the gabbro sequence show evidence of hydrous alteration at amphibolite to greenschist facies conditions (Kelley *et al.*, 1992; Gillis and Roberts, 1999). The earliest alteration is focused along amphibole-filled micro- and macroscopic fractures that were sealed at high temperatures (650–750 °C; Fig. 9.6), similar to most oceanic gabbros (Manning *et al.*, 2000). These are locally

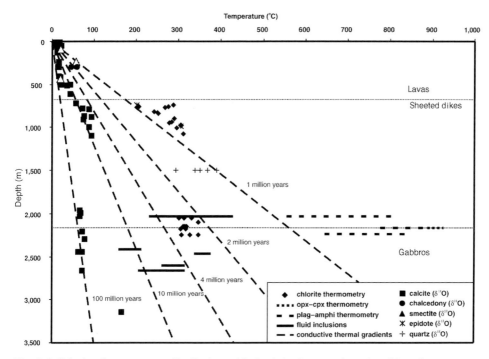

Fig. 9.6 Calculated temperature distribution with depth in the crustal section. Note that temperatures are derived from a wide range of indicators that mark different stages in the evolution of the hydrothermal systems of Troodos. Hydrothermal discharge in Troodos is localized to small areas, and the points from alteration pipes and stockworks have been omitted from the diagram. Temperature gradients (dashed lines) show calculated conductive temperature profiles for different times (in millions of years) after the formation of the ophiolite (see Chapter 10). A simple way of interpreting the results is that a calculated point may have formed in a conductive environment at the time shown by the conductive gradient line that passes through that point, or in a convectively cooled environment at an earlier time. Thus the calculated calcite $\delta^{18}O$ points in the lavas and the upper parts of the sheeted dikes may have formed in a conductive environment 10–20 M.y. after the ophiolite formed, or in a convective environment at some earlier time. The deeper points are likely to reflect convectively cooled conditions. This is consistent with the field and petrographic evidence that this calcite formed during off-axis low-temperature circulation. Oxygen isotope data: temperatures were calculated assuming unmodified Cretaceous seawater ($\delta^{18}O = -1.0\%o$); data from Schiffman *et al.* (1987); Vibetti *et al.* (1989a); Gillis and Robinson (1990); K. Gillis, unpubl. data (1994). Fluid inclusion fields for two-phase inclusions (liquid–vapor) only; data from (Schiffman *et al.*, 1987; Vibetti *et al.*, 1989b; Kelley *et al.*, 1992; Gillis and Roberts, 1999). Thermometric data: chlorite – followed method of Cathelineau (1988), data from Gillis and Robinson (1990), Gillis and Roberts (1999); plagioclase– amphibole – followed method of Holland and Blundy (1994), data from Gillis and Roberts (1999); orthopyroxene–clinopyroxene – followed method of Lindsley (1983), data from Gillis and Roberts (1999).

overprinted by lower temperature assemblages (350–550 °C). Plagiogranites are generally more altered than associated gabbros, and are locally epidotized (Kelley *et al.*, 1992). Fluids trapped in these rocks show a wide range in salinity, from less than seawater up to 60–70 wt. % NaCl (Kelley *et al.*, 1992). The characteristics of these fluid inclusions are best explained by a magmatic fluid source (Kelley *et al.*, 1992) that may have originated in part from primary water present in the andesitic magmas of the ophiolite, or perhaps from hydrous minerals present in dikes stoped into the magma chamber. Fluid inclusion data for plutonic rocks from the Oman ophiolite are very similar to Troodos (Nehlig, 1991; Juteau *et al.*, 2002). For a detailed comparison with gabbro-hosted fluid inclusions in modern ocean crust see Kelley and Früh-Green (2000).

9.3.6 *Hydrothermal alteration of the lower crust of the Oman ophiolite*

To gain a more complete view of hydrothermal alteration patterns in the lower crust it is necessary to consider the gabbroic sequence of the Oman ophiolite. There it has been shown that the extent and nature of hydrothermal alteration varies significantly on regional (kilometers) and local (meters) scales. The most pervasive alteration is found in the upper few hundred meters of the gabbro sequence or associated with fault zones and late-stage intrusions at deeper levels.

Where the gabbroic sequence has experienced only minor tectonism, there is a general trend of diminishing alteration with depth (Gregory and Taylor, 1981; Nehlig and Juteau, 1988a,b). Incipient penetration of hydrothermal fluids is marked by amphibole-filled microfractures that equilibrated at high temperatures (>700 °C) and coalesced into macroscopic vein systems at slightly lower temperatures (Manning *et al.*, 2000; Nehlig and Juteau, 1988a). In one tectonic block, a systematic study of amphibole-filled microfractures yielded temperatures of $\sim$700 °C near the sheeted dike–gabbro transition to $\sim$825 °C near the petrological Moho (Manning *et al.*, 2000). In the same area, cooling rates were calculated using closure temperatures of diffusive exchange of Ca from olivine to clinopyroxene (Coogan *et al.*, 2002). Their results document changes in cooling rates with depth, with more rapid cooling in the upper third of the gabbro sequence and much slower cooling in the remaining gabbros, trends that are consistent with conductive cooling. In another area, O- and Sr-isotope depth profiles also record the diminishing effects of hydrothermal activity toward the Moho (Gregory and Taylor, 1981; McCulloch *et al.*, 1981). The lower third of the gabbro sequence displays trace hydrous alteration and only modest shifts in igneous ^{18}O values, indicative of fluid–rock exchange at high temperatures (>600 °C) and low water/rock ratios. The upper two-thirds of the gabbros have lower whole-rock δ^{18}O values (down to 3.6‰) and are more pervasively altered. Whole-rock Sr-isotope data mimic the O-isotope depth trends in that they show little or no influence of hydrothermal fluids in the lowermost gabbros (McCulloch *et al.*, 1981). Collectively, these studies show that where tectonism is minimal, incipient penetration of hydrothermal fluids occurs at high temperatures throughout the gabbro sequence and that fluid–rock exchange is more extensive and continues to lower temperatures in the upper gabbros.

The extent and nature of hydrothermal alteration is very different where gabbroic sequences have been subjected to significant tectonism and/or complex intrusive histories. Similar to the less-tectonized blocks, in these areas, the entire gabbro sequence is cut by high-temperature amphibole veins. These early veins are cut by lower temperature zoisite–epidote, prehnite, and/or calcite vein systems interpreted to have formed off-axis (Nehlig and Juteau, 1988a). The zoisite–epidote veins primarily replace pegmatitic gabbros in lowermost gabbros whereas prehnite veins are associated with tectonized zones throughout the gabbroic sequence (Nehlig and Juteau, 1988a). O-isotope depth profiles for several areas show significant variability in comparison to $\delta^{18}O$ profiles for undisrupted gabbro sequences (Stakes and Taylor, 1992). In one location where the gabbros are highly deformed and fractured, depleted $\delta^{18}O$ values extend to very near the Moho. In another area with late-stage plagiogranite and composite wehrlite–gabbro intrusions, gabbros are enriched in ^{18}O relative to igneous values due to the effect of localized hydrothermal systems (Stakes and Taylor, 1992).

The relationships observed in the Oman ophiolite show that the extent and nature of fluid–rock interaction in the lower oceanic crust is strongly influenced by the extent of tectonism prevalent in a given place at a given time. This suggests that proximity to features such as an overlapping spreading center or transform faults contribute to the extent of mass and heat exchange between the lower and upper crust and, ultimately, perhaps, between the lower crust and the ocean.

9.4 Transition to low-temperature, off-axis circulation

9.4.1 Early off-axis mineralization

There are a number of indications in Troodos that moderate temperature hydrothermal circulation continued in the rift flanks but near the axis, based on evidence from fluid discharge zones. The best examples come from an area in the northeast of the ophiolite where silicified, gold-rich deposits, apparently the sites of fluid discharge, can be dated stratigraphically as having formed off-axis (Prichard and Maliotis, 1998). In these deposits, silicification and gold enrichment affects not only basaltic lavas, but also a thickness of at least several meters of the umbers overlying the lavas. This dates the silicification to a period after the deposition of the umbers, and hence after the crust had spread at least a few kilometers beyond the edge of the zone of lava deposition. The sediments above the umbers are not silicified, but this is not a strong age constraint, since later sediments did not start to accumulate until 20 M.y. after crustal construction. The silicified umbers contain about 90% of SiO_2, and the less mobile elements in the umbers are about a factor of ten lower in the silicified umbers than in unsilicified umbers (Prichard and Maliotis, 1998), suggesting that the umber was in general simply diluted by the addition of silica, and that very little compaction of the umber had taken place at the time of silicification.

Kokkinovounaros is the largest of these deposits. Capping the deposit, and now largely mined away, was a unit of silicified umber which was underlain by intensely leached and gossanized pillow lava. Less-gossanized outcrops at each end of the deposit show that this unit was made of basalts largely altered to clays, silicified and impregnated by pyrite, and cut by quartz–pyrite veins. In other deposits of this type, the silicification and gold enrichment of umber has taken place without precipitation of sulfides. Perhaps this difference is due to variation of fluid temperature between the two deposits, or perhaps, as Prichard and Maliotis (1998) suggest, to solutions having passed through an underlying massive sulfide deposit at Kokkinovounaros. The age of this mineralization clearly postdates that of the high-temperature circulation that produced the sulfide deposits. All but one of the sulfide deposits were covered by later lavas, and hence formed within the reach of lava flows, while the silica–gold deposits formed not only beyond the reach of the lava flows, but also after 10 m or more of umber had accumulated on top of the latest lava flow. The solutions that formed the silica–gold deposits were capable of transporting large amounts of silica, and of precipitating pyrite, but no good constraints have been reported on their temperature. Such deposits might prove to bridge the gap between the high-temperature axial deposits and the low-temperature off-axis alteration.

9.4.2 *Regional alteration of the sheeted dikes: on- or off-axis?*

One of the major problems of interpreting hydrothermal systems of ophiolites (and of the ocean crust as well) is how to understand the hydrothermal alteration that affects most of the sheeted dike complex. Away from the epidosite zones and the contact with the plutonics (see above), the sheeted dike complex is pervasively altered to greenschist facies mineral assemblages (Baragar *et al.*, 1989; Gillis and Roberts, 1999; Gillis and Robinson, 1990). Within the transition between the lavas and sheeted dikes, dikes and pillow and flow screens are altered to smectite, mixed-layer smectite/chlorite, laumontite, and quartz. Immediately below this transition, the upper part of the sheeted dikes is altered to greenschist facies. Within the sheeted dike complex, there is gradual change in the greenschist assemblage with depth such that chlorite–quartz–albite are dominant in the upper dikes whereas amphibole–andesine $\pm$ chlorite $\pm$ quartz dominate the lower dikes; there is also an increase in the degree of replacement of primary igneous phases with depth. These assemblages indicate a temperature of 200–250 °C in the upper sheeted dikes, and a steady increase in temperature to about 400 °C throughout the remainder of the sheeted dike unit (Fig. 9.6). On a regional scale, the sheeted dikes display lateral variation in their alteration mineral assemblages that reflect fluctuations in temperature and/or local permeability. Higher-temperature zones extend to shallower stratigraphic depths where the sheeted dikes are intruded by high-level and, commonly, off-axis plutons (e.g. Eddy *et al.*, 1998). Lower-temperature zones extend to deeper crustal levels in areas that have been extensively faulted (e.g. Gillis and Robinson, 1990). Using outcrop-scale fracture data, van Everdingen (1995) demonstrated that hydrothermal fluid flow in the sheeted dike complex was broadly distributed along dike-parallel fractures and faults. He also showed that mineralized fractures sealed with high-temperature assemblages are more common in the basal dikes than in the shallow dikes (where fractures commonly remain unmineralized). He concluded that high-temperature fluid flow was diffuse (broadly distributed) at depth and became focused at shallower depths.

Corresponding chemical changes with depth are illustrated in Fig. 9.4. Na_2O displays no change or enrichment in shallow dikes (albitization), whereas CaO shows some mobility but no consistent trends. The dikes are uniformly depleted in K_2O and Rb. MnO, Cu, Zn, and S show the most variability (Fig. 9.4), reflecting small sulfide-rich zones and localized zones of metal depletion which have the same degree of metal depletion as epidosite zones (see below). Sr isotope ratios fall between those of Cretaceous seawater and primary magmas (Bickle and Teagle, 1992; Bickle *et al.*, 1999), and display a much narrower range than in the overlying lavas (Fig. 9.4). These values indicate that the time-integrated fluid flux was high and that fluid flow was pervasive (i.e. not channelized) throughout the sheeted dike complex.

A similar temperature range of alteration, as shown by greenschist facies mineral assemblages and the isotopic record, is seen in the upper part of the sheeted dikes of the ocean crust, though the proportion of primary phases preserved is much higher in the oceans than in Troodos (Alt *et al.*, 1996; Gillis *et al.*, 2001; Humphris and Thompson, 1978).

Many of the authors who have studied this pervasive alteration, both in ophiolites and in the ocean, have considered that it was caused by the flow of water recharging axial black smoker systems. There are, however, problems in reconciling the temperatures within the upper parts of the sheeted dike unit from Fig. 9.6 with significant downflow of water. In a vigorous black smoker system such as TAG, the volume flow rate is close to 1 m^3 s^{-1} (Humphris and Cann, 2000). If the down-flow that feeds this is distributed over a recharge zone of, say 10 km^2, then the downward volumetric fluid flux (the fluid flow rate per unit cross-sectional area, or Darcy velocity) would be 10^{-7} m s^{-1}. Given a depth of flow of 1500 m, the corresponding Peclet number, calculated as $Pe = (qh\rho c)/\lambda$, (where q is the volume flux, h is the distance through which the fluid flows, c is the specific heat of the fluid, ρ is the fluid density, and λ is the thermal conductivity of the fluid saturated rock) for black smoker recharge would be 300. Such a high Peclet number would give a very substantial downward deflection of the isotherms (Fig. 9.7), so that temperatures as high as those estimated here would not be observed high in the sheeted dikes. To reach such high temperatures at relatively shallow depths in a recharge would require Peclet numbers of around one or less, and hence volume flow rates very much smaller than that of a vigorous black smoker.

How may this argument be reconciled with the observations? The problem is severe. Not only can the alteration of the upper part of the sheeted dikes not have formed in the recharge of a black smoker system, but its regional distribution means that the alteration cannot correspond to the discharge of a system of high-temperature flow.

One possibility is that the alteration took place at the axis during intervals between episodes of vigorous high-temperature hydrothermal flow. It is possible that, in these

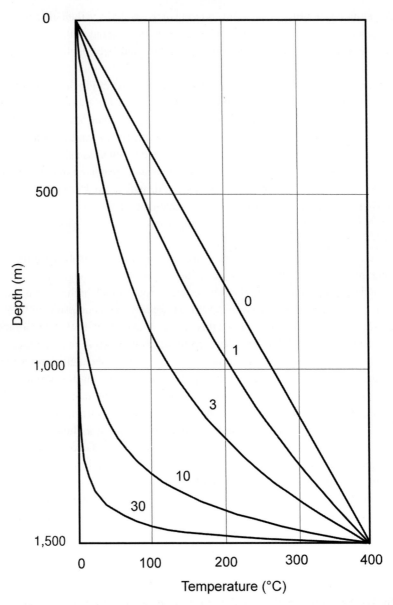

Fig. 9.7 Geothermal gradients calculated for Peclet numbers of 0, 1, 3, 10, and 30 in fluid recharging a hydrothermal system in which the heat source is at a depth of 1500 m and at a temperature of 400 °C. The fluid is assumed to flow out of the bottom of the recharge. Corresponding values for the Darcy velocity of the flow are 0, 3×10^{-10}, 9×10^{-10}, 3×10^{-9}, and 9×10^{-9} m s^{-1}. In black smokers, the geothermal gradients close to the base of the system would be different from those shown here, because fluid flow at that level is lateral rather than vertical. Adapted from Bredehoeft and Papadopulos (1965; see Chapters 11 and 12 for formulation).

intervals, cracks are sealed and a conductive regime is established, allowing high temperatures to be reached shallow in the sheeted dikes.

Another possibility is that the alteration formed off-axis, but close to the axis. In this model, as the crust spreads away from the axis, fault activity ceases, fractures are sealed by precipitation of hydrothermal minerals, and the crustal permeability decreases. Because the fluid flow velocity is closely related to permeability, reduced permeability would lead to reduced flow velocity, and hence to the possibility of higher temperatures within the upper part of the sheeted dikes. The observed temperatures might have been generated in an essentially conductive environment at crustal ages of about 1–2 Ma, as shown in Fig. 9.6. However, this model would require some high-temperature alteration of the transition zone between lavas and dikes and of the lower lavas too, which has not been identified. The problem is one that requires further attention.

9.5 Prolonged and continued off-axis hydrothermal circulation

9.5.1 Alteration zones developed off-axis in the volcanics

Away from massive sulfide deposits, the volcanic sequence is variably altered to low-temperature ($\leq 50\,°C$) mineral assemblages (Gass and Smewing, 1973; Bednarz and Schmincke, 1989; Gillis and Robinson, 1990). Two mappable alteration zones have been recognized (Fig. 9.8). The seafloor weathering zone occurs discontinuously along the sediment–lava interface (paleoseafloor) and extends downward into the lava pile for tens to a few hundred meters. It is best developed in areas exposed to cold, oxidative seawater for up to 30 M.y., as indicated by the age gap between the generation of the ophiolite and the age of the earliest sedimentary cover. The low-temperature zone underlies the seafloor weathering zone and extends down to the transition into the sheeted dike complex. Locally, the low-temperature zone extends upward to the paleoseafloor where the lavas have been covered by umbers.

The key mineralogical and geochemical characteristics of these zones are as follows. Lavas within the seafloor weathering zone are characterized by pervasive oxidative alteration and fracture-infilling (Fig. 9.9a). Volcanic glass is completely replaced by smectite, Fe-oxyhydroxides, $\pm$ zeolites; interpillow zones, breccia zones, and pillow margins are cemented by pervasive calcite veining (Fig. 9.9b,c); and vesicles and fractures are clay-lined and filled with carbonate, smectite, $\pm$ zeolites. Lavas are strongly enriched in Fe^{3+}, K_2O, Rb, Ba, Sr, CO_2, $^{87}Sr/^{86}Sr$, and $\delta^{18}O$, and depleted in SiO_2 and Na_2O (Fig. 9.4). K-enrichment and Na-depletion reflect the pervasive replacement of plagioclase by adularia. CaO is both enriched and depleted, reflecting the combined effects of plagioclase replacement and calcite infilling. Pillow margins are more oxidized and Fe-rich relative to pillow cores, whereas K_2O, Ba, and Sr are less enriched in pillow margins than pillow cores (Baragar *et al.*, 1991).

The low-temperature zone is distinguished from the seafloor weathering zone by non-pervasive alteration which varies in intensity both between and within individual lithological

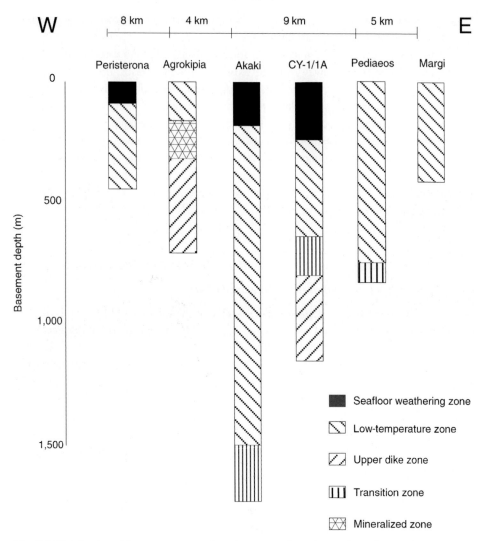

Fig. 9.8 Distribution of low-temperature alteration zones along the northern flanks of the ophiolite (modified from Gillis and Robinson, 1988).

units. Pillow margins and interpillow zones are altered to smectite, ± zeolites (analcime, phillipsite; Fig. 9.9*e*), fresh glass is locally preserved; and margins are generally uncemented (Fig. 9.9e). Vesicles and fractures are clay-lined and variably filled with smectite, ± zeolites (analcime, phillipsite), ± calcite, ± celadonite. Most massive and sheet flows are relatively fresh with minor surficial celadonite and Fe-hydroxide staining. However, zones of intense alteration occur along flow margins where volcanic glass is replaced by celadonite, smectite, zeolites (clinoptilolite, mordenite), ± calcite, ± chalcedony. Lavas display similar geochemical trends as the seafloor weathering zone, but the degree of element mobility is less pronounced (Fig. 9.4). Random distribution of celadonite and, to a lesser extent,

Fig. 9.9 Outcrop photographs of the seafloor weathering (*a–c*) and low-temperature zones (*d–f*). (*a*) Pillow outcrop showing pervasive oxidation and calcite (Unit A); (*b*) pillow margins with oxidized rims and interpillow zone cemented with smectite and calcite; (*c*) pillow breccia cemented with smectite and calcite (Unit A); (*d*) pillow outcrop, note lack of oxidation and calcite; (*e*) pillow margins and interpillow zone composed of smectite and zeolite (Unit B); and (*f*) outcrop displaying heterogeneous mixtures of pillows and sheet flows. Lithological units from Schmincke *et al.* (1983).

adularia are responsible for the scatter in K_2O, hence Na and Ca mobility is much less pronounced.

9.5.2 Inferred alteration conditions within the volcanics

Alteration conditions within the lava sequence are discussed in detail by Bednarz and Schmincke (1989) and Gillis and Robinson (1990). Here we summarize only the salient points of these studies. Oxidative alteration was most prevalent in zones of high permeability near the top of the lava pile, such as pillow margins and breccias. Reaction with chemically unmodified seawater led to the oxidation of basaltic-Fe and the formation of Fe oxyhydroxides. As seawater migrated inward toward the cores of pillows and with depth in the lava pile, progressive depletion in oxygen inhibited the formation of Fe oxyhydroxides. An increase in pH resulted, favoring the formation of saponite and the replacement of plagioclase by K-feldspar. As alteration proceeded, zeolites were the next phase to form, filling vesicles and fractures and replacing volcanic glass. The association of specific zeolites and clay minerals with either pillows or flows reflects fluid composition, such as pH and Si/Al contents. Mg-poor calcite was the last phase to precipitate from chemically modified seawater.

A particular feature of the Troodos lavas is the common association of celadonite, Si-saturated zeolites (clinoptilolite, mordenite), and chalcedony with sheet and massive flows. Gass and Smewing (1973) attributed this assemblage to an early axial alteration event whereby the intensity and grade of metamorphism increased with depth. Gillis and Robinson (1985) predicted that the distribution of celadonite reflected distinctive conditions along flow margins relative to pillows. Celadonite formation is favored by high $SiO_{2(aq)}$ contents coupled with low pH and oxidative conditions. These conditions were likely achieved by the focusing of fluid flow along flow margins where glass alteration readily releases $SiO_{2(aq)}$ and higher permeability maintained oxidative conditions. By contrast, fluid flow in pillows was more diffuse due to radial fracture patterns which resulted in fluid compositions buffered by both glass and crystalline rock (i.e. less Si-rich). It may be noted that celadonite is more abundant in the Troodos lavas than in the modern oceanic crust, due to the higher SiO_2 contents of the lavas relative to MORB. Moreover, MORB-hosted celadonite is attributed to the local alteration of glass and titanomagnetite and/or precipitation from Fe–Si-rich hydrothermal fluids mixed with seawater (Alt and Honnorez, 1984).

Temperature constraints come from oxygen isotope data for secondary minerals from the Akaki River lava section (Fig. 9.6). Geological constraints (see below) indicate that this area was a paleotopographic high that remained unsedimented and open to seawater circulation for ~30 M.y. (Gillis and Robinson, 1990). Calculated temperatures for calcite are uniformly low (<20 °C) within the seafloor weathering zone, which extends to a depth of 500 m in this section. Hence, there is no evidence for focused discharge of warm fluids at this topographic high. In the underlying low-temperature zone, the temperature range is broader and displays a slight increase with depth (up to 50 °C). Chalcedony, which forms nodules within pervasively altered flow tops, records the highest temperatures.

K–Ar and Sr-isotopic age data show that the duration of mineral precipitation ranged from <1 to 40 M.y. following crustal formation (Staudigel *et al.*, 1986; Staudigel and Gillis, 1990; Gallahan and Duncan, 1994; Booij *et al.*, 1995). Individual outcrops were sealed over a narrow timeframe but there is no systematic age progression with depth or geographic area (Gallahan and Duncan, 1994). This suggests that fluid flow and secondary mineral precipitation was controlled by local alteration conditions rather than time or space

(Gallahan and Duncan, 1994). For comparison, estimates for the duration of mineral precipitation in modern oceanic crust range from 10 to 20 M.y. (e.g. Staudigel *et al.*, 1986).

9.5.3 Relationship of alteration patterns to sedimentary cover/paleoseafloor topography

There is significant lateral variation in the distribution of alteration zones along the northern flank of the ophiolite due to local variations in permeability, paleoseafloor morphology, and the age of the conformable sedimentary cover (Fig. 9.8). The volcanic paleoseafloor had a low surface relief with fault scarp heights ≤20 m (Robertson, 1977) and low-relief seamount-like volcanic edifices (Schmincke and Bednarz, 1990) superimposed on a broader-scale long-wavelength relief. Umbers accumulated locally in small depressions and more widely in the longer wavelength seafloor lows. Other areas were first covered by sediments much later (sediments of the Lefkara Formation are >10–30 M.y. younger than basement), and must have been basement topographic highs. These highs and lows are separated by a few kilometers across strike and may have had an amplitude of as much as 100–200 m. The seafloor weathering zone is best developed in high areas, apparently because of the combined effects of the local hydrology (basement highs often act as discharge sites; see Chapter 8), and exposure to freely circulating seawater for up to 30 M.y. The relatively low permeability of the umbers (see above, Section 9.3.3) probably inhibited the free circulation of seawater locally beneath lenses of umber. Field relations show that there is significant lateral heterogeneity in the distribution of the seafloor weathering and low-temperature zones and that the seafloor weathering zone comprises <5–35% of the volcanic sequence (with an average of approximately 10%; Gillis and Robinson, 1990).

9.5.4 Evidence for off-axis fluid-flow patterns

The distribution of alteration zones and lithologic units composed of pillows, flows, hyaloclastites, and/or breccias varies laterally on a scale of tens to hundreds of meters. Outcrop-scale heterogeneity in the distribution of secondary minerals and pervasiveness of alteration suggests that fluid flow was channelized at length-scales of meters to tens of meters (i.e. the length-scale that can be traced in outcrop). This is most evident in outcrops that include more than one lithology (e.g. pillows and sheet flows, Fig. 9.9*f*) where flow tops and some, but not all, pillows are pervasively altered while retaining some porosity. Alteration in outcrops comprising only pillows or flows may also be heterogeneous, especially in the low-temperature zone. Breccia and hyaloclastite units are generally pervasively altered and are uniformly cemented in the seafloor weathering zone and are variably cemented in the low-temperature zone (Gillis and Robinson, 1990). Evidence for channelized flow also comes from whole-rock Sr-isotope data. The broad range in $^{87}Sr/^{86}Sr$ ratios for the lavas from different depths in the lava pile and with different degrees of alteration (Fig. 9.4) is indicative of kinetically limited fluid–rock exchange, which is consistent with channelized, heterogeneous flow (Bickle and Teagle, 1992; Bickle *et al.*, 1999).

Field relations give one the sense that high-volume fluid conduits were not restricted to any one location for the duration of fluid flow but rather that the location of these conduits migrated with time. This prediction is supported by K–Ar and Rb–Sr dates for celadonite such that samples from the same flow unit may have the same age whereas those from a flow unit a few tens of meters away may have very different ages (Gallahan and Duncan, 1994). Moreover, celadonite from areas with high primary permeability (e.g. breccias) are typically younger than those with lower primary permeability such as the columnar-jointed centers of sheet flows (Gallahan and Duncan, 1994), suggesting that some conduits stayed open for longer periods of time than others.

Fisher and Becker (2000) have recently proposed that fluid flow in the flank environment is focused within discrete channels that occupy small volumes of the crust. They envision that flow occurs preferentially through a network of connected conduits composed of breccia zones, and pillow and flow margins. It may be expected that such conduits would be preserved in the geological record as zones of pervasive alteration which show evidence for either high initial permeability or maintenance of some permeability. Away from these conduits, voids would be progressively filled with secondary minerals as the crust ages. These predictions are consistent with recent observations of scale- and age-dependence of upper crustal permeability (Chapter 7), and with the field and geochemical data from the Troodos lavas summarized here.

9.5.5 *Evolution of porosity and permeability structure in the volcanics*

Outcrop scale, macroscopic porosity data (Fig. 9.10) has been determined from field photographs to assess how the porosity created by crustal accretion processes is modified as a section of crust ages (Gillis and Sapp, 1997). Measurements were made for outcrops composed of either pillows or flows at different stratigraphic depths in the lavas. Porosity values for pillow outcrops account for the relative abundance of pillows and interpillow zones (i.e. the tightness of packing of the pillows), and the porosity of interpillow zones and margins, and intrapillow fractures and vesicles. Porosity values in sheet flows account for intraflow fractures and vesicles but not glassy flow tops.

Pillow outcrops within 250 m of the paleoseafloor have higher mean initial macroscopic porosity (14–17%) than pillow outcrops deeper in the volcanic sequence (6–10%) due to higher vesicularity in the shallow lavas and tighter packing of pillows with depth. Initial macroscopic porosity values for massive flows (6–9%) overlap the range for the pillow outcrops deeper in the volcanic sequence. The upper portion of the lava pile is dominated by pillows and breccias, whereas flows are most abundant in the lower portions, although the specific proportions and depth at which these variations occur is quite variable (Schmincke *et al.*, 1983; Schmincke and Bednarz, 1990). Most of the lava sequence represents the main stage of crustal construction and is composed of intercalated flow and pillow units that formed at high eruption rates (Schmincke and Bednarz, 1990). Locally, the uppermost lavas

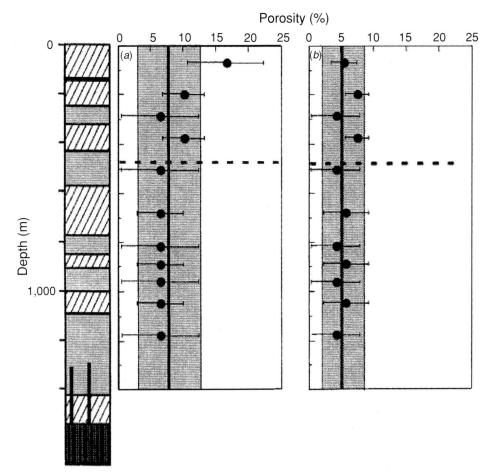

Fig. 9.10 Distribution of calculated initial and final macroscopic porosity in the Akaki River lava section. (*a*) Initial porosity (average 8.3%; standard error 4.7%) and (*b*) final porosity (average 5.2%; standard error 3.5%). Porosity data from Gillis and Sapp (1997). Stratigraphic column on left illustrates the distribution of lithological units dominated by pillows (diagonal hatch), flows (light gray), and underlying dikes (vertical hatch). Lithological units from Schmincke *et al.* (1983).

are composed of pillow volcanoes that formed during the waning stages of volcanic activity at lower eruption rates. Thus constructional processes influenced the spatial distribution of macroscopic porosity and, in particular, locally produced an upper layer of higher initial porosity.

The macroscopic porosity of the lava sequence was reduced during off-axis alteration by the infilling of primary voids with secondary minerals and by the alteration of volcanic glass within interpillow zones. Macroscopic primary porosity was reduced by 10–11% within the upper part of the seafloor weathering zone (within 250 m of the paleoseafloor). Beneath the seafloor weathering zone, pillow and sheet flow outcrops showed decreases from primary porosity of <1–3% and 2–5%, respectively. After off-axis alteration, final macroscopic porosity values (4–8%) are similar throughout the lava pile (Fig. 9.10). Thus the greatest decrease in porosity occurs where initial macroscopic porosity was high, while there is no systematic variation in the final macroscopic porosity with depth in the volcanic sequence.

Gillis and Sapp (1997) concluded that the initial macroscopic porosity of volcanic piles is influenced by the eruptive processes that control the spatial distribution and characteristics of morphologic types. Macroscopic porosity was lowest at depth due to differences between pillows that formed during the main and waning stages of crustal construction. Subsequent alteration processes caused a reduction in macroscopic porosity throughout the volcanic pile, with the greatest reduction occurring close to the paleoseafloor. The combined effects of volcanic morphology and alteration resulted in no systematic variation with depth in the final macroscopic porosity.

9.5.6 Off-axis alteration in the sheeted dike complex and plutonic sequence

In the sheeted dike complex and plutonic sequence, low-temperature assemblages composed of zeolites (laumontite, stilbite, analcime), clay minerals (smectite, mixed-layer smectite/chlorite), and calcite are generally restricted to fault zones and fractures. Calculated temperatures for calcite in the sheeted dikes (75–100 °C) are higher than in the lavas (Fig. 9.6) but are slightly lower than temperatures indicated by the zeolite and clay-mineral assemblages (100–200 °C; Gillis and Roberts, 1999). Hence, these assemblages record decreasing temperatures in the sheeted dikes as the crust moved off-axis.

A range of low temperatures has been obtained from fluid inclusions and oxygen isotopes in calcite for samples from the plutonics (Fig. 9.6). These represent different stages in the cooling of the plutonic section. Temperatures of 200–400 °C from the fluid inclusions probably record the first stages of cooling of the plutonics close to the axis. Temperatures of 80–180 °C recorded in calcite probably reflect alteration at about the same time as the calcite was being deposited in the lavas and sheeted dikes, and hence a time several million years after crustal formation. It should be noted, however, that the calcite temperatures in the deep dike and shallow plutonic parts of the section are significantly lower than would be present under conductive conditions in crust a few to a few tens of millions of years old (Fig. 9.6). The data suggest that either the calcite was precipitated very late in the history of the ophiolite, or that circulation of fluids within these units was sufficient to reduce temperatures below a conductive profile (as suggested by data from the ocean crustal Hole 504B; see Chapter 8).

9.6 Overview of evidence from ophiolites

Within the Troodos ophiolite it is possible to distinguish three sequential stages of hydrothermal circulation and hydrogeological flow. Despite the fact that there are many places where the evidence from the three stages overlaps, because of the excellent exposure over many tens of kilometers of a wide range of structural levels in the ophiolite it is possible to find areas where one or another stage dominates over the others.

The three episodes are:

1. High-temperature hydrothermal circulation close to the spreading axis. This circulation yielded black smokers, sulfide deposits, and metalliferous sediments, all closely analogous in geological settings, mineralogy, textures, and chemistry to those of modern mid-ocean ridge ventfields. Beneath these surface expressions lie stockwork zones and alteration pipes within the volcanic sequence, in which hydrothermally altered basalt is cut by sulfide–quartz veins. These pipes are concentrically zoned, and are cored by fractured and very intensely altered basalt, chemically modified at greenschist facies conditions. Close to the base of the sheeted dike complex are epidote–quartz zones, where dikes are pervasively altered at around 350 °C and characterized by pervasive metasomatism and recrystallization. These are depleted in the metals that are enriched in the sulfide deposits and metalliferous sediments, and contain abundant fluid inclusions that retain a record of fluid–rock interaction. We interpret these epidosite zones as the reaction zones in which cool seawater is transformed to hot hydrothermal fluid by heat from an underlying magma chamber and by reaction with the wall rocks. This axial circulation would have ended within 1–2 km of the spreading axis.

2. Intermediate-temperature circulation within crust a few hundred thousand to a few million years old. The discharge sites of this circulation are marked by silicification and gold enrichment of basalts and the overlying umbers. In some of the discharge sites, pyrite was precipitated during the silicification, perhaps from solutions at higher temperatures. During the discharge, the underlying extrusives were altered at low temperature to clays. The change from high-temperature axial circulation to the early off-axis circulation was probably related to the ceasing of active faulting, and the sealing of fractures. Such sealing may have slowed hydrothermal circulation in the sheeted dikes, allowing higher temperatures to develop, perhaps contributing to the pervasive alteration of that unit.

3. Long-term, off-axis circulation in the lavas, sheeted dikes, and gabbros. Temperatures in sheeted dikes and gabbros decreased from axial values and fluid flow became highly localized within fault zones. In the lavas, an oxidative alteration front migrated down into the volcanic sequence, creating a zone of pervasive alteration and carbonate cementation. The thickness and distribution of this zone was controlled by the original seafloor morphology, the rate and nature of sedimentation, and the duration of open circulation. It is poorly developed in areas where the lavas were quickly buried by synvolcanic umbers. The oxidative alteration front is best developed at inferred topographic highs due to the abundance of breccia units with high permeability and prolonged exposure to seawater (up to 30 M.y.). Beneath this region, alteration proceeded at slightly higher temperatures but at more reducing conditions for at least 20–30 M.y. Fluid flow in the lavas was both diffuse and channelized, with the location of the conduits migrating with time.

References

Adamides, N. G. 1980. The form and environment of formation of the Kalavasos ore deposits, Cyprus. In *Ophiolites; Proceedings, International Ophiolite Symposium*, ed. A. Panayiotou. Nicosia, Cyprus: Geological Survey, pp. 117–127.

Adamides, N. G. 1990. Hydrothermal circulation and ore deposition in the Troodos ophiolite, Cyprus. In *Ophiolites – Oceanic Crustal Analogues*, eds. J. Malpas, E. M. Moores, A. Panayiotou, and C. Xenophontos. Nicosia, Cyprus: Geological Survey, pp. 685–704.

Agar, S. M. and Klitgord, K. D. 1995. A mechanism for decoupling within the oceanic lithosphere revealed in the Troodos ophiolite. *Nature* **374**: 232–238.

Alt, J. C. 1994. A sulfur isotopic profile through the Troodos ophiolite, Cyprus: primary composition and the effects of seawater hydrothermal alteration. *Geochim. Cosmochim. Acta* **58**: 1,825–1,840.

Alt, J. C. and Honnorez, J. 1984. Alteration of the upper oceanic crust, DSDP Site 417: mineralogy and chemistry. *Contrib. Mineral. Petrol.* **87**: 149–169.

Alt, J. C., Laverne, C., Vanko, D. A., Tartarotti, P., Teagle, D. A. H., Bach, W., Zuleger, E., Erzinger, J., Honnorez, J., Pezard, P. A., Becker, K., Salisbury, M. H., and Wilkens, R. H. 1996. Hydrothermal alteration of a section of upper oceanic crust in the eastern equatorial Pacific: a synthesis of results from Site 504 (DSDP Legs 69, 70, and 83, and ODP Legs 111, 137, 140, and 148). In *Proceedings of the Ocean Drilling Program, Scientific Results*, eds. J. C. Alt, H. Kinoshita, L. B. Stokking, and P. J. Michael. College Station, TX: Ocean Drilling Program, pp. 417–434.

Banerjee, N. R., Gillis, K. M., and Muehlenbachs, K. 2000. Discovery of epidosites in a modern oceanic setting, the Tonga fore-arc. *Geology*. **28**: 151–154.

Baragar, W. R. A., Lambert, M. B., Baglow, N., and Gibson, I. L. 1989. Sheeted dikes from CY-4 and surface sections: Troodos Ophiolite. In *Cyprus Crustal Study Project: Initial Report, Hole CY-4*, Geological Survey Canada Paper 88–9, eds. I. L. Gibson, J. Malpas, P. T. Robinson, and C. Xenophontos. Ottawa: Geological Survey of Canada, pp. 69–106.

Baragar, W. R. A., Ludden, J. N., and Auclair, F. 1991. Alteration effects in pillow lavas from the Cy-1 drillhole, Upper Volcanic Sequence, Troodos Ophiolite. In *Cyprus Crustal Study Project: Initial Report, Holes CY-1 and 1A*, Geological Survey Canada Paper 90–20, eds. I. L. Gibson, J. Malpas, P. T. Robinson, and C. Xenophontos. Ottawa: Geological Survey of Canada, pp. 117–131.

Bear, L. M. 1960. *The Geology and Mineral Resources of the Akaki–Lythrodondha Area: Cyprus Geological Survey Department Memoir, 3*. Nicosia: Cyprus Geological Survey Department, 122 pp.

  1963. *The Mineral Resources and Mining Industry of Cyprus*. Nicosia: Cyprus Geological Survey Department, 208 pp.

Bednarz, U. and Schmincke, H.-U. 1989. Mass transfer during sub-seafloor alteration of the upper Troodos crust (Cyprus). *Contrib. Mineral. Petrol.* **102**: 93–101.

Bettison-Varga, L., Varga, R. J., and Schiffman, P. 1992. Relation between ore-forming hydrothermal systems and extensional deformation in the Solea graben spreading center, Troodos ophiolite, Cyprus. *Geology* **20**: 987–990.

Bickle, M. J. and Teagle, D. A. H. 1992. Strontium alteration in the Troodos ophiolite: implications for fluid fluxes and geochemical transport in mid-ocean ridge hydrothermal systems. *Earth Planet. Sci. Lett.* **113**: 219–237.

Bickle, M. J., Teagle, D. A. H., Beynon, J., and Chapman, H. J. 1999. The structure and controls on fluid–rock interactions in ocean ridge hydrothermal systems: constraints from the Troodos ophiolite. In *Modern Ocean Floor Processes and the Geological Record*, Geology Society Special Publication, eds. R. A. Mills and K. Harrison. London: Geology Society, pp. 127–152.

Blome, C. D. and Irwin, W. P. 1985. Equivalent radiolarian ages from ophiolitic terranes of Cyprus and Oman. *Geology* **13**: 401–404.

Bonatti, E. and Joensuu, O. 1966. Deep sea iron deposits from the South Pacific. *Science* **154**: 643–645.

Booji, E., Gallahan, W. E., and Staudigel, H. 1995. Ion-exchange experiments and Rb/Sr dating on celadonites from the Troodos Ophiolite, Cyprus. In *The Mantle–Ocean Connection*, eds. H. Staudigel, F. Albarede, D. Hilton, and T. Elliott. *Chem. Geol.* **126**: 155–167.

Boström, K. and Peterson, M. N. A. 1966. Precipitates from hydrothermal exhalations on the East Pacific Rise. *Econ. Geol.* **61**: 1,258–1,265.

Boyle, J. F. 1990. The composition and origin of oxide metalliferous sediments from the Troodos ophiolite, Cyprus. In *Ophiolites: Oceanic Crustal Analogues*, eds. J. Malpas, E. M. Moores, A. Panayioutou, and C. Xenophontos. Nicosia: Cyprus Geological Survey Department, pp. 705–717.

Boyle, J. F. and Robertson, A. H. F. 1984. Evolving metallogenesis at the Troodos spreading axis. In *Ophiolites and Oceanic Lithosphere*, eds. I. G. Gass, S. J. Lippard, and A. W. Shelton. London: Blackwell Scientific Publications, pp. 169–181.

Bredehoeft, J. and Papadopulos, I. S. 1965. Rates of vertical groundwater movement estimated from the earth's thermal profile. *Water Resources Res.* **11**: 325–328.

Cann, J. R., Prichard, H. M., Malpas, J. G., and Xenophontos, C. 2001. Oceanic inside corner detachments of the Limassol Forest area, Troodos ophiolite, Cyprus. *J. Geol. Soc. Lond.* **158**: 757–767.

Cathelineau, M. 1988. Cation site occupancy in chlorites and illites as a function of temperature. *Clay Min.* **23**: 471–485.

Chapman, H. J. and Spooner, E. T. C. 1977. ^{87}Sr/^{86}Sr enrichment of ophiolitic sulfide deposits in Cyprus confirms ore formation by circulating seawater. *Earth Planet. Sci. Lett.* **35**: 71–78.

Constantinou, G. 1976. Genesis of the conglomerate structure, porosity and collomorphic textures of the massive sulphide ores of Cyprus. In *Metallogeny and Plate Tectonics*, Geological Association of Canada Special Paper 14. Ottawa: Geological Association of Canada, pp. 187–210.

1980. Metallogenesis associated with the Troodos Ophiolite. In *Ophiolites: Proceedings, International Ophiolite Symposium*, ed. A. Panayiotou. Nicosia: Cyprus Geological Survey Department, pp. 663–674.

Constantinou, G. and Govett, G. J. S. 1972. Genesis of sulphide deposits, ochre and umber of Cyprus. *Trans. Inst. Min. Metal.* **81** *Sect. B: Appl. Earth Sci*: B34–B46.

1973. Geology, geochemistry, and genesis of Cyprus sulfide deposits. *Econ. Geol.* **68**: 843–858.

Coogan, L. A., Jenkin, G. R. T., and Wilson, R. N. 2002. Constraining the cooling rate of the lower oceanic crust: a new approach applied to the Oman ophiolite. *Earth Planet. Sci. Lett.* **199**: 127–146.

Cowan, J. G. 1989. Geochemistry of reaction zone source rocks and black smoker fluids in the Troodos ophiolite. Ph.D. thesis, University of Newcastle-upon-Tyne.

Cowan, J. and Cann, J. 1988. Supercritical two-phase separation of hydrothermal fluids in the Troodos ophiolite. *Nature* **333**: 259–261.

Cyprus Geological Survey 1982. *Mineral Resources Map of Cyprus*. Nicosia: Cyprus Geological Survey Department.

Dietrich, D. and Spencer, S. 1993. Spreading-induced faulting and fracturing of oceanic crust: examples from the sheeted dyke complex of the Troodos ophiolite, Cyprus. In *Magmatic Processes and Plate Tectonics*, Geological Society Special Publication, eds. H. M. Prichard, T. Alabaster, N. B. W. Harris, and C. R. Neary. Bath: Geological Society of London, pp. 121–140.

Eddy, C. A., Dilek, Y., Hurst, S., and Moores, E. M. 1998. Seamount formation and associated caldera complex and hydrothermal mineralization in ancient oceanic, Troodos ophiolite (Cyprus). *Tectonophysics* **292**: 189–210.

Edmond, J. M., Measures, C., Mangum, B., Grant, B., Sclater, J. C., Collier, R., Hudson, A., Gordon, L. I., and Corlisss, J. B. 1979. On the formation of metal-rich deposits at ridge crests. *Earth Planet. Sci. Lett.* **46**: 19–30.

Fisher, A. T. and Becker, K. 2000. Channellized fluid flow in oceanic crust reconciles heat flow and permeability data. *Nature* **403**: 71–74.

Gallahan, W. E. and Duncan, R. A. 1994. Spatial and temporal variability in crystallisation of celadonites within the Troodos ophiolite, Cyprus: implications for low-temperature alteration of the oceanic crust. *J. Geophys. Res.* **99**: 3,147–3,161.

Gass, I. G. 1968. Is the Troodos massif of Cyprus a fragment of Mesozoic ocean floor? *Nature* **220**: 39–42.

Gass, I. G. and Smewing, J. D. 1973. Intrusion, extrusion and metamorphism at constructive margins: evidence from the Troodos Massif, Cyprus. *Nature* **242**: 26–29.

Gass, I. G., MacLeod, C. J., Murton, B. J., Panayiotou, A., Simonian, K. O., and Xenophontos, C. 1994. *The Geology of the Southern Troodos Transform Zone*. Geological Survey Department Memoir 9. Nicosia: Cyprus Geological Survey Department.

Gibson, I. L., Malpas, J., Robinson, P. T., and Xenophontos, C., eds. 1989. *Cyprus Crustal Study Project: Initial Report, Hole CY4*, Geological Survey Canada Paper 88–9. Ottawa: Geological Survey of Canada, 393 pp.
    1991. Major and trace element analytical data: Cyprus Crustal Study Project Holes CY-1 and CY-1A. In *Cyprus Crustal Study Project: Initial Report, Holes CY-1 and CY-1A*, eds. I. L. Gibson, J. Malpas, P. T. Robinson, and C. Xenophontos, Geological Survey Canada Paper 90–20. Ottawa: Geological Survey of Canada, pp. 263–283.

Gillis, K. M. 2002. The rootzone of an ancient hydrothermal system exposed in the Troodos ophiolite, Cyprus. *J. Geol.* **110**: 57–74.

Gillis, K. M. and Coogan, L. A. 2002. Anatectic migmatites from the roof of an oceanic magma chamber. *J. Petrol.* **43**: 2,075–2,095.

Gillis, K. M. and Roberts, M. 1999. Cracking at the magma–hydrothermal transition: evidence from the Troodos ophiolite. *Earth Planet. Sci. Lett.* **169**: 227–244.

Gillis, K. M. and Robinson, P. T. 1985. Low-temperature alteration of the extrusive sequence, Troodos ophiolite, Cyprus. *Can. Mineral.* **23**: 431–441.
    1988. Distribution of alteration zones in the upper oceanic crust. *Geology* **16**: 262–266.
    1990. Patterns and processes of alteration in the lavas and dykes of the Troodos Ophiolite, Cyprus. *J. Geophys. Res.* **95**: 21,523–21,548.

Gillis, K. M. and Sapp, K. 1997. Distribution of porosity in a section of upper oceanic crust exposed in the Troodos Ophiolite. *J. Geophys. Res.* **102**: 10,133–10,149.

Gillis, K. M., Muehlenbachs, K., Stewart, M., Karson, J., and Gleeson, T. 2001. Fluid flow patterns in fast-spreading East Pacific Rise crust exposed at Hess Deep. *J. Geophys. Res.* **106**: 26,311–26,329.

Gregory, R. T. and Taylor, H. P. 1981. An oxygen isotope profile in a section of Cretaceous oceanic crust, Samail ophiolite, Oman: evidence for $\delta^{18}O$ buffering of the oceans by deep (>5 km) seawater–hydrothermal circulation at mid-ocean ridge. *J. Geophys. Res.* **86**: 2,737–2,755.

Heaton, T. H. E. and Sheppard, S. M. F. 1977. Hydrogen and oxygen isotope evidence for sea-water–hydrothermal alteration and ore deposition, Troodos complex, Cyprus. In *Volcanic Processes in Ore Genesis*, Geological Society of London Special Publication 7. London: Geological Society of London, pp. 42–57.

Holland, T. and Blundy, J. 1994. Non-ideal interactions in calcic amphiboles and their bearing on amphibole–plagioclase thermometry. *Contrib. Mineral. Petrol.* **116**: 433–447.

Humphris, S. E. and Cann, J. R. 2000. Constraints on the energy and chemical balances of the modern TAG and ancient Cyprus sea-floor sulphide deposits. *J. Geophys. Res.* **102**: 28,477–28,488.

Humphris, S. E. and Thompson, G. 1978. Hydrothermal alteration of oceanic basalts by seawater. *Geochim. Cosmochim. Acta* **42**: 107–125.

Humphris, S. E., Herzig, P. M., Miller, D. J., *et al.* 1995. The internal structure of an active sea-floor massive sulphide deposit. *Nature* **377**: 713–716.

Juteau, T., Manac'h, G., Moreau, O., Lécuyer, C., and Ramboz, C. 2000. The high temperature reaction zone of the Oman ophiolite: new field data, microthermometry of fluid inclusions, PIXE analyses and oxygen isotopic ratios. *Mar. Geol.* **21**: 351–385.

Kelley, D. S. and Früh-Green, G. L. 2000. Volatiles in mid-ocean ridge environments. In *Ophiolites and Oceanic Crust: New Insights from Field Studies and Ocean Drilling Program*, Geological Society America Special Paper, eds. Y. Dilek, E. M. Moores, D. Elthon, and A. Nicolas. Boulder, CO: Geological Society of America, pp. 237–260.

Kelley, D. S., Robinson, P. T., and Malpas, J. G. 1992. Processes of brine generation and circulation in the oceanic crust: fluid inclusion evidence from the Troodos Ophiolite, Cyprus. *J. Geophys. Res.* **97**: 9,307–9,322.

Lindsley, D. H. 1983. Pyroxene thermometry. *Am. Mineral.* **68**: 477–493.

Little, C. T. S., Cann, J. R., Herrington, R. J., and Morisseau, M. 1999. Late Cretaceous hydrothermal vent communities from the Troodos ophiolite, Cyprus. *Geology* **27**: 1,027–1,030.

Lydon, J. W. and Galley, A. G. 1986. The geochemical and mineralogical zonation of the Mathiati alteration pipe and its genetic significance. In *Metallogeny of Basic and Ultrabasic rocks*, eds. M. J. Gallagher, R. A. Ixer, C. R. Neary, and H. M. Prichard. London: London Institute of Mineralogy and Metallogy, pp. 46–68.

Malpas, J. 1990. Crustal accretionary processes in the Troodos ophiolite, Cyprus: evidence from field mapping and deep crustal drilling. In *Ophiolites: Oceanic Crustal Analogues*, eds. J. Malpas, E. M. Moores, A. Panayiotou, and C. Xenophontos. Nicosia: Cyprus Geological Survey Department, pp. 65–74.

Malpas, J., Brace, T., and Dunsworth, S. M. 1989. Structural and petrologic relationships of the CY-4 drill hole of the Cyprus Crustal Study Group. In *Cyprus Crustal Study Project: Initial Report, Hole CY-4*, Geological Survey Canada Paper 88–9, eds. I. L. Gibson, J. Malpas, P. T. Robinson, and C. Xenophontos. Ottawa: Geological Survey Canada, pp. 39–68.

Manning, C. E., MacLeod, C., and Weston, P. E. 2000. Lower crustal cracking front at fast-spreading ridges: evidence from the East Pacific Rose and the Oman ophiolite.

In *Ophiolites and Oceanic Crust: New Insights from Field Studies and Ocean Drilling Program*, Geological Society America Special Paper 349, eds. Y. Dilek, E. Moores, D. Elton, and A. Nicolas. Boulder, CO: Geological Society America, pp. 261–272.

McCulloch, M. T., Gregory, R. T., Wasserburg, G. J., and Taylor, H. P., Jr. 1981. Sm–Nd, Rb–Sr, and $^{18}O/^{16}O$ isotopic systematics in an oceanic crustal section: evidence from the Samail Ophiolite. *J. Geophys. Res.* **86**: 2,721–2,735.

Metz, S., Trefry, J. H., and Nelsen, T. A. 1988. History and geochemistry of a metalliferous sediment core from the Mid-Atlantic Ridge at 26° N. *Geochim. Cosmochim. Acta* **52**: 2,369–2,378.

Mills, R. A. and Elderfield, H. 1995. Hydrothermal activity and the geochemistry of metalliferous sediment. In *Seafloor Hydrothermal Systems*, Geophysical Monograph Series 91, eds. S. E. Humphris, R. A. Zierenberg, L. S. Mullineaux, and R. E. Thomson. Washington, DC: Geophysical Union, pp. 392–407.

Moores, E. M. and Vine, F. J. 1971. The Troodos Massif, Cyprus and other ophiolites as oceanic crust: evaluation and implications. *Phil. Trans. Roy. Soc. Lond. A* **268**: 443–466.

Mottl, M. J. 1983. Metabasalts, axial hot springs, and the structure of hydrothermal systems at mid-ocean ridges. *Geol. Soc. Am. Bull.* **94**: 161–180.

Muenow, D. W., Garcia, M. O., Aggrey, K. E., Bednarz, U., and Schmincke, H. U. 1990. Volatiles in submarine glasses as a discriminant of tectonic origin: application to the Troodos ophiolite. *Nature* **343**: 159.

Mukasa, S. B. and Ludden, J. N. 1987. Uranium–lead isotopic ages of plagiogranites from the Troodos ophiolite, Cyprus and their tectonic significance. *Geology* **15**: 825–828.

Nehlig, P. 1991. Salinity of oceanic hydrothermal fluids: a fluid inclusion study. *Earth Planet. Sci. Lett.* **102**: 310–325.

Nehlig, P. and Juteau, T. 1988a. Deep crustal seawater penetration and circulation at ocean ridges: evidence from the Oman ophiolite. *Mar. Geol.* **84**: 209–228.

1988b. Flow porosities, permeabilities and preliminary data on fluid inclusions and fossil thermal gradients in the crustal sequence of the Sumail ophiolite (Oman). *Tectonophysics* **151**: 199–221.

Pearce, J. A., Lippard, S. J., and Roberts, S. 1984. Characteristics and tectonic significance of supra-subduction zone ophiolites. In *Marginal Basin Geology: Volcanic and Associated Sedimentary and Tectonic Processes in Modern and Ancient Marginal Basins*, eds. B. P. Kokelaar and M. F. Howells. London: Geological Society London, pp. 77–94.

Prichard, H. M. and Maliotis, G. 1998. Gold mineralisation associated with low-temperature, off-axis, fluid activity in the Troodos ophiolite, Cyprus. *J. Geol. Soc. Lond.* **155**: 223–231.

Richardson, C. J., Cann, J. R., Richards, H. G., and Cowan, J. G. 1987. Metal-depleted root zones of the Troodos ore-forming hydrothermal systems, Cyprus. *Earth Planet. Sci. Lett.* **84**: 243–253.

Richards, H. G., Cann, J. R., and Jensenius, J. 1989. Mineralogical zonation and metasomatism of the alteration pipes of Cyprus sulfide deposits. *Econ. Geol.* **84**: 91–115.

Robertson, A. H. F. 1975. Cyprus umbers: basalt–sediment relationships on a Mesozoic ocean ridge. *J. Geol. Soc. Lond.* **131**: 511–531.

1977. Tertiary uplift history of the Troodos Massif, Cyprus. *Geol. Soc. Am. Bull.* **88**: 1,763–1,772.

Robertson, A. H. F. and Woodcock, N. H. 1979. Mamonia Complex, Southwest Cyprus: evolution and emplacement of a Mesozoic continental margin. *Geol. Soc. Am. Bull.* **90**: 1,651–1,665.

Robinson, P. T., Melson, W. G., O'Hearn, T., and Schmincke, H.-U. 1983. Volcanic glass compositions of the Troodos ophiolite, Cyprus. *Geology.* **11**: 400–404.

Schiffman, P. and Smith, B. M. 1988. Petrology and oxygen isotope geochemistry of a fossil seawater hydrothermal system within the Solea Graben, Northern Troodos Ophiolite, Cyprus. *J. Geophys. Res.* **93**: 4,612–4,624.

Schiffman, P., Smith, B. M., Varga, R. J., and Moores, E. M. 1987. Geometry, conditions and timing of off-axis hydrothermal metamorphism and ore-deposition in the Solea graben. *Nature* **325**: 423–425.

Schiffman, P., Bettison, L. A., and Smith, B. M. 1990. Mineralogy and geochemistry of epidosites from the Solea graben, Troodos ophiolite, Cyprus. In *Ophiolites: Oceanic Crustal Analogues.* eds. J. Malpas, E. Moores, A. Panayiotou, and C. Xenophontos. Nicosia: Cyprus Geological Survey Department, pp. 673–684.

Schmincke, H.-U. and Bednarz, U. 1990. Pillow, sheet flow, and breccia flow volcanoes and volcano-tectonic-hydrothermal cycles in the extrusive series of the northeastern Troodos ophiolite, Cyprus. In *Ophiolites: Ocean Crustal Analogues*, eds. J. Malpas, E. M. Moores, A. Panayiotou, and C. Xenophontos. Nicosia: Cyprus Geological Survey Department, pp. 185–206.

Schmincke, H.-U., Rautenschlein, M., Robinson, P. T., and Mehegan, J. M. 1983. Troodos extrusive series of Cyprus: a comparison with oceanic crust. *Geology* **11**: 405–409.

Simonian, K. O. and Gass, I. G. 1978. Arakapas fault belt, Cyprus: a fossil transform fault. *Geol. Soc. Am. Bull.* **89**: 1,220–1,230.

Smewing, J. D., Simonian, K. O., and Gass, I. G. 1975. Metabasalts from the Troodos massif, Cyprus: genetic implications deduced from petrology and trace element compositions. *Contrib. Mineral Petrol.* **51**: 49–64.

Spooner, E. T. C. and Bray, C. J. 1977. Hydrothermal fluids of seawater salinity in ophiolitic sulfide ore deposits in Cyprus. *Nature* **266**: 808–812.

Spooner, E. T. C., Chapman, H. J., and Smewing, J. D. 1977. Strontium isotopic contamination and oxidation during ocean floor hydrothermal metamorphism of the ophiolitic rocks of the Troodos massif, Cyprus. *Geochim. Cosmochim. Acta* **41**: 873–890.

Stakes, D. S. and Taylor, H. P., Jr. 1992. The Northern Samail Ophiolite: an oxygen isotope, microprobe and field study. *J. Geophys. Res.* **97**: 7,043–7,080.

Staudigel, H. and Gillis, K. M. 1990. The timing of hydrothermal alteration in the Troodos Ophiolite. In *Ophiolites: Oceanic Crust Analogues*, eds. J. Malpas, E. M. Moores, A. Panayiotou and C. Xenophontos. Nicosia: Cyprus Geological Survey Department, pp. 665–672.

Staudigel, H., Gillis, K., and Duncan, R. 1986. K/Ar and Rb/Sr ages of celadonites from the Troodos ophiolite, Cyprus. *Geology* **14**: 72–75.

van Everdingen, D. A. 1995. Fracture characteristics of the sheeted dyke complex, Troodos ophiolite, Cyprus: implications for permeability of oceanic crust. *J. Geophys. Res.* **100**: 19,957–19,972.

Varga, R. J., Gee, J. S., Bettison-Varga, L., Anderson, R. S., and Johnson, C. L. 1999. Early establishment of seafloor hydrothermal systems during structural extension: paleomagnetic evidence from the Troodos Ophiolite. *Earth Planet. Sci. Lett.* **171**: 221–235.

Vibetti, N. J., Kerrich, R., and Fyfe, W. S. 1989a. Oxygen and carbon isotope studies of hydrothermal alteration in the Troodos ophiolite complex, Cyprus. In *Cyprus Crustal Study Project: Initial Report, Hole CY-4*, Geological Survey Canada Paper 88–9, eds. I. L. Gibson, J. Malpas, P. T. Robinson, and C. Xenophontos. Ottawa: Geological Survey Canada, pp. 221–228.

    1989b. Hypersaline fluids discovered in the Troodos Ophiolite. In *Cyprus Crustal Study Project: Initial Report, Hole CY-4*, Geological Survey Canada Paper 88–9, eds. I. L. Gibson, J. Malpas, P. T. Robinson, and C. Xenophontos. Ottawa: Geological Survey Canada, pp. 229–234.

Wilson, R. A. M. 1959. *The Geology of the Xeros–Troodos Area*. Cyprus Geological Survey Memoir 1. Nicosia: Cyprus Geological Survey Department.

Part III

Heat and fluid fluxes

10

Deep-seated oceanic heat flux, heat deficits, and hydrothermal circulation

Robert N. Harris and David S. Chapman

10.1 Introduction

The Earth is losing heat at a rate of 44.2×10^{12} W with about 70% occurring through oceanic lithosphere (Pollack *et al.*, 1993). Understanding the magnitude and geographic variation of this heat loss, and the associated physical and chemical processes has been a major pursuit of earth scientists for five decades (e.g. Lee and Uyeda, 1965; Williams and Von Herzen, 1974; Chapman and Pollack, 1975; Jessop *et al.*, 1976; Sclater *et al.*, 1980).

The first-order pattern of oceanic heat flux can be explained by the gradual and progressive cooling of lithosphere continuously created at a mid-ocean ridge and rafted away from the ridge by seafloor spreading. Cooling of the lithosphere leads to subsidence through thermal contraction and thickening through the accretion of upper mantle material. Mathematical and physical descriptions of this conceptual model are both elegant and simple, providing an explanation of many observations with a limited number of free parameters.

In young oceanic lithosphere, advective heat loss due to crustal hydrothermal circulation is superimposed on the slow conductive heat loss of the lithosphere. Hydrothermal circulation dominates heat transfer where temperature and topographic gradients are sufficient to drive fluid flow, crustal permeabilities are high, and there are open pathways for seawater to enter and exit the oceanic crust. Near the ridge crest and in young seafloor, conditions are ideal for the upper oceanic crust to host vigorous ventilated hydrothermal convection. With the progression of time, temperature gradients and crustal permeability decrease, sediment accumulation isolates the permeable crust from seawater, and heat loss by fluid flow is diminished.

Other than direct venting of high-temperature fluids from the seafloor near ridge axes, direct observation of hydrothermal circulation in the crust is difficult. Instead, one infers the presence of fluid flow by measuring the physical or chemical effects of advection and comparing those effects with models. Model sensitivity studies suggest which observations are sensitive to various forms of hydrothermal convection and show that temperature fields and particularly seafloor heat flow patterns comprise the most diagnostic physical measurements for constraining the nature and vigor of circulation.

Heat flux is the product of the thermal gradient and thermal conductivity, both of which are typically measured *in situ* by inserting a gravity-driven heat flux probe into sediments. The requirement for sediment precludes probe measurements on large areas of young seafloor and other bare rock environments where advective heat loss is greatest. Seafloor drilling provides opportunities to measure sub-surface temperature fields and thermal blankets (Johnson and Hutnak, 1996) make possible the computation of heat flux in bare rock environments, but the cost of both of these techniques prohibits a spatially dense set of measurements. Once sediment ponds have accumulated on young seafloor, it is physically possible to make heat flux measurements with a probe, but the measurements are often systematically biased. Both observations and theory suggest that sediment ponds tend to be areas of anomalously low heat flux due to fluid flow (Chapter 8). Hydrologic recharge and discharge partially occurs through adjacent topographic highs, but the lack of sediments often precludes making measurements to confirm the inference, and focused ventilation in areas of discharge would probably dilute whatever thermal signal might be advected in venting fluids.

Fortunately, both recharge and discharge, do not have to be measured to estimate the magnitude of the advective heat loss. If the deep-seated heat flux is known, then spatially integrated heat flux deficits in areas affected by recharge or lateral flux must match heat flux excesses in discharge areas. Heat flux deficits, in fact, provide important information about the nature and vigor of hydrothermal convection.

While there is general consensus both about the importance of hydrothermal circulation in young oceanic crust and about first-order variations of heat flux, seafloor depth, and lithospheric thickness as functions of seafloor age, the details of these processes remain controversial. The recognition that the difference between observed and predicted values of heat flux are largely due to hydrothermal circulation provides a convenient way of estimating the magnitude of advective heat loss through the seafloor and also the flux of water through the oceanic crust (Wolery and Sleep, 1976; Sleep and Wolery, 1978; Stein and Stein, 1994b). However, magnitudes of inferred ridge flank hydrothermal circulation depend in part on the chosen model of lithospheric heat flux which remains contentious. Discussions of advective heat loss also focus on the processes controlling the magnitude and patterns of hydrothermal circulation (Chapter 11) and when fluid flux becomes thermally negligible (Chapter 13). All of these topics remain areas of active investigation.

Thus, the problem confronting the marine hydrologist is to separate the background lithospheric heat flux from the hydrothermal advective heat flux. Strategies for estimating both of these quantities are discussed below and rely on judicious combinations of heat flux and bathymetric data, as well as careful selections of environmentally favorable sites in which to make measurements. The key to these tasks is a clear understanding of each process. The purpose of this chapter is (a) to review models used to predict lithospheric or deep-seated heat flux in oceanic lithosphere, (b) to show how observations are combined with model predictions of heat flux to estimate the magnitude of seafloor hydrothermal convection, and (c) to provide examples of surveys that advance our understanding of hydrothermal circulation in the oceanic crust.

10.2 Reference models for the thermal evolution of oceanic lithosphere

Seafloor bathymetry and heat flux provide critical evidence for the thermal state of oceanic lithosphere. These data are complementary in that bathymetry, under conditions of iso-static equilibrium, constrains the integrated temperature structure through the lithosphere while observations of heat flux constrain the shallow temperature regime. Because heat flux observations depend on the thermal gradient at the surface, they are sensitive to shallow thermal processes and are therefore suited to estimating hydrothermal circulation in areas of advective heat flux. In contrast, bathymetry depends primarily on the integrated ther-mal structure and is better suited to estimating plate-scale model parameters. As a result, studies calibrating plate-scale thermal models often weight bathymetry more heavily than heat flux (e.g. Davis and Lister, 1974; Parsons and Sclater, 1977; Carlson and Johnson, 1994). A notable exception is the lithospheric model GDH1 (Stein and Stein, 1992), which jointly fits both bathymetry and heat flux (on crust older than 55 Ma). Other constraints on the thermal state of oceanic lithosphere such as surface wave velocities, intraplate earth-quakes, and flexural thicknesses, are relatively poor quantitative indicators, because the temperature dependence of rheological properties is poorly known. Thus, the combination of seafloor depth and heat flux provides the principal evidence for the thermal evolution of the lithosphere.

Figure 10.1 shows an example of the variation of seafloor bathymetry and heat flux as a function of crustal age. This example (Stein and Stein, 1992) incorporates data from the north Pacific and northwest Atlantic. The depth plot (Fig. 10.1*a*) shows that, in general, mid-ocean ridges are elevated to a depth of 2.5 km and that aging lithosphere ultimately subsides to depths of about 5.5 km. For crust younger than about 80 Ma the depth increases linearly with the square-root of age according the simple boundary layer cooling (Fig. 10.1*b*; Parker and Oldenburg, 1973; Davis and Lister, 1974). For older ocean floor, the depth increase tapers off exponentially to an asymptotic value of 5,650 m (Stein and Stein, 1992). However, note that the variability of ocean depth increases with age (Sclater and Wixon, 1986; Renkin and Sclater, 1988). Heat flux also generally decreases as a function of seafloor age, but deviates from model predictions at young ages (Fig. 10.1*c*). Figure 10.1*d* shows heat flux plotted as a function of the inverse square-root of age, the function of time predicted by simple boundary layer cooling (see Section 10.2.1). Heat flux is highly variable in young seafloor but, unlike bathymetry, the variability decreases in older seafloor (however, see Chapter 13). These measures of variability may reflect perturbations to the commonly assumed progressive cooling of oceanic lithosphere.

Two classes of models have been proposed to explain these first-order observations between bathymetry and heat flux as functions of age. In the half-space model, the litho-sphere progressively cools and thickens as it spreads away from the ridge crest (Turcotte

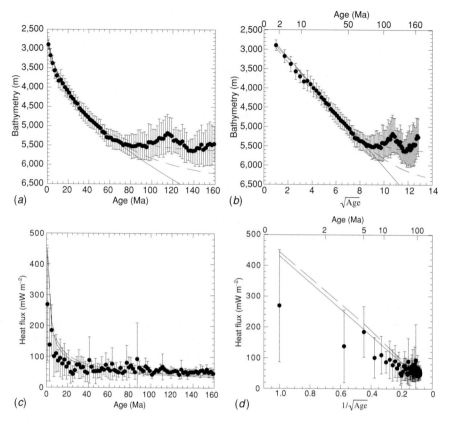

Fig. 10.1 Depth of the seafloor (*a*, *b*) and heat flux (*c*, *d*) as functions of seafloor age. Data come from the analysis of Stein and Stein (1992); data are sorted into 2-M.y. age bins and plotted as bin average (solid circles) and standard deviation (brackets). Curves show model predictions for a half-space lithosphere cooling model (solid curve), a plate model (long dashes), and a modified plate model GDH1 (short dashes). See Table 10.1. for model parameters.

and Oxburgh, 1967; Parker and Oldenburg, 1973). The predicted ocean depth is proportional to the square-root of age and heat flux decreases inversely with the square-root of age. In the plate model, the lithosphere continues to thicken until the geothermal gradient reaches equilibrium and cooling ceases (McKenzie, 1967; Sclater and Francheteau, 1970), presumably as a result of heat being supplied from the asthenosphere at a uniform rate. In this model the predicted ocean depth and heat flow vary exponentially with age. Both of these models are one-dimensional so that horizontal heat conduction is assumed negligible. This assumption holds as long as the width of the plate is large compared to its thickness so that all heat flows in the vertical direction. Additionally, both models assume that heat production is negligible as indicated by heat production measurements of basalt.

10.2.1 Half-space model

The half-space model applied to the thermal evolution of oceanic lithosphere refers to a half-space at an initial constant temperature T_i whose upper boundary temperature is changed to T_0 at time zero and maintained at that temperature. The half-space model has been described elsewhere (e.g. Davis and Lister, 1974) and detailed derivations can be found in many texts (e.g. Turcotte and Schubert, 1982; Fowler, 1990). The temperature, T_0, is the temperature at the seafloor and is commonly assumed 0 °C. The variation of temperature with depth z below the seafloor, and age t is formulated in terms of a step change in surface temperature and given by:

$$T(t, z) = T_i \text{erf}\left(\frac{z}{\sqrt{4\kappa t}}\right) \tag{10.1}$$

where erf is the error function, and κ is thermal diffusivity of the half space. The product of the thermal gradient at $z = 0$ and the thermal conductivity λ, gives the surface heat flux, f, as:

$$f = \lambda T_i / \sqrt{\pi \kappa t} \tag{10.2}$$

The cooling lithosphere contracts and is loaded by an increasing column of seawater through time. Assuming that the seafloor is regionally in isostatic equilibrium, as indicated by gravity measurements, the depth to the upper surface of unsedimented basement is given by:

$$d(t) = d_r + 2\alpha T_i \frac{\rho_m}{(\rho_m - \rho_w)}\sqrt{\frac{\kappa t}{\pi}} \tag{10.3}$$

where d_r is the depth to the ridge crest at zero age, α is the thermal expansivity of the mantle, ρ_m is the density of the mantle at $T = T_m$, and ρ_w is the density of seawater. The important feature of this model is that the thermal structure, surface heat flux, and seafloor depth vary with the square-root of age. This model was calibrated by Davis and Lister (1974) using ocean depth data (Sclater *et al.*, 1971) from the Pacific, Atlantic, and Indian Oceans (Table 10.1). Half-space cooling models successfully describe seafloor depth for ages younger than 80 Ma (Davis and Lister, 1974).

10.2.2 Plate model

The plate model (McKenzie, 1967; Sclater and Franchetau, 1970; Parsons and McKenzie, 1978) differs from half-space cooling in that the depth of cooling and the thickness of the thermal boundary layer are limited by a constant temperature lower boundary. This fixed temperature, or sometimes heat flux, is specified at a particular depth, defining the base of the plate. This model was developed to explain the approximately constant heat flux and the cessation of subsidence in old regions.

The mathematical formulation for the plate model is developed in many studies (Parsons and Sclater, 1977; Stein and Stein, 1992; Carlson and Johnson, 1994). We follow the formulation used by Carlson and Johnson (1994) that makes explicit the exponential age

Table 10.1 *Model parameters*

Model		Half-space cooling HS74[a]	Plate cooling models	
			PS77[b]	GDH1[c]
Initial temperature (°C)	T_i	1,120–1,220	–	–
Thermal diffusivity (m² s⁻¹)	κ	8×10^{-7}	8×10^{-7}	8×10^{-7}
Thermal conductivity (W m⁻¹ K⁻¹)	λ	3.3	3.1	3.1
Thermal expansivity (K⁻¹)	α	4×10^{-5}	$3.28 \pm 1.19 \times 10^{-5}$	$3.1 \pm 0.8 \times 10^{-5}$
Mantle density (kg m⁻³)	ρ_m	3,300	3,330	3,300
Water density (kg m⁻³)	ρ_w	1,000	1,000	1,000
Plate thickness (km)	a	–	125 ± 10	95 ± 15
Basal temperature (°C)	T_m	–	$1,333 \pm 274$	$1,450 \pm 250$
Ridge depth (m)	d_r	2,500	2,500	2,600

[a]HS74 (Davis and Lister, 1974).
[b]PS77 (Parsons and Sclater, 1977).
[c]GDH1 (Stein and Stein, 1992).

dependence of the geotherm, heat flux, and ocean depth. Temperature as a function of the depth z, and distance from ridge crest, x, can be expressed by:

$$T(x, z) = T_m \left[z/L + \sum_n A_n \sin(n\pi z/L) \exp(-\beta_n x/L) \right] \qquad (10.4)$$

where T_m is the temperature of the plate base, L is the thermal plate thickness, and

$$A_n = 2/(n\pi), \quad \beta_n = (Pe^2 + n^2\pi^2)^{1/2} - Pe, \quad Pe = vL/(2\kappa) \qquad (10.5)$$

Pe is the Peclet number and v is the plate velocity. Where Pe is sufficiently large ($Pe \gg n\pi$), $B_n x/L \sim (n^2\kappa\pi^2/L)t$. With this approximation, (10.4) can be given by

$$T(z, t) = T_m \left[z/L + \sum A_n \sin(n\pi z/L) \exp(-n^2 at) \right] \qquad (10.6)$$

for the first few terms of the series when $n\pi$ is sufficiently small and where $a = \kappa\pi^2/L^2$. This formulation makes the exponential dependence on seafloor age explicit. Heat flux as a function of age is given by:

$$f(t) = f_m \left[1 + 2 \sum_{n=1}^{\infty} \exp(-n^2 at) \right] \qquad (10.7)$$

where $f_m = \lambda T_m/L$ is the asymptotic heat flow for old seafloor. Depth as a function of age is given by:

$$d(t) = d_r + d_s \left[1 - 8/\pi^2 \sum_j j^{-2} \exp(-j^2 at) \right] \qquad (10.8)$$

where $j = 1, 3, 5$; and d_s is the asymptotic subsidence of the seafloor given by:

$$d_s = \frac{\alpha T_m L \rho_m}{2(\rho_m - \rho_w)} \qquad (10.9)$$

Standard reference curves for the plate model were originally given by Parsons and Sclater (1977) and more recently updated by Stein and Stein (1992). Following Stein and Stein (1992), models using parameters given by Parsons and Sclater (1977) and Stein and Stein (1992) are denoted PS77 and GDH1, respectively. PS77 was calibrated using depth data only from the North Pacific and North Atlantic, whereas GDH1 was calibrated using both depth and heat flux (from crust >50 Ma) from the North Pacific and northwest Atlantic. In order to avoid biasing the age–depth relation and use as much data as possible, both Parsons and Sclater (1977) and Stein and Stein (1992) include data close to hot-spots such as Hawaii and Bermuda. GDH1 represents a marked improvement in fit to the expanded data set relative to PS77 (Stein and Stein, 1992). The GDH1 plate is substantially thinner and hotter than PS77 as indicated by the respective model parameters (Table 10.1). Asymptotic values of heat flux are 48 and 34 mW m^{-2}, for GDH1 and PS77, respectively. These values represent the background flux of heat from the asthenosphere carried by convection. The plate models and their predictions characterize the average seafloor, but variations between and within plates are large such that best-fitting thermal parameters are different for different ocean basins (Stein and Stein, 1994a). Johnson and Carlson (1992) made a fit of the plate model to DSDP/ODP drilling data (sediment thickness and basement depth) and found best-fitting model parameters intermediate between those of PS77 and GDH1.

Figure 10.1 shows that for crustal ages younger than 80–100 Ma, the depth increases linearly with the square-root of age (e.g. Davis and Lister, 1974; Parsons and Sclater, 1977; Sclater and Wixon, 1986; Renkin and Sclater, 1988; Stein and Stein, 1992). Parsons and Sclater (1977) emphasized that because the plate cools from the top down, both the plate model and half-space cooling models are essentially the same until cooling has progressed to the point where, in the case of the plate model, further cooling is retarded by the lower boundary condition. The time taken for a thermal disturbance to have a significant effect (16% perturbation) at a given depth is defined by a thermal length calculation where:

$$t = \frac{l^2}{4\kappa}. \qquad (10.10)$$

For a thermal diffusivity κ of 8×10^{-7} m^2 s^{-1}, the thermal length, l, corresponding to 100 Ma is about 100 km. At ages older than about 100 Ma, ocean depth appears to flatten with respect to the half-space model and shows greater variability. One reason for the large variability in bathymetry is that old lithosphere such as the western half of the Pacific plate is densely populated by seamounts and oceanic plateaus (Renkin and Sclater, 1988; Wessel, 2001).

10.2.3 A preferred reference model

The choice of an appropriate reference for the thermal evolution of oceanic lithosphere remains somewhat contentious, and depends in part on the objectives at hand. The choice of reference model is not clear cut because no single model, whether it is a plate model or half-space model, adequately explains all of the data (Stein and Stein, 1992; Carlson and Johnson, 1994). Regional variations of ocean depth systematically exceed assigned uncertainties and such departures from the plate model are significant (e.g. Marty and Cazenave, 1989). Overall, plate models provide a better fit to heat flux and bathymetric data than half-space models, especially for crust of older ages (Fig. 10.1). Thus plate models, such as GDH1, provide the best average description of observed heat flux and depth as a function of age and may serve as a reference against which to measure anomalies. The principal drawback of the plate model is that it is a simple mathematical abstraction for a lithosphere consisting of a rigid mechanical and a thermal boundary layer (Parsons and McKenzie, 1978). The plate model does not address the process that limits growth such that the plate thickness and lower boundary temperature are really only parameters of convenience. McNutt (1995) summarizes various candidate processes that might limit the thickness of the lithosphere, but to date these processes remain poorly resolved.

The half-space boundary-layer cooling model provides a reasonable fit to cooling for crust as old as 80–100 Ma, but predicts deeper ocean depths and lower heat flux values in older lithosphere than are generally observed (Fig. 10.1). One possible explanation for the discrepancy involves non-uniformitarian processes, whereby major transient thermal events in the past have affected both heat flux and bathymetry for the oldest seafloor. If this is the case, no simple model (i.e. uniform seafloor spreading with constant thermal boundary conditions) could explain all the data (Davis, 1988; Lister *et al.*, 1990). One example of this class of models has been to combine half-space cooling with plate reheating (Heestand and Crough, 1981; Carlson and Johnson, 1994; Nagihara *et al.*, 1996). Proponents of the reheating model argue that when the distance from hot-spots and hot-spot swells is large the seafloor subsides in agreement with the half-space cooling model. The northwest and southwest Atlantic, two regions relatively unaffected by hot-spot activity, subside according to simple half-space cooling theory over large areas (e.g. Marty and Cazenave, 1989). Nagihara *et al.* (1996) pursued this argument by carefully selecting a series of wide flat basins of different crustal ages. In this way topographic subsidence can be used to predict heat flux, with the only model assumption being that the heat flux is related to subsidence through thermal expansivity, Eq. (10.3). They found that these basins fall closer to a boundary layer cooling curve than PS77 or GDH1. Furthermore, these basins plot on the age–depth curve younger than their crustal ages, suggesting a reheating event.

For the purpose of having a reference thermal model to evaluate heat flux deficits in young seafloor and the magnitude of hydrothermal circulation, it seems prudent to select a simple reference model appropriate for young seafloor that satisfies the following criteria: (a) the model should be physically based; (b) when compared with observations, the model should yield anomalies that also have a physical explanation; (c) the model should provide a reasonable fit to available data; and (d) the model should be simple so that observations can easily be compared to model predictions. One should also note that there are many geologic processes that can add or inject heat into the lithosphere such as stretching, plume interaction, intrusion, or magmatic underplating, but few processes extract heat. Thus, most lithospheric thermal anomalies that have a physical explanation (criterion (b)) should be expected to be positive.

A useful analogy can be made to gravity reference models and anomalies. The gravity reference is physically based, being computed for a rotating Earth and given on a reference spheroid by a simple formula. At heights above the reference spheroid one can compute a free-air gravity anomaly, a Bouguer gravity anomaly, or an isostatic gravity anomaly, each of which involves different assumptions about mass compensation in the Earth. In each case, gravity anomalies can be explained by mass excesses or deficiencies (caused by rock density distributions) relative to the model assumptions. The fact that the Bouguer gravity prediction fails to match observations at high elevations does not negate the usefulness of the simple and convenient Bouguer gravity anomaly in assessing mass excesses and deficiencies in the crust. Likewise with thermal reference models and thermal anomalies, the failure of the half-space model to match observations in seafloor older than 100 Ma does not negate the usefulness of the half-space model in determining heat flux deficits and excesses in much of the seafloor today, and especially in young seafloor.

For assessing the hydrothermal heat flux deficit, we choose as a reference thermal model, particularly applicable in seafloor up to 100 Ma old, the half-space model for which surface heat flux is given by:

$$f = \frac{C}{\sqrt{t}} \tag{10.11}$$

Average seafloor heat flux between ages t_1 and t_2 for the reference model is given by

$$f_{t_1 - t_2} = \frac{2C}{(t_2 - t_1)}(\sqrt{t_2} - \sqrt{t_1}) \tag{10.12}$$

and average heat flux from the mid-ocean ridge to an isochron of age t is:

$$f_{\text{avg}} = \frac{2C}{\sqrt{t}} \tag{10.13}$$

A comparison of model predictions with "unperturbed" heat flux data in oceanic crust is shown in Fig. 10.2. The data are restricted to well-sedimented sites known to be more than 20 km from any basement outcrop to eliminate or minimize the effects of open hydrothermal circulation (Sclater *et al.*, 1976; Davis, 1988). Although the data have considerable variance, especially in seafloor younger than 10 Ma, they show good agreement with an inverse square-root age boundary-layer cooling curve out to an age of roughly 100 Ma. A constant value of C of 500 mW m^{-2} M.y.$^{1/2}$ provides a good fit to existing data.

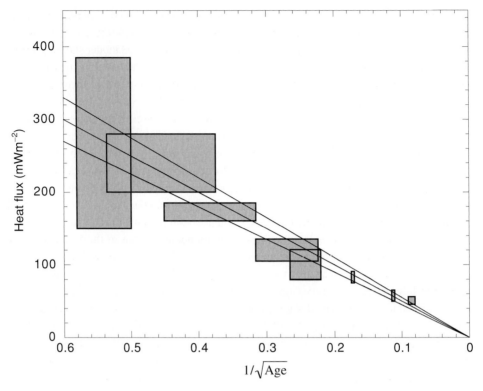

Fig. 10.2 Reliable heat flux data for young seafloor (Sclater *et al.*, 1976) plotted against the reciprocal square-root of age (in Ma) to test model predictions. Solid lines represent half-space lithosphere cooling models for values of the constant *C*, Eq. (10.11) equal to 475, 500, and 525 mW m^{-2} M.y.$^{1/2}$, respectively.

For seafloor older than 100 Ma, this reference thermal model consistently predicts heat flux that is lower than observed heat flux by about 10 mW m^{-2} just as the Bouguer gravity model prediction is too high by about 200 mGal for large expanses of elevated continents such as western North America. The physical explanation for the Bouguer gravity anomaly is a low-density crustal root that is not included in the gravity model. The physical explanation for the old seafloor heat flux anomaly is additional heat into the base of the lithosphere that is not included in the reference model as discussed above.

10.3 Hydrothermal circulation and heat flow deficits

Two common characteristics of heat flux through young seafloor are well illustrated in Fig. 10.1. First, while heat flux values are usually higher than the global average, they are substantially lower than predicted by conductive thermal models of lithospheric evolution (the reference model). Second, heat flux values for young crust almost invariably exhibit significantly more variability than observations on older seafloor. Several lines of reasoning suggest both of these characteristics are the consequence of hydrothermal circulation (Lister, 1972). As discussed previously, the vast majority of heat flux observations are made by inserting a several-meter-long heat flow probe into sediments. In areas of hydrothermal circulation, sediment ponds are typically areas of lower than expected heat flux due to the cooling effect of ventilation through surrounding outcrop (Chapters 8, 11, 12). Heat that would normally reach the surface in the absence of fluid flow is carried laterally to a discharge region. This process lowers the thermal gradient in the overlying sediments over a considerable distance from the outcrops themselves, biasing even randomly distributed heat flux observations toward lower values. This bias is made worse because measurements cannot be made in areas of outcrop where discharge normally occurs. As a result, values significantly greater than the reference level are rarely observed, and average heat flux is often well below the reference model prediction. In areas of vigorous crustal fluid circulation and lateral flow, the sediment–crust interface is nearly isothermal, and the thermal gradient is a strong function of the sediment thickness. This often accounts for much of the variability in heat flux values through young seafloor. Thermal effects of chemical reactions between the water and crust are considered negligible (Wolery and Sleep, 1976), and other sources of environmental noise, such as sedimentation, recent slumping, and thermal refraction, account for neither the heat flux deficit nor observed variability (e.g. Von Herzen and Uyeda, 1967; Langseth and Von Herzen, 1970).

Figure 10.3 is a multi-scale schematic diagram depicting a heat flux deficit determined from either an individual survey result, a heat flux compilation for a particular spreading ridge, or a compilation for a global data set. Consider a heat flux deficit ($f_{obs} - f_{ref}$) for an element of seafloor between age t_1 and t_2. The rate of heat extracted by hydrothermal circulation is:

$$F_i = (f_{obs} - f_{ref})A_i \qquad (10.14)$$

where A_i is the area of affected seafloor within the isochrons t_1 and t_2. The total rate of heat extracted for a particular spreading center or for the global ridge system is the summation of (10.14) for that ridge segment or over all the seafloor exhibiting a heat flow deficit:

$$F_{tot} = \sum F_i \qquad (10.15)$$

Sclater *et al.* (1980; Table A1, and reproduced in modified form by Stein and Stein, 1994b) provide a convenient set of area–age values for the seafloor. The heat extracted from the upper oceanic crust is delivered advectively back into the ocean by a volume flow of water Q_v given by:

$$Q_v = \frac{F}{\rho c(T_2 - T_1)} \qquad (10.16)$$

where c is the specific heat of water and T_1 and T_2 are recharge and discharge temperatures, respectively, at the seafloor (Fig. 10.3).

A summary of estimates of global hydrothermal circulation heat loss rates is given in Table 10.2. It is interesting to note that two of these estimates (Williams and Von Herzen,

Table 10.2 *Estimates of heat flow deficit based on thermal measurements*

Study	Age range (Ma)	Advective heat flow (W)	Fluid circulation temperature (°C)	Mass flow (kg yr^{-1})
Williams and Von Herzen (1974)	0–2	8.5×10^{12}		–
Wolery and Sleep (1976)	0–17	5×10^{12}	50–300	$1.3 – 8.1 \times 10^{14}$
Sclater *et al.* (1980)	0–50	10×10^{12}		–
Stein and Stein (1994b)	1–65	11×10^{12}	~50	1.3×10^{15}
Elderfield and Schultz (1996)	1–65	7×10^{12}	5–15	$3.7 – 11.0 \times 10^{15}$

1974; Wolery and Sleep, 1976) were made before 1979 when the discovery of black smokers provided the first direct evidence of seafloor hydrothermal venting. Williams and Von Herzen (1974) did not perform the summation in (10.15), but instead argued that oceanic heat flux could be partitioned into a background of 47 mW m^{-2} and a transient heat flux resulting from lithospheric cooling. Their analysis of heat flux measurements on young crust suggested that only a small part of the transient heat was released by thermal conduction in very young seafloor. They calculated that if all of the transient heat in crust out to 2 Ma (equivalent to half the total transient heat) is removed by hydrothermal circulation, the global advective heat loss rate would be 8.5×10^{12} W, or about 20% of the global heat loss.

Subsequent investigators based their estimates of hydrothermal heat loss on actual integrations of the observed seafloor heat flow deficit. Wolery and Sleep (1974) first performed the integration on a global data set but divided the globe into fast-spreading ridges and slow-spreading ridges to simplify the integration. Data analyses at that time suggested heat flow deficits to seafloor ages 17 and 23 Ma in fast- and slow-spreading ridges, respectively, corresponding to a removal of 32 and 42% of lithospheric cooling heat for those ridge systems through hydrothermal convection (Wolery and Sleep, 1976). The global heat loss rate through advection was computed to be 5×10^{12} W. The global mass flow though the seafloor to account for the advective heat removal was calculated by Wolery and Sleep (1976) to be between 8.1 and 1.3×10^{14} kg yr^{-1}, assuming discharge temperatures ranging between the limits of 50 and 300 °C, respectively.

Sclater *et al.* (1980) applied the heat flux deficit analysis in a systematic way to all oceans. Using all available data, they computed the average of marine heat flux observations for the five major ocean basins (North Pacific, South Pacific, Indian, North Atlantic, South Atlantic) for 13 age groups. For each of the age brackets, they compared observed heat flux to a reference heat flux from their lithospheric cooling model (Table 10 of Sclater *et al.*, 1980). The total advective heat loss rate out to 65 Ma seafloor was computed to be 10.3×10^{12} W, with almost 60% of that heat loss occurring in crust younger than 9 Ma. Subsequently, Stein and Stein (1994a, 1997) used a global marine heat flow data set filtered for quality (Stein and Abbott, 1991) and averaged into 2 M.y. bins, and compared the data to GDH1 to calculate the heat flow deficit as a function of age. Stein and Stein (1994b) estimated a global heat loss rate of 11×10^{12} W. This estimate is larger than previous

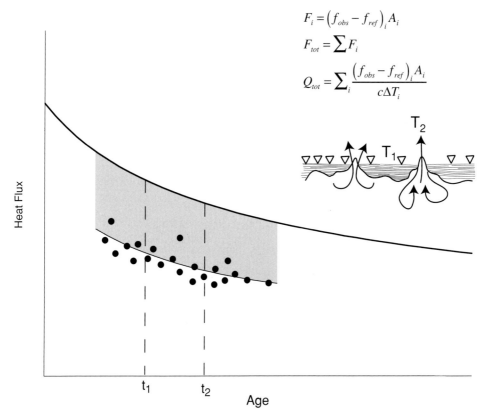

$$F_i = \left(f_{obs} - f_{ref}\right)_i A_i$$

$$F_{tot} = \sum F_i$$

$$Q_{tot} = \sum_i \frac{\left(f_{obs} - f_{ref}\right)_i A_i}{c\Delta T_i}$$

Fig. 10.3 Schematic diagram illustrating a heat flow deficit resulting from open hydrothermal circulation in the oceanic crust. A deficit (shaded region) exists when the heat flux measurements (solid circles) are systematically lower than a reference curve (solid line) for modeled lithosphere cooling. Deficits exist because most heat flux measurements are made in sediment ponds where fluid recharge in surrounding outcrop carries away some of the lithospheric heat flux. Note that for significant advective heat loss, rapid recharge also needs to take place in areas where basement crops out. The integrated heat flow deficit, F_{tot}, gives an estimate of the total heat lost through the seafloor by advection, and has been used with assumed values of average fluid circulation temperature to estimate the total mass flow of water, Q_{tot}, to extract that heat (Table 10.2). A proper estimate for Q_{tot} (equation shown) requires use of a circulation temperature that is locally determined, since basement temperatures in ridge-flank settings vary over a large range.

estimates of hydrothermal heat flow (Table 10.2) for two reasons. First, GDH1 has a hotter basal reference temperature than used by previous investigators and thus their inferred heat flux deficits are larger for all age brackets. Second, the Stein and Stein (1994a) analysis suggests that ventilated hydrothermal circulation extends to crust as old as 65 Ma for all oceans, and thus the integration is performed over a relatively large area of seafloor. The mass flow of 1.3×10^{15} kg yr^{-1} inferred from this thermal analysis is also larger than given by previous investigators; that difference arises primarily from assumptions about discharge temperatures, Eq. (10.16). Stein and Stein (1994b) assume that high-temperature (250 °C) discharge characterizes seafloor younger than 1 Ma, but that the rest of the seafloor exhibiting

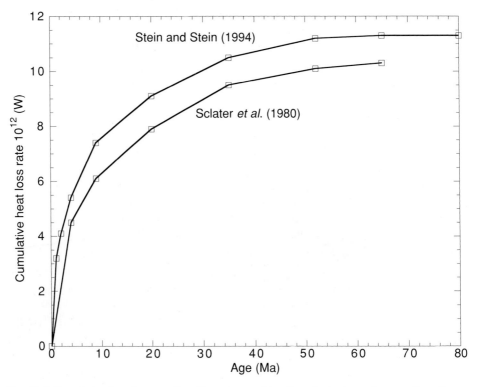

Fig. 10.4 Cumulative heat loss rate for all oceans resulting from hydrothermal convection. Curves were estimated by integrating the heat flow deficit (defined using the preferred lithospheric reference of Section 10.2.3) between seafloor isochrons over the area of the seafloor between the same isochrons.

a heat flow deficit has a lower temperature discharge of 50 °C . Even this lower assumed temperature may be too high (Wheat and Mottl, Chapter 19). Elderfield and Schultz (1996) based an estimate of off-axis mass flow on an assumed formation water temperature in the range of 5–15 °C. Using an off-axis advective heat loss estimate based on the average heat flux compilation of Stein and Stein (1994b) and a cooler plate than predicted by GDH1, they calculated a global advective flow between 3.7 and 11×10^{15} kg yr^{-1}.

Two estimates for the cumulative advective heat loss for the globe as a function of seafloor age are shown in Fig. 10.4 (Sclater *et al.*, 1980; Stein and Stein, 1994b). Although the temporal extent of this heat loss and the physical conditions that diminish and eventually eliminate the heat flux deficit (e.g. sediment sealing, crustal alteration, and permeability loss) remain uncertain, the result for young seafloor is robust. About 5×10^{12} W, or half the heat loss rate, is associated with seafloor younger than 4 Ma. Two-thirds of the advective heat loss occurs in seafloor younger than 8 Ma, and 83% in crust younger than 20 Ma. Thus the cumulative value of approximately 10^{13} W is unlikely to change significantly by the addition of new data, unless detailed surveys alter our view of the processes that cause heat flux deficits. The same robustness cannot be ascribed to the estimation of volumetric or mass flow, for the effective discharge temperature for ridge flank hydrothermal circulation systems remains poorly defined. Use of a temperature of 50 °C in (10.16) together with the cumulative heat loss yields a global hydrothermal mass flow of the order of 10^{15} kg yr^{-1}. The mass of the oceans is 1.37×10^{21} kg, so the inferred hydrothermal mass flow suggests that the entire mass of the oceans may cycle through the upper crust in about one million years. Lower effective formation fluid temperatures are likely (see Fig. 19.6), so the cycling time may be even shorter.

While the global marine heat flux data set yields broad global estimates of hydrothermal heat loss and mass flow (Sclater *et al.*, 1980; Stein and Stein, 1994b; Stein *et al.*, 1995), there are troubling discrepancies between inferences about hydrothermal circulation drawn from the global analyses and those drawn from detailed local studies. For example, numerous local studies have pointed to the importance of sediment thickness, completeness of sediment cover, and basement relief as having controlling local influences on the vigor, pattern, and degree of ventilation of hydrothermal circulation (Sclater, 1970; Sclater *et al.*, 1974, 1976; Lister, 1972; Anderson and Hobart, 1976; Davis and Lister, 1977; Anderson *et al.*, 1977; Davis *et al.*, 1989, 1999; Abbott *et al.*, 1992; Chapter 6). Several others have demonstrated that observed heat flux can equal predicted heat flux at ages that are determined by a threshold continuity and thickness of sediment accumulation. Ventilation is effectively stopped by sediment cover at 10 Ma on the southern East Pacific Rise (Villinger *et al.*, 2002), 7–8 Ma at the Galapagos Spreading Center (Sclater *et al.*, 1974), 5–6 Ma at the Costa Rica Rift (Langseth *et al.*, 1983), and 1.5 Ma at the Juan de Fuca Ridge (Davis *et al.*, 1999). In contrast, the Stein and Stein (1994b) analysis of the global data suggests a uniform sediment-sealing age of 65 Ma for all oceans, i.e. that sediment thickness plays a relatively minor role governing the longevity of hydrothermal ventilation. And in some enigmatic settings, heat flux deficits have been found to be unusually large even where sediments appear to be thick and continuous, such as in the region of generally well-sedimented seafloor off central America (Langseth and Silver, 1996; Fisher *et al.*, 2003). For the most part, these discrepancies may result from the vintage of heat flux measurements and the increasing recognition that environmental factors provide a strong control on the nature of heat transfer. Many measurements in the global heat flux data set are often 50–100 km apart and have only sparse environmental information (basement relief, sediment thickness, and cover extent), whereas modern surveys have paid more attention to environmental controls with complementing seismic surveys and extensive swath mapping. With greater ability to navigate precisely and to position ships dynamically (often to within meters), heat flux transects with penetrations only tens to hundreds of meters apart are becoming more routine. These closely spaced measurement transects coupled with environmental information are revealing new insights into the nature of hydrothermal circulation, and allowing the global heat flux data set to be interpreted with greater wisdom.

Whatever the details regarding the timing of the sealing process, it is certain that hydrothermal circulation evolves with the aging of oceanic crust (Fig. 10.5; Lister, 1972, 1983; Davis and Lister, 1977; Jacobson, 1992; Stein and Stein, 1994b; Carlson, 1998). In young seafloor near ridge axes, hydrothermal circulation is characterized by rapid flow rates and unencumbered exchange of seawater with the crystalline crust. This style of circulation is manifested in heat flux values that are significantly lower than predicted by cooling

Basement	Rough or smooth	Rough or smooth	Rough	Smooth	Rough	Smooth
Sediment cover	Sparse or none	Incomplete	Extensive and thin	Extensive and thin	Extensive and thick	Extensive and thick
f/f_{ref}	0.1–0.2	0.1–0.2	0.5	~1	~1	~1
Std. dev. mW m^{-2}	80	70	60–70	20–40	60	20

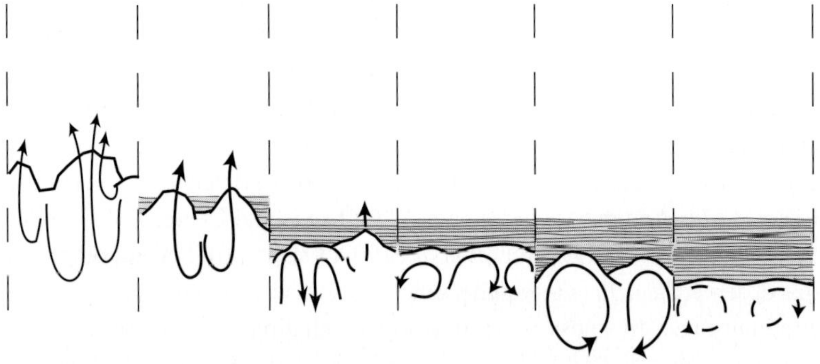

Fig. 10.5 Cartoon showing possible hydrothermal convection patterns and their heat flux conse-quences for six hydrologic environments with varying basement relief and sediment cover. Crustal age generally increases to the right.

models of oceanic lithosphere. Heat flux values of 10–20% of the reference heat flux are not uncommon, and heat flux is highly variable (Fig. 10.5). The convective regime is not radically changed with the gradual addition of sediment, especially if the sediment cover is incomplete and concentrated to sediment ponds within an otherwise bare rock environ-ment. Because measurements can be made in basement lows where sediment is present, this environment is strongly represented in heat flux compilations. Standard deviations of heat flux surveys in this regime can be 70 mW m^{-2} or more, with total variablility even greater, i.e. comparable to the mean measured heat flux (e.g. Sclater *et al.*, 1974). With greater age and the addition of hydraulically resistive sediment, the hydrothermal circula-tion evolves from an open (ventilated) to a closed (recirculating) pattern (Fig. 10.5), but until the sediment is sufficiently thick and continuous to isolate the permeable igneous crust from the ocean, there may still be significant leakage through infrequent basement ridges and partially buried seamounts. Such leakage may produce very high but geographically concentrated heat flux anomalies surrounded by broad areas with as much as 50% heat flux deficit. Variability in heat flux over rough basement topography with complete sediment drape can still be high (Fig. 10.5), but once the sediment has effectively isolated circulation within the igneous crust from exchange with the ocean, the mean heat flux values approach the reference value. Closed circulation redistributes heat within the circulation system, but the integrated surface heat flux will equal the cooling lithospheric heat flux in this regime. However, given that buried basement topography exerts a significant control on the heat flow pattern, incomplete spatial sampling can still produce a bias toward lower heat flux. As burial continues, heat flux becomes more uniform, although the signature of hydrothermal circulation may still be present. Wherever variations exceed those expected from conductive refraction effects, redistribution of heat in the crust beneath the sediments is likely to be the cause (Chapter 13), and caution must be exercised when using widely spaced measurements to estimate the regional level of heat flux.

Because many of the important variables describing the local environment at each measurement of the global heat flux data set are often poorly known, drawing conclusions about hydrologic processes at a local scale is clearly problematic, and assessing regional average heat flow can only be done with considerable uncertainty. In contrast, local surveys with dense data coverage provide greater context for understanding heat transport processes and interpreting heat flow deficits. One such example carried out at a young ridge flank site is considered next.

10.4 The eastern flank of the Juan de Fuca Ridge

The eastern flank of the northern Juan de Fuca Ridge has been the setting for numerous studies of heat flux and hydrothermal circulation in young oceanic crust. There is extensive sediment cover on seafloor as young as 0.6 Ma, a variety of basement environments (flat areas, basement ridges, and isolated outcrops referred to as permeable penetrators), and relatively easy access. With respect to sedimentation, the eastern flank of the Juan de Fuca Ridge is anomalous, but the rapidly deposited sediments allow a high density of heat flux measurements to be made and show conclusively that sediment exerts a strong influence on the age at which the heat flux deficit becomes negligible (Davis *et al.*, 1999).

The main heat flux result from this area is shown in Fig. 10.6. Heat flux is shown for a transect that starts about 20 km from the ridge axis where seafloor age is 0.6 Ma and continues to the southeast for about 100 km where the seafloor age is 3.7 Ma. Heat flux nearest the axial region of igneous rock outcrop is only 100 mW m^{-2}, about 15% of that expected from cooling lithosphere models, but increases to expected heat flux over a distance of about 20 km where the age of the seafloor is 1.2 Ma. Heat flux along the entire transect is very coherent; the local variability at the kilometer scale is closely related to buried basement topography, with high heat flux over topographic highs and low heat flux over topographic lows. The systematic increase in heat flux away from the basement outcrop can be explained by lateral flow of cool seawater from the area of extensive outcrop in the vicinity of the ridge axis into the upper igneous crust that lies under the sediment cover. The rate of fluid flow is estimated to be of the order of meters to tens of meters per year (Davis *et al.*, 1999; Stein and Fisher, 2002; Chapter 11), although the details of flow, including recharge, discharge, and local convective mixing, has not been determined.

The observation that heat flow returns to the reference value predicted from lithosphere cooling at 1.2 Ma does not mean that hydrothermal circulation ceases at that point. Quite to the contrary, there is considerable evidence to suggest the presence of vigorous closed convection within the igneous crust throughout the transect. Downward continuation of heat flux values along the transect in Fig. 10.6 (Davis *et al.*, 1989) initially suggested that the

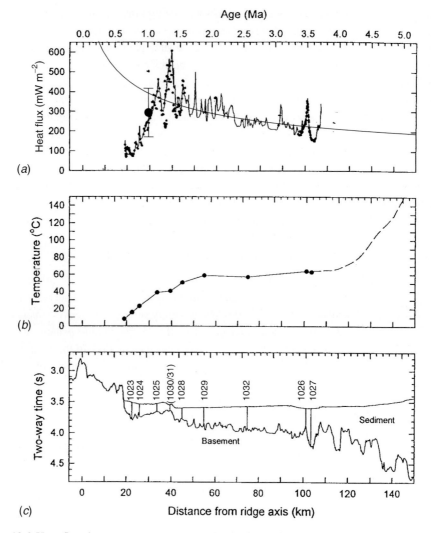

Fig. 10.6 Heat flux, basement temperature, and seismic section for a Juan de Fuca Ridge flank transect crossing through ODP Leg 168 drilling Sites 1023–1032. (*a*) Heat flux measured at the seafloor (dots) and estimated on the basis of sediment thickness and basement temperatures (line). (*b*) Basement temperatures measured at drilling sites (dots) and interpolated between drilling sites (line). (*c*) Seismic travel time to the seafloor and basement (from Davis *et al.*, 1999), and locations of Leg 168 drilling sites.

top of the basement beyond 1.5 Ma crust was roughly isothermal at a temperature between 50 and 70 °C. This inference has subsequently been confirmed by drilling and post-drilling observations along the ODP Leg 168 transect (Davis *et al.*, 1997) where temperatures determined at the top of the basement for a considerable part of the transect varied only from 60 to 64 °C in spite of considerable variations in depth to basement (Fig. 10.6). The most convincing confirmation of the isothermal uppermost basement temperature condition comes from ODP Sites 1026 and 1027 (separated by 2.2 km). Whereas sediment thickness differs by a factor of 2.5, the basement temperatures differ by only 2 °C. Such an isothermal condition requires vigorous convection (Chapters 8 and 12).

The conditions that exist on the Juan de Fuca Ridge also provide an opportunity for interpolating heat flux between measurement points and estimating continuous profiles or maps of heat flux across the seafloor. Such profiles and maps allow true spatial integration of heat flux and heat flux deficits. A continuous heat flux profile for the Juan de Fuca Ridge (Fig. 10.6) was made by combining seafloor heat flux data, sediment physical and thermal properties, seismic reflection times to basement, and ODP measurements of basement depth and temperature (Davis *et al.*, 1999). The continuous profile shows a heat flux deficit in crust younger than 1.2 Ma but the mean heat flow for seafloor older than 1.2 Ma agrees remarkably well with half-space cooling when averaged over a spatial scale that fully spans the scale of local variability (Fig. 10.6).

The heat flux deficit over the sedimented seafloor between ages 0.6 and 1.2 Ma (Fig. 10.6*c*) has been integrated to yield a cumulative heat loss of about 3 MW km^{-1} of strike length for this section of the Juan de Fuca Ridge (Davis *et al.*, 1999). If the conductive heat flux over the bare-rock section from 0 to 0.6 Ma seafloor is no more than the 100 mW m^{-2} observed at the sediment onlap, then an additional 16 MW km^{-1} strike length is being lost by advection over bare rock for a total heat loss rate of 19 MW km^{-1} strike length. Thus a value of about 20 MW km^{-1} strike-length is a robust minimum for mid-ocean ridge advective heat loss; advective heat loss on ridges with less sediment and therefore a greater area to host open-exchange circulation (Fig. 10.5) could be as high as 170 MW km^{-1} strike-length if open exchange continues to a more typical seafloor age of 65 Ma as suggested by Stein and Stein (1994b).

The advantage of detailed heat flux studies over global statistical studies in understanding hydrothermal processes in the seafloor is highlighted in Fig. 10.6, where all the 0–2 Ma heat flow data for the transect are compressed into a single bin as would be done in the global compilations. The average of 96 measurements is 389 mW m^{-2} with a standard deviation of 131 mW m^{-2}. Although this single value would be interpreted to have a deficit of 318 mW m^{-2} compared to the average expected heat flux of 707 mW m^{-2} for seafloor 0–2 Ma, neither the return to predicted heat flux at 1.2 Ma, the important correlation of heat flux with basement topography, nor the nature of the heat flux deficit recovery would have been revealed.

10.5 Discussion

Recent, high density heat flow surveys collocated with extensive seismic profiling and often accompanied by piston coring, pore-water chemistry, and ODP drilling are shedding new light on controls for hydrothermal convection in young seafloor (Davis *et al.*, 1989, 1999; Grevemeyer *et al.*, 1999; Harris *et al.*, 2000; Fisher *et al.*, 2003). Not only do these detailed studies give guidance on how marine heat flux surveys focused on hydrothermal convection

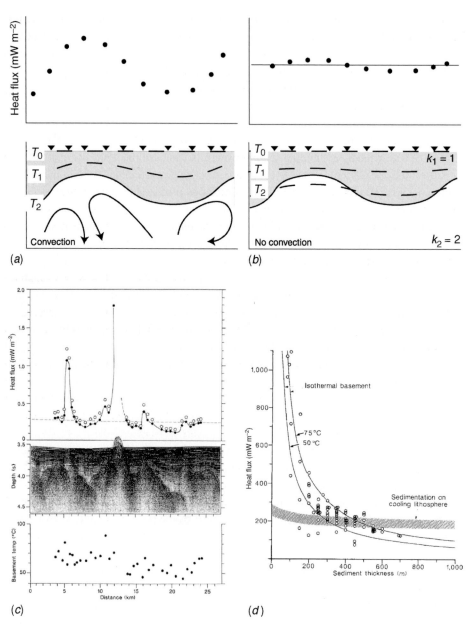

Fig. 10.7 Topographic test for hydrothermal circulation. (*a*) Schematic diagram showing heat flux patterns over basement relief with vigorous convection. (*b*) Schematic diagram showing heat flux pattern over basement relief with no convection. Variations due to the contrast in thermal conductivity between the basement and overlying sediments are normally small. (*c*) An example, from the eastern flank of the Juan de Fuca Ridge, of heat flux over basement relief with vigorous convection. In this example the basement sediment interface is approximately isothermal. Solid circles indicate raw heat flux determinations, open circles denote heat flow corrected for the thermal effects of sedimentation. Dashed line shows the theoretical heat flux expected for the age of seafloor. Basement temperatures

should be conducted in the future, but they also provide insight for interpreting the large body of data gathered under less ideal conditions, limited instrumentation, and imprecise navigation. Most critically, the detailed surveys confirm how important it is to understand the environment surrounding the heat flow sites; this includes defining sediment thickness variations, extent of sediment cover, and basement relief.

It appears that the heat flux deficit (the locally averaged observed heat flux relative to that expected from lithosphere cooling models; Fig. 10.5) remains a robust measure of the ventilation efficiency of open, hydrothermal convection beneath the seafloor. Heat is extracted by pore-water circulation systems recharging and discharging in bare-rock environments where heat flow probe measurements are precluded.

The spatially integrated heat flux deficit over the area of the seafloor hosting ventilated convection provides an estimate of the advective heat loss rate for a particular spreading ridge. Older data compilations are useful for such calculations, although more recent studies with close station spacing and/or a continuous heat flux profile inferred from determinations of basement temperature and burial depth provide a more precise estimate of local heat loss. The most uncertain parameter in converting advective heat loss into mass flow through the upper oceanic crust is basement fluid temperature. Current global estimates of mass flow computed from heat loss observations vary from 10^{14} to 10^{16} kg yr^{-1} (Table 10.2; Chapter 19), corresponding to the entire ocean mass passing though the upper oceanic crust between 1 and possibly less than 0.1 M.y respectively.

There is likely a transition in the heat flux patterns between seafloor that permits ventilated convection and seafloor with sufficient thickness and areal extent of sediment to prevent thermally significant recharge of seawater into the permeable igneous crust and discharge of hydrothermal water back to the ocean. In transition regions, the average measured conductive heat flux through the seafloor must increase from a fraction to the full expected heat flow value, and variability may fall (Fig. 10.5). In the absence of spatial biasing (e.g. where measurements are preferentially made in low-heat-flux environments) the integrated or average heat flow over closed convection systems should equal the lithospheric reference heat flux. High variability in heat flow can persist in spite of thick sediments, however, if there is considerable basement relief, especially if there is vigorous, closed hydrothermal circulation in the upper igneous crust, and measurements must be made with this possibility in mind.

A useful test for hydrothermal circulation in a sedimented environment of the seafloor is a "topo test" in which a closely spaced heat flux survey is conducted over seismically imaged rough basement topography (Fig. 10.7). A "topo test" in marine

---

are estimated based on uncorrected heat flux and the sediment thickness as determined from the seismic reflection profile (Davis *et al.*, 1989). (d) Observed heat flux as a function of sediment thickness for all values in the study area. Reprinted from Davis, E. E., Chapman D. S., Forster, C. B., and Villinger, H. 1989. Heat flow variations correlated with buried basement topography on the Juan de Fuca Ridge flank. Nature, **342**: 533–537, Copyright Nature Publishing Group, http://www.nature.com.

heat flux is loosely analogous to the fold test in paleomagnetism. In the fold test, remanent magnetizations are determined at multiple positions in a single bed (isochron) following the fold. If the remanent magnetizations are unidirectional in the folded geometry, it is inferred that the magnetization was acquired post folding. If the magnetization exhibited the rotations of the fold, and unfolding the bed restores a unidirectional remanent magnetization, then it can safely be assumed that magnetization was acquired prior to folding.

Figure 10.7 illustrates the diagnostic feature of a topo test for hydrothermal circulation. If convection is sufficiently vigorous to create a near-isothermal condition at the top of the permeable igneous basement, then this state will be manifested by heat flux that is inversely proportional to sediment thickness (Fig. 10.7a). High heat flux will occur over buried topographic ridges and seamounts and low heat flux will occur over thickly buried topographic troughs. If hydrothermal circulation does not exist, or if it is insufficiently vigorous to homogenize the upper crust thermally (see Fig. 8.2), the heat flux across rough topography will be constant except for a relatively small thermal refraction and sedimentation effect (Chapter 13). A plot of heat flux versus sediment thickness (Fig. 10.7d) illustrates how large the difference can be between convective and conductive regimes in the topo test. This plot shows data gathered over a basement ridge–trough configuration on the Juan de Fuca Ridge flank (Fig. 10.7c) and confirms the existence of hydrothermal circulation (Fig. 10.7d).

In the absence of detailed knowledge about the environment of the heat flux sites, as is the case for much of the older data, one can still exploit the variability of the heat flux results to make inferences about convection (Chapter 13). Both heat flux variability and the variability normalized by measured heat flux in age groups decrease with age of seafloor (Stein and Stein, 1994b). The decline in variability is inferred to reflect a decline in hydrothermal activity. It may also be a simple consequence of greater sediment thickness, however; detailed studies are required to resolve this ambiguity.

Where extensive marine heat flux surveys have been completed, they have provided a rich data set for examining the process of hydrothermal circulation in the oceanic crust. A continuing goal of marine heat flux studies should be to document for many ridge systems the nature and geographic extent of the three basic thermal regimes: the open circulation regime whose heat flux deficit leads to estimates of heat loss rates and mass fluxes of seawater through the oceanic crust; the closed circulation convection regime where exchange of seawater with the convection cells is precluded by hydraulically resistive sediments; and the purely conductive regime where reduced permeability of the crust and reduced driving forces in older and cooler lithosphere combine to shut down convection.

10.6 Conclusions and recommendations

1. Heat flux remains one of the most useful observational constraints for studying cooling models of oceanic lithosphere and hydrothermal circulation within oceanic crust.
2. For seafloor younger than about 100 Ma, a convenient and useful reference curve for lithospheric cooling is $C/\sqrt{\text{Age}}$. This reference curve is simple, physically based on boundary-layer cooling

theory, and yields anomalies that are also physically based. A value of $C = 500$ mW m^{-2} M.y.$^{1/2}$ provides a good fit to reliable data. For lithosphere older than 80–100 Ma, heat flux levels out at about 45–50 mW m^{-2} as a consequence of some form of chronic or episodic heating from great depth.

3. Global analyses of heat flux data grouped into regular age bins for oceanic crust have been historically important in leading to the recognition of hydrothermal convection as a mechanism to explain heat flux deficits in young seafloor. Global analyses yield average results on global heat flux deficits, and heat loss rates.

4. The advective heat flow for the entire oceanic ridge system is about 10^{13} W, with about two-thirds of the heat loss occurring in seafloor younger than 8 Ma. The mass flow necessary to produce this heat loss is probably greater than 10^{15} kg yr^{-1}, equivalent to cycling the entire mass of the oceans though the upper crust in less than a million years. The total mass flow depends on the temperature difference between recharge and discharge; this is not well defined, thus this mass flow estimate is approximate.

5. Improved estimates for total mass flow require an improved understanding of processes controlling hydrothermal circulation, and this will depend on new closely spaced heat flux surveys with particular attention paid to the environment in which the thermal measurements are made. In particular, constraints on sediment thickness are required to determine basement fluid temperatures.

6. Although carried out in an atypical seafloor environment (high sedimentation rate), experiments on the eastern flank of the Juan de Fuca Ridge provide a useful template for studying thermal aspects of hydrothermal circulation beneath and through the seafloor. The combination of closely spaced heat flux determinations with collocated seismic profiles and strategic cores for sampling pore fluids offers the greatest opportunity for understanding hydrothermal circulation.

Acknowledgments

This paper benefited from useful discussions with C. Stein and A. Fisher, and reviews by R. Von Herzen and C. Jaupart. US National Science Foundation grant OCE-0001944 supported this work.

References

Abbott, D. H., Stein, C. A., and Diachok, O. 1992. Topographic relief and sediment thickness: their effects on the thermal evolution of the oceanic crust. *Geophys. Res. Lett.* **19**: 1,975–1,978.

Anderson, R. N. and Hobart, M. A. 1976. The relation between heat flow, sediment thickness, and age in the eastern Pacific. *J. Geophys. Res.* **81**: 2,968–2,989.

Anderson, R. N., Langseth, M. G., and Sclater, J. G. 1977. The mechanisms of heat transfer through the floor of the Indian Ocean. *J. Geophys. Res.* **82**: 3,391–3,409.

Carlson, R. L. 1998. Seismic velocities in the uppermost oceanic crust: age dependence and the fate of layer 2A. *J. Geophys. Res.* **103**: 7069–7077.

Carlson, R. L. and Johnson, H. P. 1994. On modeling the thermal evolution of the oceanic upper mantle: an assessment of the cooling plate model. *J. Geophys. Res.* **99**: 3,201–3,214.

Chapman, D. S. and Pollack, H. N. 1975. Global heat flow: a new look. *Earth Planet. Sci. Lett.* **28**: 23–32.

Davis, E. E. 1989. Thermal aging of oceanic lithosphere. In *Handbook of Seafloor Heat Flow*, eds. J. A. Wright and K. E. Louden. Boca Raton, FL: CRC Press, pp. 145–168.

Davis, E. E. and Lister, C. R. B. 1974. Fundamentals of ridge crest topography. *Earth Planet. Sci. Lett.* **21**: 405–413.

1977. Heat flow measured over the Juan de Fuca Ridge: evidence for widespread hydrothermal circulation in a highly heat transportive crust. *J. Geophys. Res.* **82**: 4,845–4,860.

Davis, E. E., Chapman, D. S., Forster, C. B., and Villinger, H. 1989. Heat-flow variations correlated with buried basement topography on the Juan de Fuca Ridge flank. *Nature* **342**: 533–537.

Davis, E. E., Chapman, D. S., Villinger, H., Robinson, S., Grigel, J., Rosenberger, A., and Pribnow, D. 1997. Seafloor heat flow on the eastern flank of the Juan de Fuca Ridge: data from "Flankflux" studies through 1995. In *Proceedings of the Ocean Drilling Program, Initial Reports*, Vol. 168, eds. E. E. Davis, A. T. Fisher, J. V. Firth, *et al.* College Station, TX: Ocean Drilling Program, pp. 23–33.

Davis, E. E., Chapman, D. S., Wang, K., Villinger, H., Fisher, A. T., Robinson, S. W., Grigel, J., Pribnow, D., Stein, J., and Becker, K. 1999. Regional heat-flow variations across the sedimented Juan de Fuca Ridge eastern flank: constraints on lithospheric cooling and lateral hydrothermal heat transport. *J. Geophys. Res.* **104**: 17,675–17,688.

Elderfield, H. and Schultz, A. 1996. Mid-ocean ridge hydrothermal fluxes and the chemical composition of the ocean. *Ann. Rev. Earth Planet. Sci.* **24**: 191–224.

Fisher, A. T., Stein, C. A., Harris, R. N., Wang, K., Silver, E. A., Pfender, M., Hutnak, M., Cherkaoui, A., Bodzin, R., and Villinger, H. 2003. Abrupt thermal transition reveals hydrothermal boundary and role of seamounts within the Cocos Plate. *Geophys. Res. Lett.* **30**, doi. 10.1029/2002GLO16766.

Fowler, C. M. R. 1990. *The Solid Earth.* Cambridge: Cambridge University Press.

Grevemeyer, I., Kaul, N., Villinger, H., and Weigel, W. 1999. Hydrothermal activity and the evolution of the seismic properties of upper oceanic crust. *J. Geophys. Res.* **104**: 5,069–5,079.

Harris, R. N., Von Herzen, R. P., McNutt, M. K., Garven, G., and Jordahl, K. 2000. Submarine hydrogeology of the Hawaiian archipelagic apron. 1. Heat flow patterns north of Oahu and Maro Reef. *J. Geophys. Res.* **105**: 21,353–21,369.

Heestand, R. L. and Crough, S. T. 1981. The effect of hot spots on the oceanic age–depth relation. *J. Geophys. Res.* **86**: 6,107–6,114.

Jacobson, R. S. 1992. Impact of crustal evolution on changes of the seismic properties of the uppermost oceanic crust. *Rev. Geophys.* **30**: 23–42.

Jessop, A. M., Hobart, M., and Sclater, J. G. 1976. *World Wide Compilation of Heat Flow Data, Geothermal Series*, No. 5. Ontario: Department of Energy, Mines and Resources.

Johnson, H. P. and Carlson, R. L. 1992. Variation of sea floor depth with age: a test of models based on drilling results. *Geophys. Res. Lett.* **19**: 1,971–1,974.

Johnson, H. P. and Hutnak, M. 1996. Conductive heat flow measured in unsedimented regions of the seafloor. *Eos, Trans. Am. Geophys. Union* **77**: 321–324.

Langseth, M. G. and Silver, E. A. 1996. The Nicoya convergent margin: a region of exceptionally low heat flow. *Geophys. Res. Lett.* **23**: 891–894.

Langseth, M. G. and Von Herzen, R. P. 1970. Heat flow through the floor of the world oceans. In *The Sea*, Vol. 4, Part 1, ed. A. E. Maxwell. New York: Wiley Interscience, pp. 299–352.

Langseth, M. G., Cann, J. R., Natland, J. H., and Hobart, M. 1983. Geothermal phenomena at the Costa Rica Rift: background and objectives for drilling at Deep Sea Drilling Project Sites 501, 504, and 505. In *Initial Reports of the Deep Sea Drilling Project*, Vol. 69, eds. J. R. Cann, M. G. Langseth, J. Honnorez, *et al.* Washington, DC: US Govt. Printing Office, pp. 5–29.

Lee, W. H. K. and Uyeda, S. 1965. Review of heat flow data. In *Terrestrial Heat Flow, Geophysical Monograph*, Vol. 8, ed. W. H. K. Lee. Washington, DC: American Geophysical Union, pp. 87–190.

Lister, C. R. B. 1972. On the thermal balance of a mid-ocean ridge. *Geophys. J. Int.* **26**: 515–535.

1983. The basic physics of water penetration into hot rocks. In *Hydrothermal Processes at Seafloor Spreading Centers*, eds. by P. A. Rona, K. Bostrom, L. Laubier, and K. L. Smith Jr. New York: Plenum, pp. 141–168.

Lister, C. R. B., Sclater, J. G., Davis, E. E., Villinger, H., and Nagihara, S. 1990. Heat flow maintained in ocean basins of great age: investigations in the north-equatorial west Pacific. *Geophys. J. Int.* **102**: 603–630.

Marty, J. C. and Cazenave, A. 1989. Regional variations in subsidence rate of oceanic plates: a global analysis. *Earth Planet. Sci. Lett.* **94**: 301–315.

McKenzie, D. P. 1967. Some remarks on heat flow and gravity anomalies. *J. Geophys. Res.* **72**: 6,261–6,273.

McNutt, M. K. 1995. Marine geodynamics: depth–age revisited. *Rev. Geophys., Supp.* **33**: 413–418.

Nagihara, S., Lister, C. R. B., and Sclater, J. G. 1996. Reheating of old oceanic lithosphere: deductions from observations. *Earth Planet. Sci. Lett.* **139**: 91–104.

Parker, R. L. and Oldenburg, D. W. 1973. Thermal model of ocean ridges. *Nature Phys. Sci.* **242**: 137–139.

Parsons, B. and Mckenzie, D. P. 1978. Mantle convection and the thermal structure of the plates. *J. Geophys. Res.* **83**. 4,485–4,496.

Parsons, B. and Sclater, J. G. 1977. An analysis of the variation of ocean floor bathymetry and heat flow with age. *J. Geophys. Res.* **82**: 803–827.

Pollack, H. N., Hurter, S. J., and Johnston, J. R. 1993. Heat flow from the Earth's interior: analysis of the global data set. *Rev. Geophys.* **31**: 267–280.

Renkin, M. L. and Sclater, J. G. 1988. Depth and age in the North Pacific. *J. Geophys. Res.* **93**: 2,919–2,935.

Sclater, J. G. and Francheteau, J. 1970. The implications of terrestrial heat flow observations on current tectonic and geochemical models of the crust and upper mantle. *Geophys. J. Roy. Astron. Soc.* **20**: 509–542.

Sclater, J. G. and Wixon, L. 1986. The relationship between depth and age and heat flow and age in the western North Atlantic. In *The Western North Atlantic Region*, eds. P. R. Vogt and B. E. Tucholke. Boulder, CO: Geological Society of America, pp. 257–270.

Sclater, J. G., Anderson, R. N., and Bell, M. L. 1971. Elevation of ridges and evolution of the central eastern Pacific. *J. Geophys. Res.* **76**: 7,888–7,915.

Sclater, J. G., Von Herzen, R. P., Williams, D. L., Anderson, R. N., and Klitgord, K. 1974. The Galapagos spreading center: heat-flow in the north flank. *Geophys. J. Roy. Astron. Soc.* **38**: 609–626.

Sclater, J. G., Crowe, J., and Anderson, R. N. 1976. On the reliability of ocean heat flow averages. *J. Geophys. Res.* **81**: 2,997–3,006.

Sclater, J. G., Jaupart, C., and Galson, D. 1980. The heat flow through oceanic and continental crust and the heat loss of the Earth. *Rev. Geophys. Space Phys.* **18**: 269–311.

Sleep, N. H. and Wolery, T. J. 1978. Egress of hot water from the midocean ridge hydrothermal systems: some thermal constraints. *J. Geophys. Res.* **93**: 5,913–5,922.

Stein, C. A. and Abbott, D. H. 1991. Heat flow constraints on the South Pacific Superswell. *J. Geophys. Res.* **96**: 16,083–16,099.

Stein C. A., and Stein, S. 1992. A model for the global variation in oceanic depth and heat flow with lithospheric age. *Nature* 359: 123–129.

1994a. Comparison of plate and asthenospheric flow models for the thermal evolution of oceanic lithosphere. *Geophys. Res. Lett.* **21**: 709–712.

1994b. Constraints on hydrothermal heat flux through the oceanic lithosphere from global heat flow. *J. Geophys. Res.* **99**: 3,081–3,095.

1997. Sea-floor depth and the Lake Wobegon effect. *Science* **275**: 1,613–1,614.

Stein C. A., Stein, S., and Pelayo, A. M. 1995. Heat flow and hydrothermal circulation. In *Seafloor Hydrothermal Systems*, Geophysical Monograph 91, eds. S. E. Humphris, R. A. Zierenberg, L. S. Mullineaux, and R. E. Thompson. Washington, DC: pp. 425–455.

Stein, J. S. and Fisher, A. T. 2002. Multiple scales of hydrothermal circulation in Middle Valley, northern Juan de Fuca Ridge: physical constraints and geologic models. *J. Geophys. Res.* **106**: 8,563–8,580.

Turcotte, D. L. and Oxburgh, E. R. 1967. Finite amplitude convective cells and continental drift. *J. Fluid Mech.* **28**: 29–42.

Turcotte, D. L. and Schubert, G. 1982. *Geodynamics: Applications of Continuum Physics to Geologic Problems*. New York: Wiley.

Villinger, H., Grevemeyer, I., Kaul, N., Hauschild, J., and Pfender, M. 2002. Hydrothermal heat flux through aged oceanic crust: where does the heat escape? *Earth Planet. Sci. Lett.* **202**: 159–170.

Von Herzen, R. P. and Uyeda, S. 1967. Heat flow through the eastern Pacific ocean floor. *J. Geophys. Res.* **68**: 4,219–4,250.

Wessel, P. 2001. Global distribution of seamounts inferred from gridded Geosat/ERS-1 altimetry. *J. Geophys. Res.* **106**: 19,431–19,441.

Williams, D. L. and Von Herzen, R. P. 1974. Heat loss from the Earth: new estimate. *Geology* **2**: 327–330.

Wolery, T. J. and Sleep, N. H. 1976. Hydrothermal circulation and geochemical flux at mid-ocean ridges. *J. Geol.* **84**: 249–275.

11

Rates of flow and patterns of fluid circulation

Andrew T. Fisher

11.1 Significance of fluid circulation rates and patterns

The rates and patterns of fluid circulation in seafloor hydrothermal systems control the efficiency of lithospheric heat extraction, the nature of fluid–rock interaction, and the extent of seafloor and sub-seafloor biospheres supported by fluid, energy, and solute fluxes. It has been challenging to determine the patterns and rates of fluid flow within seafloor hydrothermal systems for several reasons. Active systems are remote and measurements are often difficult to make. Although newly developed tools have led to important advances in *in situ* characterization (Chapter 8), considerable challenges remain in determining where and how quickly fluids move through the sub-seafloor environment. Fossil systems provide an integrated record of fluid flow (Chapter 9), but system heterogeneity and overprinting of multiple generations of activity make quantitative interpretation difficult. Exposures are limited within both active and fossil systems, with the former often restricted to widely spaced or isolated boreholes and seafloor outcrops, and the latter comprising mainly ophiolites and rare seafloor exposures (for example, along fracture zones or rifts, e.g. Karson, 2002). Fluid flow within seafloor hydrothermal systems is transient on a range of time scales, and some of these systems operate over lateral and vertical distances of kilometers or more. Additional complexities arise due to the scaling of rock properties within heterogeneous, fractured crust (Chapter 7). A more subtle difficulty is the influence of assumptions and ideas that became embedded in the literature at an early stage of research. Even after some of these ideas have been found to conflict with observations, they continue to influence interpretations.

In this chapter, rates of fluid flow in hydrothermal systems, the lateral extent of circulation, preferred directions of fluid flow, and the "shapes" of fluid pathways are discussed. Techniques applied to resolve these parameters include use of thermal and chemical tracers, seafloor and borehole observations, and coupled modeling. Considerably more is known about apparent rates of fluid flow than is known about overall patterns of fluid circulation, but even estimates of rates should be viewed with caution because estimates are generally based on fundamental assumptions regarding the primary direction of flow, nature of

Hydrogeology of the Oceanic Lithosphere, eds. E. E. Davis and H. Elderfield. Published by Cambridge University Press.
© Cambridge University Press 2004.

the flow pathways, and other critical parameters. The delineation of fluid flow pathways remains a topic of intense debate within land-based hydrogeologic systems, where there is often extensive access to data and samples; the paucity of available seafloor data is bound to result in considerable uncertainty. Although the focus of this book is on ridge flank hydrothermal systems, limited observations and inferences from studies of ridge crest systems are also presented because crustal properties acquired at the ridge (Chapters 3, 9, and 15) are likely to have continued influence as the seafloor evolves (Chapter 5).

11.2 Rates of fluid circulation

11.2.1 Flow-rate concepts and nomenclature

Researchers concerned with global thermal and geochemical budgets often refer to quantities of volume flow [L^3/T] or mass flow [M/T], but these are difficult to determine by direct measurement, even at known exit points of hydrothermal venting (Converse *et al.*, 1980; Schultz *et al.*, 1992). Instead, estimates of volume flow from these systems are often made at a larger scale (that of the vent field, ridge segment, or global ocean) based on a tracer such as heat or a conservative solute (e.g. Baker *et al.*, 1996; Elderfield and Schultz, 1996; Mottl *et al.*, 1998).

Rates of fluid flow in hydrogeologic systems are often discussed in terms of either fluid velocity (u) or volume discharge per area (q), both of which have dimensions of [L/T]. In this text we use the term "flux" to describe the latter, although in the hydrogeologic literature the terms "specific discharge" or "Darcy velocity" are used. Average fluid velocity relates to flux through the effective porosity, $u = q/n_e$, where n_e is the volume fraction of rock occupied by pores or fractures that contribute significantly to fluid flow. As a practical matter, determining the effective porosity independent of flux and velocity is difficult. The average velocity is a lower limit to the local particle velocity because hydrogeologic flow paths tend to be tortuous. In fact, it is not possible to determine an upper limit on particle velocities within hydrothermal systems because the rate of internal mixing (that not contributing to net flow through the system) cannot be measured once it becomes sufficiently rapid so as to homogenize the fluid in storage.

A reservoir is a region within which fluid, energy, or solutes are stored. Distinct reservoirs (i.e. for fluid and heat) may coincide in location and shape, or they may be physically separated. Reservoirs may change in location and shape with time or may be entirely conceptual. Even if it is not possible to define reservoir boundaries, the reservoir concept provides a framework for discussion of water–rock interaction (i.e. the reservoir is the place where emerging sub-surface water acquires a characteristic chemistry, etc.).

Fluid flow rate and reservoir concepts can be combined through calculation of a residence time, t_r. The residence time can be calculated on the basis of fluid reservoir size and mass (or volume) rate of fluid flow: t_r = mass storage/mass flow rate (or volume storage/volume flow rate) or on the basis of energy, t_r = energy storage/energy flow rate. Of the two fluid flow options, the mass calculation makes more sense for high-temperature hydrothermal systems because of large differences in fluid density associated with changes in pressure and enthalpy. The residence time concept strictly applies only to systems at dynamic steady state, where there is no change in flow rate or the quantity in storage. Under these conditions, for a system that is perfectly well mixed (every particle has an equal probability of appearing anywhere in the reservoir at any time), the residence time represents the average amount of time that each particle will spend in the reservoir. For an idealized "single-pass" system, in which fluid moves as a plug from one end to the other, the residence time is the same as the travel time calculated on the basis of average fluid velocity. For more complex systems, in which there is considerable recirculation, mixing, or exchange between reservoirs, particle residence times could be considerably different from those calculated under the assumption of single-pass, plug flow.

Diffusion and dispersion are discussed in this chapter with regard to solute transport. Diffusion is a process by which matter is transported from one part of a system to another through random molecular (Brownian) motion, with net transport occurring from areas of high concentration to areas of low concentration (Crank, 1975). Solutes have a coefficient of diffusion in pure water, a term that is independent of concentration and fluid flow rate but depends on temperature, species, and interaction between species. The effective diffusion coefficient within a porous medium is usually lower than the value in pure water because diffusion is considerably more rapid within the fluid of a pore network than through solid matrix material, and the pore network generally occupies a small fraction of the rock volume. Hydrodynamic dispersion comprises a combination of diffusion and mechanical dispersion, a process that results from differences in solute transport rates within and between pore channels. Rather than being constant like the diffusion coefficient, mechanical dispersion varies with fluid velocity, as described with a coefficient called dispersivity. Dispersivity is known to vary with measurement length (Neuman, 1990; Gelhar *et al.*, 1992); its value must be determined at the scale of interest.

11.2.2 Estimating vertical fluxes

One method for estimating vertical fluid fluxes in groundwater systems was introduced by Bredehoeft and Papadopulos (1965). A horizontal confining layer (a geological unit having a permeability that is low relative to that of surrounding materials) is bounded at the top and bottom by interfaces held at constant temperatures, and flow across the confining layer is entirely vertical and occurs at steady state (Fig. 11.1). The assumption of vertical flow through a sub-horizontal confining layer is generally reasonable, as fluids tend to take the shortest possible path through units having low permeability (see Section 11.3). The assumption that upper and lower boundaries have fixed temperatures is most appropriate when the flux through the confining layer is modest relative to the size of overlying and underlying reservoirs. In the case of seafloor hydrothermal systems, this model has been applied most commonly to marine sediments, bounded at the top by the ocean (essentially an infinite sink for heat), and at the base by a hydrothermal reservoir (allowing application of either a constant temperature or a constant heat flux boundary condition, as discussed

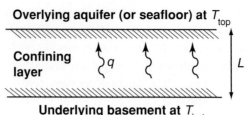

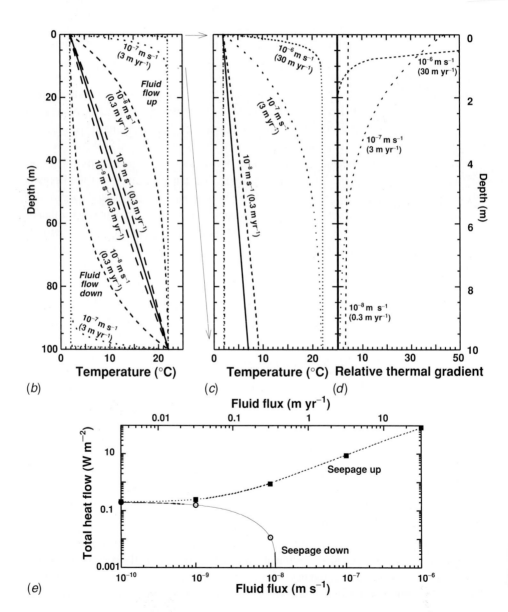

later). The same theory can be applied to basement layers, in principle, but it has proven difficult to determine equilibrium formation temperatures within hard rock.

For the case of constant-temperature boundaries, temperatures at depth, z, below the upper boundary may be calculated (Bredehoeft and Papadopulos, 1965) as:

$$T(z) = T_{\text{top}} + (T_{\text{bot}} - T_{\text{top}}) \left[\frac{\exp\left(\beta \frac{z}{L}\right) - 1}{\exp(\beta) - 1} \right] \tag{11.1}$$

where $\beta = (\rho_w c_w q L)/\lambda$, $\rho_w c_w$ = volumetric heat capacity of water, L = layer thickness, and λ = thermal conductivity of the fluid-saturated matrix. Example calculations illustrate several important characteristics of this solution. Imagine a sediment layer 100 m thick, with fixed temperature boundaries ($T_{\text{bot}} - T_{\text{top}} = 20\,°\text{C}$) and a constant thermal conductivity of $1\ \text{W m}^{-1}°\text{C}^{-1}$. In the absence of fluid flow, conductive heat flux would be $0.2\ \text{W m}^{-2}$ (Fig. 11.1). A vertical seepage flux in excess of $\sim 10^{-9}\ \text{m s}^{-1}$ ($30\ \text{mm yr}^{-1}$) is needed to detect advection based on the non-uniformity of heat flux determined from thermal data extending to ~ 100 mbsf (Fig. 11.1b). Conventional oceanic heat flow probes are generally ≤ 5 m long, and up-flow rates $> 10^{-8}\ \text{m s}^{-1}$ ($0.3\ \text{m yr}^{-1}$) are required to cause a resolvable curvature in plots of temperature versus depth over this depth range. In addition, surface probes can only detect curvature due to up-flow; thermal gradients resulting from down-flow at equivalent rates will appear to be conductive when measured with short probes, although they will be anomalously low (Fig. 11.1b,c). A plot of thermal gradient versus depth (Fig. 11.1d) is often helpful in evaluating the consistency of deviations from conductive conditions.

Equation (11.1) does not explicitly include advection of heat across the confining layer boundaries. The boundary temperatures are fixed and the thermal gradients immediately adjacent to the boundaries indicate the total heat flux through the layer. For upward flow,

---

Fig. 11.1 Influence of vertical fluid flow on thermal state of a confining layer with fixed-temperature boundary conditions and various flow rates (modified from Bredehoeft and Popadopulos, 1965). (*a*) Cartoon of system configuration, with layer thickness, L, thermal conductivity, λ, upper boundary held at T_{top} and bottom boundary held at T_{bot}. (*b*) Calculated thermal structure within a 100-m-thick layer, with $T_{\text{top}} = 2\,°\text{C}$, $T_{\text{bot}} = 22\,°\text{C}$, and $\lambda = 1\ \text{W m}^{-1}°\text{C}$, based on Eq. (11.1). (*c*) Detail plot of upper 10 m of the layer. Upward fluid flux of $0.3\ \text{m yr}^{-1}$ or more is generally needed to detect fluid flow within marine sediments collected in this depth range, based on the non-linearity induced by advective heat transport. Downward fluid flow may be manifested as low (but apparently conductive) heat flow. (*d*) Detail plot of upper 10 m of the confining layer showing variations in thermal gradient with depth. Relative gradient is calculated as apparent values measured at 10-cm intervals divided by the conductive gradient across the 100-m-thick layer. Upward flow increases the gradient near the upper boundary, and reduces the gradient near the lower boundary. (*e*) Total heat flux (conductive and advective) through a 100-m-thick layer with fixed-temperature boundary conditions. Total seafloor heat flux increases without limit with increasingly rapid upward seepage, and rapidly goes to zero for the case of rapid downward seepage. The two boundaries function as infinite heat sources or sinks. An alternative formulation limits total heat flow (Sleep and Wolery, 1978).

the upper boundary serves as an infinite heat sink; the faster the seepage rate, the steeper the thermal gradient at the upper boundary and the greater the total heat flux (Fig. 11.1*e*). In the same way, the lower boundary serves as an infinite heat source for the case of upward fluid flow. For the case of downward seepage, conductive heat flux at the seafloor goes to zero as the seepage rate increases (Fig. 11.1*e*).

Several workers have introduced modifications to this mathematical model. Sleep and Wolery (1978) solved the same fundamental one-dimensional heat flow equation as Bredehoeft and Papadopulos (1965), but applied a fixed heat flux boundary condition, which may be more representative in some environments. Lu and Ge (1996) modified (11.1) by relaxing the requirement that all fluid and heat flows vertically, and allowed lateral (sub-horizontal) heat transport within the confining layer to be represented through use of a heat source or sink term. As a result, temperatures within the confining layer can be greater than or less than those on both of the fixed-temperature upper and lower boundaries. This occurs if the rate at which heat is transported laterally along the layer is large relative to the rate at which heat is transported vertically. The highly layered nature of volcanic oceanic crust (Chapters 3 and 9) suggests that some ridge flank systems may host horizontal and vertical flow components of similar magnitudes, particularly if there are large lateral thermal gradients (for example, close to a spreading center or a recently emplaced seamount).

Systematic changes in thermal conductivity with depth can induce significant curvature in a conductive, steady-state, thermal gradient, and such variations must be considered before seepage rates can be inferred with confidence from thermal gradient observations. Thermal conductivity variations in shallow nature sediments on the order of $\times 2$ are common. Transient response to recent changes in bottom-water temperatures can also lead to curved thermal gradients within shallow sediments (e.g. Beck *et al.*, 1985; Barker and Lawver, 2000; Von Herzen *et al.*, 2001; Davis *et al.*, 2003). This occurs most often within shallow water, but it can also occur in deep oceanic environments where bottom water comes from multiple sources having different temperatures. Unlike vertical fluid flow through sediments, which tends to be localized above basement highs and other features (e.g. Mottl, 1989; Giambalvo *et al.*, 2000), changes in bottom-water temperatures should influence heat flow measurements made over a relatively broad area.

As described elsewhere (Chapter 16), pore-water geochemical data can also be used to estimate rates of vertical seepage. The primary differences in working with chemical data are: (i) chemical diffusivity is about three orders of magnitude less than thermal diffusivity, making chemically based flow-rate estimates commensurately more sensitive (see Fig. 6.8); and (ii) many chemical species are non-conservative, requiring that reactions be accounted for in estimating seepage rates (e.g. Mottl and Wheat, 1994; Wheat and McDuff, 1994; Giambalvo *et al.*, 2002). The concentration profiles that result from non-conservative behavior may be similar to those derived by Lu and Ge (1996); steady-state concentrations may be greater than the larger of the two boundary conditions, or smaller than the lesser of the two. Because of the greater sensitivity of geochemical tracers, it is often valuable to compare estimates of seepage based on both thermal and geochemical methods. Chemical diffusivity is so low that there is little overlap in sensitivity between these methods, but if vertical flow occurs at thermally detectible rates, fluid chemistry should be dominated by advection rather than diffusion or reaction. Conversely, pore-fluid profiles in which seepage fluxes fall within a chemically quantifiable range should coincide with thermally conductive conditions.

Several published studies have shown curved thermal gradients measured in shallow seafloor sediments. Estimates of vertical seepage fluxes from these analyses are generally 10^{-8}–10^{-6} m s^{-1} (0.3–30 m yr^{-1}). Anderson *et al.* (1979) documented curved thermal gradients within the upper 5 m of sediments from the central Indian Ocean, in an area with thin sediment cover and considerable basement relief. Thermal data were collected with outrigger probes on a large lance, and thermal conductivity measurements made on cores from the area indicated only modest changes with depth, too small to account for the non-uniform gradients. Abbott *et al.* (1981) measured permeabilities of core samples recovered from this region, and concluded that measured permeabilities were too low to allow the rates of flow suggested by thermal data with reasonable pressure gradients. Geller *et al.* (1983) documented curved thermal gradients in the upper few meters of the thick sediments of the Bengal Fan using outrigger probes. Thermal conductivity variations in this area were also insufficient to account for the observed curvature in thermal gradients, and Geller *et al.* concluded that sediment compaction and deformation may be responsible for fluid seepage through the seafloor in this region. Fisher and Becker (1991) estimated fluid seepage rates from curvature in thermal gradients in Guaymas Basin, a sedimented spreading center in the Gulf of California, using a thin, 2-m-long probe. This study did not include thermal conductivity measurements, but earlier work in the region suggested little change with depth in the upper 2–3 m. All of these studies were made where changes in bottom-water temperatures are unlikely to explain the curved thermal gradients, and all showed considerable spatial variability in apparent seepage rates. Non-uniform gradients have also been reported in the equatorial Pacific (Mayer *et al.*, 1985), but with no definitive explanation other than the possibility of upward seepage of pore water.

Evidence for downward seepage has also been found. Abbott *et al.* (1984) studied core samples, pore-fluid chemistry, and heat flow from the Guatemala Basin, and found evidence for downward fluid flow into the sediments at low velocities. Similar interpretations of slow downflow have been made on the basis of chemical, thermal, and pore-pressure measurements in other ridge flank settings (McDuff, 1981; Langseth *et al.*, 1984; Mottl, 1989; Langseth *et al.*, 1992; Fisher *et al.*, 2001).

While these results are interesting, curved thermal gradients in shallow sediments indicative of vertical fluid seepage are relatively rare. In most areas, low sediment permeability (Chapter 6) and the modest hydrothermal forces available to drive fluid flow (Chapter 8) generally do not result in thermally significant seepage through shallow sediments. For example, hundreds of heat flow measurements have been made over the sedimented northern Juan de Fuca Ridge and flank (Davis and Villinger, 1992, Davis *et al.*, 1992, 1999; Stein and Fisher, 2001). This is a setting where fluid seepage might be thought likely, since pressure and thermal gradients are relatively high and some sediment (turbidite) layers are coarse grained. Instead, thermal conditions in the shallow sub-surface throughout this region are conductive (except immediately adjacent to hydrothermal vent sites; e.g. Stein

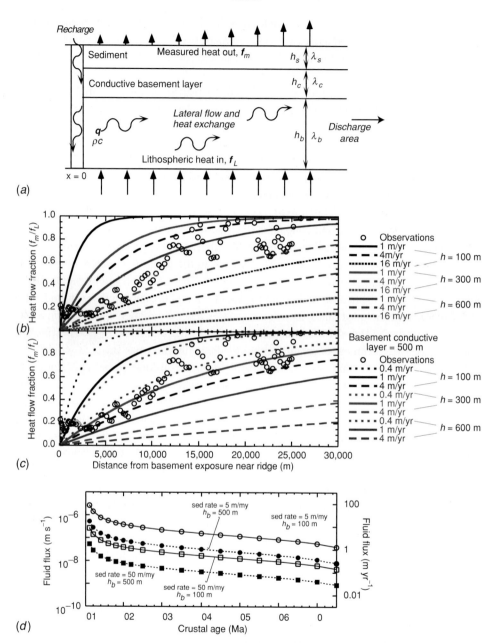

Fig. 11.2 The well-mixed aquifer model in concept and application. (*a*) Cartoon showing configuration for the well-mixed aquifer model (Langseth and Herman, 1981), including addition of a conductive layer within upper basement (Rosenberg *et al.*, 2000). Fluid recharges the basement aquifer at 0 °C and flows laterally toward a discharge region, the location of which is unspecified, gaining heat from the underlying lithosphere along the way. The fluid within the aquifer is defined to be vertically isothermal; this emulates efficient local mixing. (*b*) Comparison between seafloor heat flow observations from

and Fisher, 2001). The permeability of fine-grained terrigenous sediments appears to be sufficiently low (Chapter 6) to keep seepage rates well-below thermal detectability (e.g. Davis *et al.*, 1992; Wheat and Mottl, 1994).

In summary, while it is possible to detect curvature in thermal gradients associated with vertical fluid seepage and to use this information to estimate seepage rates, other possible explanations should be evaluated before fluid seepage is inferred and rates are estimated. Corrections should be made to account for thermal conductivity variations with depth. It is also important to verify that seepage rates apparent from curved thermal gradients are consistent with sediment properties and available driving forces. In several studies cited above, analysis of sediment properties suggests that unrealistically high driving forces would be required to sustain upward seepage at rates inferred from thermal data. Giambalvo *et al.* (2000) looked for evidence that upward seepage may modify sediment properties (in particular, increase permeability) through diagenesis or measurable physical processes, but could find no evidence that this occurred within fine-grained turbidites or hemipelagic sediments. This does not mean that such processes do not occur in some environments, but it does suggest that researchers should be cautious in using thermal data alone to infer fluid seepage rates. It is ideal to collect and analyze pore-fluid profiles from the same sites where curved thermal gradients are found, since thermally significant fluid flow rates should result in pore-fluid chemistry that is dominated by advection. Although there are several published examples of curved thermal gradients that could result from fluid flow, there are few data sets that combine thermal, chemical, and hydrogeologic observations so as to allow confident estimation of seepage rates.

11.2.3 Lateral rates of flow

In a study of heat flow in the Brazil Basin, western Atlantic Ocean, Langseth and Herman (1981) introduced the well-mixed aquifer (WMA) model for estimating rates of lateral fluid flow. The model is based on a one-dimensional, steady-state consideration of coupled heat and fluid flow (Fig. 11.2), with water flowing laterally and capturing heat entering the system from below. Close to the point of fluid recharge, fluid temperatures will be relatively low;

---

the eastern flank of Juan de Fuca Ridge (Davis *et al.*, 1999) and calculations based on the well-mixed aquifer model (Langseth and Herman, 1981), assuming various flow rates and thicknesses for the hydrothermal layer. Heat flow values have been normalized by dividing by values predicted using a one-dimensional conductive cooling model (Parsons and Sclater, 1977), and plotted versus distance from the area of significant basement exposure to the west (assumed recharge area). (*c*) Comparison between seafloor heat flow observations and calculations based on the modified well-mixed aquifer model (Rosenberg *et al.*, 2000), assuming various flow rates and thicknesses for the hydrothermal layer, and a conductive basement layer 500 m thick. The inclusion of a conductive basement layer allows slower fluid velocities to explain the general pattern of seafloor heat flow observations. (*d*) Calculated rates of mean fluid flow based on consideration of the global heat flow anomaly for seafloor of various ages, different depths of circulation, and a range of sedimentation rates (modified from Fisher and Becker, 2000).

as the fluid moves away from the recharge area, it acquires a temperature consistent with heat input from below, the depth of circulation, and the thermal properties of the fluid–rock system.

The upper oceanic crust is idealized as a horizontally layered system, with a conductive boundary layer (sediments), underlain by an aquifer within which thermally significant transport occurs (shallow basement). Water is supplied to one end of the layer at the temperature of bottom-water, and heat is exchanged during lateral flow (Fig. 11.2*a*). The aquifer layer is assumed to be well-mixed vertically, having a temperature determined by the balance between heat entering from below and leaving through the top of the system. Temperature is thus a function only of horizontal position. The steady-state heat balance for this system is:

$$h_b\lambda_b\frac{\mathrm{d}^2T}{\mathrm{d}x^2} - h_b q\rho c\frac{\mathrm{d}T}{\mathrm{d}x} + f_L - f_m = 0 \tag{11.2a}$$

where the first term is lateral heat conduction, the second term is lateral heat advection, the third term is heat flux in at the base of the aquifer, and the fourth term is heat flux out at the top (measured conductive heat flux at the seafloor). If bottom-water temperature is assumed to be $0\,°\mathrm{C}$ and sediment thermal conductivity is constant with depth, measured seafloor heat flux is defined as $f_m = K_sT/h_s$. If lateral heat conduction is small relative to the other terms (a reasonable assumption if the lateral flow rate is large and there are no local magmatic heat sources to generate large lateral thermal gradients), the first term in (11.2a) can be ignored and the solution is:

$$\frac{f_m}{f_L} = 1 - \mathrm{e}^{-Ax} \tag{11.2b}$$

where $A = K_s/(h_b h_s q\rho c)$.

The critical information required to compare predictions to observations using this model is: measured seafloor heat flux, the distance from the measurement points to the site of recharge, the thicknesses of the confining layer and the permeable rock layer through which fluid flows laterally, estimated heat input from below the aquifer, and thermal properties of the sediment and rock–fluid system. In the Brazil Basin, application of the well-mixed aquifer model yielded estimated lateral fluid flux on the order of 10^{-9}–10^{-8} m s^{-1} (0.03 –0.3 m yr^{-1}) for an aquifer 1 km thick (Langseth and Herman, 1981). Assuming an effective porosity in basement of 1–10%, this range of fluxes indicates average velocities on the order of 10^{-8}–10^{-6} m s^{-1} (0.3–30 m yr^{-1}). This approach was subsequently used to estimate lateral flow rates in the eastern equatorial Pacific Ocean (Baker *et al.*, 1991), below a sediment pond on the western flank of the Mid-Atlantic Ridge (Langseth *et al.*, 1984; Langseth *et al.*, 1992), on the eastern flank of the Juan de Fuca Ridge (Davis *et al.*, 1999), and in the Alarcon Basin, Gulf of California (Fisher *et al.*, 2001).

The well-mixed aquifer model has been extended and applied in additional ways. For example, Rosenberg *et al.* (2000) relaxed the requirement that thermally significant lateral flow occur immediately below the sediment layer in uppermost basement (Fig. 11.2*a*).

Because it is assumed that fluid enters the lateral flow layer at bottom-water temperature, deeper circulation allows more efficient extraction of heat from the upper lithosphere. The solution is the same as that derived by Langseth and Herman (1981), except that the parameter, A, is redefined to include the thickness and thermal conductivity of the conductive basement layer above the aquifer. In another example, Davis *et al.* (1999) included a high-conductivity proxy for the effects of lateral convective mixing as well as local variations in sediment thickness (not included in the well-mixed aquifer formulation), although this was done via a numerical model, not with the analytic solution of (11.2).

Calculations based on the well-mixed aquifer model are compared to field observations in Fig. 11.3. Seafloor heat flow data are from the eastern flank of Juan de Fuca Ridge (see Chapter 8 for additional discussion of the setting). Analytical solutions were generated based on models that exclude and include a conductive boundary within upper basement (Figs. 11.2*b* and *c*, respectively). Lateral fluxes on the order of 1–4 m yr^{-1} are required by the model if fluid flow occurs within the upper 100–600 m of basement. Values ×10 lower are consistent with observations if the top of the hydrothermal layer is at a depth of 500 m into basement. This efficiency arises from the flow and heat exchange taking place with a greater temperature difference. Fisher and Becker (2000) applied the well-mixed aquifer model on a global basis to estimate typical rates of circulation and available driving forces, and thus basement permeabilities, required to explain the global ridge flank thermal anomaly (see Chapter 10). These calculations were based on linking available driving forces to the thickness of the sediment section (varies as a function of basement age, using a range of sediment accumulation rates), thickness of the basement aquifer, temperature of the aquifer (varies with sediment thickness, aquifer thickness, and basement age), and distances between fluid recharge and discharge sites. Even with a large range in likely sediment and basement properties and system geometries, these calculations yielded a relatively narrow range of characteristic lateral fluid fluxes of 10^{-8}–10^{-6}m s^{-1} (0.3–30 m yr^{-1}) for seafloor younger than 20 Ma, decreasing to 10^{-9}–10^{-7}m s^{-1} (0.03–3 m yr^{-1}) for seafloor between 40 and 65 Ma (Fig. 11.2*e*).

Baker *et al.* (1991) adapted the well-mixed aquifer model for use with geochemical data in the Eastern Equatorial Pacific Ocean, reasoning that the loss of solutes from basement fluids to sediments was mathematically equivalent to conductive loss of heat. They recognized that pore-fluid compositions within thick sediment layers (400–500 m) were bounded by bottom-water compositions at both the top and the bottom of the sediment layer. Qualitatively, this required that basement pore fluids be relatively young, since these fluids had not had an opportunity to react significantly with basement. Quantitatively, Sr and ^{87}Sr/^{86}Sr ratios allowed estimates of basement residence times on the order of 10,000–30,000 years. Assuming typical distances between recharge and discharge sites of 20–200 km (a weakly constrained estimate of the characteristic outcrop spacing in the area), fluxes on the order of 10^{-7}m s^{-1} (3 m yr^{-1}) are suggested.

Elderfield *et al.* (1999) use a similar approach to estimate average lateral fluid flow velocities within uppermost basement along a transect of sites on the eastern flank of the Juan de Fuca Ridge. Pore fluids were collected from sediments just above basement, and assumed to be in equilibrium with underlying basement fluids. Based on (i) sulfate, chloride, and radiocarbon analyses, (ii) the spacing between sites, and (iii) a plug-flow (no dispersion) model for solute transport, lateral fluid flow velocities in upper basement are on the order of 2 m yr^{-1}, with net transport from west to east, a direction consistent with seafloor heat flow data (Davis *et al.*, 1999). Assuming an effective porosity in basement of 1–10%, the lateral flux is 0.02–0.2 m yr^{-1}, significantly lower than estimated from thermal constraints.

One difficulty with applying geochemical age estimates to fluids in oceanic basement is that corrections are needed to account for solute loss to surrounding, hydrologically inactive regions by diffusion and mixing (e.g. Sanford, 1997; Goode, 1996; Alt-Epping and Smith, 2001; Bethke and Johnson, 2002; Park *et al.*, 2002). Similar corrections generally are not required for flow rates estimated using heat as a tracer because of the relative efficiency of thermal conduction; this also helps to explain why dispersion is generally neglected in numerical models of coupled heat and fluid flow (Chapter 12).

There are two distinct approaches that have been used to estimate the magnitude of tracer losses in heterogeneous systems, both of which suggest that age corrections appropriate for seafloor hydrothermal systems are likely to be large, on the order of ×10 to ×100 or more (Sanford, 1997; Bethke and Johnson, 2002). The analysis of Bethke and Johnson (2002), based on the "age mass" concept of Goode (1996), yields the smallest age correction. Imagine that a packet of circulating water within an aquifer suffers diffusive loss of "youth" to surrounding, water-bearing but less permeable rocks during transport. The fluid velocity within an aquifer sandwiched between confining (stagnant) layers, estimated from apparent ages at two locations along a flow path, should be calculated as:

$$u = \left[\frac{(1 + F)\,\Delta x}{\Delta t_a} \right] \tag{11.3}$$

where $F = (n_{con}h_{con})/(n_{aqf}h_{aqf})$, n = porosity, h = layer thickness, subscripts aqf and con refer to the aquifer and confining layers, respectively; Δx is the lateral spacing between sample locations; and Δt_a is the apparent age difference between fluids at each location. It is initially surprising that the magnitude of the correction does not depend on either the fluid velocity or the diffusivity of the aquifer or confining layer. This results from the assumption that, after a sufficiently long time following recharge, the loss of age mass from the primary flow channels reaches steady state. The correction necessary to account for this loss is based entirely on F, the ratio of the volume of water within the stagnant regions to the volume of water within the aquifer. Note that (11.3) reduces to the plug-flow approximation if $F = 0$. The correction required by this equation would be appropriate for both conservative and reactive tracers.

Figure 11.3 illustrates the magnitude of the correction appropriate for the eastern flank of Juan de Fuca Ridge. The system is somewhat more complex than that described by Bethke and Johnson (2002), in that lateral flow is thought to occur mainly within shallow basement, so the confining layers above the aquifer (marine sediments) and below the aquifer (less permeable basement) have different properties. If we assume reasonable values for aquifer effective porosity (1–10%), aquifer thickness (10–500 m), confining layer porosity (1–5% for underlying basement, 50–60% for overlying sediments), and confining layer thickness (1,000 m for underlying basement, 200 m for overlying sediments), the appropriate correction factor for this hydrogeologic system is 2–50 (Fig. 11.3*b*). This is likely to be a low-end estimate of the necessary correction because (i) the confining layer below the shallow basement aquifer is many kilometers thick (making the volume of water in the confining layer even larger than assumed); and (ii) these calculations neglect the influence of flow channeling within the aquifer itself (isolation of most flow within a small fraction of the most permeable regions), which would reduce effective porosity.

The theory developed by Sanford (1997) for interpretation of radiocarbon ages from aquifer fluids, rooted in concepts introduced by Sudicky and Frind (1981), is also based on the idea that most fluid flow is focused within regions that are surrounded by stagnant layers in which transport is dominantly diffusive. A stagnant zone "width factor" is defined as $w = \tanh(h_s \sqrt{\lambda/K_m}/2)$, where h_s = thickness of stagnant layers, K_m = molecular diffusion within this zone, and λ = decay constant for radiocarbon. The rate of diffusive loss is calculated as:

$$\lambda' = \left[2w\sqrt{\lambda K_m}/(nh_a) \right] \tag{11.4}$$

where n = aquifer porosity and h_a = aquifer thickness. The relative importance of diffusive to radiometric loss is expressed by the ratio λ'/λ. Typical values of λ'/λ noted for "volcanic rock layers" and "fractured rocks" are on the order of $10^2 - 10^5$ (Sanford, 1997). Corrected fluid velocities (u_c) based on radiocarbon measurements are calculated from uncorrected age values as: $u_c = (\Delta x/\Delta t_a)(\lambda + \lambda'/\lambda)$. As the value of λ' increases, the correction factor goes to λ'/λ (Fig. 11.3*c*).

The rate of diffusive loss predicted by (11.4) depends on the thicknesses of both aquifer and stagnant layers, values that can be estimated from core and borehole observations. For example, lithologic and electrical resistivity logs from DSDP/ODP Hole 395A were compared by shipboard scientists after an initial phase of drilling and experiments and interpreted to indicate a series of vertically distinct, basaltic flows (Matthews *et al.*, 1984). Each flow unit is characterized by greater electrical resistivity at the base and lower electrical resistivity at the top, interpreted to indicate an increase in porosity in rocks deposited during the final stages of each effusive event. Temperature logs collected soon after drilling (Becker *et al.*, 1984) indicated that bottom water was being drawn down Hole 395A, and additional geophysical data were collected during a subsequent visit (Becker *et al.*, 2001). The spontaneous potential (SP) log was used to locate intervals within Hole 395A into which borehole water flowed. Deflections in the SP log clearly correlate with the tops of individual resistivity sequences, suggesting that these thin layers, independently interpreted based on

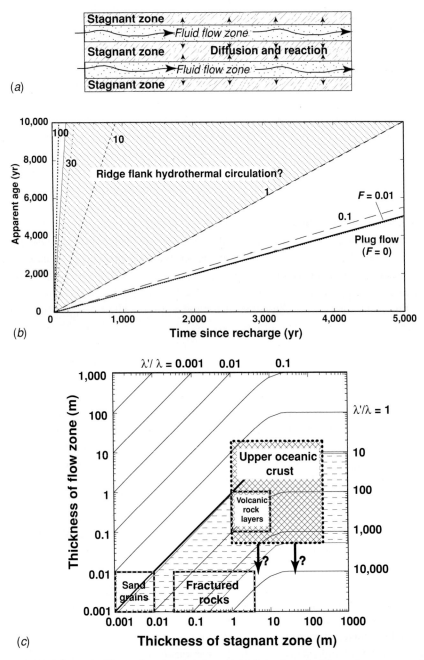

Fig. 11.3 (*a*) Schematic illustration showing how fluid flow through thin regions surrounded by confining (stagnant) zones can lead to loss of tracers. (*b*) Calculations of apparent age of water versus actual time since recharge, with different values of F, the ratio of the volume of water within confining layer(s) to the volume of water within the aquifer (Bethke and Johnson, 2002). The region indicated

lithologic and resistivity data to have higher porosity, are also the most hydrologically active intervals. The typical thicknesses of the most and least hydrologically active sections of this borehole are on the order of 1–10 and 10–100 m, respectively (Fisher, 2003). Similar scales of layering, on the basis of combined geological, geochemical, and geophysical observations, have been noted in numerous other DSDP and ODP drill-holes (Larson *et al.*, 1993; Alt, 1995; Bach *et al.*, 2003).

Initial consideration of the Sanford (1997) analysis suggests that correction factors on the order of ×10 to ×1,000 should be applied to fluid velocities estimated from plug-flow interpretations of geochemical tracer experiments (Fig. 11.3*c*), but these corrections may be too low. In kilometer-scale tracer studies within the Mirror Lake fractured rock system, Becker and Shapiro (2000) and Shapiro (2001) have shown that effective chemical diffusivity of the rock matrix is at least 3×10^{-8} m^2 s^{-1}, a full order of magnitude *greater* than the diffusivity of the tracer in water. This result contrasts with the standard concept of diffusive transport within a porous medium, in which the effective diffusivity is lower than that for the solute in water, often by an order of magnitude or more (depending on the effective porosity and tortuosity of the pore network). The documented range in fracture transmissivity at the Mirror Lake field site is six orders of magnitude (Shapiro and Hsieh, 1998), and the high effective diffusivity is thought to result from preferential migration of tracer along the most permeable fractures, which are generally well connected over relatively short distances. Thus the high effective diffusivity essentially results from a component of hydrodynamic dispersion that is independent of fluid flow velocity. If the effective diffusivity of geochemical tracers in ridge flank hydrothermal circulation systems is also greater than that for free water, a distinct possibility within a heterogeneous, fractured aquifer system, then the diffusive loss correction that needs to be applied to ^{14}C data would result in λ'/λ values considerably greater than ×1,000 (Fig. 11.3*c*).

Thus fluid velocities and fluxes estimated from plug-flow interpretation of geochemical observations on the eastern flank of Juan de Fuca Ridge are likely to be underestimates. Correction by a factor of ×10 to ×100 would make geochemical estimates similar to those based on thermal data, and would indicate average fluid velocities on the order of hundreds to thousands of meters per year. This is a very rapid fluid velocity in geological terms and suggests that both passive and active tracer experiments conducted at a large scale should be possible within ridge flank hydrothermal systems. This result also has important

---

by the diagonal shading is that which seems likely for several ridge flank hydrothermal systems, requiring corrections on the order of 2–50. (*c*) Illustration of corrections needed for ^{14}C data in order to estimate actual fluid ages, based on the thickness of flow zones and stagnant zones (modified from Sanford, 1997). The value λ'/λ is essentially the ratio of diffusive loss to radioactive decay, and is the magnitude of the correction that should be applied to estimates of water age based on ^{14}C analyses of basement pore fluid. Observations of crustal lithostratigraphy and *in situ* measurements of crustal properties indicate that λ'/λ is at least 10–1,000, and field estimates of dispersivity from fractured aquifers (discussed in the text) suggest that even larger corrections may be needed.

implications for the nature of water–rock interactions and the cycling of nutrients within ridge flank reservoirs, since it would imply relatively short fluid residence times.

11.2.4 Hydrothermal residence times

Researchers have attempted to estimate hydrothermal fluid residence times using both thermal and chemical tracers. Kadko and Moore (1988) and Kadko and Butterfield (1998) estimated hydrothermal residence times on the Juan de Fuca Ridge using radium. These studies suggest that fluids spend on the order of several years interacting with host rock at elevated temperatures. Other researchers have based their estimates on energy or geochemical mass-balance considerations (Davis and Fisher, 1994; Humphris and Cann, 2000). Fisher (2003) combined these results with additional estimates of hydrothermal residence times, yielding estimates in general agreement with earlier calculations.

Residence time calculations require estimates of both reservoir size and steady-state flux. Geochemical mass balance considerations led to estimates of the volume of fluid required to produce ore deposits in Middle Valley, northern Juan de Fuca Ridge (Davis and Fisher, 1994) and at the main hydrothermal field at TAG (Humphris and Cann, 2000), 5×10^8 and 4×10^8 m^3, respectively. Instantaneous heat budget estimates for these regions, the Broken Spur segment of the Mid-Atlantic Ridge, and the main field of the Endeavour Segment, Juan de Fuca Ridge, were provided by these studies and by Baker *et al.* (1996), Murton *et al.* (1999), and references therein. Heat flow was converted to equivalent mass and volume flow by assuming characteristic fluid properties (density $= 675$ kg m^{-3}, heat capacity $= 6400$ J kg^{-1} K^{-1}, and temperature drop from reaction zone to seafloor $= 350\,^\circ$C).

Independent estimates of hydrothermal reservoir sizes as a function of mean spreading rate were based on considerations of the likely depth extent, width off axis, and effective porosity of the crust hosting the primary circulation system (Fisher, 2003). It was assumed that the depth extent of the reservoir scales with spreading rate ($z_{max} = 5$ km, slow spreading; $z_{max} = 1$ km, fast spreading), based on the observed inverse relation between spreading rate and depths to both magma lens reflectors and measurable earthquakes (e.g. Huang and Solomon, 1988; Kong *et al.*, 1992; Purdy *et al.*, 1992; Phipps Morgan and Chen, 1993; Chen and Phipps Morgan, 1996; Wilcock and Fisher, 2003). It was also assumed that the hydrothermal reservoir is wider at slower spreading ridges, based on the presence of a wider neovolcanic zone and more extensive axis-parallel faulting in these settings. Finally, it was assumed that effective porosities within the hydrothermal reservoir are between 1 and 10%, a reasonable range for fractured rock.

Given these parameter values, estimated reservoir sizes are on the order of 2×10^7 to 2×10^8 m^3 km^{-1} of ridge for a fast-spreading ridge (full spreading rate $= 160$ mm yr^{-1}), and 5×10^8 to 5×10^9 m^3 km^{-1} of ridge for a slow-spreading ridge (full spreading rate 20 mm yr^{-1}), values that straddle the geochemical estimates described earlier for Middle Valley and TAG. Additional estimates of steady-state heat flow (and thus mass and volume flow) come from consideration of the global distribution of known vent sites (e.g. Baker, 1996; Baker *et al.*, 1996). These estimated reservoir volumes and hydrothermal budgets

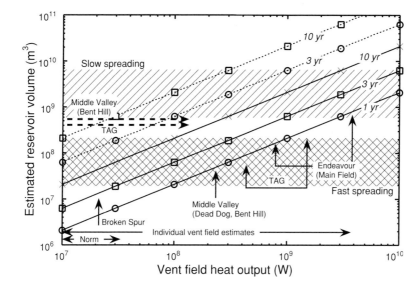

Fig. 11.4 Calculations and estimates of ridge crest hydrothermal reservoir volumes and heat output, and resulting fluid residence times in the sub-surface (Fisher, 2003). Estimated reservoir volumes are shown per kilometer of ridge crest, based on seafloor spreading full rates of 20–160 mm yr^{-1} ("slow spreading" and "fast spreading," respectively). Estimates of reservoir volumes needed to produce massive sulfide deposits are labeled with horizontal arrows for TAG and Middle Valley–Bent Hill. Vent field heat flow estimates are from Baker *et al.* (1996), Murton *et al.* (1999), Humphris and Cann (2000), and references therein, for Broken Spur, Middle Valley–Dead Dog and Bent Hill, TAG, and the Endeavour Segment – Main Field. The heat output range labeled "Norm" indicates normalized high-temperature vent field output (corrected for the fraction of ridge hosting high-temperature circulation at any time (Baker and Urabe, 1996). Diagonal lines are steady-state fluid residence times based on these values and effective porosities of >1% (solid lines) or 10% (dashed lines). Ten-year residence time with effective porosity of 1% is equivalent to one-year residence time with effective porosity of 10%, since the same amount of water–rock interaction is assumed. These calculations show that residence times on the order of years (as opposed to weeks or decades) are consistent with a wide range of geochemical and geophysical considerations.

are consistent with hydrothermal residence time, t_r = 1–10 years (Fig. 11.4). There is considerable uncertainty in many of the parameters used to estimate these residence times, but ongoing field programs focusing on fluxes, hydrogeological properties, and flow scales will help to refine these calculations.

These high-temperature hydrothermal residence times are considerably shorter than estimates derived for ridge flank hydrothermal systems, on the order of 10^4 years (Baker *et al.*, 1991; Elderfield *et al.*, 1999). However, as noted previously, residence time estimates for ridge flanks based on geochemical estimates and a plug-flow approximation almost certainly require correction by at least a factor of ×10, and perhaps by a factor of ×100 to ×1,000 or more. Interestingly, such corrections would result in hydrothermal residence time estimates for ridge flank hydrothermal fluids that are about the same as those from spreading centers, although there is no particular reason why this might be expected.

11.3 Patterns of fluid circulation

11.3.1 Heat flow observations and the occurrence of "cellular convection"

Elder (1965) was among the first to speculate that widespread hydrothermal circulation could advect significant quantities of heat from young oceanic lithosphere. Although there were few direct observations allowing quantification of this process at the time, inferences were drawn from related studies of buoyancy-driven convection. Lapwood (1948), Wooding (1960), and Nield (1968) evaluated conditions necessary for the onset of convection, including the imposition of various pressure and temperature boundary conditions. These and other studies also assessed the shape of convection cells within homogeneous, porous systems heated from below.

Lister (1972) analyzed heat flow measurements from near the Juan de Fuca Ridge, attributed the variations in observed values to hydrothermal circulation, and drew several sketches illustrating hypothetical flow paths (see Chapter 2). The convection cells in these drawings generally had an aspect ratio (cell width/cell height) near one, and showed the convection cells as two-dimensional in a ridge-perpendicular plane, parallel to the direction of spreading. Limited observational data available at the time were consistent with these ideas.

The first study that combined detailed seafloor heat flow observations with bench-top models of buoyancy driven convection, and drew explicit inferences regarding the geometry of circulation, was that of Williams *et al.* (1974). This paper and companion studies of young seafloor north and south of the Galapagos Spreading Center (Detrick *et al.*, 1974; Klitgord and Mudie, 1974; Sclater *et al.*, 1974), documented fundamental characteristics of young oceanic crust formed at a moderate-rate spreading center: a median valley having topographic relief of several hundred meters, abyssal hills and associated faults (generally inward dipping) that form sub-parallel to the ridge with spacing of several kilometers, and patchy sediment cover that thickens and becomes more continuous with distance from the ridge. Williams *et al.* (1974) also noted a critical pattern in seafloor heat flow: values measured along transects oriented perpendicular to the ridge (parallel to the direction of spreading) were generally suppressed below that expected based on models of conductive lithospheric cooling (e.g. Sclater and Francheteau, 1970), but showed systematic variations, with alternating regions of relatively high and low values. The amplitude of these variations was one order of magnitude, and the wavelength of variations was 5–10 km, about the same as the abyssal hill spacing.

The regularity of heat flow highs and lows led Williams *et al.* (1974) to consider the occurrence within oceanic crust of "cellular convection" organized in two-dimensional rolls having an aspect ratio close to one (Fig. 11.5). Seafloor heat flow observations could be explained by cellular convection having rising limbs below heat flow highs (sometimes

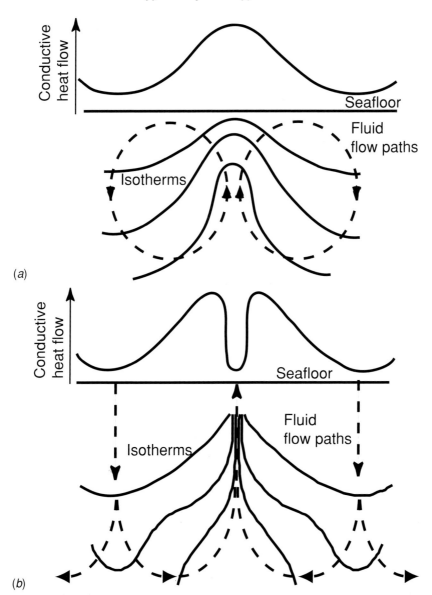

Fig. 11.5 Cartoon of hypothesized fluid convection cells and resulting seafloor heat flow (modified from Williams *et al.*, 1974). Note low aspect ratio of convection cells, with depth of circulation being approximately equal to the width of the cells. (*a*) Fluid flow pathways, isotherms, and seafloor heat flow resulting from convection within basement rocks covered with a low-permeability sediment layer. (*b*) Fluid flow pathways, isotherms, and seafloor heat flow resulting from convection within basement rocks, where water can readily pass through incomplete sediment cover. Heat flow is suppressed immediately above the hydrothermal up-flow zone because conditions close to the seafloor are essentially isothermal.

associated with basement and seafloor highs), and downgoing limbs below seafloor heat flow lows (sometimes associated with basement and seafloor lows), as discussed by Lister (1972). Williams *et al.* (1974) suggested that convection "cells are approximately equidimensional." Thus it was concluded, "[w]ith our observed modulation [of seafloor heat flow values] of wavelength of 6 km, equidimensional Rayleigh cells would penetrate approximately 3 km and greater depths are certainly possible." Subsequent field studies by Green *et al.* (1981), Langseth *et al.* (1988), and others documented swaths of high and low heat flow on ridge flanks that were elongated sub-parallel to the spreading center. These observations were consistent with the idea that the primary direction of fluid circulation was perpendicular to the ridge, and that these systems could be modeled using cross-section simulations.

The first numerical studies of two-dimensional, porous media convection within mainly isotropic crustal systems were thus oriented perpendicular to the ridge (Ribando *et al.*, 1976; Fehn and Cathles, 1979; Fehn *et al.*, 1983). Permeability in these studies was generally made homogeneous or smoothly varying (for example, decreasing exponentially) with depth. Results of this work reinforced earlier interpretations of ridge flank (and, to some degree, ridge crest) hydrothermal systems: (i) convection cells tend to have aspect ratios near one and thus a wavelength of circulation similar to the depth of circulation; and (ii) conductive heat flow at the seafloor tends to be elevated above rising convection limbs and suppressed above downgoing limbs, allowing both the wavelength and depth extent of circulation to be inferred from measurements of seafloor heat flow. However, there are good reasons to reconsider both of these concepts, as discussed in the following section.

11.3.2 Challenges to the cellular convection concept

There are several reasons why we should not expect hydrothermal convection in the ocean floor to form cells similar to those that form in an isotropic, homogeneous porous medium: (i) the oceanic crust is strongly layered, (ii) the crustal aquifer is heterogeneous and anisotropic, and (iii) convection geometry is likely to vary over time. Oceanic crust is constructed in irregular layers (e.g. Houtz and Ewing, 1976; Purdy and Detrick, 1986; Karson, 1998, 2002; Jacobson, 1992; Smith and Cann, 1999). The uppermost few hundred meters are composed mainly of pillows, flows, and breccia, with proportions of extrusive components varying with spreading rate, proximity to fracture zones and the ends of ridge segments, and the overall robustness of magmatic processes. Deeper sections comprise mainly dikes and other intrusive rocks. Diking is most common within and near the neovolcanic zone (e.g. Smith and Cann, 1993; Delaney *et al.*, 1998), but continues to some distance off-axis. The crust is modified tectonically, beginning close to the ridge and extending well off-axis (e.g. Macdonald *et al.*, 1996; Blackman *et al.*, 1998), providing a ridge-parallel fabric of faults and fractures that is superimposed on the sub-horizontal, layered structure established during initial construction (see Chapters 3 and 9 and references therein).

We should expect that the combination of horizontal layering and sub-horizontal to sub-vertical faulting and fracturing imposes considerable heterogeneity and anisotropy

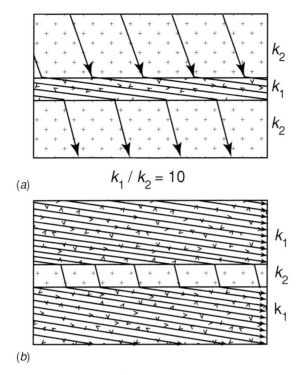

(a)

$$k_1 / k_2 = 10$$

(b)

Fig. 11.6 Illustration of the importance of formation anisotropy on the dominant direction of fluid flow (modified from Hubbert, 1940). In both drawings, there is a permeability ratio of $k_1/k_2 = 10$, and the gradient driving flow is directed from the upper left to the lower right. (a) A thin region of high permeability between layers of lower permeability. Flow lines are refracted to flow along the high-permeability layer, remaining within this layer for as long as possible. (b) A thin region of low permeability between layers having higher permeability. Flow lines are refracted so as to cross the region of low permeability as rapidly as possible. In both of these cases, there would be no refraction of flow only if the gradient were oriented perpendicular or parallel to the layering.

to crustal hydrogeologic properties. Permeability anisotropy within layered sedimentary aquifers commonly exceeds ×10–50 (horizontal to vertical, e.g. Bair and Lahm, 1996); anisotropy within brecciated and fractured oceanic basement, associated with either sub-horizontal layering or sub-vertical faulting, may be ×100 or more. Anisotropy (even in the absence of heterogeneity) leads to refraction of flow lines such that the dominant fluid flow direction can be highly oblique to the gradient in driving forces (Fig. 11.6). Individual layers, fractures, and faults within natural hydrogeological systems can act as conduits for flow, barriers to flow, or both at the same time (e.g. Caine *et al.*, 1996), and it is generally not possible to predict how an individual structure will influence fluid flow patterns or intensity without direct testing. Hydrothermal circulation and associated reactions impose additional heterogeneity through differential rock alteration (e.g. Alt, 1995), leading to the development of hydrogeologically distinct regions within the crust. One example of this hydrothermal "compartmentalization" in oceanic basement comes from ODP Hole

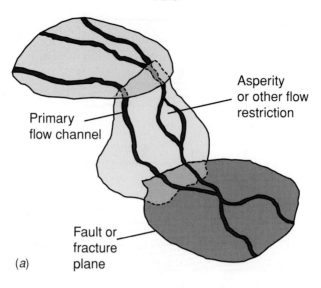

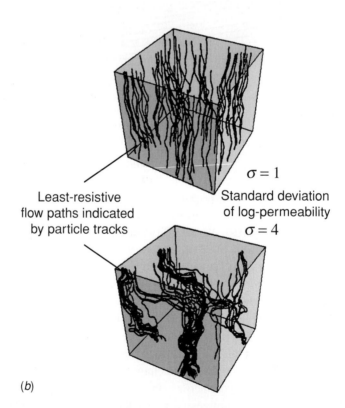

Fig. 11.7 Illustration of two processes that can lead to flow channeling (modified from Wilcock and Fisher, 2002). (*a*) At the scale of individual, intersecting fracture planes, most flow occurs along a small fraction of each fracture surface, away from asperities or other blockages (Tsang *et al.*, 1991).

801C in the western Pacific Ocean, drilled into one of the oldest remaining sections of the seafloor. This crust is layered in terms of both primary lithology and secondary modification, including adjacent, alternating zones of oxidized and reduced hydrothermal alteration, glassy pillows, and massive Fe–Si precipitates (e.g. Larson *et al.*, 1993; Alt and Teagle, 2003).

Heterogeneous rock systems, particularly those within fractured rock, are widely recognized to experience flow channeling (e.g. Tsang and Neretnieks, 1998), a process by which the vast majority of the fluid flow (and associated solute and heat transport) is concentrated to occur within a small fraction of the total rock volume. This process occurs on many scales (Fig. 11.7). Within individual fault and fracture surfaces, asperities block fluid flow along most of the surface (e.g. Tsang *et al.*, 1991). Similarly within heterogeneous systems at a larger scale, fluid circulation tends to follow the paths that offer the least resistance for a given driving force magnitude and direction (e.g. Tsang and Tsang, 1989; Clemo and Smith, 1997). Thus the dominant flow paths are not fixed in a physical sense, but will vary as a function of the interplay between heterogeneity and driving forces. In stochastic numerical studies of three-dimensional fluid flow at a larger scale (Moreno and Tsang, 1994), it was shown that systems having a large standard deviation in permeability will develop preferential flow paths that utilize a small volume of the formation porosity. It is these variations in permeability, rather than the average value, that determine the extent of flow channeling. Because oceanic basement rocks are highly heterogeneous, particularly within the upper few hundred meters, fluid flow must be highly channeled. This interpretation is consistent with the common association between seafloor structure and hydrothermal venting (e.g. Karson and Rona, 1990; Delaney *et al.*, 1992; Wright *et al.*, 1995; Kleinrock and Humphris, 1996) and with the rock record of heterogeneous alteration (Chapter 15).

Hydrothermal circulation patterns within many crustal systems are also likely to vary in shape with time (to be oscillatory or otherwise unstable), given sufficiently high driving forces and permeability (Kimura *et al.*, 1986; Davis *et al.*, 1997), preventing delineation of characteristic shapes of convection cells. This complexity is exacerbated near the critical point for water, as within many ridge crest hydrothermal systems (Goldfarb and Delaney, 1988; Von Damm, 1995), where phase separation and the formation of vapor and/or brine can result in fluid segregation and divergence of flow paths and fluid properties, but instability is also expected under conditions documented within many ridge flank circulation systems.

Studies of the eastern flank of the Juan de Fuca Ridge identified one area that was thought to have a flat-topped basaltic aquifer buried below several hundred meters of

---

(*b*) At a larger scale, within heterogenous rock systems having the same mean permeability, flow is much more highly focused when there is a high standard deviation in log-permeability (Moreno and Tsang, 1994). This process does not depend on the absolute magnitude of mean (or effective system) permeability; even extremely permeable systems experience flow channeling, because fluids prefer to travel along the paths of least resistance.

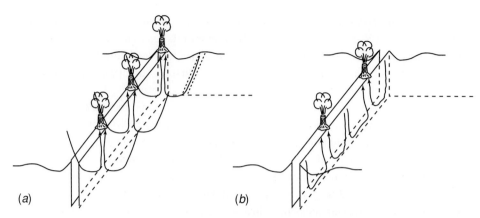

Fig. 11.8 Cartoons showing two end-member circulation geometries for both ridge crest and ridge flank hydrothermal circulation (inspired by Haymon *et al.*, 1991; Delaney *et al.*, 1992; Sohn *et al.*, 1997). Central slot indicates either spreading axis (for ridge crest systems) or a high-angle fault (for ridge flank systems). Drawings are not to scale. Actual flow geometries are likely to be more complicated than either end member. (*a*) Across-strike flow. (*b*) Along-strike flow. Note that there are sites of fluid recharge located between up-flow zones for this circulation geometry.

low-permeability sediments. This seemed to be an ideal place to test the for the occurrence of near-unity-aspect-ratio convection cells within hydrologically isolated basement, as hypothesized to occur by Lister (1972) and Williams *et al.* (1974). Field and numerical analyses were interpreted initially to indicate that near-unity-aspect-ratio convection cells having a width and height of about 600 m were responsible for small-scale variations in seafloor heat flow (Davis *et al.*, 1992, 1996; Snelgrove and Forster, 1996). However, subsequent numerical and seismic studies (Fisher and Becker, 1995; Davis and Chapman, 1996) showed that subtle variations in relief at the aquifer top could control the geometry of underlying convection and lead to the observed heat flow pattern. In a similar way, the locations of isolated basement outcrops, which provide fluid entry and exit points for many ridge flank systems, are likely to influence convection geometry (e.g. Lister, 1972; Fisher *et al.*, 2002; Villinger *et al.*, 2002). Although the concept of cellular convection remains rooted in many descriptive models of hydrothermal circulation, it has yet to be shown that near-unity-aspect-ratio cells are actually favored within natural seafloor systems.

11.3.3 Along-strike versus across-strike fluid flow

There is a tendency when drawing cartoons of seafloor hydrothermal circulation to collapse the crust and the associated fluid pathways into a two-dimensional cross-section oriented perpendicular to the ridge (Fig. 11.8*a*). This results, in part, from the way in which researchers developed and tested ideas about crustal formation and evolution. Early views of seafloor structure were based largely on seismic studies (e.g. Raitt, 1963) and comparison with ophiolites (e.g. Moores and Vine, 1971; Cann, 1974). Conventional reflection and refraction studies are inherently two dimensional, and commonly profiles are oriented so as to cross perpendicular to primary structures (cross-strike) because lines sub-parallel to structure (along-strike) are usually more difficult to interpret. Also, many early studies of ophiolites focused on relations between crustal structure and spreading processes (nature of magma chamber, orientation and mode of emplacement of dikes, etc.), emphasizing cross-strike relations.

Observations of fluid circulation at spreading centers illustrate the importance of along-strike flow (Fig. 11.8*b*). Haymon *et al.* (1991) noted the alignment and consistent spacing between sites of high-temperature hydrothermal venting on the East Pacific Rise (EPR) at $9-10°$ N, and suggested that this could result from along-axis convection. Seafloor obser-vations along the EPR show that axial fissures are strongly aligned sub-parallel to the ridge axis (e.g. Wright *et al.*, 1995), likely imparting a hydrogeologic fabric to the upper crust. Seismic refraction studies have also indicated anisotropic (along-strike) crack distributions extending to depths of several kilometers (Dunn and Toomey, 2001). Collectively, these observations suggest that the dominant orientation of hydrothermal circulation cells on the EPR may be along axis (Haymon, 1996).

Delaney *et al.* (1992) similarly observed the association of hydrothermal vent sites on the Endeavour Segment of the Juan de Fuca Ridge with along-strike structures, an observation that has proven robust as new vent sites have been discovered (Kelley *et al.*, 2001). Wilcock and McNabb (1996) noted that Endeavour Segment vent fields are elongated sub-parallel to the spreading axis, and suggested that permeability anisotropy could favor along-axis circulation cells. Hydrothermal vent fluids on the Endeavour Segment of the Juan de Fuca Ridge have chemistries indicative of substantial interaction with sediments at elevated temperatures, even though this part of the ridge is sediment free at the seafloor (Lilley *et al.*, 1993; Butterfield *et al.*, 1994, 1997). One explanation is that fluids interact with sediments interbedded with lavas at depth, but another option is that along-ridge transport brings hydrothermal fluids tens of kilometers along strike from the north, where the ridge is sedimented (Wilcock and Fisher, 2003). As at the EPR, seismic studies along the Juan de Fuca Ridge also detect large-scale seismic anisotropy within the upper crust (e.g. Sohn *et al.*, 1997), with greater velocities in the along-axis direction, consistent with a crustal fabric established during or soon after crustal formation. Of course, actual flow geometries in these systems are likely to be complex, even three-dimensional (e.g. Travis *et al.*, 1991), so delineation of individual vector components (along-strike or across-strike) may not indicate the dominant direction of fluid flow within any particular system.

There are considerably fewer studies of ridge flank hydrothermal systems that provide an indication of the importance of along-strike versus across-strike flow. As described earlier, the occurrence of elongate heat flow anomalies sub-parallel to the spreading center has been interpreted to indicate a predominantly across-strike fluid circulation direction. However, once vigorous circulation largely homogenizes temperatures in upper basement, the pattern of seafloor heat flow will mimic the distribution of basement relief and sediment thickness no matter what the dominant flow direction (Chapter 8). Although geochemical and geothermal data on the western end of the ODP Leg 168 transect (Davis *et al.*, 1999; Elderfield *et al.*, 1999) are consistent with a dominant flow direction from west to east (across-strike), basement outcrops to the north and south of this transect could play an important role in guiding fluid to and from the basement aquifer. Along the eastern end of this transect, fluid exiting the top of the Baby Bare basement outcrop appears to be focused along one or more ridge-parallel faults (Mottl *et al.*, 1998; Becker *et al.*, 2000). In addition, pore-fluid data from uppermost basement around Baby Bare suggest an evolution in an along-strike direction, from south-southwest to north-northeast (Wheat *et al.*, 2000). Heat flow data from a large basement outcrop 52 km south-southwest of Baby Bare, along structural strike, suggest that this feature may allow recharge of seawater into this part of the ridge flank hydrothermal system (Fisher *et al.*, 2003). Seismic anisotropy studies on ridge flanks (Stephen, 1981, 1985) are also consistent with preferential orientation of fractures in an along-strike direction.

Several recent studies have also indicated hydrogeologic connections between ridge crests and ridge flanks. For example, Johnson *et al.* (2000) documented changes in ridge crest venting temperatures (and possibly flow rates) associated with off-axis seismicity. Davis *et al.* (2001) showed that borehole pressures on ridge flanks can respond to largely aseismic spreading at the axis. These studies suggest that there are important connections between stress state, seismic activity, and crustal properties over distances of kilometers or more, but they do not require significant fluid flow or heat flow between ridge crest and ridge flank areas. In each of these cases, strain in the plate is probably responsible for the long-distance transport of energy, with the initial hydrogeologic response being a local phenomenon (Chapter 8). The extent of heat, fluid, and solute exchange between ridge crest and ridge flank hydrothermal systems remains largely unconstrained. Quantifying these fluxes is important for understanding numerous processes, including mechanisms of heat extraction and crustal construction, and the nature of thermal rebound following the cessation of high-temperature circulation at the ridge (e.g. Chen and Phipps Morgan, 1996; Fisher, 2002)

11.3.4 Directions of fluid flow

It would seem to be a simple matter to determine the direction of fluid flow between field sites by measuring pressures in pairs of sealed boreholes, but this has been challenging because pressure differences are small (Davis and Becker, 2002). A more fundamental difficulty is that coupled heat and fluid flow problems in which fluid properties vary with pressure and temperature often cannot be solved based on a static consideration of differences in absolute pressure. Pressure is a form of potential energy, but the direction of fluid flow will follow the gradient in pressure only if: (i) permeability is either homogeneous or is anisotropic with the primary permeability aligned with the steepest gradient; and (ii) variations in fluid density occur only in the vertical direction, and surfaces of equal density correspond to surfaces of equal pressure (e.g. Hubbert, 1956; Hickey, 1989; Oberlander, 1989). Particularly in cases where pressure differences are small and there are lateral changes in fluid properties, a fully coupled solution is required to determine the fluid flow direction (e.g. Ingebritsen and Sanford, 1998).

Numerical studies of ridge flank circulation may also yield non-unique results, with pre-dicted flow directions that depend on grid geometry and boundary and initial conditions. Fisher *et al.* (1994, 1990) showed numerical simulations of ridge flank hydrothermal cir-culation enhanced by basement and seafloor relief. In these studies, fluid rose from depth below the peak of basement and topographic highs and then flowed laterally down the slop-ing flanks of local ridges, consistent with bench-top models (Hartline and Lister, 1981). These numerical studies used a transient, integrated finite-difference model of coupled heat and fluid flow, but the first analysis was completed with a rectangular grid, similar in form to earlier studies (e.g. Fehn and Cathles, 1979), while the second analysis employed a curvilinear (non-rectangular) grid that allowed more efficient fluid and advective heat flow. A later analysis that included finer grid spacing on the flank of a buried basement ridge demonstrated that, if the basement aquifer is sufficiently permeable and is isotropic, wide convection cells will break up into smaller cells (Fisher and Becker, 1995). At one level the use of wider cells could be interpreted to introduce a numerical "mesh effect," but if the actual system is highly anisotropic, then the use of wide cells (which favors sub-horizontal flow) may actually be geologically realistic. This possibility requires field observatories that can distinguish between wide and narrow cell geometries, or allows an independent assessment of permeability anisotropy.

Steady-state modeling by Davis *et al.* (1997) and Wang *et al.* (1997) quantified the influence of topography and basement relief on convection in the upper oceanic basement of a ridge flank, and determined relations between crustal properties, convection geometry, and thermal and pressure homogenization within a basement aquifer. In one set of simulations, it was shown that the direction of flow relative to the hydrologic structure depends on the starting state of the numerical model. In addition, once convection becomes sufficiently vigorous so as to homogenize basement temperatures (generally a result of permeabilities so great as to also make lateral pressure gradients very small), flow in either direction can result in similar pressure and thermal conditions within shallow basement. Spinelli and Fisher (2004) simulated coupled heat and fluid transport between a buried basement ridge and trough on the eastern flank of the Juan de Fuca Ridge, and compared modeling results to date from long-term, sub-seafloor observatories (Chapter 8). New models show that the preferred flow direction at dynamic "steady state" (convection is unstable and oscillatory) depends on geological and model initial conditions when basement permeability is homogeneous. However, when most of the permeability in upper basement was concentrated within thin channels, the flow directions was always upward in basement below the buried basement high. These solutions also gave a better match to sub-seafloor observations than did models having downward flow within the buried basement ridge.

Stein and Fisher (2003) obtained a related result when they modeled a 26-km-wide section of the eastern flank of the Juan de Fuca Ridge (Fig. 11.9). Fluid could be forced to enter basement close to the ridge crest to the west and then to flow to the east, but when transient models were started with a hydrostatic initial condition (either cold or warm, based on a conductive thermal solution), the final fluid flow direction was from east to west, opposite to that inferred from heat flow and geochemical observations (Davis *et al.*, 1999;

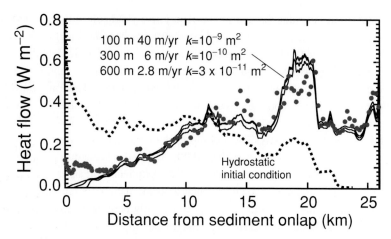

Fig. 11.9 Comparison between heat flow observations and results of transient numerical models of heat and fluid flow along the western end of the ODP Leg 168 transect (Fig. 8.3*b*). Dots show oceanic heat flow values measured with a seafloor probe (Davis *et al.*, 1999). Lines indicate output from numerical models (Stein and Fisher, 2003). Models were cast using a two-dimensional domain oriented perpendicular to the ridge to the west, with sediment thickness and seafloor and basement relief based on seismic data. Fluid entered and exited the seafloor through the side boundaries. The thickness of the permeable aquifer in uppermost basement was varied from 100 to 600 m, and permeability was adjusted to allow sufficiently rapid lateral flow so as to match the general pattern of seafloor heat flow. Heat input at the base of the domain is greatest on the left, where the seafloor is youngest, and decreases to the east, with greater distance from the ridge. Dotted line shows heat flow resulting from a simulation in which the initial pressure condition was "warm" hydrostatic, based on a conductive solution. All simulations started with this initial condition resulted in flow from east to west, a direction opposite to that indicated by seafloor heat flow measurements. However, when fluid flow was initially forced from west to east, and then the forcing was discontinued, fluid continued to flow in this direction because a "hydrothermal siphon" had been established. Basement permeability was then adjusted to allow a good match between model results and observations, with the fluid flux and permeability values shown for various thicknesses of the basement aquifer.

——→

lateral-flow layer is 10^{-11} m^2, and temperature contour interval is 5 °C. Because the system is allowed to convect freely it is thermally more homogeneous than in the case with no vertical mixing. (*c*) Temperature field for same case, but with isotropic permeability in lateral-flow layer of 10^{-13} m^2. Temperature contour interval is 20 °C. Convection still occurs, but mixing is less vigorous than in the case with higher permeability. (*d*) heat flow fraction (f_m/f_L) as a function of distance from the recharge area for all three cases. The case with no vertical mixing is most efficient in suppressing seafloor heat flow, and the case with the most vigorous vertical mixing is the least efficient in suppressing seafloor heat flow.

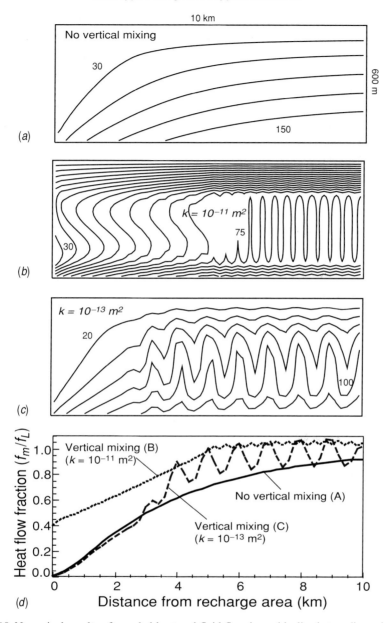

Fig. 11.10 Numerical results of coupled heat and fluid flow in an idealized, two-dimensional system, illustrating the influence of mixed convection on heat transport efficiency (Rosenberg *et al.*, 2000). All models have the same geometry, and fluid is forced to flow from left to right ($q = 10^{-8}$ m s^{-1}, 0.3 m yr^{-1}) along a 400-m-thick basement aquifer placed below low-permeability sediments. (*a*) Temperature field for numerical solution of large-scale lateral flow with vertical mixing prevented through use of permeability anisotropy in basement. Temperature contour interval is 30 °C. (*b*) Temperature field for same case but with vertical mixing allowed; isotropic permeability in

Elderfield *et al.*, 1999). This occurred because heat flow into the base of the grid was greatest at the western end of the domain, where the lithosphere was youngest, leading to up-flow in this area. However, once rapid flow was initiated from west to east in the transient models, through use of temporary, forced-flow boundary conditions, a "hydrothermal siphon" was established, and the small difference in basement pressure below cool and warm columns of water (recharge and discharge sites, respectively) continued to move fluid from west to east even after forcing was discontinued.

These studies illustrate a fundamental limitation in using numerical models to infer fluid flow directions: the results of both steady-state and transient models often depend on the starting condition. In the case of transient simulations, this starting state should not be a numerical convenience; it should represent realistic geological conditions. Representing such conditions within a system that evolves over geological times poses unique challenges; taken to an extreme, the correct initial condition is established when magma upwells below the spreading center and intrusive and extrusive layers are formed.

Another complexity in characterizing the patterns and rates of fluid flow in ridge flank hydrothermal systems is the occurrence of mixed convection, i.e. local convection superimposed on rapid net throughput of fluid. Numerical analyses have shown that mixed convection occurs within a lateral-flow system when permeability is sufficiently large and isotropic (Fig. 11.10). Convective mixing helps to homogenize basement temperatures locally, but reduces the efficiency with which heat is extracted by crustal-scale lateral flow (Davis *et al.*, 1999; Rosenberg *et al.*, 2000; Stein and Fisher, 2003). Numerical models can be made to allow or exclude mixed convection, depending on the absolute magnitude of permeability, the degree of permeability anisotropy, aquifer geometry, and the magnitude of lateral driving forces, but it is not clear how one would test for the occurrence of this process within natural systems.

11.4 Summary and recommendations

Enormous progress has been made in the last several decades in quantifying the rates and directions of fluid flow within seafloor hydrothermal systems. Continued characterization is essential for determining the dynamics and impacts of fluid, heat, solute, and biological fluxes. There are several thermal and geochemical techniques that allow estimation of vertical fluid fluxes through sediments, and lateral fluid fluxes through basement aquifers. We should take advantage of apparent discrepancies between flow rates estimated using different methods, as these differences may help us to find flaws in analytical techniques or assumptions, or may tell us something fundamental about the nature of flow systems. Although assumptions about the isotropic, homogeneous nature of sub-seafloor hydrothermal systems allow us to apply simple models, we should continue to explore the anisotropy and heterogeneity that is to be expected in complex, geological environments. Many of these complexities result from the way in which oceanic lithosphere is created and evolves, so we should make all possible use of geological and other information that can help to guide development of increasingly sophisticated models.

We must continue to develop and apply tools that allow us to map thermal, structural, and hydrogeologic properties spatially and temporally. Technology is available that will allow us to test directly the nature of crustal anisotropy and heterogeneity, the effective porosity of basement aquifers within which transport of fluids occurs over distances of tens of kilometers, and the continuity (or isolation) of thin regions with the crust separated by only a few meters. Fundamental questions regarding the nature of flow processes within heterogeneous, fractured rock systems can be tested within the seafloor to an extent that is not possible in many aquifer systems on land, because of the consistency and simplicity of boundary conditions and our ability to conduct crustal-scale experiments that last for years and extend for vast distances. On the other hand, there may be basic differences between the hydrogeology of oceanic and continental crustal systems that will be delineated by future studies. Thus marine hydrogeologists must design experiments that will test fundamental assumptions, discriminate between competing conceptual models, and reconcile seemingly contradictory data sets.

References

Abbott, D., Menke, W., Hobart, M., and Anderson, R. 1981. Evidence for excess pore pressures in Southwest Indian Ocean sediments. *J. Geophys. Res.* **86**: 1,813–1,827.

Abbott, D. H., Menke, W., Hobart, M., Anderson, R. N., and Embley, R. W. 1984. Correlated sediment thickness, temperature gradient, and excess pore pressure in a marine fault block basin. *Geophys. Res. Lett.* **11**: 485–488.

Alt, J. C. 1995, Subseafloor processes in mid-ocean ridge hydrothermal systems. In *Seafloor Hydrothermal Systems: Physical, Chemical, Biological and Geological Interactions*, eds. S. E. Humphris, R. A. Zierenberg, L. S. Mullineaux, and R. E. Thompson. Washington, DC: American Geophysical Union, pp. 85–114.

Alt, J. C. and Teagle, D. A. H. 2003. Hydrothermal alteration of upper oceanic crust formed at a fast spreading ridge: mineral, chemical, and isotopic evidence from ODP Site 801. *Chem. Geol.* **201**(3–4): 191–211.

Alt-Epping, P. and Smith, L. 2001. Computing geochemical mass transfer and water/rock ratios in submarine hydrothermal systems: imlications for estimating the vigour of convection. *Geofluids* **1**(3): 163–182.

Anderson, R. N., Hobart, M. A., and Langseth, M. G. 1979. Geothermal convection through oceanic crust and sediments in the Indian Ocean. *Science* **204**: 828–832.

Bach, W., Humphris, S. E., and Fisher, A. T. in press. Fluid flow and fluid–rock interaction within oceanic crust: reconciling geochemical, geological and geophysical observations. In *Subseafloor Biosphere at Mid-ocean Ridges*, eds. C. Cary, E. Delong, D. Kelley, and W. S. D. Wilcock. Washington, DC: American Geophysical Union.

Bair, E. S. and Lahm, T. D. 1996. Variations in capture-zone geometry of a partially penetrating pumping well in an unconfined aquifer. *Ground Water* **34**(5): 842–852.

Baker, E. T. 1996. Geological indexes of hydrothermal venting. *J. Geophys. Res.* **101**(B6): 13,741–13,753.

Baker, E. T. and Urabe, T. 1996. Extensive distribution of hydrothermal plumes along the superfast spreading East Pacific Rise, 13° 30′–18° 40′ S. *J. Geophys. Res.* **101**(B4): 8,685–8,695.

Baker, E. T., Chen, Y. J., and Morgan, J. P. 1996. The relationship between near-axes hydrothermal cooling and the spreading rate of mid-ocean ridges. *Earth. Planet. Sci. Lett.* **142**: 137–145.

Baker, P., Stout, P., Kastner, M., and Elderfield, H. 1991. Large-scale lateral advection of seawater through oceanic crust in the central equatorial Pacific. *Earth. Planet. Sci. Lett.* **105**: 522–533.

Barker, P. and Lawver, L. 2000. Anomalous temperatures in central Scotia Sea sediments: bottom water variation or pore water circulation in old crust. *Geophys. Res. Lett.* **27**(1): 13–16.

Beck, A. E., Wang, K., and Shen, P. Y. 1985. Sub-bottom temperature perturbations due to temperature variations at the boundary of inhomogeneous lake or oceanic sediments, *Tectonophysics* **121**: 11–24.

Becker, K., Langseth, M. G., and Hyndman, R. D. 1984. Temperature measurements in Hole 395A, Leg 78B. In *Initial Reports of the Deep Sea Drilling Project*, eds. R. Hyndman and M. Salisbury. Washington, DC: US Govt. Printing Office, pp. 689–698.

Becker, K., Bartetzko, A., and Davis, E. E. 2001. Leg 174B synopsis: revisiting Hole 395A for logging and long-term monitoring of off-axis hydrothermal processes in young oceanic crust. In *Proceedings of the Ocean Drilling Program, Scientific Results*, eds. K. Becker and M. J. Malone. College Station, TX: Ocean Drilling Program, pp. 1–12.

Becker, M. W. and Shapiro, A. M. 2000. Tracer transport in fractured crystalline rock: evidence of nondiffusive breakthrough tailing. *Water Resources Res.* **36**(7): 1,677–1,686.

Becker, N. C., Wheat, C. G., Mottl, M. J., Karsten, J., and Davis, E. E. 2000. A geological and geophysical investigation of Baby Bare, locus of a ridge flank hydrothermal system in the Cascadia Basin. *J. Geophys. Res.* **105**(B10): 23,557–23,568.

Bethke, C. M. and Johnson, T. M. 2002. Paradox of groundwater age. *Geology* **30**(2): 107–110.

Blackman, D. K., Cann, J. R., Janssen, B., and Smith, D. K. 1998. Origin of extensional core complexes: evidence from the Mid-Atlantic Ridge at Atlantis fracture zone. *J. Geophys. Res.* **103**(B9): 21,315–21,333.

Bredehoeft, J. D. and Papadopulos, I. S. 1965. Rates of vertical groundwater movement estimated from the earths thermal profile. *Water Resources Res.* **1**: 325–328.

Butterfield, D. A., McDuff, R. E., Mottl, M. J., Lilley, M. D., Lupton, J. E., and Massoth, G. J. 1994. Gradients in the composition of hydrothermal fluids from the Endeavour segment vent field: phase separation and brine loss. *J. Geophys. Res.* **99**: 9,561–9,583.

Butterfield, D. A., Jonasson, I. R., Massoth, G. J., Feely, R. A., Roe, K. K., Embley, R. E., Holden, J. F., McDuff, R. E., Lilley, M. D., and Delaney, J. R. 1997. Seafloor eruptions and evolution of hydrothermal fluid chemistry. *Phil. Trans. Roy. Soc. Lond. A* **355**(1,723): 369–386.

Caine, J. S., Evans, J. P., and Forster, C. B. 1996. Fault zone architechure and permeability structure. *Geology* **24**: 1,025–1,028.

Cann, J. R. 1974. A layered model for oceanic crust developed. *Geophys. J. Roy. Astron. Soc.* **39**: 169–187.

Chen, Y. J. and Phipps Morgan, J. 1996. The effects of spreading rate, the magma budget, and the geometry of magma emplacement on the axial heat flux at mid-ocean ridges. *J. Geophys. Res.* **101**(B5): 11,475–11,482.

Clemo, T. and Smith, L. 1997. A hierarchical approach to simulation of flow and transport in fractured media. *Water Resources Res.* **33**(8): 1,763–1,784.

Converse, D. R., Holland, H. D., and Edmund, J. M. 1980. Flow rates in the axial hot springs of the East Pacific Rise (21° N): implications for the heat budget and the formation of massive sulfide deposits. *Earth. Planet. Sci. Lett.* **69**: 159–175.

Crank, J. 1975, *The Mathematics of Diffusion*. Oxford: Oxford University Press.

Davis, E. E. and Becker, K. 2002. Observations of natural-state fluid pressures and temperatures in young oceanic crust and inferences regarding hydrothermal circulation. *Earth. Planet. Sci. Lett.* **204**: 231–248.

Davis, E. E. and Chapman, D. S. 1996. Problems with imaging cellular hydrothermal convection in oceanic crust. *Geophys. Res. Lett.* **24**: 3,551–3,554.

Davis, E. E. and Fisher, A. T. 1994. On the nature and consequences of hydrothermal circulation in the Middle Valley sedimented rift: inferences from geophysical and geochemical observations, Leg 139. In *Proceedings of the Ocean Drilling Program, Scientific Results*, eds. E. E. Davis, M. J. Mottl., A. T. Fisher, and J. F. Slack. College Station, TX: Ocean Drilling Program, pp. 695–717.

Davis, E. E. and Villinger, H. 1992. Tectonic and thermal structure of the Middle Valley sedimented rift, northern Juan de Fuca Ridge. In *Proceedings of the Ocean Drilling Program, Initial Results*, eds. E. E. Davis, M. Mottl, and A. T. Fisher. College Station, TX: Ocean Drilling Program, pp. 9–41.

Davis, E. E., Chapman, D. S., Mottl, M. J., Bentkowski, W. J., Dadey, K., Forster, C., Harris, R., Nagihara, S., Rohr, K., Wheat, G., and Whiticar, M. 1992. FlankFlux: an experiment to study the nature of hydrothermal circulation in young oceanic crust. *Can. J. Earth Sci.* **29**(5): 925–952.

Davis, E. E., Chapman, D. S., and Forster, C. B. 1996. Observations concerning the vigor of hydrothermal circulation in young volcanic crust. *J. Geophys. Res.* **101**: 2,927–2,942.

Davis, E. E., Wang, K., He, J., Chapman, D. S., Villinger, H., and Rosenberger, A. 1997. An unequivocal case for high Nusselt-number hydrothermal convection in sediment-buried igneous oceanic crust. *Earth. Planet. Sci. Lett.* **146**: 137–150.

Davis, E. E., Chapman, D. S., Wang, K., Villinger, H., Fisher, A. T., Robinson, S. W., Grigel, J., Pribnow, D., Stein, J., and Becker, K. 1999. Regional heat flow variations across the sedimented Juan de Fuca Ridge eastern flank: constraints on lithospheric cooling and lateral hydrothermal heat transport. *J. Geophys. Res.* **104**(B8): 17,675–17,688.

Davis, E. E., Wang, W., Thomson, R. E., Becker, K., and Cassidy, J. F. 2001. An episode of seafloor spreading and associated plate deformation inferred from crustal fluid pressure transients. *J. Geophys. Res.* **106**(B10): 21,953–21,963.

Davis, E. E., Wang, K., Becker, K., Thomson, R. E., and Yashayaev, I. 2003. Deep-ocean temperature variations and implications for errors in seafloor heat flow determinations. *J. Geophys. Res.* **108**(B1): 2,034.

Delaney, J. R., Robigou, V., McDuff, R., and Tivey, M. 1992. Geology of a vigorous hydrothermal system on the Endeavor segment, Juan de Fuca Ridge. *J. Geophys. Res.* **97**: 19,663–19,682.

Delaney, J. R., Kelley, D. S., Lilley, M. D., Butterfield, D. A., Baross, J. A., Wilcock, W. S. D., Embley, R. W., and Summit, M. 1998. The quantum event of oceanic crustal accretion: impacts of diking at mid-ocean ridges. *Science* **281**(5374): 222–230.

Detrick, R. S., Williams, D. L., Mudie, J. D., and Sclater, J. G. 1974. The Galapagos Spreading Centre: bottom-water temperatures and the significance of geothermal heating. *Geophys. J. Roy. Astron. Soc.* **38**: 627–636.

Dunn, R. A. and Toomey, D. R. 2001. Crack-induced seismic anisotropy in the oceanic crust across the East Pacific Rise (9° 30′ N). *Earth Planet. Sci. Lett.* **189**: 9–17.

Elder, J. W. 1965, Physical processes in geothermal areas. In *Terrestrial Heat Flow*, ed. W. H. K. Lee. Washington, DC: American Geophysical Union, pp. 211–239.

Elderfield, H. and Schultz, A. 1996. Mid-ocean ridge hydrothermal fluxes and the chemical composition of the ocean. *Ann. Rev. Earth Planet. Sci.* **24**: 191–224.

Elderfield, H., Wheat, C. G., Mottl, M. J., Monnin, C., and Spiro, B. 1999. Fluid and geochemical transport through oceanic crust: a transect across the eastern flank of the Juan de Fuca Ridge. *Earth. Planet. Sci. Lett.* **172**: 151–165.

Fehn, U. and L. Cathles, 1979. Hydrothermal convection at slow-spreading midocean ridges. *Tectonophysics* **55**: 239–260.

1986. The influence of plate movement on the evolution of hydrothermal convection cells in the oceanic crust. *Tectonophysics* **125**: 289–312.

Fehn, U., Green, K., Von Herzen, R. P., and Cathles, L. 1983. Numerical models for the hydrothermal field at the Galapagos Spreading Center. *J. Geophys. Res.* **88**: 1,033–1,048.

Fisher, A. 2003. Geophysical constraints on hydrothermal circulation: observations and models. In *Energy and Mass Transfer in Submarine Hydrothermal Systems*, eds. P. Halbach, V. Tunnicliffe, and J. Hein. Berlin: Dahlem University Press, pp. 29–52.

Fisher, A. and Becker, K. 2000. Channelized fluid flow in oceanic crust reconciles heat-flow and permeability data. *Nature* **403**: 71–74.

Fisher, A., Giambalvo, E., Sclater, J., Kastner, M., Ransom, B., Weinstein, Y., and Lonsdale, P. 2001. Heat flow, sediment and pore fluid chemistry, and hydrothermal circulation on the east flank of Alarcon Ridge, Gulf of California. *Earth. Planet. Sci. Lett.* **188**: 521–534.

Fisher, A. T. and Becker, K. 1991. Heat flow, hydrothermal circulation and basalt intrusions in the Guaymas Basin, Gulf of California. *Earth. Planet. Sci. Lett.* **103**: 84–99.

1995. The correlation between heat flow and basement relief: observational and numerical examples and implications for upper crustal permeability. *J. Geophys. Res.* **100**: 12,641–12,657.

Fisher, A. T., Becker, K., Narasimhan, T. N., Langseth, M. G., and Mottl, M. J. 1990. Passive, off-axis convection on the southern flank of the Costa Rica Rift. *J. Geophys. Res.* **95**: 9,343–9,370.

Fisher, A. T., Becker, K., and Narasimhan, T. N. 1994. Off-axis hydrothermal circulation: parametric tests of a refined model of processes at Deep Sea Drilling Project/Ocean Drilling Program site 504. *J. Geophys. Res.* **99**: 3,097–3,121.

Fisher, A. T., Davis, E. E., Hutnak, M., Spiess, V., Zühlsdorff, L., Cherkaoui, A., Christiansen, L., Edwards, K. M., Macdonald, R., Villinger, H., Mottl, M. J., Wheat, C. G., and Becker, K. 2003. Hydrothermal recharge and discharge across 50 km guided by seamounts on a young ridge flank. *Nature* **421**: 618–621.

Gelhar, L., Welty, C., and Rehfeldt, K. 1992. A critical review of data on field-scale dispersion in aquifers. *Water Resources Res.* **28**(7): 1,955–1,974.

Geller, G. A., Weissel, J. K., and Anderson, R. N. 1983. Heat transfer and intraplate deformation in the central Indian Ocean. *J. Geophys. Res.* **88**(B2): 1,018–1,032.

Giambalvo, E., Fisher, A. T., Darty, L., Martin, J. T., and Lowell, R. P. 2000. Origin of elevated sediment permeability in a hydrothermal seepage zone, eastern flank of Juan de Fuca Ridge, and implications for transport of fluid and heat. *J. Geophys. Res.* **105**(B1): 913–928.

Giambalvo, E., Steefel, C., Fisher, A., Rosenberg, N., and Wheat, C. G. 2002. Effect of fluid–sediment reaction on hydrothermal fluxes of major elements, eastern flank of the Juan de Fuca Ridge. *Geochim. Cosmochim. Acta* **66**(10): 1,739–1,757.

Goldfarb, M. S. and Delaney, J. R. 1988. Response of two-phase fluids to fracture configurations within submarine hydrothermal systems. *J. Geophys. Res.* **93**: 4,585–4,594.

Goode, D. J. 1996. Direct simulation of groundwater age. *Water Resources Res.* **32**: 289–296.

Green, K. E., Von Hergen, R. P., and Williams, D. L. 1981. The Galapagos spreading center at 86° W: a detailed geothermal field study. *J. Geophys. Res.* **86**: 979–986.

Hartline, B. K. and Lister, C. R. B. 1981. Topographic forcing of super-critical convection in a porous medium such as the oceanic cost. *Earth Planet. Sci. Lett.* **55**: 75–86.

Haymon, R. M. 1996, The response of ridge-crest hydrothermal systems to segmented episodic magma supply. In *Tectonic, Magmatic, Hydrothermal, and Biological Segmentation of Mid-ocean Ridges*, ed. C. L. Walker. Washington, DC: American Geophysical Union, pp. 157–168.

Haymon, R. M., Fornari, D. J., Edwards, M. H., Carbotte, S., Wright, D., and Macdonald, K. C. 1991. Hydrothermal vent distribution along the East Pacific Rise crest (9° 09′−54′ N) and its relationship to magmatic and tectonic processes on fast-spreading mid-ocean ridges. *Earth. Planet. Sci. Lett.* **104**: 513–534.

Hickey, J. J. 1989. An approach to the field study of hydraulic gradients in variable-salinity ground water. *Ground Water* **27**(4): 531–539.

Houtz, R. and Ewing, J. 1976. Upper crustal structure as a function of plate age. *J. Geophys. Res.* **81**: 2,490–2,498.

Huang, P. Y. and Solomon, S. C. 1988. Centroid depths of mid-ocean ridge earthquakes: dependence on spreading rate. *J. Geophys. Res.* **93**: 13,445–13,477.

Hubbert, M. K. 1940. The theory of groundwater motion. *J. Geol.*, **48**: 785–944.
  1956. Darcy's law and the field equivalent of the flow of underground fluids. *Trans. Am. Inst. Mining Metal. Eng.* **207**: 222–239.

Humphris, S. E. and Cann, J. R. 2000. Constraints on the energy and chemical balances of the modern TAG and ancient Cyprus seafloor sulfide deposits. *J. Geophys. Res.* **105**(B12): 28,477–28,488.

Ingebritsen, S. E. and Sanford, W. E. 1998. *Groundwater in Geologic Processes*. New York: Cambridge University Press.

Jacobson, R. S. 1992. Impact of crustal evolution on changes of the seismic properties of the uppermost oceanic crust. *Rev. Geophys.* **30**: 23–42.

Johnson, H. P., Hutnak, M., Dziak, R. P., Fox, C. G., Urcuyo, I., Cowen, J. P., Nabelek, J., and Fisher, C. 2000. Earthquake-induced changes in a hydrothermal system on the Juan de Fuca mid-ocean ridge. *Nature* **407**: 174–176.

Kadko, D. and Butterfield, D. A. 1998. The relationship of hydrothermal fluid composition and crustal residence time to maturity of vent fields on the Juan de Fuca Ridge. *Geochim. Cosmochim. Acta* **62**(9): 1,521–1,533.

Kadko, D. and Moore, W. 1988. Radiochemical constraints on the crustal residence time of submarine hydrothermal fluids: Endeavour Ridge. *Geochim. Cosmochim. Acta* **52**: 659–668.

Karson, J. A. 1998. Internal structure of oceanic lithosphere: a perspective from tectonic windows. In *Faulting and Magmatism at Mid-ocean Ridges*, eds. W. R. Buck, P. T. Delaney, J. A. Karson, and Y. Lagabrielle. Washington, DC: American Geophysical Union, pp. 177–218.

2002. Geologic structure of the uppermost oceanic crust created at fast- to intermediate-rate spreading centers. *Ann. Rev. Earth Planet. Sci.* **30**: 347–384.

Karson, J. A. and Rona, P. A. 1990. Block-tilting, transfer faults, and structural control of magmatic and hydrothermal processes in the TAG area, mid-Atlantic 26° North. *Geol. Soc. Am. Bull.* **102**: 1,635–1,645.

Kelley, D. S., Delaney, J. R., and Yoerger, D. R. 2001. Geology and venting dynamics of the Mothra hydrothermal field, Endeavour Segment, Juan de Fuca Ridge. *Geology* **29**(10): 959–962.

Kimura, S., Schubert, G., and Straus, J. M. 1986. Route to chaos in porous-medium thermal convection. *J. Fluid Mech.* **166**: 305–324.

Kleinrock, M. C. and Humphris, S. E. 1996. Structural control on sea-floor hydrothermal activity at the TAG active mound. *Nature* **382**: 149–153.

Klitgord, K. D. and Mudie, J. D. 1974. The Galapagos Spreading Center: a near-bottom geophysical study. *Geophys. J. Roy. Astron. Soc.* **26**: 623–626.

Kong, L. S. L., Solomon, S. C., and Purdy, G. M. 1992. Microearthquake characteristics of a mid-ocean ridge along-axis high. *J. Geophys. Res.* **97**(B2): 1,659–1,685.

Langseth, M. G. and Herman, B. 1981. Heat transfer in the oceanic crust of the Brazil Basin. *J. Geophys. Res.* **86**: 10,805–10,819.

Langseth, M. G., Hyndman, R. D., Becker, K., Hickman, S. H., and Salisbury, M. H. 1984. The hydrogeological regime of isolated sediment ponds in mid-oceanic ridges. In *Initial Reports of the Deep Sea Drilling Project*, eds. R. H. Hyndman and M. H. Salisbury. Washington, DC: US Govt. Printing Office, pp. 825–837.

Langseth, M. G., Mottl, M. J., Hobart, M. A. and Fisher, A. T. 1988. The distribution of geothermal and geochemical gradients near site 501/504: implications for hydrothermal circulation in the oceanic crust. In *Proceedings of the Ocean Drilling Program, Initial Reports*, Vol. 111, eds. K. Beaker, H. Sakai, *et al.* College Station, TX: Ocean Drilling Program, pp. 23–32.

Langseth, M. G., Becker, K., Von Herzen, R. P., and Schultheiss, P. 1992. Heat and fluid flux through sediment on the western flank of the Mid-Atlantic Ridge: a hydrogeological study of North Pond. *Geophys. Res. Lett.* **19**: 517–520.

Lapwood, E. 1948. Convection of a fluid in a porous media. *Proc. Camb. Phil. Soc.* **44**: 508–521.

Larson, R. L., Fisher, A. T., and Jarrard, R. 1993. Highly layered and permeable Jurassic oceanic crust in the western Pacific. *Earth. Planet. Sci. Lett.* **119**: 71–83.

Lilley, M. D., Butterfield, D. A., Olson, E. J., Lupton, J. E., Macko, S. A., and McDuff, R. E. 1993. Anomalous CH4 and NH4$^+$ concentrations at an unsedimented mid-ocean ridge hydrothermal system. *Nature* **364**(6432): 45–47.

Lister, C. R. B. 1972. On the thermal balance of a mid-ocean ridge. *Geophys. J. Roy. Astron. Soc.* **26**: 515–535.

Lu, N. and Ge, S. 1996. Effect of horizontal heat and fluid flow on the vertical temperature distribution in a semiconfining layer. *Water Resources Res.* **32**(5): 1,449–1,453.

Macdonald, K. C., Fox, P. J., Alexander, R. T., Pockalny, R., and Gente, P. 1996. Volcanic growth faults and the origin of Pacific abyssal hills. *Nature* **380**: 125–129.

Matthews, M., Salisbury, M., and Hyndman, R. 1984. Basement logging on the Mid-Atlantic Ridge, Deep Sea Drilling Project Hole 395A. In *Initial Reports of the Deep Sea Drilling Project*, eds. R. D. Hyndman and M. H. Salisbury. Washington, DC: US Govt. Printing Office, pp. 717–730.

Mayer, L., Theyer, F., *et al.* 1985. Site 571. *Initial Reports of the Deep Sea Drilling Project*, Vol. 85, eds. L. Mayer, F. Theyer, *et al.* Washington, DC: US Govt. Printing Office, pp. 23–32.

McDuff, R. E. 1981. Major cation gradients in DSDP interstitial waters: the role of diffusive exchange between seawater and upper oceanic crust. *Geochim. Cosmochim. Acta* **45**: 1,705–1,713.

Moores, E. M. and Vine, F. J. 1971. The Troodos massif, Cyprus, and other ophiolites as oceanic crust: evaluations and implications. *Phil. Trans. Roy. Soc. Lond. Series A* **268**: 443–466.

Moreno, L. and Tsang, C.-F. 1994. Flow channeling in strongly heterogeneous porous media: a numerical study. *Water Resources Res.* **30**: 1,421–1,430.

Mottl, M. J. 1989. Hydrothermal convection, reaction and diffusion in sediments on the Costa Rica Rift flank, pore water evidence from ODP Sites 677 and 678. In *Proceedings of the Ocean Drilling Program, Scientific Results*, Vol. 111, eds. K. Becker and H. Sakai. College Station, TX: Ocean Drilling Program, pp. 195–214.

Mottl, M. J. and Wheat, C. G. 1994. Hydrothermal circulation through mid-ocean ridge flanks: fluxes of heat and magnesium. *Geochim. Cosmochim. Acta* **58**: 2,225–2,237.

Mottl, M. J., Wheat, C. G., Baker, E., Becker, N., Davis, E., Feeley, R., Grehan, A., Kadko, D., Lilley, M., Massoth, G., Moyer, C., and Sansone, F. 1998. Warm springs discovered on 3.5 Ma oceanic crust, eastern flank of the Juan de Fuca Ridge. *Geology* **26**: 51–54.

Murton, B. J., Redbourn, L. J., German, C. G., and Baker, E. T. 1999. Sources and fluxes of hydrothermal heat, chemicals and biology within a segment of the Mid-Atlantic Ridge. *Earth. Planet. Sci. Lett.* **171**: 301–317.

Neuman, S. P. 1990. Universal scaling of hydraulic conductivities and dispersivities in geologic media. *Water Resources Res.* **26**(8): 1,749–1,758.

Nield, D. A. 1968. Onset of thermohaline convection in a porous medium. *Water Resources Res.* **4**: 553–560.

Oberlander, P. L. 1989. Fluid density and gravitational variations in deep boreholes and their effect on fluid potential. *Ground Water* **27**(3): 341–350.

Park, J., Bethke, C. M., Torgersen, T., and Johnson, T. M. 2002. Transport modeling applied to the interpretation of groundwater 36Cl age. *Water Resources Res.* **38**: 1–15.

Parsons, B. and Sclater, J. G. 1977. An analysis of the variation of ocean floor bathymetry and heat flow with age. *J. Geophys. Res.* **82**: 803–829.

Phipps Morgan, J. and Chen, Y. J. 1993. The genesis of oceanic crust: magma injection, hydrothermal circulation, and crustal flow. *J. Geophys. Res.* **98**: 6,283–6,297.

Purdy, M. and Detrick, R. 1986. Crustal structure of the Mid-Atlantic Ridge at 23° N from seismic refraction studies. *J. Geophys. Res.* **91**: 3,739–3,762.

Purdy, M., Kong, L. S. L., Christeson, G. L., and Solomon, S. C. 1992. Relation between spreading rate and the seismic structure of mid-ocean ridges. *Nature* **355**: 815–817.

Raitt, R. W. 1963, The crustal rocks. In *The Sea*, ed. M. N. Hill. New York: Wiley-Interscience, pp. 85–102.

Ribando, R., Torrence, K., and Turcotte, D. 1976. Numerical models for hydrothermal circulation in the oceanic crust. *J. Geophys. Res.* **81**: 3,007–3,012.

Rosenberg, N., Fisher, A., and Stein, J. 2000. Large-scale lateral heat and fluid transport in the seafloor: revisiting the well-mixed aquifer model. *Earth. Planet. Sci. Lett.* **182**: 93–101.

Sanford, W. E. 1997. Correcting for diffusion in Carbon-14 dating of ground water. *Ground Water* **35**(2): 357–361.

Schultz, A., Delaney, J. R., and McDuff, R. E. 1992. On the partitioning of heat flux between diffuse and point source seafloor venting. *J. Geophys. Res.* **97**: 12,299–12,315.

Sclater, J. G. and Francheteau, J. 1970. The implications of terrestrial heat flow observations on current tectonic and geochemical models of the crust and upper mantle of the earth. *Geophys. J. Roy. Astron. Soc.* **20**: 509–542.

Sclater, J. G., Von Herzen, R. P., Williams, D. L., Anderson, R. N., and Klitgord, K. 1974. The Galapagos Spreading Center: heat flow on the north flank. *Geophys. J. Roy. Astron. Soc.* **38**: 609–626.

Shapiro, A. M. 2001. Effective matrix diffusion in kilometer-scale transport in fractured rock. *Water Resources Res.* **37**(3): 507–522.

Shapiro, A. M. and Hsieh, P. A. 1998. How good are estimates of transmissivity from slug tests in fractured rocks? *Ground Water* **36**(1): 37–48.

Sleep, N. and Wolery, T. 1978. Egress of hot water from mid-ocean ridge hydrothermal systems, some thermal constraints. *J. Geophys. Res.* **83**: 5,913–5,922.

Smith, D. K. and Cann, J. R. 1993. Building the crust at the Mid-Atlantic Ridge. *Nature* **365**: 707–715.

  1999. Constructing the upper crust of the Mid-Atlantic Ridge: a reinterpretation based on the Puna Ridge, Kilauea Volcano. *J. Geophys. Res.* **104**: 25,379–25,399.

Snelgrove, S. H. and Forster, C. B. 1996. Impact of seafloor sediment permeability and thickness on off-axis hydrothermal circulation: Juan de Fuca Ridge eastern flank. *J. Geophys. Res.* **101**: 2,915–2,925.

Sohn, R. A., Webb, S. C., Hildebrand, J. A., and Cornuelle, B. C. 1997. Three-dimensional tomographic velocity structure of upper crust, CoAxial segment, Juan de Fuca Ridge: implications for on-axis evolution and hydrothermal circulation. *J. Geophys. Res.* **102**: 17,679–17,695.

Spinelli, G. A. and Fisher, A. T. 2004. Hydrothermal circulation within rough basement on the Juan is Fuca Ridge Flank. *Geochem. Geophys. Geosys.* **5**(2): Q02001, doi: 10.1029/2003GC000616.

Stein, J. S. and Fisher, A. T. 2001. Multiple scales of hydrothermal circulation in Middle Valley, northern Juan de Fuca Ridge: physical constraints and geologic models. *J. Geophys. Res.* **106**(B5): 8,563–8,580.

  2003. Observations and models of lateral hydrothermal circulation on a young ridge flank: reconciling thermal, numerical and chemical constraints. *Geochem. Geophys. Geosys.* doi:10.1029/200290000415.

Stephen, R. 1981. Seismic anisotropy observed upper oceanic crust. *Geophys. Res. Lett.* **8**: 865–868.

Stephen, R. A. 1985. Seismic anisotropy in the upper oceanic crust. *J. Geophys. Res.* **90**: 11,383–11,396.

Sudicky, E. A. and Frind, E. O. 1981. Carbon 14 dating of groundwater in confined aquifers: implications of aquitard diffusion. *Water Resources Res.* **17**(4): 1,060–1,064.

Travis, B. J., Janecky, D. R., and Rosenberg, N. D. 1991. Three-dimensional simulation of hydrothermal circulation at mid-ocean ridges. *Geophys. Res. Lett.* **18**: 1,441–1,444.

Tsang, C.-F. and Neretnieks, I. 1998. Flow channeling in heterogeneous fractured rocks. *Rev. Geophys.* **36**(2): 275–298.

Tsang, C.-F. and Tsang, Y. W. 1989. Flow channeling in a single fracture as a two-dimensional strongly heterogeneous permeable medium. *Water Resources Res.* **25**(9): 2,076–2,080.

Tsang, C.-F., Tsang, Y. W., and Hale, F. V. 1991. Tracer transport in fractures: analysis of field data based on a variable-aperature channel model. *Water Resources Res.* **27**(12): 3,095–3,106.

Villinger, H., Grevemeyer, I., Kaul, N., Hauschild, J., and Pfender, M. 2002. Hydrothermal heat flux through aged oceanic crust: where does the heat escape? *Earth Planet. Sci. Lett.* **202**(1): 159–170.

Von Damm, K. L. 1995. Controls on the chemistry and temporal variability of seafloor hydrothermal fluids. In *Seafloor Hydrothermal Systems: Physical, Chemical, Biological and Geological Interactions*, eds. S. E. Humphris, R. A. Zierenberg, L. S. Mullineaux, and R. E. Thompson. Washington, DC: American Geophysical Union, pp. 222–247.

Von Herzen, R., Ruppel, C., Molnar, P., Nettles, M., Nagihara, S., and Ekstrom, G. 2001. A constraint on the shear stress at the Pacific–Australian plate boundary from heat flow and seismicity at the Kermadec Forearc. *J. Geophys. Res.* **106**(B4): 6,817–6,833.

Wang, K., He, J., and Davis, E. E. 1997. Influence of basement topography on hydrothermal circulation in sediment-buried oceanic crust. *Earth Planet. Sci. Lett.* **146**: 151–164.

Wheat, C. G. and McDuff, R. 1994. Hydrothermal flow through the Mariana Mounds: dissolution of amorphous silica and degredation of organic mater on a mid-ocean ridge flank. *Geochim. Cosmochim. Acta* **58**: 2,461–2,475.

Wheat, C. G. and Mottl, M. J. 1994. Hydrothermal circulation, Juan de Fuca Ridge eastern flank: factors controlling basement water composition. *J. Geophys. Res.* **99**: 3,067–3,080.

Wheat, C. G., Elderfield, H., Mottl, M. J., and Monnin, C. 2000. Chemical composition of basement fluids within an oceanic ridge flank: implications for along-strike and across-strike hydrothermal circulation. *J. Geophys. Res.* **105**(B6): 13,437–13,447.

Wilcock, W. S. D. and Fisher, A. T. in press. Geophysical constraints on the sub-seafloor environment near mid-ocean ridges. In *Subseafloor Biosphere at Mid-ocean Ridges*, eds. C. Cary, E. Delong, D. Kelley, and W. S. D. Wilcock. Washington, DC: American Geophysical Union.

Wilcock, W. S. D. and McNabb, A. 1996. Estimates of crustal permeability on the Endeavour segment of the Juan de Fuca mid-ocean ridge. *Earth Planet. Sci. Lett.* **138**: 83–91.

Williams, D. L., Von Herzen, R. P., Sclater, J. G., and Anderson, R. N. 1974. The Galapagos Spreading Centre, lithospheric cooling and hydrothermal circulation. *Geophys. J. Roy. Astron. Soc.* **38**: 587–608.

Wooding, R. A. 1960. Instability of a viscous liquid of variable density in a vertical Hele-Shaw cell. *J. Fluid Mech.* **7**: 501–515.

Wright, D. J., Haymon, R. M., and Macdonald, K. C. 1995. Breaking new ground: estimates of crack depth along the axial zone of the East Pacific Rise ($9°12' - 54'$ N). *Earth Planet. Sci. Lett.* **134**: 441–457.

12

Applying fundamental principles and mathematical models to understand processes and estimate parameters

Kelin Wang

12.1 Introduction

12.1.1 Modeling approaches in marine hydrogeology

The structure of oceanic lithosphere is relatively simple compared to that of the continents (see Chapter 3), but marine hydrogeology presents its own challenges. (i) The flow system beneath the ocean is much less accessible, and flow patterns usually have to be inferred from indirect observations such as heat flow measurements, borehole temperatures, seismic wave velocities, and geochemical indicators. (ii) The predominant driving force for fluid flow, thermal buoyancy, is typically small. In contrast, groundwater flow on land is usually driven by much larger potential differences due to surface topography. (iii) Flow caused by the deformation of the hosting formation is relatively important. Flow may be caused by permanent deformation such as sediment accretion at subduction zones and by seismic or aseismic faulting. Significant pressure fluctuation or fluid flow may also accompany elastic deformation due to tidal loading and other types of periodic stress changes.

Fluids flow through interconnected pore spaces between solid grains in seafloor sediment, and mostly through interconnected fractures in the underlying igneous formation. Depending on the scale of interest and hence the level of complexity, three types of models can be used to describe the flow systems, namely, the pipe model, the fracture model, and the porous-medium model. The pipe model, in which the fluid flows along a channel through an impermeable rock formation, provides the least amount of detail and is designed to simulate the average thermal and chemical effects of a flow system (Lowell, 1991). Models of flow through fractured rocks provide a realistic description of the flow system in the igneous formation, although the overwhelming degree of heterogeneity of a fracture network usually requires stochastic modeling approaches. Fracture-flow models have important theoretical implications and are being increasingly used in the studies of solute transport in groundwater systems (Tsang and Neretnieks, 1998) but have yet to find practical applications to seafloor hydrogeology. Intermediate between these two models in terms of accounting for details is the most widely used porous-medium model, which is the focus of this chapter.

Hydrogeology of the Oceanic Lithosphere, eds. E. E. Davis and H. Elderfield. Published by Cambridge University Press.
© Cambridge University Press 2004.

This chapter is designed to provide a summary of the fundamental physical principles frequently encountered in the study of oceanic lithosphere hydrogeology. An effort is made to explain some conceptual (and occasionally numerical) details that are not always explained in standard textbooks and technical papers, because experience with frequently asked questions indicates that some seemingly trivial points do require clarification. After a comparison of different forms of Darcy's Law for porous medium flow in the rest of the first section, the interaction of the temperature field with fluid flow and the effects of elastic medium deformation on the pressure and flow fields will be discussed. Near the end of the chapter is a brief discussion on fluid flow in deforming sediments that provides simple examples of how to deal with porous medium flow in the presence of permanent and large matrix deformation. Equations that govern geochemical processes are not discussed in this chapter, although some similarities between heat transport and chemical transport are mentioned in Section 12.2.1.

12.1.2 Darcy's Law for flow in porous media

Flow in a porous medium is governed by Darcy's Law:

$$\boldsymbol{q} = -\frac{k}{\mu} \cdot (\nabla P - \rho_f \boldsymbol{g}) \tag{12.1a}$$

where $\boldsymbol{q}$ is the volume flux of the fluid (average rate of volumetric flow through a unit area), P is fluid pressure, k is the permeability of the porous medium, ρ_f and μ are the density and viscosity of the fluid, respectively, and $\boldsymbol{g}$ is the acceleration of gravity. The permeability is anisotropic if the rock structure and fabric allow fluid to flow more easily in one direction than in other directions. In this chapter, a vector is represented by a bold-face character or a character with an index subscript. For example, in a three-dimensional Cartesian coordinate system, if the x_3 axis points upward, $\boldsymbol{g} = (0, 0, -g)$, $g_3 = -g$, $|\boldsymbol{g}| = g$, and $\rho_f g_j = -\rho_f g \partial x_3 / \partial x_j = -\rho_f g \partial_{3j}$. In this case, (12.1a) in index notation is:

$$q_i = -\frac{k_{ij}}{\mu} \left(\frac{\partial P}{\partial x_j} + \rho_f g \frac{\partial x_3}{\partial x_j} \right) \tag{12.1b}$$

If the x_3 axis points downward, the plus sign in the parentheses becomes a minus sign. Darcy's Law describes a macroscopic diffusion process, being analogous to heat conduction or chemical diffusion in that flux is proportional to a gradient. The volumetric fluid flux is called the Darcy flux, Darcy velocity, or specific discharge. An individual fluid particle must follow a tortuous route. The average linear velocity of fluid particles relative to the solid matrix is the Darcy flux divided by porosity n, $\boldsymbol{q}/n$.

The validity of using a porous medium to approximate a fractured rock depends on the spatial scale of interest (Chapter 7). A formation with densely and evenly spaced fractures can be well approximated using a porous medium, as is numerically verified by Yang *et al.* (1998) for the case of hydrothermal circulation. If fluid flow is controlled by relatively few large fractures, Darcy's Law is invalid at scales equivalent to the fracture scale, but the formation-scale permeability as defined by the average fluid flux and pressure gradient using Darcy's Law is a useful concept that allows the effective average hydrologic characteristics of a geological formation to be quantified (Chapter 8).

Equation (12.1) states that fluid flow in the porous medium is driven by the non-hydrostatic part of the pressure gradient. In a hydrostatic state, $\nabla P = \rho_f \mathbf{g}$, and there is no fluid flow. If a density change (e.g. due to its temperature dependence) causes $\rho_f \mathbf{g}$ to be different from the local hydrostatic state, the fluid is said to be buoyant. While $\rho_f \mathbf{g}$ is strictly a local quantity, ∇P depends on the behavior of every other part of the entire flow system. In some cases, a background (or reference) pressure field such as the hydrostatic component is of no interest and is understood to have been subtracted from the total pressure, and (12.1a) becomes:

$$ \boldsymbol{q} = -\frac{k}{\mu} \cdot \nabla P \tag{12.2} $$

Symbol P represents the total fluid pressure in (12.1) but the incremental part of the fluid pressure in (12.2). If fluid density is uniform $\rho_f = \rho_o$, as is often assumed in land hydrogeology, and the x_3 axis is oriented upward, one can define a hydraulic head $h = P/\rho_o g + x_3$, so that (12.1a) becomes:

$$ \boldsymbol{q} = -\frac{k\rho_o g}{\mu} \cdot \nabla h = -K \cdot \nabla h \tag{12.3} $$

The composite parameter $K = k\rho_o g/\mu$ is termed the "hydraulic conductivity." Integrated over the thickness of a confined aquifer, it becomes the "transmissivity" of the aquifer. If the fluid density varies with temperature (T), the head must be defined using a reference density at a given temperature $\rho_o = \rho_f(T_o)$. Equation (12.1) then becomes:

$$ \boldsymbol{q} = -\frac{k\rho_o g}{\mu} \cdot \left[\nabla h - \left(1 - \frac{\rho_f}{\rho_o} \right) \nabla x_3 \right] \tag{12.4} $$

Equation (12.4) seems to have separated the Darcy flow into head-driven and buoyancy-driven components, but this separation does not contribute to our physical intuition. The problem is that the second term of (12.4) is not a correct mathematical description of buoyancy. If ρ_f is temperature dependent, this term exists even at a hydrostatic state where there is no flow. "Head" has a clear physical meaning only when fluid density is independent of temperature. For seafloor hydrogeology, in which thermal buoyancy is the primary driving force, it is advisable to avoid basing discussions on the concept of head.

12.2 Fluid flow and heat transfer

12.2.1 Convective and conductive heat transfer

Thermal buoyancy drives sub-seafloor fluid flow (Chapter 8). Fluid flow, in turn, transports thermal energy and distorts the thermal field (Chapter 10). Temperature and conservative chemical species can both be good tracers of fluid flow (Chapter 11), and with some modification, the theory describing convective heat transfer is to some degree applicable to chemical transport (Chapters 16 and 20).

A unit volume of a fluid increases its temperature from T_o, an arbitrary reference temperature, to T by absorbing the amount of heat $\rho_f c_f (T - T_o)$, where c_f is specific heat. The value of c_f depends on whether the heat is added at constant pressure or constant volume. It takes slightly less energy, thus a smaller c_f, to heat up the fluid to the same temperature at constant volume than at constant pressure, because the fluid does not need to spend energy to do work to expand itself during the heating process. For the ensuing equations to be strictly correct, c_f at constant pressure should be used, but it makes little practical difference if the wrong one is used. When the fluid moves in a porous medium with Darcy flux $\boldsymbol{q}$, it gives rise to an advective (or convective) heat flux:

$$\boldsymbol{f}_a = \rho_f c_f \boldsymbol{q} (T - T_o) \tag{12.5}$$

The composite parameter $\rho_f c_f$ is called the (volumetric) thermal capacity. The $\rho_f c_f$ of water at room temperature is about 4.2×10^6 J m^{-3} K^{-1}, larger than that of oceanic crustal rocks by about 25%. Therefore water can transport significant amounts of heat.

Heat is also transferred by conduction, due purely to a temperature gradient. The conductive heat flux is:

$$\boldsymbol{f}_c = -\lambda \cdot \nabla T \tag{12.6}$$

where λ is the thermal conductivity of the solid–fluid mixture, generally a tensor but in most practical cases assumed to be isotropic. Thus, λ can be estimated from the solid matrix conductivity λ_m, fluid conductivity λ_f, and porosity n. One widely used empirical formula is the geometric mean:

$$\lambda = \lambda_m^{(1-n)} \lambda_f^n \tag{12.7}$$

There are many other formulae (Beck, 1988; Zimmerman, 1989) that may have better physical basis but have lengthier mathematical expressions. The differences between different formulae are generally within experimental uncertainties. Most dry rocks have conductivity values of 2–4 W m^{-1} K^{-1}, but the value for water is only about 0.6 W m^{-1} K^{-1}. Therefore water-saturated high-porosity rocks or sediments may have rather low conductivities. The minus sign in (12.6) indicates that heat is always conducted from hotter to colder regions. Equations (12.2) and (12.6) describe diffusion processes at very different scales. In Darcy flow, diffusion is the volumetric average effect of viscous flow along tortuous pore spaces, but in heat conduction, kinetic energy is transferred between molecules.

The total heat flux including both $\boldsymbol{f}_c$ and $\boldsymbol{f}_a$ is:

$$\boldsymbol{f} = \boldsymbol{f}_c + \boldsymbol{f}_a = -\lambda \cdot \nabla T + \rho_f c_f \boldsymbol{q} (T - T_o) \tag{12.8}$$

When we determine seafloor heat flux by measuring the vertical temperature gradient $\Delta T / \Delta z$, we only determine $\boldsymbol{f}_c$. Fluid flow may distort the gradient in different ways and cause heat flux measurements to be highly variable. In young oceanic crust, the overall effect of water circulation through the seafloor is to make the temperature, and thus the average gradient, at shallow depths lower than what would be for a purely conductive thermal regime (Chapter 10). The missing part of f, i.e. f_a, is not detected by gradient measurements.

Conservation of energy leads to:

$$\nabla \cdot \lambda \cdot \nabla T - \rho_f c_f \boldsymbol{q} \cdot \nabla T + H = \rho c \frac{\partial T}{\partial t} \tag{12.9}$$

where t is time, and:

$$\rho c = n \rho_f c_f + (1 - n) \rho_m c_m \tag{12.10}$$

is the thermal capacity of the solid–fluid mixture, with ρ_m and c_m being the density and specific heat of the solid matrix. H is a heat source/sink term, which may include heat due to the decay of unstable isotopes, chemical reaction, viscous energy dissipation, and rapid pressurization of the fluid by faulting or other non-linear processes. The importance of the advection term (second term) relative to the conduction term (first term) in (12.9) depends not only on the magnitude of the Darcy flux but also on its direction. Where $\boldsymbol{q}$ and ∇T are perpendicular to each other, advective heat transfer has no local thermal effect. In (12.6) through (12.9), we assume that the fluid and solid phases are in local thermal equilibrium. This assumption is valid for a steady-state process (i.e. $\partial T / \partial t = 0$) but not always valid for a time-dependent process. The thermal conductivity generally depends on temperature (Clauser and Huenges, 1995), but such dependence is often ignored in hydrogeological problems, mainly because its effects are overshadowed by the spatial variations of and/or uncertainties in the rock permeability.

The thermal diffusivity of the solid–fluid mixture is defined as:

$$\kappa = \frac{\lambda}{\rho c} \tag{12.11}$$

The thermal conductivity controls the conductive heat flux, but the thermal diffusivity controls how fast a temperature perturbation propagates by heat conduction. At any given λ, a larger ρc leads to slower diffusion because more time is needed to change the temperature of the material along the path of the temperature perturbation. An appreciation of the time and distance scales of one-dimensional diffusion can be had by considering a sudden temperature change ΔT_o at the surface of a uniform half-space at $t = 0$. The resultant temperature change at distance L from the boundary is:

$$\Delta T(t, L) = \Delta T_o \operatorname{erfc}\left(\frac{L}{2\sqrt{\kappa t}}\right) \tag{12.12}$$

At time

$$\tau(L) = \frac{L^2}{\kappa} \tag{12.13}$$

ΔT reaches 48% of ΔT_o. $\tau(L)$ is called the thermal time constant for distance L. Note that the one-dimensional time constant is not applicable to general three-dimensional diffusion. As an example of three-dimensional diffusion, we consider a ΔT_o that occurs within a spherical region of radius a. Outside the source region at radial distance L from the surface of the sphere, ΔT at time $\tau(L)$ is that of (12.12) multiplied by a factor of $a/(L + a)$, (Carslaw and Jaeger, 1959; p. 247).

Chemical diffusion is governed by a law similar to (12.6), with concentration in the place of temperature. If chemical diffusion occurs only in the fluid phase, solid–fluid equilibrium is not a relevant issue. The chemical diffusivity is usually smaller than the thermal diffusivity by a factor of 10^3, and therefore chemical advection becomes predominant much more easily (Chapter 11). However, velocity-dependent hydrodynamic dispersion due to tortuous fluid paths, an effect usually unimportant in the heat equation (12.9), acts to the same effect as diffusion (Bear, 1972). Compared with heat transport, in which conduction in the solid phase tends to smooth the temperature field, solute transport is much more sensitive to the heterogeneity of the flow system.

12.2.2 Fluid flow with prescribed velocity

We first consider the thermal effects of fluid flow with a given velocity, ignoring buoyancy and other driving mechanisms for the flow system. Very often, we can estimate flow velocity using proxies such as temperatures (Chapters 7, 8, and 11), and we then infer the differential pressure required to drive this flow. Four examples are given in this section. The first two are steady-state analytical solutions for very simple geometries. For more complex geometries, some simplified analytical solutions of (12.9) are available (e.g. Domenico and Palciauskas, 1973), but numerical solutions are generally required. The last two examples are solutions obtained using the finite element method.

Uniform vertical flow in a horizontal formation

If the basement is overpressured or underpressured relative to the hydrostatic state, fluid flow through the overlying sediment section is nearly vertical. The thermal effect of the one-dimensional system can be described by the following solution. Assume the upper ($z = 0$) and lower ($z = L > 0$) boundaries of the formation are kept at temperatures T_o and $T_L > T_o$, respectively, and the vertical fluid flow has a Darcy flux q (Fig. 12.1a). The normalized temperature T' as a function of $z' = z/L$ is then (Bredehoeft and Papadopulos, 1965):

$$T'(z') = \frac{T(z') - T_o}{T_L - T_o} = \frac{\exp(Pe\, z') - 1}{\exp(Pe) - 1} \tag{12.14}$$

where Pe is the Peclet number of the system, defined as:

$$Pe = \frac{\rho_f c_f q L}{\lambda} = \frac{\rho_f c_f q (T_L - T_o)}{\lambda (T_L - T_o)/L} = \frac{f_a(L)}{f_c'} \tag{12.15}$$

where $f_a(L)$ is the advective heat flux at $z = L$, and f_c' is the conductive heat flux of the system in the absence of fluid flow. Pe is the ratio of advective to conductive heat transfer. When $|Pe| \gg 1$, advection dominates; and when $|Pe| \ll 1$, conduction dominates. T' is

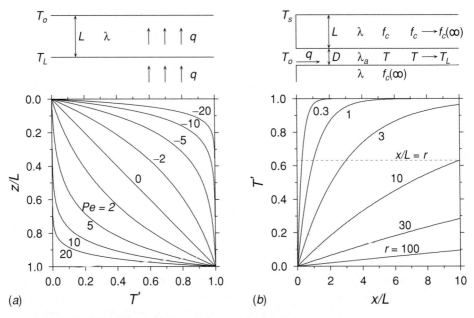

Fig. 12.1 (*a*) Thermal effects of vertical Darcy flux q in a horizontal layered formation. z is depth below the upper surface. T' is defined in (12.14), and Pe in (12.15). (*b*) Thermal effects of net Darcy flow q in a thin, well-mixed (i.e. zero vertical thermal gradient) horizontal aquifer. x is distance from the point where the aquifer temperature is T_o. T' is defined in (12.17), and r in (12.18).

shown in Fig. 12.1*a* for a number of Pe values. Chapter 11 provides examples of applying this solution to infer fluid flux from temperature measurements.

If $f_c(0)$ and $f_c(L)$ are the conductive heat fluxes at $z = 0$ and L, respectively, it can be easily verified that $f_c(z')/f_c(0) = e^{Pez'}$ and that

$$\frac{f_c(z') - f_c(L)}{f_c(0) - f_c(L)} = 1 - T'(z') \tag{12.16}$$

The total heat flux through the system is $f = f_c(0) = f_c(L) + f_a(L)$. Because we have used T_o as the reference temperature for advective heat flux, $f_c(0)$ equals the total heat flux. Upward flow ($Pe < 0$) increases the total energy flux by bringing in warm fluid across the lower boundary; the fluid loses heat conductively on its way up, increasing f_c and diminishing f_a.

Flow in a thin horizontal aquifer

Under certain conditions, fast lateral fluid flow may occur in the permeable igneous crust covered by low-permeability sediments. The thermal effect of such flow may be described by the following solution. We consider a thin horizontal aquifer of thickness D at depth L (Fig. 12.1*b*). The temperature at this depth in the absence of fluid flow is T_L. A constant flux q enters from one end of the aquifer ($x = 0$) at an initial temperature of T_o. The aquifer is assumed to be well mixed, i.e. to have a constant temperature across its thickness, and is also assumed to be sufficiently thin such that the effect of mixing in the horizontal direction, when compared to horizontal heat conduction and advection, can be neglected. It is also possible to use a larger thermal conductivity for the aquifer to approximate the enhanced horizontal as well as vertical heat transfer by mixing (see discussions of Nusselt number, p. 388). The normalized temperature T' of the aquifer as a function of $x' = x/L$ is then (Langseth and Herman, 1981):

$$T'(x') = \frac{T(x') - T_o}{T_L - T_o} = 1 - \exp\left(-\frac{x'}{r}\right) \qquad (12.17)$$

in which:

$$\frac{1}{r} = \frac{\rho_f c_f q L}{2\lambda_a}\left[\sqrt{1 + \frac{4\lambda_a \lambda}{(\rho_f c_f q)^2 DL}} - 1\right] \qquad (12.18)$$

where λ_a and λ are the conductivities of the aquifer and surrounding impermeable rock formations, respectively. Parameter r is a dimensionless length scale of the temperature disturbance caused by the flow. At $x' = r$, the disturbance has decreased to e^{-1} (Fig. 12.1*b*). If the flow is fast such that the second term under the square-root is small, e.g. $q > 10^{-8}\,\mathrm{m\,s^{-1}}$ for an DL of 1,000 m^2, r becomes a Peclet number $Pe = \rho_f c_f q D/\lambda$ for this system. A time-dependent solution for the same system assuming large q was given by Ziagos and Blackwell (1981). At steady state, T' also represents the change of surface heat flux $f_c = \lambda(T - T_s)/L$ with distance:

$$T'(x') = \frac{f_c(x') - f_c(0)}{f_c(\infty) - f_c(0)} \qquad (12.19)$$

where $(T_s < T_L)$ is the temperature of the upper surface, and $f_c(0)$ and $f_c(\infty)$ are heat flux values at $x' = 0$ and $x' \gg r$, respectively. Defining heat flux from $T - T_s$ neglects the effect of horizontal heat conduction, but the approximation is fairly accurate if $r \gg 1$. An application of (12.19) assuming $T_s = T_o$ is discussed in Chapter 11.

Constant flow in a vertical conduit

Figure 12.2 shows steady-state temperature fields associated with upward Darcy flux q in a 100-m wide vertical conduit (shaded zone) obtained using a finite element model. The upper and lower boundaries of the model are kept at constant temperatures 0 and 280 °C, respectively. The model emulates fluid "leakage" through a confining sediment layer on top of a permeable basement that is kept isothermal by vigorous hydrothermal circulation (see Chapter 8). The two-dimensional (infinite in the strike dimension) conduit represents a linear discharge zone, and the cylindrical (axisymmetric) conduit represents a pipe-like focused discharge area. The left-side boundary is the plane or line of symmetry of the system, and the right-side boundary (too far to be included in the display box of Fig. 12.2) is assigned zero

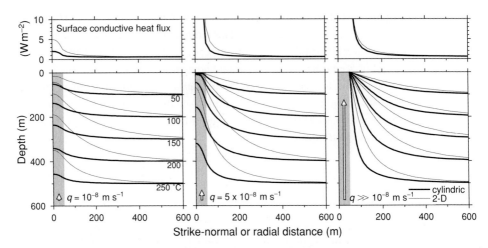

Fig. 12.2 Temperature field in the presence of vertical Darcy flux q along a permeable conduit of 100 m width (for two-dimensional geometry) or diameter (for axisymmetric cylindrical geometry). The symmetrical left half of the model domain is not shown. See text for parameters and boundary conditions.

horizontal heat flux. A thermal conductivity of $1.2 \text{ W m}^{-1} \text{ K}^{-1}$ is used, appropriate for high-porosity deep-sea sediments. With the same flow rate, a cylindrical conduit allows much less advective heat than a two-dimensional conduit and therefore causes much smaller temperature perturbations. The large difference between the $q = 10^{-8} \text{ m s}^{-1}$ and $q = 5 \times 10^{-8}$ m s^{-1} cases illustrates the sensitivity of the near-field thermal regime to small differences in the flow rate. The solution for the $q \gg 10^{-8} \text{ m s}^{-1}$ case was obtained by assuming an isothermal conduit with the same temperature as the lower boundary. It shows the maximum perturbation that the flow can cause to the thermal field. Five conduit widths (or diameters) away from the conduit boundary, the thermal perturbation is negligibly small, especially for the cylindrical case. In natural systems, the conduit width may change with depth and produce different thermal anomalies at the surface. For example, a broader observed thermal anomaly would suggest a conduit that widens with depth.

Transient flow along a dipping fault

Transient flow along dipping faults has been proposed to explain the displacement of the bottom simulating reflector (BSR), which represents the stability field of methane hydrates, in the Cascadia accretionary prism off Oregon (Davis *et al.*, 1995). Faults may act as high-permeability channels connecting regions of different non-hydrostatic pressures. The flow may be turned on and off by permeability changes due to fault motion (such as earthquake rupture) or mineral precipitation, and by pressure changes in the fault zone. Figure 12.3 shows the temperature field around a dipping fault (thin dashed line) along which fluid starts to flow updip at a constant rate at time zero. The actual calculation domain is much larger than the display box. The model has a uniform basal heat flux of 70 mW m^{-2}, and a 0 °C

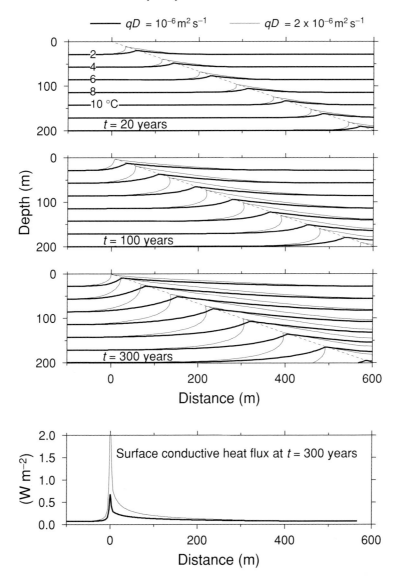

Fig. 12.3 Temperature field in the presence of updip flow q along a thin fault zone of width D. t is time after the flow is started. See text for parameters and boundary conditions.

upper surface. Two vertical boundaries, far from the region of interest, are assigned a zero horizontal heat flux. The thermal conductivity is uniformly 1 W m^{-1} K^{-1}, representative of high-porosity sediments near the toe of an accretionary prism. For a thin flow channel, the thermal effect depends only on the product of the Darcy flux and channel thickness, qD, which is the volume of flow per unit time through a unit along-strike length of the fault. A comparison between the $qD = 10^{-6}$ m^2 s^{-1} and the $qD = 2 \times 10^{-6}$ m^2 s^{-1} results shows the sensitivity of the thermal field to the fault flow. The upper wall of the fault is much more strongly affected than the lower wall.

12.2.3 Buoyancy-driven convection

Equations and fluid properties

Conservation of fluid mass for saturated single-phase flow in a non-deforming matrix leads to:

$$\nabla \cdot (\rho_f \boldsymbol{q}) + \frac{\partial (n\rho_f)}{\partial t} = 0 \tag{12.20}$$

This, combined with Darcy's Law (12.1), and (12.9) are the coupled equations governing thermal-buoyancy driven flow in a porous medium. It is often assumed that the density variation is important only to the buoyancy term $\rho_f \boldsymbol{g}$ (the Boussinesq approximation) and that the density of the solid phase does not change with pressure and temperature. With these assumptions, (12.20) becomes:

$$\nabla \cdot \boldsymbol{q} = 0 \tag{12.21}$$

The Boussinesq approximation degrades the accuracy of the mathematical formulation, but the damage is probably marginal for many systems considering the other assumptions underlying the application of Darcy's Law to geological formations.

In addition to the Boussinesq approximation, parameters such as the thermal expansivity α and viscosity μ are often assumed to be constants. A constant α means that fluid density is a linear function of temperature. With numerical methods, it is not difficult to model these parameters more accurately. Figure 12.4 shows the thermodynamic properties of pure water as presented in Keenan *et al.* (1978). Fluid properties are complex near the critical point ($T = 374$ °C and $P = 22$ MPa for pure water; and $T \approx 405$ °C and $P \approx 30$ MPa for seawater), but in most applications of off-ridge-axis seafloor hydrogeology, T is much less than 300 °C. Near-critical temperatures need to be considered only for such flows as associated with magma intrusion (e.g. at ridge axes) or fast fault slip (friction). At sub-critical conditions the properties of seawater are very similar to those of the pure water (Bischoff and Rosenbauer, 1985). Because of salinity, seawater is denser than pure water by less than 3%, but the density change as a function of temperature or pressure is similar to that of pure water.

Below the critical point and at boiling pressure/temperature conditions (e.g. the points of discontinuity of the 20 MPa curves in Fig. 12.4), liquid and vapor coexist. At super-critical conditions, there is no distinction between fluid and vapor for pure water, but a two-phase boundary still exists for seawater. As seawater is heated to approach this super-critical two-phase boundary, a small amount of dense brine instead of vapor is formed (e.g. Bischoff and Rosenbauer, 1984). Sub-critical two-phase flow is important in terrestrial hydrothermal systems (Ingebritsen and Hayba, 1994), and super-critical two-phase flow may be important

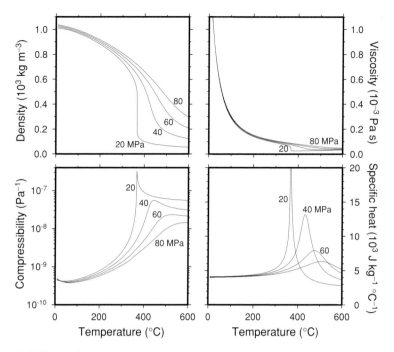

Fig. 12.4 Thermodynamic properties of pure water. Data are from Keenan *et al.* (1978).

in ridge-axis hydrothermal processes (e.g. Lowell and Germanovich, 1997). Two-phase flow is less a concern for off-ridge-axis sub-seafloor systems.

Cellular convection in a flat formation

For a homogeneous, isotropic flat-lying porous layer of thickness L with its impervious top and bottom boundaries maintained at temperatures T_o and $T_o + \Delta T$, respectively, free convection takes place with a cell width L (i.e. wavelength $2L$) if the Rayleigh number

$$Ra = \frac{kg\alpha L \Delta T}{\kappa \mu} \tag{12.22}$$

reaches a critical value Ra_c. With constant fluid properties, $Ra_c = 4\pi^2 \approx 40$, but with pressure- and temperature-dependent fluid properties, Ra_c can be higher or lower than that defined by average properties (Straus and Schubert, 1977). The Rayleigh number is a measure of thermal buoyancy versus viscous resistance. Thermal buoyancy drives the fluid upward and downward, and it has to overcome viscous resistance to form convection cells before the thermal anomaly causing it is dissipated by conduction. If the convection wavelength is very small, interaction between the closely spaced downward and upward flow paths adds viscous resistance. If the wavelength is very large, there will be much resistance to horizontal flow. The cell width L is the best compromise that requires the lowest Ra_c. Convection takes place at this wavelength before other wavelengths can materialize. If heat is supplied conductively to the base of the permeable layer through thermally resistive material (e.g. low-permeability rock), the cell width will be larger than L, because the thermal structure of the convection will extend into the low-permeability region. If the top boundary is pervious and kept at a constant pressure, the ventilated convection will have a Ra_c of about 27 and cell width of $1.35L$ (Nield, 1968).

Cellular convection models relevant to ridge and ridge flank hydrothermal circulation studies are of the above two types: close-top and open-top systems. The close-top models, with an impermeable or much less permeable layer on top, simulate the condition of a sediment-buried basement (Elder, 1967; Kvernvold, 1979; de la Torre Juarez and Busse, 1995; Davis *et al.*, 1996). The open-top models simulate the condition of ventilated convection such as usually seen at the ridge crest (Wilcock, 1998; Cherkaoui and Wilcock, 1999, 2001). Numerical models do not have the luxury of an infinite horizontal dimension. Lateral boundaries in these models affect the aspect ratio of convection cells because an integer number of cells must fit in the model box. If the boundary-enforced cell width is not optimal, the onset of convection should occur at a $Ra > Ra_c$.

Much effort has been made in detecting and modeling off-axis cellular convection in the permeable igneous oceanic crust, with the hope that the cell width constrained by heat flow measurements could tell us something about the circulation depth. As reviewed in Chapter 11, the real oceanic crust is much too heterogeneous to be described in any detail by the above porous-layer models. Analog and numerical models of cellular convection have met little success when it comes to fitting observations point by point, and as of today we still know very little about the circulation depth. However, these models are important in helping us understand the fundamental principles of hydrothermal circulation. Among the important lessons we have learned from simple cellular convection models, the following two are of particular interest.

1. Convection patterns at high Ra. Analog and numerical models of cellular convection have revealed that beyond Ra_c, there are other types of instabilities that transform convection into various time-dependent states. For Ra between Ra_c and $\sim 10Ra_c$, convection is steady, with the Darcy flux increasing roughly linearly with Ra. For Ra above $\sim 10Ra_c$, convection pattern fluctuates periodically, and at yet higher Ra the pattern becomes chaotic (Kimura *et al.*, 1986; Graham and Steen; 1994; Cherkaoui and Wilcock, 1999). In both the periodic and chaotic states, the time scale of the convection pattern variability is a tiny fraction of the thermal time constant $\tau(L)$ see (12.13), and the heat transfer characteristics vary by only a small amount about the average state.

2. Efficiency of heat transfer at high Ra. The efficiency of convective heat transfer for a saturated porous layer heated from below is represented by the Nusselt number, defined as the total heat transferred through the system to the heat that would be transferred by conduction alone in the absence of convection:

$$Nu = \frac{f}{\lambda \Delta T / L} \qquad (12.23)$$

The relation of Nu with Ra at high Rayleigh numbers is of fundamental importance. If Nu can be estimated, one can infer Ra and thus permeability from the Nu–Ra relationship

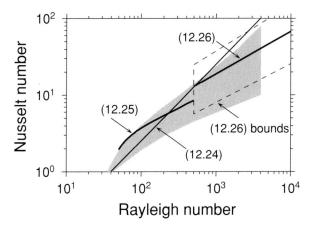

Fig. 12.5 Nusselt number–Rayleigh number relationship for convection in a porous layer heated from below. The gray zone is the range of experimental and model results as summarized in Chen (1979) and Davis *et al.* (1997).

(Davis *et al.*, 1997; Chapter 7). For Ra values moderately above Ra_c in close-top systems, theoretical analyses and laboratory experiments show (e.g. Bejan, 1984):

$$Nu = \frac{Ra}{Ra_c} \qquad (12.24)$$

but deviation from this relation toward lower Nu values is observed in nearly all experiments and numerical models for high Ra values. The flattening of the Nu–Ra curve is thought to indicate that at very high Rayleigh numbers porous-medium convection behaves more like free convection of a viscous fluid (Elder, 1967). Many Nu–Ra scaling relations for isotropic porous media have been proposed, based on various theoretical and empirical arguments (see review by Cherkaoui and Wilcock, 1999). In addition, sudden changes in the Nu–Ra slope occur when convection changes between different time-dependent states. Summaries of experimental results can be found in Chen (1979) and Lister (1990), and Fig. 12.5 shows the range of the Nu–Ra relationships reported by many authors. Lister (1990), on the basis of experiments using porous media composed of beads and curled fiber and through boundary layer analysis, proposed the following relation for $50 < Ra < 500$:

$$Nu = 0.159(Ra + 708)^{0.25}(Ra - 4\pi^2)^{0.36} \qquad (12.25)$$

The behavior at higher Ra values depended on the type of matrix he used. Viewing cracked rock to be something between matrices formed by beads (flow in a porous medium) and curled fiber (free convection in the presence of obstacles), Lister (1990) further suggested the following relationship for high Rayleigh numbers:

$$Nu = 0.425(\pm 40\%)Ra^{0.55 \pm 0.05} \qquad (12.26)$$

Equations (12.25) and (12.26) are listed here as examples out of many candidates. They are not meant to be the only, or even an accurate, description of the Nu–Ra relationship. When these formulae are used, potentially large uncertainties must be recognized.

Convection in the presence of seafloor or basement topography

If the vertical thermal gradient is greater than adiabatic, fluid has a tendency to convect. In a homogeneous flat layer, whether convection actually takes place depends on Ra, the quantification of the ability of thermal buoyancy to overcome viscous resistance, as discussed above. In the presence of lateral thermal heterogeneity such as due to seafloor or basement topography, the fluid will convect even if Ra is less than Ra_c. In reality, there is little reason for fluid not to convect beneath the seafloor, because there is always some topographic relief. Lister (1972) and Lowell (1980) discussed such topographically controlled sub-critical convection. If the crust is covered by a thin veneer of sediment (just enough to prevent ventilation; Chapter 6), sub-seafloor temperature at the same depth below sealevel should be higher under topographic peaks than under topographic troughs, and buoyancy should drive water to ascend under the peaks. If the crust is covered by thick sediment such that basement troughs are warmer than peaks, water should ascend under the troughs. The pattern of this sub-critical convection is regulated entirely by topography. The rate of flow is thermally insignificant but may be geochemically significant (see examples in Chapter 8). The flow may occur to large depths where the diminishing permeability would inhibit convection without topography, and it may play an important role in hydrating the oceanic plate prior to subduction (Chapter 21).

Super-critical convection at Ra up to about $5Ra_c$ tends to have wavelengths associated with the Rayleigh instability, i.e. weakly influenced by topography (Hartline and Lister, 1981). For even higher Rayleigh numbers up to about $10Ra_c$, in two-dimensional steady-state convection models, the pattern is controlled mostly by topography, except for small topographic wavelengths and extremely low topographic relief (Wang *et al.*, 1997). Beyond $10Ra_c$, the convection pattern changes with time but is expected to be strongly affected by topography. As expected, larger scale topographic relief has a stronger influence on the flow pattern.

Figure 12.6 shows a model of low-Ra steady state hydrothermal convection originally reported in Wang *et al.* (1997). The model has a permeable basement of 600 m thickness with a sinusoidal topography and a sediment cover of 200–400 m. The sediment and the deeper, less permeable basement are assumed to have permeabilities of 10^{-16} and 10^{-17} m^2, respectively. The upper boundary (flat seafloor beneath 2 km of water) is assigned a 0 °C temperature, and the lower boundary, 3.3 km below the seafloor (setting it deeper makes no difference), is assigned a uniform heat flux of 250 mW m^{-2}, appropriate for very young ridge flank crust. The two vertical boundaries are lines of symmetry. Thermal conductivities (λ) for the sediment and basement are 1 and 2 W m^{-1} K^{-1}, respectively. All water properties are as illustrated in Fig. 12.4. For the permeable middle layer to reach Rayleigh instability ($Ra = Ra_c$), a permeability of about $k = 5 \times 10^{-15}$ m^2 is required. With a constant heat flux instead of constant temperature basal boundary condition, Ra increases with k not in a simple linear fashion, because the average ΔT across the layer thickness decreases with increasing heat transfer efficiency. The results for $k = 10^{-15}$ m^2 shows the topographically controlled pattern and the insignificant thermal effect of sub-critical convection. The

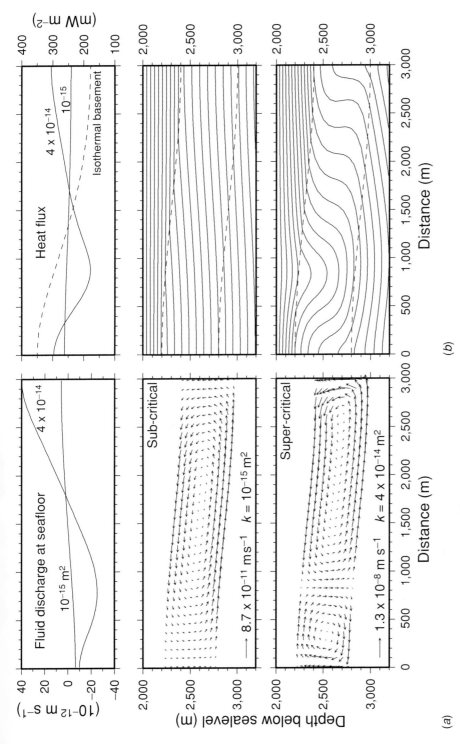

Fig. 12.6 Hydrothermal convection in permeable basement (dashed lines in (b)) with topography. (a) Flow regime. Arrows are Darcy fluxes. Flow in the sediment and less permeable deeper basement is not shown. (b) Thermal regime. The flat seafloor, 2,000 m below sealevel, is kept at 0 °C. Temperature contour interval is 10 °C. Seafloor conductive heat flux that would be caused by an isothermal sediment–basement interface is shown as a dashed line in the top panel. See text for parameters and boundary conditions.

formation temperature and the seafloor heat flux are nearly identical to those of the purely conductive regime (results for pure conduction are not shown). Super-critical convection at $k = 4 \times 10^{-14}$ m^2 tends to have cell widths that are at least partly controlled by Rayleigh instability. The rate of fluid discharge at the seafloor correlates with flow direction in the basement. Upward flow in the basement creates a higher fluid pressure at the sediment–basement interface that tends to drive the fluid to flow up through the sediment. Downward flow in the basement has the opposite effect. In addition to the pressure variation along the interface, the entire basement may for other reasons have a higher or lower background pressure field that will uniformly increase or decrease the seafloor fluid discharge rate. With increasing basement permeability and convection rate, the pressure and temperature variations along the sediment–basement interface will eventually become very small, and the fluid flow through the sediment section and seafloor reverses direction (Chapter 8). In the limiting case of extremely vigorous convection, the interface will become isothermal (or more strictly, adiabatic). The seafloor heat flux that would be produced by an isothermal interface is shown in Fig. 12.6b for comparison purpose.

In the sediment-buried young oceanic crust of the eastern flank of the Juan de Fuca Ridge, the sediment–basement interface indeed appears to be isothermal even in the presence of a basement topographic relief of over 350 m (Davis *et al.*, 1997; Wang and Davis, 2003; Chapter 12.7). A Nu of at least 25 is inferred to keep a nearly isothermal sediment–basement interface. Ra is inferred from Nu to be in the thousands (Fig. 12.5). Therefore, the formation-scale permeability is inferred to be very high. If Darcy's Law is still applicable, the flow is expected to be three-dimensional, chaotic, and strongly influenced by topography, but no experiment to date can constrain the maximum depth and exact circulation pattern. However, the case for very fast flow is unequivocal. With such a large basement topography, less vigorous convection or the absence of convection would both lead to rather large temperature variations along the sediment–basement interface, as illustrated by the sub-critical case of Fig. 12.6b.

Convection due to other externally imposed thermal heterogeneities

Topography introduces relatively modest thermal heterogeneities to the sub-sea formation. Large thermal anomalies occur for other reasons. Models have been developed for buoyancy-driven flow at and near a mid-ocean ridge axis (e.g. Fehn *et al.*, 1983; Wilcock, 1998), around ridge magma intrusions (e.g. Travis *et al.*, 1991; Cherkaoui *et al.*, 1997), within and around faults (Yang *et al.*, 1996), between high-permeability recharge and discharge zones (e.g. Fisher and Becker, 2000), and near a volcanic edifice (e.g. Harris *et al.*, 2000). In some cases, such as with magma intrusion, thermal heterogeneities arise because a heat source is introduced. In other cases, such as with faults, water is driven by buoyancy to flow along a high-permeability channel, and the flow in turn creates a thermal anomaly to drive fluid flow in the surrounding rock. In more general situations, both the heat source and high-permeability channels can be introduced by a common process. The patterns of these flows, at least near the thermal anomalies, are strongly controlled by the externally imposed thermal heterogeneities.

12.3 Oceanic crust as a poroelastic medium

12.3.1 A review of the theory

Interpretations of many seafloor hydrogeological observations involve the theory of poroelasticity. Aspects of the theory have often been presented with cumbersome mathematical expressions and potentially confusing definitions and naming of parameters (see discussions by Kümpel, 1991). The following review is designed to present a concise version of the theory. A comprehensive treatment is given by Wang (2000). The sediment or, at the formation scale, the fractured igneous crust can be considered to consist of an elastic matrix frame with compressibility β_m hosting a fluid with compressibility β_f in its pore space. Here the effect of temperature is ignored. In what follows, stresses and pressures should be understood as being incremental to a background field unless otherwise specified. Stress σ_{ij} that deforms the fluid–solid mixture is called the total stress. The total stress satisfies the stress equilibrium equation:

$$\frac{\partial \sigma_{ji}}{\partial x_j} = 0 \qquad (12.27)$$

Following the rock-mechanics convention, compressive stress is defined to be positive, such that total pressure σ_t is the average of the three normal stresses without a minus sign (i.e. $\sigma_t = \sigma_{ii}/3 = (\sigma_{11} + \sigma_{22} + \sigma_{33})/3$). The gravitational force ρg_j has been excluded from (12.27) because we only consider incremental stresses. Also excluded is the inertial force due to acceleration, which we must include when studying seismic wave propagation. The part of the total stress that deforms the solid is called the effective stress and is defined as:

$$\sigma'_{ij} = \sigma_{ij} - \alpha_m P \delta_{ij} \qquad (12.28)$$

where $\alpha_m = 1 - \beta_s/\beta_m$, with β_s being the compressibility of the solid constituents of the matrix frame. Therefore the equilibrium equation for the effective stress is:

$$\frac{\partial \sigma'_{ji}}{\partial x_j} + \alpha_m \frac{\partial P}{\partial x_i} = 0 \qquad (12.29)$$

The second term shows that the fluid pressure gradient acts on the matrix like a body force. If the matrix is isotropic and linearly elastic, it obeys Hooke's Law:

$$\varepsilon_{ij} = \frac{1 + v_m}{E_m} \sigma'_{ij} - \frac{3 v_m}{E_m} \sigma'_t \delta_{ij} \qquad (12.30)$$

where ε_{ij} is strain (tensor) and $\sigma'_t = \sigma'_{ii}/3 = \sigma_t - \alpha_m P$ is the effective pressure. The relation between the matrix frame's Young's modulus E_m, Poisson's ration v_m, and β_m is $E_m = 3(1 - 2v_m)/\beta_m$.

Now let us consider how fluid flows in this system. Porosity n, compressibilities β_f, β_m, and β_s constitute a storage compressibility Kümpel (1991):

$$\zeta = \alpha_m \beta_m + n\alpha_f \beta_f \tag{12.31}$$

where $\alpha_f = 1 - \beta_s/\beta_f$. The storage behavior of a porous medium is the result of these elastic properties. If head, instead of fluid pressure, is used for the analysis as in (12.3) and (12.4), then the storage behavior depends also on the weight of the fluid, and this is why the definition of the (three-dimensional) specific storage in hydrogeology is $S = \rho_f g \zeta$. Note that the most commonly used "specific storage" for aquifer analysis is the one-dimensional version of S; one-dimensional poroelastic parameters will be explained in Section 12.3.3. A change in total pressure σ_t affects both the matrix and pore fluid. The fraction of the pressure change instantaneously (elastically) taken up by the fluid is called the loading efficiency (γ) (or the "Skempton coefficient"). It can be shown that:

$$\gamma = \frac{\alpha_m \beta_m}{\zeta} = \frac{1}{1 + n\alpha_f \beta_f / \alpha_m \beta_m} \tag{12.32}$$

A more compressible matrix leads to a larger loading efficiency because the fluid has to take up a greater share of the incremental load. Conservation of fluid mass leads to (Rice and Cleary, 1976):

$$-\nabla \cdot \boldsymbol{q} = \zeta \frac{\partial}{\partial t}(P - \gamma \sigma_t) \tag{12.33}$$

where $\gamma \sigma_t$ is the instantaneous (i.e. undrained) fluid pressure response due to changes in total stress and can be understood as a transient fluid source term. In deriving (12.33), a non-linear term accounting for fluid compressibility along the flow path (Mase and Smith, 1987) has been ignored. This term is part of the non-Boussinesq term ignored in (12.21), but the Boussinesq approximation and the assumption that $\beta_m = 0$ also lead to $\zeta = 0$ for (12.21). Additional fluid source/sink terms, such as due to mineral dehydration (Bekins et al., 1995) and plastic (as opposed to elastic) volumetric strain (Moore and Vrolijk, 1992) can be added. Thermal expansion due to intense heating events such as from fault friction and magma implacement can also contribute to the fluid source (e.g. Mase and Smith, 1987). However, some of these processes are accompanied by permeability changes and make (12.33) very non-linear. In conjunction with Hooke's Law and Darcy's Law, (12.29) and (12.33) are the coupled governing equations of linear poroelasticity. Parameters involved are listed in Table 12.1. For numerical calculation, it is often convenient to write σ_t in terms of P and volumetric strain θ using the following relation:

$$\theta = \beta_m \sigma'_t = \beta_m(\sigma_t - \alpha_m P) \tag{12.34}$$

This gives

$$-\nabla \cdot \boldsymbol{q} = (\alpha_m \beta_s + n\alpha_f \beta_f)\frac{\partial P}{\partial t} - \alpha_m \frac{\partial \theta}{\partial t} \tag{12.35}$$

Table 12.1 *Definition of poroelastic parameters*

Symbol	Definition[a]	Name	Other names
v_m		Matrix frame Poisson's ratio	Drained Poisson's ratio
β_m		Matrix frame compressibility	Drained compressibility
β_f		Fluid compressibility	
β_s		Solid compressibility	Grain compressibility
α_m	$1 - \beta_s/\beta_m$		Effective-stress coefficient
α_f	$1 - \beta_s/\beta_f$		
ζ	$\alpha_m\beta_m + n\alpha_f\beta_f$	Storage compressibility	
γ	$\alpha_m\beta_m/\zeta$	Loading efficiency	Skempton coefficient
ϕ	$2\alpha_m(1 - 2v_m)/[3(1 - v_m)]$		
β'_m	$\beta_m(1 + v_m)/[3(1 - v_m)]$	one-dimensional frame compressibility	Confined frame compressibility
ζ'	$\zeta - \alpha_m\beta_m\phi$	one-dimensional storage compressibility	
γ'	$\alpha_m\beta'_m/\zeta'$	one-dimensional loading efficiency	Tidal loading efficiency

[a] In these expressions, n is porosity.

This is Eq. (12.2) of Kümpel (1991). The sign of the last term of (12.35) is different from Kümpel's because, in this chapter, an increase in θ represents contraction (i.e. density increase), as is obvious in (12.34). If σ_t is a known function of space and time, (12.33) can be solved without using the stress equation. If σ_t is not known, (12.33) or (12.35) and (12.29) are coupled and have to be solved simultaneously.

From (12.1) and (12.33), a (three-dimensional) hydraulic diffusivity, analogous to the thermal diffusivity defined in (12.11), can be defined:

$$\eta = \frac{k}{\mu\zeta} = \frac{K}{S} \tag{12.36}$$

Hydraulic conductivity K has been discussed following (12.3), and specific storage S has been discussed after (12.31). The ratio k/μ in (12.1) controls the rate of Darcy flow, but diffusivity η characterizes the "speed" and propagation distance of a pressure signal through the porous medium.

12.3.2 *Further discussions of poroelastic parameters*

Four independent elastic moduli are required to describe a saturated isotropic poroelastic medium mathematically, but using the five parameters β_s, β_f, β_m, v_m, and n as above appears to be conceptually simple and to result in the most concise expressions. Parameters such as α_m, α_f, ζ, and γ are derived from these basic parameters but have their own clear physical meaning. Other combinations of parameters can be used to present the same theory. For example, we can define an "undrained" compressibility of the solid–fluid mixture $\beta = \beta_m(1 - \alpha_m\gamma)$, such that under undrained conditions $\theta = \beta\sigma_t$ (compare this with (12.34)). It can be shown that β defined here is identical to that derived by Gassmann (1951). Similarly, one can define other undrained elastic moduli. Undrained parameters like β usually give rise to lengthier expressions for (12.33) or (12.35), but they are needed to describe propagation of seismic waves in porous formations.

Many parameters can be experimentally constrained. Seismic wave speeds constrain undrained elastic moduli. The loading efficiency and storage compressibility can be determined in laboratories on rock samples (with scaling effects in fractured formations kept in mind). The parameters thus determined are often used as basic parameters, with numerous naming conventions. Investigators new to this subject are advised to work with the four elastic moduli used in this chapter plus porosity, permeability, and viscosity, and to try to write other parameters seen in the literature in terms of this group. Kümpel (1991) provides a detailed review and clarification of all the parameters.

In an anisotropic poroelastic material, instantaneous fluid pressure is affected by shear stress, as well as total pressure, but there are laboratory results indicating the presence of a shear-stress influence in some isotropic materials (e.g. Skempton, 1954; Wang, 1997; Lockner and Stanchits, 2002). It is found that the effect of shear stress may result in a smaller response of instantaneous fluid pressure to loading. It may result from a change in pore-space geometry thus porosity in response to shear deformation or from a reversible stress-induced matrix anisotropy due to closure of crack-like pore spaces. If necessary, a shear stress term can be added to (12.33), e.g. Wang (1997).

The fluid compressibility β_f, and consequently ζ and γ, can be greatly affected by the presence of gas. For water containing a fraction n_g of gas, $\beta_f = \beta_{w+g}$ can be shown to be (Wang *et al.*, 1998):

$$\beta_f = (1 - n_g)\beta_w + n_g\beta_g + (1 - n_g)\xi(\beta_g - \beta_\xi - \beta_w) \qquad (12.37)$$

where β_w and β_g are the compressibilities of water and gas, respectively, and ξ is the volumetric solubility of gas in water, i.e. the amount of dissolved gas per volume of water expressed as the equivalent volume of free gas under the in situ pressure and temperature condition. The parameter

$$\beta_\xi = -\frac{1}{\xi}\frac{\partial\xi}{\partial P} \qquad (12.38)$$

represents the solubility change with pressure; it can be imagined to be the compressibility of the "space" in water holding dissolved gas. For an ideal gas, $\beta_\xi = 0$. The third term of (12.37) results from the fact that dissolution of gas in water contributes to the gas volume change. Within this term, β_g controls how the gas is driven into and out of the space ξ in water, β_ξ controls how this space changes with fluid pressure, and β_w reflects the effect of water volume change. The solubility effect is negligible for high-frequency loading such as the passing through of seismic waves but can be important for slower processes such as tidal loading. β_f with and without the solubility effect are called steady-state and instantaneous compressibilities of the gas-bearing fluid, respectively. The steady-state and instantaneous β_f (normalized by β_w) for methane-bearing water are compared in Fig. 12.7. The solubility effect is relatively important for $n_g < 2\%$. Relevant to some ridge-axis hydrothermal vent systems, behavior of vapor-bearing water would be fundamentally similar.

12.3.3 Pressure and flow due to seafloor loading

The strongest signal in seafloor borehole pressure records is often due to tidal loading (Chapter 8). When studying loading by ocean tides, we usually assume a "confined" one-dimensional situation in which there is no horizontal deformation. Because of the Poisson's effect in elastic deformation (stress in one direction affects strain in another direction), definitions of some parameters in the one-dimensional situation are different from the three-dimensional version (Table 12.1). For example, γ becomes the one-dimensional or "tidal" loading efficiency γ' (van der Kamp and Gale, 1983), and ζ becomes the one-dimensional storage compressibility ζ' (Kümpel, 1991). In the one-dimensional version of (12.33), γ and ζ are replaced by γ' and ζ', respectively. To understand the seafloor loading process, it is useful to consider the formation response in the following two theoretical examples. Because we only need to consider the loading-induced incremental pressure, we ignore the background pressure and use Darcy's Law in the form of (12.2), in conjunction with (12.33).

The first example involves a sudden pressure increase (a step change) σ_b at the bounding surface ($z = 0$) of a uniform poroelastic half-space (Fig. 12.8). Thus the fluid pressure at the surface and the total pressure σ_t in the medium both suddenly increase to σ_b. The instantaneous fluid pressure increase $P = \gamma'\sigma_b$ occurs elastically everywhere in the pore space of the medium. The rest of the surface pressure perturbation $(1 - \gamma')\sigma_b$ propagates into the medium by diffusion (Darcy flow). The solution for the one-dimensional version of (12.33) therefore consists of an elastic term and a diffusive term:

$$P(z, t) = \sigma_b \left[\gamma' + \left(1 - \gamma'\right) \operatorname{erfc} \left(\frac{z}{2\sqrt{\eta' t}} \right) \right] \qquad (12.39)$$

where η' is the one-dimensional hydraulic diffusivity defined by (12.36) but using ζ' (Table 12.1). After the initial stepwise loading, the total pressure $\sigma_t = \sigma_b$ no longer changes; the subsequent diffusion process only changes the partitioning of σ_t between the fluid (P) and the solid matrix (σ_t'). As in (12.13), a one-dimensional diffusion time constant $\tau'(L) = L^2/\eta'$ can be defined for length scale L. At $t \ll \tau'$, the diffusive signal does not reach L, and $P = \gamma'\sigma_b$ (Fig. 12.8). At this stage, the $(1 - \gamma')\sigma_b$ part of the total pressure perturbation is entirely taken up by the matrix and its constituent, but with increasing time, this pressure will be gradually transferred to the fluid phase. At $t \gg \tau'$, the diffusive term of

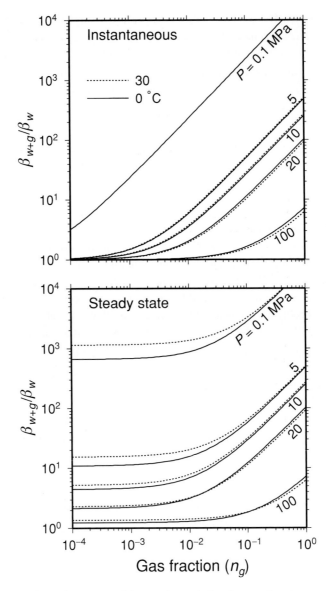

Fig. 12.7 Compressibility of water containing a volumetric fraction n_g of methane gas, normalized by the compressibility of gas–free water. The steady-state compressibility includes the effect of methane solubility in water.

P increases to $(1 - \gamma')\sigma_b$, and hence $P = \sigma_b$ (Fig. 12.8). At this stage, the effective pressure (on the matrix frame) is $\sigma'_t = (1 - \alpha_m)\sigma_b$. The effective pressure is not zero because of the compressibility of the solid constituent of the matrix frame (β_s). A more complex surface loading history can be considered a series of step changes, and combining the above solution for all different steps gives the resultant system response.

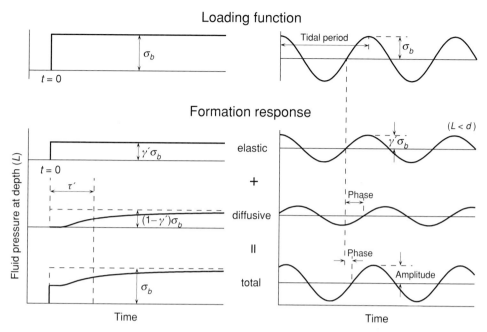

Fig. 12.8 Formation fluid pressure at depth L in a uniform poroelastic half-space in response to stepwise (left) and sinusoidal (right) surface loading. For step loading, $\tau' = L^2/\eta'$ is the diffusion time constant at depth L. For sinusoidal loading, the formation response is shown for a depth less than the penetration length of the diffusion wave d (defined in (12.41)).

The second example involves a continuous sinusoidal pressure variation at the surface with period F, $\sigma_B(t) = \sigma_b \cos(2\pi t/F)$, (Fig. 12.8). Thus the total pressure in the medium is $\sigma_t(t) = \sigma_B(t)$. With this boundary condition, the solution for the one-dimensional version of (12.33) is:

$$P(z, t) = \sigma_b \left[\gamma' \cos\left(\frac{2\pi t}{F}\right) + (1 - \gamma')e^{-\frac{\pi z}{d}} \cos\left(\frac{2\pi t}{F} - \frac{\pi z}{d}\right) \right] \qquad (12.40)$$

where:

$$d = \sqrt{\pi \eta' F} \qquad (12.41)$$

The first term of (12.40) is the elastic component which is always in phase with the loading function (Fig. 12.8). The second term is the diffusive component which propagates into the medium as a diffusion wave with a wavelength $2d$, a phase lag $\pi z/d$, and an amplitude that decreases exponentially with z. The phase lag φ of the total pressure signal is the combined effect of the elastic and diffusive components (Fig. 12.8). The amplitude and phase lag of P as functions of depth are shown in Fig. 12.9 for a few γ' values. The maximum φ value and its depth of occurrence are uniquely determined by γ' (Wang and Davis, 1996) with the relation illustrated by the crossing points of the dashed line and the phase curves in Fig. 12.9.

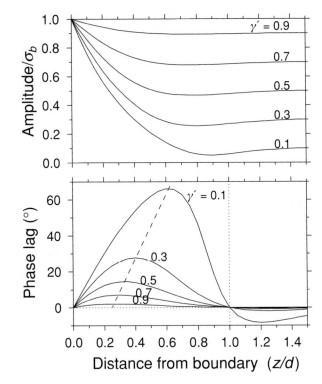

Fig. 12.9 Amplitude and phase lag of fluid pressure in a uniform poroelastic half-space under periodic surface loading for five one-dimensional loading efficiency values. The dashed line is the trajectory of the maximum phase lag. The maximum phase lag and its depth are uniquely determined by the loading efficiency (after Wang and Davis, 1996).

Distance d defined in (12.41) is called the penetration length of the diffusion wave. At this distance the amplitude of the diffusion wave attenuates to $e^{-\pi}(1 - \gamma')\sigma_b$. The penetration length of a diffusion wave is smaller for a shorter wave period. For most seismic wave frequencies, d is practically zero. With a 12-hour tidal period, d is $\sim$1–10 m for typical seafloor sediments having a frame compressibility of 2×10^{-9} Pa^{-1} and a permeability of 10^{-17}–10^{-15} m^2. For young igneous oceanic crust, the formation-scale permeability can be of the order of 10^{-10} m^2, and application of the half-space solution to the Juan de Fuca Ridge flank gives a d value as large as 10 km for the same loading period (Davis *et al.*, 2000; Chapter 7). In the latter case, the diffusion wave propagation is in the horizontal direction. A more complex loading function can be expressed as a number of sinusoidal functions of different frequencies (i.e. discrete Fourier transform), and combining the above solution for all frequencies gives the resultant system response.

Wang and Davis (1996) derived a solution for a plane-layered poroelastic medium subject to periodic surface loading. A loading-efficiency contrast across any layer interface causes an instantaneous differential fluid pressure which generates Darcy flow across the interface

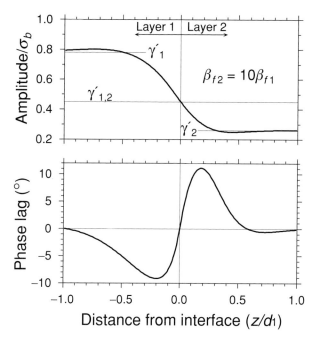

Fig. 12.10 Amplitude and phase lag of fluid pressure around an isolated boundary of loading-efficiency contrast. Parameter values are given in Table 12.2.

and diffusion waves that propagate away from the interface in both directions. The pressure response right at the interface is elastic and is a weighted average of the elastic responses of the two layers. A loading efficiency for the interface between layer j and layer $j + 1$ can be derived (Wang *et al.*, 1998):

$$\gamma'_{j,j+1} = \frac{\gamma'_j + r\gamma'_{j+1}}{1 + r} \tag{12.42}$$

where $r = (k_{j+1}\mu_j d_j)/(k_j\mu_{j+1}d_{j+1})$. At a given loading period, a layer with a thickness greater than $2d$ can be approximately treated as a half-space. Boundaries spaced at least $2d$ apart are called isolated boundaries. An isolated boundary acts like the loading surface of a half-space, with the loading function given by $\sigma_B(t)$ times the boundary loading efficiency of (12.42). This often greatly simplifies the problem. Estimates of tidal energy dissipation by fluid flow in sub-sea formations were made by Wang *et al.* (1999) using the approximation of isolated boundaries. An example of the isolated boundary is given in Fig. 12.10, where the interface $z = 0$ is sufficiently far away from the seafloor and other layer interfaces. The two sediment layers have identical material properties, except that the fluid compressibility of layer 2 is greater than that of layer 1 by a factor of 10, as is the case if the pore water is gas-free in layer 1 but contains a few percent gas in layer 2 (Fig. 12.7). From the parameter values given in Table 12.2 and the definition of γ' in Table 12.1, the one-dimensional loading efficiencies for the two layers are $\gamma'_1 = 0.78$ and $\gamma'_2 = 0.26$. The loading efficiency

Table 12.2 *Parameter values used in the example of Fig. 12.10*

Symbol	Name	Value[a]
ν_m	Matrix frame Poisson's ratio	0.1
β_m	Matrix frame compressibility	2×10^{-9} Pa^{-1}
β_{f1}	Layer 1 fluid compressibility	0.45×10^{-9} Pa^{-1}
β_s	Solid compressibility	0.02×10^{-9} Pa^{-1}
n	Porosity	0.5
μ	Fluid viscosity	1.31×10^{-3} Pa s

[a]The two layers have identical properties except $\beta_{f2} = 10\beta_{f1}$. Permeability need not be specified if distance is normalized as z/d. d is defined in (12.41).

of their interface is $\gamma'_{1,2} = 0.45$, according to (12.42). The diffusion wave propagating from the interface into layer 1 is a negative perturbation to the elastic component, and that is why the total fluid pressure has a phase lead instead of a phase lag relative to the loading function.

Any internal boundaries where there are loading-frequency contrasts or gradients (oriented in any direction, and two-dimensional or three-dimensional) will generate fluid flow and pressure waves in response to loading. Elastic properties, porosity, and thus the loading efficiency vary with lithology and fracture density over a wide range of spatial scales, and thus tidally induced multi-scale fluid flow must be a very common phenomenon in permeable formations. The tidally driven, cyclic motion of the fluid particles is superimposed on the buoyancy driven flows discussed in Section 12.2.

12.3.4 Faulting induced flow

Faulting induces strain in the lithosphere surrounding the rupture area. The volumetric component of this strain field (and possibly shear strain; see Section 12.3.2) contributes to total pressure σ_t and thus fluid pressure P. Pressure change and associated fluid low in response to sub-seafloor faulting events have been detected by borehole monitoring (Davis *et al.*, 2001) and an example is discussed in Chapter 8 (Fig. 8.14). For earthquake faulting or fast aseismic slip, the coevent deformation is usually considered undrained. With given fault geometry and slip vector, we can readily calculate the three-dimensional strain field θ, using either an analytical dislocation solution (e.g. Okada, 1992) or, if structural heterogeneity is to be considered, a numerical solution. From

$$\theta = \beta_m(\sigma_t - \alpha_m P_o) = \beta_m \left(\frac{1}{\gamma} - \alpha_m \right) P_o \qquad (12.43)$$

θ can be converted into the instantaneous fluid pressure response P_o. Note that the second equality is valid only for the undrained situation. With P_o as the initial condition, (12.33) or (12.35) can be solved to determine the resultant Darcy flow field. The fluid pressure will eventually evolve back to its unperturbed state, unless the faulting causes a permanent hydrological change such as opening an escape channel for an initially overpressured formation.

Unlike the one-dimensional situation of the preceding section where the total pressure σ_t is entirely determined by the boundary condition, Darcy flow in three dimensions will modify the total stress σ_{ij} and hence the total pressure. The changing σ_t in (12.33) or θ in (12.35) in turn affects the fluid pressure. Therefore, the flow and stress equations are coupled. As a first-order simplification, it is often assumed that the effect of changing P on σ_t can be ignored, so that if there is no subsequent loading such as continuing slip of the rupture zone or adjacent segments of the fault, σ_t is constant with time (Ge and Stover, 2000; Davis *et al.*, 2001). If $\partial \sigma_t / \partial t = 0$, (12.33) becomes the standard equation used in groundwater hydrology and is a pure diffusion equation analogous to that of heat conduction. However, including and ignoring flow-stress coupling have been shown to lead to qualitatively different predictions of post-seismic ground deformation for a specific fault geometry and set of parameters (Bosl and Nur, 1998). Hence, the importance of the coupling is still an issue worth exploring.

Other post-seismic processes may change the total pressure, too. Viscoelastic stress relaxation in the underlying mantle following a crustal earthquake or strain event has time scales of years to hundreds of years. If its time scale is much longer than the Darcy diffusion time constant defined for the spatial scale of interest, its contribution to $\partial \sigma_t / \partial t$ can be neglected. In studying Darcy flows induced by plate boundary events such as at subduction zones, short-term fault and mantle response may have to be considered. Geodetic observations have recorded continuing crustal deformation of the time scale of months to years after the main shock, and this may be commensurate with the hydrologic diffusion time scale of thick sediment accumulations on ridge flanks or of accretionary and non-accretionary prism sections.

In calculating the spatial distribution of P_o, one should be aware of potential artifacts due to the simplification of fault geometry and motion. To model the fluid response correctly, one must appreciate to what degree the real strain field is represented by the model strain field and be cautious of potential uncertainties. (i) Fault edges are particularly likely to generate artifacts, and some published results suffer from these artifacts. The edges are often modeled as displacement discontinuities, giving rise to huge strains and spurious P_o values. Comparison of model results with data should be made in places sufficiently far away from these edges, because the far-field deformation is not sensitive to details of the fault geometry and motion. (ii) Near the rupture zone, local curvature of the fault surface and heterogeneous slip may cause large strain variations over short distances. A smooth model fault with well-defined edges is not designed to represent these near-field effects. (iii) Small fractures due to mechanical failure near the primary rupture, particularly in materials directly above the trace of a shallowly buried rupture zone, may create or destroy pore space and render relation (12.43) invalid.

12.4 Fluid flow in deforming rocks and sediments

12.4.1 Approximations and assumptions

The elastic deformation discussed in the preceding section is considered infinitesimal. Seafloor sediments may experience severe, permanent deformation, especially in accretionary prisms. Crustal rocks must suffer severe deformation in ridge-axis settings. Faulting is a form of localized large permanent deformation but has been treated above as a boundary condition to generate elastic deformation around the rupture zone. However, if the average dimension of pervasively distributed individual fractures is orders of magnitude smaller than the spatial scale of the process under investigation, the permanent deformation can be considered diffuse. For example, the simple model of a critically tapered Coulomb wedge well describes the large-scale behavior of locally complex processes of mountain building and accretionary prism deformation (Davis *et al.*, 1983).

The Darcy flux q (or the average linear velocity q/n) is relative to the solid matrix. If the matrix itself is moving and/or deforming with a velocity field u_m relative to a fixed coordinate system, the total fluid velocity in this system should be:

$$u_f = u_m + \frac{q}{n} \tag{12.44}$$

For this class of models, the solid constituent of the matrix is often assumed incompressible ($\beta_s = 0$). With this assumption, conservation of both fluid and solid masses leads to an equation similar to (12.35):

$$-\nabla \cdot q = n\beta_f \frac{\partial P}{\partial t} + \nabla \cdot u_m \tag{12.45}$$

This is Eq. (6.2.1.21) of Huyakorn and Pinder (1983). The matrix contraction rate

$$-\nabla \cdot u_m = \frac{\partial \theta}{\partial t} \tag{12.46}$$

as a fluid source term includes both permanent and elastic strain. Again, other fluid source terms can be added as discussed in Section 12.2.1. Note that β_m does not appear explicitly in this formulation and the effect of the solid matrix compressibility is in the $\nabla \cdot u_m$ term. Another often seen formulation has a $\nabla \cdot u_m$ appended directly to (12.33), such that the coefficient for $\partial P/\partial t$ is ζ instead of $n\beta_f$ as in (12.45) or $\alpha_m\beta_s + n\alpha_f\beta_f$ as in (12.35), e.g. Saffer and Bekins (1998). In this latter practice, u_m is understood to represent permanent deformation only, and the effect of β_m is accounted for by ζ. Which formulation to use depends on how the solid velocity u_m is calculated. If u_m includes both permanent and elastic deformation, then (12.45) should be used.

Fluid pressure has a first-order impact on the brittle strength of rocks, and rock failure may significantly modify permeability (e.g. Nathenson, 2000). Therefore, permanent deformation and Darcy flow are non-linearly coupled processes. Fully coupled models are rarely attempted except for engineering applications. The main difficulties are the limited knowledge of in situ rock or sediment rheology in the presence of fluid and the very poor knowledge of the permeability creation and reduction associated with permanent deformation. There is also a non-trivial challenge of getting meaningful numerical results out of a highly non-linear calculation that involves both material non-linearity (plasticity and stress-dependent permeability) and geometrical non-linearity (large deformation). The theory of critical state soil mechanics with finite strain has been used to explain qualitatively characteristic behavior of accretionary prism deformation (Karig, 1990). Numerical modeling of the accretionary prism fluid regime using critical state theory has been attempted by assuming infinitesimal deformation (Stauffer and Bekins, 2001). If the focus is on the Darcy flow and its thermal and geochemical consequences, it is a common practice to assume that (i) certain kinematic aspects of the deformation pattern and thus the strain field are known, (ii) the porosity is a known function of space or effective pressure, and (iii) the permeability is known or is a known function of porosity or effective pressure.

One-dimensional sediment compaction

The simplest example of permanent deformation is vertical compaction of sediment. Some details of sediment properties and factors affecting the compaction process are discussed in Chapter 6. Instead of modeling the mechanical process of porosity reduction with increasing sediment load, we usually use an observationally constrained porosity–depth function $n(z)$, assumed to be constant with time, and obtain a kinematic solution. The model is useful in correcting seafloor heat flow observations for the effect of sedimentation and in estimating the rate of fluid expulsion. One most commonly used $n(z)$ function is:

$$n(z) = n_o \exp\left(-\frac{z}{b}\right) \tag{12.47}$$

For seafloor sediments, surface porosity n_o is around 0.5–0.7, and the length scale b is around 1–2 km. If we set up the depth coordinates such that $z = 0$ is fixed at the sediment surface (Hutchison, 1985), the solid velocity at $z = 0$ is just the sedimentation rate u_o, and the basement moves to greater depths at the sediment accumulation rate. The accumulation rate is less than the sedimentation rate because of compaction. If both the solid and fluid are assumed incompressible, the solid flux $(1 - n(z))u_m(z)$ and fluid flux $n(z)u_f(z)$ relative to the fixed coordinate system must both be constant with depth at any given time, i.e.:

$$[1 - n(z)]u_m(z) = (1 - n_o)u_o \tag{12.48a}$$

$$n(z)u_f(z) = n(z_B)u_f(z_B) = n(z_B)u_m(z_B) \tag{12.48b}$$

where z_B is the current depth of the sediment–basement interface which is determined by solving $dz_B/dt = u_m(z_B)$, (Wang and Davis, 1992). Although constant with depth, both fluxes may change with time because of time-dependent u_o and increasing z_B. Relative to the seafloor, fluid velocity u_f is downward but is slower than that of the deforming matrix. The Darcy flux from (12.48b) and (12.44),

$$q = n(z_B)u_m(z_B) - n(z)u_m(z) \tag{12.49}$$

is of course everywhere upward relative to the deforming matrix. For the porosity function of (12.47), if $n_o = 0.7$, $b = 1500$ m, and $z_B = 500$ m, q at the seafloor is $-0.4u_o$, and the average linear fluid velocity relative to sediment q/n is $-0.57u_o$. The flow rate can be comparable to that of topographically driven sub-critical convection in low-permeability sediment. Except for very high sedimentation rates, the flow is thermally insignificant, but it can be chemically significant.

The above solution says nothing about pore fluid pressure P, although by assuming a permeability structure and using Darcy's Law, a P distribution can be defined by integrating q. In cases of high sedimentation rate and low permeability, higher than hydrostatic pore pressure can develop, but rigorous modeling of the pore pressure development requires a better understanding of the basic physics of the consolidation process.

Two-dimensional accretionary prism deformation

The toe area of an accretionary prism is among the most hydrologically active tectonic settings. While such settings are outside the primary focus of this volume, some discussion of the hydrogeology is justified. Fluid flow influences the thermal and geochemical budgets of the prism, and fluid pressure controls the mechanical behavior of the sediment and fault zones and may affect the occurrence of great earthquakes. In accretionary prism models, fluid flow may be channelized along highly permeable conduits, such as the decollement and other faults, and diffuse flow may take place everywhere in the deforming sediment (Fig. 12.11*a*; Saffer and Bekins, 1999; Henry, 2000). It is likely, at least where young lithosphere is subducted, that the subducting oceanic basement hosts hydrothermal convection as in the ridge flank environment (Chapters 7 and 8) and/or serves as a high-permeability conduit. For the deforming sediment, the fluid source term (dewatering rate) in (12.45) is important.

To calculate dewatering and fluid flow rates in the prism, we need to know $\boldsymbol{u}_m$, the velocity of the deforming solid matrix. For reasons explained in Section 12.4.1, the calculation of $\boldsymbol{u}_m$ in prism fluid flow models has usually been based on kinematic sediment accretion models. Ignoring fluid compressibility, two different two-dimensional kinematic models of sediment deformation have been assumed: (A) sediment flow lines diverge uniformly landward (Bekins and Dreiss, 1992), or (B) a vertical sediment column stays vertical as it travels landward relative to the deformation front and is stretched vertically (pure-shear deformation; Le Pichon *et al.*, 1990). Wang (1994) generalized these two models by allowing curved upper and lower surfaces and showed that the two assumptions led to similar dewatering rates and pore-fluid pressures. Figure 12.11*b* defines the coordinate system and geometrical parameters. In this system, $z = s(x) - y$ is the depth below seafloor. The horizontal component of the solid matrix velocity u_x is determined as follows.

For Model A, u_x is a function of both x and y:

$$\frac{u_x}{u_o} = \frac{D_o[1 - n(z_o)]}{D[1 - n(z)]} \tag{12.50}$$

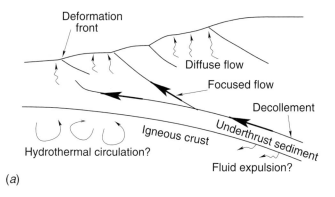

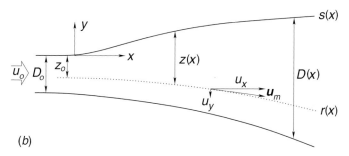

Fig. 12.11 (*a*) Schematic illustrations of the types of fluid flow in an accretionary prism setting. (*b*) Simplified model of the prism defining the coordinate system and geometrical parameters for (12.50) through (12.54).

where porosity n can vary with x as well as z. For Model B, u_x is a function of x only:

$$\frac{u_x}{u_o} = \frac{D_o[1 - N(D_o)]}{D[1 - N(D)]} \tag{12.51}$$

where

$$N(z) = \frac{1}{z}\int_0^z n(z)\,\mathrm{d}z \tag{12.52}$$

is the average porosity from the seafloor to depth z. For both models, the vertical component of the solid velocity u_y is given by:

$$u_y = u_x\frac{\mathrm{d}r}{\mathrm{d}x} \tag{12.53}$$

Expressions for the slope of the sediment flow line $\mathrm{d}r/\mathrm{d}x$ are given by Wang (1994). The dewatering rate $-\nabla \cdot \boldsymbol{u}_m$ determined from u_x and u_y can then be used to determine Darcy flux. For example, if we assume that the horizontal component of Darcy flux is negligible and that the basement is impermeable, the vertical Darcy flux is readily determined by integrating $-\nabla \cdot \boldsymbol{u}_m$ over a vertical column of the prism (Wang, 1994):

$$q_y(x, z) = -\int_z^D \nabla \cdot \boldsymbol{u}_m \mathrm{d}z \tag{12.54}$$

These simple kinematic solutions demonstrate that sediment velocity depends strongly on porosity, and the calculation of fluid flow rate should not be based on a kinematic sediment velocity model that is independent of porosity. Discussions made in the last paragraph of Section 12.4.2 regarding pore-fluid pressures also apply to these two-dimensional cases. In this class of models, the pore-fluid pressure is not used to obtain Darcy flux but is inferred from the kinematically derived Darcy flux by assuming a permeability structure. The limitation of this approach should be recognized when the calculated pore pressures are referred to.

12.4.2 Thermal effects of deformation and deformation-induced flow

To model the thermal effects of sediment deformation and associated fluid flow, we must consider both the matrix velocity and pore-fluid velocity. Assuming local thermal equilibrium between the solid matrix and the pore fluid, the equation of heat transfer is (e.g. Wang *et al.*, 1993):

$$\nabla \cdot \lambda \cdot \nabla T - [n\rho_f c_f \boldsymbol{u}_f + (1 - n)\rho_m c_m \boldsymbol{u}_m] \cdot \nabla T + H = \rho c \frac{\partial T}{\partial t} \tag{12.55}$$

Using (12.44) and (12.10), the expression in the square brackets can also be written as $\rho c \boldsymbol{u}_m + \rho_f c_f \boldsymbol{q}$. This equation differs from (12.9) by including the solid velocity, both directly and through the fluid velocity (12.44), in the advection term.

There are two ways of representing solid deformation. What has been used for (12.55) is the Eulerian representation, i.e. all velocities and fluxes are referenced to a coordinate system fixed in space. In grid-based numerical calculation using the Eulerian representation, the grid does not deform except for special purposes. Each grid point is like a police car parked at a street corner, watching normal traffic ($\boldsymbol{u}_m$) and deranged vehicles ($\boldsymbol{u}_f$). The alternative is the Lagrangian representation, in which the grid deforms and traces sediment flow. Each grid point is then like a police car traveling with the normal traffic ($\boldsymbol{u}_m$), watching deranged vehicles ($\boldsymbol{q}/n$). With the Lagrangian representation, the advection of heat by sediment flow is automatically accounted for by the deforming grid, and therefore (12.55) becomes (12.9), as if there is no solid deformation. For a steady-state problem in which $\boldsymbol{u}_m$ is constant with time, one is forced to use the Eulerian representation, because otherwise the grid deformation would never stop.

12.5 Concluding remarks: objectives of modeling

Mathematical models are used in Earth science to perform two tasks: to provide quantitative tests for conceptual models and to constrain parameter values. Both tasks require fitting model-generated numbers to data, but their objectives are distinctly different. By performing the first task, we admit our ignorance and thus develop mathematical models to evaluate various possible processes that we infer from observations, reasoning, and intuition. For this objective, models are designed to investigate certain aspects of a complicated system. The greatest challenge in developing these models is to decide, in order to isolate the most important process, what can and cannot be neglected. The decision involves thought experiments, back-of-the-envelope calculations, and perhaps some numerical model testing. The second task usually implies that the processes are understood, and all we need is to determine values for the parameters involved. The most efficient way of performing the second task involves the use of inversion methods.

In modeling, a better fit to data is a means not a purpose. The complexity of the Earth system and paucity of observations leave vast room for ambiguities, and therefore every model is only an expression of the scientist's limited understanding of the natural processes. An inversion of data is no exception, because it requires additional constraints to stabilize the inverse solution as well as the assumption that the forward model is correct. It is up to the modeler to decide how much effort should go into scrutinizing the validity of the models in representing natural processes and how much should go into refining various measures of model fit. For the purpose of understanding processes, the acceptance of a model may not be based only on how well it fits data (contrary to the belief of many manuscript reviewers), for numerous models may provide an equally good fit. The acceptance is often based on whether the concept of the model is consistent with other evidence and knowledge that is not part of the modeling equations.

With fast computers and many freely or commercially available computer codes, one may quickly discover the pleasure of being able to develop elaborate models with neither an understanding of the natural processes nor a knowledge of the mathematics. This has fostered a peculiar trend of using mathematical models as black-boxes, with the misconception that adding fine details makes the model more accurate. Models developed this way, with the help of colorful and animated computer graphics, may impress a less-informed readership, but for modeling to become an integral part of research, one must make an effort to understand how the physical problem is translated into a mathematical one.

References

Bear, J. 1972. *Dynamics of Fluids in Porous Media*. New York: American Elsevier.

Beck, A. E. 1988. Methods for determining thermal conductivity and thermal diffusivity. In *Handbook of Terrestrial Heat-Flow Density Determination*, eds. R. Haenel, L. Rybach, and L. Stegena. Dordrecht: Kluwer Academic, pp. 87–124.

Bejan, A. 1984. *Convection Heat Transfer*. New York: Wiley.

Bekins, B. A. and Dreiss, S. J. 1992. A simplified analysis of parameters controlling dewatering in accretionary prisms. *Earth Planet. Sci. Lett.* **109**: 275–287.

Bekins, B. A., McCaffrey, A. M., and Dreiss, S. J. 1995. Episodic and constant flow models for the origin of low-chloride waters in a modern accretionary complex. *Water Resources Res.* **31**: 3,205–3,215.

Bischoff, J. L. and Rosenbauer, R. J. 1984. The critical point and two-phase boundary of seawater, 200–500 °C. *Earth Planet. Sci. Lett.* **68**: 172–180.

   1985. An empirical equation of state for hydrothermal seawater (3.2 percent NaCl). *Am. J. Sci.* **285**: 725–763.

Bosl, W. J. and Nur, A. 1998. Numerical simulation of postseismic deformation due to pore fluid diffusion. In *Poromechanics: A Tribute to Maurice A. Biot*, ed. J.-F. Thimus. Rotterdam: Balkema, pp. 23–28.

Bredehoeft, J. D. and Papadopulos, I. S. 1965. Rates of vertical groundwater movement estimated from the earth's thermal fields. *Water Resources Res.* **1**: 325–328.

Carslaw, H. S. and Jaeger, J. C. 1959. *Conduction of Heat in Solids*, 2nd Edn. New York: Oxford University Press.

Chen, P. 1979. Heat transfer in geothermal systems. *Adv. Heat Transfer* **14**: 1–105.

Cherkaoui, A. S. M. and Wilcock, W. S. D. 1999. Characteristics of high Rayleigh number two-dimensional convection in an open-top porous layer heated from below. *J. Fluid Mech.* **394**: 241–260.

   2001. Laboratory studies of high Rayleigh number circulation in an open-top Hele-Shaw cell: an analog to mid-ocean ridge hydrothermal systems. *J. Geophys. Res.* **106**: 10,983–11,000.

Cherkaoui, A. S. M., Wilcock, W. S. D., and Baker, E. T. 1997. Thermal fluxes associated with the 1993 diking event on the CoAxial segment, Juan de Fuca Ridge: a model for the convective cooling of a dike. *J. Geophys. Res.* **102**: 24,887–24,902.

Clauser, C. and Huenges, E. 1995. Thermal conductivity of rocks and minerals. In *Rock Physics and Phase Relations: A Handbook of Physical Constants*. ed. T. J. Ahrens. Washington, DC: American Geophysical Union.

Davis, D. M., Suppe, J., and Dahlen, F. A. 1983. Mechanics of fold-and-thrust belts and accretionary wedges. *J. Geophys. Res.* **88**: 1,153–1,172.

Davis, E. E., Becker, K., Wang, K., and Carson, B. 1995. Long-term observations of pressure and temperature in hole 892B, Cascadia accretionary prism. In *Proceedings of the Ocean Drilling Program, Scientific Results*, Vol. 146, eds. B. Carson, G. K. Westbrook, R. J. Musgrave, and E. Suess. College Station, TX: Ocean Drilling Program, pp. 299–311.

Davis. E. E., Chapman, D. S., and Forster, C. B. 1996. Observations concerning the vigor of hydrothermal circulation in young oceanic crust. *J. Geophys. Res.* **101**: 2,927–2,942.

Davis, E. E., Wang, K., He, J., Chapman, D. S., Villinger, H., and Rosenberger, A. 1997. An unequivocal case for high Nusselt-number hydrothermal convection in sediment-buried igneous oceanic crust. *Earth Planet. Sci. Lett.* **146**: 137–150.

Davis, E. E., Wang, K., Becker, K., and Thomson, R. E. 2000. Formation-scale hydraulic and mechanical properties of oceanic crust inferred from pore-pressure response to periodic seafloor loading. *J. Geophys. Res.* **105**: 13,423–13,435.

Davis, E. E., Wang, K., Thomson, R. E., Becker, K., and Cassidy, J. F. 2001. An episode of seafloor spreading and associate plate deformation inferred from crustal fluid pressure transients. *J. Geophys. Res.* **106**: 21,953–21,963.

de la Torre Juarez, M. and Busse, F. H. 1995. Stability of two-dimensional convection in a fluid-saturated porous medium. *J. Fluid Mech.* **292**: 305–323.

Domenico, P. A. and Palciauskas, V. V. 1973. Theoretical analysis of forced convective heat transfer in regional ground-water flow. *Geol. Soc. Am. Bull.* **84**: 3,803–3,814.

Elder, J. W. 1967. Steady free convection in a porous medium heated from below. *J. Fluid Mech.* **27**: 29–48.

Fehn, U., Green, K. E., Von Herzen, R. P., and Cathles, L. M. 1983. Numerical models for the hydrothermal field at the Galapagos spreading center. *J. Geophys. Res.* **88**: 1,033–1,048.

Fisher, A. T. and Becker, K. 2000. Channelized fluid flow in oceanic crust reconciles heat flow and permeability data. *Nature* **403**: 71–74.

Gassmann, F. 1951. Über die Elastizität Poröser Medien. *Vierteljahrsschr. Naturforsch. Ges. Zürich* **96**: 1–23.

Ge, S. and Stover, S. C. 2000. Hydrodynamic response to strike- and dip-slip faulting in a half-space. *J. Geophys. Res.* **105**: 25,531–25,524.

Graham, M. D. and Steen, P. H. 1994. Plume formation and resonant bifurcations in porous-media convection. *J. Fluid Mech.* **272**: 67–89.

Harris, R. N., Garven, G., Georgen, J., McNutt, M. K., Christiansen, L., and Von Herzen, R. P. 2000. Submarine hydrogeology of the Hawaiian Archipelagic Apron. Part 2, Numerical simulations of flow. *J. Geophys. Res.* **105**: 21,371–21,385.

Hartline, B. K. and Lister, C. R. B. 1981. Topographic forcing of super-critical convection in a porous medium such as the oceanic crust. *Earth Planet. Sci. Lett.* 55: 75–86.

Henry, P. 2000. Fluid flow at the toe of the Barbados accretionary wedge constrained by thermal, chemical, and hydrogeologic observations and models. *J. Geophys. Res.* **105**: 25,855–25,872.

Hutchison, I. 1985. The effects of sedimentation and compaction on oceanic heat flow. *Geophys. J. Roy. Astron. Soc.* **83**: 439–459.

Huyakorn, P. S. and Pinder, G. F. 1983. *Computational Methods in Sub-surface Flow*. San Diego, CA: Academic Press.

Ingebritsen, S. E. and Hayba, D. O. 1994. Fluid flow and heat transport near the critical points of H_2O. *Geophys. Res. Lett.* **21**: 2,199–2,202.

Karig, D. E. 1990. Experimental and observational constraints on the mechanical behavior in the toes of accretionary prisms. In *Deformation Mechanisms, Rheology and Tectonics*, Geology Society of London Special Publication 54, eds. R. J. Knipe and E. H. Rutter. London: Geology Society, pp. 383–398.

Keenan, J. H., Keyes, F. G., Hill, P. G., and Moore, J. G. 1978. *Steam Tables*. New York: Wiley.

Kimura, S., Schubert, G., and Straus, J. M. 1986. Route to chaos in porous-medium thermal convection. *J. Fluid Mech.* **166**: 305–324.

Kvernvold, O. 1979. On the stability of non-linear convection in a Hele-Shaw cell. *J. Heat Mass Transfer* **22**: 395–400.

Kümpel, H.-J. 1991. Poroelasticity: parameters reviewed. *Geophys. J. Int.* **105**: 783–799.

Langseth, M. G. and Herman, B. M. 1981. Heat transfer in the oceanic crust of the Brasil Basin. *J. Geophys. Res.* **86**: 10,805–10,819.

Le Pichon, X., Henry, P., and Lallemant, S. 1990. Accretion and erosion in subduction zones: the role of fluids. *Ann. Rev. Earth Planet. Sci.* **21**: 307–331.

Lister, C. R. B. 1972. On the thermal balance of a mid-ocean ridge. *J. Geophys. Roy. Astron. Soc.* **26**: 515–535.

  1990. An explanation for the multivalued heat transport found experimentally for convection in a porous medium. *J. Fluid Mech.* **214**: 287–320.

Lockner, D. A. and Stanchits, S. A. 2002. Undrained poroelastic response of sandstones to deviatoric stress change. *J. Geophys. Res.* **107**(B12): 2,353.

Lowell, R. P. 1980. Topographically driven sub-critical hydrothermal convection in the oceanic crust. *Earth Planet. Sci. Lett.* **49**: 21–28.

1991. Modeling continental and submarine hydrothermal systems. *Rev. Geophys.* **29**: 457–476.

Lowell, R. P. and Germanovich, L. N. 1997. Evolution of brine-saturated layer at the base of a ridge-crest hydrothermal sytem. *J. Geophys. Res.* **102**: 10,245–10,255.

Mase, C. W. and Smith, L. 1987. Effects of frictional heating on the thermal, hydrologic, and mechanical response of a fault. *J. Geophys. Res.* **92**: 6,249–6,272.

Moore, J. C. and Vrolijk, P. 1992. Fluids in accretionary prisms. *Rev. Geophys.* **30**: 113–135.

Nathenson, M. 2000. The dependence of permeability on effective stress for an injection test in the Higashi–Hachimantai geothermal field. *Geophys. Res. Lett.* **27**: 589–592.

Nield, D. A. 1968. Onset of thermohaline convection in a porous medium. *Water Resources Res.* **4**: 553–560.

Okada, Y. 1992. Internal deformation due to shear and tensile faults in a half space. *Bull. Seismol. Soc. Am.* **82**: 1,018–1,040.

Rice, J. R. and Cleary, M. P. 1976. Some basic stress diffusion solutions for fluid-saturated elastic porous media with compressible constituents. *Rev. Geophys. Space Phys.* **14**: 227–241.

Saffer, D. M. and Bekins, B. A. 1998. Episodic fluid flow in the Nankai accretionary complex: timescale, geochemistry, flow rate, and fluid budget. *J. Geophys. Res.* **103**: 30,351–30,370.

1999. Fluid budgets at convergence plate margins: implications for the extent and duration of fault-zone dilation. *Geology* **27**: 1,095–1,098.

Skempton, A. W. 1954. The pore pressure coefficients A and B. *Geotechnique* **4**: 143–147.

Stauffer, P. and Bekins, B. A. 2001. Modeling consolidation and dewatering near the toe of the northern Barbados accretionary complex. *J. Geophys. Res.* **106**: 6,369–6,383.

Straus, J. M. and Schubert, G. 1977. Thermal convection of water in a porous medium: effects of temperature- and pressure-dependent thermodynamic and transport properties. *J. Geophys. Res.* **82**: 325–333.

Travis, B. J., Janecky, D. R., and Rosenberg, N. D. 1991. Three-dimensional simulation of hydrothermal circulation at mid-ocean ridges. *Geophys. Res. Lett.* **18**: 1,441–1,444.

Tsang, C.-F. and Neretnieks, I. 1998. Flow channeling in heterogeneous fractured rocks. *Rev. Geophys.* **36**: 275–298.

van der Kamp, G. and Gale, J. E. 1983. Theory of earth tide and barometric effects in porous formations with compressible grains. *Water Resources Res.* **19**: 538–544.

Wang, H. F. 1997. Effects of deviatoric stress on undrained pore pressure response to fault slip. *J. Geophys. Res.* **102**: 17,943–17,950.

2000. *Theory of Linear Poroelasticity with Applications to Geomechanics.* Princeton, NJ: Princeton Press.

Wang, K. 1994. Kinematic models of dewatering accretionary prisms. *J. Geophys. Res.* **99**: 4,429–4,438.

Wang, K. and Davis, E. E. 1992. Thermal effects of marine sedimentation in hydrothermally active areas. *Geophys. J. Int.* **110**: 70–78.

1996. Theory for the propagation of tidally induced pore pressure variations in layered sub-seafloor formations. *J. Geophys. Res.* **101**: 11,483–11,495.

2003. High permeability of young oceanic crust constrained by thermal and pressure observations. In *Land and Marine Hydrogeology*, eds. M. Taniguchi, K. Wang, and T. Gamo. Amsterdam: Elsevier, pp. 165–188.

Wang, K., Hyndman, R. D., and Davis, E. E. 1993. Thermal effects of sediment thickening and fluid expulsion in accretionary prisms: model and parameter analysis. *J. Geophys. Res.* **98**: 9,975–9,984.

Wang, K., He, J., and Davis, E. E. 1997. Influence of basement topography on hydrothermal circulation in sediment buried igneous oceanic crust. *Earth Planet. Sci. Lett.* **146**: 151–164.

Wang, K., Davis, E. E., and van der Kamp, G. 1998. Theory for the effects of free gas in subsea formations on tidal pore pressure variations and seafloor displacements. *J. Geophys. Res.* **103**: 12,339–12,353.

Wang, K., van der Kamp, G., and Davis, E. E. 1999. Limits of tidal energy dissipation by fluid flow in subsea formations. *Geophys. J. Int.* **139**: 763–768.

Wilcock, W. S. D. 1998. Cellular convection models of mid-ocean ridge hydrothermal circulation and the temperatures of black smoker fluids. *J. Geophys. Res.* **103**: 2,585–2,596.

Yang, J., Edwards, R. N., Molson, J. W., and Sudicky, E. A. 1996. Fracture-induced hydrothermal convection in the oceanic crust and the interpretation of heat-flow data. *Geophys. Res. Lett.* **23**: 926–932.

Yang, J., Latychev, K., and Edwards, R. N. 1998. Numerical computation of hydrothermal fluid circulation in fractured Earth structures. *Geophys. J. Int.* **135**: 627–649.

Ziagos, J. P. and Blackwell, D. D. 1981. A model for the effect of horizontal fluid flow in a thin aquifer on temperature–depth profiles. *Geotherm. Resources Co. Trans.* **5**: 221–223.

Zimmerman, R. W. 1989. Thermal conductivity of fluid-saturated rocks. *J. Petrog. Sci. Eng.* **3**: 219–227.

13

Geothermal evidence for continuing hydrothermal circulation in older (>60 M.y.) ocean crust

Richard P. Von Herzen

13.1 Introduction

Hydrothermal circulation is now widely recognized as a nearly ubiquitous process for young (<60 M.y.) seafloor, affecting the chemistry of ocean water and the chemical and mineralogical alteration of a significant fraction of the ocean crust (e.g. Wolery and Sleep, 1976; Chapters 14–21). Early detailed marine geothermal investigations (e.g. Lister, 1972; Williams *et al.*, 1974) showed that hydrothermal circulation in the uppermost ocean crust (i.e. possibly to several kilometers below the seafloor) strongly controls the magnitude and detailed patterns of heat flow through young seafloor. The transfer of heat by pore-water advection was inferred to dominate over conduction in many areas, and the spatial distribution of seafloor heat flux was inferred to indicate the geometry of the sub-seafloor pore-water circulation (i.e. generally high values above upwellings, low values above downwellings).

As the lithosphere created at mid-ocean ridges cools and migrates laterally with seafloor spreading, the heat flux decreases and the driving force for hydrothermal circulation diminishes. At the same time, the relatively impermeable seafloor sediment layer thickens and increasingly covers the topography of the relatively permeable (but decreasing with age) basement rock. Both of these processes promote a gradual change in the mechanism of seafloor heat transfer from primarily advection to conduction with increasing lithospheric age. This evolution is seen in the distribution of heat flow values with seafloor age, which is characterized by mean values that are significantly lower than expected (the heat flow deficit) from lithospheric cooling models for young (<60 Ma) seafloor (Stein and Stein, 1992; Stein *et al.*, 1995; Chapter 10; Fig. 13.1) and by large statistical scatter. The low mean values are believed to be the result of unmeasured heat advected through the seafloor. Lateral scales of hydrothermal circulation patterns are typically a few kilometers to a few tens of kilometers, such that multiple heat flow measurements spaced at distances of 1–2 km or less are required to reveal their lateral geometries (e.g. Williams *et al.*, 1974; Green *et al.*, 1981).

Theory, as well as laboratory and numerical modeling, suggest that the hydrothermal circulation patterns are spatially controlled by topography of the seafloor or of the permeable

Hydrogeology of the Oceanic Lithosphere, eds. E. E. Davis and H. Elderfield. Published by Cambridge University Press.
© Cambridge University Press 2004.

Heat flow discrepancy due to hydrothermal flux

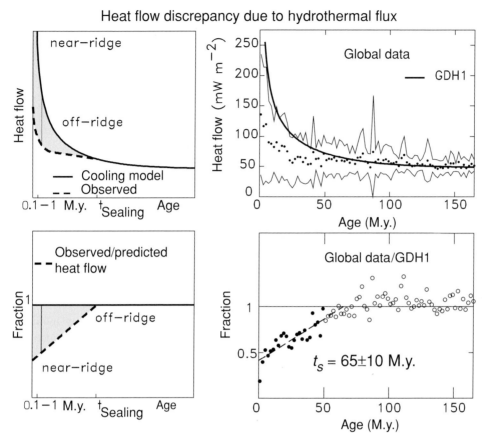

Fig. 13.1 Estimating hydrothermal heat flux from heat flow data (from Stein *et al.*, 1995). Left: The heat flow discrepancy, the difference between the heat flow predicted by a lithospheric cooling model and the lower values observed (shaded), is assumed to indicate the heat transported by water flux. Data are compared to a lithospheric cooling model in raw form (top) and as the fraction of the predicted heat flow that is observed (bottom). The water flux is divided somewhat arbitrarily into high-temperature near-ridge flow and low-temperature off-ridge flow. Beyond a "sealing age," defined as that where the observed and predicted heat flow approximately coincide, hydrothermal heat transfer across the seafloor is presumed to have largely ceased. Right: Observed heat flow versus age for the global data set and predictions of the GDH1 thermal model, shown in raw form (top) and fraction (bottom). Data means and standard deviations are shown for 2-M.y. bins. The sealing age estimated by a least squares fit to the heat flow fraction for ages <50 M.y. (closed circles) is 65 $\pm$ 10 M.y. Further details are discussed in Chapter 10. (Reprinted from Stein, C. A., Stein, S., and Pelayo, A. 1995. Heat flow and hydrothermal circulation. In *Seafloor Hydrothermal Systems: Physical, Chemical, Biological, and Geological Interactions*, Monograph 91, eds. S. E. Humphris, R. A. Zierenberg, L. S. Mullineaux, and R. E. Thomson. Washington, DC: American Geophysical Union, Fig. 2, copyright 1995 American Geophysical Union.)

upper basement layer as it becomes covered by sediments (e.g. Lister, 1972; Hartline and Lister, 1981; Wang *et al.*, 1997). Hydrothermal discharge through the seafloor is generally found above basement highs, although seawater recharge can also occur (e.g. Davis *et al.*, 1999; Fisher *et al.*, 2003; Chapter 8). The topographic roughness of the upper boundary of igneous ocean crust is generally a function of the rate of separation between lithospheric plates at mid-ocean ridges, being smoother (lower amplitude) for higher rates (>several centimeters per year) compared to slower rates (e.g. Bird and Pockalny, 1994). This relationship may cause the heat lost by hydrothermal processes to extend to greater ages for slower spreading ridges (e.g. Atlantic, Indian) compared to faster spreading ones (e.g. Pacific; Anderson and Langseth, 1977) as a result of both the lesser degree to which rugged basement topography can be buried by sediments, and the greater driving force generated in the presence of large-amplitude relief of the permeable igneous layer. Of course, the rate of sedimentation, which is a complex function of geographic location in the oceans (e.g. Chapter 6), may well affect such a relationship. Furthermore, perhaps independent of basement topography, the distribution of basement rock permeability is probably the most critical parameter that controls the existence and nature of hydrothermal circulation, particularly its magnitude, homogeneity, and degree of anisotropy (e.g. Fisher and Becker, 1995, 2000).

The lithospheric age at which hydrothermal circulation ceases or becomes insignificant is poorly understood, mostly because it probably occurs gradually and measurements to detect it on older seafloor are relatively sparse. The statistics of worldwide marine heat flow values (Stein and Stein, 1992; Stein *et al.*, 1995) indicate that the unmeasured advective heat loss through the seafloor becomes small (or at least thermally negligible) at ages $\gtrsim$50–65 M.y. Although the theoretical conductive lithospheric heat loss is approached by measured heat flow means at this age, considerable scatter still exists among values obtained at similar ages for older seafloor (Fig. 13.1). Some of this scatter may be caused by the variations associated with thermal refraction of rough basement topography beneath a sediment cover, but circulation of pore water may also continue in the permeable basement beneath a nearly continuous sediment layer that inhibits significant advective heat loss through it. This paper reviews the evidence for such continuing circulation from existing detailed geothermal investigations that were implemented for other purposes and thus are not ideal. It should be noted that pore-water circulation below the seafloor, whatever its cause, has the potential to alter the distribution of seafloor heat flow. A likely driving force for any such circulation in older seafloor far from tectonic boundaries is the ubiquitous heat flux from below, hence hydrothermal circulation, although other mechanisms may need to be considered in particular settings.

13.2 Relevant detailed geothermal investigations

Over the past several decades, the methodologies for seafloor heat flow measurements have been improved in several important ways. Geothermal probes deployed from research vessels now include the violin-bow Lister design (Hyndman *et al.*, 1979) in addition to the cylindrical (Bullard) and outrigger (Ewing) probes (e.g. Louden and Wright, 1989). All designs have also incorporated *in situ* measurement of sediment thermal conductivity at the same locations as thermal gradient measurements (Sclater *et al.*, 1969; Lister, 1970, 1979; Von Herzen, 1987), resulting in a more accurate measure of the local heat flux. Most significantly, techniques for multiple measurements ("pogo" methodology) as part of a single deployment of instrumentation to the seafloor, combined with acoustic telemetry and extended duration instrumentation operations, have resulted in more efficient measurements and acquisition of accurate and detailed data (e.g. Von Herzen, 1987; Davis *et al.*, 1997). Most of the investigations described and referenced below have utilized these recent improvements in methodologies.

Table 13.1 lists detailed heat flow investigations that may be relevant to discerning the possibility of hydrothermal circulation in older (>60 Ma) seafloor. Most of these surveys were not carried out for this purpose, but generally for other objectives such as determining the thermal influence of hot-spots, the relationship of heat flow versus seafloor age, and the hazards of waste containment in seafloor sediments. The selected investigations comprise those made in the older seafloor of the main ocean basins including the Gulf of Mexico (Fig. 13.2). They do not include detailed investigations of active subduction (trench or fore-arc) regions, which usually have more complex thermal signatures as a result of their special tectonic situation (e.g. Von Herzen *et al.*, 2001), nor of back-arc regions, particularly in the western Pacific (e.g. Yamano and Uyeda, 1989), which include younger (but sometimes unknown age) seafloor created at back-arc spreading centers with likely hydrothermal circulation. The closely spaced sites are commonly associated with hot-spot surveys, e.g. Hawaii (Von Herzen *et al.*, 1982; Von Herzen *et al.*, 1989; Harris *et al.*, 2000), Bermuda (Detrick *et al.*, 1986), Cape Verde (Courtney and White, 1986), and Crozet (Courtney and Recq, 1986).

The 58 sites (Table 13.1) are selected because of the high quality and relatively close spacing of their heat flow measurements, usually a few kilometers or less, which are required to discern potentially subtle lateral variability associated with crustal hydrothermal circulation. The measurements commonly are distributed spatially in the form of linear profiles, or can be projected to linear segments within each survey. Although it is impractical here to describe all the profiles in detail, the following sections present criteria for the partially subjective designation as to whether hydrothermal circulation may be thermally significant at each site, with some example profiles presented and discussed in detail. All the selected linear profiles or segments are illustrated as cross-sections in the CD-ROM appendix for those readers interested in more detail.

13.3 Interpretation and modeling

13.3.1 Diagnostic criteria

The primary objective of this paper is to determine whether the heat flow patterns for each of the selected profiles can be explained mainly by conduction through the seafloor, or whether advective processes (e.g. hydrothermal circulation) are required. Criteria that may

Table 13.1 *Summary of selected detailed heat flow (HF) surveys*

Ocean	Lat. (°)	Long. (°)	Age[a]	Circ?[b]	N[c]	Reference	Max./Min. (% of mean)[d]
NW Atl.	23.75 N	60 W	~85	Y	47	Embley *et al.* (1983), Sta. 9, 10, 11, 13	+330, +18.6, +51.2, +70.8
Four short profiles in same region. Turbidite sediments up to 200-m thick above local sediments up to 400-m thick. Basement topography up to 500 m. HF × 10 normal over 0.5-km distance.							
NW Atl.	23.5 N	59 W	~85	Y?	11	Embley *et al.* (1983), Sta. 12	+42.9
HF varies ±15 mW m^{-2} systematically over 7-km-long profile; large errors on some values. Not correlated with topography.							
NW Atl.	29 N	71 W	~150	N?	11	Detrick *et al.* (1986), Site 1	−13.9
200–300 m basement undulating beneath ~1 km sediment thickness (prof. mean: 1.09 ± 0.04s). Small (±10 mW m^{-2}) systematic HF variations over 20-km-long profile?							
NW Atl.	30.5 N	70 W	148	N	27	Detrick *et al.* (1986), Site 2	−12.0, +14.9
Sediments 500–800-m thick completely cover basement; mean (two profiles) 0.67 ± 0.07s. Most HF values within mutual error bars.							
NW Atl.	31 N	68 W	139	N	24	Detrick *et al.* (1986), Site 3	+8.1, −7.4
Sediments 750–1,000-m thick completely cover basement; mean (two profiles) 0.87 ± 0.07s. All HF values within mutual error bars.							
NW Atl.	31 N	67 W	127	N	11	Detrick *et al.* (1986), Site 4	+13.6
Sediments (turbidites, flat seafloor) 1.0–1.5-km thick completely cover undulating basement (prof. mean 1.13 ± 0.16s).							
NW Atl.	31.5 N	66 W	121	Y?	26	Detrick *et al.* (1986), Site 5	+41.9, +19.4, +14.7
Systematic HF variation (±15 mW m^{-2}) across Profile 1, associated with a generally systematic variation of ~600–900 m in basement depth across profile. Large errors for some HF values. Sediment thickness (three profiles) = 0.77 ± 0.13s.							
NW Atl.	32 N	64.5 W	105	Y?	23	Detrick *et al.* (1986), Site 6	+24.5, +14.7
Generally systematic HF variation (±10 mW m^{-2}) across both W–E and N–S profiles, wavelength 15–20 km. Sediments 0.5–1.2-km thick completely covering basement; mean (two profiles) 0.82 ± 0.20s).							
NW Atl.	33.5 N	67.5 W	141	N	28	Detrick *et al.* (1986), Site 7	−19.4
Mostly uniform HF (within individual measurement error) along NW–SE profile. Sediments 500–700-m thick completely cover basement (prof. mean: 0.61 ± 0.06s.							

418

(cont.)

NW Atl.	27 N	71–72 W	~150	Y?	58	Davis et al. (1984), Site 2 (Sta. 4, 6, 11)	−14.5, +42.3, −13.5, +28.1
NW Atl.	27.5 N	73.8 W	~160	Y	24	Davis et al. (1984), Site 3	+29.6, +24.8
NW Atl.	27.9 N	74.7 W	~165	N?	24	Davis et al. (1984), Site 4	−13.1
NW Atl.	25 N	68 W	120	N	41	Galson and Von Herzen (1981)	−14.3, +7.8, +5.6, −21.3
NW Atl.	32–33 N	75–76 W	~175	N?	50	Ruppel et al. (1995)	+41.1, −31.2, +51.8, +45.4 (−14.7)
NW Atl.	40.5 N	60 W	~160	N	21	Louden et al. (1987)	−21.2
NE Atl.	15.5 N	24.5 W	~124	N?	11	Courtney and White (1986), Site A	+28.7
NE Atl.	16 N	23.5 W	~135	N	6	Courtney and White (1986), Site B	+15.0
NE Atl.	18 N	24.5 W	~124	Y	21	Courtney and White (1986), Site C	+32.7
NE Atl.	19 N	24.5 W	~125	N	31	Courtney and White (1986), Site D	−11.9, +6.6
NE Atl.	20 N	24.5 W	~127	Y?	23	Courtney and White (1986), Site E	−5.4, −12.8

Notes:

NW Atl., 27 N: Small (±10 mW m^{-2}) but systematic HF variations over most of E–W profile, wavelength 15–20 km. Sediment thickness 15–20 km. Sediment thickness ~1 ± 0.1 km. Significant HF variations in NE–SW direction (two profiles), not NW–SE (two profiles)?

NW Atl., 27.5 N: Systematic HF variation (±15–20 mW m^{-2}) along SW–NE profile. Sediment thickness ~1.8 km.

NW Atl., 27.9 N: Mostly random variability (±5 mW m^{-2}) over 22-km profile, except lower values near ends? Sediment thickness ~1.8 km.

NW Atl., 25 N: Four short (<6-km) profiles without obvious systematic variation (but may not be long enough?).

NW Atl., 32–33 N: Apparent significant variability along most profiles, but BWT variations, sediment slumping, and salt diapirs may be responsible. Sediment thickness >4 km.

NW Atl., 40.5 N: No significant systematic HF variation over 20-km NNW–SSE profile of 21 values. ±1 SD error bars of most HF values overlap. Sediment thickness >2 km (profile mean: 2.58 ± 0.12s).

NE Atl., 15.5 N: Two contiguous lower HF values over ~15-km profile. 700–800-m sediment cover over all basement (profile mean: 0.69 ± 0.06s).

NE Atl., 16 N: Some variability (45–60 mW m^{-2}), but only six values over a short (4-km) profile. 250–300-m thick sediments apparently cover all basement (profile mean: 0.29 ± 0.01s).

NE Atl., 18 N: Systematic variation from ~50 to ~70 mW m^{-2} over 30-km profile. 650–800-m-thick sediment cover (profile mean: 0.76 ± 0.05s).

NE Atl., 19 N: Values randomly(?) vary between ~45 and 55 mW m^{-2} over two profiles each ~15-km long. 0.8 to >1.2 km sediment over all basement, systematically varying along profiles (mean (two profiles): 1.20 ± 0.17s).

NE Atl., 20 N: Small (5–10 mW m^{-2}) but systematic variability over two profiles 2.5 and 7 km long, resp. ~0.85 to >1 km sediment thickness (mean (two profiles): 0.85 ± 0.09s).

Table 13.1 (cont.)

Ocean	Lat. (°)	Long. (°)	Age[a]	Circ?[b]	N[c]	Reference	Max./Min. (% of mean)[d]
NE Atl.	20.5 N	24.5 W	~127	N	27	Courtney and White (1986), Site F	+18.6, −10.1

No apparent systematic variation over two profiles ~7 and 12 km long. 0.9–1.3 km sediment covers basement (mean (two profiles): 1.09 ± 0.15s).

Ocean	Lat. (°)	Long. (°)	Age[a]	Circ?[b]	N[c]	Reference	Max./Min. (% of mean)[d]
NE Atl.	22.5 N	24.5 W	~126	N	8	Courtney and White (1986), Site G	+11.2, −7.2

Only a few values (5, 3) over two short (8-, 5-km) profiles, resp. 1.0–1.3 km sediment over basement (mean (two profiles): 1.16 ± 0.12s).

NE Atl.	31.5 N	25 W	106	Y?	55	Noel and Hounslow (1988), Noel (1985)	−19.5, +18.6, +11.8, +51.1

Three (of four) profiles show systematic HF variability, correlated with basement topography (amplitude a few hundred meters, but completely sediment covered). 0.2–1.3 km sediment over all basement (mean (four profiles): 0.41 ± 0.26s).

W Pac	12 N	151 E	~170	N?	32	Lister et al. (1990), Site 1	−4.1, −2.9

Very small (±1 $mW\,m^{-2}$) systematic HF variation along two profiles 22 and 11 km length.

W Pac	13 N	156 E	~165	N	18	Lister et al. (1990), Site 2	+1.9

Very uniform HF over 13-km profile.

W Pac	8 N	164 E	~160	Y?	14	Lister et al. (1990), Site 3	+8.5

Systematic HF variation along 20-km profile from 48 to 42 $mW\,m^{-2}$. Uniformly thick sediments to >500 m depth, basement not visible.

W Pac	9.5 N	180	~150	Y?	13	Lister et al. (1990), Site 4	+23.5

Systematic HF variation >15 $mW\,m^{-2}$ over 12-km profile; high HF correlates with basement high (or shallowing of uppermost reflectors).

W Pac	30.3 S	175.3 W	~100	N	10	Von Herzen et al. (2001), S profile ref.	+19.1

Small (±5 $mW\,m^{-2}$) apparently random variations about mean over 16-km profile. Sediment thickness 50–200 m (profile mean: 0.14 ± 0.06s).

W Pac	28.2 S	174.7 W	~100	Y?	10	Von Herzen et al. (2001), N profile ref.	+19.8

Systematic HF variations 5–10 $mW\,m^{-2}$, wavelengths 6–8 km, superimposed on increase from 50 to 65–70 $mW\,m^{-2}$ over 14-km profile from N to S. Sediment thickness ~150 m (profile mean: 0.15 ± 0.02s).

Pacific	17.5 N	158.5 W	95	N	11	Von Herzen et al. (1982), Site A	−6.1, +12.8

Almost all values for each of two short (10–12-km) profiles within ±1 SD of each other. Total sediment thickness not determined.

Pacific	20 N	159 W	96	N	8	Von Herzen et al. (1982), Site B	+17.6

All but one (seven of eight) value within ±1 SD of others over 25-km-length profile. Seamount near-anomalous value. Total thickness not determined.

Pacific	19 N	160.5 W	98	N?	18	Von Herzen et al. (1982), Site C	16.8, −13.2	Some systematic variation along two 20-km-long profiles. Highest anomaly coincident with steep topographic slope of 100-m-high bump. Total sediment thickness not determined.
Pacific	20 N	160 W	86	N	4	Von Herzen et al. (1982), Site D	−10.3	All four values over 11-km-long profile within ±1 SD of one another. Total sediment thickness not determined.
Pacific	20.5 N	161.5 W	88	N	8	Von Herzen et al. (1982), Site E	+7.9, +5.1	All values along two profiles within 5–6 mW m^{-2} of each other. Total sediment thickness not determined.
Pacific	21.5 N	166 W	94	N	10	Von Herzen et al. (1982), Site F	−20.1, +11.0	Appears to be an about random distribution over two profiles 12- and 16-km length with five and six measurements each, resp. Total sediment thickness not determined.
Pacific	23N	169.5 W	97	N	17	Von Herzen et al. (1982), site G	−17.9, +11.6	Most values over two 16-km-long profiles within ±1 SD of one another. Total sediment thickness not determined.
Pacific	24.5 N	174.5 W	109	N?	17	Von Herzen et al. (1982), Site H	+20.0, −15.1	Values vary somewhat systematically from 50 to 70 mW m^{-2} over 16-km-length of one profile, but all values within ±2 SD. Total sediment thickness not determined.
Pacific	31 N	168 W	99	N?	18	Von Herzen et al. (1989), Site 1	+11.3, −12.7	Small (5–10 mW m^{-2}) systematic HF increase over both 16-km-long profiles (but mostly within error bars). Total sediment thickness not determined.
Pacific	29 N	169 W	101	N	16	Von Herzen et al. (1989), Site 2	+17.2, +7.3	Values over two profiles 12 and 18 km long mostly within error bars, with one higher value at top of seafloor slope (W–E profile). Total sediment thickness not determined.
Pacific	28 N	169.75 W	102	N	19	Von Herzen et al. (1989), Site 3	+7.0, +14.2	All values over two profiles 14 and 24 km in length are within error bars, no systematic trends. Total sediment thickness not determined.
Pacific	26.75 N	170.5 W	103	Y	19	Von Herzen et al. (1989), Site 4	−13.5, +16.4	Values over two profiles 12 and 22 km long vary systematically by ±10 mW m^{-2}, wavelengths 7–10 km. Nearly same location as the "Maro southern disturbed zone" of Harris et al. (2000) below. Total sediment thickness not determined.
Pacific	24.5 N	171.5 W	99	N	11	Von Herzen et al. (1989), Site 5	−10.9	Small (±5 mW m^{-2}) variations over 8-km-long profile, but most values are within error bars. Total sediment thickness not determined.
Pacific	23.5 N	171.75 W	100	N	22	Von Herzen et al. (1989), Site 6	+4.1, −6.1, +10.6	Small (±5 mW m^{-2}) HF variations over two longer profiles 12 and 16 km long, somewhat systematic on "middle" profile (but within error bars). Values for short (3.5-km) W–E profile not as reliable (thousands of variations?). Total sediment thickness not determined.

(cont.)

Table 13.1 (cont.)

Ocean	Lat. (°)	Long. (°)	Age[a]	Circ?[b]	N[c]	Reference	Max./Min. (% of mean)[d]
Pacific	22.5 N	173 W	102	N	19	Von Herzen et al. (1989), Site 7	+12.5, −8.3
Small (±5 mW m^{-2}) and seemingly random variations with distance over 8- and 18-km-long profiles. Total sediment thickness not determined.							
Pacific	21 N	173 W	104	N	20	Von Herzen et al. (1989), Site 8	+7.8, −9.2
Random HF variability (±5 mW m^{-2}) over two profiles ~13 km long? Total sediment thickness not determined.							
Pacific	26.5 N	170.5 W	103	Y	19	Harris et al. (2000) – Maro southern disturbance zone	+19.5
HF varies systematically over ±10 mW m^{-2} with 15–20 km wavelength. Modest correlation with basement topography Sediment cover 0.5–1.5 km over rough basement topography (profile mean: 1.11 ± 0.24s). Nearly same location as Von Herzen et al. (1989), Site 4, above.							
Pacific	27 N	170 W	103	Y	15	Harris et al. (2000) – Maro northern disturbance zone	−19.6
Systematic HF variability over ±7–10 mW m^{-2}, wavelength 10–15 km. No apparent correlation to basement topography. Sediment cover ~300–900 m over moderately rough basement (profile mean: 0.57 ± 0.19s).							
Pacific	27.25 N	170 W	~103	N	24	Harris et al. (2000) – N end Maro profile	+28.4 (−9.0)
HF values mostly within measurement error bars, with one value ~20 mW m^{-2} higher near 200 km. Sediment cover 500–700 m over fairly smooth basement topography (profile mean: 0.46 ± 0.06s).							
Pacific	22.5 N	157.5 W	~86	Y	36	Harris et al. (2000) – Oahu southern disturbance zone	+40.0
Systematic HF variability over ±10–15 mW m^{-2}, wavelengths 10–30 km. No obvious correlation to basement topography. Sediment thickness decreases systematically from ~2 to 1 km thickness from S to N (profile mean: 1.41 ± 0.31s).							
Pacific	23 N	157 W	~86	Y	33	Harris et al. (2000) – Oahu northern disturbance zone	−48.1
Systematic HF variability over ±10–15 mW m^{-2}, wavelengths 10–30 km. Some correlation with basement topography? Sediment thickness decreases from ~1 to 0.5 km over profile from S to N (profile mean: 0.60 ± 0.13s).							
Pacific	23.25 N	157.25 W	~86	N	13	Harris et al. (2000) – between Oahu disturbance zones	−26.9
Relatively uniform HF over 20 km profile, with one lower value; between two regions interpreted as showing circulation. Sediment thickness decreases slightly from ~1 to 0.8 km over profile from S to N (profile mean: 0.90 ± 0.06s).							

Location	Lat	Long	Age[a]	Designation[b]	N[c]	Reference	Difference[d]
SW Indian	49 S	54 E	~66	N?	10	Courtney and Recq (1986), Site M6	−7.9

Small (±5 mW m^{-2}) but possibly systematic variation over 8-km profile of projected measurements. Sediment thickness several hundred meters.

| SW Indian | 48 S | 54 E | ~66 | N? | 10 | Courtney and Recq (1986), Site M5 | +15.4 |

Scattered HF values ~57–70 mW m^{-2} over 3-km profile. 400–500-m sediment thickness.

| SW Indian | 47 S | 53 E | ~70 | N | 7 | Courtney and Recq (1986), Site M4 | +13.4 |

HF values over ~13-km profile mostly within their resp. error bars. 400–500-m sediment thickness.

| W Indian | 16 S | 57 E | 66 | Y | 24 | Bonneville et al. (1997), N profile W of fracture zone | +42.7 |

Systematic HF variation ±20 mW m^{-2} over ~50 km wavelength, not correlated with basement topography. Sediment thickness varies 0.5–1 km, mostly dependent on basement topography (profile mean: 0.71 ± 0.12s).

| W Indian | 17 S | 58 E | 77 | N | 25 | Bonneville et al. (1997), N profile E of fracture zone | +44.1 |

Except for three individually separate high values, all other values along profile are within error bars. Sediment 0.7–2.1-km thick completely covers basement (profile mean: 1.25 ± 0.32s).

| W Indian | 17 S | 56 E | 64 | Y? | 12 | Bonneville et al. (1997), S profile W of fracture zone | +37.7 |

Mean trend of values increasing from W to E along profile, from ~40 to 70 mW m^{-2}, anticorrelated with basement topography. Sediments 350–700-m thick completely cover basement (profile mean: 0.48 ± 0.10s).

| Gulf of Mexico | 24 N | 94 W | 120–170? | N | 10 | Nagihara et al. (1992) | +5.3 |

Most site values (multiple penetrations) within error bars of each other, although small (2–4 mW m^{-2}) systematic minimum at center of profile. Sites separated by 10–20 km along profile. Sediment thickness >5 km (profile mean: 5.59 ± 0.36s).

[a] Lithospheric age (M.y.) determined primarily from seafloor magnetic anomalies (e.g. Muller et al., 1997).

[b] Designation (see text) as to whether survey data indicate hydrothermal circulation (Y = yes, Y? = perhaps, N? = uncertain, N = no).

[c] Number of heat flow measurements in survey.

[d] Difference between maximum (+) or minimum (−) value and mean profile value (% of mean value). Numbers in parentheses ignore most extreme value.

[e] Sediment thickness estimates assume seismic velocity of 2 km s^{-1}, so that 1 s TWTT = 1 km.

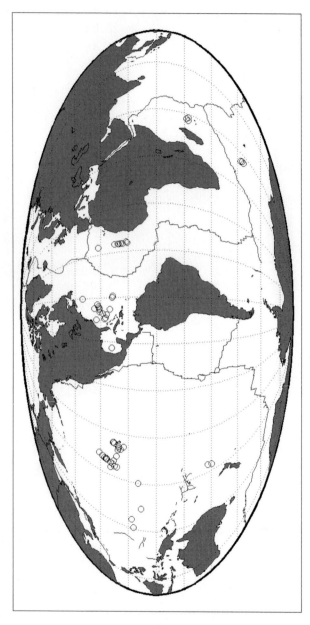

Fig. 13.2 Geographic locations of selected marine geothermal surveys on older seafloor (circles, from Table 13.1). The continents (shaded) and active mid-ocean ridge plate boundaries are also shown for reference.

be used to identify active hydrothermal circulation from seafloor heat flow profiles should be based on known physical principles and realistic models and parameters. In addition to the spatial variability of heat flow, the relative seafloor depth and in some cases the thickness of sediments above acoustic basement are available along the profiles. Variations in both seafloor depth and sediment thickness may cause lateral variability in vertical heat flux, as a result of focusing and defocusing of the vertically conducted heat by the non-horizontal configurations of the layer boundaries, combined with the contrasts in thermal conductivity between layers (e.g. Birch, 1967; Lachenbruch, 1968; Sclater *et al.*, 1970).

The main objective for most of the geothermal surveys over older ocean floor was to obtain an accurate mean value, either for purposes of establishing the heat flux versus age relationship (Parsons and Sclater, 1977; Stein and Stein, 1992) or to deduce large-scale anomalies associated with tectonic processes (e.g. Von Herzen *et al.*, 1982; Courtney and White, 1986). Usually any variability in heat flux along the profiles was ignored, or considered as either random measurement variability or caused by local processes for which the ratio of the perturbation relative to the mean value over the profiles was believed negligible. As may be deduced from some examples shown here, these assumptions may not have been always correct, particularly for surveys where systematic variations of heat flux were of a scale similar to the length of measured profiles.

For purposes of this discussion, the main objective is to discriminate the sources of any profile heat flux variability between purely conductive anomalies and those that may be the result of fluid advection, either in the seafloor sediments or basement rock. In most situations, any thermally significant pore-water flow will probably occur in shallow basement rock, rather than sediments, because the fluid permeability of the latter is normally much less than that of the former (e.g. Von Herzen *et al.*, 1992; Chapters 6 and 7). Fortunately, much of the older seafloor has relatively complete sediment cover that tends to fill in the basement topography, so that the relatively smooth or gently dipping seafloor is probably not a significant source of heat flow variability. Unfortunately, many of the profiles lack determinations of sediment thickness variations, especially for the earlier surveys. Thus we are unable to calculate directly the effects of thermal conductivity contrasts between sediments and basement rock on the heat flow variability for such profiles, assess whether heat flow variability is correlated with sediment thickness variations, or determine whether the absence of variations may occur where sediment thickness variations are absent.

For those profiles with co-located seismic data, the maximum conductive anomalies caused by topography of the sediment–basement interface can be estimated using simple models. Comparison of those with observed variations provides a test of whether fluid advection is likely to be significant. In cases where the available seismic data indicate relatively subdued basement topography, we may be able to infer the existence of fluid circulation from small but systematic heat flux variations over the profiles. No detailed modeling has been done to estimate the fluid flux rates or the possible geometries of any fluid flows (e.g. Chapter 8), which was beyond the scope of this initial analysis of the relevant geothermal surveys. Hopefully these important quantitative objectives may be accomplished with additional research and surveys.

An estimate of the maximum conductive heat flow anomalies caused by topography of the interface between sediments and basement rock depends mostly on (i) the ratio of the height to wavelength of the topography, (ii) the mean depth of basement burial compared to the amplitude of its topography, and (iii) the contrast in thermal conductivity between basement rock and overlying sediment. Conductive anomalies will increase with the height-to-wavelength ratio and conductivity contrast, but decrease with burial depth of the topography. For profiles where seismic data are lacking, an approximate limiting estimate of the possible effects of basement topography can be made by assuming burial of rough basement by sediments with a maximum contrast in thermal conductivity.

Some relevant analytical models have been developed, one in which basement topography is approximated by oblate spheroid shapes and isotropic thermal conductivity is assumed (Carslaw and Jaeger, 1959). However, this model lacks exact analogy with the sub-seafloor configurations of sediment and basement rock. It applies to an isolated conductive anomaly (rock) in an infinite medium of contrasting conductivity (sediment), whereas the sub-seafloor basement topographic anomalies are almost always connected to a horizontally continuous basement layer below. Also, a uniform-temperature, and in cases considered here, approximately horizontal boundary (seafloor) exists some distance above, which is a function of sediment thickness. Both of these factors will result in an overestimate of conductive heat flow anomalies and, more importantly, the basement geometry may not be readily approximated with such an analytical model. Other analytical models have similar shortcomings.

Numerical models have the flexibility to simulate steady-state conductive anomalies produced by more realistic basement rock configurations. The results discussed below are obtained using a two-dimensional finite element model (namely, Bethke *et al.*, 1988) without pore-water flow and with two layers of uniform and isotropic thermal conductivity separated by an interface with variable geometry to simulate sediment covering basement rock. Although three-dimensional models can be employed, the volcanic and tectonic processes at seafloor spreading centers result in a strongly two-dimensional fabric of the volcanic seafloor (e.g. Goff, 1991), and three-dimensional modeling is rarely justified. For most of the surveys, heat flow measurements were located at sufficient distance (e.g. greater than 10–20 km) from obvious outcropping basement topography to minimize complications in the geothermal field from either conductive refraction or pore-water advection. However, particularly for surveys lacking seismic data, the possible effects on seafloor heat flow of conduction refraction from basement topography buried by sediment need to be estimated to establish a threshold to identify surveys having larger anomalies that may be ascribed to advection.

For the selected surveys with seismic data (Table 13.1), except for several located on some of the oldest seafloor having increased sediment thickness near passive continental margins, the mean sediment thickness (0.76 ± 0.34 (SD) km, $n = 29$) is relatively uniform for seafloor created 60–150 Ma (Fig. 13.3). The somewhat surprising result that sediment thickness does not increase significantly over this age range is consistent with more comprehensive data over all ocean regions (Chapter 6). Beneath the sediment, basement topography is created

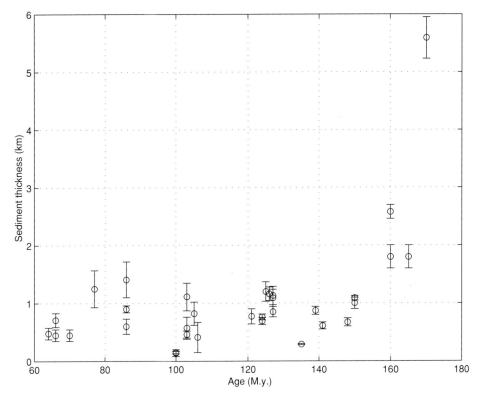

Fig. 13.3 Sediment thickness above basement versus seafloor age for selected heat flow surveys (Table 13.1). Mean sediment seismic velocity of 2 km s^{-1} is assumed at the 33 sites with seismic data. Sedimentation at older (>160 M.y.) sites is probably affected by proximity to passive continental margins.

initially near spreading centers mainly as abyssal hills having significant lineations oriented approximately normal to the spreading directions (e.g. Goff, 1991). These lineations are interrupted by fracture zones (FZs) composed of lineated topography trending approximately orthogonal to the abyssal hills. As the seafloor ages, seamounts apparently created by hot-spot volcanism from mantle convection become more common.

Abyssal hill topography may be described statistically as a spectrum of two-dimensional sinusoids having amplitude to wavelength ratios that range up to about 0.05 (e.g. Fig. 10 of Goff, 1991), and with characteristic lengths of several to >10 times their widths. The larger FZs that interrupt them have widths that range up to several tens of kilometers with lineated steep-sloped ridges that range in height up to several kilometers (e.g. Vema FZ in the North Atlantic, van Andel, *et al.*, 1971). The heat flow survey sites selected here (Table 13.1) mostly lack sufficient sediment to bury such large FZs, although several of these surveys (e.g. Von Herzen *et al.*, 1989; Bonneville *et al.*, 1997; Harris *et al.*, 2000) may be near or over smaller FZs. Thus the potential basement surface beneath sites without seismic

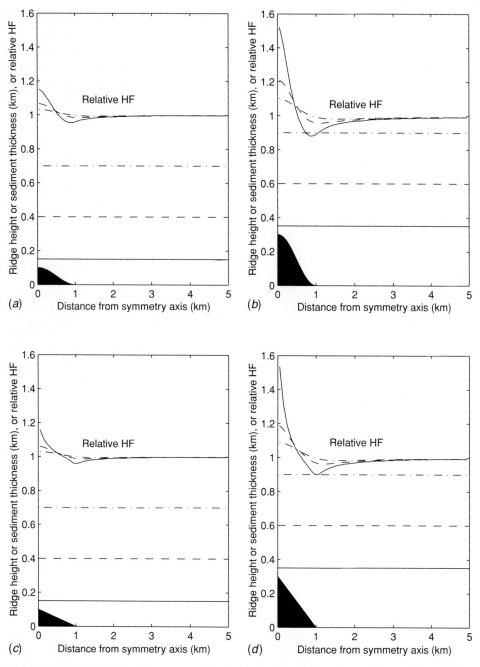

Fig. 13.4 Results of finite element simulations of steady-state heat flow above two-dimensional basement ridges of either sinusoidal (*a*, *b*) or triangular (*c*, *d*) shapes buried by sediment. The ridge cross-sections are represented by the opaque shapes in the lower left of each panel, with vertical to horizontal aspect ratios of 0.025 (*a*, *c*) and 0.075 (*b*, *d*) as defined by Goff (1991). They are buried by

data is modeled as either two-dimensional sinusoidal-shaped topography (abyssal hills) or triangular ridges (FZs) with appropriate vertical to horizontal aspect ratios, all buried by sediment.

The steady-state conductive heat flow anomalies over isolated two-dimensional basement ridges having either sinusoidal or triangular shapes are shown in Fig. 13.4. Results are given for aspect ratios of 0.025 and 0.075, thereby bracketing Goff's (1991) maximum value, with sediment thickness ranging between factors of 1.07 and 7.0 of the modeled basement ridge heights. The calculated heat flow anomalies can attain about 50% for the largest ridge aspect ratios and thinnest sediment cover for both ridge shapes, with an apparently sharper (more focused) anomaly pattern for the triangular ridge shape compared to the sinusoid. Other than the large positive near-axis anomalies, the anomaly is negative over approximately the lower half of both ridge shapes for the thinnest sediment cover. For these models, the mean anomaly over the width of the ridge structures ranges between ~5 and 10% for both the triangular and sinusoidal shapes. Other modeled ridge shapes (e.g. flat-topped wedge) give comparable results. Thus whereas relatively large conductive heat flow anomalies may occur at the peaks of such ridge structures, they decrease rapidly with distance from the peak, such that the mean anomalies are more modest. The focused anomaly patterns emphasize the need for closely spaced heat flow measurements in such surveys.

The models apply to two-dimensional basement ridges, whereas seamounts are mostly three-dimensional conical-shaped structures (e.g. Smith, 1988). Although the model results do not directly apply to seamounts, a similar pattern would be expected. Probably the maximum conductive anomalies will be even higher for three-dimensional versus two-dimensional structures having the same aspect ratios, although they should decrease more rapidly with distance from the peak of a three-dimensional structure. The probability of a heat flow measurement being directly over a seamount peak along a randomly located profile (as for many of the selected surveys) is quite low. Thus although such high values seem possible from conductive focusing by basement structure, they would be quite rare statistically. Even if a profile did cross a seamount peak, only one or two measurements would detect the highest part of the anomaly for widely spaced (>1 km) measurements. Thus the continuity of any anomalies is also an important criterion for evaluating the possibility of advection at the selected sites (Table 13.1).

The conductive anomalies over two-dimensional sinusoidal basement topography were also modeled (Fig. 13.5). For an aspect ratio of 0.05, corresponding to Goff's (1991) maximum ratio for abyssal hills, the maximum heat flow anomaly over the peaks ranges between about 35 and 3%, and the minimum over the troughs between -20 and -3%, depending on

---

a sediment layer of three different thickness values (50, 300, and 600 m above the top of the ridges) shown as horizontal lines. The heat input is a uniform 2 km beneath the base of each panel, and no heat flows through the sides, so that only the ridge half-width is shown. A ratio of 2 is assumed for the uniform and isotropic thermal conductivity of basement and sediment, respectively. The heat flow anomaly patterns shown as curves near the value of 1 are relative to the basal heat flux.

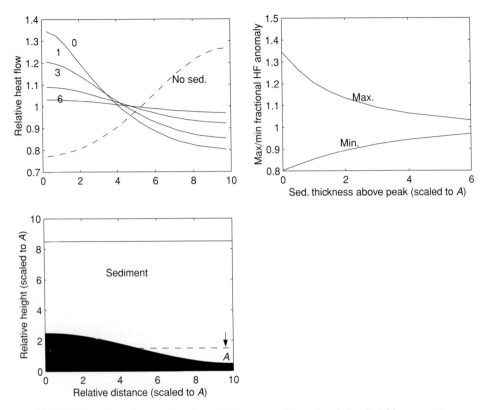

Fig. 13.5 Calculated steady-state heat flow (HF) over two-dimensional sinusiodal basement topography covered by sediment. The basement sinusoid, shown in the lower left panel, has an aspect ratio of 0.05, as defined by Goff (1991). A ratio of 2 is assumed for the uniform and isotropic thermal conductivity of basement and sediment, respectively. The heat flow (HF) patterns, relative to the uniform flux at depth, are shown in the upper left panel for different sediment thickness above the basement high relative to the amplitude (A) of the sinusoid. The right panel shows the maximum and minimum heat flow anomalies as a function of sediment thickness.

sediment thickness. The extreme maximum and minimum are for the case with sediment just overlapping the highs, and the maximum anomalies particularly are reduced rapidly with increasing sediment cover (e.g. to ∼25% for sediment thickness reaching 0.5 of the sinusoidal amplitude above the basement highs). Thus for sinusoidal basement topography with a characteristic amplitude of 100 m and wavelength of 2 km buried by only 100 m of sediment over the peaks, the maximum calculated heat flow anomaly is about 21%; it can exceed 25% if the sediment thickness above the peaks is only about 50 m or less.

More realistic estimates of conductive anomalies may be calculated using constraints from seismic reflection data where they are available. An example is shown in Fig. 13.6, where the heat flow variability is calculated with a two-dimensional model for the relatively

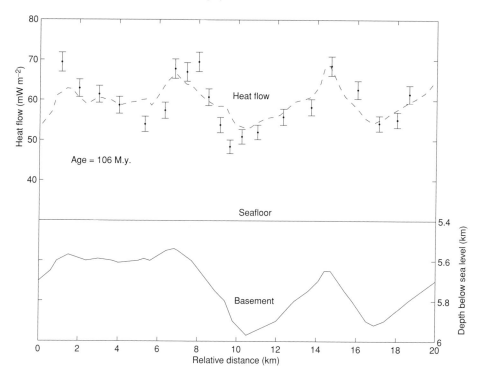

Fig. 13.6 Heat flow and seafloor structure for a profile across the Madeira abyssal plain, northeast Atlantic. Seafloor and basement topographies were obtained from Fig. 5 of Noel and Hounslow (1988), and heat flow data (dots with unified error estimates normalized to mean value) from profile CD9B/30 of their Table 1. Effects of refraction through the structure (dashed curve) are calculated using a finite element model, assuming two-dimensional structure and a sediment–basement thermal conductivity contrast of 2. The calculated heat flux is normalized to the mean of the measured values.

rough basement topography of the Madeira abyssal plain (Noel and Hounslow, 1988), with ~160 m sediment thickness above the highest basement peak. The result gives maximum heat flow anomalies of about ±10% of the mean value. The modeled heat flux based on the basement topography shows excellent correlation with the pattern of the measured heat flux, although the amplitude of the variation in measured values appears to exceed somewhat the modeled variability. This apparent contrast in measured versus modeled heat flux amplitudes is even greater if thermal conductivities of both sediment and rock are modeled as depth dependent (A. Fisher, pers. comm., 2002). Thermally significant pore-water circulation at this site is probably the cause of the anomalous variability.

An assumption of a uniform thermal conductivity contrast between rock and sediment maximizes the model heat flow anomalies; the actual contrast may be expected to vary from the assumed 2 : 1 ratio by perhaps ±25%, depending on sediment type, thickness, consolidation state, and basement porosity. Beneath older seafloor with generally thick (greater than a few hundred meters) sediments, the assumed conductivity contrast may generally be significantly less than 2 : 1 (e.g. Hyndman, 1976), especially for increasing sediment thickness. Data from some ocean drill sites even suggests that the mean conductivity of old basalt can be even lower than that of overlying pelagic sediment (e.g. Table 10 and Fig. 33 of Shipboard Scientific Party, 1993), although that is undoubtedly exceptional.

Some reduction of basement thermal conductivity is bound to occur with alteration, however, and older basement rocks altered by seawater circulation are likely to have significantly lower conductivity than fresh basalt. Both of these factors (i.e. increase of sediment thermal conductivity with depth, decrease of basalt conductivity with age or alteration) will generally tend to reduce the conductive anomalies caused by basement topography. However, a simple model test, in which the mean conductivity contrast was reduced (1.5 : 1 rather than 2 : 1) for a flat-top ridge model showed that the reduction in the mean heat flow anomaly over the ridge is significantly less than the conductivity contrast reduction (29 versus 50%, respectively). Hence it seems that the heat flow anomalies are not proportional to the mean conductivity contrast, and this test suggests that the shape or aspect ratio of the two-dimensional ridges is more important than the conductivity contrast in causing steady-state heat flow anomalies for at least some ranges of the modeling parameters.

On the basis of the numerical modeling described above, it seems possible, but unlikely, that thermal conductivity contrasts between basement topography and overlying sediments would result in steady-state surface heat flow anomalies that exceed about 25% of the regional value. For this purpose, Table 13.1 lists the maximum difference ($\pm$) between any heat flow values from their mean for each selected profile. It seems reasonable to assume that systematic heat flow anomalies greater than about $\pm25\%$ of the profile means may result from the effects of basement hydrothermal circulation, rather than (or in addition to) anomalies caused by refraction of conductive heat flux. Based on the above discussion, the heat flow variations due to basement structure for most older seafloor environments seem likely to be less than this threshold anomaly value unless significant advective transport is present. However, an anomaly of this magnitude, or greater, consisting of only a single value has the potential to be caused by a steady-state conductive effect of a nearby basement structure (e.g. Fig. 13.4), even if statistically improbable. In several such cases among the profiles considered (Table 13.1), a question mark is assigned to the evaluation as to whether the anomalous heat flux may be the result of conduction, or whether advection of pore water is more likely.

Isolated anomalous heat flow values for some profiles (e.g. Fig. 13.7) generally tend to be positive. Particularly where the sediments are relatively thick (>1 km), such isolated anomalies seem unlikely to reflect the conductive effects of hydrothermal circulation in rocks beneath the sediments, but could be associated with fluid flow through fractures in the sediments. Where seismic data exist along the profiles considered here, their relatively low resolution does not normally allow visualization of faults in either sediments or basement. The generally positive (rather than negative) single-valued heat flow anomalies along many of the profiles suggest that upward (rather than downward) flow may be associated with

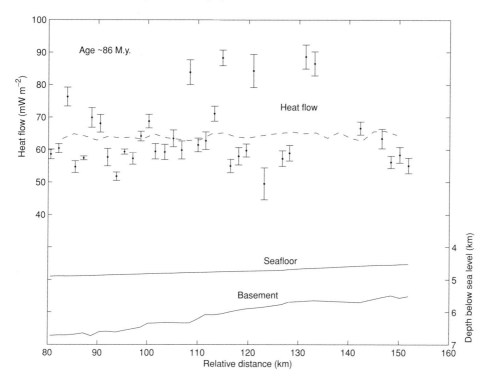

Fig. 13.7 Measured heat flow (above, dots with error bar (±1 SD) symbols) and simplified seafloor structure (below, solid lines) in the southern disturbed zone of the profile north of Oahu, Hawaii, from Harris *et al.* (2000). Calculated heat flux (dashed line) is from a finite element model based on an assumed two-dimensional structure as shown with a ratio of 2 between the assumed uniform and isotropic thermal conductivities of basement and sediment, respectively, normalized to the mean of the measured values. Relative distances are the same as those designated in the published data table. Seafloor (lithospheric) age from Table 13.1.

relatively high-angle faults, at least in the sediments. High-resolution multi-channel seismic (MCS) reflection instrumentation may be capable of detecting such faults in sediments (e.g. Figs. 2 and 3 of Harris *et al.*, 2000), although such data are rarely available along heat flow profiles.

In contrast, several of the selected profiles (e.g. Fig. 13.8) show generally systematic trends in heat flow across them, either with or without more localized anomalies. The most likely cause of such trends would also seem to be basement hydrothermal circulation, even where the total variation across such profiles is less than 25% of the mean value; several of these sites are also assigned question marks in Table 13.1. Where the wavelength of any heat flow pattern is greater than the apparent associated topographic wavelengths, the profile azimuth may be oblique to the principal horizontal axis of a two-dimensional pattern of basement hydrothermal circulation, or reflect lateral advective heat transport in basement that is not directly related to basement topography. Neither of these possibilities is readily

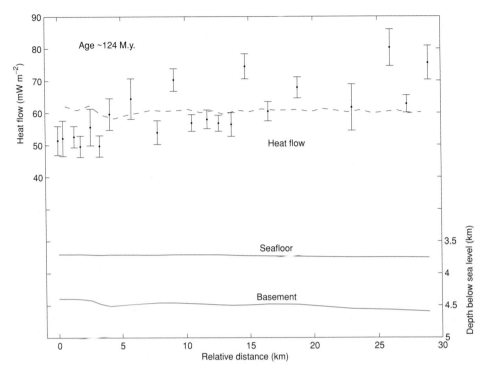

Fig. 13.8 Measured heat flow and seafloor structure versus distance plotted along a linear profile from the Cape Verde hot-spot, northeast Atlantic, Site C of Courtney and White (1986). Symbols and lines have the same meanings as in Fig. 13.7.

evaluated for the selected heat flow profiles, for randomly oriented profiles cannot provide clear evidence for circulation having a significant trend.

13.3.2 Examples of profile characteristics

Space does not permit a detailed description of all the surveys included in Table 13.1, but a few examples serve to show the typical variability that provides the evidence (or lack thereof) for hydrothermal circulation. The data acquired at Site 3 on the Bermuda Rise, North Atlantic (Detrick *et al.*, 1986), is an example of a location where the heat flow values are quite uniform (Fig. 13.9). Two profiles 15–20 km long show no systematic variations in heat flux from the mean values of 51.4 ± 1.8 (SD) and 50.0 ± 2.2 mW m^{-2}, respectively, for the northern and southern profiles. In fact, almost all of the individual values are identical within the estimated measurement uncertainties at this site. The sediment thickness is relatively uniform, up to ~ 1 km (1 s TWTT after the seafloor reflection), and appears to cover the relatively smooth basement completely. Thus there is no evidence for significant hydrothermal circulation at this site having a seafloor age of ~ 140 M.y. at least none that perturbs the heat flux.

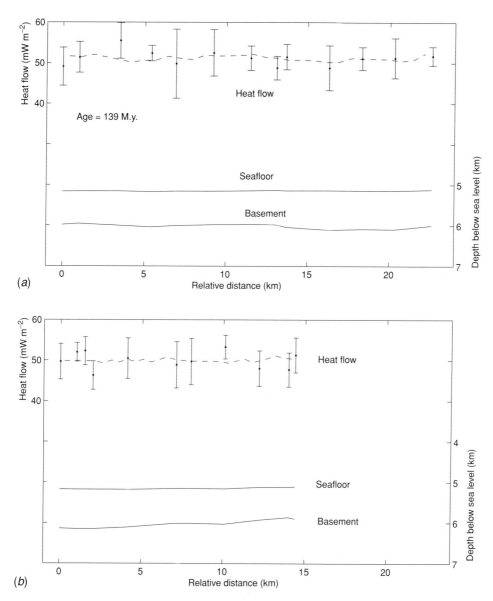

Fig. 13.9 Measured heat flow and seafloor structure versus distance along two profiles in the northwest Atlantic, Bermuda Rise, at Site 3 of Detrick *et al.* (1986): (*a*) northern profile and (*b*) southern profile of the Bermuda Rise. Symbols and lines have the same meanings as in Fig. 13.7.

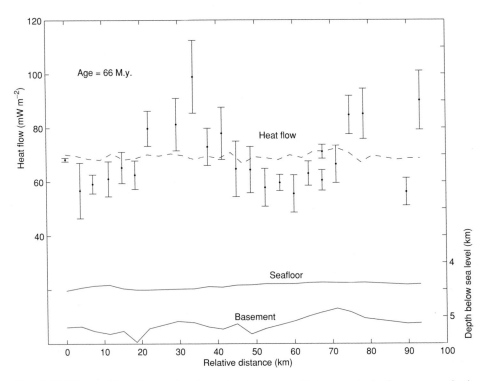

Fig. 13.10 Measured heat flow and seafloor structure versus distance west of a fracture zone in the Indian Ocean (Reunion hot-spot, northern profile west of fracture zone) from profile 1 of Bonneville *et al.* (1997). Symbols and lines have the same meanings as in Fig. 13.7.

On the other hand, the large systematic variations (up to ~43% above the mean value) over the long profile west of an Indian Ocean fracture zone measured by Bonneville *et al.* (1997; Fig. 13.10) seem best explained by active hydrothermal circulation. The eastern end (at the largest relative distance) of this profile abuts the fracture zone. The heat flow pattern appears to have a modest correlation with basement topography, although sediments greater than several hundred meters thick completely cover basement along the profile. The approximate wavelength (40–50 km) of the topography and primary heat flow variation is unusually large; this may be a result of the non-orthogonality of the profile (~125°) relative to the strike of the basement structure. Active hydrothermal circulation would seem to characterize this 66 Ma site in the western Indian Ocean.

The profile of Courtney and White (1986) at their Site C on the Cape Verde Rise (Fig. 13.8) may represent a similar situation. In this case the heat flow varies generally systematically over the profile length of ~30 km, punctuated by single-valued maxima up to >30% above the mean. Sediment thickness is ~700–800 m, and basement topography is apparently relatively flat over the profile. The isolated heat flow maxima may be caused by upward migration of pore water along faults that extend close to the seafloor, although no independent evidence exists for them. The apparent wavelength of the systematic variation over

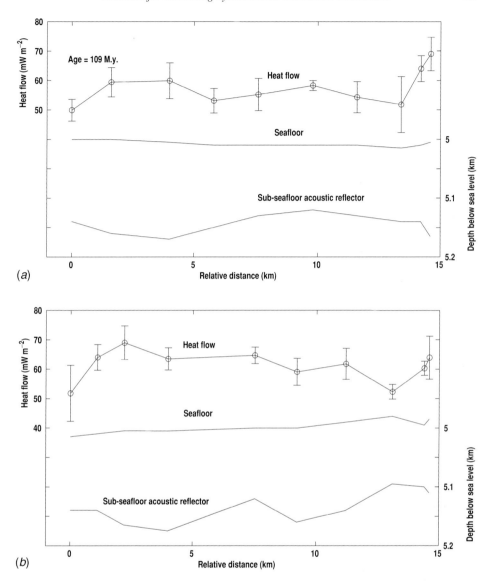

Fig. 13.11 Measured heat flow and seafloor depth versus distance along two profiles on the Hawaiian swell at Site H of Von Herzen *et al.* (1982): (*a*) NW–SE profile and (*b*) S–N profile. Basement depth not determined.

the profile may exceed 50–60 km, again with a possibility that the orientation of the profile azimuth is not normal to the structural trends that control the hydrology.

More questionable profiles in terms of an explanation by hydrothermal circulation are those of Site H of Von Herzen *et al.* (1982) near the Hawaiian ridge (Fig. 13.11). Although some coherent variability may exist, and the mean values along the profiles may differ (57.5 $\pm$ 5.9 versus 61.0 $\pm$ 5.4 mW m^{-2}), most of the values lie within their mutual measurement errors. Total sediment thickness is undetermined, but may be several kilo-meters near the Hawaiian ridge (e.g. Harris *et al.*, 2000). The maximum heat flow value of Profile 1 is 20% greater than the mean, and the minimum for Profile 2 is 15% below its profile mean. Although it seems unlikely that basement topography is responsible, local sedimentation effects at this site, i.e. near the Hawaiian Islands, may cause some of the variability.

A somewhat stronger case for a continuing hydrothermal process appears in the measure-ments at Site 3 of Davis *et al.* (1984) on old ($\sim$160 Ma) seafloor in the northwest Atlantic basin (Fig. 13.12). The sediments are relatively thick ($>$1.5 km) and flat lying above unde-termined (but probably rough) basement topography. Profiles oriented north–south and southwest–northeast have maximum heat flows that are nearly 30 and 25% greater than the respective profile means (45.2 ± 6.4 and 47.4 ± 7.1 mW m^{-2}). The north–south profile has more punctuated single-valued maxima, whereas the southwest–northeast profile is charac-terized by a wider (3-km) heat flow maximum defined by three contiguous measurements 10–15 mW m^{-2} higher than the other values near the center of the 10-km-long profile. Particularly with the relatively thick sediments covering basement, both types of variability seem unlikely to be the results of conductive heat flow anomalies. Pore-water advection would seem probable at this site.

One of the strongest cases for active pore-water advection among all the selected sites is provided by a study on $\sim$85 Ma seafloor of the Nares abyssal plain in the northwest Atlantic by Embley *et al.* (1983). Each of the five short (1–7 km) heat flow profiles shows significant heat flow anomalies in a region of rough basement topography partially covered by sediments up to several hundred meters thick. The main part of the sediment cover consists of turbidites, probably both locally and regionally derived, although a more recent veneer of pelagic ooze up to 50 m thick allowed heat flow measurements to be made on elevated seafloor as well. The highest heat flow anomaly over an apparent basement elevation nearly outcropping at the seafloor is about an order of magnitude greater than the regional mean value (Fig. 13.13). Unfortunately, except for relatively shallow ($\sim$100 m) seafloor-penetrating 3.5-kHz echo sounding, co-located seismic profiles are lacking. But this rather narrow (few tenths of a kilometer) heat flow anomaly has such a large amplitude that it seems almost certain to be caused by active pore-water advection, not simply the result of focusing of conductive heat flux by an unusually shaped basement hill or ridge.

13.4 Discussion

The foregoing types of data were used to identify the processes (conduction versus advec-tion) that are likely to cause the heat flow patterns along each of the selected profiles. As each profile is associated with variable distributions of data, the assignments are partially subjective. The primary criteria are the magnitude of the extreme (maximum or minimum) heat flow value compared to the mean (Table 13.1), and the continuity of that extreme value along the profiles. The anomalies are compared to those expected from conduction alone, as estimated with case-specific (Figs. 13.6–13.10) or generic (Fig. 13.4) modeling; if the

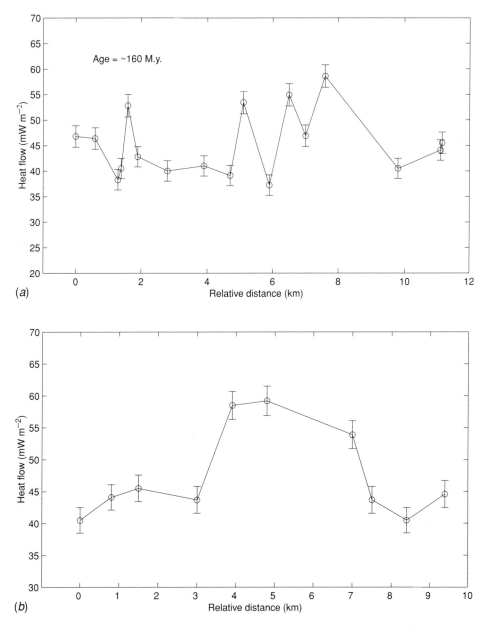

Fig. 13.12 Measured heat flow versus distance along two profiles in the northwest Atlantic basin at Site 3 of Davis *et al.* (1984): (*a*) S–N profile and (*b*) SW–NE profile. Seafloor depths and structure not available.

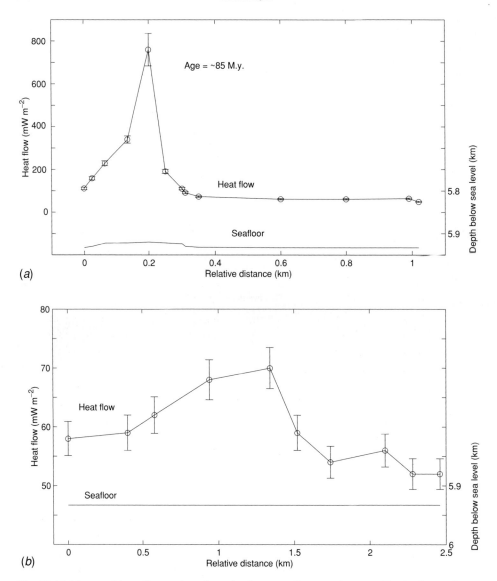

Fig. 13.13 Measured heat flow and seafloor depth versus distance along profiles in the northwest Atlantic at Stations 9 (*a*) and 10 (*b*) of Nares abyssal plain of Embley *et al.* (1983).

variability significantly exceeds that expected by conductive refraction, it is assumed that advection is a significant heat transfer mechanism. Where the profile data are marginal, the classification in Table 13.1 is given a question mark, either favoring circulation or not; these assignments involve some degree of subjectivity, and may be most susceptible to different interpretations. A brief description of the data for each selected site is given in Table 13.1, and the profiles for each site are plotted in the CD-ROM appendix for interested readers.

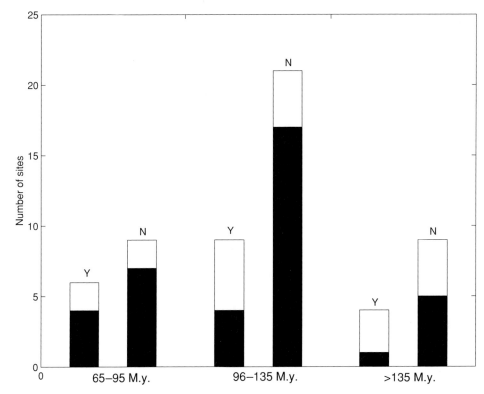

Fig. 13.14 Bar diagram showing numbers of sites in three different seafloor-age bins with or without heat flow evidence for hydrothermal circulation (Table 13.1). "Y" indicates sites that possess variations indicative of hydrothermal circulation, "N" indicates those where signs of advective heat transport are absent, and white parts of bars show the number of sites having more questionable data.

In many cases, the available surveys are relatively closely grouped (Fig. 13.2) as a result of a focus on particular geological or geophysical problems. The fact that many of these surveys are located on or near hot-spots may bias the evidence for the existence of hydrothermal circulation for the entire seafloor. However, the anomalous conductive heat flow associated with most hot-spots appears to exceed the normal heat flux versus age relationship by only 10–25% or less (e.g. Von Herzen *et al.*, 1989). The main effect of hot-spots is to elevate the seafloor over broad regions by deep thermal or dynamic uplift without significant reheating or thinning of the lithosphere. It seems unlikely that a threshold heat flux value controls the existence of hydrothermal circulation for most of ocean crust; rather, changes in permeability and of boundary conditions (e.g. thickness and continuity of the relatively impermeable sediment cover) of the ocean crust are probably more important. Additional measurements and modeling are needed to clarify these issues.

To summarize the available data, the results of Table 13.1 are grouped into three different seafloor-age bins: 65–95, 96–135, and >135 M.y. (Fig. 13.14). The compilation suggests at least two general deductions: (i) hydrothermal circulation probably continues at some locations on the oldest seafloor, and (ii) the fraction of sites with likely or probable continuing circulation to the total number of sites decreases with increasing age. This fraction as determined by all compiled data varies from 0.4 for the 65–95 M.y. age range, to about 0.3 for both the 96–135 and >135 M.y. sites. For the clearest results (without question marks, Table 13.1), the fractions vary systematically as 0.36, 0.19, and 0.17 over the three increasing age ranges, respectively. Although these statistics are based on a relatively small number of sites that are probably not representative of the entire seafloor, they suggest that there may be no sharp cutoff in hydrothermal circulation with age of the seafloor or completeness of the sediment cover. The observation that the heat flux approaches the theoretical conductive value for seafloor ages >50–60 M.y. (Stein and Stein, 1992) suggests that most of the circulation after this age may be confined to a permeable basement beneath sediments, without significant venting or advection of heat directly through the seafloor. Obviously more detailed geothermal data in other older seafloor regions would be useful to confirm this suggestion, as well as more numerous and systematically distributed data from other relevant disciplines such as pore-water and basement rock geochemistry.

Although quantitative modeling of pore-water flow has not been emphasized here, some heuristic estimates of flow velocities may be obtained from dimensional analysis of the heat flow patterns where they exist. At least two physical models may be appropriate for such estimates. One is to assume that, as has been done for heat flow anomalies on young crust (e.g. Davis and Lister, 1977), they result from the variations in depth below the seafloor to an essentially isothermal basement. In this case, the lateral pore-water flow velocity (V) must be sufficient to dominate the loss of heat by vertical diffusion through the sediments, or approximately $V > \alpha/h$ (α is thermal diffusivity, h is sediment thickness). Obviously the greatest velocities are required for the thinnest sediment layer. For $h_{min} < 100$ m on old crust, $V_{max} > 0.1$ M.y.$^{-1}$ for $\alpha = 3 \times 10^{-7}$ m^2 s^{-1}. For this model, heat flow variations should correlate (inversely) with sediment thickness.

Another estimate is based on the lateral distances between heat flow anomalies on the seafloor, typically 10–20 km, assuming that they represent cellular convection in the basement rock. In this case, V_{max} must not exceed values that would significantly attenuate or eliminate the basement thermal anomalies before they are conducted to the seafloor through sediments, over an approximate time scale $t = h^2/\alpha$. For a typical $h = 1$ km for old seafloor, $t = 10^5$ years, resulting in $V_{max} < 0.1$ m yr^{-1}. For greater lateral distances between heat flow anomalies, V_{max} is correspondingly increased. It is interesting, but probably coincidental, that similar limiting maximum pore-water velocities are obtained for both models, although it is a lower bound for the isothermal basement model and an upper bound for that based on typical lateral distances between seafloor heat flow anomalies. Some of the selected surveys with the largest anomalies appear to have co-located basement highs, seemingly favoring an isothermal basement model, although most of the significant anomalies for the selected profiles are not obviously associated with basement elevations. The results of both estimates suggest that fluid velocities of order 0.1 m yr^{-1} could cause many of the anomalies, but these are probably uncertain by ±1 order of magnitude, depending on the actual model, for which more information is needed for most of the surveys.

13.5 Conclusions

Compared to expected conductive anomalies, detailed marine heat flow surveys suggest that hydrothermal circulation may continue for some sites in the main ocean basins in old (>60 Ma) lithosphere at thermally significant rates. A partially subjective but conservative analysis shows that the fraction of sites with evidence for circulation decreases from about 0.4 over the 65–95 M.y.-seafloor-age range to about 0.2–0.3 for seafloor created 96–135 and >135 Ma. It is possible that the conservative modeling and assumptions made will cause these fractions to be underestimated. Although the data suggest that hydrothermal circulation may continue but decrease in intensity with age for ocean crust created ~60–175 Ma, the geographic distribution of the existing measurements is inadequate to determine the extent and exact nature of this seafloor-age variation. Most of these sites are completely covered with thick (more than a few hundred meters) sediments, so that the circulation is probably confined to the permeable upper basement rocks without significant hydrological connection to the ocean. If basement topography is lacking or subdued, it may be difficult to use lateral heat flux variations to detect or estimate the magnitude of hydrothermal circulation (e.g. Davis *et al.*, 1996). However, evidence is generally inadequate at the selected sites to determine why some sites indicate continuing circulation, but not others. Multi-disciplinary investigations, particularly including co-located seismic profiles and detailed heat flow measurements over buried basement topography of significant relief, should optimize the opportunities to quantify hydrothermal circulation for older seafloor sites. If possible, the seismic surveys should be sufficiently detailed to determine the three-dimensional configuration of at least the sediment–basement interface to correlate with the surface heat flow distribution.

The evidence in this paper for such continuing circulation is based almost entirely on geophysical data, primarily detailed geothermal measurements and seismic profiles. Few, if any, geochemical measurements exist at the selected sites, and geochemical data have the potential to determine anomalies that may result from lower advective velocities, primarily as a result of the smaller diffusion coefficients of chemical species compared to thermal diffusion coefficients for seafloor rock and sediment. This fact reinforces the desirability of including multi-disciplinary studies to detect the possibility of relatively weak advection that may continue in older seafloor environments. Downhole measurements and samples from drilling at selected locations for these objectives at older seafloor sites should also provide important data on the sub-seafloor depths at which continuing circulation exists.

Acknowledgments

I am grateful for the initial stimulation toward this topic by the Ocean Hydrology Workshop, organized by Earl Davis and Harry Elderfield under the International Lithosphere

Panel and supported by JOI, and hosted by Andy Fisher at University of California, Santa Cruz (UCSC), in December 1998. I am also indebted to Andy Fisher for initial numerical models of the thermal effects of two-dimensional basement ridges covered by sediments, particularly of the Madeira abyssal plain, and for steering me toward other relevant modeling code. Rob Harris and Keith Louden provided initial insightful reviews, and Earl Davis gave valuable guidance and detailed comments on several drafts of the paper. I appreciate the permission of Carol Stein and the American Geophysical Union to reproduce Fig. 13.1. The Earth Sciences Department at UCSC provided hospitality during visits in 2002–2003 when this paper was mostly prepared and revised. The Woods Hole Oceanographic Institution provided travel and other support.

References

Anderson, R. N. and Langseth, M. G. 1977. The mechanisms of heat transfer through the floor of the Indian Ocean. *J. Geophys. Res.* **82**: 3,391–3,409.

Bethke, C. M., Harrison, W. J., Upson, C., and Altaner, S. P. 1988. Supercomputer analysis of sedimentary basins. *Science* **239**: 261–267.

Birch, F. 1967. Low values of oceanic heat flow. *J. Geophys. Res.* **72**: 2,261–2,262.

Bird, R. T. and Pockalny, R. A. 1994. Late Cretaceous and Cenozoic seafloor and oceanic basement roughness: spreading rate, crustal age and sediment thickness correlations. *Earth Planet. Sci. Lett.* **123**: 239–254.

Bonneville, A., Von Herzen, R. P., and Lucazeau, F. 1997. Heat flow over Reunion hot-spot track: additional evidence for thermal rejuvenation of oceanic lithosphere. *J. Geophys. Res.* **102**: 22,731–22,747.

Carslaw, H. S. and Jaeger, J. C. 1959. *Conduction of Heat in Solids*, 2nd Edn. Oxford: Oxford University Press, 510 pp.

Courtney, R. C. and Recq, M. 1986. Anomalous heat flow near the Crozet Plateau and mantle convection. *Earth Planet. Sci. Lett.* **79**: 373–384.

Courtney, R. C. and White, R. S. 1986. Anomalous heat flow and geoid across the Cape Verde Rise: evidence for dynamic support from a thermal plume in the mantle. *Geophys. J. Roy. Astron. Soc.* **87**: 815–867.

Davis, E. E. and Lister, C. R. B. 1977. Heat flow measured over the Juan de Fuca Ridge: evidence for widespread hydrothermal circulation in a highly heat transportive crust. *J. Geophys. Res.* **82**: 4,845–4,860.

Davis, E. E., Lister, C. R. B., and Sclater, J. G. 1984. Toward determining the thermal state of old ocean lithosphere: heat flow measurements from the Blake–Bahama outer ridge, north-western Atlantic. *Geophys. J. Roy. Astron. Soc.* **78**: 507–545.

Davis, E. E., Chapman, D. S., and Forster, C. B. 1996. Observations concerning the vigor of hydrothermal circulation in young oceanic crust. *J. Geophys. Res.* **101**: 2,927–2,942.

Davis, E. E., Chapman, D. S., Villinger, H., Robinson, S., Grigel, J., Rosenberger, A., and Pribnow, D. 1997. Seafloor heat flow on the eastern flank of the Juan de Fuca Ridge: data from "FLANKFLUX" studies through 1995. In *Proceeding of the Ocean Drilling Program, Initial Reports*, Vol. 168, eds. E. E. Davis, A. T. Fisher, J. V. Firth, *et al.* College Station, TX: Ocean Drilling Program, pp. 23–33.

Davis, E. E., *et al.* 1999. Regional heat flow variations across the sedimented Juan de Fuca Ridge eastern flank: constraints on lithospheric cooling and lateral hydrothermal heat transport. *J. Geophys. Res.* **104**: 17,675–17,688.

Detrick, R. S., Von Herzen, R. P., Parsons, B., Sandwell, D., and Dougherty, M. 1986. Heat flow observations on the Bermuda Rise and thermal models of mid-plate swells. *J. Geophys. Res.* **91**: 3,701–3,723.

Embley, R. W., Hobart, M. A., Anderson, R. N., and Abbott, D. 1983. Anomalous heat flow in the northwest Atlantic: a case for continued hydrothermal circulation in 80-M.y. crust. *J. Geophys. Res.* **88**: 1,067–1,074.

Fisher, A. T. and Becker, K. 1995. Correlation between seafloor heat flow and basement relief: observational and numerical examples and implications for upper crustal permeability. *J. Geophys. Res.* **100**: 12,641–12,657.

2000. Channelized fluid flow in oceanic crust reconciles heat-flow and permeability data. *Nature* **403**: 71–74.

Fisher, A. T., Davis, E. E., Hutnak, M., *et al.* 2003. Hydrothermal recharge and discharge across 50 km guided by seamounts on a young ridge flank. *Nature* **421**: 618–621.

Galson, D. A. and Von Herzen, R. P. 1981. A heat flow survey on anomaly M0 south of the Bermuda Rise. *Earth Planet. Sci. Lett.* **53**: 296–306.

Goff, J. A. 1991. A global and regional stochastic analysis of near-ridge abyssal hill morphology. *J. Geophys. Res.* **96**: 21,713–21,737.

Green, K. E., Von Herzen, R. P., and Williams, D. L. 1981. The Galapagos spreading center at 86 degrees W: a detailed geothermal field study. *J. Geophys. Res.* **86**: 979–986.

Harris, R. N., Von Herzen, R. P., McNutt, M. K., Garven, G., and Jordahl, K. 2000. Submarine hydrogeology of the Hawaiian archipelagic apron, 1. Heat flow patterns north of Oahu and Maro Reef. *J. Geophys. Res.* **105**: 21,353–21,369.

Hartline, B. K. and Lister, C. R. B. 1981. Topographic forcing of supercritical convection in a porous medium such as the oceanic crust. *Earth Planet. Sci. Lett.* **55**: 75–86.

Hyndman, R. D. 1976. Seismic structure of the oceanic crust from deep drilling on the mid-Atlantic ridge. *Geophys. Res. Lett.* **3**: 201–204.

Hyndman, R. D., Davis, E. E., and Wright, J. A. 1979. The measurement of marine geothermal heat flow by a multipenetration probe with digital acoustic telemetry and insitu thermal conductivity. *Mar. Geophys. Res.* **4**: 181–205.

Lachenbruch, A. H. 1968. Rapid estimation of the topographic disturbance to superficial thermal gradients. *Rev. Geophys.* **6**: 365–400.

Lister, C. R. B. 1970. Measurement of in situ sediment conductivity by means of a Bullard-type probe. *Geophys. J. Roy. Astron. Soc.* **19**: 521–532.

1972. On the thermal balance of a mid-ocean ridge. *Geophys. J. Roy. Astron. Soc.* **26**: 515–535.

1979. The pulse-probe method of conductivity measurement. *Geophys. J. Roy. Astron. Soc.* **57**: 451–461.

Lister, C. R. B., Sclater, J. G., Davis, E. E., Villinger, H., and Nagihara, S. 1990. Heat flow maintained in ocean basins of great age: investigations in the north-equatorial west Pacific. *Geophys. J. Int.* **102**: 603–630.

Louden, K. E., and Wright, J. A. 1989. Marine heat flow data: a new compilation of observations and brief review of its analysis. In *Handbook of Seafloor Heat Flow*, eds. J. A. Wright and K. E. Louden. Boca Raton, FL: CRC Press, pp. 3–67.

Louden, K. E., Wallace, D. O., and Courtney, R. C. 1987. Heat flow and depth versus age for the Mesozoic northwest Atlantic Ocean: results from the Sohm abyssal plain and implications for the Bermuda Rise. *Earth Planet. Sci. Lett.* **83**: 109–122.

Muller, R. D., Roest, W. R., Royer, J.-Y., Gahagan, L. M., and Sclater, J. G. 1997. Digital isochrons of the world's ocean floor. *J. Geophys. Res.* **102**: 3,211–3,214.

Nagihara, S., Sclater, J. G., Beckley, L. M., Behrens, E. W., and Lawver, L. A. 1992. High heat flow anomalies over salt structures on the Texas continental slope, Gulf of Mexico. *Geophys. Res. Lett.* **19**: 1,687–1,690.

Noel, M. 1985. Heat flow, sediment faulting and porewater advection in the Madeira abyssal plain. *Earth Planet. Sci. Lett.* **73**: 398–406.

Noel, M. and Hounslow, M. 1988. Heat flow evidence for hydrothermal convection in Cretaceous crust of the Madeira abyssal plain. *Earth Planet. Sci. Lett.* **90**: 77–86.

Parsons, B. and Sclater, J. G. 1977. An analysis of the variation of ocean floor bathymetry and heat flow with age. *J. Geophys. Res.* **82**: 803–827.

Ruppel, C., Von Herzen, R. P., and Bonneville, A. 1995. Heat flux through an old (~175 Ma) passive margin: offshore southeastern USA. *J. Geophys. Res.* **100**: 20,037–20,057.

Sclater, J. G., Corry, C. E., and Vacquier, V. 1969. In situ measurement of the thermal conductivity of ocean-floor sediments. *J. Geophys. Res.* **74**: 1,074–1,081.

Sclater, J. G., Jones, E. J. W., and Miller, S. P. 1970. The relationship of heat flow, bottom topography and basement relief in Peake and Freen deeps, northeast Atlantic. *Tectonophysics* **10**: 283–300.

Shipboard Scientific Party 1993. Site 872. In *Proceeding of the Ocean Drilling Program, Initial Reports*, Vol. 144, eds. I. Premoli Silva, J. Haggerty, F. Rack, *et al.* College Station, TX: Ocean Drilling Program, pp. 105–144.

Smith, D. K. 1988. Shape analysis of Pacific seamounts. *Earth Planet. Sci. Lett.* **90**: 457–466.

Stein, C. A. and Stein, S. 1992. A model for the global variation in oceanic depth and heat flow with lithospheric age. *Nature* **359**: 123–129.

Stein, C. A., Stein, S., and Pelayo, A. 1995. Heat flow and hydrothermal circulation. In *Seafloor Hydrothermal Systems: Physical, Chemical, Biological, and Geological Interactions*, Monograph 91, eds. S. E. Humphris, R. A. Zierenberg, L. S. Mullineaux, and R. E. Thomson. Washington, DC: American Geophysical Union, pp. 425–445.

van Andel, Tj. H., Von Herzen, R. P., and Phillips, J. D. 1971. The VEMA fracture zone and the tectonics of transverse shear zones in oceanic crustal plates. *Mar. Geophys. Res.* **1**: 261–283.

Von Herzen, R. P. 1987. Measurement of oceanic heat flow. In *Methods of Experimental Physics–Geophysics*, Vol. 24B, eds. C. Sammis and T. Henyey. San Diego, CA: Academic Press, pp. 227–263.

Von Herzen, R. P., Detrick, R. S., Crough, S. T., Epp, D., and Fehn, U. 1982. Thermal origin of the Hawaiian swell: heat-flow evidence and thermal models. *J. Geophys. Res.* **87**: 6,711–6,723.

Von Herzen, R. P., Cordery, M. J., Detrick, R. S., and Fang, C. 1989. Heat flow and the thermal origin of hot spot swells: the Hawaiian swell revisited. *J. Geophys. Res.* **94**: 13,783–13,799.

Von Herzen, R. P., Goldberg, D., and Manghnani, M. 1992. Physical properties and logging of the lower oceanic crust: Hole 735B. In *The Indian Ocean: A Synthesis of*

Results from the Ocean Drilling Program, Monograph **70**, eds. R. A. Duncan, *et al.* Washington, DC: American Geophysical Union, pp. 41–56.

Von Herzen, R. P., Ruppel, C., Molnar, P., Nettles, M., Nagihara, S., and Ekstrom, G. 2001. A constraint on the shear stress at the Pacific–Australian plate boundary from heat flow and seismicity at the Kermadec forearc. *J. Geophys. Res.* **106**: 6,817–6,833.

Wang, K., He, J., and Davis, E. E. 1997. Influence of basement topography on hydrothermal circulation in sediment-buried igneous oceanic crust. *Earth Planet. Sci. Lett.* **146**: 151–164.

Williams, D. L., Von Herzen, R. P., Sclater, J. G. and Anderson, R. N. 1974. The Galapagos spreading center: lithospheric cooling and hydrothermal circulation. *Geophys. J. Roy. Astron. Soc.* **38**: 587–608.

Wolery, T. J. and Sleep, N. H. 1976. Hydrothermal circulation and geochemical flux at mid-ocean ridges. *J. Geol.* **84**: 249–275.

Yamano, M. and Uyeda, S. 1989. Heat flow in the western Pacific. In *Handbook of Seafloor Heat Flow*, eds. J. Wright and K. Louden. Boca Raton, FL: CRC Press, pp. 277–303.

Part IV

Geochemical state and water–rock reactions

14

Alteration and mass transport in mid-ocean ridge hydrothermal systems: controls on the chemical and isotopic evolution of high-temperature crustal fluids

W. E. Seyfried, Jr. and Wayne C. Shanks, III

14.1 Introduction

Marine hydrothermal systems represent one of the most dynamic and diverse environments on Earth. Fueled by the crystallization and cooling of sub-seafloor magma, these systems provide a key mechanism by which heat and mass are exchanged between the oceanic crust and underlying mantle and overlying hydrosphere. The rate of magmatic heat supply that drives seawater advection is ultimately constrained by crustal production, which is a sensitive function of spreading rate, and involves both latent heat (heat released when new crust is formed) and heat associated with lithospheric cooling (Mottl and Wheat, 1994; Pelayo *et al.*, 1994; Stein and Stein, 1994; Stein *et al.*, 1995). Magmatic processes (intrusion and crystallization) are largely responsible for hydrothermal circulation at the ridge axis, while off-axis low-temperature circulation and venting is more likely sustained by conductive cooling of the crust. Lister (1982) termed these two very different hydrothermal systems, "active" and "passive," respectively. Although it is increasingly clear that the passive system dominates total power output, it is the active system that contributes more to the large compositional changes in seawater chemistry manifested by black smoker fluids at mid-ocean ridges.

Owing to their extreme compositional variability, axial vent fluids must be associated with corresponding changes in the composition of the oceanic crust. Thus, it is not surprising that numerous studies of hydrothermally altered rocks recovered from the ocean crust and ophiolite outcrops on land have revealed large-scale changes in mineralogy, chemistry, and isotopic composition as a function of the inferred depth in the fossil hydrothermal systems (Humphris and Thompson, 1978; Ito *et al.*, 1983; Harper *et al.*, 1988; Kimball, 1988; Gillis and Thompson, 1993; Gillis *et al.*, 1993; Alt, 1995; Alt and Teagle, 2000; Bach *et al.*, 2001). The challenge, however, has long been to link vent fluids to their counterpart in the sub-seafloor from which the fluids were derived; a region that is often referred to as the "hydrothermal reaction zone."

The hydrothermal reaction zone is simply that portion of the fluid flow system where rock and mineral components are able to buffer the fluid, and factors of state, such as

---

Hydrogeology of the Oceanic Lithosphere, eds. E. E. Davis and H. Elderfield. Published by Cambridge University Press.
© Cambridge University Press 2004.

temperature and pressure, play an increasingly important role in controlling mass transfer processes. Even chloride, which is entirely fluid derived, is sensitive to temperature and pressure change due to constraints imposed by phase equilibria in the $NaCl–H_2O$ system. The combination of increasing temperature with attendant effects on reaction rates, together with the existence of thermally and/or tectonically induced crack networks, tend to facilitate mass transfer and enhance approach to equilibrium. The modest shifts of strontium isotopes and apparent lack of Mg enrichment in bulk mineral assemblages in high-temperature reaction zones (Alt *et al.*, 1996; Teagle *et al.*, 1998), indicate rock-dominated conditions with increasing depth in sub-seafloor hydrothermal systems. Moreover, whole rock $\delta^{18}O$ values decrease, as one would expect for a system at high temperatures (>325 °C), where the oxygen isotope inventory is controlled by reactions between alteration minerals and compositionally evolved seawater. These conditions (high temperatures, low fluid/rock ratio) are generally consistent with the relative enrichment of incompatible trace elements in vent fluids, in keeping with results of trace element partitioning studies (Seyfried *et al.*, 1984; Berndt and Seyfried, 1990).

This chapter will attempt to build on recent advances in experimental, theoretical, isotopic, and field-based studies that have resulted in an improved understanding of the complex feedback between fluids and minerals in submarine geothermal systems. Where it was once assumed that vent fluids remained largely unchanged on decadal scales in terms of temperature and composition, it is now well recognized that episodic tectonic and magmatic events can trigger processes that profoundly affect the temporal evolution of vent fluids and the composition of rocks and minerals with which they coexist. Phase separation processes, in particular, which can be linked to such events, can have a dominating influence on vent fluid chemistry (Von Damm *et al.*, 1997; Von Damm, 2000). These processes respond to a rich array of variables that interrelate in still uncertain ways and can be elucidated only by an integration of geological, geochemical, geophysical, and, increasingly, microbiological studies.

The approach we have taken is to focus on selective species, including isotopes, which have been shown to play dominant roles in hydrothermal alteration processes at mid-ocean ridges. Chloride, sodium, calcium, iron, H_2, and H_2S, and isotopes of hydrogen, carbon, boron, lithium, oxygen, and sulfur, for example, are exceedingly sensitive indicators of rock–fluid interaction processes at elevated temperatures and pressures. Taken together, these parameters allow constraints to be imposed on pH and redox (reduction–oxidation) conditions, which are, along with temperature and pressure are mineral and fluid source terms: the master variables in all geochemical systems. Although our efforts are largely restricted to basaltic and ultramafic rock systems at mid-ocean ridges, the highly integrated approach we emphasize can be applied to virtually all hydrothermal systems.

14.2 Compositional characteristics of hot spring vent fluids at mid-ocean ridges

14.2.1 Halide variability

The most conspicuous variation in mid-ocean ridge hydrothermal vent fluid chemistry is the wide range of dissolved halogen (chloride, bromide) concentrations (Fig. 14.1). Considering

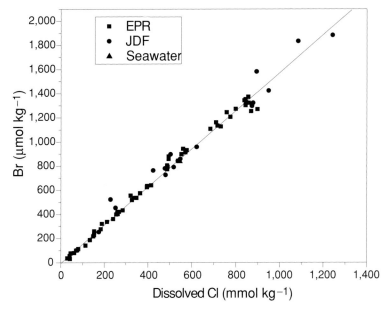

Fig. 14.1 Dissolved chloride versus bromide for hydrothermal vent fluids on the East Pacific Rise and Juan de Fuca Ridge (Michard *et al.*, 1984; Von Damm *et al.*, 1985; Von Damm and Bischoff, 1987; Bowers *et al.*, 1988; Campbell *et al.*, 1988a; Evans *et al.*, 1988; Campbell and Edmond, 1989; Palmer, 1992; Butterfield and Massoth, 1994; Butterfield *et al.*, 1994; You *et al.*, 1994; Trefry *et al.*, 1994; Von Damm, 1995; Von Damm *et al.*, 1995; Charlou *et al.*, 1996b; Oosting and Von Damm, 1996; Von Damm *et al.*, 1997; Von Damm, 2000). The large and systematic range in concentration of both species is suggestive of phase separation processes, where owing to the effects of temperature and pressure, seawater separates into vapor and brine components (see text). Berndt and Seyfried (1990) and Berndt and Seyfried (1997) showed that phase separation has little effect on Br/Cl ratio. Thus, that these data reveal a linear correlation that passes through seawater is not surprising. Mineral–fluid reaction and/or analytical effects can cause departures from the linear relation (see text). This is especially important with respect to halite dissolution, which has been suggested as a means of lowering Br/Cl in vapor-rich vents at EPR 9–10° N (Oosting and Von Damm, 1996).

that the source fluid driving hydrothermal circulation in sub-seafloor hydrothermal systems is seawater, the variability of chloride and bromide concentrations has long been of interest (Edmond *et al.*, 1979). The large chloride variability and relatively constant Br/Cl ratio in vent fluids (Fig. 14.1) is most consistent with phase separation, where temperature–pressure changes cause seawater to separate into vapor and brine components. At temperatures below the critical point of seawater (408 °C) the process of phase separation is equivalent to boiling; a lower-density vapor separates from a higher-density liquid. For the two component NaCl–H$_2$O system, however, phase separation is also possible at temperatures above the critical point, in which case a small amount of brine separates from the liquid, causing the liquid to become more vapor like. This process is referred to as super-critical phase separation (Bischoff and Pitzer, 1989; Bischoff, 1991), while the actual separation mechanism is often referred to as brine condensation (Fournier, 1987; Bischoff and Rosenbauer, 1987). The composition of the coexisting vapor and brine is an explicit function of temperature,

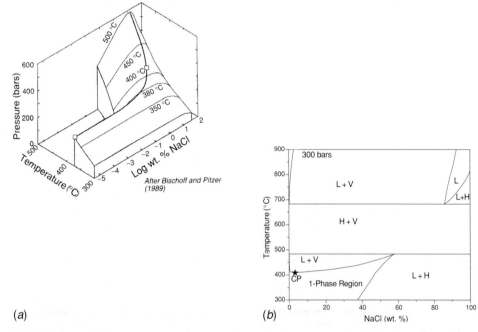

(a)

(b)

Fig. 14.2 Phase relations in the NaCl–H$_2$O system at elevated temperatures and pressures. The effect of temperature, pressure, and total dissolved chloride on two-phase behavior and criticality for the range of conditions applicable to hydrothermal fluids at mid-ocean ridges (Bischoff and Pitzer, 1989) is shown in (a), while phase relations as a function of temperature and at a specific pressure (300 bars) are depicted in (b). These data show that at 300 bars halite has a broad stability field in temperature–composition space. Conditions such as depicted in (b) may be encountered during diking events in sub-seafloor systems. Data for construction of (b) are from Anderko and Pitzer (1993).

pressure, and bulk fluid composition (Bischoff and Pitzer, 1989), which gives rise to a three-dimensional saddle-shaped topology in P–T–X space (Fig. 14.2a). For a given NaCl composition, increasing temperature and decreasing pressure can result in extreme variability in composition of vapors and conjugate brines, which eventually may intersect the halite stability field. For example, phase equilibria in the NaCl–H$_2$O system show that at 300 bars, temperatures between 500 and 700 °C can result in halite formation (Fig. 14.2b). At these low pressures, the coexisting vapor contains less than 0.01% NaCl, whereas just prior to the elimination of two liquid phases, the brine is predicted to contain approximately 40% NaCl, which gives some sense of the compositional variability that is possible due to phase separation in sub-seafloor hydrothermal systems. At 300 bars and temperatures greater than 700 °C, two phases again dominate, as halite is restricted to a relatively small region in NaCl–H$_2$O space (Fig. 14.2b). Although pressures and temperatures this low and high, respectively, may not be typical, it is likely that conditions sufficient to generate highly saline brines and render halite stable do exist, especially following diking events that impact hydrothermal systems at relatively shallow regions of the ocean crust. Indeed, Oosting and

Von Damm (1996), Berndt and Seyfried (1997) and Von Damm (2000) suggested halite dissolution to account for the low Br/Cl ratios of vapor-rich fluids at EPR 9–10° N.

The relatively high temperatures required for phase separation and chloride variability of vent fluids is often at odds with the measured temperatures of these fluids. This suggests that sub-seafloor mixing between vapors and brines and evolved seawater subsequent to phase separation processes, and/or conductive cooling effects, are required (Von Damm, 1988; Berndt and Seyfried, 1990; Berndt *et al.*, 1996b).

14.2.2 Dissolved cations

Because seawater is the primary agent of hydrothermal alteration in sub-seafloor reaction zones, it can be expected that phase-separation-induced changes in chloride will result in equally significant changes in most other species due to mass and charge balance constraints. Such would certainly be the case for fluid in the absence of rock components. For example, based on experiments in the super-critical region, Berndt and Seyfried (1990) showed that Na and Cl partition similarly between vapors and brines. In sub-seafloor reaction zones, however, phase separation occurs in the presence of rock components, and thus key elements in the fluid are constrained by mineral solubility effects. Indeed, representative vent fluid data for Na, Ca, Fe, and H_2S from EPR and JDF localities show that the dissolved concentrations of these species express dramatic changes with dissolved chloride (Fig. 14.3).

Dissolved Na is taken up by primary and secondary minerals in the host rock as indicted by the divergence from the Na/Cl ratio in seawater, especially for high-chloride vent fluids (Fig. 14.3*a*). This becomes even clearer when Na data are normalized to chloride and plotted against Cl (Fig. 14.3*b*). This normalization procedure has been emphasized as a means to distinguish better vapor–brine partitioning from rock–fluid interaction effects (Von Damm *et al.*, 1997; Von Damm, 2000). Although there is considerable scatter in the data for the low-Cl fluids, leaching of Na from the rock or halite dissolution is indicated by the Na/Cl ratios being greater than seawater (Von Damm, 2000). The generally lower Na/Cl ratios for the JDF relative to EPR vent fluids may be caused by slight differences in plagioclase composition between the two ridge segments and/or constraints imposed by temperature and pressure variability.

The relatively high Ca concentrations in vent fluids (Fig. 14.3*c*) are largely a function of the Ca-rich nature of the oceanic crust (plagioclase, diopside), dissolved chloride concentration (phase separation effects), as well as the relative abundances of Na and Mg in seawater, which effectively exchange with Ca during hydrothermal alteration. In particular, Mg removal from seawater largely takes place in the form of chlorite precipitation, which generates acidity and enhances dissolution of Ca-bearing minerals, especially in the recharge zone of sub-seafloor hydrothermal systems (Seyfried and Bischoff, 1981; Seyfried, 1987). At sufficiently high temperatures, however, Ca released to the fluid at relatively low temperatures is effectively taken up by calcic secondary phases, such as epidote solid solutions and amphibole, and provides a primary control on pH in the coexisting fluid at greater crustal depths. As before, vent fluids considered here (Fig. 14.3*c*) are largely from eastern

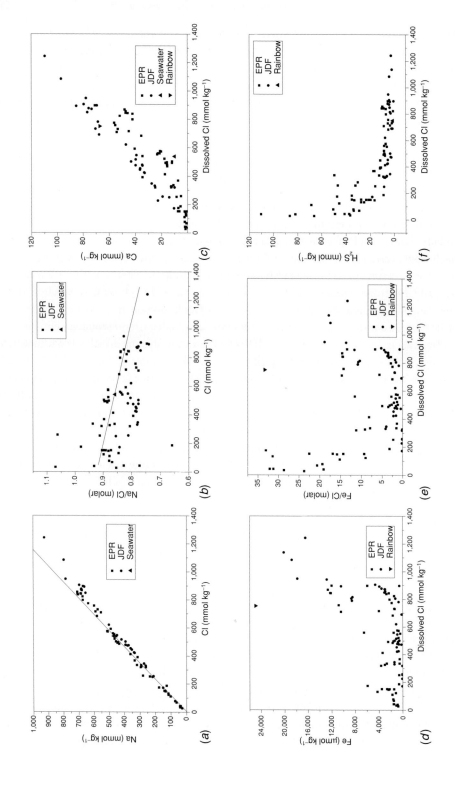

Pacific localities, although the ultramafic-hosted Rainbow system (MAR) is also included. In the Rainbow fluids, high Ca concentrations are almost certainly derived from dissolution of clinopyroxene components in the host rocks.

Other alkaline earth elements, such as Sr, Ba, and beryllium are moderately soluble in aqueous fluids at elevated temperatures and pressures, and tend to be enriched relative to seawater in hot spring vent fluids (Butterfield *et al.*, 1994; Von Damm, 1995; Seyfried *et al.*, 2003).

Very few species in hydrothermal vent fluids at mid-ocean ridges have been investigated as thoroughly as dissolved Fe. Reasons for this include the relative abundance of Fe in chimney and mound structures through which hot spring fluids vent at the seafloor and the importance of iron sulfides in formation of analogous ancient massive sulfide ore deposits. Fe-sulfide mineralization in chimney structures, however, is due mainly to pH and redox changes induced by seawater mixing (Janecky and Seyfried, 1984). Because dissolved Fe in end-member hydrothermal fluids is a sensitive function of temperature, pressure, pH, dissolved chloride, and redox conditions, however, it can provide important clues to phase equilibria controls in sub-seafloor reaction zones (Ding and Seyfried, 1992).

Dissolved Fe from representative vent fluids increases with increasing chloride, much the same as Ca, Sr, and other species that form strong chloride complexes (Fig. 14.3*d*). Looking more closely at these data, however, two obvious anomalies are apparent: the relatively high Fe concentration for the Rainbow vent fluids (Douville *et al.*, 2002); and, the high dissolved Fe in the low chloride vents from EPR, 9–10° N (Von Damm, 2000). Both of these observations become even clearer when Fe/Cl ratio is plotted against chloride (Fig. 14.3*e*), which clearly shows that processes other than chloride complexing are

←——

Fig. 14.3 Dissolved Na, Ca, Fe, and H_2S versus chloride (*a, c, d, f*) in vent fluids from eastern Pacific localities and the Rainbow hydrothermal system on the Mid-Atlantic Ridge (see Fig. 14.1 for specific references). Data for the Rainbow system are from Douville *et al.* (2002). In general, the decrease in dissolved Na with increasing chloride (*a*) likely results from mineralization effects related to plagioclase recrystallization reactions. The solid line in (*a*) is drawn from the origin and through seawater to provide a better measure of the magnitude of the Na decrease with increasing dissolved chloride. This can be illustrated better, however, by a chloride normalization plot, which allows rock–fluid interaction effects to be distinguished from those caused by phase separation (*b*); see text and Von Damm (2000). A linear regression line is shown for comparison. Although Na fixation is observed for both EPR and JDF vent fluids, the extent of this appears to be slightly greater for the JDF system. Na fixation can be linked to Ca release (*c*), which is entirely consistent with plagioclase recrystallization reactions. Normalization of the Fe data (*e*) strongly indicates that a process other than chloro-complexing is responsibility for the relatively high Fe concentrations in vapor-rich fluids; see text and Von Damm (2000). Fe concentrations in vent fluids from Rainbow are conspicuously high and strongly suggest the existence of an acid generating mechanism in the sub-seafloor reaction zone, the ultramafic lithology notwithstanding. Dissolved H_2S concentrations are also greatly elevated in the vapor-rich vent fluids (*f*), which is best accounted for by partitioning of H_2S and other volatile species into low-density vapor phases coexisting with fresh basalt, which initially maintains a high reductive capacity. Seawater composition is from the GERM database, Geochemical Earth Reference Model, http://earthref.org/GERM/inded.html.

responsible for Fe mobility at EPR 9–10° N. Von Damm (2000) emphasized this point earlier for the EPR data, while Seyfried *et al.* (2003) noted it more recently for vent fluids from the Main Endeavour Field, JDF. In both instances, factors such as pH, redox, and temperature were proposed to account for high Fe contents. In the case of the Rainbow system, however, the high Fe (highest of all vent fluids sampled to date) takes on added significance when one considers that this vent system is hosted in ultramafic rocks (Douville *et al.*, 2002). Bulk compositional constraints notwithstanding, an acid generating process must be active at Rainbow to account for the unusually high Fe/Cl ratio.

The chemical variations of hydrothermal vent fluids at mid-ocean ridges are, of course, much more complex than indicated by the few key species selected for review here. The excellent studies by Campbell *et al.* (1988a), Von Damm (1990, 1995, 2000), Butterfield and Massoth (1994), and Butterfield *et al.* (1994), in particular, should be consulted to gain a better understand of the full range of dissolved cation concentrations that are possible.

14.2.3 *Redox constraints in sub-seafloor hydrothermal systems*

Mid-ocean ridge vent fluids exhibit a wide range of dissolved H_2 concentrations (Welhan and Craig, 1983; Merlivat *et al.*, 1987; Evans *et al.*, 1988; Charlou *et al.*, 1993, 1996a, 1998; Lilley *et al.*, 1993; Von Damm *et al.*, 1997; Douville *et al.*, 2002; Seyfried *et al.*, 2003). Some of this is due to differences in temperature, pressure, redox controls, and phase separation effects in sub-seafloor reaction zones, but subsequent cooling, mixing, and mineralization can account for this as well. Most vent fluids from locations along the East Pacific Rise and Juan de Fuca Ridge with near seawater dissolved chloride concentrations, have relatively uniform (0.2–0.4 mmol kg^{-1}) H_2 concentrations (Lilley *et al.*, 1993). Vapor-rich fluids, which have clearly undergone phase separation, tend to have relatively high H_2 concentrations (Lilley personal communication). This accounts for the relatively high H_2 concentrations in vent fluids from the Main Endeavour Field (MEF), JDF (Seyfried *et al.*, 2003) and at EPR 9–10° N (Von Damm, 2000). Although H_2-rich vapors can be generated in part by temperature- and pressure-dependent changes in Henry's Law coefficients, which characterizes gas–aqueous fluid partitioning, the early eruptive stages sub-seafloor magmatic activity is inherently reducing, which enhances H_2 generation during hydrothermal alteration.

The high H_2 concentration (14 mmol kg^{-1}) in Rainbow vent fluids (36° 14′ N, MAR) is anomalous for fluids with dissolved chloride in excess of seawater values. This, however, is clearly related to the effects of oxidation by H_2O of ferrous silicate components in ultramafic rocks, which host the Rainbow hydrothermal system (Barriga *et al.*, 1997; Fouquet *et al.*, 1997; Douville *et al.*, 2002), as follows:

$$\text{Fayalite} + 0.6667\,H_2O = 0.6667\,\text{Magnetite} + SiO_{2(aq)} + 0.6667\,H_{2(aq)} \quad (14.1)$$

In this reaction, fayalite represents the fayalite component of olivine solid solution. The ferrosilite component of orthopyroxene, however, is an equally effective agent for H_2 generation in ultramafic systems.

Moreover, the high dissolved CH_4 in Rainbow vent fluids strongly suggests Fischer–Tropsch synthesis, which underscores the level of reduction that is possible in ultramafic systems (Berndt *et al.*, 1996a; Holm and Charlou, 2001):

$$HCO_3^- + H^+ + 4H_{2(aq)} = CH_{4(aq)} + 3H_2O \qquad (14.2)$$

Dissolved H_2S is also sensitive to redox conditions and is often present in relatively high concentrations in vent fluids (Fig. 14.3*f*). As with H_2, the relatively high H_2S in vent fluids having low dissolved chloride can be linked to reducing conditions at depth during early stages of hydrothermal alteration, but also to the generation of low-density vapors, which can enhance the solubility of dissolved gases. Indeed, the highest H_2S concentrations (up to 120 mm kg^{-1}) are manifested by the vapor-rich fluids at Aa vent (EPR 9° 46.5′ N), which were sampled shortly after a volcanic eruption in 1991 (Von Damm *et al.*, 1997). Temperatures of fluids from this vent achieved values as high as 403 °C at the seafloor, and likely higher at depth, which would enhance phase separation effects. Similar arguments apply for the MEF, where unusually high H_2S concentrations (20–30 mm kg^{-1}) and low chloride concentrations are also observed (Butterfield *et al.*, 1994; Seewald *et al.*, 2003; Seyfried *et al.*, 2003).

In contrast to the anomalously high H_2S-rich vapors, vent fluids with dissolved chloride in excess of seawater tend to have low H_2S contents (<5 mm kg^{-1}). This, in part, reflects phase separation effects, and the loss of H_2S to the vapor phase. Mineralization processes also may be important because the fluids typically have high Fe, which may precipitate Fe-sulfide on ascent to the seafloor. Finally, it is likely that many of the high-chloride fluids may have been derived during alteration at relatively high fluid/rock mass ratios where relatively oxidizing conditions can stabilize the bornite–magnetite redox assemblage rather than pyrite–magnetite, precluding high H_2S concentrations in coexisting fluid.

14.2.4 Fluid/rock mass ratio

It has long been recognized that hot spring vent fluids at mid-ocean ridges are characterized by relatively high concentrations of mobile trace elements (Li, B, Rb, Cs) and isotopes of some of these elements, which allow calculation of the relative abundance of rock and fluid components during hydrothermal alteration processes, assuming knowledge of the composition of the fluid and rock prior to alteration (Von Damm *et al.*, 1985; Spivack and Edmond, 1987; Bowers *et al.*, 1988; Campbell *et al.*, 1988a,b). Hydrothermal systems having undergone phase separation, however, present added complications due to element partitioning between brines and vapors (Berndt and Seyfried, 1990; Von Damm *et al.*, 1997). Thus, chloride normalization procedures are often necessary to constrain mass transfer

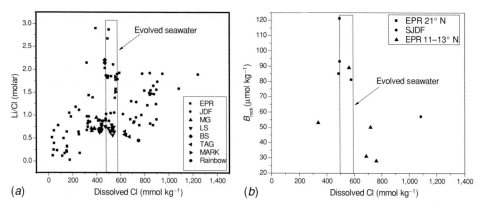

(a)

(b)

Fig. 14.4 Li/Cl ratio (molal) versus chloride for vent fluids from Pacific and Atlantic hydrothermal systems (*a*). The relatively low ratio for low-chloride vent fluids likely reflects short fluid-residence times and/or high fluid/rock mass ratios of low-density, vapor-rich fluids during rock–fluid inter-action. The moderately high Li/Cl ratios for fluids at or above seawater-dissolved chloride likely reflect a more stable state of heat and mass transfer associated with crack propagation into pre-viously unaltered crust (see text). Data sources are the same as noted in Fig. 14.1. In addition, however, data for MAR vent sites are from Campbell *et al.* (1988b), Edmond *et al.* (1995), James *et al.* (1995), Von Damm *et al.* (1998), and Charlou *et al.* (2000). LS, BS, MG designations in the legend refer to Lucky Strike, Broken Spur, and Menez Gwen, respectively. Boron isotope data for selected vent fluids (*b*) allow seawater and rock-derived boron sources to be distinguished (Spivack and Edmond, 1987; Berndt and Seyfried, 1990). The lower B_{rock} value for vent fluids with dis-solved chloride that departs significantly from seawater is either due to lower values of boron in the basaltic substrate or alteration at relatively high fluid/rock mass ratios. The relatively high B_{rock} values for 21° N EPR vents corroborate inferences drawn from Li/Cl data depicted in (*a*) (see text), as well as oxygen isotope data reported by Spivack and Edmond (1987), Bowers (1989), and Shanks (2001). Sources of data for (*b*) are described in Spivack and Edmond (1987) and Berndt and Seyfried (1990).

during rock–fluid interaction processes in systems where phase separation is known to have occurred (Von Damm, 2000).

Alkali element/Cl ratios of mid-ocean ridge vent fluids reveal a complex pattern char-acterized by relatively low values for fluids having low dissolved chloride concentrations and higher values for vent fluids having near or above seawater chloride concentrations. For example, we can illustrate this for lithium by making use of vent systems characterized by varying degrees of phase separation, as indicated by chloride variability, and lithologic controls covering a wide compositional range, as indicated by tectonic setting (Fig. 14.4*a*). The strikingly low Li/Cl ratios for vapor-rich fluids issuing from vents at EPR 9–10° N tend to be associated with active magmatism as was the case in the aftermath of the 1991 eruptive event (Von Damm, 2000). Initial eruptive stages of magmatically active systems are characterized by particularly high thermal gradients, which can cause large-scale changes in the transport properties of the aqueous fluid, vapor phase buoyancy, and short residence times for fluids at the peak hydrothermal condition where rock–fluid interaction effects tend to be greatest. Sub-seafloor dike intrusions are accompanied by considerable faulting and fracturing, which can be expected to induce formation of highly permeable up-flow zones (Wilcock and McNabb, 1996; Wilcock, 1997), which may further decrease the opportunity for chemical exchange between the fluid and coexisting crustal rocks. Although these factors, individually or in combination, limit fluid access to rock, trace alkali element/Cl ratios, such as Li/Cl, are seldom less than seawater (Von Damm, 2000). It is unclear, however, whether or not the low Li/Cl ratio (or any other similar indicator) is more a function of kinetic effects imposed by factors related to residence time or to an increase in the mass of fluid relative to rock caused by changes in permeability. Either effect will give rise to the observed low alkali Li/Cl ratios. Penetrative stages of reaction, where fluid propagates into recently crystallized, but still hot magmatic bodies, in contrast, can be expected to increase alkali element/Cl ratios. It is likely that this stage dominates processes of heat and mass transfer at mid-ocean ridges as originally emphasized by Lister (1982, 1995), and thus, it is not surprising that most vent fluids are characterized by relatively high alkali element/Cl ratios, as depicted in Fig. 14.4*a* by Li/Cl ratios.

Fluid propagation into fresh rock may be enhanced by seismic activity at or near the conductive boundary layer at the interface of the magmatic–hydrothermal system. Sohn *et al.* (1998), for example, reported a change in temperature of fluids issuing from Bio9 vent at 9° 50.2′ N on the East Pacific Rise following a microearthquake swarm in 1995. Although compositional changes in fluid chemistry were not determined simultaneously with temperature measurements, the first sampling opportunity seven months after the seismic event revealed increases in dissolved chloride and silica relative to pre-seismic levels. At Bio9, dissolved chloride and silica increased from $\sim$300 to 500 mmol kg^{-1} and from 14 to 15 mmol kg^{-1} immediately before and after the seismic activity, respectively. Changes of this sort are consistent with a deepening of the hydrothermal reaction zone and greater interaction with fresh basalt (Fornari *et al.* 1998; Sohn *et al.*, 1998). Corresponding increases in Li/Cl, other alkali element/Cl ratios, and rock-derived boron are to be expected, which for alkali element/Cl ratios is in keeping with actual observations (Bray and Von Damm, 2003; Fig. 14.4*a*).

The relative constancy of Li/Cl versus chloride as depicted in Fig. 14.4*a* can be interpreted to indicate phase equilibria control (Von Damm, pers. comm., 2002). Although this is not likely in a strict sense thermodynamically, Li (and other trace alkali elements) can be described by bulk fluid/rock partition coefficients. Seyfried *et al.* (1984) showed that at 400 °C approximately 70% of the Li initially in fresh basalt partitions into the aqueous fluid. It is this partitioning behavior that accounts for the relatively high dissolved Li in hydrothermal vent fluids when access to fresh rock in unimpeded by direct magmatic events. The residual Li in the rock, however, likely occupies sites in the crystal lattice of a number of primary and secondary minerals, typically Mg-bearing phases due to the similar ionic radii of Mg and Li. Results of a Li–Li isotope study of altered basalts from the deep root zone of ODP site 504B (Chan *et al.*, 2002) confirm extensive, but not quantitative leaching of Li during high-temperature hydrothermal alteration – in keeping with experimental data. It is important to stress, however, that reversible exchange between Li in the fluid and coexisting minerals in the rock has not been experimentally demonstrated at temperatures, pressures, and for bulk compositions applicable to high-temperature reaction zones at mid-ocean ridges (You *et al.*, 1996; James *et al.*, 2003). This, together with the significant amount of Li dissolution that does take place during fluid–rock reaction, however, indicates that the capacity of the rock to buffer Li in solution is ineffective and the relative constancy of Li and other so-called soluble elements in vent fluids on an absolute and chloride-normalized basis is more likely related to physical controls linked to the relative masses of fluid and rock during hydrothermal alteration than to any solubility limiting chemical control. This is the major difference between phase relations involving trace incompatible elements and major elements, such as Na, Ca, Fe, and $SiO_{2(aq)}$, which have long been recognized as solubility controlled. Although we have emphasized Li in this discussion, similar arguments can be made for virtually all trace elements that exhibit incompatible behavior during fluid–rock interaction at elevated temperatures and pressures, chief among these are B, Rb, Cs, and to a lesser degree K.

Although it may be largely coincidental, fluids having dissolved chloride near seawater values tend to be characterized by the highest Li/Cl ratios (Fig. 14.4*a*). For typical MORB lithologies, data from 21° N illustrate this best, although fluids issuing from vents at Broken Spur and MARK also reveal relatively high Li/Cl ratios and near-seawater dissolved chloride concentrations. For reasons which may be unique to the EPR 21° N hydrothermal system, however, the dissolved concentrations of a wide range of incompatible elements in vent fluids from this area suggest exceedingly low fluid/rock mass ratios, which have remained relatively constant, since the time of initial discovery in 1979 (Von Damm *et al.*, 1985; Campbell *et al.*, 1988b). In addition to simple alkali element/Cl data, such as Li/Cl (Fig. 14.4*a*), low fluid/rock mass ratios for the 21° N system are also indicated by boron isotope data. Application of boron isotopes to vent fluids to constrain hydrothermal alteration processes involving boron is needed due to high and low boron concentrations in seawater ($\sim$430 μmol kg^{-1}) and MORB ($\sim$20–40 μmol kg^{-1}), respectively (Spivack and Edmond, 1987). Thus, using data and equations from Spivack and Edmond (1987) and Berndt and Seyfried (1990), rock-derived boron concentration in vent fluids for which data are available can be determined (Fig. 14.4*b*). In comparison with other vent fluids, the vents at EPR 21° N reveal unusually high B_{rock} concentrations, indicative of fluid/rock mass ratios as low as 0.3–0.5, assuming 40 μmol kg^{-1} rock for MORB boron. Moreover, assuming a constant MORB composition for the range of vent fluids considered indicates fluid/rock mass ratios of 1–2 for these other vent systems. Unambiguous interpretation of the correlation between B_{rock} and vent fluid chloride is not obvious. Since chloride variability can be clearly linked to phase separation processes, however, suggests phase separation is a contributing factor. In general, the range of fluid/rock mass ratios predicted from boron isotope data is in good agreement with what can be inferred from Li/Cl (Fig. 14.4*a*; assuming MORB Li), as well as oxygen isotope systematics (Bowers and Taylor, 1985; Shanks, 2001; see below.)

A model for EPR 21° N involving fluid interaction with fresh rock in response to crack propagation seems to be the most likely explanation to account for the steady-state time series observations, although this places constraints on the mechanism by which this occurs, since to the best of our knowledge, only modest changes have taken place in vent fluid temperatures and dissolved chloride and silica concentrations with time (approximately ten year interval), indicating that deepening of the hydrothermal reaction zone has not been overly dramatic (Von Damm *et al.*, 1985; Campbell *et al.*, 1988b). The 21° N vent system was again visited in 2002 (Von Damm *et al.*, 2002), and it will be of interest to examine changes, if any, in the dissolved incompatible element concentrations in association with other indicators of sub-seafloor temperature and depth (pressure). It would, of course, be particularly helpful in terms of constraining reaction zone processes, if time integrated data were ultimately available for both the flux of fluid and chemicals so that a better sense of reaction zone volume could be estimated.

Estimates of fluid/rock mass ratio based on the dissolved inventory of trace incompatible elements in vent fluids need to take explicit account of lithologic constraints, as suggested earlier. This is true as well whenever isotope data are used for the purpose of determining fluid/rock mass ratio (Shanks, 2001). This is not a serious concern, however, for MORB-hosted hydrothermal systems, such as for sites on the EPR and on the Mid-Atlantic Ridge at TAG, but this is not always the case. For example, hydrothermal venting at Lucky Strike (Langmuir *et al.*, 1997) and Menez Gwen (Charlou *et al.*, 2000), south of approximately 37° N on the Mid-Atlantic, suggests an enriched basaltic substrate possibly influenced by near proximity to the Azores hot-spot, which appears to contain anomalously high barium and potassium, but low lithium. Thus, for these hydrothermal systems, the relatively low Li/Cl ratio (Fig. 14.4*a*) may not allow simple conversion to fluid/rock ratio without additional information on the chemistry of the fresh rock. Moreover, many of the MAR vent systems are unusually tectonically active (Wilcock and Delaney, 1996), which may enhance fluid flow through reactivated crack networks contributing to a highly altered substrate and unusually low inventories of trace alkali elements, lowering further Li/Cl ratios of hydrothermal fluids. Indeed, tectonically enhanced permeability effects may be the primary cause of the unusually low Li/Cl ratios of vent fluids at TAG (Fig. 14.4*a*), since substrate composition (N-MORB) is well constrained. The relatively high fluid/rock mass ratios estimated for TAG vent fluids are consistent with strontium isotope data at TAG where values from approximately three to seven can be estimated using data from Berndt *et al.* (1988) and Gamo *et al.* (1996).

Fluids venting from the ultramafic-hosted hydrothermal Rainbow system reveal unusually low Li/Cl ratios (Fig. 14.4*a*). This may be due to unusually low lithium in the peridotite involved in alteration at this site, but could also be caused by a relatively low fluid/rock partition coefficient for Li during hydrothermal alteration. Data are not available to test the latter possibility unambiguously. A third possibility involving alteration at high fluid/rock mass ratio can likely be ruled out by other aspects of the fluid chemistry, including relatively high potassium and rubidium concentrations (Douville *et al.*, 2002).

Fluid/rock mass ratios estimated from chloride-normalized concentrations of largely incompatible elements and imposed constraints can provide important insight on a wide range of chemical and physical factors in sub-seafloor hydrothermal systems. As with other parameters that reflect the end result of complex reaction-path-dependent processes, non-unique interpretations can often result. Part of this may be caused by uncertainties in the composition of the rock protolith as well as lack of information on trace element partitioning data. Although most trace alkali elements and boron are largely incompatible in hydrothermal alteration phases, other than perhaps boron, subtle differences still exist. Chloride normalized concentrations of trace alkali elements in vent fluids need to be considered in combination with other information, such as isotopic data, to achieve the most accurate representation of the relative masses of fluid and rock in hydrothermal alteration zones at mid-ocean ridges.

14.3 Silicate phase equilibria in basalt-hosted hydrothermal systems

14.3.1 Metastable plagioclase–fluid equilibria

Berndt and Seyfried (1993) showed that Ca and Na exchange between chloride-bearing fluid and plagioclase coexisting with quartz is reversible at 400 °C, 400 bars, and highly dependent on plagioclase composition. Plagioclase remained homogeneous during the experiments despite the fact that intermediate plagioclase is metastable with respect to end-member phases at these conditions. In fact, the data suggest that a solid solution model for plagioclase under hydrothermal conditions approaches ideal behavior. Although the absolute concentrations of Ca and Na changed greatly with plagioclase composition and chloride, $m\text{Ca}^{++}/m^2\text{Na}^+$ ratios remained constant. This has important implications for Ca and Na systematics for MOR vent fluids, especially considering the relative abundance of plagioclase in oceanic crustal rocks.

The model can be tested by comparing dissolved Ca and Na concentrations in MOR vent fluids with analogous data predicted from plagioclase solubility constraints. Data show that vent fluids from JDF and EPR, which constitute a statistically meaningful subset of all vent fluids, closely follow a trend intermediate between that predicted for An_{82} and An_{60} (Fig. 14.5). For comparison, dissolved concentrations of Ca and Na predicted assuming the coexistence of anorthite and albite (as might be expected for a fully equilibrated system at 400 °C; Bowers *et al.*, 1985), are also shown. Vent fluid chemistry is not consistent with such a trend. Unequivocally, the enhanced concentrations of Ca provided by metastable plagioclase–fluid equilibria account best for Ca/Na systematics of vent fluid data.

The relatively high dissolved Ca concentration in fluids coexisting with metastable igneous plagioclase is particularly significant from the standpoint of pH control. Experimental and theoretical data show that Ca metasomatism is likely the primary pH controlling reaction in sub-seafloor hydrothermal systems at elevated temperatures and pressures (Berndt *et al.*, 1989; Seyfried *et al.*, 1991), which can be depicted by the following reaction:

$$\text{Anorthite} + 1.5\text{H}_2\text{O} + \text{Ca}^{++} + \text{SiO}_{2(\text{aq})} + 0.5\,\text{Hematite} = \text{Epidote} + 2\text{H}^+ \quad (14.3)$$

In this reaction, anorthite and epidote are end-member phases, whereas in nature, both are solid solutions. Explicit account of mineral solid solution effects can have a dramatic influence on the composition of the coexisting aqueous fluid, in particular, pH. To illustrate

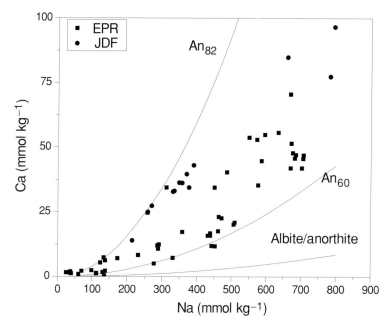

Fig. 14.5 Dissolved Ca versus Na for hydrothermal vent fluids from sites on the Juan de Fuca (JDF) Ridge and East Pacific Rise (EPR; see Fig. 14.1) in comparison with solubility constraints imposed by plagioclase–fluid equilibria (An_{82}, An_{60}), as well as end-member anorthite and albite–fluid equilibria. All mineral solubility calculations were performed at 400 °C, 400 bars. Data indicate that the Ca and Na concentrations of vent fluids are in good agreement with plagioclase–fluid equilibria where the mCa^{++}/m^2Na^+ ratio has been shown to be an explicit function of the composition of plagioclase solid solution (Berndt and Seyfried, 1993).

this, we constructed an ion-activity diagram in the $CaO–Na_2O–Al_2O_3–SiO_2–FeO–Fe_2O_3–H_2O–HCl–H_2S$ system at 400 °C, 500 bars depicting the effect of redox and plagioclase composition (Fig. 14.6). In comparison with constraints imposed by the solubility of end-member minerals (anorthite, albite, clinozoisite), the addition of Fe to the system in the form of various Fe-bearing redox buffers results in a dramatic decrease in aCa^{++}/a^2H^+ and aNa^+/aH^+ ratios due to the enhanced stability of epidote and plagioclases solid solutions. Log aCa^{++}/a^2H^+ and aNa^+/aH^+ ratios decrease with increasing oxidative intensity and anorthite component of plagioclase solid solution. For example, assuming constant chloride (e.g. 0.55 mmol kg^{-1}) and plagioclase composition (An_{70}), plagioclase–epidote–quartz–fluid equilibria depicted in Fig. 14.6 indicate a pH decrease of approximately 0.5 units in going from the more reducing (QFM) to oxidizing (HMP) conditions. At elevated temperatures and pressures, a pH change of this magnitude would have a profound effect on the solubility of all coexisting minerals in the system.

Increasing dissolved Cl has the capacity as well to lower the pH of a redox buffered aqueous fluid coexisting with plagioclase and epidote at elevated temperatures and pressures (Fig. 14.7a). As shown, pH is predicted to decrease by approximately an order of magnitude with increasing Cl from very low values (0.03 mmol kg^{-1}) to concentrations as high as

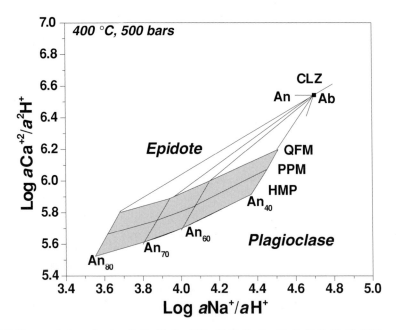

Fig. 14.6 Phase relations in the $CaO–Al_2O_3–SiO_2–FeO–Fe_2O_3–H_2O–H_2S–NaCl–HCl$ system at 400 °C, 500 bars, which show the effect of redox variability on changes in log aCa^{++}/a^2H^+ and log aNa^{++}/aH^+ in chloride-bearing fluid coexisting with plagioclase and epidote solid solutions. The following buffer assemblages provide redox constraints: hematite–magnetite–pyrite (HMP), pyrite–pyrrhotite–magnetite (PPM), and quartz–fayalite–magnetite (QFM). Explicit treatment of plagioclase and epidote solid solutions in Fe-bearing systems is essential to understand phase relations in subseafloor hydrothermal systems at mid-ocean ridges better. Calculations were made using SUPCRT92 (Johnson *et al.*, 1992).

1.5 mol kg^{-1}. This is caused by phase equilibria effects linked to the increase in dissolved Ca^{++} that occurs with increasing chloride for fluids coexisting with plagioclase and epidote solid solution (Fig. 14.7*b*). As a consequence of pH lowering, Fe concentrations are predicted to increase dramatically, while dissolved Al decreases from ~10 to 2 µmol kg^{-1}, which is

---

Fig. 14.7 Phase equilibria in the $CaO–Na_2O–FeO–Fe_2O_3–Al_2O_3–SiO_2–H_2S–H_2O–HCl$ system at 400 °C, 500 bars depicting the effect of dissolved chloride on the composition of the fluid phase. Mineral phases in the system include: plagioclase solid solution (An$_{80}$), epidote solid solution, quartz, pyrite, pyrrhotite, and magnetite, which represents an assemblage of phases in keeping with incipient alteration of the oceanic crust by seawater. Owing to the tendency of Ca and Fe to partition into a brine phase during hydrothermal alteration processes, the predicted increase in the concentration of these species with increasing chloride is not surprising (*a*). Vent fluids reveal similar trends, especially for Ca (*b*), while Fe is not only affected by chloride, but also other variables, which account for the conspicuous increase in Fe in low chloride vent fluids (*c*). Predicted changes in dissolved Al (*a*) indicate a slight negative correlation with chloride. Calculations were performed using the EQ3/6 software package taking explicit account of recent upgrades in the SUPCRT92 database (see text). Portions of this figure have been published elsewhere (Butterfield *et al.*, 2003) and reproduced here with permission of Dahlen University Press.

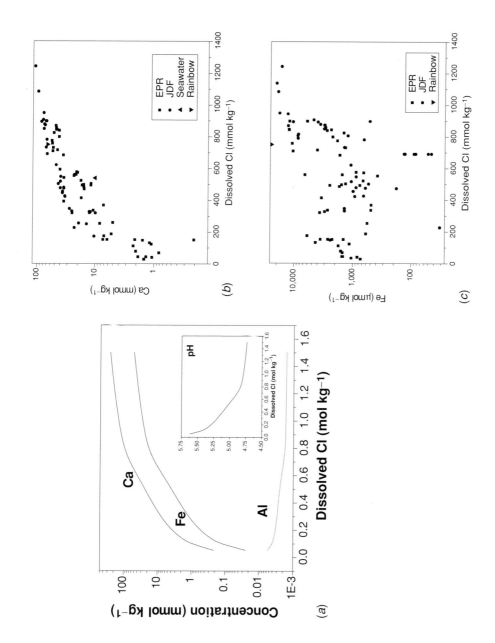

likely due to weakening of neutral hydroxo-complexes of Al (Tagirov and Schott, 2001) in the more acidic, chloride-rich fluids.

In general, chloride-dependent changes in fluid chemistry predicted agree reasonably well with measured concentrations of similar species in vent fluids (Von Damm, 1995), although this is truer for Ca than Fe. Vent fluid Al data are generally insufficient to allow rigorous comparison over the complete range of dissolved chloride considered. To illustrate this better for Ca and Fe, however, vent data were plotted on a log axis, which can be more easily compared with predicted trends of these species (Figs. 14.7*b*,*c*). Vent fluids reveal surprisingly high Fe concentrations at relatively low chloride (Fig. 14.7*c*), while the Ca data closely follow the predicted trend. Any of a number of mechanisms may account for the divergence between measured and predicted Fe in the low chloride fluids, including: lower pH conditions in the vent systems; the existence of aqueous complexes of Fe not included in the thermodynamic database used to model mineral solubility, but present in the natural system; or higher temperatures in sub-seafloor reaction zones. Additional experimental and field data are needed to resolve this apparent inconsistency unambiguously.

14.3.2 Constraints from the geologic record

It is important to note that the geochemical considerations presented to this point implicitly assume epidote coexistence with calcic plagioclase in the hydrothermal reaction zone that feeds black smoker fluids, in keeping with reaction (14.3) and Fig. 14.6. The existence of these phase relations have been questioned, however, based on results of mineralogic and petrologic studies of fossil hydrothermal systems (Gillis *et al.*, 1993; Gillis, 2002). For example, from petrologic investigation of a suite of hydrothermally altered rocks from the Kane Fracture Zone (23–24° N, MAR), Gillis and Thompson (1993) recognized a distinct grade of alteration characterized by albitic plagioclase (An$_{10-30}$), coexisting with amphibole, chlorite, sphene, and minor epidote. This assemblage, which apparently formed at temperatures from 250 to 450 °C, was observed to be depleted in trace transition metals, likely originated near the base of the sheeted dike complex, and was inferred to represent a fossil hydrothermal "reaction zone" for vent fluids in the MARK region. The relatively low abundance of epidote coexisting with plagioclase in these rocks, however, may be linked to a number of factors including, temperature, redox conditions, and activity of dissolved SiO_2. For example, at temperatures significantly greater than 425 °C, distinctly reducing redox conditions (QFM equivalent), and SiO_2 concentrations below quartz saturation, epidote is unstable relative to calcic amphibole (McCollom and Shock, 1998). Although the "epidote-out" consequences of this may be appealing, the coexisting aqueous phase is predicted to look nothing like a vent fluid. Indeed, reaction calculations by McCollum and Shock (1998), for an NaCl fluid at 425 °C coexisting with gabbro and amphibole-rich alteration products, reveal dissolved Ca and Fe concentrations of 0.1 and 0.0001 mmol kg^{-1}, respectively, and pH of approximately 7.5. In effect, the exceedingly low Ca is largely responsible for the relatively high pH, which precludes Fe mobility. Moreover, owing to the reducing

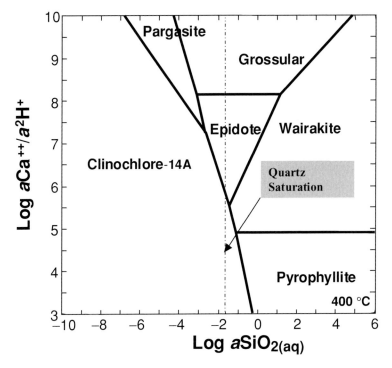

Fig. 14.8 Phase equilibria in the $CaO–Na_2O–FeO–Fe_2O_3–Al_2O_3–SiO_2–H_2S–H_2O–HCl$ system at 400 °C, 500 bars emphasizing the role of dissolved $SiO_{2(aq)}$ on the relative stability of epidote. As noted in Fig. 14.6, constraints imposed by the coexistence of plagioclase–epidote–quartz and a Cl-bearing aqueous fluid buffer log aCa^{++}/a^2H^+ values at approximately 5.5–5.8 depending on redox intensity and anorthite content of plagioclase solid solution. Clearly, quartz saturated conditions would be expected to stabilize epidote solid solution in sub-seafloor hydrothermal systems.

conditions, dissolved H_2 in the seawater chlorinity fluids can be predicted to be very high, which again is in sharp contrast with field data from many vents (Seyfried and Ding, 1995).

Vent fluid chemistry and results from experimental and theoretical models are consistent with metastable coexistence of calcic plagioclase and a chloride-bearing aqueous fluid under moderately oxidizing conditions and temperatures and pressures in the vicinity of 400–425 °C, 300–500 bars, respectively, which requires formation of epidote solid solution as well, provided dissolved SiO_2 concentrations are at or near quartz saturation (Fig. 14.8).

In contrast, the base of the hydrothermal up-flow zone may be easily recognized in the geologic record due to the focused nature of fluid flow through initially high-permeability conduits and the large compositional changes likely as minerals in the rock attempt to equilibrate with potentially very large volumes of fluid. This is especially true if the fluids are sufficiently hot such that decompression effects during fluid ascent trigger or enhance phase separation. For example, although it may appear counterintuitive, phase separation is an inherently oxidative process. Indeed, the production and removal of H_2 in the fluid necessitate increases in the oxygen fugacity of the bulk system, reflected by the production of ferric iron (Bischoff and Rosenbauer, 1987), which resides in minerals such as hematite, magnetite, or epidote. One form of expression of this may be epidosites, which are well developed in intensively hydrothermally altered basal portions of sheeted intrusive complexes in many ophiolites (Richardson *et al.*, 1987; Nehlig *et al.*, 1994; Gillis, 2002), and in at least one modern oceanic setting (Banerjee *et al.*, 2000). Fluid inclusion data from epidosites from different ophiolites terrains (Nehlig *et al.*, 1994; Juteau *et al.*, 2000; Gillis, 2002) indicate temperatures as high as 430 °C and range of salinities that confirm phase separation effects. Moreover, in the case of the Oman epidosites, Raman spectroscopic data reveal hematite and anhydrite daughter minerals in fluid inclusions (Juteau *et al.*, 2000), providing additional support for the existence of oxidative conditions. Although it is likely that a wide range of other chemical and physical factors may play a role in the formation of these interesting rocks, which may, in fact, limit their existence in mid-ocean ridge hydrothermal systems (Gillis, 2002), phase separation induced oxidation may be an overarching requirement. Finally, it is of interest to note the very existence of epidote at the base of the up-flow zone provides support for the notion that fluids entering this region from the underlying hydrothermal system are at or close to epidote saturation, which strengthens the connection between epidote and at least the upper portions of the reaction zone responsible for the composition of black smoker fluids.

Sub-seafloor drilling into an active, high-temperature reaction zone, which is precluded at present by technological limitations, will probably be required to resolve the apparent discrepancy between the fluid and rock records. It is becoming increasingly clear, however, that the sub-seafloor reaction zone at mid-ocean ridges needs to be viewed incrementally, both in space and time. The physical and chemical conditions in one part of the system are not the same as in all other parts, or even in the same part at a different point in time. Moreover, magmatic and tectonic variability can cause large-scale changes in the composition of fluids on relatively short time scales (Von Damm *et al.*, 1995; Seyfried *et al.*, 2003), which must influence mineral compositions at depth, but on a scale that may be difficult to resolve from petrologic studies of the rock record.

Thus, a multi-faceted approach with due consideration of experimental, theoretical, isotopic, and field constraints is the most effective way to assess sub-seafloor hydrothermal alteration processes – whether viewed from the perspective of the fluid or the rock record. For multi-component systems, it can often be misleading to overexpress the significance of one variable, such as temperature, in lieu of a broad range of other chemical and physical factors. Clearly, a holistic approach is the preferred means of studying both fossil and modern hydrothermal systems, and it is this approach that offers the greatest chance for success in linking one system to the other.

14.4 Silicate phase equilibria in ultramafic-hosted hydrothermal systems

It has long been known that ultramafic bodies outcrop at mid-ocean ridges, especially at the intersection of the Mid-Atlantic Ridge with offsetting fracture zones from approximately 15° through 36° N latitude (Charlou *et al.*, 1998). The involvement of these rocks in sub-seafloor hydrothermal alteration processes has largely been inferred from studies of dredged samples (Bonatti *et al.*, 1984), ODP drilling (Alt and Shanks, 1998), and from anomalies of methane, manganese, and other species in the water column overlying the oceanic crust in these areas (Charlou *et al.*, 1991, 1998; Rona *et al.*, 1992). With the recent discovery of the Rainbow hydrothermal system at 36° 14′ N latitude, however, and, more recently, the Lost City vent field on the Atlantis Fracture Zone (Kelley *et al.*, 2001), the conditions and implications of hydrothermal alteration of ultramafic rocks in a marine setting are finally becoming clear.

The Rainbow system is characterized by venting of black smoker fluids at temperatures as high as 362 °C (Donval *et al.*, 1997; Douville *et al.*, 2002). In addition to high temperatures, however, these fluids are characterized by H_2 and CH_4 concentrations of approximately 14 and 2 mmol kg^{-1}, respectively, values significantly higher than observed for other non-sedimented ridge systems venting hydrothermal fluid with dissolved chloride in excess of seawater (Donval *et al.*, 1997). Moreover, Rainbow vent fluid reveals dissolved Fe concentrations as high as 24 mmol kg^{-1}, 7 mmol kg^{-1} $SiO_{2(aq)}$, and nearly 750 mmol kg^{-1} dissolved chloride (Douville *et al.*, 2002). Although the high H_2 and CH_4 are strongly indicative of serpentinization processes (Janecky and Seyfried, 1986; Berndt *et al.*, 1996a), in keeping with theoretical predictions of the composition of fluids coexisting with ultramafic assemblages at temperatures as high as 400 °C (Janecky and Seyfried, 1986; Wetzel and Shock, 2000), other aspects of the fluid chemistry, such as, the relatively high Fe and $SiO_{2(aq)}$, are very much at odds with the theoretical predictions assuming a fully equilibrated system. As noted previously, high Fe concentrations suggest low pH, in sharp contrast with high pH values predicted for fluid–mineral equilibria in the MgO–FeO–SiO_2–H_2O–HCl system at elevated temperatures and pressures (Wetzel and Shock, 2000). Indeed, at 400 °C, 500 bars, these investigators reported dissolved Fe and $SiO_{2(aq)}$ concentrations that do not exceed 0.1 mmol kg^{-1}. In the case of Fe, this is approximately 250 times lower than actually measured in vent fluids at Rainbow. Clearly, alteration processes at Rainbow are more complex than predicted by a fully equilibrated reaction path model.

To shed light on hydrothermal alteration processes at Rainbow, Allen and Seyfried (2003) conducted a series of mineral solubility experiments at 400 °C, 500 bars. In general, these investigators suggested that the relatively slow rates of olivine hydrolysis at relatively high temperatures, together with non-stoichiometric dissolution of enstatite, allow sufficient $SiO_{2(aq)}$ to enter solution to render talc and tremolite stable. Even in experiments dominated by an overwhelming abundance of olivine, in keeping with the bulk composition of abyssal peridotites, dissolved $SiO_{2(aq)}$ exceeds that in equilibrium with olivine by approximately an order of magnitude (Fig. 14.9), and high Fe concentrations are also observed. In the experiment, as at Rainbow, the high Ca concentrations are derived from Mg (seawater) for Ca (diopside) exchange, which enhances tremolite formation, as follows:

$$2\,\text{Diopside} + 2\,\text{Enstatite} + Mg^{++} + 2.0\,H_2O$$
$$\rightarrow Ca^{++} + 0.25\,\text{Tremolite} + 0.667\,\text{Chrysotile} + 0.1667\,\text{Talc} \qquad (14.4)$$

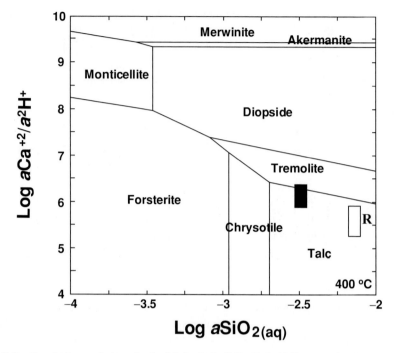

Fig. 14.9 Predicted phase relations in the MgO–CaO–SiO$_2$–H$_2$O–HCl system at 400 °C, 500 bars in comparison with experimental mineral solubility data (hatched symbol) from Allen and Seyfried (2003). Non-stoichiometric dissolution of SiO$_2$ from orthopyroxene and clinopyroxene during the experiments results in the formation of tremolite and talc, which together with release of Ca from diopside (clinopyroxene) causes the pH of the coexisting aqueous fluid to achieve relatively low values. In effect, the slow rate of olivine hydrolysis at temperatures greater than approximately 375 °C precludes titration of the low pH (Allen and Seyfried, 2003). This is in sharp contrast to theoretical predictions assuming full equilibrium (see Fig. 14.10). The experimental data provide a mechanism to account for H$^+$ generation in the ultramafic-hosted Rainbow hydrothermal system. The "R"-marked symbol defines the approximate location of Rainbow vent fluid (see Allen and Seyfried (2003). Portions of this figure have been published elsewhere and reproduced here with permission of Elsevier Press.

Thus, it is tremolite coexisting with Ca-rich fluids from the dissolution of clinopyroxene (diopside) that provides the key to the acidity needed to account for the relatively high Fe in the ultramafic-hosted Rainbow system (Fig. 14.9). Interestingly, the relatively low pH values at Rainbow are similar to values calculated for basaltic systems at similar temperatures and pressures. Acid generation by tremolite formation in the CaO–MgO–FeO–SiO$_2$–H$_2$O–HCl system at elevated temperature and pressures is very effective. It is indeed enlightening that an alkaline fluid (seawater) reacted with an alkaline mineral assemblage (ultramafic rock) can produce such an acidic fluid.

The "peridotite" – seawater experiments of Allen and Seyfried (2003) produced a fluid that is similar in many ways to the Rainbow vent fluid, but contrasts sharply with that

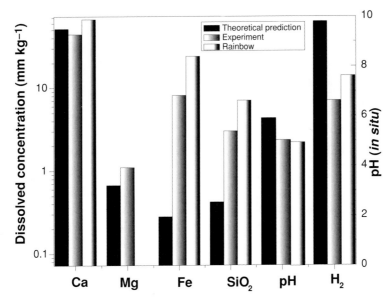

Fig. 14.10 Dissolved concentrations of Ca, Mg, Fe, SiO$_2$, and H$_2$ and pH in aqueous chloride-bearing fluid coexisting with olivine and pyroxenes (orthopyroxene and clinopyroxene) at elevated temperatures and pressures. Experimental data are from Allen and Seyfried (2003); theoretical data are from Wetzel and Shock (2000): while field data are from the ultramafic-hosted Rainbow system (Douville *et al.*, 2002). Field and experimental systems are in good agreement likely because of common silica release mechanisms, which stabilize H$^+$-generating hydrous silicates. Disequilibria effects are not considered in theoretical modeling studies, which then fail to account for the dissolved concentration of species especially sensitive to these.

predicted assuming full thermodynamic equilibrium (Fig. 14.10). It is the relatively high dissolved SiO$_2$ concentrations in the experimental and natural systems that are entirely responsible for this. In time, however, the bulk composition of the ultramafic assemblage at Rainbow and in the experiments will cause dissolved SiO$_{2(aq)}$ to decrease and pH to increase, which will result in quantitative removal of Fe from solution, in a manner consistent with theoretical predictions. The value of experiments, however, is to show that instability and metastability often need to be considered, even for reactions thought to proceed rapidly at elevated temperatures and pressures.

The relatively high H$_2$ concentrations measured in the experiments of Allen and Seyfried (2003), predicted from theoretical data (Wetzel and Shock, 2000), and observed at Rainbow (Douville *et al.*, 2002; Fig. 14.10), result from Fe oxidation of unstable ferrous silicates in the ultramafic assemblages, as noted previously. Orthopyroxene, in particular, has virtually no field of stability coexisting with an aqueous fluid at temperatures and pressures that can be inferred for Rainbow, suggesting that the Fe in this phase serves as the primary source of H$_2$, when oxidized by water. Although the high H$_2$ in vent fluids from ultra-mafic systems (Rainbow) provides an important clue to the mechanism of hydrothermal alteration involving ultramafic rocks, it takes on added significance when one considers its potential as an energy source for chemolithoautotrophic microbial populations hosted in the sub-seafloor at mid-ocean ridges (Alt and Shanks, 1998). In a sense, microbial ecosystems in the sub-seafloor at ridges provides yet another type of reaction zone, which can contribute to the fate of chemical species initially released into fluids separated in space and time from the site at which the final reaction step takes place.

14.5 Applications of stable isotopes to mid-ocean ridge hydrothermal systems

Stable isotopes have been important tools for understanding the origin of seafloor massive sulfide deposits, hydrothermal vent fluids, and hydrothermal alteration processes from the earliest discoveries. For example, oxygen isotope studies of greenstones dredged from the seafloor and metabasaltic rocks from ophiolites indicated seawater involvement long before vent fluids were discovered on the mid-ocean ridges (Muehlenbachs and Clayton, 1972; Spooner *et al.*, 1974). Initial stable isotope studies of the 21° N sulfide deposits and vent fluids showed unequivocally that seawater sulfate reduction had occurred to produce sulfide deposits with $\delta^{34}S$ values of 3–5‰ (Hekinian *et al.*, 1980). Similarly, the first study of oxygen isotopes of black smoker fluids showed that they were dominated by seawater that had reacted with basalt/gabbro at high temperature (Welhan and Craig, 1983). Recently, Shanks *et al.* (1995) and Shanks (2001) have provided overviews and summaries of hydrogen, carbon, oxygen, and sulfur isotopic data for vent fluids, hydrothermal deposits, and hydrothermally altered rocks from mid-oceanic ridges, back-arc basins, and island-arc settings. These data show more similarities than differences between the diverse geologic settings and will be discussed here in terms of processes. These include water–rock reactions, phase separation, and contributions of magmatic volatiles. The discussion will also be extended to include other stable isotope systems (B, Li) and emerging stable isotope studies of some transition elements (Fe, Cu).

14.5.1 Hydrogen and oxygen

Water–rock interaction

Hydrogen and oxygen isotope values of hydrothermal vent fluids have been studied from most of the known high-temperature vent sites on the seafloor (Shanks *et al.*, 1995; Shanks, 2001). Most of the data are from mid-ocean ridges (Fig. 14.11) and range from δD of -2 to 3‰ and $\delta^{18}O$ of 0.2–2‰ relative to local bottom water. Several investigators have shown that vent fluid oxygen and hydrogen isotope data are consistent with reaction of evolved seawater with silicate rocks at high temperatures (350–400 °C) and fairly low (<1) water/rock mass ratios (Bowers and Taylor, 1985; Bowers, 1989; Bohlke and Shanks, 1994; Shanks *et al.*, 1995; Shanks, 2001). Furthermore, the vent fluid isotope data (Fig. 14.11) appear to be most consistent with water–rock reactions that produce chlorite (or some other hydrous sheet silicate) in the temperature range 300–400 °C, rather than higher temperature reactions that produce amphibole rather than chlorite (see Shanks *et al.* (1995)

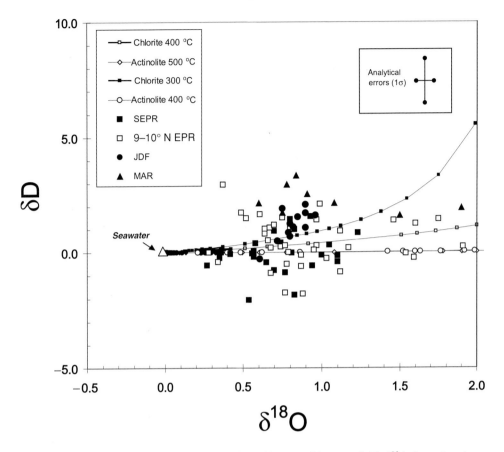

Fig. 14.11 Hydrogen and oxygen isotope data for mid-ocean ridge vent fluids (‰). Isotopic values have been normalized to local bottom waters, so observed isotopic effects are due to hydrothermal processes. Data are from Shanks (2001). Curves for seawater reaction with chlorite–anorthite (70) and actinolite–anorthite (70) at decreasing water/rock ratios as fluids progress away from unaltered seawater. Most of the data correspond best to the chlorite–anorthite (70) curves for 300–400 °C indicating water–rock reactions dominate stable isotope systematics of vent fluids. Anomalously high δD values for MAR vent fluids and anomalously low δD values for some low chloride EPR vent fluids are outside the range of analytical error and may be attributed to serpentinization reactions (MAR) and phase separation (EPR).

for detailed discussion). However, the relatively large analytical error (± 1‰ 1σ) of the hydrogen isotope analyses of these fluids precludes conclusive differentiation of chlorite from amphibole reactions.

Three conclusions can be stated with certainty based on the data in Fig. 14.11: (1) most lower temperature reactions (<300 °C) are unimportant in controlling the final isotope values of black smoker vent fluids; (2) water/rock mass ratios are typically in the range from 5 to 0.5 and reactions at ratios of <0.1, which would produce $\delta^{18}O$ values >2‰, are not indicated; and (3) the most positive and most negative δD values fall outside the range of analytical uncertainty (even at 2σ) relative to the calculated fractionation pathways and therefore are likely due to processes other than simple reaction with unaltered silicate rocks.

What processes could produce the anomalous δD values? The strongly positive or negative δD values might be produced by reaction with rocks that were previously altered at higher or lower temperatures, by phase separation processes, or by contributions of magmatic water (for the low δD values). Reaction with rocks altered at very high temperatures (450–600 °C) might account for high δD vent fluids, such as those from the Mid-Atlantic Ridge (Fig. 14.11), if hydrous phases in the high-temperature rocks had relatively high δD values. In this case, fluids in the reaction zone might exchange or produce additional alteration minerals with more negative δD values, thus producing fluids with positive δD values. However, this is unlikely because available evidence suggests that δD fractionation does not vary much over this temperature range (Chacko et al., 2001) and actual δD values for ophiolitic and seafloor chlorites and amphiboles (Stakes and O'Neil, 1982; Stakes, 1991) indicate that amphiboles have more negative δD values than chlorites; the opposite of the required trend. Similarly, fluids with negative δD values might result from reaction with sediments that formed at low temperatures. Data indicate that hydrous minerals in marine sediments commonly have δD values in the -70 to $-100‰$ range (Shanks, 2001) and hydrothermal reaction calculations with sediments predict that fluids will have negative δD values (Bohlke and Shanks, 1994). However, the vent fluids with the most negative δD values are from 9 to 10° N on the East Pacific Rise or from the southern EPR; two areas of extremely fast-spreading rate and slow sedimentation. It is extremely unlikely that sediments are involved in the reaction zones of these areas.

Magmatic water

Magmatic water from a variety of igneous rocks has δD values of -40 to $-80‰$ and $\delta^{18}O$ values from -6 to $-8‰$. Thus addition of a small amount of magmatic water to the EPR vent fluids could produce the negative δD values observed. However, crystallizing basalt or gabbro is a poor source of magmatic water because of low water contents (~ 0.2 wt. %) and an unusual circumstance would be required to separate significant amounts of magmatic water from mid-ocean ridge basalts (MORBs).

Although CO_2 in most vent fluids can be derived by leaching of rocks during hydrothermal alteration, extremely high concentrations (up to ~ 120 mm kg^{-1}) in low Cl vents following the 9° 46–52′ N EPR eruption in 1991 probably require direct magmatic degassing. Assessing direct degassing is complicated by phase separation processes, which are an inevitable consequence of hydrothermal circulation near magmatic systems.

Phase separation

Stable isotope fractionation between coexisting liquid water and water vapor in NaCl solutions at temperatures of sub-critical and super-critical phase separation have been studied experimentally (Horita et al., 1995; Berndt et al., 1996b). Both studies found that water vapor is D-enriched relative to liquid water in the critical region, with fractionations generally

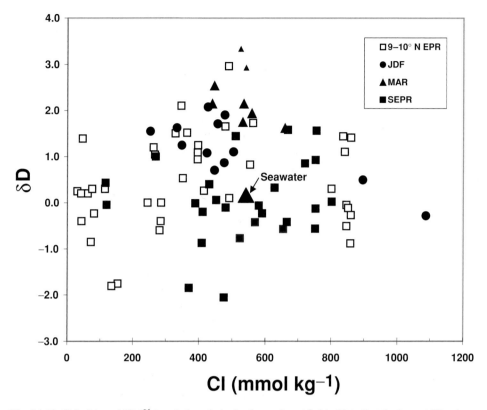

Fig. 14.12 Chloride and δD (‰) variations in hydrothermal vent fluids. Note that the lowest δD values are for low chloride vents and the highest δD values are for samples with chloride near seawater.

between 5 and 10‰. Thus a simple equilibrium exchange would produce vapors with δD values 5–10‰ higher than the residual brine. Clearly this does not explain the negative values from the EPR, which are all from low-salinity vent fluids. However, phase separation is a dynamic process, so Rayleigh distillation effects must be considered. Furthermore, phase separation processes are quite different in the sub-critical and super-critical regions, as describe above. Berndt *et al.* (1996b) have described a possible process that involves loss of much of the evolved water vapor during phase separation, as might occur during an isobaric heating event related to dike emplacement into the near surface of the mid-ocean ridge. Under these conditions, the residual brine will attain very negative δD values and the vapor evolved may also have negative δD values. Thus, phase separation may produce significant positive or negative δD variations depending on the exact conditions of separation and the fraction of fluid phase separated. This is consistent with the general lack of correlation of vent fluid δD with Cl concentration (Fig. 14.12). However, when specific sets of vent fluids are examined, for example the 9–10° N vent fluids with Cl less than seawater (Fig. 14.12), a rough positive correlation of δD with Cl is observed that is consistent with the phase separation model proposed by Berndt *et al.* (1996b). In contrast, the D-enriched

MAR fluids are close to normal seawater salinity and probably cannot be explained by phase separation processes alone.

14.5.2 Carbon and sulfur

Carbon and sulfur isotope variations are taken together here because they are both strongly affected by redox reactions, which are mediated by microbial processes at low temperatures (<114 °C), and because both carbon and sulfur are important components of magmatic volatiles (Welhan and Craig, 1979, 1983; Shanks *et al.*, 1995; Urabe *et al.*, 1995; Alt and Shanks, 1998, 2002; Shanks, 2001). Carbon and sulfur exist in both oxidized (SO_4 and CO_2) and reduced forms (H_2S and CH_4) in hydrothermal fluids. In the hydrothermal reaction zone, water–rock reaction with iron-bearing minerals causes reduction of most, but not all, SO_4 to H_2S. Simultaneously, iron–monosulfide solid-solution minerals (mss, approximately pyrrhotite) in basalts are quite soluble and they dissolve during hydrothermal alteration until saturation with pyrite or pyrrhotite (mss) is attained (Fig. 14.13). CH_4 also coexists with CO_2, but is much less abundant.

At magmatic conditions, H_2S and CO_2 are the stable species in gases evolved from MORB, but SO_2 may be important in some island-arc related intermediate volcanic systems. Direct degassing of sub-seafloor magma is thought to be an insignificant contributor of sulfur to black smoker vent fluids because water–rock reaction experiments, without the presence of externally derived magmatic gases, accurately reproduces H_2S contents observed in vent fluids. Because sulfur is a reactive component, any H_2S (or SO_2) released from cooling magma into the hydrothermal system will be mixed with sulfur derived from water–rock reactions and levels will be controlled by reactions outlined in this paper.

Stable isotope studies of $\delta^{13}C$ and $\delta^{34}S$ provide important constraints on possible reaction processes in hydrothermal vent fluids. Sulfur isotopes have been studied extensively because of the importance of these systems as analogs to ore-forming systems. In general, $\delta^{34}S$ values of high-temperature (>250 °C) vent fluid H_2S range from 2.9 to 8.6‰, reflecting a mixture of sulfur derived directly from the rock by mss dissolution and from reduction of seawater sulfate. Primary sulfide in MORB has $\delta^{34}S$ values close to 0‰ (±0.5) and seawater sulfate is quite uniform at 21‰. Reduction of seawater sulfate by water–rock reaction is near quantitative (<1 mmol kg^{-1} SO_4 in end-member hydrothermal vent fluids) but most sulfate is removed as anhydrite due to retrograde solubility and precipitation during seawater ingress and heat up. Thus, sulfate that is reduced has $\delta^{34}S$ values close to 21‰. Addition of reduced sulfide to igneous sulfide derived from mss dissolution gives the observed range of isotopic values of vent fluids (2.9–8.6‰), indicating 14–41% reduced seawater sulfate in the vent fluid H_2S.

Vent fluid sulfur isotope data also may provide insight into reaction conditions and processes during phase separation, as inferred from Na and Li systematics in vent fluids. H_2S contents of vent fluids are highest in low Cl fluids that have undergone extensive phase separation, due to progressive partition of gases in the vapor phase. Most vent fluids show no correlation of $\delta^{34}S$ values with H_2S content (Fig. 14.14). However, samples collected from

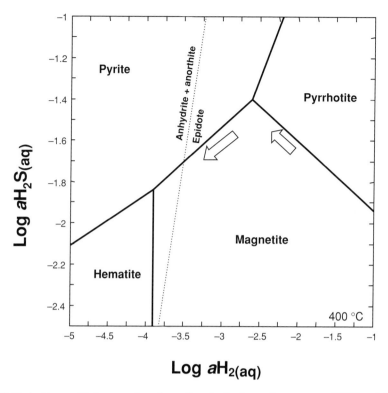

Fig. 14.13 Reduction–oxidation reactions in seafloor hydrothermal systems at 400 °C and 500 bars. Low water–rock reducing systems are buffered near pyrite–pyrrhotite–magnetite or along the pyrrhotite–magnetite join. More oxidizing systems, including most vent fluids, plot along the pyrite–magnetite join at a position controlled by the *anhydrite + anorthite (plagioclase) = clinozoisite (epidote)* reaction. Phase separation processes partition gases into the escaping vapor phase and produce more oxidizing conditions in the reaction zone. This may lead to increased pyrite dissolution and oxidation of some H_2S to SO_4, leading to anhydrite precipitation.

9 to 10° N on the EPR immediately following the 1991 ridge-crest eruption (Haymon *et al.*, 1991) provide a unique data set that dramatically confirms our assertion that phase separation is an inherently oxidizing process. The 1991 samples with extremely enriched H_2S follow a trend to lower $\delta^{34}S$ values with increasing degrees of phase separation (Fig. 14.14). The cause of this trend can be understood in terms of degassing, which removes both H_2 and H_2S into the vapor phase, and mineral reactions that attempt to compensate for this oxidative affect. If, immediately prior to phase separation, vent fluid redox chemistry is controlled by pyrite–magnetite equilibria on the anhydrite + anorthite–clinozoisite boundary (Fig. 14.13), as demonstrated from experimental studies (Seyfried and Ding, 1995), then H_2 and H_2S loss will cause pyrite to dissolve and some of the H_2S produced will be oxidized and precipitated as anhydrite. Pyrite exists in altered basalts because of reaction with seawater SO_4 to oxidize pyrrhotite (mss) to pyrite. This results in incorporation of seawater-derived

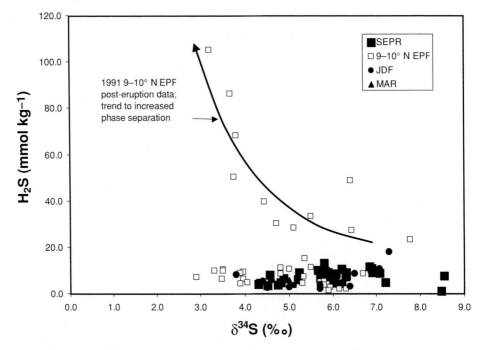

Fig. 14.14 H$_2$S and δ^{34}S variations in hydrothermal vent fluids. Data are from Shanks (2001). Samples with H$_2$S less than 20 mmol kg^{-1} form an elongate cluster that shows no systematic variation with δ^{34}S. However, low chloride samples from a high-temperature vent sampled in 1991 and 1992 following the volcanic eruption on the EPR at 9° 46–52′ N correlate to a distinct and dramatic exponential curve. As phase separation progressed to produce higher and higher H$_2$S fluids, δ^{34}S values decreased from near 8‰ to about 3‰. This change in δ^{34}S values is due to pyrite dissolution (see Fig. 14.13) in the deep reaction zone contributing isotopically light (<3‰) H$_2$S to the evolving hydrothermal system.

SO$_4$ and produces pyrite with δ^{34}S values of about 1.0–1.5‰ (Woodruff and Shanks, 1988). Thus, pyrite dissolution in response to vapor loss adds isotopically light sulfide to the vent fluids, causing the trend to δ^{34}S values of 3‰ or less in fluids with highest H$_2$S. This oxidative reaction mechanism during phase separation may also explain the long-observed but enigmatic presence of anhydrite with δ^{34}S values much lower than seawater in hydrothermally altered rocks recovered by the Ocean Drilling Program (Alt and Shanks, 1998, 2002; Alt *et al.*, 1998).

Carbon isotope systematics are less well understood and are currently the subject of intensive studies relating to abiogenic synthesis reactions and the possible origin of life. In general, vent fluid CO$_2$ related to MORB systems has δ^{13}C values in the −5 to −7‰ range, consistent with inferred values of CO$_2$ in basalt. Methane (CH$_4$) is a persistent minor component in vent fluids and has δ^{13}C values of −15 to −21‰ in most MOR systems. These values are not consistent with isotopic equilibrium with CO$_2$ at vent fluid temperatures but indicate formation at considerably higher temperatures (∼620–770 °C) than indicated for reaction zone conditions. Vent fluid CH_4 may have formed at higher temperature during cooling of the MORB and remained trapped in the rock until mobilized by hydrothermal processes, or CO_2–CH_4 may simply have failed to achieve equilibrium isotopic exchange during hydrothermal reactions. In systems with significant sedimentary components in the hydrothermal reaction zone, such as Guaymas Basin and Middle Valley and, apparently, the Main Endeavour field on the JDF Ridge, CH_4 attains much lower $\delta^{13}C$ values, ranging down to $-55‰$. These values are consistent with microbiological reduction processes at temperatures well below vent fluid temperatures.

In serpentine-hosted systems (Alt and Shanks, 1998, 2002) very negative $\delta^{34}S$ values (to $-45‰$) are found in hydrothermal sulfide minerals. These negative values are interpreted as due to thermophilic microbial sulfate-reduction processes that probably utilize H_2 as an energy source to produce CH_4. Methane is abundant in serpentine-related vent fluids (Charlou *et al.*, 1998) and $\delta^{13}C$ evidence from carbonate veins in serpentine at the MAR Kane Fracture Zone site (ODP Hole 920) also suggests methane production (Alt and Shanks, 2002) related to serpentinization reactions.

14.5.3 Boron and lithium

Boron and lithium stable isotopes are useful because they have significant isotopic fractionation during hydrothermal processes and well-understood chemical reaction processes (Seyfried *et al.*, 1984; Spivack and Edmond, 1987; Berndt and Seyfried, 1990; Palmer, 1991, 1996; Chan *et al.*, 1993; Campbell *et al.*, 1994; James *et al.*, 1995). Both elements are quite soluble at high-temperature conditions and thus provide tracers of the amount of rock reacted in the hydrothermal system. The isotopic systematics can constrain key compositional variables owing to the distinctly different Li and B isotopic composition of seawater, basalt, and its alteration products. For example, the δ^6Li and $\delta^{11}B$ of seawater are -32.3 ± 0.5 and $39.5‰$, respectively. Fresh MORB has δ^6Li from -3.5 to -4.8 and $\delta^{11}B$ from -2.2 to $-3.8‰$ (Spivack and Edmond, 1987; Chan and Edmond, 1988). The isotopic compositions of these elements in altered basalt tend to be intermediate between these limits, being very much a function of mineral composition, the relative abundance of seawater and basaltic components during fluid/rock alteration, and temperature, which governs the magnitude of the fluid–mineral isotopic fractionation factors (Spivack and Edmond, 1987; Chan *et al.*, 1993). In spite of these variables, hydrothermal vent fluids at mid-ocean ridges reveal a relatively narrow range of δ^6Li and $\delta^{11}B$ values, once the seawater component of these species in accounted for, which for B involves a non-trivial adjustment owing to the relatively high B concentration of seawater and its distinct isotopic composition relative to fresh basalt (Spivack and Edmond, 1987).

The range of δ^6Li values for vent fluids from Pacific and Atlantic hydrothermal systems is similar, which suggests similar conditions of hydrothermal alteration. In effect, these data rule out the previously suggested existence of weathered basalt as a source of trace alkalis for MAR vent fluid (Palmer and Edmond, 1989; Chan *et al.*, 1993). Such a process would clearly be reflected by a shift to more negative δ^6Li values, a result confirmed by experimental results (Seyfried *et al.*, 1998), but not observed in the vent fluid δ^6Li values. The δD data for vent fluids also do not show evidence for reactions with weathered rocks, which would produce negative δD values in the vent fluids. In fact, the δD values of MAR vent fluids measured to date are among the highest observed (Fig. 14.11). The δ^6Li compositions of vent fluids, however, are best modeled by isotopic exchange processes involving dissolution of Li from fresh basalt at elevated temperatures and pressures and small but significant re-precipitation in hydrous alteration phases. The isotopic fractionation factor for this is on the order of 3 per mil at 350 °C, which contrasts sharply with analogous processes at lower temperatures where mineral–water fractionation can be as high as 19 per mil for seafloor weathering (Chan *et al.*, 1992). It is in fact the relatively large enrichment of the heavy isotope of Li in rocks altered at low-temperature that precludes weathered basalt as a source of Li for hot spring vent fluids. Finally, it is important to note, that time series data for Li isotopes in vent fluids at EPR 21° N, reveal a time invariance (Chan *et al.*, 1993) that is best accounted for by penetration of seawater into fresh rock together with dissolution re-precipitation processes discussed earlier, at conditions (*T*, *P*, *X*) that have remained relatively, if not entirely constant, over the five-year sampling interval.

In contrast with Li, boron isotopes in vent fluids indicate no evidence of fractionation because boron is quantitatively leached from basalt during alteration at high temperatures (Seyfried *et al.*, 1984). Thus, the isotopic composition of boron is virtually identical to that of fresh MORB (Spivack and Edmond, 1987), and the addition of rock-derived boron to the hydrothermal fluid has been used to infer seawater/rock mass ratios for a number of vent sites in the eastern Pacific, as noted previously. Moreover, Berndt and Seyfried (1990) used experimental constraints together with boron isotope data from the eastern Pacific vent fluids to show that phase separation occurred at temperatures as high as 425 °C.

Boron isotopes in vent fluids from Mid-Atlantic Ridge sites often reveal evidence of boron depletion relative to seawater (Spivack and Edmond, 1987). Indeed, James *et al.* (1995) estimated that approximately 57% of seawater boron has been removed in the low temperature of the hydrothermal convection cell to account for the boron concentration and isotopic composition in high-temperature vent fluids at Broken Spur, 29° N Mid-Atlantic Ridge. Similar analysis for other MAR hydrothermal fluids indicate 23, 34, and 61% boron removal at low temperatures in MARK 1990, MARK 1986, and TAG 1990 fluids, respectively. Although the specific mechanism to account for this is not entirely clear, boron uptake during serpentinization is certainly one possibility. Seyfried and Dibble (1980) showed that boron is taken up from seawater during peridotite alteration at relatively low temperatures (150–200 °C). Thus, considering the availability of peridotite along fracture zones associated with the tectonically active MAR, the observed loss of seawater-derived boron in many of the MAR vent fluids is consistent with a serpentinization mechanism, as emphasized by Palmer (1996).

Returning to the enigmatic high δD vent fluids from the MAR (Fig. 14.11), we note that these are also B-depleted fluids and we speculate that the enriched δD values of these fluids may be due to low-temperature serpentinization of ultramafic rocks during recharge of seawater into the hydrothermal system prior to high-temperature basalt/gabbro reactions in the deep reaction zone. Low-temperature serpentinization, if reaction occurs at sufficiently low water/rock ratios, could result in significant B-loss, δD increase (~ 1–2%), and little change or actual decrease in $\delta^{18}O$ in the fluid. Subsequent high-temperature reaction with basalt would increase B and both δD and $\delta^{18}O$ along trajectories like those plotted in Fig. 14.11, except the starting fluid will no longer have seawater isotopic values. Low-temperature $\delta^{18}O$ shifts in fluid composition will be eliminated by high-temperature re-equilibration in the reaction zone, but δD shifts at low temperature will likely persist because of the low H content of rocks and because high-temperature reactions produces shifts in the same direction as low-temperature reactions. Thus, a two-step reaction process could account for both the B depletion and the relative δD enrichment of the MAR fluids.

14.5.4 Iron and copper

The first data on iron and copper isotopes in vent fluids and sulfide structures have recently been published (Zhu *et al.*, 2000; Sharma *et al.*, 2001). Iron isotope ($\delta^{56}Fe$) variations in vent fluids indicate small but significant variations (-0.30 to -0.77%) relative to the Canyon Diablo iron meteorite and the US Geological Survey basalt standard BCR-2. These results suggest non-biological fractionation of iron isotopes in hydrothermal vent systems, possibly due to precipitation of isotopically heavy sulfide minerals.

Copper isotope values (expressed as $\varepsilon^{65}Cu$, which is similar to the δ scale but uses 0.1% units) of chalcopyrite from seafloor chimneys from a variety of sites vary from about -5 to 12 ε units relative to the National Institutes of Standards Technology (NIST) 976 Cu standard (Zhu *et al.*, 2000). Variations are significant within and between hydrothermal vent fields. Variations are most likely due to fractionation processes in the hydrothermal systems.

14.6 Conclusions

Homogenous and heterogeneous equilibria involving stable and metastable minerals and fluids at a range of temperatures and pressures, govern the chemical and isotopic evolution of mid-ocean ridge hot spring vent fluids. Phase relations in the $NaCl$–H_2O system play a key role in accounting for halide variability, which, in turn, influences the dissolved concentrations of virtually all other species. Indeed, at sufficiently high temperatures and low pressures, segregation of dilute vapors and concentrated brines is predicted, while at relatively extreme conditions halite formation is possible, although slight retrograde effects can induce halite dissolution, as suggested by low Br/Cl ratios of vent fluids at EPR 9–10° N.

Mixing of vapors and brines with evolved seawater in sub-seafloor reaction zones accounts best for the compositional variability of hot spring vent fluids. The correspondence between dissolved Ca and Na with chloride, for example, indicates phase equilibria control by metastable calcic plagioclase. Na fixation, however, is enhanced with increasing dissolved chloride, while release of Na is more likely for vapor-dominated systems. In both cases, plagioclase dissolution/precipitation reactions are likely, as opposed to formation of end-member components of plagioclase solid solution. Plagioclase–fluid equilibrium not only controls dissolved Ca and Na concentrations in vent fluids, but also buffers pH. At temperatures and pressures in the vicinity of 400 °C and 500 bars, plagioclase is predicted to coexist with epidote solid solution over a wide range of redox conditions. More oxidizing conditions enhance epidote stability, which, for a given dissolved chloride concentration, lowers pH. Experimental and theoretical data also reveal that increasing dissolved chloride causes pH to decrease, and dissolved Ca and Fe to increase exponentially while Al is affected less by this.

Results of experimental and theoretical models account well for vent fluid chemistry, but are not in full agreement with petrologic studies of fossil hydrothermal systems where the coexistence of calcic plagioclase and epidote solid solution is seldom observed. Considering the dynamic nature of rock–fluid interaction processes at mid-ocean ridges, on temporal and spatial scales, which are just now being fully explored, together with constraints imposed by existence of metastable minerals, it may be difficult to resolve unambiguously the apparent discrepancy between fluid and rock records without actually drilling into an active high-temperature reaction zone.

There are cases, however, where the extent of mass transfer is sufficiently large to be accurately preserved in the rock record. One such example may involve systems where phase separation effects dominate the local lithology. Indeed, the partitioning of H_2 and H_2S into the vapor phase during phase separation/segregation necessitates oxidation of the rock. In effect, vapor release is an inherently oxidizing process and the evidence of this in the form of ferric iron-rich mineralization likely exists in the region of the sub-seafloor reaction zone from which the vapors are derived. Epidosites, which are thought to represent fossil hydrothermal up-flow zones situated at the base of the sheeted intrusive complex in ophiolites, may owe their origin to just such a process.

In contrast to basalt/gabbro-hosted hydrothermal systems, the recently discovered Rainbow system at 36° 14′ N latitude on the Mid-Atlantic Ridge is situated in ultramafic rocks, yet the vent fluids are characterized by unusually high Fe concentrations, which suggest low fluid pH – not what one typically associates with hydrothermal alteration processes involving ultramafic mineral assemblages. Recent results of experimental studies, however, indicate that kinetic effects involving non-stoichiometric release of SiO_2 from pyroxene minerals, together with sluggish rates of olivine hydrolysis, allow formation of relatively SiO_2-rich hydrous secondary minerals, such as talc and tremolite, which accounts for the observed acidity and high Fe of the Rainbow vent fluids.

Stable isotope studies of modern hydrothermal vent fluids, seafloor massive sulfide deposits, and hydrothermally altered mineral assemblages recovered from the ocean crust provide important constraints on sub-seafloor hydrothermal alteration processes. Some isotopic systems have long been relied on to accomplish this (oxygen, hydrogen, carbon sulfur, boron, lithium), while others (Cu, Fe) offer great promise for the future.

Oxygen isotope data from vent fluids show that water–rock reactions at peak hydrothermal condition dominate earlier stages of reaction at lower temperatures along the fluid flow path. Moreover, these data indicate water/rock mass ratios from 5 to 0.5, which are in good agreement with ratios determined by boron and strontium isotope data, and with constraints imposed by the dissolved concentration of mobile trace alkali elements in vent fluids. Hydrogen isotope data reveal that most fluids follow mineral–fluid fractionation trends, but that the most positive and negative δD values fall outside this range, suggesting processes other than simple reaction with unaltered silicates at peak hydrothermal conditions. Alternative explanations entail reaction with rocks previously altered at higher or lower temperatures, phase separation processes, or by contributions of magmatic water (for the low δD values). We speculate that high δD values may be linked to formation of low-temperature serpentinized peridotite with the downwelling limb of seawater derived hydrothermal fluid, which subsequently reacts with basalt/gabbro at higher temperatures.

Stable isotope studies of $\delta^{13}C$ and $\delta^{34}S$ provide important constraints on sources of carbon and sulfur in vent fluids, as well as on the nature of the reduction–oxidation processes in sub-seafloor reaction zones. Carbon isotope systematics are consistent with a mantle source of carbon for vent fluids from unsedimented ridges. These data, however, are not consistent with isotopic equilibration of CO_2 and CH_4 at vent fluid temperatures, but rather indicate formation at much higher temperatures, which implicates leaching of the gases from magmatic components in the rock, where equilibration ceased at temperature below approximately 620 °C. Carbon isotope data have been used with great success to identify sedimentary sources of carbon in hydrothermal systems.

Vent fluid sulfur data provide a powerful means for not only assessing sulfur sources in sub-seafloor hydrothermal systems, but also constraining reaction processes that confirm the previously established linkages between phase separation, fluid/rock mass ratio, and oxidation effects. These data have also been relied on to constrain mixing reactions between hydrothermal fluid and seawater resulting in seafloor sulfide mineralization.

Owing largely to the significant fractionation that is possible between isotopes of lithium and boron during hydrothermal alteration processes, studies using these isotopes have helped to constrain fluid/rock mass ratios and fluid and mineral source terms in sub-seafloor reaction zones, as well as to link recharge and reaction zone processes more effectively, with collateral benefit for development of reaction path models.

Chemical and isotopic data from vent fluids together with constraints imposed by experimental and theoretical studies have enhanced greatly our understanding of the temporal, spatial, and compositional dimensions of hydrothermal reactions zones at mid-ocean ridges. Where once it was thought that the sub-seafloor reaction zone was a region defined largely by relatively constant conditions, it is now recognized that the source region for vent fluid chemistry is a dynamic environment and may change, both chemically and physically, in space and time, giving rise to compositional variability in fluids and coexisting minerals on time scales unimagined earlier.

Acknowledgments

The manuscript benefited greatly from thoughtful reviews by Jeff Alt and Karen Von Damm. Editorial comments and scientific recommendations by Harry Elderfield and Earl Davis also improved the paper. The authors would also like to thank Ms. Sharon Kressler, Department of Geology and Geophysics, University of Minnesota, for assisting with preparation of the final version of the manuscript. Funding from the National Science Foundation through research grants OCE-0117117, OCE-9911471, and OCE-9818908 to the senior author made possible many of the concepts developed in the paper.

References

Allen, D. E. and Seyfried, W. E., Jr. 2003. Alteration and mass transfer in the MgO–CaO–FeO–Fe$_2$O$_3$–SiO$_2$–Na$_2$O–H$_2$O–HCl system at 400 °C and 500 bars: implications for pH and compositional controls on vent fluids from ultramafic-hosted hydrothermal systems at mid-ocean ridges. *Geochim. Cosmochim. Acta* **67**: 1,531–1,542.

Alt, J. C. 1995. Subseafloor processes in mid-ocean ridge hydrothermal systems. In *Seafloor Hydrothermal Systems: Physical, Chemical, Biological, and Geological Interactions*, eds. S. E. Humphris, R. A. Zierenberg, L. S. Mullineaux, and R. E. Thomson. Washington, DC: American Geophysical Union, pp. 85–114.

Alt, J. C. and Shanks, W. C. III 1998. Sulfur in serpentinized oceanic peridotites; serpentinization processes and microbial sulfate reduction. *J. Geophys. Res. B, Solid Earth Planets* **103**(5): 9,917–9,929.

  2002. Serpentinization of abyssal peridotites from the MARK area, Mid-Atlantic Ridge: sulfur geochemistry and reaction modeling. *Geochim. Cosmochim. Acta* **67**(4): 641–653.

Alt, J. C. and Teagle, D. A. H. 2002. Hydrothermal alteration and fluid fluxes in ophiolites and oceanic crust. In *Penrose Conference on Ophiolites and Oceanic Crust: New Insights from Field Studies and the Ocean Drilling Program*, eds. Y. Dilek, E. M. Moores, D. Elthon, and A. Nicolas. Marshall, CA, pp. 273–282.

Alt, J. C., Laverne, C., Vanko, D. A., Tartarotti, P., Teagle, D. A. H., Bach, W., Zuleger, E., Erzinger, J., Honnorez, J., *et al.* 1996. Hydrothermal alteration of a section of upper oceanic crust in the eastern equatorial Pacific: a synthesis of results from Site 504 (DSDP Legs 69, 70, and 83, and ODP Legs 111, 137, 140, and 148). In *Proceedings of the Ocean Drilling Program: Scientific Results*, Vol. 148, ed. L. B. Stokking. College Station, TX: Ocean Drilling Program, pp. 417–434.

Alt, J. C., Teagle, D. A. H., Brewer, T., Shanks, W. C., III, and Halliday, A. 1998. Alteration and mineralization of an oceanic forearc and the ophiolite–ocean crust analogy. *J. Geophys. Res. B, Solid Earth* **103**(B6): 12,365–12,380.

Anderko, A. and Pitzer, K. S. 1993. Equation of state representation of phase equilibria and volumetric properties of NaCl–H$_2$O above 573 °C. *Geochim. Cosmochim. Acta* **57**: 1,657–1,680.

Bach, W., Alt, J. C., Niu, Y., Humphris, S. E., Erzinger, J., and Dick, H. J. B. 2001. The geochemical consequences of late-stage low-grade alteration of lower ocean crust at the SW Indian Ridge: results from ODP Hole 735B (Leg 176). *Geochim. Cosmochim. Acta* **65**(19): 3,267–3,287.

Barriga, F. J. A. S., Costa, I. M. A., Relvas, J. M. R. S., Ribeiro, A., Fouquet, Y., Ondreas, H., Parson, L. M., and Anonymous 1997. The Rainbow serpentinites and serpentinite–sulphide stockwork (Mid Atlantic Ridge, AMAR segment): a preliminary report of the FLORES results. *Eos, Trans. Am. Geophys. Union* **78**(46, Suppl.): 832–833.

Berndt, M. E. and Seyfried, W. E., Jr. 1990. Boron, bromine, and other trace elements as clues to the fate of chlorine in mid-ocean ridge vent fluids. *Geochim. Cosmochim. Acta* **54**(8): 2,235–2,245.

1993. Calcium and sodium exchange during hydrothermal alteration of calcic plagioclase at 400 °C and 400 bars. *Geochim. Cosmochim. Acta* **57**(18): 4,445–4,451.

1997. Calibration of Br/Cl fractionation during subcritical phase separation of seawater: possible halite at 9 to 10° N East Pacific Rise. *Geochim. Cosmochim. Acta* **61**(14): 2,849–2,854.

Berndt, M. E., Seyfried, W. E., Jr., and Beck, J. W. 1988. Hydrothermal alteration processes at mid-ocean ridges: experimental and theoretical constraints from Ca and Sr exchange reactions and Sr isotopes ratios. *J. Geophys. Res.* **93**(5): 4,573–4,583.

Berndt, M. E., Seyfried, W. E., Jr., and Janecky, D. R. 1989. Plagioclase and epidote buffering of cation ratios in mid-ocean ridge hydrothermal fluids: experimental results in and near the supercritical region. *Geochim. Cosmochim. Acta* **53**(9): 2,283–2,300.

Berndt, M. E., Allen, D. E., and Seyfried, W. E., Jr. 1996a. Reduction of CO_2 during serpentinization of olivine at 300 °C and 500 bar. *Geology* **24**(4): 351–354.

Berndt, M. E., Seal, R. R., II, Shanks, W. C., III, and Seyfried, W. E., Jr. 1996b. Hydrogen isotope systematics of phase separation in submarine hydrothermal systems: experimental calibration and theoretical models. *Geochim. Cosmochim. Acta* **60**(9): 1,595–1,605.

Bischoff, J. L. 1991. Densities of liquids and vapors in boiling $NaCl$–H_2O solutions: a *PVTX* summary from 300 to 500 °C. *Am. J. Sci.* **291**(4): 309–338.

Bischoff, J. L. and Pitzer, K. S. 1989. Liquid–vapor relations for the system $NaCl$–H_2O: summary of the *P–T–X* surface from 300 to 500 °C. *Am. J. Sci.* **289**: 217–248.

Bischoff, J. L. and Rosenbauer, R. J. 1987. Phase separation in seafloor geothermal systems: an experimental study on the effects of metal transport. *Am. J. Sci.* **287**: 953–978.

Bohlke, J. K. and Shanks, W. C., III 1994. Stable isotope study of hydrothermal vents at Escanaba Trough: observed and calculated effects of sediment–seawater interaction. In *Geologic, Hydrothermal, and Biologic Studies at Escanaba Trough, Gorda Ridge, Offshore Northern California*, eds. J. L. Morton, R. A. Zierenberg, and C. A. Reiss. Denver, CO: US Geological Survey, pp. 223–240.

Bonatti, E., Lawrence, J. R., and Morandi, N. 1984. Serpentinization of oceanic peridotites: temperature dependence of mineralogy and boron content. *Earth Planet. Sci. Lett.* **70**: 88–94.

Bowers, T. S. 1989. Stable isotope signatures of water–rock interaction in mid-ocean ridge hydrothermal systems: sulfur, oxygen, and hydrogen. *J. Geophys. Res.* **94**(B5): 5,775–5,786.

Bowers, T. S. and Taylor, H. P. 1985. An integrated chemical and stable isotope model of the origin of mid-ocean ridge hot spring systems. *J. Geophys. Res.* **90**: 12,583–12,606.

Bowers, T. S., Von Damm, K. L., and Edmond, J. M. 1985. Chemical evolution of mid-ocean ridge hot springs. *Geochim. Cosmochim. Acta* **49**: 2,239–2,252.

Bowers, T. S., Campbell, A. C., Measures, C. I., Spivack, A. J., Khadem, M., and Edmond, J. M. 1988. Chemical controls on the composition of vent fluids at 13–11° N and 21° N, East Pacific Rise. *J. Geophys. Res.* **93**(5): 4,522–4,536.

Bray, A. and Von Damm, K. L. 2004. Controls on the alkali metal compostion of mid-ocean ridge hydrothermal fluids: constraints from 9–10° N, East Pacific Rise. *Geochim. Cosmochim. Acta* in press.

Butterfield, D. A. and Massoth, G. J. 1994. Geochemistry of North Cleft Segment vent fluids – temporal changes in chlorinity and their possible relation to recent volcanism. *J. Geophys. Res.* **99**(B3): 4,951–4,968.

Butterfield, D. A., McDuff, R. E., Mottl, M. J., Lilley, M. D., Lupton, J. E., and Massoth, G. J. 1994. Gradients in the composition of hydrothermal fluids from the Endeavor segment vent field: phase separation and brine loss. *J. Geophys. Res.* **99**: 9,561–9,583.

Campbell, A. C. and Edmond, J. M. 1989. Halide systematics of submarine hydrothermal vents. *Nature* **342**: 168–170.

Campbell, A. C., Bowers, T. S., and Edmond, J. M. 1988a. A time-series of vent fluid compositions from 21° N, EPR (1979, 1981, and 1985) and the Guaymas Basin, Gulf of California (1982, 1985). *J. Geophys. Res.* **93**: 4,537–4,549.

Campbell, A. C., Palmer, M. R., Klinkhammer, G. P., Bowers, T. S., Edmond, J. M., Lawrence, J. R., Casey, J. F., Thompson, G., Humphris, S., Rona, P., and Karson, J. A. 1988b. Chemistry of hot springs on the Mid-Atlantic Ridge. *Nature* **335**: 514–519.

Campbell, A. C., German, C. R., Palmer, M. R., Gamo, T., and Edmond, J. M. 1994. Chemistry of hydrothermal fluids from Escanaba Trough, Gorda Ridge. In *Geologic, Hydrothermal, and Biologic Studies at Escanaba Trough, Gorda Ridge, Offshore Northern California*, eds. J. L. Morton, R. A. Zierenberg, and C. A. Reiss. Denver, CO: US Geological Survey, pp. 201–222.

Chacko, T., Cole, D. R., and Horita, J. 2001. Equilibrium oxygen, hydrogen, and carbon isotope fractionation factors applicable to geologic systems. In *Stable Isotope Geochemistry*, eds. J. W. Valley, and D. R. Cole. Washington, DC: Mineralogical Society of America, pp. 1–81.

Chan, L. H. and Edmond, J. M. 1988. Variation of lithium isotope composition in the marine environment: a preliminary report. *Geochim. Cosmochim. Acta* **52**: 1,711–1,717.

Chan, L. H., Edmond, J. M., Thompson, G., and Gillis, K. 1992. Lithium isotopic composition of submarine basalts: implications for the lithium cycle in the oceans. *Earth Planet. Sci. Lett.* **108**: 151–160.

Chan, L. H., Edmond, J. M., and Thompson, G. 1993. A lithium study of hot springs and metabasalts from mid-ocean ridge hydrothermal systems. *J. Geophys. Res.* **98**: 9,653–9,659.

Chan, L. H., Alt, J. C., and Teagle, D. A. H. 2002. Lithium and lithium isotope profiles through the upper oceanic crust: a study of seawater–basalt exchange at ODP Sites 504B and 896A. *Earth Planet. Sci. Lett.* **201**(1): 187–201.

Charlou, J. L., Bougault, H., Fouquet, Y., Collette, B., Rona, P., *et al.* 1991. North and south 15° 20′ N fracture zone intersection: hydrothermal activity involves outcropping mantle in the inner floor of the rift valley. *Terra Abstracts* **3**(1): 311–312.

Charlou, J. L., Donval, J. P., Jean-Baptiste, P., Mills, R., Rona, P., Von Herzen, D., and Anonymous 1993. Methane, nitrogen, carbon dioxide, and helium isotopes in vent fluids from TAG hydrothermal field, 26° N-MAR. *Eos, Trans. Am. Geophys. Union* **74**(43, Suppl.): 99.

Charlou, J. L., Donval, J. P., Jean-Baptiste, P., Dapoigny, A., and Rona, P. A. 1996a. Gases and helium isotopes in high temperature solutions sampled before and after ODP Leg 158 drilling at TAG hydrothermal field (26° N, MAR). *Geophys. Res. Lett.* **23**(23): 3,491–3,494.

Charlou, J. L., Fouquet, Y., Donval, J.-P., Auzende, J.-M., Jean-Baptiste, P., Stievenard, M., and Michel, S. 1996b. Mineral and gas chemistry of hydrothermal fluids on an ultrafast spreading ridge: East Pacific Rise, 17° to 19° S (Naudur cruise, 1993) phase separation processes controlled by volcanic and tectonic activity. *J. Geophys. Res.* **101**(7), 15,899–15,919.

Charlou, J. L., Fouquet, Y., Bougault, H., Donval, J. P., Etoubleau, J., Jean-Baptiste, P., Dapoigny, A., Appriou, P., and Rona, P. A. 1998. Intense CH$_4$ plumes generated by serpentinization of ultramafic rocks at the intersection of the 15° 20′ N fracture zone and the Mid-Atlantic Ridge. *Geochim. Cosmochim. Acta* **62**(13): 2,323–2,333.

Charlou, J. L., Donval, J. P., Douville, E., Jean-Baptiste, P., Radford-Knoery, J., Fouquet, Y., Dapoigny, A., and Stievenard, M. 2000. Compared geochemical signatures and the evolution of Menez Gwen (37° 50′ N) and Lucky Strike (37° 17′ N) hydrothermal fluids, south of the Azores triple junction on the Mid-Atlantic Ridge. *Chem. Geol.* **171**(1–2): 49–75.

Ding, K. and Seyfried, W. E., Jr. 1992. Determination of Fe–Cl complexing in the low pressure supercritical region (NaCl fluid) – iron solubility constraints on pH of subseafloor hydrothermal fluids. *Geochim. Cosmochim. Acta* **56**(10): 3,681–3,692.

Donval, J. P., Charlou, J. L., Douville, E., Knoery, J., Fouquet, Y., Ponsevera, E., Jean-Baptiste, P., Stievenard, M., German, C. R., and Anonymous 1997. High H_2 and CH_4 content in hydrothermal fluids from Rainbow site newly sampled at 36°14′ N on the AMAR segment, Mid-Atlantic Ridge (diving FLORES cruise, July 1997): comparison with other MAR sites. *Eos, Trans. Am. Geophys. Union* **78**(46, Suppl.): 832.

Douville, E., Charlou, J. L., Oelkers, E. H., Bienvenu, P., Jove Colon, C. F., Donval, J. P., Fouquet, Y., Prieur, D., and Appriou, P. 2002. The Rainbow vent fluids (36° 14′ N, MAR): the influence of ultramafic rocks and phase separation on trace metal content in Mid-Atlantic Ridge hydrothermal fluids. *Chem. Geol.* **184**, 37–48.

Edmond, J. M., Measures, C., McDuff, R. E., Chan, L., Collier, R., Grant, B., Gordon, L. I., and Corliss, J. 1979. Ridge crest hydrothermal activity and the balances of the major and minor elements in the ocean: the Galapagos data. *Earth Planet. Sci. Lett.* **46**: 1–18.

Edmond, J. M., Campbell, A. C., Palmer, M., Klinkhammer, G. P., German, C. R., Edmonds, H. N., Elderfield, H., Thompson, G., and Rona, P. 1995. Time series studies of vent fluids from the TAG and MARK sites (1986, 1990) Mid-Atlantic Ridge: a new solution chemistry model and mechanism for Cu/Zn zonation in massive sulfide ore deposits. In *Hydrothermal Vents and Processes*, eds. L. M. Parson, C. L. Walker, and D. R. Dixon. London: Geological Society, pp. 411.

Evans, W. C., White, L. D., and Rapp, J. B., 1988. Geochemistry of some gases in hydrothermal fluids from the southern Juan de Fuca Ridge. *J. Geophys. Res.* **93**: 15,305–15,313.

Fornari, D. J., Shank, T., Von Damm, K. L., Gregg, T. K. P., Lilley, M., Levai, G., Bray, A., Haymon, R. M., Perfit, M. R., and Lutz, R. 1998. Time-series temperature measurements at high-temperature hydrothermal vents, East Pacific Rise 9 degrees 49′–51′ N: evidence for monitoring a crustal cracking event. *Earth Planet. Sci. Lett.* **160**: 419–431.

Fouquet, Y., Charlou, J. L., Ondreas, H., Radford-Knoery, J., Donval, J. P., Douville, E., Appriou, R., Cambon, P., Pell, H., Landure, J. Y., Normand, A., Ponsevera, E., German, C. R., Parson, L. M., Barriga, F. J. A. S., Costa, I. M. A., Relvas, J. M. R. S., and Ribeiro, A. 1997. Discovery and first submersible investigations on the Rainbow hydrothermal field on the MAR (36° 14′ N). *Eos, Trans. Am. Geophys. Union* **78** (46, Suppl.): 832.

Fournier, R. O. 1987. *Conceptual Models for Brine Evolution in Magmatic-Hydrothermal Systems*, US Geological Survey, Professional Paper 13,502. Boulder, CO: US Geological Survey.

Gamo, T., Chiba, H., Masuda, H., Edmonds, H. N., Fujioka, K., Kodama, Y., Nanba, H., and Sano, Y. 1996. Chemical characteristics of hydrothermal fluids from the TAG mound of the mid-Atlantic Ridge in August 1994: implications for spatial and temporal variability of hydrothermal activity. *Geophys. Res. Lett.* **23**: 3,483–3,486.

Gillis, K. M. 2002. The rootzone of an ancient hydrothermal system exposed in the Troodos ophiolite, Cyprus. *J. Geol.* **110**(1): 57–74.

Gillis, K. M. and Thompson, G. 1993. Metabasalts from the mid-Atlantic Ridge: new insights into hydrothermal systems in slow spreading crust. *Contrib. Mineral. Petrol.* **113**: 502–523.

Gillis, K. M., Thompson, G., and Kelley, D. S. 1993. A view of the lower crustal component of hydrothermal systems at the Mid-Atlantic Ridge. *J. Geophys. Res.* **98**(B11): 19,597–19,619.

Harper, G. D., Bowman, J. R., and Kuhns, R. J. 1988. A field, chemical, and stable isotope study of subseafloor metamorphism of the Josephine Ophiolite, California–Oregon. *J. Geophys. Res.* **93**(5): 4,625–4,656.

Haymon, R., Fornari, D., *et al.*, 1991. Eruption of the EPR crest at 9° 45′–52 ′N since late 1989 and its effects on hydrothermal venting: results of the ADVENTURE program, an ODP site survey with Alvin. *Eos, Trans. Am. Geophys. Union* **72**: 480.

Hekinian, R., Fevrier, M., Bischoff, J. L., Picot, P., and Shanks, W. C. 1980. Sulfide deposits from the East Pacific Rise near 21° N. *Science* **207**(4,438): 1,433–1,444.

Holm, N. G. and Charlou, J. L. 2001. Initial indications of abiotic formation of hydrocarbons in the Rainbow ultramafic hydrothermal system, Mid-Atlantic Ridge. *Earth Planet. Sci. Lett.* **191**: 1–8.

Horita, J., Cole, D. R., and Wesolowski, D. J. 1995. The activity–composition relationship of oxygen and hydrogen isotopes in aqueous salt solutions: III, Vapor–liquid water equilibration of NaCl solutions to 350 °C. *Geochim. Cosmochim. Acta* **59**(6): 1,139–1,151.

Humphris, S. and Thompson, G. 1978. Hydrothermal alteration of oceanic basalts by seawater. *Geochim. Cosmochim. Acta* **42**: 127–136.

Ito, E., Harris, M. D., and Anderson, F. T., Jr. 1983. Alteration of the oceanic crust and geologic cycling of chlorine and water. *Geochim. Cosmochim. Acta* **47**: 1,613–1,624.

James, R. H., Elderfield, H., and Palmer, M. R. 1995. The chemistry of hydrothermal fluids from the Broken Spur site 29° N Mid Atlantic Ridge. *Geochim. Cosmochim. Acta* **59**(4): 651–661.

James, R. H., Allen, D. E., and Seyfried, W. E., Jr. 2003. An experimental study of alteration of oceanic crust and terrigenous sediments at moderate temperatures (51–351 °C): insights as to chemical processes in near-shore ridge-flank hydrothermal systems. *Geochim. Cosmochim. Acta* **67**(4): 681–691.

Janecky, D. R. and Seyfried, W. E., Jr. 1984. Formation of massive sulfide deposits on oceanic ridge crests: incremental reaction models for mixing between hydrothermal solutions and seawater. *Geochim. Cosmochim. Acta* **48**: 2,723–2,738.

   1986. Hydrothermal serpentinization of peridotite within the oceanic crust: experimental investigations of mineralogy and major element chemistry. *Geochim. Cosmochim. Acta* **50**: 1,357–1,378.

Johnson, J. W., Oelkers, E. H., and Helgeson, H. C. 1992. SUPCRT92 – A software package for calculating the standard molal thermodynamic properties of minerals, gases, aqueous species, and reactions from 1-bar to 5000-bar and 0 °C to 1000 °C. *Computers and Geosci.* **18**(7): 899–947.

Juteau, T., Manac'h, G., Moreau, O., Lecuyer, C., and Ramboz, C. 2000. The high temperature reaction zone of the Oman ophiolite: new field data, microthermometry of fluid inclusions, PIXE analyses and oxygen isotopic ratios. *Mar. Geophys. Res.* **21**(3–4): 351–385.

Kelley, D. S., Karson, J. A., Blackman, D. K., Fruh-Green, G. L., Butterfield, D. A., Lilley, M. D., Olson, E. J., Schrenk, M. O., Roe, K. K., Lebon, G. T., and Rivizzigno, P. 2001. An off-axis hydrothermal vent field near the Mid-Atlantic Ridge at 30° N. *Nature* **412**(6843): 145–149.

Kimball, K. L. 1988. High-temperature hydrothermal alteration of ultramafic cumulates from the base of the sheeted dikes in the Josephine Ophiolite, NW California. *J. Geophys. Res.* **93**(5): 4,675–4,687.

Langmuir, C., Humphris, S., Fornari, D., Van Dover, C., Von Damm, K. L., Tivey, M. K., Colodner, D., Charlou, J. L., Desonie, D., Wilson, C., Fouquet, Y., Klinkhammer, G., and Bougault, H. 1997. Hydrothermal vents near a mantle hot spot: the Lucky Strike vent field at 37 degrees N on the Mid-Atlantic Ridge. *Earth Planet. Sci. Lett.* **148**(1–2): 69–91.

Lilley, M. D., Butterfield, D. A., Olson, E. J., Lupton, J. E., Macko, S. A., and McDuff, R. E., Anomolous 1993. CH_4 and NH_4 concentrations at an unsedimented mid-ocean ridge hydrothermal system. *Nature* **364**: 45–47.

Lister, C. R. B. 1982. "Active" and "passive" hydrothermal systems in the oceanic crust: predicted physical conditions. In *The Dynamic Environment of the Ocean Floor*, eds. K. A. Fanning and F. T. Manheim. Lexington Books, Gomer Publishing: pp. 441–470.

  1995. Heat transfer between magmas and hydrothermal systems, or, six lemmas in search of a theorem. *Geophys. J. Int.* **120**: 45–59.

McCollom, T. and Shock, E. L. 1998. Fluid–rock interactions in the lower oceanic crust: thermodynamic models of hydrothermal alteration. *J. Geophys. Res.* **103**(1): 547–575.

Merlivat, L., Pineau, F., and Javoy, M. 1987. Hydrothermal vent waters at 13° N on the East Pacific Rise: isotopic composition and gas concentration. *Earth Planet. Sci. Lett.* **84**(1): 100–108.

Michard, G., Albarede, F., Michard, A., Minster, J. F., Charlou, J. L., and Tan, N. 1984. Chemistry of solutions from the 13° N East Pacific Rise hydrothermal site. *Earth Planet. Sci. Lett.* **67**(3): 297–307.

Mottl, M. J. and Wheat, C. G. 1994. Hydrothermal circulation through mid-ocean ridge flanks – fluxes of heat and magnesium. *Geochim. Cosmochim. Acta* **58**(10): 2,225–2,237.

Muehlenbachs, K. and Clayton, R. N. 1972. Oxygen isotope geochemistry of submarine greenstones. *Can. J. Earth Sci.* **9**: 471–478.

Nehlig, P., Juteau, T., Bendel, V., and Cotten, J. 1994. The root zones of oceanic hydrothermal systems – constraints from the Samail Ophiolite (Oman). *J. Geophys. Res.* **99**(B3): 4,703–4,713.

Oosting, S. E. and Von Damm, K. L. 1996. Bromide/chloride fractionation in seafloor hydrothermal fluids from 9–10° N East Pacific Rise. *Earth Planet. Sci. Lett.* **144**(1–2): 133–145.

Palmer, M. R. 1991. Boron-isotope systematics of the Halmahera arc (Indonesia) lavas: evidence for involvement of the subducted slab. *Geology* **19**: 215–217.

  1992. Controls over the chloride concentration of submarine hydrothermal vent fluids: evidence from Sr/Ca and Sr^{87}/Sr^{86} ratios. *Earth Planet. Sci. Lett.* **109**: 37–47.

  1996. Hydration and uplift of the oceanic crust on the Mid-Atlantic Ridge associated with hydrothermal activity: evidence from boron isotopes. *Geophys. Res. Lett.* **23**(23): 3,479–3,482.

Palmer, M. R. and Edmond, J. M. 1989. Cesium and rubidium in submarine hydrothermal fluids: evidence for recycling of alkali elements. *Earth Planet. Sci. Lett.* **95**: 8–14.

Pelayo, A. M., Stein, S., and Stein, C. A. 1994. Estimation of oceanic hydrothermal heat flux from heat flow and depths of mid-ocean ridge seismicity and magma chambers. *Geophys. Res. Lett.* **21**(8): 713–716.

Richardson, C. J., Cann, J. R., Richards, H. G., and Cowan, J. G. 1987. Metal-depleted root zones of the Troodos ore-forming hydrothermal systems, Cyprus. *Earth Planet. Sci. Lett.* **84**: 243–253.

Seewald, J. S., Cruse, A. M., and Saccocia, P. J. 2003. Aqueous volatiles in hydrothermal fluids from the main Endeavour Field northern Juan de Fuca Ridge: temporal variability following earthquake activity. *Earth Planet. Sci. Lett.* **216**(4): 575–590.

Seyfried, W. E., Jr. 1987. Experimental and theoretical constraints on hydrothermal alteration processes at mid-ocean ridges. *Ann. Rev. Earth Planet. Sci.* **15**, 317–335.

Seyfried, W. E., Jr. and Bischoff, J. L. 1981. Experimental seawater–basalt interaction at 300 °C, 500 bars, chemical exchange, secondary mineral formation and implications for the transport of heavy metals. *Geochim. Cosmochim. Acta* **45**: 135–149.

Seyfried, W. E., Jr. and Dibble, W. E., Jr. 1980. Seawater–peridotite interaction at 300 degrees C and 500 bars: implications for the origin of oceanic serpentinites. *Geochim. Cosmochim. Acta* **44**(2): 309–322.

Seyfried, W. E., Jr. and Ding, K. 1995. Phase equilibria in subseafloor hydrothermal systems: a review of the role of redox, temperature, pH and dissolved Cl on the chemistry of hot spring fluids at mid-ocean ridges. In *Seafloor Hydrothermal Systems: Physical, Chemical, Biologic and Geological Interactions*, eds. S. E. Humphris, R. A. Zierenberg, L. S. Mullineaux, and R. E. Thompson. Washington, DC: American Geophysical Union, pp. 248–273.

Seyfried, W. E., Jr., Janecky, D. R., and Mottl, M. J. 1984. Alteration of the oceanic crust: implications for geochemical cycles of lithium and boron. *Geochim. Cosmochim. Acta* **48**: 557–569.

Seyfried, W. E., Jr., Ding, K., and Berndt, M. E. 1991. Phase equilibria constraints on the chemistry of hot spring fluids at mid-ocean ridges. *Geochim. Cosmochim. Acta* **55**: 3,559–3,580.

Seyfried, W. E., Jr., Chen, X., and Chan, L.-H. 1998. Trace element mobility and lithium isotope exchange during hydrothermal alteration of seafloor weathered basalt: an experimental study at 350 °C, 500 bars. *Geochim. Cosmochim. Acta* **62**: 949–960.

Seyfried, W. E., Jr., Seewald, J. S., Berndt, M. E., Ding, K., and Foustoukos, D. 2003. Chemistry of hydrothermal vent fluids from the Main Endeavour Field, Northern Juan de Fuca Ridge: geochemical controls in the aftermath of June 1999 seismic events. *J. Geophys. Res.* **108**: 2,429–2,449.

Shanks, W. C., III 2001. Stable isotopes in seafloor hydrothermal systems: vent fluids, hydrothermal deposits, hydrothermal alteration, and microbial processes. In *Stable Isotope Geochemistry*, eds. J. W. Valley and D. R. Cole. Washington, DC: Mineralogical Society of America, pp. 469–526.

Shanks, W. C., III, Bohlke, J. K., and Seal, R. R., II 1995. Stable isotopes in mid-ocean ridge hydrothermal systems: interactions between fluids, minerals, and organisms. In *Seafloor Hydrothermal Systems: Physical, Chemical, Biological, and Geological Interactions*, eds. S. E. Humphris, R. A. Zierenberg, L. S. Mullineaux, and R. E. Thomson. Washington, DC: American Geophysical Union, pp. 194–221.

Sharma, M., Polizzotto, M., and Anbar, A. D. 2001. Iron isotopes in hot springs along the Juan de Fuca Ridge. *Earth Planet. Sci. Lett.* **194**: 39–51.

Sohn, R. A., Fornari, D. J., Von Damm, K. L., Hildebrand, J. A., and Webb, S. C. 1998. Seismic and hydrothermal evidence for a cracking event on the East Pacific Rise crest at 9 degrees 50′ N. *Nature (London)* **396**(6707): 159–161.

Spivack, A. J. and Edmond, J. M. 1987. Boron isotope exchange between seawater and the oceanic crust. *Geochim. Cosmochim. Acta* **51**: 1,033–1,045.

Spooner, E. T. C., Beckinsale, R. D., Fyfe, W. S., and Smewing, J. D. 1974. O-18 enriched ophiolitic metabasic rocks from E. Liguria (Italy), Pindos (Greece), and Troodos (Cyprus). *Contrib. Mineral. Petrol.* **47**: 41–62.

Stakes, D. S. 1991. Oxygen and hydrogen isotope compositions of oceanic plutonic rocks: high-temperature deformation and metamorphism of oceanic layer 3. In eds. H. P. Taylor, Jr., J. R. O'Neil, and I. R. Kaplan, *Stable Isotope Geochemistry*. University Park, PA: The Geochemical Society, pp. 77–90.

Stakes, D. S. and O'Neil, J. R. 1982. Mineralogy and stable isotope geochemistry of hydrothermally altered oceanic rocks. *Earth Planet. Sci. Lett.* **57**: 285–304.

Stein, C. A. and Stein, S. 1994. Constraints on hydrothermal heat flux through the oceanic lithosphere from global heat flow. *J. Geophys. Res.* **99**: 3,081–3,095.

Stein, C. A., Stein, S., and Pelayo, A. M. 1995. Heat flow and hydrothermal circulation. In *Seafloor Hydrothermal Systems: Physical, Chemical, Biological and Geological Interactions*, eds. S. E. Humphris, R. A. Zierenberg, L. S. Mullineaux, and R. E. Thomson. Washington, DC: American Geographical Union, pp. 425–445.

Tagirov, B. and Schott, J. J. 2001. Aluminum speciation in crustal fluids revisited. *Geochim. Cosmochim. Acta* **65**: 3,965–3,990.

Teagle, D. A. H., Alt, J. C., and Halliday, A. N. 1998. Tracing the chemical evolution of fluids during hydrothermal recharge: constraints from anhydrite recovered in ODP Hole 504B. *Earth Planet. Sci. Lett.* **155**(3–4): 167–182.

Trefry, J. H., Butterfield, D. B., Metz, S., Massoth, G. J., Trocine, R. P., and Feely, R. A. 1994. Trace metals in hydrothermal solutions from Cleft Segment on the Southern Juan de Fuca Ridge. *J. Geophys. Res.* **99**(B3): 4,925–4,935.

Urabe, T., Baker, E. T., Ishibashi, J., Feely, R. A., Marumo, K., Massoth, G. J., Maruyama, A., Shitashima, K., Okamura, K., Lupton, J. E., Sonoda, A., Yamazaki, T., Aoki, M., Gendron, J., Greene, R., Kaiho, Y., Kisimoto, K., Lebon, G., Matsumoto, T., Nakamura, K., Nishizawa, A., Okano, O., Paradis, G., Roe, K., Shibata, T., Tennant, D., Vance, T., Walker, S. L., Yabuki, T., and Ytow, N. 1995. The effect of magmatic activity on hydrothermal venting along the superfast-spreading East Pacific Rise. *Science* **269**(5227): 1,092–1,095.

Von Damm, K. L. 1988. Systematics of and postulated controls on submarine hydrothermal solution chemistry. *J. Geophys. Res.* **93**: 4,551–4,562.

1990. Seafloor hydrothermal activity: black smoker chemistry and chimneys. *Ann. Rev. Earth Planet. Sci.* **18**: 173–205.

1995. Controls on the chemistry and temporal variability of seafloor hydrothermal fluids. In *Seafloor Hydrothermal Systems: Physical, Chemical, Biologic and Geologic Interactions*, eds. S. E. Humphris, R. A. Zierenberg, L. S. Mullineaux, and R. E. Thompson. Washington, DC: American Geophysical Union, pp. 222–248.

2000. Chemistry of hydrothermal vent fluids from 9°–10° N, East Pacific Rise: "time zero," the immediate posteruptive period. *J. Geophys. Res.* **105**(5): 11,203–11,222.

Von Damm, K. L. and Bischoff, J. L. 1987. Chemistry of hydrothermal solutions from the Southern Juan de Fuca Ridge. *J. Geophys. Res.* **92**: 11,334–11,346.

Von Damm, K. L., Edmond, J. L., Grant, B., Measures, C. I., Walden, B., and Weiss, R. F. 1985. Chemistry of submarine hydrothermal solutions at 21° N, East Pacific Rise. *Geochim. Cosmochim. Acta* **49**: 2,221–2,237.

Von Damm, K. L., Oosting, S. E., Kozlowski, R., Buttermore, L. G., Colodner, D., Edmonds, H. N., Edmond, J. M., and Grebmeir, J. M. 1995. Evolution of East Pacific Rise hydrothermal vent fluids following a volcanic eruption. *Nature* **375**: 47–50.

Von Damm, K. L., Buttermore, L. G., Oosting, S. E., Bray, A. M., Fornari, D. J., Lilley, M. D., and Shanks, W. C., III 1997. Direct observation of the evolution of a seafloor "black smoker" from vapor to brine. *Earth Planet. Sci. Lett.* **149**: 101–111.

Von Damm, K. L., Bray, A. M., Buttermore, L. G., and Oosting, S. E. 1998. The geochemical controls on vent fluids from the Lucky Strike vent field, Mid-Atlantic Ridge. *Earth Planet. Sci. Lett.* **160**: 521–536.

Von Damm, K. L., Parker, C. M., Gallant, R. M., and Loveless, J. P. 2002. Chemical evolution of hydrothermal fluids from EPR 21° N: 23 years later in a phase separating world. *Eos, Trans. Am. Geophys. Union* **83**(47): V61B–1365.

Welhan, J. A. and Craig, H. 1979. Methane and hydrogen in East Pacific Rise hydrothermal fluids. *American Geophysical Union; 1979 fall annual meeting* **60**(46): 863.

     1983. Methane, hydrogen and helium in hydrothermal fluids. In *Hydrothermal Processes at Seafloor Spreading Centers*, eds. P. A. Rona, K. Bostrom, L. Laubier, and K. L. Smith, Jr. New York: Plenum Press, pp. 391–411.

Wetzel, L. R. and Shock, E. L. 2000. Distinguishing ultramafic- from basalt-hosted submarine hydrothermal systems by comparing calculated vent fluid compositions. *J. Geophys. Res.* **105**(4): 8,319–8,340.

Wilcock, W. S. 1997. A model for the formation of tranisent event plumes above mid-ocean ridge hydrothermal systems. *J. Geophys. Res.* **102**(B6): 12,109–12,121.

Wilcock, W. S. D. and Delaney, J. R. 1996. Mid-ocean ridge sulfide deposits: evidence for heat extraction from magma chambers or cracking fronts? *Earth Planet. Sci. Lett.* **145**(1–4): 49–64.

Wilcock, W. S. and McNabb, A. 1996. Estimates of permeability on the Endeavour segment of the Juan de Fuca mid-ocean ridges. *Earth Planet. Sci. Lett.* **138**: 83–91.

Woodruff, L. G. and Shanks, W. C., III 1988. Sulfur isotope study of chimney minerals and vent fluids from 21° N, East Pacific Rise: hydrothermal sulfur sources and disequilibrium sulfate reduction. *J. Geophys. Res.* **93**: 4,562–4,572.

You, C.-F., Butterfield, D. A., Spivack, A. J., Gieskes, J. M., Gamo, T., and Campbell, A. J. 1994. Boron and halide systematics in submarine hydrothermal systems – effects of phase separation and sedimentary contributions. *Earth Planet. Sci. Lett.* **123**(1–4): 227–238.

You, C.-F., Castillo, P. R., Gieskes, J. M., Chan, L. H., and Spivack, A. J. 1996. Trace element behavior in hydrothermal experiments: implications for fluid processes at shallow depths in subduction zones. *Earth Planet. Sci. Lett.* **140**: 41–52.

Zhu, X.-K., O'Nions, R. K., Guo, Y., Belshaw, N. S., and Rickard, D. 2000. Determination of natural Cu-isotope variation by plasma-source mass spectrometry: implications for use in geochemical tracers. *Chem. Geol.*, **163**: 139–149.

15

Alteration of the upper oceanic crust: mineralogy, chemistry, and processes

Jeffrey C. Alt

15.1 Introduction

Convection of seawater fluids within the oceanic crust, driven by heat from formation of the crust and lithosphere, has significant effects on the chemistry and physical properties of the crust and on the composition of seawater (Edmond *et al.*, 1979a). At the axes of seafloor spreading, hydrothermal systems driven by magmatic heat sources give rise to black smoker hydrothermal vents and massive sulfide deposits, and cause hydrothermal metamorphism of ocean crust (sub-greenschist up to amphibolite grade), mainly in the sheeted dike complex and uppermost gabbros (Fig. 15.1). As the crust moves away from the spreading axis, cooling of the crust and lithosphere drives continuing off-axis convection for up to tens of millions of years. Temperatures are much lower, however, and circulation is essentially restricted to the volcanic section, becoming increasingly isolated from the ocean by accumulating sediment. The lavas are generally only slightly recrystallized at low temperatures (0–150 °C), but chemical and physical effects are nevertheless significant.

This chapter focuses mainly on low-temperature alteration of the volcanic section of oceanic crust. High-temperature axial hydrothermal metamorphism is briefly discussed here, as this has recently been reviewed elsewhere (Alt, 1995a, 1999a; Kelley *et al.*, 2002; Alt and Bach, 2003; Gillis, 2003). Data documenting alteration processes and their effects in the crust are based mainly on sampling by the Ocean Drilling Program (ODP), which has provided many sections of the uppermost few hundred meters of ocean crust (Fig. 15.2). Several deep basement holes (~400–600 m) show that processes can change with depth, and varying alteration effects in crust of different ages reveal a sequence of alteration processes that continue for up to several tens of million years. The origin and evolution of these temporal and spatial variations are discussed in this chapter in terms of mineralogy, chemistry, and hydrothermal processes.

Hydrogeology of the Oceanic Lithosphere, eds. E. E. Davis and H. Elderfield. Published by Cambridge University Press.
© Cambridge University Press 2004.

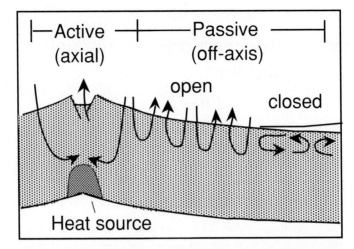

Fig. 15.1 Illustration of convective regimes in oceanic crust (after Lister, 1982). Active high-temperature convection at mid-ocean ridges is driven by a magmatic heat source, whereas lower temperature passive circulation off-axis is driven by cooling of the crust and lithosphere.

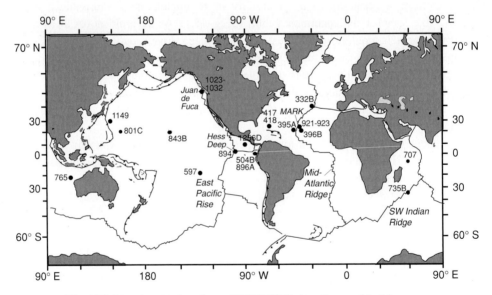

Fig. 15.2 Map showing locations of DSDP/ODP sites discussed in this paper.

15.2 Low-temperature alteration of upper oceanic crust

Because the volcanic section is more porous and permeable than the underlying sheeted dikes by one to four orders of magnitude (Fisher, 1998; Becker, this volume), much greater volumes of fluid can circulate more freely through the heterogeneous volcanic section than through the more uniform, massive sheeted dike section. Thus, the volcanic section is

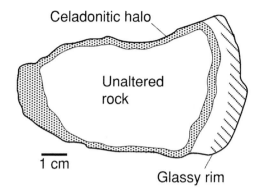

Fig. 15.3 Sketch of dredged basalt, illustrating an example of dark celadonitic alteration halos along fractures and exposed surfaces. Glassy pillow rim, impermeable microcrystalline/glassy area immediately beneath glassy rim, and host rock remain unaffected. See text for further description and discussion. (Basalt dredged from the Siqueiros fracture zone, after Natland and Rosendahl, 1980)

efficiently cooled, even at the spreading axis, where low-temperature processes occur in the lavas while higher temperature hydrothermal reactions leading to black smoker fluids occur at greater depths (Rosenberg *et al.*, 1993). Alteration processes in the uppermost oceanic crust (volcanic section) occur at low temperatures (0–150 °C), but a sequence of processes with characteristic chemical and mineralogical effects can be identified. These include: (1) early formation of dark celadonitic alteration halos in basalts by low-temperature hydrothermal fluids at the spreading axis; (2) oxidation and formation of reddish iron-oxyhydroxide-rich alteration halos along fractures by cold seawater solutions; (3) pervasive formation of saponite and pyrite during restricted flow of more evolved seawater solutions; and (4) precipitation of late carbonates filling fractures.

15.2.1 Initial low-temperature hydrothermal processes

Alteration effects in the youngest lavas (<0.1–2.5 Ma) consist of dark bands or halos, a few millimeter up to 5 cm wide, sub-parallel to fractures and exposed surfaces (Fig. 15.3; Scott and Hajash, 1976; Humphris *et al.*, 1980; Laverne and Vivier, 1983; Bohlke *et al.*, 1984; Adamson and Richards, 1990; Laverne, 1993; Hunter *et al.*, 1998). These dark halos result from filling of pore spaces with celadonite and Fe-oxyhydroxides, and the partial replacement of olivine phenocrysts by these phases. In contrast, the interior portions of the rocks are unaltered and pore spaces remain open. These same dark halo features are observed in older rocks (up to 170 Ma), but with other alteration effects superimposed, indicating that this early stage is a ubiquitous process in the uppermost crust (Andrews, 1977; Bohlke *et al.*, 1980; Alt and Honnorez, 1984; Alt *et al.*, 1986a, 1992; Gillis *et al.*, 1992; Alt, 1993; Shipboard Scientific Party, 1990, 2000).

Although typically referred to as celadonite (a ferric mica), the phyllosilicates exhibit a range of compositions from true celadonite to glauconite (Al-rich ferric mica) and nontronite

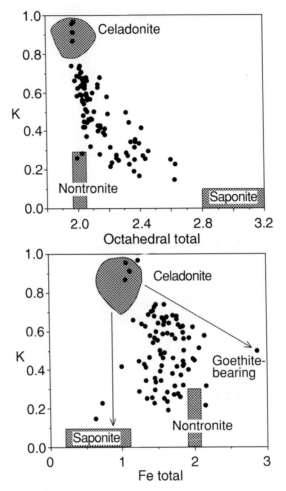

Fig. 15.4 Compositions of celadonitic phyllosilicates from oceanic basalts. What is commonly referred to as celadonite in oceanic basalts is typically a mixture of celadonite with nontronite and Fe-oxyhydroxides, but some analyses from older rocks include saponite. High Al contents of some analyses (not shown) fall in the field of glauconite rather than celadonite. See text for further discussion. Representative analyses taken from various sources (Kempe, 1974; Andrews, 1977; Scheidegger and Stakes, 1977; Buckley *et al.*, 1978; Seyfried *et al.*, 1978; Bohlke *et al.*, 1980; Humphris *et al.*, 1980; Laverne and Vivier, 1983; Alt and Honnorez, 1984; Alt *et al.*, 1986a, 1992; Peterson *et al.*, 1986; Adamson and Richards 1990; Alt, 1993; Laverne, 1993; Teagle *et al.*, 1996). All Fe as Fe^{3+}, formulae per $O_{10}(OH)_2$.

(ferric smectite; Fig. 15.4). Compositional trends toward saponite (Mg-smectite) and Fe-oxyhydroxide result from mixtures with these phases. Saponite is present in these mixtures only in the older rocks that contain superimposed alteration effects and abundant discrete saponite. Thus, ferric-iron-rich phases celadonite ± Fe-oxyhydroxides ± nontronite form initially in very young rocks, which undergo subsequent alteration with formation of abundant saponite.

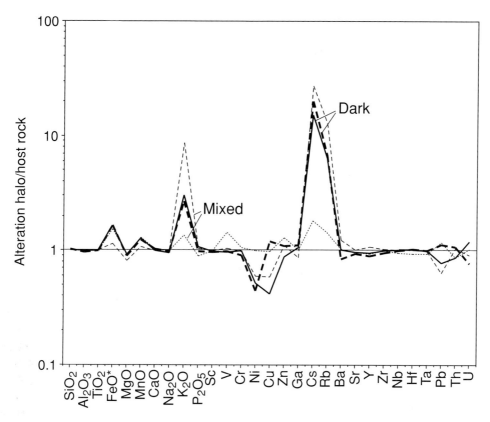

Fig. 15.5 Spidergram showing bulk-rock chemical changes resulting from the formation of alteration halos along fractures (see text). Data for alteration halos are normalized to constant Ti and compared to the composition of the adjacent less-altered host rock. Elevated iron and alkali contents result from addition of celadonite to dark alteration halos along veins (see text). Subsequent oxidation and superimposition of red or brown alteration halos on early dark halos results in formation of mixed (brown and dark) halos, which exhibit chemical changes similar to those of the dark halos. Unpublished data for ODP Site 843B (J. Alt and D. Teagle, unpublished data).

Compared to unaltered basalts, the dark alteration halos have gained Fe, K, Rb, Cs, H_2O and exhibit elevated Fe^{3+}/Fe^T, $\delta^{18}O$, and $^{87}Sr/^{86}Sr$ (Fig. 15.5; Humphris *et al.*, 1980; Laverne and Vivier, 1983; Bohlke *et al.*, 1984; Park and Staudigel, 1990; Alt, 1993; Alt and Teagle, unpublished data). Iron contents typically increase by 10–20% (1–2 wt.% FeO^T) over the amount present in unaltered rocks, but iron may increase by more than 40% (Humphris *et al.*, 1980; Laverne and Vivier, 1983). These changes reflect addition of 5–20 vol.% celadonite and iron oxide to the rock with little alteration of primary phases. Local losses of Ni and mobility of Cu are related to breakdown of olivine and recrystallization of igneous sulfides. High K, Rb, B, and Cs contents of celadonites compared to basalts contribute to the enrichment of these elements in altered basalts (Staudigel *et al.*, 1981a,b; Hart and Staudigel, 1986; Berndt and Seyfried 1986; Teagle *et al.*, 1996).

Quantifying the amount of basalt affected by this process is difficult because of the lack of deep holes (>100 m) in young crust, the heterogeneity of alteration effects in the crust (Fig. 15.6), and the presence of superimposed alteration in older rocks. In Hole 504B these effects are limited to the upper 300 m of the volcanic section (Fig. 15.7). At some sites celadonite abundance decreases below about 150–250 m into basement (Pertsev and Rusinov, 1979; Alt and Honnorez, 1984; Gillis *et al.*, 1992), but at others there are clear peaks in abundance of celadonite at shallow depths and again at depths of 400–500 m sub-basement (Holes 801C and 1256D in Fig. 15.6). Despite uncertainties, it is estimated that on the order of 2–10% of the volcanic section is affected by this process. Because primary minerals are essentially unaltered and secondary minerals fill up primary pore space, this process would result in slight increases in density and decreased porosity in the alteration halos. It is unlikely that this would affect larger scale porosity, however, except in extreme cases of more focused hydrothermal up-flow. At 170 Ma Site 801, the strong effects of axial up-flow in the volcanic section (celadonite and quartz in veins) sealed many fractures and may have affected formation physical properties, as well as affecting pathways for later fluid flow (Shipboard Scientific Party, 2000; Alt and Teagle, 2003).

Celadonites from submarine basalts have $\delta^{18}O$ values of 16.5–23.4‰, which indicate formation temperatures of less than about 80 °C (Kastner and Gieskes, 1976; Seyfried *et al.*, 1978; Stakes and O'Neil, 1982; Bohlke *et al.*, 1984; Alt and Teagle, 2003). The alkali enrichment of the dark halos can be accounted for by seawater, but an external source of iron is required. The most likely source is distal, mixed hydrothermal fluids at the spreading axis. High-temperature (~350 °C) hydrothermal fluids upwelling from the base of the low-permeability sheeted dike complex mix with cooler seawater in the highly permeable volcanic section, causing precipitation of sulfides and the resultant cooler (~10–100 °C), mixed fluids remain enriched in Fe and Si and contain elevated alkali concentrations compared to seawater (Edmond *et al.*, 1979a,b). It is this type of fluid that vents from the seafloor at the Galapagos Spreading Center and which results in the formation of hydrothermal celadonite/nontronite and Fe–Mn oxide deposits on the seafloor (Edmond *et al.*, 1979b; Alt, 1988). These fluids also comprise the diffuse venting that surrounds black smokers and that occurs sporadically along the spreading axis between high-temperature sulfide deposits and black smoker vents. The patchiness of such diffuse flow contributes to the variability in the thickness of the dark celadonitic alteration halos and in their distribution within the crust.

In the oldest oceanic crust drilled by ODP (Hole 801C), the link between distal hydrothermal fluids and the celadonite-rich dark alteration halos is clear. This hole penetrates 415 m into tholeiitic basement formed at a fast-spreading rate. A 20-m thick low-temperature (30–50 °C) hydrothermal silica + iron deposit lies atop the tholeiitic basement, which for several meters beneath is pervasively and intensely altered to celadonite, glauconite, beidellite, and K-feldspar (Fig. 15.6; Alt *et al.*, 1992; Alt and Teagle, 2003). A similar but thinner deposit occurs 125 m deeper. Veins of celadonite and quartz/chalcedony occur to the base of the drilled section, as do celadonite-bearing alteration halos (Shipboard Scientific Party, 2000). Oxygen isotope data for quartz and celadonite indicate temperatures of 40–110 °C, increasing downward (Fig. 15.8; Alt and Teagle, 2003). Here the link is

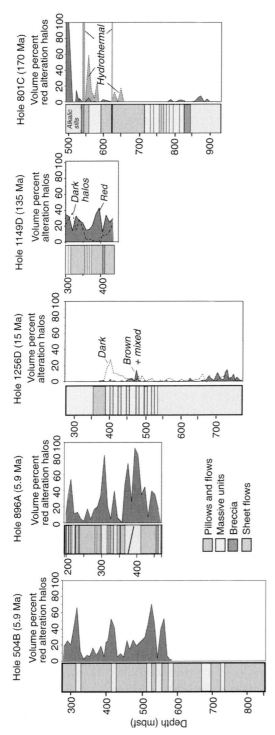

Fig. 15.6 Proportion of alteration halos versus depth in five ODP drillcores. The red (or brown) oxidation halos result from circulation of relatively cold, oxidizing seawater, giving indications of the distribution and amount of this flow with depth at the different sites (see text). Common dark celadonitic halos are indicated for Sites 1149 and 1256, but dark halos are mostly obscured by superimposed red halos at these and other sites, leading to "mixed" halos and to underestimation of the abundance of dark alteration halos. Oxidation halos tend to be better developed in massive units at Sites 504 and 896 because coarser grain size results in greater porosity (see text). Low-temperature hydrothermal silica–iron deposits (hachure) in Hole 801C and associated intense alteration (light gray shading) are indicated. Lithostratigraphy is highly simplified and schematic, and is based on core recoveries of ~30% for Holes 504B and 896A, 17% for Hole 1149D, 47% for Hole 801C, and ~50% for Hole 1256D. Geophysical logs indicate that recovered cores are representative of alteration processes in the cored crust at those sites (Alt and Teagle, 2003). Breccias are generally not indicated, except for a thick unit in Hole 801C. The top 60 m of Hole 801C consists of alkaline sills intruded into sediments. Data from Alt et al. (1996a), Alt and Teagle (2000, 2003), and Shipboard Scientific Party (2000, 2003).

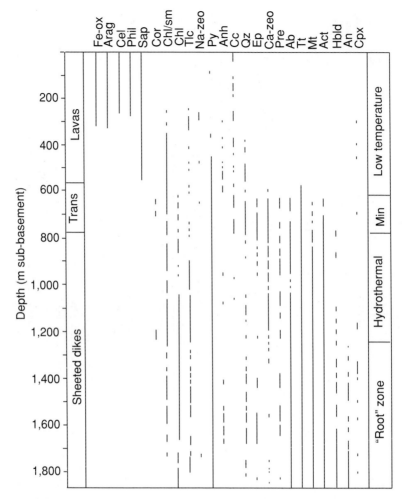

Fig. 15.7 Distribution of secondary minerals with depth in the upper oceanic crust penetrated by ODP Hole 504B (after Alt *et al.*, 1986a, 1996b). Lines show general distributions of secondary minerals, but mineral occurrences are typically heterogeneous. Min = mineralized with disseminated pyrite; Fe-ox, Fe-oxyhydroxide; Arag, aragonite; Cel, celadonite; Phil, phillipsite; Sap, saponite; Cor, corrensite; Chl/sm, chlorite/smectite; Chl, chlorite; Tlc, talc; Zeo, Zeolite (Na = analcite, natrolite; Ca = laumontite > scolecite ≫ heulandite); Py, pyrite; Anh, anhydrite; Cc, calcite; Qz, quartz; Ep, epidote; Pre, prehnite; Ab, albite; Tt, titanite; Mt, magnetite; Act, actinolite; Hbld, hornblende; An, secondary anorthite; Cpx, secondary clinopyroxene.

clear: the large amount of silica required for the thick hydrothermal deposit must be derived from higher-temperature hydrothermal fluids at depth, and the intense celadonitic alteration beneath the deposit is related to its formation. The underlying veins and alteration halos are less-intense features resulting from the same fluids flowing generally upward through the uppermost crust. Similar effects are common at other sites, but without the hydrothermal

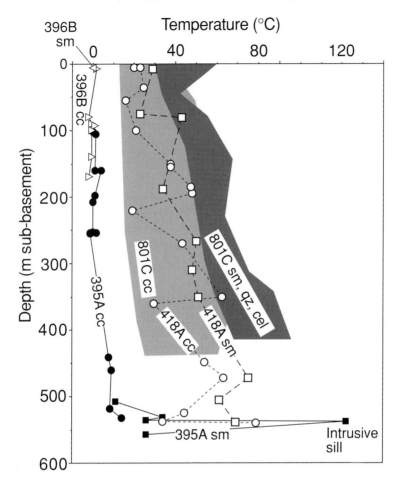

Fig. 15.8 Alteration temperatures versus depth for several deep ODP/DSDP basement holes. Temperatures calculated based on oxygen isotopic analyses of secondary vein minerals (cc, carbonates; sm, smectite; cel, celadonite, qz, quartz) and assuming equilibrium with normal seawater. Smectites form early while the crust is relatively warm, whereas carbonates form at lower temperatures after further cooling of the crust. Compare with temperatures for Hole 896A in Fig. 15.14 and for Hole 504B in Fig. 15.17. Data from Muehlenbachs (1979), Lawrence and Drever (1981), Staudigel *et al.* (1981*b*), Bohlke *et al.* (1984), Alt *et al.* (1986b, 1992), and Alt (unpublished data for Hole 801C). Data for 80 carbonate and 16 other mineral separates from Hole 801C are shown as shaded fields for clarity. Water–rock interactions at low temperature can lead to ^{18}O-depletions of fluids and slight overestimation of temperatures, whereas glacial periods can lead to ^{18}O-enrichment of seawater and slight underestimation of temperatures (Lawrence and Drever, 1981).

deposits and the associated intense alteration. Iron and Si can be enriched in ridge flank fluids by reaction with basalt at temperatures <100 °C (Wheat and Mottl, 2000), but alkalis are depleted in these fluids and the concentrations of Fe and Si are about five times lower than in low-temperature fluids derived from higher-temperature axial hydrothermal reactions at depth (Edmond *et al.*, 1979a). The latter fluids are also enriched in alkalis (Edmond *et al.*, 1979a), making these fluids more efficient at producing the intense alteration associated with hydrothermal deposits at Site 801 and the more typical celadonitic alteration halos throughout the uppermost crust (Alt and Teagle, 2003). The Si–Fe hydrothermal deposits at Site 801 formed directly atop tholeiitic basement and celadonitic alteration halos occur elsewhere in very young crust having little or no sediment cover, indicating that the source of Fe for this alteration process is not sedimentary.

The formation of dark celadonitic halos may not always be related to sustained diffuse venting of hydrothermal fluids, but could also result from episodes such as diking events or sill intrusions that disrupt hydrothermal vents and cause widespread diffuse flow of mixed hydrothermal fluids and partly evolved seawater (Von Damm *et al.*, 1995). Some celadonite may also form during oxidation (see below), but this requires that circulating seawater was sufficiently modified to transport significant amounts of iron, or that iron was derived locally (from within the rock). The small to non-existent amount of primary phase alteration in the youngest lavas plus the addition of several weight percent FeO to these rocks suggest that the low-temperature, mixed hydrothermal fluids with deep sources are the dominant process of celadonite formation in the uppermost crust.

15.2.2 *Low-temperature seawater–rock interactions (oxidation) in the volcanic section*

Lavas from the uppermost crust greater than 2.7 Ma in age retain dark celadonitic halos but also exhibit superimposed alteration effects, commonly in millimeter- to centimeter-scale zonations around veins and fractures, indicating continued hydrothermal reactions during passive off-axis circulation (Bass, 1976; Andrews, 1977, 1980; Bohlke *et al.*, 1980; Alt and Honnorez, 1984; Gillis *et al.*, 1992; Alt, 1993; Alt *et al.*, 1986a, 1996a,b; Teagle *et al.*, 1996; Hunter *et al.*, 1998; Shipboard Scientific Party, 2000; Alt and Teagle, 2003). A typical zonation is shown in Fig. 15.9, where a celadonite vein is surrounded by a dark celadonitic halo and a reddish Fe-oxyhydroxide halo, and where the remainder of the rock consists of a dark gray saponite zone. In the Fe-oxyhydroxide halo olivine is partly to totally replaced and pores filled by Fe-oxyhydroxide and Fe-oxide (goethite, hematite, and x-ray amorphous Fe-oxyhydroxide), ± saponite ± celadonite. Titanomagnetite is intensively maghemitized and primary sulfides are oxidized. The mineralogy of the dark celadonitic halo is similar, but the abundance of Fe-oxyhydroxides is much lower and celadonite greater. In the saponite zone olivine is partly to totally replaced, and pores are filled by saponite ± talc ± pyrite ± calcite. Titanomagnetite is less-intensively maghemitized and primary and secondary sulfides are disseminated in the groundmass. Secondary minerals typically comprise 5–20% of the rocks, although abundances may be higher locally, especially in interpillow areas, breccias, or fault zones (up to 40–90%; Andrews, 1977; Bohlke *et al.*, 1980; Alt *et al.*, 1996b; Hunter *et al.*, 1998; Shipboard Scientific Party, 2000, 2003).

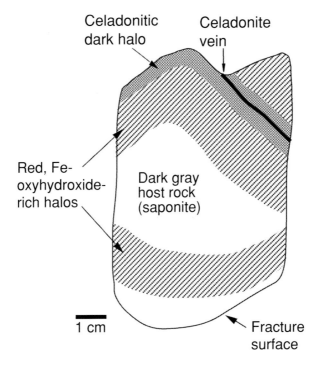

Fig. 15.9 Sketch of a hand specimen from the upper volcanic section of ODP Hole 504B, illustrating typical occurrence of early dark celadonitic halo and superimposed reddish Fe-oxyhydroxide-rich alteration halo in dark gray host rock characterized by saponite and pyrite. This is but one of many relationships between the alteration halos. See text for further description and discussion.

Figure 15.9 shows a typical alteration zonation, but relationships among the different zones vary: the Fe-oxyhydroxide zone may be superimposed upon or exterior or interior to the celadonite zone; or the celadonite zone may be absent (Figs. 15.9 and 15.10; Alt *et al.*, 1996b). These geometries indicate that the Fe-oxyhydroxide zone is the result of a later, superimposed alteration process reflecting interaction of rock with relatively cold, oxidizing seawater (Andrews, 1977, 1980; Bohlke *et al.*, 1980, 1981, 1984; Staudigel *et al.*, 1981b; Alt and Honnorez, 1984; Alt *et al.*, 1986a, 1992, 1996a,b; Alt, 1993; Alt and Teagle, 2003). The concentric alteration zonation within individual rock fragments (e.g. Fig. 15.3) and the observed chemical changes in the rocks (see below) indicate that these alteration zones form by diffusion into and out of the rock perpendicular to fractures along which seawater fluids circulate. Oxidizing conditions are maintained in the fracture and alteration halo, and the inner edge of the alteration halo is an oxidation front that migrates inward. Secondary pyrite is commonly concentrated in the more reducing host rock outside the oxidation front (Andrews, 1979; Alt *et al.*, 1989; Hunter *et al.*, 1998).

Saponite (Mg-smectite) in submarine basalts is typically iron-bearing, and compositions range to low-charge Mg-smectite (stevensite) and talc (Alt *et al.*, 1986a; Hunter *et al.*, 1998; Alt, 1999b; Giorgetti *et al.*, 2001). FeO/(FeO + MgO) is variable, ranging from 0.1 to 0.5 for saponites associated with oxidation zonations. Measured values of Fe^{3+}/Fe^T for

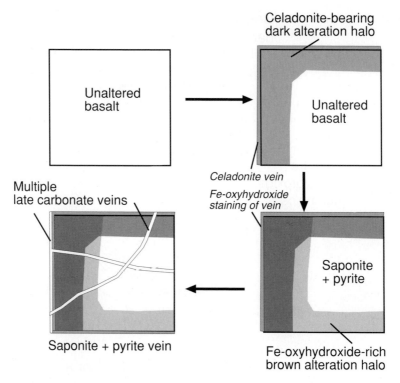

Fig. 15.10 Schematic illustration of development of alteration halos and superimposition of alteration effects in submarine basalts (after Alt and Teagle, 2003). Initial alteration consists of the formation of dark celadonitic alteration halos along fractures, from deeply sourced, low-temperature hydrothermal fluids. This is followed by circulation of cold oxidizing seawater and development of iron-oxyhydroxide-rich oxidation halos in the rock, either superimposed on earlier dark celadonitic halos or on unaltered basalt. At the same time, saponite replaces olivine and fills pores in the host rock, along with trace pyrite. Finally, late carbonate veins form in multiple stages. See text for further discussion.

saponites range up to 0.8 (Alt and Honnorez, 1984), but these high values reflect post-sampling oxidation, and ~30% of the iron in *in situ* saponite is ferrous, consistent with formation at reducing conditions (Andrews *et al.*, 1983). Thus alteration conditions can vary significantly on the scale of centimeters, even within an individual rock fragment, with ferric oxyhydroxides forming at one site and pyrite and saponite forming in the adjacent rock under more reducing conditions. Chemical changes in the rocks vary accordingly.

   Alteration temperatures estimated from oxygen isotopic data for smectites and carbonates are typically 0–80 °C (Fig. 15.8; Muehlenbachs, 1979; Lawrence, 1979, 1991; Lawrence and Drever, 1981; Staudigel *et al.*, 1981b; Bohlke *et al.*, 1984; Alt *et al.*, 1986b, 1992; Alt and Teagle, 2003). This oxidizing seawater alteration has also been referred to as low-temperature alteration, the seawater zone, seafloor weathering, oxidative diagenesis or alteration, and oxic alteration, but the term oxidizing seawater alteration is preferred. The term seafloor weathering is better reserved for rocks that remain exposed to seawater at the seafloor for millions of years (Thompson, 1973, 1983; see below).

Until recently, the effects of oxidizing seawater alteration were thought to occur mainly in the upper half of the volcanic section, where access of cold seawater to the crust is less restricted (Andrews, 1977; Alt and Honnorez, 1984; Alt *et al.*, 1986a,b; Gillis and Robinson, 1988). Hole 504B is the only basement hole to penetrate through the volcanic section, so it has been used as a reference site for the upper ocean crust. Figures 15.7 and 15.6 show that oxidation effects in Hole 504B are restricted to the upper half of the volcanic section, but alteration is heterogeneous at the meter to centimeter scale throughout this interval. Some other deep basement sites exhibit a general decrease in the frequency and intensity of oxidation effects with depth below a few hundred meters (Andrews, 1977; Pertsev and Rusinov, 1979; Alt and Honnorez, 1984; Gillis and Robinson, 1988). In contrast, other deep basement holes exhibit partitioning of oxidation effects into discrete depth intervals. Basement from 170 Ma Hole 801C exhibits only slight oxidation, which is concentrated at the top of basement and below ~300 m sub-basement (Fig. 15.6). The proportion of oxidation effects at this site (~2 vol. %) is an order of magnitude less than in the upper 300 m of Holes 504B and 896A (20–30 vol. %). Also in contrast to Hole 504B, oxidation effects in Hole 1256D, formed at a superfast-spreading rate in the eastern Pacific, are partitioned into two zones, at ~150–250 and 400–500 m sub-basement, and even increase slightly in abundance with depth (Fig. 15.6). These two examples emphasize the critical role that the local lithology and basement structure play in controlling fluid flow and alteration of the upper oceanic crust.

Critical factors for fluid flow in partly sedimented crust are the location and abundance of basement outcrops that provide sites for fluid ingress and egress (Fisher *et al.*, 2003). Sites 801 and 1256 are in crust formed at fast- to superfast-spreading rates where basement has low relief (<100 m; Carbotte and Sheirer, this volume), whereas Sites 504 and 806 are in crust formed at an intermediate spreading rate where basement relief is greater (~50–200 m; Swift *et al.*, 1998; Carbotte and Scheirer, this volume). Thus basement relief appears to have a significant effect on seawater access and oxidation of the upper crust, implying a possible spreading rate effect. In apparent conflict with this, however, are the abundant oxidation effects at Site 1149 in crust formed at a fast-spreading rate (Fig. 15.6).

The partitioning of oxidation effects into different zones at some sites (Fig. 15.6) is consistent with inferences from heat flow data for fluid flow along discrete high-permeability horizons (Becker *et al.*, 1989; Fisher *et al.*, 1990, 1994; Davis *et al.*, 1996; Fisher and Becker, 2000). An important point shown by Fig. 15.6 is that there is not a simple decrease in alteration effects with depth in the lavas. Finally, Fig. 15.6 illustrates the heterogeneity of alteration effects in the uppermost crust, showing that, with the exception of increased C contents, older crust is not necessarily more altered than younger crust. Nor is there a clear distinction between crust formed at fast versus slow- or intermediate-spreading rates (compare Holes 801C and 1149D formed at fast-spreading rates in Fig. 15.6). Clearly, the local structure and hydrothermal history are more important than the age of the crust.

Locally elevated K_2O and H_2O contents occur at spacings of ~50–100 m in the 110 Ma volcanic basement of Holes 417D and 418A (366 and 544 m deep, respectively; Fig. 15.11; Flower *et al.*, 1979). These coincide with changes in igneous chemistry and magnetic

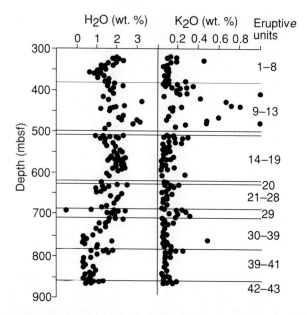

Fig. 15.11 Whole-rock H_2O and K_2O contents versus depth in 110 Ma Hole 418A. Horizontal lines show major stratigraphic breaks, which are defined on the basis of chemical eruptive units and changes in magnetic inclination (after Flower *et al.*, 1979). There is a general downward decrease in alteration intensity as indicated by decreasing H_2O and K_2O contents, with zones of more intense alteration (high H_2O and K_2O) possibly related to preferential fluid flow and alteration along lithologic boundaries (see text). This figure represents least altered "background" type alteration and does not include more highly altered samples (alteration halos, breccias, pillow rims, etc.).

inclination, suggesting that lithostratigraphic boundaries may represent preferential flow pathways along which seawater solutions circulated in the buried basement. Basalts having the most altered oxygen isotopic compositions are also from breccias and unit boundaries at this site, suggesting that these are enhanced flow pathways (Muehlenbachs, 1979).

Glass at pillow rims is altered to "palagonite" (hydrated glass), smectite (mainly saponite and Al-saponite), and zeolites (mainly phillipsite, but also analcite and rare chabazite, gismondine, and faujasite). Glass fragments are cemented by these same minerals ± celadonite and Fe-oxyhydroxides, and pyrite may be present (Melson and Thompson, 1973; Andrews, 1977; Juteau *et al.*, 1978; review in Honnorez, 1981; Staudigel and Hart, 1983; Teagle *et al.*, 1996; Alt and Mata, 2000). Chemical changes during palagonitization involve hydration, gains of K, Rb, and Cs, and losses of other cations (e.g. Si, Al, Mn, Mg, Ca, Na, P, Zn, Cu, Ni, REE), although some of the material lost from altered glass may be retained locally in replacement products and cements (see review in Honnorez, 1981; Staudigel and Hart, 1983; Alt and Mata, 2000). Microbial activity may also be important in contributing to alteration of volcanic glass (Torsvik *et al.*, 1998; Alt and Mata, 2000; Staudigel, this volume). Because non-crystalline glass is highly reactive, such altered material can contain

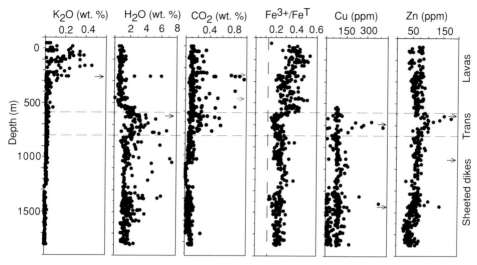

Fig. 15.12 Whole-rock geochemical data for ODP Hole 504B illustrating some of the chemical effects of different alteration processes in the upper crust. Depth in meters sub-basement (beneath 274.5 m sediment). Arrows indicate samples that plot off-scale. Low-temperature effects in the volcanic section range from more open seawater circulation (oxidation, alkali uptake) in the upper portion, to restricted seawater circulation and smaller chemical changes in the lower lavas. The transition zone and upper dikes reflect sub-greenschist to greenschist grade hydrothermal metamorphism, with mineralization in the transition zone (trans). The lower dikes were affected by greenschist to amphibolite grade hydrothermal metamorphism in the high-temperature ($\sim$400 °C) deep "root" zone of the hydrothermal system, and have lost alkalis, metals, and sulfur. After Alt *et al.* (1996b). Dashed line indicates unaltered MORB Fe^{3+}/Fe^{T}.

a large seawater chemical signature and can contribute significantly to geochemical budgets for altered ocean crust.

Basement fluids sampled from ridge flank systems have been divided into two regimes: (1) open flow of large volumes of cold ($<$30 °C) oxidizing seawater; and (2) more restricted flow of evolved fluids (low Mg) at higher temperatures (40–150 °C; Mottl and Wheat, 1994). The alteration described above is consistent with the lower temperature, high water/rock ratio regime. Bulk-rock chemical changes include variable enrichments of Fe^{T}, H_2O, K, Li, Rb, Cs, B, and U contents; losses of S and Ca; increased Fe^{3+}/Fe^{T}, $^{87}Sr/^{86}Sr$, $\delta^{11}B$, $\delta^{7}Li$, and $\delta^{18}O$; and decreased δD and $\delta^{34}S$ (e.g. Figs. 15.5, 15.12, 15.13; Andrews, 1977, 1980; Muehlenbachs, 1979; Humphris *et al.*, 1980; Hart and Staudigel, 1982; Barrett and Friedrichsen, 1983; Laverne and Vivier, 1983; Thompson, 1983; Alt and Honnorez, 1984; Bohlke *et al.*, 1984; Gillis and Robinson, 1988; Adamson and Richards, 1990; Ishikawa and Nakamura, 1992; Alt, 1993, 1995a,b; Staudigel *et al.*, 1995; Alt *et al.*, 1996a; Teagle *et al.*, 1996; Chan *et al.*, 2002; Alt and Teagle, 2003). The Fe-oxyhydroxide and celadonite zones typically exhibit the greatest changes for Fe^{T}, Fe^{3+}/Fe^{T}, H_2O, the alkalis, S, and Ca, reflecting alteration of primary phases and abundances of secondary minerals. Although

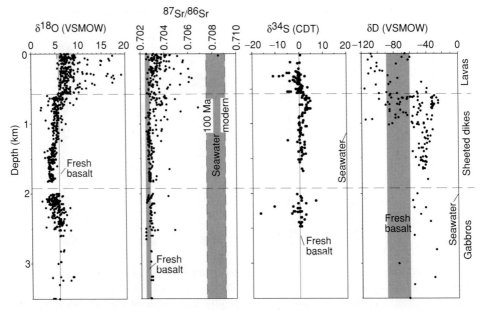

Fig. 15.13 Composite profiles of oxygen, strontium, sulfur, and hydrogen isotopic compositions of oceanic crust. Low-temperature alteration and ^{18}O-enrichment of the volcanic section give way to sub-greenshist to greenschist hydrothermal alteration and ^{18}O-depletion in the upper dikes. Mostly primary values below a few hundred meters into the plutonic section indicate little seawater penetration. After Alt and Teagle (2000). Data for submersible sampled rocks from Hess Deep and MARK, and from DSDP/ODP sites 417D, 418A, 504B, 735B, 894F-G, 896A, 801C, and 921–923 compiled in Alt and Teagle (2000), with additional data from Site 332A-C (Gray *et al.*, 1977; Muehlenbachs, 1977; O'Nions and Pankhurst, 1977; Yamaguchi *et al.*, 1977), site 395 (Hoernes *et al.*, 1978; von Drach *et al.*, 1978), and new data for Hole 735B from Bach *et al.* (2001).

Mg may be lost from individual rock samples, it is retained and even gained by the crust as smectite in veins (Alt *et al.*, 1996a).

Cumulative seawater/rock mass ratios range from near zero up to 900 for the volcanic section of Hole 504B (Alt *et al.*, 1986b). Water/rock ratios based on K uptake are 1–3 orders of magnitude lower than those based on oxidation of iron. This is in part due to post-sampling of iron in the rocks, but it also points out some of the problems with such water/rock ratio calculations. This case assumes reaction with unaltered seawater, so it does not take into account flow and reaction pathways, which certainly alter the composition of the fluid. Complete reaction of a fluid component is also unlikely, except perhaps for uptake of Mg into the crust. Thus such ratios are minima.

Differences in the apparent intensity or amount of alteration for bulk upper crustal sections among many sites can be real, but can also result from loss of material during drilling (e.g. Staudigel *et al.*, 1995; Alt *et al.*, 1996a). Interpillow material and breccias are commonly highly altered and contain abundant secondary mineral cements but are preferentially lost during drilling, which can lead to underestimation of alteration where recovery is low

(Brewer *et al.*, 1998, 1999; Haggas *et al.*, 2002; Révillon *et al.*, 2003). Elsewhere, significant variations in the temperature and intensity of seawater circulation lead to differences in alteration effects in the upper crust. In 13 Ma Atlantic crust (DSDP Hole 396B), intensive circulation of cold seawater in the upper 100 m of the basement resulted in the formation of wide (2 cm) Fe-oxyhydroxide zones, where olivine is leached and pseudomorphed by goethite, causing loss of olivine constituents from these zones (Si, Mg, Co, and Ni) and significant gain of P (Bohlke *et al.*, 1980, 1981). Minimum cumulative seawater/rock mass ratios of 3,000 are calculated for these zones assuming complete uptake of P from seawater (Bohlke *et al.*, 1981).

DSDP Hole 417A, in 110 Ma Atlantic crust, penetrates a basement hill that remained uncovered by sediment for 10 Ma and served as an outflow site for basement fluids (Donnelly *et al.*, 1979). Focusing of fluid flow and entrainment of seawater led to extreme alteration at temperatures of up to 50 °C (Donnelly *et al.*, 1979; Muehlenbachs, 1979; Alt and Honnorez, 1984; Bohlke *et al.*, 1984). The rocks contain early celadonitic halos and later Fe-oxyhydroxide zones, but rather than saponite, the rocks contain abundant K-feldspar and aluminous smectites (beidellite, Fe-beidellite, Al-saponite; Alt and Honnorez, 1984). The extensive breakdown of plagioclase and glass provided Al for formation of these aluminous phases. The breakdown of primary phases and formation of secondary minerals led to whole-rock gains of K, Rb, Cs, Li, B, P, and H_2O; increased Fe^{3+}/Fe^T, $^{87}Sr/^{86}Sr$, and $\delta^{18}O$; and loss of Ca, Mg, Si, and locally Al and Na (Donnelly *et al.*, 1979; Staudigel *et al.*, 1981a,b; Alt and Honnorez, 1984). Chemical changes are locally extreme: e.g. gains of >5 wt. % K_2O, losses of 4 wt. % MgO, and total oxidation of iron for some samples. Minimum cumulative seawater/rock mass ratios of 2,000–4,500 are calculated based on oxidation of Fe or assuming quantitative uptake of P from seawater. Nearby Holes 417D (450 m) and 418A (5 km) also exhibit rare zones of similar intense alteration and development of K-feldspar locally in the uppermost 100–200 m of basement (Pertsev and Rusinov, 1979; Alt and Honnorez, 1984).

The most extreme examples of this type of low-temperature seawater alteration are in rocks exposed to seawater at the seafloor for millions of years and which has been termed "seafloor weathering" (Thompson, 1973, 1983). Secondary mineralogy is generally poorly documented in these early studies, which focused more on bulk geochemical changes, including increases in alkalis, P, LREE, U, $^{87}Sr/^{86}Sr$, and $\delta^{18}O$, and losses of Mg, Ca, Si, Co, and Ni (Muehlenbachs and Clayton, 1972; Thompson, 1973, 1983; Ludden and Thompson, 1979; Hart 1973; Hart *et al.*, 1974).

In contrast to these more extreme cases, lavas on the east flank of the Juan de Fuca Ridge were buried by sediment and mostly sealed from free access of cold seawater relatively early (Davis *et al.*, 1992a,b). Here the effects of seawater oxidation are minimal (narrow, millimeter-wide oxidation halos), and the rocks mainly reflect early celadonite formation and later more restricted circulation and alteration (see following section; Hunter *et al.*, 1998, 1999). Basement from 170 Ma Hole 801C exhibits only slight oxidation (Fig. 15.6), which in part reflects early sealing of the crust by precipitation of celadonite and silica from upwelling low-temperature hydrothermal fluids at the spreading axis, but may also result from primary lithological controls on fluid circulation (Alt and Teagle, 2003).

Based on the occurrences of alteration effects in rocks of different ages, it can be concluded that the above alteration processes take place within a few million years (~2–13 Ma) of crustal formation. Subsequent formation of later carbonates and minor zeolites in veins also occurs, however (see following sections), and seawater fluids can circulate within basement for several tens of million years, suggesting that interactions can continue for longer periods. Available evidence, however, suggests that such circulation results in continued precipitation of carbonates in fractures but no significant alteration of the host rocks (Alt and Teagle, 1999). Taking into account uncertainties in crustal age and in Rb–Sr isochrons for smectite and celadonite, these minerals can form anytime up to about 23 Ma after crustal formation, although the best estimate from isotopic dating is generally within 10–15 Ma (Staudigel and Hart, 1985; Hart and Staudigel, 1986; Waggoner, 1993). Similar ages of smectite formation within 10–18 Ma of crustal formation are obtained by K–Ar dating (Peterson *et al.*, 1986). All these dates rely on the assumption that these minerals behave as closed systems, however, and the timing of alteration processes likely varies from site to site and even within a single site. Celadonite and silica precipitation and associated alteration at 170 Ma Site 801 are clearly related to formation of the 20-m-thick hydrothermal deposit at the spreading axis (Alt and Teagle, 2003), yet Rb/Sr isochrons for celadonite indicate its formation ~30 Ma younger than the crust (Bourrasseau, 1996). This suggests that these minerals can remain open to exchange with seawater and that such dating is not always meaningful.

Alteration can produce competing effects on physical properties at different scales (Jarrard *et al.*, 2003). On the grain scale, alteration of primary minerals (mainly olivine, but also plagioclase in more extreme cases) to phyllosilicates and Fe-oxyhydroxides should cause a decrease in density and an increase in porosity and compressibility. At the formation scale, cementation of fractures and breccias will increase density and decrease porosity, compressibility, and permeability (Wilkens *et al.*, 1991). Primary lithology and grain size also influence alteration: alteration halos are wider in coarser grained massive units, where intergranular spaces provide pathways for fluids and chemical reaction (see Fig. 15.6; Alt *et al.*, 1996a; Teagle *et al.*, 1996; Hunter *et al.*, 1999). This in turn can lead to greater alteration and possibly enhanced effects on physical properties (Jarrard *et al.*, 2003). Because of the smaller numbers of fractures (fluid pathways) in massive units, however, these rocks are on average less altered chemically than more highly fractured pillow basalts. Volcanic glass is more highly reactive than crystalline basalt (especially plagioclase and pyroxene). Thus, alteration of glass at pillow margins and flow tops releases material to solution that can be precipitated locally, cementing fractures and interpillow material and thereby contributing to increasing density and decreasing porosity and permeability at the formation scale.

15.2.3 Restricted low-temperature hydrothermal interactions in the volcanic section (reducing alteration)

Seawater fluids evolve as they percolate through and react with basalts, so oxidation effects can diminish with depth in the volcanic section or vary according to partitioning of seawater flow within the basement (Figs. 15.6 and 15.7; Andrews, 1977; Alt and Honnorez, 1984; Alt *et al.*, 1986a; Gillis and Robinson, 1988; Alt and Teagle, 2003; Shipboard Scientific Party, 2003). Seawater circulation can also evolve with time: fluids evolve to more reacted compositions as the crust is buried by sediments and fractures are sealed with secondary minerals. Thus, the effects of restricted seawater circulation can be superimposed on previous oxidation effects in the upper crust, or may occur in rocks where early seawater oxidation was lacking.

Where early celadonite and Fe-oxyhydroxide are present in veins and lining pores they are generally followed by saponite veins or pore fillings, representing later restricted seawater alteration and more reducing conditions (Seyfried *et al.*, 1978; Andrews, 1980; Bohlke *et al.*, 1980; Alt and Honnorez, 1984; Alt *et al.*, 1986a, 1992; Gillis *et al.*, 1992; Alt, 1993; Giorgetti *et al.*, 2000; Alt and Teagle, 2003). The portions of the volcanic section affected by restricted seawater circulation are generally <20% recrystallized, but locally rocks are more intensely altered (up to 40–100% secondary minerals in interpillow areas, breccias, and fault zones). Olivine is replaced, and pores and fractures are filled by saponite ± calcite ± pyrite and local talc. Trace disseminated secondary pyrite is common, and plagioclase may be partly altered to saponite or Al-saponite. Glass at pillow rims is altered to palagonite, saponite, Al-saponite, and zeolites (mainly phillipsite), and cemented by saponite, zeolites, calcite, and local pyrite.

Saponite resulting from restricted seawater circulation has generally higher FeO/(FeO + MgO) values (0.2–0.7) than saponite associated with oxidation, reflecting incorporation of greater amounts of ferrous iron under the more reducing conditions (Andrews, 1980; Alt *et al.*, 1986a). Bulk-rock FeO/(FeO + MgO) can also affect saponite compositions, with higher values in saponite within more iron-rich rocks (Porter *et al.*, 2000; Hunter *et al.*, 1999). Minor chlorite interlayers are present locally in saponite if alteration temperatures are sufficiently high, such as in the lower volcanic section of Hole 504B (>100 °C; Alt *et al.*, 1986a) and in some saponite-rich pillow breccias from the EPR (150–170 °C; Stakes and Scheidegger, 1981; Stakes and O'Neil, 1982). The trace amounts of chlorite reported locally from other volcanic sections formed very early, during initial penetration of seawater into the cooling lavas immediately after their extrusion (e.g. Bohlke *et al.*, 1980; Alt and Honnorez, 1984; Hunter *et al.*, 1998). The overall volcanic sections at the latter sites underwent restricted seawater alteration at temperatures less than about 70–90 °C (Muehlenbachs, 1979; Staudigel *et al.*, 1981b; Bohlke *et al.*, 1984; Hunter *et al.*, 1998; Alt and Teagle, 2003). The temperatures of restricted seawater alteration can thus vary from near 0 up to ∼150 °C (e.g. Fig. 15.8), depending upon the age at which the crust becomes insulated by a blanket of sediment, which is a function of basement relief and sedimentation rate.

Trace secondary pyrite is characteristic of restricted seawater alteration (Andrews, 1979; Alt *et al.*, 1986a, 1989; Hunter *et al.*, 1998). Andrews (1979) suggested that alteration results in dissolution of igneous sulfides under limited oxidation conditions and that secondary pyrite having negative δ^{34}S values precipitates. Alt and Shanks (1998), however, show that microbial reduction of seawater sulfate is important during low-temperature alteration of peridotite, and it has been suggested that microbial sulfate reduction may also be important within basaltic basement (Alt and Mata, 2000; Alt *et al.*, 2003).

Carbonates are commonly associated with saponite, but also form later (see below). Calcite is the most common carbonate, although aragonite is present in some cases and siderite and dolomite are rare (Bass, 1976; Andrews, 1977; Bohlke *et al.*, 1980; Lawrence and Drever, 1981; Lawrence, 1979, 1991; Alt and Honnorez, 1984; Teagle *et al.*, 1996; Alt *et al.*, 1992; Burns *et al.*, 1992; Yatabe *et al.*, 2000). Fibrous carbonate ± saponite veins are common in upper oceanic basement. In some cases these may reflect crystal growth simultaneous with crack opening (Bohlke *et al.*, 1980), but in others these have been interpreted as crack-seal veins indicative of fluid overpressures (Harper and Tartarotti, 1996).

Alteration of primary phases results in decreased density and a volume increase, but primary rock textures are maintained because the volume increase is accommodated by secondary material filling pore spaces and fractures in the rock. If silica is maintained constant, alteration of olivine to saponite requires a source of Al and Ca, and releases Fe and significant Mg to solution, resulting in a net volume increase of only 2%. Alteration of basalt glass at pillow rims and of silicic interstitial glass to saponite are significant sources of Si, however (Honnorez, 1981; Zhou *et al.*, 2001). Assuming input of silica and that Mg is constant for the olivine to saponite reaction results in a 110% volume increase. Thus at least some sealing of space in pores and fractures can occur with little net chemical change other than hydration. This suggests that such reactions could seal portions of the basement.

Chemical changes for bulk rocks affected by restricted seawater circulation are generally small, reflecting the evolved fluid compositions. Changes include slight increases of Fe^{3+}/Fe^T, H_2O, CO_2, $^{87}Sr/^{86}Sr$, and $\delta^{18}O$ (Figs. 15.12 and 15.13; Andrews, 1977, 1980; Muehlenbachs, 1979; Stakes and Scheidegger 1981; Staudigel *et al.*, 1981b; Alt and Honnorez, 1984; Bohlke *et al.*, 1984; Alt *et al.*, 1986a,b, 1992; Alt, 1993; Teagle *et al.*, 1996; Hunter *et al.*, 1999). Some of the oxidation, however, is the result of post-sampling oxidation of iron in saponite, which oxidizes readily upon exposure to air (Andrews *et al.*, 1983). Li substitutes for Mg in saponite leading to high Li contents of saponites and concomitant slight increases in bulk-rock concentrations and high Li contents of saponite-rich breccias (Berndt *et al.*, 1986; Chan *et al.*, 2002). Although K contents of saponites are typically low, they are greater than tholeiitic basalts, leading to slightly elevated whole-rock K contents. Ba, Rb, Th, and U contents of saponite are also greater than basalt (Porter *et al.*, 2000). In the more intensely recrystallized rocks, Mg may be gained and Ca lost. Even though individual rocks may not gain Mg, the abundant saponite in veins and cementing breccias results in a net uptake of Mg throughout the upper crust during restricted seawater circulation (Andrews, 1980; Stakes and Scheidegger 1981; Alt *et al.*, 1996a).

For the lower volcanic section of Hole 504B, the slight uptake of alkalis in the most altered breccias indicates seawater/rock ratios of about 1–2, while ratios calculated for most rocks are much less than one. This suggests highly evolved, rock-dominated solution compositions producing this type of alteration. In 155 Ma Hole 765D, early celadonite and oxidation halos are present but saponite is rare and the host rocks are only very slightly altered (Gillis *et al.*,

1992). This is interpreted to reflect relatively early sealing of the basement to circulation, inhibiting subsequent restricted seawater alteration. Sealing of basement may also have led to reheating of the crust at Site 765, which is inferred from the presence of secondary albite and titanite in the basement (Gillis *et al.*, 1992; J. Alt and P. Mata, unpublished data). In contrast, basement on the eastern flank of the Juan de Fuca Ridge is blanketed by sediment at a relatively young age, leading to elevated basement temperatures but the uppermost basement remains permeable (Davis *et al.*, 1992a,b). This leads to enhanced alteration and a greater percentage recrystallization of basalts under restricted seawater circulation conditions than at some other sites (Hunter *et al.*, 1998, 1999).

15.2.4 Aging of the volcanic section: late carbonates and zeolites

Carbonates and zeolites are typically the last phases to form in veins, vugs, and breccias of submarine basalts, although they also form earlier in the sequence (Bass, 1976; Andrews, 1977; Bohlke *et al.*, 1980; Stakes and Scheidegger, 1981; Staudigel *et al.*, 1981a; Alt and Honnorez, 1984; Alt *et al.*, 1986a, 1992; Gillis *et al.*, 1992; Alt, 1993; Teagle *et al.*, 1996; Alt and Teagle, 1999, 2003; Hunter *et al.*, 1999). The change from formation of Fe-oxyhydroxides and phyllosilicates to carbonates and zeolites reflects a general change to more alkaline conditions. Extraction of OH^- from solution with Mg or Fe to form clays or oxyhydroxides maintains slightly acidic conditions, but as Mg is depleted in solution hydrolysis reactions become more important, pH increases and carbonates and zeolites precipitate (Bass, 1976; Bohlke *et al.*, 1980; Stakes and Scheidegger 1981; Alt and Honnorez, 1984; Alt and Teagle, 2003). This interpretation is consistent with the generally low-Mg calcites that are observed filling fractures (Bass, 1976; Bohlke *et al.*, 1980, 1984; Alt and Honnorez, 1984). As reaction of seawater with basalt proceeds, Ca can be released to solution, further decreasing the Mg/Ca ratio of circulating seawater fluids. Late carbonates may also contain elevated Fe and Mn or be associated with pyrite, consistent with formation from more evolved, reducing fluids (Alt and Honnorez, 1984; Alt *et al.*, 1992; Burns *et al.*, 1992; Teagle *et al.*, 1996; Hunter *et al.*, 1999; Yatabe *et al.*, 2000).

Oxygen and carbon isotopic analyses of carbonate veins in seafloor basalts generally indicate formation from fluids having normal seawater carbon isotopic compositions at temperatures of 0–70 °C (Fig. 15.8; Lawrence, 1979, 1991; Muehlenbachs, 1979; Lawrence and Drever, 1981; Bohlke *et al.*, 1984; Alt *et al.*, 1986b, 1992; Alt, 1993; Alt and Teagle, 2003). Much less common low-δ^{13}C carbonates reflect incorporation of sedimentary organic carbon (Alt *et al.*, 1992; Alt and Teagle, 2003), or fractionation during closed system carbonate precipitation (Lawrence, 1991). Teagle *et al.* (1996) documented two generations of carbonates in veins of 6 Ma Hole 896A (Fig. 15.14). The Sr isotopic compositions of the two generations are similar, indicating that water–rock reactions did not cause significant differences in the oxygen isotopic compositions of the fluids and that the two generations of carbonate precipitated from similar solutions. Lower temperature (25–35 °C) carbonates formed during early open circulation of seawater; and a higher temperature (50–70 °C) generation formed later, at temperatures similar to the present "reheated" ridge flank conditions.

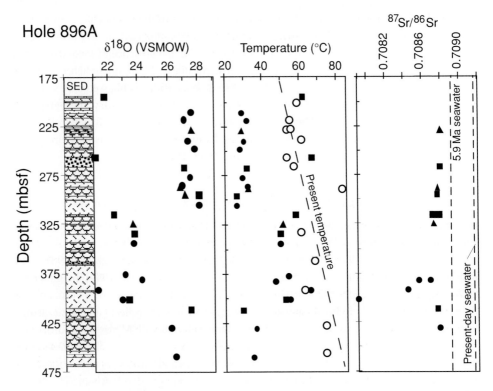

Fig. 15.14 Oxygen and strontium isotopic compositions of vein carbonates and smectites from ODP Hole 896A, with temperatures calculated assuming equilibrium with seawater (from Teagle *et al.*, 1996, 1998; J. Alt, unpublished data). The high-δ^{18}O carbonates formed early, when seawater circulation maintained lower temperatures in basement, whereas the higher-δ^{18}O carbonates formed later, as the crust was sealed by sediments and temperatures increased due to conductive heating from below. Present-day thermal gradient shown for comparison. These data suggest that later flow may have been mostly partitioned into the interval 310–400 mbsf (meters below seafloor). Smectites formed before carbonate minerals and record generally higher temperatures than carbonates.

The latter are mostly restricted to 135–220 m in the 290-m-deep core, suggesting that later flow may have been partitioned into this interval.

At 13 Ma Site 395, smectites formed while the volcanic section was still warm (10–30 °C, and up to 125 °C near an intrusive sill), whereas carbonates formed later after further cooling of the uppermost basement, at temperatures of 0–15 °C (Fig. 15.8; Lawrence and Drever, 1981). A similar relationship is shown at older sites (418 and 801; Fig. 15.8), but with wider variation in carbonate formation temperatures, which can be attributed to the long history of those sites.

Staudigel *et al.* (1981b) identified two different regimes of carbonate formation in 110 Ma Site 417/418: most carbonates there formed from seawater-like solutions, but carbonates intergrown with smectites and in breccias formed from more highly reacted solutions. The coincidence of initial ^{87}Sr/^{86}Sr ratios of vein carbonates from Sites 417 and 418 with that of 110 Ma seawater was interpreted to indicate that carbonates precipitated within a few million years of formation of the crust (Hart and Staudigel, 1978; Richardson *et al.*, 1980; Staudigel *et al.*, 1981a,b). Subsequent revisions of the magnetic anomaly time scale and of the curve for the Sr isotopic composition of seawater through time expand the uncertainty and the constraint on precipitation to within 40 Ma of crustal formation (Staudigel and Hart, 1985). Calcites from these sites have Sr/Ca ratios lower than predicted for equilibrium with seawater, however, requiring input of basaltic Ca into circulating seawater solutions, but without input of basaltic Sr, in order to lower the fluid Sr/Ca ratio without changing its Sr isotopic ratio (Hart and Staudigel, 1978; Staudigel *et al.*, 1981a,b).

Using these arguments, Sr isotopic data for one other MOR site and one on the Ninetyeast Ridge (a hot-spot trace) in the Indian Ocean indicate precipitation of carbonates within about 10–15 Ma of crustal formation (Hart and Staudigel, 1986; Hart *et al.*, 1994). At two other sites in the Indian Ocean, carbonates formed for up to about 27–40 Ma after creation of the crust (Burns *et al.*, 1992). These durations are minima, however, because a large basaltic component in solution (lowering the Sr isotopic ratio of the fluid) is indicated by the widely varying Sr contents and high Fe and Mn of the carbonates. At several other MOR and hot-spot sites the carbonates have initial $^{87}Sr/^{86}Sr$ ratios lower than seawater for any time between the formation of the crust at those sites and the present, clearly indicating a significant component of basaltic Sr in solution (e.g. Fig. 15.14; Hart and Staudigel, 1986; Waggoner, 1993; Hart *et al.*, 1994; Teagle *et al.*, 1996; Alt and Teagle, 2003). These data cast doubt on the assumption that Sr isotope ratios of seawater fluids remain constant at varying Sr/Ca ratios, which is required for the Sr dating technique for carbonate veins.

Alt and Teagle (1999) showed that the numbers of carbonate veins and the carbon content of the upper crust increase with crustal age (Fig. 15.15). The abundance of carbonate veins decreases with depth, so the temporal carbon increase occurs mostly in the upper 100 m of basement. Multiple carbonate veins in old crust document repeated creation and destruction of permeability during aging of the crust (e.g. Fig. 15.16). Carbon, oxygen, and strontium isotopic and trace element analyses of many sets of multiple cross-cutting carbonate veins in Hole 801C basalts reveal no systematic evolution of fluid compositions, however (Alt and Teagle, 2003). Carbonates are common as cements of breccias and hyaloclastites, and certainly contribute to reduction of permeability and increases of density at the formation scale. The amount of C taken up annually during low-temperature alteration of oceanic basalts is greater than the amount of C outgassed at mid-ocean ridges each year, and should be important for global C budgets (Staudigel *et al.*, 1989; Alt and Teagle, 1999).

Zeolites are also typically late forming phases, although minor in abundance. They occur in and along veins and cementing and replacing glass at pillow rims and in hyaloclastite breccias. Phillipsite is the most common, but analcite and natrolite are not uncommon and mesolite, thompsonite, chabazite, gismondine, faujasite, and stilbite also occur (Bass, 1976; Andrews, 1977; Juteau *et al.*, 1978; Bohlke *et al.*, 1980; Honnorez *et al.*, 1983; Alt and Honnorez, 1981; Alt *et al.*, 1986a; Gillis *et al.*, 1992; Teagle *et al.*, 1996). The breakdown of glass releases Al required for zeolite formation, so the association with altered glass is typical. Analcites are enriched in K, Rb, and Cs relative to basalt, contributing locally

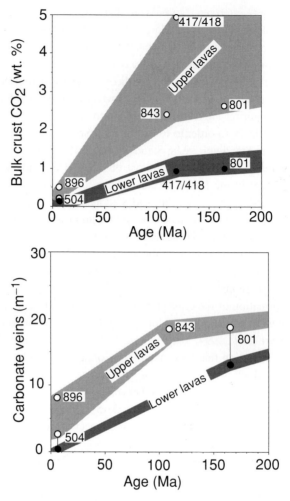

Fig. 15.15 Concentrations of CO_2 in bulk-upper oceanic crust and abundances of carbonate veins versus age of crust for different ODP cores (after Alt and Teagle, 1999). Bulk-crust CO_2 contents were calculated by mass balance combining measurements of vein abundances and thicknesses with whole-rock CO_2 contents. Bulk-crust CO_2 contents are greater in the upper volcanic section where carbonate veins are more abundant. Upper lavas defined as upper half of drilled section, where total basement penetration is 400–600 m. Estimates of Alt and Teagle (1999) for Hole 801C have been revised downward taking into account new data showing decreased bulk-rock CO_2 contents and carbonate vein abundances with depth at that site (Shipboard Scientific Party, 2000).

to the elevated alkali contents of the rocks (Staudigel *et al.*, 1981a,b). These minerals are associated with smectites and carbonates which formed at temperatures of about 0–100 °C, implying similar temperatures for the zeolites (Andrews, 1977; Juteau *et al.*, 1978; Bohlke *et al.*, 1980, 1984; Honnorez *et al.*, 1981; Alt and Honnorez, 1984; Alt *et al.*, 1986a; Teagle *et al.*, 1996). The distribution of zeolites in submarine basalts reflects a combination of the effects of varying temperature and evolution of fluid compositions (K/Na/Ca). In Hole

1 cm

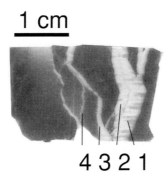

4 3 2 1

Fig. 15.16 Sample of basalt from ODP Hole 801C illustrating multiple stages of carbonate vein formation. At least four different stages of vein formation can be determined. Each vein has distinct oxygen isotopic and trace element composition, but no systematic evolution in fluid composition is documented by such veins (J. Alt and D. Teagle, unpublished data).

504B zeolites are zoned with depth: phillipsite occurs in the upper volcanic section; minor natrolite in the lower lavas; then laumontite, scolecite, and trace analcite and heulandite are late minerals in rocks of higher metamorphic grade in the transition zone and upper dike section (Fig. 15.7).

15.2.5 Late, pre-subduction alteration

ODP Site 1149 penetrates 135 Ma basement on the outer rise of the Izu-Bonin trench (Fig. 15.2). Jarrard *et al.* (2003) suggest that the very low matrix densities of these rocks may be the result of renewed circulation and alteration prior to subduction. These authors point out that crustal extension and normal faulting occur as the lithosphere flexes, which can provide vertical permeability connecting the already existing horizontal permeability (Fisher and Becker, 2000). Interstitial water profiles at Site 1149 indicate ongoing alteration in the basement, but with the exception of abundant late carbonate, there is no clear evidence for a late stage of alteration in the recovered basalts (Shipboard Scientific Party, 2000). The basalts at Site 1149 are more altered than at most older and younger sites (e.g. Fig. 15.6), but there is no evidence in the rocks to indicate that this is the result of a late or ongoing processes.

Stakes and Scheidegger (1981) sampled ~50 Ma basalts from fault scarps on the subducting slab in the Peru–Chile Trench. When altered compositions are normalized to constant Al and Ti the rocks exhibit losses of Si and Mg, gain of K, and significant oxidation of iron. These authors interpret this as the result of exposure of the rocks to cold seawater upon faulting in the last 0.5 Ma. The chemical and mineralogical changes, however, are the same as those in many younger rocks that remain buried by sediments. Some oxidation and leaching of Mg probably occurred during exposure, but it is difficult to distinguish the late exposure effects from those produced during earlier alteration at the spreading center and on ridge flanks.

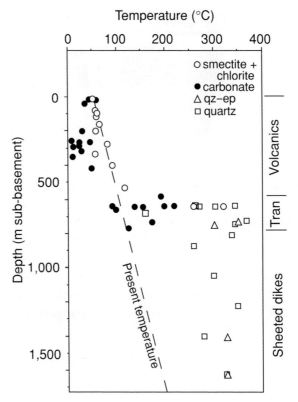

Fig. 15.17 Fossil and present temperatures in the upper oceanic crust, ODP Hole 504B. Metamorphic temperatures estimated from oxygen isotopic data for secondary minerals, assuming equilibrium and using various experimentally and empirically calibrated mineral–water (quartz, smectite, chlorite, calcite) and mineral–mineral (quartz–epidote) oxygen isotopic fractionations. (Data from Alt *et al.*, 1986b, 1989, 1995, 1996a,b; Alt, unpublished). The mineral data define the steeply stepped fossil thermal gradient that was present during axial hydrothermal metamorphism. This step coincides with a metamorphic boundary and the lithological transition from lavas to dykes, (see Fig. 15.7). Elsewhere this step in the geothermal gradient may be somewhat more gradual and lie deeper, within the upper sheeted dike complex (Gillis, 1995). Carbonates in the volcanic section are generally later, and record lower temperatures prior to basement burial and conductive reheating.

15.3 Axial hydrothermal metamorphism of the sheeted dike complex and upper plutonic section

As temperatures increase downward in the crust, phyllosilicates change from saponite (at 0 to ∼100 °C) to mixed-layer chlorite-smectite (∼100 to ∼200 °C) and then to chlorite at temperatures above about 200 °C (Figs. 15.7 and 15.17). At temperatures above about 300 °C, actinolite and actinolitic hornblende form. The upper ∼500 m of the dike complex are characterized by greenschist to sub-greenschist facies minerals, with partial reaction to albite, chlorite, actinolitic amphiboles, titanite, and trace sulfide, magnetite, and secondary apatite (Alt *et al.*, 1986a, 1996a; Gillis and Thompson, 1993). Permeability, water/rock ratio, and reaction kinetics exert important controls on recrystallization. At these conditions (up to about 350 °C) olivine and basalt glass are the least stable phases, followed by plagioclase and then pyroxene, and rocks are typically only partly recrystallized. Centimeter-size halos along veins and isolated high-porosity patches are more intensely recrystallized (50–100%) than most of the rock (10–30%). Where permeability (fracture abundance) is low, water/rock ratio is very low and reaction progress is small. In this case, olivine is the only phase altered and insufficient Al is released from plagioclase to form chlorite, so talc or chlorite-smectite form instead (Alt *et al.*, 1986a; Alt, 1999a).

The root zones of hydrothermal systems lie within the lower sheeted dike complex and the uppermost plutonic section, where the intensity of hydrothermal alteration reaches a maximum (Gillis and Thompson 1993; Alt, 1995a,b; Gillis 1995; Alt *et al.*, 1996b; Gillis and Roberts, 1999). Recrystallization is heterogeneous, and in Hole 504B is characterized by the replacement of plagioclase by calcic plagioclase (anorthite), and of pyroxene by magnesio-hornblende in alteration halos along fractures and in isolated patches, at temperatures >400 °C (Fig. 15.7; Gillis, 1995; Alt *et al.*, 1996b; Vanko *et al.*, 1996; Vanko and Laverne, 1998). The depth of these rocks, the presence of secondary calcic plagioclase, the high temperatures, partial recrystallization, and losses of Cu, Zn, and S (Fig. 15.12) are consistent with experimental and theoretical predictions of reactions at ∼1–2 km below the seafloor (Berndt and Seyfried, 1993). Metal and sulfur losses result from breakdown of sulfide minerals and titanomagnetite at high alteration temperatures (>350 °C), where the solubilities of metals and sulfide in hydrothermal fluids increase significantly (see Seyfried *et al.*, 1999). Rock-dominated fluids are indicated by low $^{87}Sr/^{86}Sr$ values of anhydrite, amphiboles, and bulk rocks, and otherwise generally small bulk chemical changes: e.g. slight increases of H_2O and CO_2, losses of K and Li, and decreased $\delta^{18}O$ (Figs. 15.12 and 15.13; Alt *et al.*, 1996b; Teagle *et al.*, 1998; Chan *et al.*, 2002).

Seawater hydrothermal effects (number of veins, amount of recrystallization, and isotopic exchange) decrease significantly below a few hundred meters into the plutonic section (e.g. Fig. 15.13). This reflects intense hydrothermal alteration of the uppermost gabbros at the spreading axis in the hydrothermal root zone immediately overlying the axial melt lens. The base of these root zones probably moves up and down within the crust, depending on the balance between spreading effects (fracturing and hydrothermal cooling) and magmatic or melt supply effects (10–100 year versus 10^5 year time scales, respectively; Hooft *et al.*, 1997; Gillis and Roberts, 1999).

Beneath the intensely altered root zone are the less altered gabbros of the solidified melt lens and underlying rocks, which were altered very near to the axis after the rocks had crystallized and cooled (Gillis 1995). Fluid penetration occurs along microcrack networks, initially at temperatures of up to 650 to 800 °C, penetrating deeper as the crust moves off axis and cools further (Manning *et al.*, 1996). The locally elevated $\delta^{18}O$ and $^{87}Sr/^{86}Sr$ values in the gabbro section of Fig. 15.13 are the result of late, low-temperature effects related to tectonic unroofing of the lower crust (e.g. Bach *et al.*, 2001). Profiles similar to those in Fig. 15.13 are exhibited by the Troodos ophiolite, but in contrast, the Oman ophiolite shows evidence for greater amounts of seawater exchange to greater depths (see Alt and Teagle, 2000).

The deep roots of discharge zones in ophiolites consist of epidosites, granular epidote + quartz + titanite rock (Richardson *et al.*, 1987; Schiffman *et al.*, 1987; Schiffman and Smith, 1988; Bickle and Teagle, 1992; Alt, 1994; Nehlig *et al.*, 1994). Epidosites are enriched in Ca, Sr, and ^{34}S; have increased Fe^{3+}/Fe^{T} and $^{86}Sr/^{86}Sr$; have lost Mg, Na, Zr, K, Cu, Zn, and S, and are depleted in ^{18}O. The rocks formed by hydrothermal reaction with Mg-depleted, Ca-enriched hydrothermal fluids at 350–440 °C, similar to the black-smoker-type vent fluids on the seafloor. Volume reduction during epidosite formation from diabase produces increased porosity that may provide pathways for flow of the large amounts of fluid required (fluid/rock ratios ~500–1000). Although epidosites have been described from suprasubduction oceanic settings (Banerjee *et al.*, 2000) they have not yet been observed in MOR crust.

In ophiolites focused up-flow zones extend upward from epidosites into quartz + epidote + sulfide veins and into shallow stockwork feeder zones beneath seafloor massive sulfide deposits (Nehlig *et al.*, 1994). Such deep up-flow zones have not yet been observed on the seafloor, however. Mineralized stockworks beneath massive sulfide deposits in ophiolites and on the seafloor record processes that include chloritization, silicification, sulfidization, sericitization, and sulfate reduction, resulting from reaction of basalt with upwelling hydrothermal fluids plus a component of seawater entrained into the feeder zone (Embley *et al.*, 1988; Janecky and Shanks, 1988; Zierenberg *et al.*, 1988; Richards *et al.*, 1989; Teagle *et al.*, 1998; Humphris and Tivey, 2000).

15.4 Seawater–crustal chemical fluxes at mid-ocean ridges and cycling at subduction zones

For some elements the effects of low-grade alteration on ridge flanks are opposite to those of high-grade axial processes and, while the chemical fluxes for both are large, these effects may offset each other. For example, the alkalis, B, and $\delta^{18}O$ decrease in the middle crust via axial hydrothermal metamorphism, but increase in the uppermost crust in passive systems on ridge flanks and the effects essentially cancel each other (e.g. Fig. 15.13; Chan *et al.*, 2002). It is the balance between these processes that buffers the oxygen isotopic composition of the oceans (Muehlenbachs and Clayton, 1976). For S isotopes the trends are reversed but the net effect on the crust is similar (Alt, 1994). Much of the alkali uptake in the uppermost crust may actually occur at the spreading axis during formation of celadonite in veins and dark alteration halos in basalt, from distal, mixed hydrothermal fluids associated with diffuse venting. For many other elements, the chemical changes are essentially unidirectional during both axial and flank processes, so the effects are cumulative (e.g. H_2O, $^{87}Sr/^{86}Sr$, Ca, Si, Mg, and U; Fig. 15.13). The combined effects of active axial hydrothermal metamorphism and lower temperature alteration in passive ridge flank systems exert significant controls on the Mg content of seawater and on the Sr isotopic compositions of the oceans (Bickle and Teagle, 1992; Mottl and Wheat, 1994). Despite the very small chemical changes in passive ridge flank systems, the enormous fluid fluxes make these processes important (Mottl and Wheat, 1994). The volcanic section, affected by low-temperature alteration, is the main carrier of the seawater signature (e.g. elevated alkalis, B, δ^{18}O, ^{87}Sr/^{86}Sr, CO$_2$), so it is this portion of the crust that contributes most to subduction zone metasomatic fluids (Staudigel *et al.*, 1995; Alt *et al.*, 1996b).

15.5 Future directions

Perhaps the biggest problem relates to the heterogeneity of the upper oceanic crust. Drilling additional basement sections is important to understand the heterogeneities in lithology and alteration effects, and how these are related. Such basement sections must be longer than the typical 100–200 m penetrations, because significant seawater effects can be encountered at depths of several hundred meters.

Detailed logging of lithology and alteration effects in basement drillcores must be coupled with geophysical logs of the boreholes in order to quantify alteration effects in the upper crust and understand their relationships to primary structure. These data are also essential to obtain more reliable geochemical budgets for the upper crust.

Because of the heterogeneity of the upper crust, it is desirable to develop a better understanding of how measurements made at the surface (e.g. sediment thickness, crustal age, basement topography, seismic velocities, heat flow, pore-fluid chemistry) relate to alteration effects at depth. Coupling programs of such surface measurements in different settings with deep (at least 400–500 m sub-basement) drilling will contribute to this goal and may ultimately enable predictive models to be developed.

Linking alteration effects and processes in different parts of submarine hydrothermal systems is also important. How does the transition from low-temperature effects in the upper crust relate to higher temperature hydrothermal alteration in the sheeted dikes and gabbros? The transition from dikes to gabbros has also not yet been cored in ocean crust, and this critical zone is important for understanding deep reactions at the source for hydrothermal fluids that vent at the surface.

Acknowledgments

Preparation of this chapter was supported by NSF OCE-9911901. Reviews by Mike Mottl and Damon Teagle, and comments by Earl Davis helped to improve the manuscript. This research used samples and/or data provided by the Ocean Drilling Program (ODP). The ODP is sponsored by the US National Science Foundation (NSF) and participating countries under management of Joint Oceanographic Institutions (JOI), Inc.

References

Adamson, A. C. and Richards, H. G. 1990. Low-temperature alteration of very young basalts from ODP Hole 648B: Serocki volcano, Mid-Atlantic Ridge. In *Proceedings of the Ocean Drilling Program, Scientific Results*, Vol. 106/109, eds. R. Detrick, J. Honnorez, W. B. Bryan, and T. Juteau. College Station, TX: Ocean Drilling Program, pp. 181–196.

Alt, J. C., 1988. Hydrothermal oxide and nontronite deposits on seamounts in the eastern Pacific. *Marine Geol.* **81**: 227–239.

1993. Low-temperature alteration of basalts from the Hawaiian Arch, ODP Leg 136. In *Proceedings of the Ocean Drilling Program, Scientific Results*, Vol. 136, eds. R. H. Wilkens, J. Firth, and J. Bender. College Station, TX: Ocean Drilling Program, pp. 133–146.

1994. A sulfur isotopic profile through the Troodos Ophiolite, Cyprus: primary composition and the effects of seawater hydrothermal alteration. *Geochim. Cosmochim. Acta* **58**: 1,825–1,840.

1995a. Subseafloor processes in mid-ocean ridge hydrothermal systems. In *Seafloor Hydrothermal Systems, Physical, Chemical, and Biological Interactions*, Geophysical Monograph 91, eds. S. Humphris, J. Lupton L. Mullineaux, and R. Zierenberg. Washington, DC: American Geophysical Union, pp. 85–114.

1995b. Sulfur isotopic profile through the oceanic crust: sulfur mobility and seawater–crustal sulfur exchange during hydrothermal alteration. *Geology* **23**: 585–588.

1999a. Hydrothermal alteration of the oceanic crust. Mineralogy, geochemistry and processes. In *Volcanic Associated Massive Sulfide Deposits, Reviews In Economic Geology*, Vol. 8, eds. T. Barrie and M. Hannington. Chelsea, MI: Society of Economic Geology, pp. 133–155.

1999b. Very low grade hydrothermal metamorphism of basic igneous rocks. In *Very Low Grade Metamorphism*, eds. M. Frey and D. Robinson. Oxford: Blackwell Scientific, pp. 169–201.

Alt, J. C. and Bach, W. 2003. Alteration of oceanic crust: subsurface rock–water interactions. In *Dahlem Workshop Report on Energy and Mass Transfer in Marine Hydrothermal Systems*, eds. P. Halbach, V. Tunnicliffe, and J. Hein. Berlin: Freie Universitat.

Alt, J. C. and Honnorez, J. 1984. Alteration of the upper oceanic crust DSDP Site 417: mineralogy and chemistry. *Contrib. Mineral. Petrol.* **87**: 149–169.

Alt, J. C. and Mata, P. 2000. On the role of microbes in the alteration of submarine basaltic glass: a TEM study. *Earth Planet. Sci. Lett.* **181**: 301–313.

Alt, J. C. and Shanks, W. C. 1998. Sulfur in serpentinized oceanic peridotites: serpentinization processes and microbial sulfate reduction. *J. Geophys. Res.* **103**: 9,917–9,929.

Alt, J. C. and Teagle, D. A. H. 1999. Uptake of carbon during alteration of oceanic crust. *Geochim. Cosmochim. Acta* **63**: 1,527–1,535.

2000. Hydrothermal alteration and fluid fluxes in ophiolites and oceanic crust. In *Ophiolites and Oceanic Crust: New Insights from Field Studies and the Ocean Drilling Program*, Geological Society of America Special Paper 349, eds. Y. Dilek, E. Moores, D. Elthon, and A. Nicolas. Boulder, CO: Geological Society of America, pp. 273–282.

2003. Hydrothermal alteration of upper oceanic crust formed at a fast spreading ridge: mineral, chemical, and isotopic evidence from ODP Site 801. *Chem. Geol.* **201**: 191–211.

Alt, J. C., Honnorez, J., Laverne, C., and Emmermann, R. 1986a. Hydrothermal alteration of a 1 km section through the upper oceanic crust, DSDP Hole 504B: the mineralogy, chemistry, and evolution of seawater–basalt interactions. *J. Geophys. Res.* **91**: 10,309–10,335.

Alt, J. C., Muehlenbachs, K., and Honnorez, J. 1986b. An oxygen isotopic profile through the upper kilometer of oceanic crust, DSDP Hole 504B. *Earth Planet. Sci. Lett.* **80**: 217–229.

Alt, J. C., Anderson, T. F., and Bonnell, L. 1989. The geochemistry of sulfur in a 1.3 km section of hydrothermally altered oceanic crust, DSDP Hole 504B. *Geochim. Cosmochim. Acta* **53**: 1,011–1,023.

Alt, J. C., Lanord, C. F., Floyd, P. A., Castillo, P., and Galy, A. 1992. Low-temperature hydrothermal alteration of Jurassic ocean crust, Site 801. In *Proceedings of the Ocean Drilling Program, Scientific Results*, Vol. 129, eds. R. L. Larson and Y. Lancelot. College Station, TX: Ocean Drilling Program, pp. 415–427.

Alt, J. C., Teagle, D. A. H., Laverne, C., Vanko, D., Bach, W., Honnorez, J., Becker, K., Ayadi, M., and Pezard, P. A. 1996a. Ridge flank alteration of upper ocean crust in the eastern Pacific: a synthesis of results for volcanic rocks of holes 504B and 896A. In *Proceedings of the Ocean Drilling Program, Scientific Results*, Vol. 148, eds. J. C. Alt, H. Kinoshita, L. B. Stokking, and P. J. Michael. College Station, TX: Ocean Drilling Program, pp. 434–452.

Alt, J. C., Laverne, C., Vanko, D., Tartarotti, P., Teagle, D. A. H., Bach, W., Zuleger, E., Erzinger, J., Honnorez, J., Pezard, P. A., Becker, K., Salisbury, M. H., and Wilkens, R. H. 1996b. Hydrothermal alteration of a section of upper oceanic crust in the eastern equatorial Pacific: a synthesis of results from Site 504DSDP legs 69, 70, and 83, and ODP legs 111, 137, 140, and 148. In *Proceedings of the Ocean Drilling Program, Scientific Results*, Vol. 148, eds. J. C. Alt, H. Kinoshita, L. Stokking, and P. Michael. College Station, TX: Ocean Drilling Program, pp. 417–434.

Alt, J. C., Davidson, G., Teagle, D. A. H., and Karson, J. A. 2003. The isotopic composition of gypsum in the Macquarie Island ophiolite: implications for the sulfur cycle and the subsurface biosphere in oceanic crust. *Geology* **31**: 549–552.

Andrews, A. J. 1977. Low temperature fluid alteration of oceanic Layer 2 basalts. *Can. J. Earth Sci.* **14**: 911–926.

1979. On the effect of low-temperature seawater–basalt interaction on the distribution of sulfur in oceanic crust, Layer 2. *Earth Planet. Sci. Lett.* **46**: 68–80.

1980. Saponite and celadonite in layer 2 basalts, DSDP Leg 37. *Contrib. Mineral. Petrol.* **73**: 323–340.

Andrews, A. J., Dollase, W. A., and Fleet, M. E. 1983. A Mossbauer study of saponite occurring in Layer 2 basalts, DSDP Leg 69. In *Initial Reports of the Deep Sea Drilling Project*, Vol. 69, eds. J. R. Cann, M. G. Langseth, J. Honnorez, R. P. Von Herzen, and S. M. White. Washington, DC: US Govt. Printing Office, pp. 585–588.

Bach, W., Alt, J. C., Niu, Y., Humphris, S. E., Erzinger, J., and Dick, H. J. B. 2001. The chemical consequences of late-stage hydrothermal circulation in an uplifted block of lower ocean crust at the SW Indian Ridge: results from ODP Hole 735B Leg 176. *Geochim. Cosmochim. Acta* **65**: 3,267–3,287.

Banerjee, N. R., Gillis, K. M., and Muehlenbachs, K. 2000. Discovery of epidosites in a modern oceanic setting, the Tonga forearc. *Geology* **28**: 151–154.

Barrett, T. J. and Friedrichsen, H. 1983. Strontium and oxygen isotopic composition of some basalts from Hole 504B, Costa Rica Rift, DSDP Legs 69 and 70. *Earth Planet. Sci. Lett.* **60**: 27–38.

Bass, M. N. 1976. Secondary minerals in oceanic basalts, with special reference to Leg 34, DSDP. In *Initial Reports of the Deep Sea Drilling Project*, Vol. 34, eds.

R. S. Yeats and S. R. Hart. Washington, DC: US Govt. Printing Office, pp. 393–431.

Becker, K., Sakai, H., Adamson, A., Alexandrovich, J., Alt, J. C., Anderson, R. N., Bideau, D., Gable, R., Herzig, P., Houghton, S., Ishizuka, H., Kawahata, H., Kinoshita, H., Langseth, M. G., Lovell, M. A., Malpas, J., Masuda, H., Merril, R. B., Morin, R. H., Mottl, M. J., Pariso, J. E., Pezard, P., Phillips, J., Sparks, J., and Uhlig, S. 1989. Drilling deep into young oceanic crust, Hole 504B, Costa Rica Rift. *Rev. Geophys.* **27**: 79–102.

Berndt, M. E. and Seyfried, W. E. 1986. B, Li, and associated trace element chemistry of alteration minerals, Holes 597B and 597C, In *Initial Reports of the Deep Sea Drilling Project*, Vol. 92, eds. M. Leinen and D. K. Rea. Washington, DC: US Govt. Printing Office, pp. 491–497.

Berndt, M. E. and Seyfried, W. E. 1993. Calcium and sodium exchange during hydrothermal alteration of calcic plagioclase at 400 °C and 400 bars. *Geochim. Cosmochim. Acta* **57**: 4,445–4,451.

Bickle, M. J. and Teagle, D. A. H. 1992. Strontium alteration in the Troodos ophiolite: implications for fluid fluxes and geochemical transport in mid-ocean ridge hydrothermal systems. *Earth Planet. Sci. Lett.* **113**: 219–237.

Bohlke, J. K., Honnorez, J., and Honnorez-Guerstein, M. B. 1980. Alteration of basalts from Site 396B, DSDP: petrographic and mineralogic studies. *Contrib. Mineral. Petrol.* **73**: 341–364.

Bohlke, J. K., Honnorez, J., Honnorez-Guerstein, B. M., Muehlenbachs, K., and Petersen, N. 1981. Heterogeneous alteration of the upper oceanic crust: correlation of rock chemistry, magnetic properties, and O isotope ratios with alteration patterns in basalts from Site 396B, DSDP. *J. Geophys. Res.* **86**: 7,935–7,950.

Bohlke, J. K., Alt, J. C., and Muehlenbachs, K. 1984. Oxygen isotope–water relations in altered deep-sea basalts: low-temperature mineralogical controls. *Can. J. Earth Sci.* **21**: 67–77.

Bourasseau, I. 1996. L'alteration de la croute oceanique Jurasique et la genese des sediments metaliferes associes (Bassin de Pigafetta, Ocan Pacifique NW). Ph.D. thesis, Universite Louis Pasteur, Strasbourg.

Brewer, T. S. *et al.* 1998. Ocean floor volcanism: constraints from the integration of core and downhole logging measurments. In *Core–Log Integration*, eds. P. K. Harvey and M. A. Lovell. London: Geological Society, pp. 341–362.

Brewer, T. S., Harvey, P. K., Haggas, S., Pezard, P., and Goldberg, D. 1999. Borehole images of the ocean crust: case histories from the Ocean Drilling Program. In *Borehole Imaging: Applications and Case Histories*, eds. M. A. Lovell, G. Williamson, and P. K. Harvey. London: Geological Society, pp. 283–294.

Buckley, H. A., Bevan, J. C., Brown, K. M., Johnson, L. R., and Farmer, V. C. 1978. Glauconite and celadonite: two separate mineral species. *Min. Mag.* **42**: 373–382.

Burns, S. J., Baker, P. A., and Elderfield, H. 1992. Timing of carbonate mineral precipitation and fluid flow in seafloor basalts, northwest Indian Ocean. *Geology* **20**: 255–258.

Carbotte, S. and Scheirer, D. 2004. Variability of ocean crustal structure created along the global midocean ridge. In *Hydrogeology of the Oceanic Lithosphere*, eds. H. Elderfield and E. Davis. Cambridge: Cambridge University Press.

Chan, L. H. Alt, J. C., and Teagle, D. A. H. 2002. Lithium and lithium isotope profiles through the upper oceanic crust: a study of seawater–basalt exchange at ODP Sites 504B and 896A. *Earth Planet. Sci. Lett.* **201**: 187–201.

Davis, E. E., Chapman, D. S., Mottl, M. J., Bentkowski, W. J., Dadey, K., Forster, C., Harris, R., Nagihara, S., Rohr, K., Wheat, G., and Whiticar, M. 1992a. FlankFlux: an experiment to study the nature of hydrothermal circulation in young oceanic crust. *Can. J. Earth Sci.* **29**: 925–952.

Davis, E. E., Mottl, M. J., and Fisher, A. T., eds. 1992b. *Proceedings of the Ocean Drilling Program, Initial Reports*, Vol. 139. College Station, TX: Ocean Drilling Program.

Davis, E. E., Chapman, D. S., and Forster, C. B. 1996. Observations concerning the vigor of hydrothermal circulation in young oceanic crust. *J. Geophys. Res.* **101**: 2,927–2,942.

Donnelly, T. W., Pritchard, R. A., Emmermann, R., and Puchelt, H. 1979. The aging of oceanic crust: synthesis of mineralogical and chemical results of DSDP Legs 51–53. In *Initial Reports of the Deep Sea Drilling Project*, eds. T. W. Donnelly *et al.*, Vol. 51–53 part 2. Washington, DC: US Govt. Printing Office, pp. 1563–1577.

Edmond J. M., Measures, C., McDuff, R. E., Chan, L. H., Collier, R., Grant, B., Gordon, L. I., and Corliss, J. B. 1979a. Ridge crest hydrothermal activity and the balances of the major and minor elements in the ocean: the Galapagos data. *Earth Planet. Sci. Lett.* **46**: 1–18.

Edmond J. M., Measures, C., Magnum, B., Grant, B., Sclater, F. R., Collier, R., Hudson, A., Gordon, L. I., and Corliss, J. B. 1979b. On the formation of metal-rich deposits at ridge crests. *Earth Planet. Sci. Lett.* **46**: 19–30.

Embley, R. W., Jonasson, I. R., Perfit, M. R., Franklin, J. M., Tivey, M. A., Malahoff, A., Smith, M. F., and Francis, T. J. G. 1988. Submersible investigation of an extinct hydrothermal system on the Galapagos Ridge: sulfide mounds, stockwork zone, and differentiated lavas. *Can. Mineral.* **26**: 517–540.

Fisher, A. T. 1998. Permeability within basaltic oceanic crust. *Rev. Geophys.* **36**: 143–182.

Fisher, A. and Becker, K. 2000. Channelized fluid flow in oceanic crust reconciles heat-flow and permeability data. *Nature* **403**: 71–74.

Fisher, A. T., Becker, K., Narasimhan, T. N., Langseth, M. G., and Mottl, M. J. 1990. Passive off-axis convection through the southern flank of the Costa Rica Rift. *J. Geophys. Res.* **95**: 9,343–9,370.

Fisher, A. T., Becker, K., and Narasimhan, K. 1994. Off-axis hydrothermal circulation: parametric tests of a refined model of processes at DSDP/ODP site 504. *J. Geophys. Res.* **99**: 3,097–3,123.

Fisher, A. T., Davis, E. E., Hutnak, M., Spiess, V., Zuhlsdorff, L., Cherkaoui, A., Christiansen, L., Edwards, K., Macdonald, R., Villinger, H., Mottl, M. J., Wheat, C. G., and Becker, K. 2004. Hydrothermal recharge and discharge across 50 km guided by seamounts on a young ridge flank. *Nature* **421**: 618–621.

Flower, M. F. J., Ohnmacht, W., Robinson, P. T., Marriner, G., and Schmincke, H. U. 1979. Lithological and chemical stratigraphy at DSDP Sites 417 and 418. In *Initial Reports of the Deep Sea Drilling Project*, eds. T. Donnelly, J. Francheteau, W. Bryan, P. Robinson, M. Flower, and M. Salisbury, Vol. 51–53. Washington, DC: US Govt. Printing Office, pp. 939–956.

Gillis, K. M. 1995. Controls on hydrothermal alteration in a section of fast-spreading oceanic crust. *Earth Planet. Sci. Lett.* **134**: 473–489.

2003. Subseafloor geology of hydrothermal systems at oceanic spreading centers. In *Dahlem Workshop Report on Energy and Mass Transfer in Marine Hydrothermal Systems*, eds. P. Halbach, V. Tunnicliffe, and J. Hein. Berlin: Freie Universitat.

Gillis, K. M. and Roberts, M. 1999. Cracking at the magma–hydrothermal transition: evidence from the Troodos ophiolite. *Earth Planet. Sci. Lett.* **169**: 227–244.

Gillis, K. M. and Robinson, P. T. 1988. Distribution of alteration zones in the upper oceanic crust. *Geology* **16**: 262–266.

Gillis, K. M. and Thompson, G. 1993. Metabasalts from the Mid-Atlantic Ridge: new insights into hydrothermal systems in slow-spreading crust. *Contrib. Mineral. Petrol.* **113**: 502–523.

Gillis, K. M., Ludden, J. N., Plank, T., and Hoy, L. D. 1992. Low temperature alteration and subsequent reheating of shallow oceanic crust at Hole 765D, Argo abyssal plain. In *Proceedings of the Drilling Program, Scientific Results*, Vol. 123, eds F. M. Gradstein, J. N. Ludden, *et al.* College Station, TX: Ocean Drilling Program, pp. 191–199.

Giorgetti, G., Marescotti, P., Cabella, R., and Lucchetti, G. 2001. Clay mineral mixtures as alteration products in pillow basalts from the eastern flank of Juan de Fuca Ridge: a TEM–AEM study. *Clays Clay Mineral.* **36**: 75–91.

Gray, J. Cummings, G. L. and Lambert, R. S. St. J. 1977. Oxygen and strontium isotopic compositions and thorium and uranium contents of basalts from DSDP Leg 37 cores. In *Initial Reports of the Deep Sea Drilling Project*, Vol. 37, eds. F. Aumento, W. G. Melson, *et al.* Washington, DC: US Govt. Printing Office, pp. 607–612.

Haggas, S. L., Brewer, T. S., and Harvey, P. K. 2002. Architecture of the volcanic layer from the Costa Rica Rift, constraints from core–log integration. *J. Geophys. Res.* **107** (B2 ECV2): 1–14.

Harper, G. D. and Tartarotti, P. 1996. Structural evolution of upper layer 2, Hole 896A. In *Proceedings of the Ocean Drilling Program, Scientific Results*, Vol. 148, eds. J. C. Alt, H. Kinoshita, L. Stokking, and P. J. Michael. College Station, TX: Ocean Drilling Program, pp. 245–259.

Hart, R. 1973. Chemical exchange between sea water and deep ocean basalts. *Earth Planet. Sci. Lett.* **9**: 269–279.

Hart, S. R. and Staudigel, H. 1978. Oceanic crust: age of hydrothermal alteration. *Geophys. Res. Lett.* **5**: 1,009–1,012.

1982. The control of alkalis and uranium in seawater by ocean crust alteration. *Earth Planet. Sci. Lett.* **58**: 202–212.

1986. Ocean crust vein mineral deposition: Rb/Sr ages, U–Th–Pb geochemistry, and duration of circulation at DSDP Sites 261, 462, and 516: *Geochim. Cosmochim. Acta* **50**: 2,751–2,761.

Hart, S. R., Erlank, A. J., and Kable, E. J. D. 1974. Sea floor basalt alteration: some chemical and Sr isotopic effects. *Contrib. Mineral. Petrol.* **44**: 219–230.

Hart, S. R., Blusztajn, J., Dick, H. J. B., and Lawrence, J. R. 1994. Fluid circulation in the oceanic crust: contrast between volcanic and plutonic regimes. *J. Geophys. Res.* **99**: 3,163–3,174.

Honnorez, J. 1981. The aging of the oceanic crust at low temperature. In *The Sea*, Vol. 7. *The Oceanic Lithosphere*, ed. C. Emiliani. New York: Wiley, pp. 525–587.

Honnorez, J., *et al.* 1981. Hydrothermal mounds and young ocean crust of the Galapagos: preliminary deep sea drillling results, Leg 70. *Geol. Soc. Am. Bull.* **92**: 457–472.

Hooft, E. E., Detrick R. S., and Kent, G. M. 1997. Seismic structure and indicators of magma budget along the southern East Pacific Rise. *J. Geophys. Res.* **102**: 27,319–27,340.

Hoernes, S., Friedrichsen, H. and Schock, H. H. 1978. Oxygen and hydrogen and trace element investigations on rocks of DSDP Hole 395A, Leg 45. In *Initial Reports of the Deep Sea Drilling Project*, Vol. 45, eds. W. G. Melson, P. Rabinowitz, *et al.* Washington, DC: US Govt. Printing Office, pp. 541–550.

Humphris, S. E. and Tivey, M. K. 2000. A synthesis of geological and geochemical investigations of the TAG hydrothermal field: insights into fluid-flow and mixing processes in a hydrothermal system. In *Ophiolites and Oceanic Crust: New Insights from Field Studies and the Ocean Drilling Program*. Geological Society America Special Paper 349, eds. Y. Dilek, E. Moores, D. Elthon and A. Nicolas. Boulder, CO: Geological Society of America, pp.213–235.

Humphris, S. E., Melson, W. G., and Thompson, R. N. 1980. Basalt weathering on the East Pacific Rise and Galapagos Spreading Center, DSDP Leg 54. In *Initial Reports of the Deep Sea Drilling Project*, Vol. 54, eds. B. R. Rosendahl and R. Hekinian. Washington, DC: US Govt. Printing Office, pp. 773–787.

Hunter, A. G. and ODP Leg 168 Scientific Party 1998. Petrological investigations of low-temperature hydrothermal alteration of the upper crust, Juan de Fuca Ridge, ODP Leg 168. In *Modern Ocean Floor Processes and the Geological Record*, Special Publication 148, eds. R. A. Mills and K. Harrison. London: Geological Society, pp. 99–125.

Hunter, A. G., Kempton, P. D., and Greenwood, P. 1999. Low temperature fluid-rock ineraction – an isotopic and mineralogical perspective of upper crustal evolution, eastern flank of the Juan de Fuca Ridge, JdFR, ODP Leg 168. *Chem. Geol.* **155**: 3–28.

Ishikawa, T. and Nakamura, E. 1992. Boron isotope geochemistry of the oceanic crust form DSDP/ODP Hole 504B. *Geochim. Cosmochim. Acta* **56**: 1,633–1639.

Janecky, D. A. and Shanks, W. C. 1988. Computational modeling of chemical and sulfur isotopic reaction processses in seafloor hydrothermal systems: chimneys, massive sulfides, and subjacent alteration zones. *Can. Mineral.* **26**: 805–825.

Jarrard, R. D., Abrams, L. J., Pockalny, R., Larson, R. L., and Hirono, T. 2004. Physical properties of upper oceanic crust: ODP Hole 801C and the waning of hydrothermal circulation. *J. Geophys. Res.* **108**(D7), 2188, doi: 10-1029/2001JB001727.

Juteau, T., Bingol, F., Noack, Y., Whitechurch, H., Hoffert, M., Wirrmann, D., and Courtois, C. 1979. Preliminary results: mineralogy and geochemistry of alteration products in Leg 45 basement samples. In *Initial Reports of the Deep Sea Drilling Project*, Vol. 45, eds. W. G. Melson and P. D. Rabinowitz. Washington, DC: US Govt. Printing Office, pp. 613–645.

Kastner, M. and Gieskes J. M. 1976. Interstitial water profiles and sites of diagenetic reactions, Leg 35, Bellingshausen abyssal plain. *Earth Planet. Sci. Lett.* **33**: 11–20.

Kelley, D. S., Baross, J. A., and Delaney, J. R. 2002. Volcanoes, fluids, and life at mid-ocean ridge spreading centers. *Ann. Rev. Earth Planet. Sci.* **30**: 385–491.

Kempe, D. R. C. 1974. The petrology of the basalts, Leg 26. In *Initial Reports of the Deep Sea Drilling Project*, Vol. 26, eds. T. A. Davies and B. P. Luyendyk. Washington, DC: US Govt. Printing Office, pp. 465–504.

Laverne, C. 1993. Occurence of siderite and ankerite in young basalts from the Galapagos Spreading Center, DSDP Holes 605G and 507B. *Chem. Geol.* **106**: 27–46.

Laverne, C. and Vivier, G. 1983. Petrographical and chemical study of basement from the Galapagos Spreading Center, Leg 70. In *Initial Report of the Deep Sea Drilling Project*, Vol. 70, eds. R. P. Von Herzen, J. Honnorez, *et al.* Washington, DC: US Govt. Printing Office, pp. 375–390.

Lawrence, J. R. 1979. Temperatures of formation of calcite veins in the basalts from DSDP Holes 417A and 417D. In *Initial Reports of the Deep Sea Drilling Project*,

Vol. 51–53, eds. T. Donnelly, *et al.* Washington, DC: US Govt. Printing Office, pp. 1183–1184.

1991. Stable isotopic composition of pore waters and calcite veins. In *Proceedings of the Ocean Drilling Program, Scientific Results*, Vol. 121, J. Weissel, J. Pierce, E. Taylor, and J. Alt. College Station, TX: Ocean Drilling Program, pp. 1–6.

Lawrence, J. R. and Drever, J. I. 1981. Evidence for cold water circulation at DSDP Site 395: isotopes and chemistry of alteration products. *J. Geophys. Res.* **86**: 5,125–5,133.

Lister, C. R. B. 1982. "Active" and "passive" hydrothermal systems in the ocean crust. Predicted physical conditions. In *The Dynamic Environment of the Ocean Floor*, eds. K. A. Fanning and F. T. Manheim. Lexington, MA: Heath, pp. 441–470

Ludden, J. N. and Thompson, G. 1979. An evaluation of the behavior of the rare earth elements during the weathering of sea floor basalt. *Earth Planet. Sci. Lett.* **43**: 85–92.

Manning, C. E., Weston, P. E., and Mahon, K. I. 1996. Rapid high-temperature metamorphism of the East Pacific Rise gabbros from Hess Deep. *Earth Planet. Sci. Lett.* **144**: 123–132.

Melson, W. G. and Thompson, G. 1973. Glassy abyssal basalts, Atlantic sea floor near St. Paul's Rocks: petrography and composition of secondary clay minerals. *Geol. Soc. Am. Bull.* **84**: 703–716.

Mottl, M. J. and Wheat, C. G. 1994. Hydrothermal circulation through mid-ocean ridge flanks: fluxes of heat and magnesium. *Geochim. Cosmochim. Acta* **58**: 2,225–2,237.

Muehlenbachs, K. 1977. Oxygen isotope geochemistry of DSDP Leg 37 rocks. In *Initial Reports of the Deep Sea Drilling Project*, Vol. 37, eds. F. Aumento, W. G. Melson, *et al.* Washington, DC: US Govt. Printing Office, pp. 617–620.

1979. The alteration and aging of the basaltic layer of the seafloor: oxygen isotope evidence from DSDP/IPOD Legs 51, 52, and 53. In *Initial Reports of the Deep Sea Drilling Project*, Vol. 50–50, eds. T. Donnelly, J. Francheteau, W. Bryan, P. Robinson, M. Flower, and M. Salisbury. Washington, DC: US Govt. Printing Office, pp. 1,159–1,167.

Muehlenbachs, K. and Clayton, R. N. 1972. Oxygen isotope studies of fresh and weathered submarine basalts. *Can. J. Earth Sci.* **9**: 172–184.

1976. Oxygen isotope composition of the oceanic crust and its bearing on seawater. *J. Geophys. Res.* **81**: 4,365–4,369.

Natland, J. H. and Rosendahl, B. H. 1980. Driing difficulties in basement during DSDP Leg 54. In *Initial Reports of the Deep Sea Drilling Project*, Vol. 54, eds. B. R. Rosendahl and R. Hekinian. Washington, DC: US Govt. Printing Office, pp. 593–603.

Nehlig, P., Juteau, T., Bendel, V., and Cotten, J. 1994. The root zones of oceanic hydrothermal systems: constraints from the Samail ophiolite Oman. *J. Geophys. Res.* **99**: 4,703–4,713.

O'Nions, R. K. and Pankhurst, R. J. 1977. Sr isotope and rare earth element geochemistry of DSDP Leg 37 basalts. In *Initial Reports of the Deep Sea Drilling Project*, Vol. 37, eds. F. Aumento, W. G. Melson, *et al.* Washington, DC: US Govt. Printing Office, pp. 599–602.

Park, K. H. and Staudigel, H. 1990. Radiogenic isotope ratios and initial seafloor alteration in submarine Serocki Volcano basalts. In *Proceedings of the Ocean Drilling Program, Scientific Results*, Vol. 106/109, eds. R. Detrick, J. Honnorez, W. B. Bryan, and T. Juteau. College Station, TX: Ocean Drilling Program, pp. 117–122.

Pertsev, N. N. and Rusinov, V. L. 1979. Mineral assemblages and processes of alteration in basalts at DSDP Sites 417 and 418. In *Initial Reports of the Deep Sea Drilling Project*, Vol. 51–53, eds. T. Donnelly, J. Francheteau, W. Bryan, P. Robinson,

M. Flower, and M. Salisbury. Washington, DC: US Govt. Printing Office, pp. 1,219–1,242.

Peterson, C., Duncan, R., and Scheidegger, K. F. 1986. The sequence and longevity of basalt alteration at Deep Sea Drilling Project Site 597. In *Initial Reports of the Deep Sea Drilling Project*, Vol. 92, eds. M. Leinen and D. K. Rea. Washington, DC: US Govt. Printing Office, pp. 491–497.

Porter, S., Vanko, D. A., and Ghazi, M. 2000. Major and trace element compositions of secondary clays in basalts altered at low temperature, easter flank of the Juan de Fuca Ridge. In *Proceedings of the Ocean Drilling Program*, Vol. 168, eds. A. Fisher, E. E. Davis, and C. Escutia. College Station, TX: Ocean Drilling Program, pp. 149–157.

Révillon, S., Barr, S. R., Brewer T. S., Harvey P. K., and Tarney J., 2004. An alternative approach using integrated gamma-ray and geochemical data to estimate the inputs to subduction zones from ODP Leg 185, Site 801. *G-cubed* **3**(12): 8902, doi: 10.1029/2002GC000344.

Richards, H. G., Cann, J. R., and Jensenius, J. 1989. Mineralogical zonation and metasomatism of the alteration pipes of Cyprus sulfide deposits. *Econ. Geol.* **84**: 91–115.

Richardson, C. J., Cann, J. R., Richards, H. G., and Cowan, J. G. 1987. Metal-depleted root zones of the Troodos ore-forming hydrothermal systems, Cyprus. *Earth Planet. Sci. Lett.* **84**: 243–253.

Richardson, S. H., Hart, S. R., and Staudigel, H. 1980. Vein mineral ages of old oceanic crust. *J. Geophys. Res.* **85**: 7,195–7,200.

Rosenberg, N. D., Spera, F. J., and Haymon, R. M. 1993. The relationship between flow and permeability field in seafloor hydrothermal systems. *Earth Planet. Sci. Lett.* **116**: 135–153.

Scheidegger, K. F. and Stakes, D. S. 1977. Mineralogy, chemistry, and crystallization sequence of clay minerals in altered tholeiitic basalts from the Peru Trench. *Earth Planet. Sci. Lett.* **36**: 413–422.

Schiffman, P. and Smith, B. M. 1988. Petrology and oxygen isotope geochemistry of a fossil seawater hydrothermal system within the Solea Graben, northern Troodos ophiolite, Cyprus. *J. Geophys. Res.* **93**: 4,612–4,624.

Schiffman, P., Smith, B. M., Varga, R. J., and Moores, E. M., 1987. Geometry, conditions and timing of off-axis hydrothermal metamorphism and ore deposition in the Solea Graben. *Nature* **325**: 423–425.

Scott, R. B. and Hajash, A. 1976. Initial submarine alteration of basaltic pillow lavas: a microprobe study. *Am. J. Sci.* **276**: 480–501.

Seyfried, W. E., Shanks, W. C., and Dibble, W. E. 1978. Clay mineral formation in DSDP Leg 34 basalts. *Earth Planet. Sci. Lett.* **41**: 265–276.

Seyfried, W. E., Ding, K., Berndt M. E. and Chen X., 1999. Experimental and theoretical controls on the composition of mid-ocean ridge hydrothermal fluids. In *Volcanic Associated Massive Sulfide Deposits, Reviews In Economic Geology*, Vol. 8, eds. T. Barrie and M. Hannington. Chelsea, MI: Society for Economic Geology, pp. 181–200.

Shipboard Scientific Party 1990. Site 765. In *Proceedings of the Ocean Drilling Program, Initial Results*, Vol. 123, eds. J. N. Ludden, F. M. Gradstein, *et al.* College Station, TX: Ocean Drilling Program, pp. 63–267.

　　2000. Site 801. In eds. T. Plank, J. N. Ludden, C. Escutia, *et al.*, *Proceedings of the Ocean Drilling Program, Initial Results*. **185**: 1–222 [online]. Available from World

Wide Web: http://www-odp.tamu.edu/publications/185_IR/VOLUME/CHAPTERS/ IR185_03.PDF

2003. Site 1256. In eds. D. Wilson, D. A. H. Teagle, G. Acton, *et al.*, *Proceedings of the Ocean Drilling Program, Initial Results*, **206** [online]. Available from World Wide Web: http://www/odp.tamu.edu/publications.

Stakes, D. S. and O'Neil, J. R. 1982. Mineralogy and stable isotope geochemistry of hydrothermally altered oceanic rocks. *Earth Planet. Sci. Lett.* **57**: 285–304.

Stakes, D. S. and Scheidegger, K. F. 1981. Temporal variations in secondary minerals from Nazca Plate basalts. In *Nazca Plate: Crustal Formation and Andean Convergence*, ed. L. D. Kulm. Boulder, CO: Geological Society of America, pp. 109–130.

Staudigel, H. and Hart, S. R. 1983. Alteration of basaltic glass: mechanisms and significance for the oceanic crust–seawater budget. *Geochim. Cosmochim. Acta* **47**: 337–350.

1985. Dating of ocean crust hydrothermal alteration: strontium isotope ratios from Hole 504B carbonates and the re-interpretation of Sr isotope data from Deep Sea Drilling Project Sites 105, 332, 417, and 418. In *Initial Reports of the Deep Sea Drilling Project*, Vol. 83, eds. R. N. Anderson, J. Honnorez, and K. Becker. Washington, DC: US Govt. Printing Office, pp. 297–303.

Staudigel., H., Hart, S. R., and Richardson, S. H. 1981a. Alteration of the oceanic crust: processes and timing. *Earth Planet. Sci. Lett.* **52**: 311–327.

Staudigel, H., Muehlenbachs, K., Richardson, S. H., and Hart, S. R. 1981b. Agents of low temperature ocean crust alteration. *Contrib. Mineral. Petrol.* **77**: 150–157.

Staudigel, H. R., Hart, S. R., Schmincke, H. U., and Smith, B. M. 1989. Cretaceous ocean crust at DSDP Sites 417 and 418: carbon uptake from weathering versus loss by magmatic outgassing. *Geochim. Cosmochim. Acta* **53**: 3,091–3,094.

Staudigel, H., Davies, G. R., Hart, S. R., Marchant, K. M., and Smith, B. M. 1995. Large scale Sr, Nd, and O isotopic anatomy of altered oceanic crust: DSDP sites 417/418. *Earth Planet. Sci. Lett.* **130**: 169–185.

Swift, S. A., Kent, G. M., Detrick, R. S., Collins, J. A., and Stephen, R. A. 1998. Oceanic basement structure, sediment thickness, and heat flow near Hole 504B. *J. Geophys. Res.* **103**: 15,337–15,391.

Teagle, D. A. H., Alt, J. C., Bach, W., Halliday, A. N., and Erzinger, J. 1996. Alteration of upper ocean crust in a ridge-flank hydrothermal upflow zone: mineral, chemical and isotopic constraints from ODP Hole 896A. In *Proceedings of the Ocean Drilling Program, Scientific Results*, Vol. 148, eds. J. C. Alt, H. Kinoshita, and L. Stokking. College Station, TX: Ocean Drilling Program, pp. 119–150.

Teagle, D. A. H., Alt, J. C., Chiba, H., Humphris, S., and Halliday, A. N. 1998. Strontium and oxygen isotopic constraints on fluid mixing, alteration and mineralization in the TAG hydrothermal deposit. *Chem. Geol.* **149**: 1–24.

Thompson, G. 1973. A geochemical study of the low-temperature interaction of seawater and oceanic igneous rocks. *Eos, Trans. Am. Geophys. Union* **54**: 1,015–1,019.

1983. Basalt–seawater interaction. In *Hydrothermal Processes at Seafloor Spreading Centers*. eds. P. A. Rona, K. Bostrom, and K. L. Smith. New York: Plenum, pp. 225–278.

Torsvik T., Furnes, H., Muehlenbachs, K., Thorseth, I. H., and Tumyr, O. 1998. Evidence for microbial activity at the glass–alteration interface in oceanic basalts. *Earth Planet. Sci. Lett.* **162**: 165–176.

Vanko, D. A. and Laverne, C. 1998. Hydrothermal anorthitization of plagioclase within the magmatic/hydrothermal transition at mid-oceanic ridges: examples from deep sheeted dikes (Hole 504B, Costa Rica Rift) and a sheeted dike root zone (Oman ophiolite). *Earth Planet. Sci. Lett.* **162**: 27–43.

Vanko, D. A., Laverne, C., Tartarotti, P., and Alt, J. C. 1996. Chemistry and origin of secondary minerals from the deep sheeted dikes cored during ODP Leg 148, Hole 504B. In *Proceedings of the Ocean Drilling Program, Scientific Results*, Vol. 148, eds. J. C. Alt, H. Kinoshita, L. Stokking, and P. Michael. College Station, TX: Ocean Drilling Program, pp. 71–86.

Von Drach, V., Muller-Sohnius, D., Kohler, H., and Huckenholz, H. G. 1978. Sr isotope ratios on whole rock samples of Leg 45 basalts. In *Initial Reports of the Deep Sea Drilling Project*, Vol. 45, eds. W. G. Melson, P. Rabinowitz, *et al.* Washington, DC: US Govt. Printing Office, pp. 535–540.

Von Damm, K. L., Oosting, S. E., Kozlowski, R., Buttermore, L. G., Colodner, D. C., Edmonds, H. N., Edmond, J. M., and Grebmeier, J. M. 1995. Evolution of East Pacific Rise hydrothermal vent fluids following a volcanic eruption. *Nature* **375**: 47–50.

Waggoner, D. G. 1993. The age and alteration of central Pacific oceanic crust near Hawaii, Site 843. In *Proceedings of the Ocean Drilling Program*, Vol. 136, eds. R. H. Wilkens, J. Firth, and J. Bender. College Station, TX: Ocean Drilling Program, pp. 119–132.

Wheat, C. G. and Mottl, M. J. 2000. Composition of pore and spring waters from Baby Bare: global implications of geochemical fluxes from a ridge flank hydrothermal system. *Geochim. Cosmochim. Acta* **64**: 629–642.

Wilkens, R. H., Fryer, G. J. and Karsten, J. 1991. Evolution of porosity and seismic structure of upper oceanic crust: importance of aspect ratios. *J. Geophys. Res.* **96**: 17,891–17,995.

Yamaguchi, M., Armstrong, R. L., Russell, R. D., and Slawson, W. F. 1977. Strontium and lead isotopic investigations of igneous rocks from the mid Atlantic Ridge, DSDP Leg 37. In *Initial Reports of the Deep Sea Drilling Project*, Vol. 37, eds. F. Aumento, W. G. Melson, *et al.* Washington, DC: US Govt. Printing Office, pp. 613–616.

Yatabe, A., Vanko, D. A., and Ghazi, M. 2000. Petrography and chemical compositions of secondary calcite and aragonite in Juan de Fuca Ridge basalts altered at low temperature. In *Proceedings of the Ocean Drilling Program, Scientific Results*, Vol. 168, eds. A. Fisher, E. E. Davis, and C. Escutia. College Station, TX: Ocean Drilling Program, pp. 137–148.

Zhou, W., Peacor, D. R., Alt, J., Van der Voo, R., and Kao, L.-S. 2001. TEM study of the alteration of glass in MORB by inorganic processes. *Chem. Geol.* **174**: 365–376.

Zierenberg, R. A., Shanks, W. C., Seyfried, W. E., Koski, R. A., and Strickler, M. D. 1988. Mineralization, alteration and hydrothermal metamorphism of the ophiolite-hosted Turner–Albright sulfide deposits, southwest Oregon. *J. Geophys. Res.* **93**: 4,657–4,674.

16

Ridge flank sediment–fluid interactions

Miriam Kastner and Mark D. Rudnicki

16.1 Introduction

16.1.1 Background

Sediment and pore-fluid geochemical records provide insights into the processes that regulate Earth's environment and their spatial and temporal evolution. These sediments are an integral part of ridge flank fluid flow systems and provide the thermal blanket that traps heat and allows the upper oceanic basement and circulating fluids to warm up and become chemically reactive. Depending on their lithology (primarily permeability) and thickness, the sediments may permit the passage of downwelling and upwelling fluids (i.e. Kastner *et al.*, 1986; Giambalvo *et al.*, 2000; Fisher *et al.*, 2001), or may be in diffusional communication with upper oceanic basement fluids (Baker *et al.*, 1991; Elderfield *et al.*, 1999; Silver *et al.*, 2000). The reactions between these fluids and the sediments impact diagenesis, and thus the composition of the sediments and of the fluids that may modify paleoceanographic records. Given sufficient mass flux, the reactions between circulating fluids and sediments are likely to act as sources and/or sinks for some elements and isotopes, in addition to those provided through basement–seawater reactions alone, and to affect seawater chemistry.

Despite the importance of ridge flank hydrothermal circulation ($\sim$70% of global hydrothermal fluids remove heat from the lithosphere from >1 Ma, until$\sim$65 Ma, Stein and Stein, 1994), little is known about the detailed hydrology. The overall relations between key ridge flank variables such as sediment thickness, composition, basement age, temperature, and basement topography, have been described (e.g. Anderson and Hobart, 1976; Davis *et al.*, 1989; Stein and Stein, 1992, 1994; Snelgrove and Foster, 1996; Schultz and Elderfield, 1997; Fisher and Becker, 2000; Giambalvo *et al.*, 2000; Fisher *et al.*, 2001, Chapter 6). But the relations between these variables and fluid and geochemical fluxes, as well as how much of the fluid flow is channeled through outcrops or faults, and how much is advecting through the sediment cover, are as yet inadequately quantified. Therefore, the relative roles fluid–sediment versus fluid–oceanic basement reactions play in determining ridge flank chemical and isotopic fluxes are also, thus far, poorly known. The amount of convective

Hydrogeology of the Oceanic Lithosphere, eds. E. E. Davis and H. Elderfield. Published by Cambridge University Press.
© Cambridge University Press 2004.

fluid flow decreases as the crust ages, cools, and is covered by sediments (Sclater *et al.*, 1976; Stein and Stein, 1994; see also Chapter 8). Qualitatively, at higher heat flow sites on younger oceanic basement that are blanketed by thinner sediment sections, the fluid chemical signatures are more intense. However, lower heat flow sites with thicker sediment sections, and hence less-intense fluid chemical signatures, extend over significantly larger areas of the seafloor. Because of the relatively small number of sites that have been studied to date, and the more subtle geochemical indicators of ridge flank processes, the geochemical significance of fluid circulation through the oceanic ridge flanks is still uncertain (see also Chapter 20). However, as suggested by Elderfield and Schultz (1996), because of the huge fluxes of fluid through the flanks small concentration anomalies will have a major impact on ocean chemistry.

Key unresolved questions with implications for off-axis fluid and solute fluxes, and thus chemical composition of the ocean and mantle are:

- What are the characteristic relations between fluid and heat fluxes, the thickness and type of sediment cover, basement topography, and the nature of fluid flow pathways at ocean crust >1 to ~10 Ma and >10 to ~65 Ma?
- What is the magnitude of flow which is focused through basement outcrops and/or faults compared with diffuse flow at ridge flanks? How much of the diffuse flow is through the sediment cover?
- What are the chemical, isotopic, microbial, and hydrological implications of fluid circulation through the sediment cover and/or in the upper oceanic basement, and the consequences for ocean–basement fluid exchange?
- What are the differences in the rates of typical diagenetic reactions in diffusive versus advective fluid regimes at ridge flanks?
- What is the significance of ridge flank hydrothermal circulation for global water, chemical, and isotopic fluxes?

Except for the young thickly sedimented eastern flank at the Juan de Fuca Ridge that has been studied extensively across a transect spanning from well-ventilated to sediment-sealed conditions, the database on ridge flank fluid recharge and discharge through sediments is very limited, therefore answers to only some of these questions are provided in this chapter. Our focus accordingly is on available data analysis and synthesis and recommendations for future work.

16.1.2 *Significance of pore-fluid chemical and isotopic profiles*

Critical information on fluid–sediment reactions and their impact on the physical and chemical properties of the sediments, microbial activity, material and solute fluxes may be inferred from high-resolution pore-fluid chemical and isotopic depth profiles, both in diffusion and advection controlled ridge flank regimes. For example, smectite dehydration and transformation to illite (between ~50 and 120 °C) produces fluids that have lower chlorinity and salinity, higher silica concentrations, and a distinct combination of Cl, O, H, B, and $^{87}Sr/^{86}Sr$ isotope ratios, characteristics that might be discernable in moderate temperature fluids.

The chemistry and isotopic composition of pore fluids can be used to constrain *in situ* mineral–fluid reactions, the temperature range within which these reactions occur, and the general composition of the reacting solid phases. At the low- to moderate-temperature fluid flow regimes prevailing at ridge flanks, pore fluids may not be in chemical and/or isotopic equilibrium, kinetic processes are of greater importance. Kinetic considerations imply that fluid advective velocity may exert a control on pore-fluid chemistry that may not be discernable from single profile measurements. In contrast, at higher temperature fluid flow regimes, typical of ridge crest flow systems, fluid–rock reactions rapidly reach thermodynamic equilibrium.

Thus, obtaining high-resolution and/or continuous time-series sediment pore-fluid chemical and isotopic data (e.g. Wheat *et al.*, 2000b) can provide the following important information about ridge flank sediment and upper oceanic basement hydrologic regimes and associated solute and water fluxes.

On fluid–solid reactions and fluid flow

As noted by Wheat and McDuff (1995), pore-fluid geochemical data are considerably more sensitive tracers of pore-fluid velocities than heat flow data, and geochemical fluxes can be obtained by advection–diffusion modeling of the geochemical profiles. Hence, geochemical data provide (1) a measure of the degree of water–basement and water–sediment reactions, (2) the means to calculate chemical and isotopic fluxes, and (3) a gauge to evaluate crustal aging.

On the hydrology downwards and/or upwards fluid flow (recharge and discharge)

It is possible to model: (1) at low flow rates (<1 cm yr^{-1}) the position of diagenetic reaction maxima or minima; (2) fluid advection that shifts chemical signals attributed to chemical paleoceanographic events that altered seawater composition, for example, the change in the $\delta^{18}O$, and Cl concentration of seawater at the last glacial maximum (LGM; Schrag *et al.*, 2002); and (3) at higher flow rates the chemical gradients at boundaries (sediment–water, sediment–basement) or shifts in chemical maxima or minima.

On the influence of fluid–sediment reactions on sediment physical properties, thus hydrology

High heat flow may quicken the pace of diagenesis such as cementation that affects permeability. For example, at the Costa Rica Rift: ODP Site 505, heat flow is low, and the calcareous ooze is only slightly altered to chalk in the lower 22 m of the sediment section (sediment section is 232 m thick). At the adjacent high heat flow ODP Sites 501/504, however, there are two distinct diagenetic boundaries, from ooze to chalk at 143 m, and from chalk to cherty limestone at 235 m (sediment section is 264–274 m thick; Mottl *et al.*, 1983).

On the alteration of basement formation fluids by interactions with sediment pore fluids

This is the path by which seawater reaches oceanic basement influencing the composition of the basement pore fluid (Wheat and Mottl, 1994). If seawater enters basement directly through faults and outcrops or recharges through the sediment column so rapidly that reaction with the sediment does not significantly alter the composition of the recharging bottom water, the basement fluid composition will resemble that produced by seawater–basalt reaction at a specified temperature and water/rock ratio, and depending on the residence time may be modified by diffusive exchange across the sediment–basement interface. If seawater enters basement by slow recharge through the sediment section, the composition of the seawater in the basement will reflect reactions with both sediment and basalt. The basement formation fluid composition may thus be modified either by interaction with sediments during recharge, or by diffusive exchange between sediment pore fluids and the basement aquifer. The ratio of sediment and basaltic components in basement formation fluids will reflect the relative amounts of reaction in each environment and provide insight into the hydrology (Baker *et al.*, 1991; Silver *et al.*, 2000). The composition of fluids discharging through the sediment cover (cold seeps) may also experience sub-surface diagenetic reactions, conductive cooling, and mixing with bottom seawater. Higher temperature fluids discharging at sediment covered axial vent sites carry similar mixed chemical signatures.

16.2 Sites of off-axis fluid flow

In the last two decades, extensive geophysical, thermal, chemical, and isotopic studies have been conducted at the following ridge flank discharge and recharge regions: the Galapagos Mounds, Costa Rica Rift, central and eastern equatorial Pacific, the Juan de Fuca Ridge at 48° N, Alarcon Basin, and west flank of the Mid-Atlantic Ridge (Fig. 16.1). Important gains have been made in understanding processes of fluid circulation and reactions, as briefly summarized below.

16.2.1 Galapagos Mounds

The Galapagos Mounds, located at ~86° W, 18–32 km south of the Galapagos Rift, on oceanic basement 0.5–0.9 Ma, were discovered in the mid seventies during a Deep-Tow study (Lonsdale, 1977). Heat flow data indicated lower than expected values for the age of the plate (Williams *et al.*, 1974). Follow-up detailed studies involved sampling with submersibles, coring, and drilling on two DSDP Legs 54 and 70, in 1979 and 1981, respectively. The fluid in the basement originates from the recharge of seawater through the sediment, as indicated by the high silica concentrations, saturated with respect to opal-A solubility (~800–1,000 μmol kg^{-1}). In contrast, fluid flow into the basement through basement outcrops and/or faults would contain silica concentrations similar to bottom-water, which is ~15% of the opal-A equilibrium solubility concentration at *in situ* temperature and pressure.

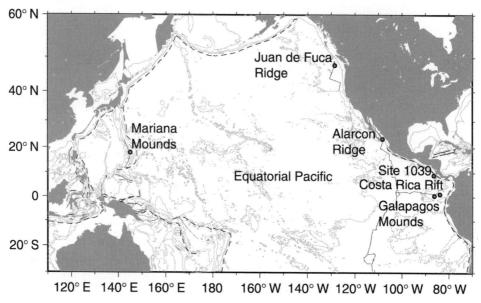

Fig. 16.1 Location map of the sites discussed in the text.

In areas of recharge, oxygen and nitrate are utilized and some Fe and Mn mobilized. Opal-A is efficiently dissolved and transported to the upper oceanic basement. This silica-rich fluid upwells through basement outcrops, faults, and the thin, ~30 m, sediment cover. The elevated silica content enhances the preservation of biogenic opal-A, in particular where authigenic opal-A is precipitated. However, because the upwelling fluid also transports dissolved Fe and Mn, nontronite and Mn-Fe-oxyhydroxides precipitate at the oxidation front. The Galapagos Mounds consist primarily of nontronite covered with Mn-Fe-oxyhydroxides.

Near the nontronite mounds, biogenic opal-A dissolution is enhanced because nontronite precipitation lowers the silica concentration in the pore fluids relative to opal-A solubility. However, the equilibrium Si concentration of nontronite solubility is about four times higher than bottom-seawater Si concentration. Therefore, a net flux of Si must occur into the ocean equal to the fluid flux multiplied by nontronite solubility Si concentration. Based on pore-water Ca, Mg, and Fe profiles, fluid advection through the mounds is vigorous at 15–35 cm yr^{-1}, and in the sediments near the mounds it is ~1 cm yr^{-1} (Bender, 1983; Maris *et al.*, 1984).

The fluid flow through the sediment modifies the paleoceanographic record of the biogenic silica productivity and preservation. Microbial activity is most likely also enhanced by the fluid flow regime, thus also affecting the original organic C paleoceanographic record. This is accompanied by a flux of old dissolved organic carbon (DOC) into seawater. The influence of ridge flank DOC flux into seawater on the average age of the DOC is unknown.

16.2.2 Costa Rica Rift

The transition from low (10–20% of the predicted conductive heat flow) to normal heat flow at the southern flank of the Costa Rica Rift near 84° W, has attracted much research since the early eighties (Langseth *et al.*, 1983; Hobart *et al.*, 1985). This ridge flank was extensively drilled on three DSDP legs, i.e. 68, 69, and 111. Sites 501–505 and 677–678, are located on oceanic crust 3.9–5.9 Ma. The basement topography in the area near Site 505 (3.9 Ma) is rough and, although covered by 130–240 m of sediments, has numerous basement outcrops and heat flow is low and variable. The heat flow at Sites 501, 504, 677, and 678 (on ~5.9 Ma basement) is, in contrast, close to the predicted value (Langseth *et al.*, 1988). The sediment cover here is 170–310 m thick and continuous. Generally, heat flow is higher at topographic highs, where sediment cover is thinner. These are sites of fluid discharge through the sediments. Recharge occurs through the thicker sediment over areas of basement lows (Langseth *et al.*, 1988), at 1–6 mm yr^{-1}, as indicated by pore-fluid chemical gradients (Mottl, 1989). The fluid flow through the sediment enhances sediment diagenesis, particularly of biogenic opal-A dissolution, carbonate precipitation, and microbially mediated organic matter oxidation. These reactions have locally altered and even erased the original paleoceanographic records, such as biogenic silica and organic C productivities and preservations.

16.2.3 Mariana Mounds

The Mariana Mounds are located in a knoll field 50–150 m high on 3 Ma oceanic basement on the western flank of the Mariana back-arc spreading center, west of the arc, at ~18° N; DSDP Sites 453–454 are situated nearby. Along a transect from 0 to 5.8 Ma the oceanic basement is blanketed by sediments 3–100 m thick and heat flow varies between 30–50% of the theoretically predicted value; the knolls have higher heat flow than the adjacent topographically lower areas. Pore-water chemical data, especially the Ca and Mg profiles along this transect indicate that only a small fraction of the heat is lost by fluid flow through the sediment cover. The Mounds area, covered by ~40 m of sediment, was studied more extensively by piston coring and submersible sampling (Dadey, 1991; Wheat and McDuff, 1994).

Most of the recharge in the Mounds area occurs through a distinct fault scarp in the vicinity, thus the basement fluid is chemically similar to bottom seawater, except for the loss of oxygen, nitrate, and slightly elevated silica concentrations (220~280 μmol kg^{-1}). Along the transect flow velocities are high, between 2–160 cm yr^{-1} (Abott *et al.*, 1983). Fluid flow through the mounds ranges from <1 cm yr^{-1} to >100 m yr^{-1} (Dadey, 1991; Wheat and McDuff, 1994; Wheat and McDuff, 1995). At such upwelling rates sediment diagenesis is enhanced, especially the dissolution of silica. Particularly, biogenic opal-A has been completely dissolved in the basal few meters in the Mounds area, and enhanced microbially mediated oxidation of organic matter has been documented (Wheat and McDuff, 1995). However, because recharge is mostly through faults and basement outcrops the dissolved silica flux into the ocean is lower than at the Galapagos Mounds where much of the recharge is through the sediment.

16.2.4 Central and eastern equatorial Pacific

The three East Pacific Rise (EPR) ridge flank regions with well-documented fluid circulation systems are: an area $>6 \times 10^6$ km^2, situated in the central equatorial Pacific from 110 to 160° W, about 5° S to 8° N; at the western flank at $\sim$19° S; and at the eastern flank at $\sim$9.4° N, 86° W, $\sim$2 km west of the Costa Rica trench.

The central equatorial Pacific region spans crustal ages from $\sim$15 to 70 Ma; is covered by thick, up to $>$500 m, sediment, and has about 50–75% of the predicted heat flow (Sclater *et al.*, 1976). Sediment pore-fluid chemical profiles, particularly of Ca, Sr, Mg, sulfate, and ^{87}Sr/^{86}Sr, obtained during DSDP Leg 85 (Stout, 1985), revealed the existence of a large-scale lateral fluid advection system in the upper oceanic basement (Baker *et al.*, 1991). The location of DSDP/ODP drill holes and examples of sulfate profiles are shown in Fig. 16.2. The chemical and isotopic components which are more susceptible to diagenetic reactions in the sediment column – Sr, sulfate, and ^{87}Sr/^{86}Sr – show concentration reversals near basement. At the sediment/basement boundary, their concentrations and isotope ratio resemble those of modern bottom seawater. The concentrations of components less susceptible to sediment diagenetic reactions such as Ca and Mg, however, remain nearly constant to basement depth. Modeling of the basal profile reversals suggests a rapid lateral circulation of seawater through a high-permeability upper oceanic basement. The geochemical profile reversals indicate diffusive communication between the low permeability basal sediments and the seawater basement "river." Modeling the profiles provides a residence time of the basement fluid of $\sim$20,000 years and inferred pore-fluid flow rates between 1 and 10 m yr^{-1}. Similar pore-fluid concentration reversal were observed at the Costa Rica Rift flank, DSDP Site 505, 3.9 Ma basement age (Mottl *et al.*, 1983), and near the Costa Rica trench ODP Site 1039, $\sim$24 Ma basement age (Chan and Kastner, 2000; Kastner *et al.*, 2000; Silver *et al.*, 2000). Such chemical reversals are likely to modify the *in situ* diagenetic reactions.

Extremely low heat flow values, 8–13 mW m^{-2}, significantly lower (10–12% of lithospheric value) than the theoretically predicted value for 24 Ma crust, of 95–105 mW m^{-2} (Stein and Stein, 1994) were measured on the EPR ridge flank near the boundary with the Cocos plate, seaward of the Middle America trench, at $\sim$9.4° N, 86° W (Langseth and Silver, 1996; Barckhausen, *et al.*, 2001). ODP Site 1039 was drilled on ODP Leg 170 to explore the origin of this very low heat flow. The site is covered by $\sim$400 m hemipelagic and pelagic sediment. The regionally depressed heat flow has been interpreted as evidence for vigorous fluid flow within the upper oceanic basement which effectively refrigerates the plate by fluid advection (Langseth and Silver, 1996). The pore-fluid concentration and isotopic gradients in the basal section of the overlying sediment show reversals, and near basement the concentration and isotope ratios approach modern-day seawater values (see Fig. 16.3), indicating that the heat loss at this site is not by flow through the thick sediment cover, but by vigorous lateral fluid flow of bottom seawater in the underlying permeable upper oceanic basement. This is most clearly manifested in the Ca, Mg, Sr, Li, and Si

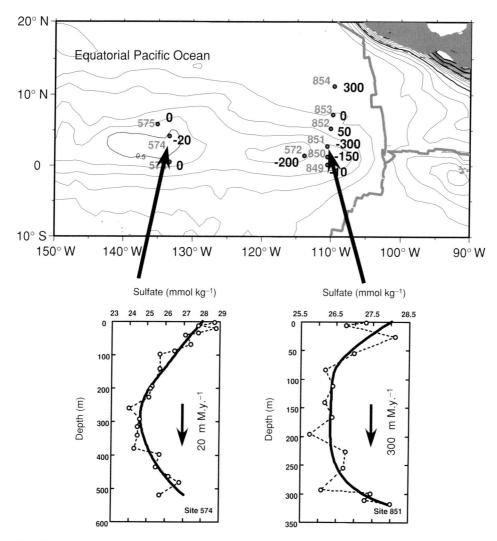

Fig. 16.2 Location of sites drilled in sediments from the equatorial Pacific sediment bulge, together with advective velocities (volumetric flux) obtained by modeling pore-water sulfate concentrations. Advective velocities are given as bold numbers, in m Ma^{-1}. Negative numbers signify downwelling of fluids. Samples obtained during DSDP Leg 85 (Sites 572–575) were modeled by Richter (1996). Samples obtained during ODP Leg 138 (Sites 849–854) were modeled by Rudnicki (unpublished data).

concentration and Sr and Li isotope ratio depth profiles (Chan and Kastner, 2000; Kastner *et al.*, 2000, pers. comm; Silver *et al.*, 2000). The concentration and isotope ratio reversals for Li and Sr are shown in Fig. 16.3. The chemically unaltered nature of the basement fluid indicates that recharge must be through deep faults or basement outcrop(s) at seamounts not as yet located (Chan and Kastner, 2000; Silver *et al.*, 2000). Modeling the pore-fluid chemical and isotopic gradients in the basal sediments indicates that the *in situ* diagenetic reactions

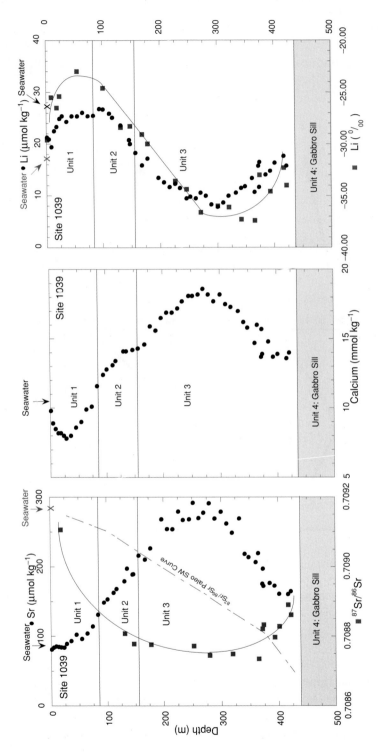

Fig. 16.3 Pore-fluid Li concentrations and δ^6Li data, and Sr concentrations and ^{87}Sr/^{86}Sr data from ODP Site 1039, eastern flank of the East Pacific Rise near the boundary with the Cocos plate, seaward of the Middle America trench at ~9.4° N, 86° W. Both concentration and isotope ratio profiles show reversals in the basal section, and near basement approach modern-day seawater values, caused by diffusional communication with a fluid flow regime in the upper oceanic basement. The seawater Sr isotope curve is also shown.

are altered by diffusional communication with the basement fluid (seawater) flow regime, having a residence time of 15,000 to <20,000 years and a lateral vigorous fluid flow rate of a minimum of 2–5 m yr^{-1}. These flow rates are also sufficient to explain the suppressed heat flow observations (Silver *et al.*, 2000).

At the 19° S region, an E–W transect of Sites 597–602 was drilled on DSDP Leg 92 (130–117° W) on young oceanic basement, 1–28 Ma, having thin sediment cover. The heat flow across the transect is generally lower than predicted except for several high heat flow zones, as at Site 600C, with a surface heat flow value of 580 mW m^{-2} instead of the expected value of 245 mW m^{-2}. Advection through the outcrops and/or faults in this region are suggested. The pore-fluid and sediment geochemical data (Leinen *et al.*, 1986), and in particular the intense basal sediment carbonate diagenesis to hard chalk, at a burial depth of just ~8 m, at DSDP Site 600C, also indicate advection through the thin sediment cover at this site. The chalk is directly overlying oceanic basement of 4.6 Ma (Kastner *et al.*, 1986). An *in situ* pore fluid study along this transect, near DSDP Sites 600–602 indicated both discharge and recharge of fluid through the thin sediment cover of up to 50 m thick, with flow velocities of 1–5 cm yr^{-1} (Bender *et al.*, 1985/1986). They also observed a positive correlation between basement topography, heat flow, and fluid flow; accordingly, upwelling occurs at the high heat flow sites. Similar correlations were documented at the Costa Rica Rift flank and eastern Juan de Fuca Ridge flank.

16.2.5 The Juan de Fuca Ridge at 48° N

The eastern flank of the Juan de Fuca Ridge, at 48° N, has been studied extensively across a transect spanning from well-ventilated to sediment sealed conditions. Crustal ages range from 0.6 to 3.5 Ma (Davis *et al.*, 1992; Wheat and Mottl, 1994). During ODP Leg 168, four hydrological observatories (CORKS) were installed for monitoring temperature and pressure, and for sampling continuous fluid chemistry using Osmo-Samplers. Two Osmo-Samplers were successfully recovered after three years operation (Wheat *et al.*, 2000b) at ODP Sites 1024 and 1027.

Because of its proximity to the tectonically active North American margin, ocean basement older than 0.6–1.0 Ma becomes heavily sedimented by turbidites, up to 700 m thick, that largely seal the sediment–basement boundary for advective flow through the sediment. The basement topography is not smooth, but has ridges parallel to the ridge axis, as in the Alarcon Basin (discussed below). Fluid circulation in the very sparsely sedimented young basement (<0.6 Ma) is intense; heat flow measured is only ~15% of the predicted value. Fluid flow is through basement outcrops and faults and the basement fluid chemistry thus has bottom-seawater composition (Elderfield *et al.*, 1999). Sediment thickness at the young end of this transect varies from 60–180 m, and heat flow increases from ~15 to 100% of the expected value. Similar to the Costa Rica flank, the heat flow is locally higher at basement highs than at adjacent basement lows. At the basement highs, pore-fluid chemical gradients indicate fluid advection through the sediments where the thickness does not exceed ~160 m; with the exception of a rapid (up to 60 mm yr^{-1}) upwelling area at local seafloor outcrops

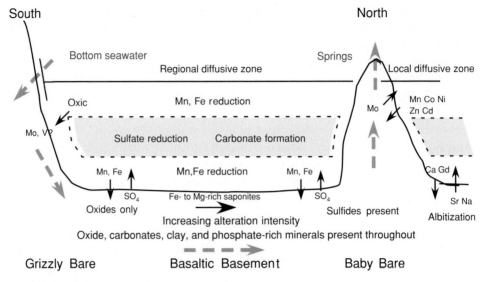

South   North

Fig. 16.4 A cartoon of basement fluid–sediment interactions on the Eastern flank of the Juan de Fuca Ridge at 48° N, from Wheat *et al.* (Fig. 6, 2002a). Bottom seawater enters basement at basement outcrops such as Grizzly Bare – a partially buried seamount – and flows to the northeast, parallel-ing the ridge axis. The fluids vent at two outcrops known as Baby Bare and Mama Bare (dashed arrows). Along this flow path the seawater becomes increasingly more altered and warms from ~2 to ~63 °C. Alteration occurs via reactions within basaltic basement as well as the result of diffusive exchange between the overlying sediment, where diagenetic reactions alter pore-water compositions, and basement. This diffusive flux impacts the composition of formation waters on regional (e.g. sulfate) and local scales (e.g. Mn, Co, Ni, Zn, Cd, and Mo).

known as the "Three Bares," the calculated flow rates through the sediments typically reach only a few millimeters per year. However, the geothermal profiles through the sediments are unaffected by the flow. The upwelling fluid is seawater altered from reactions with basement at temperatures that increase with increasing distance from the unsedimented seafloor at the young end of the transect, from <10 to 60 °C. It is significantly enriched in Ca and depleted in Mg. Wheat *et al.* (2002) have produced a summary cartoon of pore-fluid–basement-water reactivity based on pore-fluid compositions and at 3–4 Ma oceanic basement age (Fig. 16.4).

16.2.6 Alarcon Basin

The Alarcon Basin is a moderate rate (50 mm yr^{-1}) spreading basin, located in the mouth of the Gulf of California, at the northern end of the East Pacific Rise. It was first surveyed in 1998 using swath mapping, magnetics, gravity, and single-channel seismics (Lonsdale, pers. comm.). A coordinated heat flow, coring, and pore-fluid geochemistry study of the east flank of this spreading ridge, on a seismic line along 0.30–0.65 Ma oceanic basement,

9–20 km from the ridge axis, immediately followed, using the new bathymetric map (Fisher *et al.*, 2001). The sediment cover is ~40 m thick. The best evidence for hydrothermal circulation in the basement was derived from drilling of DSDP Site 482C, located ~12 km from the ridge axis. The drill-hole through 143 m of sediment and 47 m into the basement, turned into a "man-made" hot spring soon after drilling (Dunnebier and Blackinton, 1980). Heat flow values are 15–55% of the expected conductive value for this age of oceanic basement (Fisher *et al.*, 2001). Heat flow is lowest along the trough margins and highest in its center, suggesting that basement exposures and trough-bounding faults focus the hydrothermal recharge. Sediment and pore-fluid geochemical data, together with reactive transport modeling, identified localized recharged flux areas through the sediment at 2–10 mm yr^{-1}, most likely driven by a basement that is underpressured beneath the sediment. These data also indicate that the dominant fluid regime within the sediments is diffusive. Thus, most of the measured heat flow deficit is through focused basement recharge and intense fluid flow in the basement; discharge areas have yet to be identified.

16.2.7 Mid-Atlantic Ridge west flank, DSDP Site 395A

A similar hydrologic regime to the Alarcon Basin was documented on the west flank of the Mid-Atlantic Ridge, DSDP Site 395A, on a sediment pond on 7.3 Ma oceanic basement (Becker *et al.*, 1984; Langseth *et al.*, 1992). The heat flow is suppressed by 75–80% below the expected value of the crustal age. Pore-fluid chemical data together with negative pressure recorded with pore-pressure instruments deployed in the pond (Fang *et al.*, 1993), are consistent with bottom-seawater recharge at this pond. The basement outcrops on the sides of the pond are probably the sites for most seawater recharge.

16.3 Effects of diagenesis on geochemistry of pore fluids and sediments

16.3.1 Key processes

Ridge flank sediments affected by advective fluid flow may contain geochemical signatures of the flow superimposed on the geochemical characteristics of the sediments involved. The diagenetic characteristics of the sediments – carbonates, siliceous oozes, hemipelagic, or terrigenous – will determine the extent to which chemical alteration due to fluid flow can be recognized. Off-axis fluid convection may enhance the normal progress of diagenetic reactivity through an increase in temperature or transport of solutes out of the reaction zones, hence enhancing dissolution of some phases, or may disrupt the diagenetic sequence by introducing pore fluids of a composition unlike those which would otherwise evolve. The amplitudes and depths of pore-fluid concentration maxima or minima reflecting paleoceanographic events are modified by ridge flank fluid advection as well.

The most pronounced hydrothermal inputs are indicated by elevated concentrations of Ca, Si, B, F; the transition elements Fe, Cu, Zn, and Pb; and Pb isotopes. Alkali element concentrations are often depleted at low temperatures but enriched at moderate and high temperatures, and Sr isotope ratios, if altered, reflect the fluid source. Additional components are those scavenged by Fe-oxyhydroxides, for example, the rare earth elements.

16.3.2 Carbonate diagenesis

Carbonate diagenesis is ubiquitous above the carbonate compensation depth (CCD), and is most intense in organic-rich sediments (>0.5% C), where sedimentation rates exceed 20 cm kyr^{-1} and bacterial sulfate reduction is clearly observed in the pore fluids. The main diagenetic processes which can occur are recrystallization and authigenic carbonate formation.

Recrystallization of biogenic carbonate via dissolution–precipitation leads to an increase of pore-fluid Sr concentration. The Sr isotopic composition of the pore fluids may be affected, depending on the rate of recrystallization. Authigenic carbonate formation will decrease pore-fluid Ca (and possibly Mg, Mn, and/or Fe) and increase Sr concentrations but not affect the Sr isotope ratios. The decrease in Ca concentration may not be recognizable except where carbonate precipitation is driven by the rapid oxidation of organic matter through the processes of bacterial sulfate reduction. Carbonate precipitation, often as cement, may also strongly influence the hydrological regime of the associated sediments by decreasing permeability, and thus indirectly affect the rates of most fluid–sediment reactions including the potential of ridge flank fluid flow through the cemented sections. The cementation, ooze/chalk, and/or chalk/limestone transformations are enhanced at elevated temperatures and fluid advection. Therefore, at young ridge flank systems, some chalk or limestone (and even dolomite) may directly overly oceanic basement, as for example in the vicinity of Site 504B, and at the western flank of the East Pacific Rise, 18° 55.70′ S, DSDP Site 600C. At this site the oceanic basement is 4.6 Ma, and limestone occurs at just ~8 mbsf; its formation was enhanced by convective flow of relatively unaltered seawater through the thin sediment cover (Kastner *et al.*, 1986). This type of flow attenuates variations in temperature and pore-water chemistry in the upper basement and overlying sediments, as for example at the area surrounding DSDP Sites 501/504, at the flank of the Costa Rica Rift (Langseth *et al.*, 1988). Limestone formation may eventually seal the basal sediment, restricting further convective fluid exchange between seawater and the oceanic basement.

16.3.3 Siliceous sediment diagenesis

Transformation of opal-A to opal-CT

Between 30 and 80 °C, opal-A dehydrates and recrystallizes to opal-CT. In the pore fluids, Si concentrations increase significantly and are controlled by opal-A or -CT solubility with minor dilution by the released H_2O (on average, diatoms and radiolaria contain ~10 wt. % H_2O). There may be small increases in B and ^{10}B. If intense, this reaction may result in the formation of opal-CT chert that affects the physical properties of the sediments, hence the hydrology. The rate of this reaction is temperature dependent, approximately doubling per 10 °C temperature increase (e.g. Kastner, 1981). Therefore, at young ridge flanks the formation of opal-CT can be enhanced. Depending on the hydrology, convective flow at ridge flanks may either enhance or decrease the preservation or precipitation of opal-A or -CT, in the form of silica cement, authigenic opal-A or -CT, or opal-CT and chert. For example, at the Mariana Mounds, the basement fluid (bottom-seawater) is undersaturated in dissolved Si with respect to the solubility of both opal-A and -CT; as the fluid advects through the sediment it enhances Si dissolution. However, at the Galapagos Mounds where fluid downwells through the sediment and dissolves opal-A, the basement fluid is saturated with respect to opal-A. Upon upwelling, this Si-rich fluid enhances opal preservation accompanied by opal-A and/or opal-CT precipitation (Wheat and McDuff, 1994) and combines with the elevated dissolved Fe concentrations to form nontronite (Bender, 1983; Maris *et al.*, 1984). The remaining dissolved Si is returned to the ocean.

Transformation of opal-CT to quartz

Between 50 and >80 °C, Si concentrations decrease and at >150 °C are controlled by quartz solubility; some H_2O, B, and ^{10}B are released into the fluid. In contrast to the high-porosity and permeability diatomaceous sediment, massive quartz chert is the most impermeable sediment, and may act as a hydrological seal even if only a few meters thick. Similar to the opal-A to opal-CT reaction, the opal-CT to quartz transformation reaction rate also approximately doubles for each 10 °C temperature increase. At ridge flanks with fluid recharge and discharge through sediments containing siliceous ooze, such as at the Galapagos, the discharged basement fluid may be saturated with respect to opal-CT solubility, and hence could precipitate authigenic opal-CT or quartz in the sediment. The remaining Si will be returned to the ocean.

16.3.4 Clay mineral diagenesis

The potentially important clay mineral transformations and formation reactions are discussed below.

Smectite to illite transformation

The transformation reaction occurs mostly between ~50 and 120–150 °C (e.g. Perry and Hower, 1970; Środoń, 1999, and references therein). The pore fluid is diluted by the release of H_2O; however, Si and Sr concentrations somewhat increase; whereas K, B, Li and Al concentrations and ^6Li and ^{10}B decrease. The $^{87}Sr/^{86}Sr$ ratio depends on the precursor smectite; if terrigenous the pore-fluid Sr will become more radiogenic than the contemporary seawater ratio, but if volcanic it will become less radiogenic. Because of the temperature dependence of this reaction it may be significant on very young (1–2 Ma) ridge flanks with a thin cover of clay-rich sediment, or at young (~1–5 Ma) thickly sedimented flanks, such as the eastern flank of the Juan de Fuca Ridge.

Smectite ion exchange

This reaction is important only in thicker sediment sections where sedimentation rates exceed 10–20 cm yr^{-1}, with >0.5% organic C content. Exchange with ammonium, produced by *in situ* bacterially mediated organic C metabolic reactions, increases pore-fluid alkali metals (Li, K, and even Na) and Mg concentrations; ^{6}Li and ^{10}B are released into the pore fluid as well.

Ridge flank authigenic nontronite formation

At young ridge flanks with thin sediment cover, fluid advection through the sediment (as well as basement outcrops) is possible. Si–Fe-rich advecting fluids, from opal-A dissolution and Fe mobilization after oxygen and nitrate utilization at recharge sites, produce a typical ridge flank sediment that consists primarily of authigenic nontronite and/or opal-A, as observed at Galapagos (Bender, 1983; Maris, *et al.*, 1984).

Illite to chlorite reaction

Because temperatures of >200–250 °C are required, this is not a typical ridge flank reaction. It is likely more important at axial sedimented ridge crest hydrothermal systems.

Serpentinization

Near ridge crests in advective hydrothermal regimes, oceanic basement serpentinization may affect the overlying sediment pore-fluid chemistry, in particular depleting it in Mg, Li, B, and Cl concentrations, ^{6}Li, ^{10}B, ^{37}Cl, and H$_2$O. Oceanic basement serpentinization may only indirectly influence ridge flank fluid chemistry and fluxes, and is likely an important reaction controlling the H$_2$O budget in subduction zones (Peacock, 2001).

16.3.5 Terrigenous sediments

Feldspar dissolution

At moderate temperatures ($\sim$50–150 °C), plagioclase dissolution enriches pore fluids in Ca, Al, Si, Sr, and its non-radiogenic isotopes. Plagioclase albitization that involves utilization of Na and Si is important near the ridge crest, not on ridge flanks.

Volcanic ash diagenesis

Volcanic ash either dissolves or alters to smectite and/or zeolites.

Dissolution of volcanic ash enriches the pore fluids in appropriate components, particularly in Ca, Li, B, Si, and in non-radiogenic Sr isotopes.

Volcanic ash alteration to authigenic smectite: the reaction decreases pore-fluid concentrations of Li and Mg, increases Ca and Sr concentrations, and consumes H$_2$O. Pore-fluid Sr isotopes will become less radiogenic and depleted in H$_2$O, ^{6}Li, ^{10}B, and ^{18}O. This reaction proceeds at all temperatures and its rate approximately doubles per 10 °C increase in temperature.

Volcanic ash diagenesis to zeolites: at low <70–80 °C , the K and Na zeolites (phillipsite and heulandite) form. The more Si-rich K zeolites, primarily clinoptilolite, also form in siliceous ooze-rich sediments. At higher sedimentation rate young ridge flanks, Ca zeolites (e.g. laumontite) become dominant at >80 °C. Zeolite formation thus depletes pore fluids of K, Na, or Ca, Al, Si, and some H_2O, and possibly B and ^{10}B (e.g. Kastner and Stonecipher, 1978).

16.3.6 Organic matter reactions

Bacterial and thermogenic reactions

The hierarchy of bacterial utilization of organic matter has been thoroughly documented, and is applicable to ridge flanks with high sedimentation rates, such as the Juan de Fuca and Guaymas Basin flanks. After reduction of O_2, NO_3^-, and Fe^{3+} (e.g. Claypool and Kaplan, 1974). The most important pore-fluid characteristic of this environment is that sulfate becomes limited and redox sensitive elements become mobile. Other related signatures are sulfide, ammonium, and alkalinity production, and, when sulfate has been reduced, methane production and fermentation (e.g. Kvenvolden, 1993; Borowski *et al.*, 1999; Boetius *et al.*, 2000; Chapter 17, and references therein). The bacterial activity in the sub-surface, particularly the extent of sulfate reduction and methane escape through the seafloor may affect seawater chemistry, especially ocean alkalinity, and thus the partitioning of CO_2 between the atmosphere and ocean (D'Hondt *et al.*, 2002). In this type of environment the pore fluids are also enriched in dissolved Fe, Mn, Ba, Br, and I, and may be depleted in K, Ca and/or Mg, and Sr. In addition to organic matter content, the intensity of the above concentration gradient depends on the amount of terrigenous matter available and the sedimentation rate. At temperatures >80 °C, the liquid hydrocarbon window is reached and an increase in dissolved higher hydrocarbons is observed. This and the above reactions do not occur in the open ocean ridge flank setting, because young flanks have a thin cover of sediments, are poor in organic C, and are thus sulfate-rich. Here, fluids advected from the altered oceanic basement that carry nutrients, such as Fe and Mn, could stimulate microbial activity in the sediment–pore-fluid environment, enhancing the bacterial degradation of organic matter in both sulfate-rich and sulfate-poor environments, and may alter the original microbial hierarchy also.

16.4 The important proxies for ridge flank seawater–sediment interactions and their application

16.4.1 Chemical proxies for pore-fluid advection

The application of chemical proxies to determine the advective velocity of pore fluid necessarily involves constraints due to sampling considerations. Samples retrieved by coring fall into two categories: conventional coring from surface ships and research submersibles (push cores, piston and gravity cores, box and multi-cores) produce relatively short core lengths, typically <10 m, dependent on sediment type; and second, cores obtained through scientific drilling, which can sample hundreds of meters down to and into the oceanic basement.

For rates of advection greater than $\sim$1 cm yr^{-1}, up-flow will create large gradients close to the sediment–water interface. Similarly, strong down-flow will produce strong gradients close to the sediment–basalt contact. These gradients must be adequately sampled to allow the calculation of flow rates. If unimpeded flow (e.g. through outcrops or faults) recharges the basement aquifer, seawater may pervade the sediment column with no chemical contrast at the sediment–basement contact. Under these conditions it would be impossible to determine the rate or direction of pore-fluid advection from geochemical analyses alone. At flow rates >1 m yr^{-1}, temperature profiles show curvature, which can be modeled to calculate flow rates.

Strong basement fluid upwelling brings fluids close to the sediment–water interface. These fluids can be analyzed to determine the composition in the aquifer immediately underlying the sediment column. However, this fluid composition may not be representative of the majority of the fluid in the basement aquifer. It is important to stress that fluids deeper in the aquifer may not have been modified by fluid–sediment interactions. Although basement fluid compositions cannot be determined directly at low flow rates, the calculation of slower flow rates and the direction of advection are important for understanding the geometry of circulation cells as the pore fluids move in response to basement under or overpressures.

Proxies for slow (<1 cm yr^{-1}) fluid advection in DSDP and ODP cores

Slow fluid advection may be sensitively quantified by one-dimensional models, for species such as sulfate, Ca^{2+}, Mg^{2+}, and alkalinity (although sediment reactions must be considered: in fact, they can be used in the proxy). The sensitivity comes from *in situ* reactions producing minima or maxima that can be shifted by pore-fluid advection. In carbonate sediments, Sr and $^{87}Sr/^{86}Sr$ can be used as tracers for advection.

A good strategy is to model multiple tracers to assess the consistency of any velocity estimate. Advection, diffusion, and reaction can be expressed by the general diagenetic equation (Berner, 1980):

$$\frac{\partial \eta \, (C)}{\partial t} = \frac{\partial}{\partial z} \left(\eta D_C \frac{\partial \, (C)}{\partial z} \right) - \frac{\partial \boldsymbol{q} \, (C)}{\partial z} + \eta \sum R \qquad (16.1)$$

Where (C) is the concentration of a solute, η is porosity, D_C is the diffusion coefficient corrected for temperature and tortuosity, $\boldsymbol{q}$ (volumetric fluid flux) is the Darcy velocity, t is time, z is depth, and ΣR represents the sum of any reaction terms.

Sulfate

The high concentration of sulfate in seawater (28 mmol kg^{-1}) compared to dissolved oxygen (<400 μ mol kg^{-1}), makes sulfate reduction the dominant biochemical process in sediments where there is an excess of organic matter ($\geq$0.5 wt. %). The depletion in seawater sulfate by bacterial sulfate reduction results in an increase in carbonate alkalinity:

$$2CH_2O + SO_4^{2-} \leftrightarrow H_2S + 2HCO_3^- \qquad (16.2)$$

this ultimately drives carbonate precipitation:

$$Ca^{2+} + 2HCO_3^- \leftrightarrow CaCO_3 + H_2O + CO_2 \qquad (16.3)$$

The rate law for sulfate reduction has been found to be approximately dependent on the square of the sedimentation rate (r; Toth and Lerman, 1977; Berner, 1978; Tromp *et al.*, 1995):

$$k = 0.057r^{1.94} \qquad (16.4)$$

where k is measured per anum and r is in centimeters per year.

The reaction term for sulfate in (16.5) is given (Berner, 1978; Boudreau and Westrich, 1984) by:

$$\sum R = -\frac{\Delta_s(1-\eta)}{\Delta_f \eta} Lk \left[\frac{(SO_4^{2-})}{K_{SO_4} + (SO_4^{2-})} \right](G) \qquad (16.5)$$

In this equation, Δ_s and Δ_f are the solid and fluids densities, L ($=0.5$) is a stoichiometric constant for (16.2), and (G) is the concentration of organic carbon. The rate of organic matter degradation is controlled using Monod kinetics, where K_{SO_4} (~ 1 mmol kg^{-1}) is the saturation constant (Monod, 1949). When $(SO_4^{2-}) \gg K_{SO_4}$, the rate of organic matter degradation is independent of the sulfate ion concentration, but becomes first order in sulfate when $(SO_4^{2-}) \ll K_{SO_4}$. This prevents negative sulfate concentrations, which can occur in the Berner model if the oxidant (sulfate) is exhausted before all the organic matter is oxidized (Boudreau and Westrich, 1984). The scheme is to simultaneously fit the Darcy velocity $q = \eta v$ and the initial organic carbon content of the sediment (G_0).

Figure 16.2 shows calculated pore-fluid velocities at the eastern equatorial Pacific sediment bulge, based on sulfate data from Leg 85 and 138 pore fluids (Stout, 1985).

Strontium

Strontium is a reactive element in carbonate sediments – diagenetic recrystallization of biogenic carbonate results in a loss of strontium to the pore fluid, increasing pore-fluid strontium concentrations. In some cases, the precipitation of SrSO$_4$ (celestite) may occur if celestite saturation is exceeded (Baker and Bloomer, 1988). In hemipelagic sediments, the reactivity of non-carbonate sediment components may mask the effects of carbonate recrystallization at low carbonate contents.

The rate of bulk carbonate recrystallization is denoted by R_{age}, and can be modeled as a simple analytic function (Richter and DePaolo, 1987):

$$R_{age} = R_\alpha + R_\beta \exp(-age/R_\gamma) \qquad (16.6)$$

where R_α, R_β, and R_γ are three modeled constants. This relationship has been chosen to represent decreasing recrystallization rates with time since deposition, with long-term steady-state recrystallization determined by R_α.

For strontium, the reaction term in (16.1) is given by (Richter, 1996):

$$\sum R = R_{\text{age}} \frac{\Delta_s(1-\eta)}{\Delta_f \eta} \left[(\text{Sr}) - K_{\text{Sr}} \frac{(\text{Ca})}{(\text{Ca}^{2+})} (\text{Sr}^{2+}) \right] \tag{16.7}$$

K_{Sr} is a distribution coefficient ($0.02 \sim 0.04$; Baker *et al.*, 1982; Richter, 1996).

The model may also be applied to the $^{87}\text{Sr}/^{86}\text{Sr}$ isotopic ratio, a method that is especially relevant at sites on older crust where the $^{87}\text{Sr}/^{86}\text{Sr}$ isotopic ratio of the carbonate is significantly different than modern and may be reflected in the pore-fluid $^{87}\text{Sr}/^{86}\text{Sr}$ isotopic ratio. For example, in the eastern equatorial Pacific (Baker *et al.*, 1991), fluid down-flow at a rate of 2 mm yr^{-1} has been identified at ODP Site 572, on a crustal age of 15 Ma (Richter, 1993; Richter, 1996).

These processes may also occur to a limited extent in hemipelagic sediments where carbonate is a minor component, but any effect on the pore-fluid strontium concentration due to carbonate recrystallization is more likely to be swamped by the presence of additional reactive sediment components such as clays and volcanic fragments. Isotope mixing plots ($^{87}\text{Sr}/^{86}\text{Sr}$ versus 1/Sr) are useful in identifying the presence of reactive layers provided that the $^{87}\text{Sr}/^{86}\text{Sr}$ ratios of such layers are distinct. For example, pore-water Sr and $^{87}\text{Sr}/^{86}\text{Sr}$ concentrations from ODP Leg 168, Site 1025 Juan de Fuca Ridge, show multiple sources and sinks in the sediment column, some of which appear to be related to feldspar-rich turbidite layers (Fig. 16.5).

Problems associated with fitting advective velocities using reactive transport models fall into three categories: (1) adequate measurement of sediment physical properties such as porosity η, formation factor f, and temperature; (2) the formulation of the appropriate reaction rate laws; and (3) adequate knowledge of sediment–pore fluid reactions at *in situ* temperatures.

The effect of sediment physical properties on calculated pore-fluid profiles and advective velocities is due to the effect on the diffusion coefficient. This arises because the diffusion coefficient is dependent both on the temperature and also on the fluid pathway between closely packed grains, termed the sediment tortuosity, or ϑ. The bulk sediment diffusivity for solute C is determined by:

$$D_C = D^0/\vartheta^2 \tag{16.8}$$

where D^0 is the diffusion coefficient at infinite dilution (e.g. Boudreau, 1997). The tortuosity can be calculated as $\vartheta^2 = \eta f$, where the formation factor (f) is the ratio of the electrical resistivity of the bulk sediment to that of the pore fluid alone (McDuff and Ellis, 1979). Alternatively, tortuosity has been empirically calibrated in terms of the sediment porosity alone by some authors, e.g. $\vartheta^2 = 1 - 2 \ln \eta$ (Boudreau, 1997).

The second issue involves the formulation of the rate law for any given reaction. A good example of this is Richter's rate of carbonate recrystallization (R_{age}), which implicitly

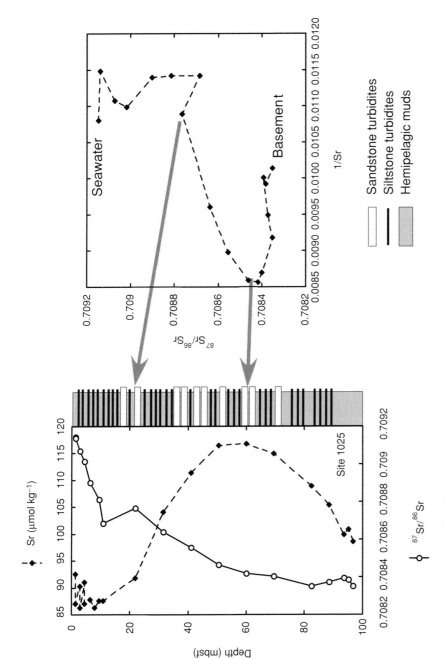

Fig. 16.5 Pore-fluid Sr and ^{87}Sr/^{86}Sr data from Site 1025, eastern flank of the Juan de Fuca Ridge at 48° N (Elderfield *et al.*, 1999; Hunter and Elderfield, 2001). Modeling pore-fluid sulfate profiles indicates weak up-flow at 100 m Ma^{-1} (Rudnicki *et al.*, 2001). Sr concentrations show a sediment sink at ∼10 mbsf, and a source at ∼60 mbsf. However, the ^{87}Sr/^{86}Sr versus 1/Sr mixing plot shows multiple sources.

describes decreasing recrystallization rates since burial. This has been found to be appropriate for near-pure carbonate sediments. However, the formulation may break down if, for example, recrystallization rates increase due to enhanced thermal gradients or changes in pore-fluid chemistry due to the influence of basement flow (see below). In these cases, applying the model to $^{87}Sr/^{86}Sr$ may serve as a validation or indication as to the appropriateness of the model, especially in older carbonate sediments where there are significant variations in $^{87}Sr/^{86}Sr$ downcore (a function of sediment age and the secular variation of seawater $^{87}Sr/^{86}Sr$), and reactivity continues downcore such that the isotopic ratio does not depend solely on the boundary conditions and transport.

The third complication is that on ridge flanks covered by hemipelagic and terrigenous sediments, pore fluids and sediments interact at 30–60 °C, which is outside the temperature range generally considered for most diagenetic studies. There is a mismatch between the high-carbon, high-temperature setting of ridge flank hemipelagic/terrigenous sediments and diagenesis in either low-carbon, low-temperature pelagic sediments or high-carbon, low-temperature coastal marine sediments. Also, the temperatures involved are lower than those considered at sedimented axial hydrothermal sites, e.g. Escanaba Trough, and in modeling studies which consider interactions at temperatures 100~350 °C.

Proxies for fluid advection in short (<10 m) cores

If the fluid velocity is sufficiently rapid, reaction between the pore fluids and the sediment will be insignificant and a pure transport (advection–diffusion) equation can be applied. Therefore, any species which is reactive in basement will show a pore-water anomaly which can be modeled. Commonly used species include Ca^{2+}, Mg^{2+}, and F^-. An exception occurs when fluid/rock ratios are high and basement temperatures are low, in which case basement fluids may be indistinguishable from seawater. Under these circumstances it would be impossible to calculate the rate and direction of flow from chemical measurements alone. The transport model used is a simplification of the one-dimensional steady-state advection–diffusion-reaction equation making the assumption of constant porosity and diffusivity (Craig, 1969; Goloway and Bender, 1982). The equation is:

$$\frac{dC}{dt} = D_b \frac{d^2C}{dz^2} + q \frac{dC}{dz} + J_0 e^{-az} = 0 \tag{16.9}$$

where D_b is the apparent sediment diffusion coefficient ($cm^2\ s^{-1}$) calculated assuming constant porosity and tortuosity, C is the concentration of the solute ($\mu mol\ kg^{-1}$), J_0 is the reaction rate at the sediment–water interface ($\mu mol\ kg^{-1}\ s^{-1}$), a is the e-folding distance characterizing the downcore decrease in the reaction rate (cm^{-1}), q is the vertical advection rate of the pore water ($cm\ s^{-1}$), and z is depth in the sediment (cm).

The equation has the solution (Maris *et al.*, 1984):

$$C = C_{max} + \left[C_0 - C_{max} - \frac{J_0}{Da(a - q/D)} \right] e^{-zq/D} + \frac{J_0}{Da(a - q/D)} e^{-az} \tag{16.10}$$

Here, C_0 is the concentration at the seabed, and C_{max} is the asymptotic concentration of the solute. For the case of no reaction, $J_0 = 0$ and the equation becomes:

$$C = C_0 + (C_{max} - C_0)(1 - e^{-zq/D}) \qquad (16.11)$$

For cores where profiles do not reach an asymptote, the following equation may be used:

$$C = C_0 + (C_{max} - C_0)\frac{(1 - e^{-zq/D})}{(1 - e^{-z_{max}q/D})} \qquad (16.12)$$

where C_{max} and Z_{max} are the concentration and depth of the deepest sample. Maris *et al.* (1984) have applied these equations to pore fluid Ca^{2+}, Mg^{2+}, Mn^{2+}, F^-, phosphate, and silicate in cores taken around the Galapagos Mounds. For phosphate, typical values of the model parameters were $a \sim 0.02\,cm^{-1}$, $J_0 \sim -2 \times 10^{-9}\,\mu mol\,s^{-1}$, $D \sim 4 \times 10^{-6}\,cm^2\,s^{-1}$ given concentrations in $\mu mol\,kg^{-1}$. The calculated advection rates were $\sim 3 \times 10^{-8}\,cm\,s^{-1}$ (or $\sim 1\,cm\,yr^{-1}$). A comparison of advective velocities calculated for calcium, magnesium, and fluoride showed good agreement, with better agreement found when reaction was considered in modeling fluoride profiles. Advection rates calculated using phosphate and Mn^{2+} were not consistent, perhaps due to sampling artifacts and additional reaction terms not considered in the model.

The assumption of constant porosity (η) and tortuosity (ϑ) in these models can be justified on the basis of the short core length: porosity and tortuosity are taken into account by reducing diffusion coefficients by a factor of $1/\vartheta^2$ (~ 2). Over longer cores, significant variations of porosity, tortuosity, and temperature may occur requiring a more complete model. Detailed analytic solutions to diagenetic equations involving varying porosity and tortuosity with depth have been published (e.g. Wheat and McDuff, 1994), but multi-species diagenetic models are best solved using numerical solutions (e.g. Fisher *et al.*, 2001; Giambalvo *et al.*, 2002).

Semi-conservative tracers: Cl, $\delta^{18}O$

Temporal changes in the composition of deep ocean water can drive changes in the composition of pore fluids in the underlying sediment. Changes in the chemistry of seawater due to the formation of ice during glacial times lead to an increase in the overall salinity of seawater (through brine exclusion) and an increase in the $\delta^{18}O$ through the formation of ice enriched in ^{16}O. Interactions with volcanic ash can alter $\delta^{18}O$, Mg, and Ca in pore fluids (Lawrence *et al.*, 1979; Gieskes and Lawrence, 1981; Lawrence and Gieskes, 1981). Also, carbonate recrystallization at elevated temperatures has been shown to affect sediment and pore-water $\delta^{18}O$ in carbonates (Lawrence *et al.*, 1976). At DSDP Site 501/504, Costa Rica Rift, an increase in pore-water $\delta^{18}O$, has been attributed to a diagenetic front marked by carbonate recrystallization and chert formation at a temperature of $\sim 56\,°C$ (Mottl *et al.*, 1983). Both $\delta^{18}O$ and Cl pore-fluid compositions can be affected by the presence of gas hydrates; however, hydrates only form in organic-rich sediments, such as in the Guaymas Basin and on continental margin slopes with low to moderate geothermal gradients, but not in typical ridge flank pelagic settings.

Pore-water chlorinity is a more reliable indicator of advective velocity than $\delta^{18}O$ because the ratio of analytical precision to expected signal is less for Cl. Modeling of chloride profiles is appropriate for fluid advective velocities <2 cm yr^{-1} (Wheat and McDuff, 1995). However, there may be difficulties in setting the lower boundary condition if $\delta^{18}O$ is altered during basement reactivity.

16.4.2 Temperature sensitive indicators

The temperature sensitive indicators presently available are important to interpret the observations uniquely, especially to identify and quantify the fluid–sediment reactions that control pore-fluid chemical and isotopic signatures, and to detect mixing between basement and sediment fluids. To substantiate reaction temperatures, redundant temperature proxies need to be used.

Hydrocarbons

At low temperatures in sulfate-depleted sediments methane is produced by microbially mediated organic matter oxidation following sulfate reduction, and at elevated temperatures (>80–$100\,°C$) the higher hydrocarbons (C_3 to C_6) form. The relative abundance of the higher hydrocarbons and C isotopes yield information about reaction temperatures and fluid migration. This is typical of sediments with $>0.5\%$ organic matter, where high sedimentation rates of >10–20 cm yr^{-1} prevail, thus are likely to occur at young thickly sedimented ridge flanks, such as at the eastern flank of the Juan de Fuca Ridge.

Alkali metals and δ^6Li

Li and K concentrations (and their ratios) are presently most useful as approximate sensitive geothermometers. At low temperatures both Li and K are known to partition preferentially into the solid phases, whereas at elevated temperatures both are preferentially released into the fluid phase. Rubidium and Cs should behave similarly; their threshold temperatures are poorly known. The threshold temperature for release into the fluid phase for Li and K is not as yet well constrained, but occurs certainly between 100–$250\,°C$ (Seyfried *et al.*, 1984; You *et al.*, 1997; Seyfried *et al.*, 1998; You and Gieskes, 2001). Experimental work on constraining the threshold temperatures are in progress (by one of us, MK, in collaboration with R. Rosenbauer, at the hydrothermal laboratory at USGS, Menlo Park).

Lithium isotope ratios and concentrations provide important insights on the hydrology, fluid–solid reactions, and approximately constrain reaction temperature. They are affected by clay mineral (especially those rich in Mg) and zeolite diagenesis. Both processes preferentially fractionate ^6Li from low to greenschist facies temperatures, enriching the fluid in ^7Li. Although the reactivity of Li and fractionation factor of δ^6Li has been studied in laboratory experiments at elevated temperatures (100–$250\,°C$), the behavior of Li in sediments at lower temperatures is less well known. Recent pore-fluid data from the eastern flank of the Juan de Fuca Ridge (ODP Leg 168), Fig. 16.6, shows that below approximately

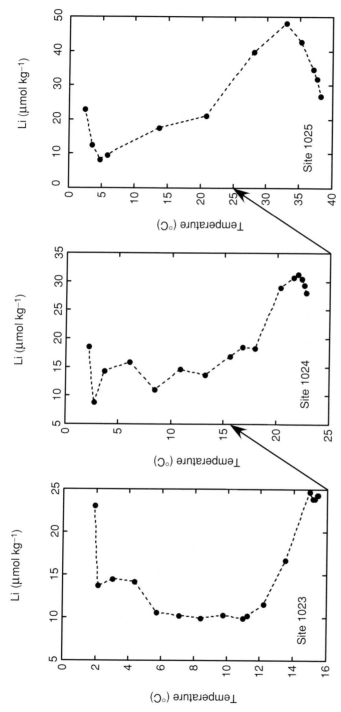

Fig. 16.6 Example profiles of Li versus *in situ* temperature for Sites 1023, 1024, and 1025, ODP Leg 168. All sites show a depletion in pore-fluid Li from a seawater concentration of 26 μmol kg^{-1}, at temperatures <12 °C. The sites show a progressive Li enrichment at higher temperature as basement temperatures increase from 15.5 ~ 38.6 °C. At site 1025, Li concentrations decrease above ~34 °C, indicating that the sediment, rather than the basement, is the source of the pore-fluid Li. Sediment thickness: Site 1023, 192.9 m; Site 1024, 167.8 m; Site 1025, 97.5 m (Davis *et al.*, 1997; Elderfield *et al.*, 1999).

20 °C, Li is removed into the sediment (e.g. Site 1023). At higher temperatures, the sediment becomes a source for Li to the pore fluids. At Site 1024, Li is added to the pore fluids at a temperature of ≈20 °C, whereas at Site 1025 the pore-fluid Li maximum occurs at >30 °C. The evidence from δ^6Li reveals a more complicated picture – Li removal within 5~10 m of the seafloor is accompanied by an *increase* in δ^6Li with depth in the pore fluids, implying, net removal of the heavier ^7Li isotope; this is thermodynamically unfeasible. James and Palmer (2000) have suggested that ^6Li is added instead. The most likely source of the ^6Li to the pore fluids is Li quantitatively expelled (i.e. without fractionation) by cation exchange with NH_3, as documented by Chan and Kastner (2000) at ODP Site 1039 on the eastern flank of the EPR. This indicates that within the top 5–10 m of sediment on the flank of the Juan deFuca Ridge, there is both net uptake of Li, required to decrease the overall concentration but enrich the fluid in ^7Li, and simultaneous release of lithium enriched in ^6Li, required to increase the δ^6Li isotopic composition. These reactions and their interplay have to be adequately described in order to interpret uniquely the observations and quantify the reactions involved.

Below the Li minimum in the sediment, and at higher temperatures, lithium is added to the pore fluids, accompanied by an increase in ^6Li, which is consistent with a clay source.

In the basal section of ODP Site 1039, the reversals in Li isotope ratios and concentrations, shown in Fig. 16.3, provided supporting evidence for a lateral seawater flow system in the upper oceanic basement underlying these sediments (Chan and Kastner, 2000).

The other cation geothermometers are empirical, and are calibrated by fitting curves to concentration–temperature data from various fluid–sediment systems, including non-marine systems, that were determined at different laboratories. The Na^+ and K^+ concentrations (Truesdell, 1976; Fournier, 1979) are useful for geothermometry in higher-temperature fluid-solid systems (>150–200 °C). The $Na^+ - K^+ - Ca^{2+}$ concentrations geo-thermometer, however, is applicable to non-carbonate fluid–solid systems at temperatures <150 °C. After successfully predicting temperature by utilizing over 300 Gulf Coast pore-fluid Mg^{2+} and Li^+ concentration data, the empirical Mg^{2+} and Li^+ geo-thermometer was advocated by Kharaka and Mariner (1989). The Na^+ and Li^+ geo-thermometer, suggested by Fouillac and Michard (1981), provided very similar temperatures to those obtained by the Mg^{2+} and Li^+ system when applied to several mud volcanoes at Barbados (Martin and Kastner, 1996).

Boron and $\delta^{11}B$

B isotope ratios and concentrations provide evidence for clay alteration and serpentine formation at depth. B is released into the fluid phase at all temperatures, and at about 150 °C most of the B is in the fluid. In clay alteration and transformation reactions, B is released and the isotope ratio of the residual solution decreases. When serpentine forms, the opposite occurs: B uptake is high and isotope ratios in the residual solution increase (Seyfried *et al.*, 1984; Spivack *et al.*, 1987; You *et al.*, 1997). Illite is particularly enriched in B and has the most negative B isotopes among the clays; thus the smectite → illite reaction could be recognized using this method. Combining the B isotopes and concentration data with other major and minor chemical data involved in the smectite/illite reaction (for example, K or Rb which are enriched in illite) should provide unequivocal evidence for this reaction.

Chloride and bromide

Cl (Br) concentrations, together with Cl isotope ratios, can be used to distinguish between simple dehydration reactions and dehydration plus formation of high-temperature hydrous minerals such as serpentine, talc, chlorite, and amphiboles (Magenheim *et al.*, 1994; Ransom *et al.*, 1995; Spivack *et al.*, 2002). The higher temperature silicates contain hundreds or more ppm Cl, whereas smectite contains $\sim$100 ppm. Thus, higher temperature reactions near the ridge crest have greater impact than moderate to lower temperature reactions on the Cl mass balance and isotope ratio in the fluids. When smectite dehydrates, much H_2O and some Cl and ^{37}Cl are released into the fluid phase; thus the Cl isotope ratio of the fresher fluid is enriched in ^{37}Cl. However, the formation of high-temperature hydrous minerals at or close to the ridge crest (at >200–250 °C), preferentially consumes ^{37}Cl, enriching the residual fresher fluid in ^{35}Cl. Ash alteration to smectite increases the Cl concentration in the pore fluid and also preferentially consumes ^{37}Cl, thus enriching the residual more saline fluid in ^{35}Cl.

$\delta^{18}O$ and deuterium

Oxygen together with hydrogen isotopes of the pore fluids are sensitive to reaction temperatures, assuming the reaction is constrained by other parameters, such as those discussed above.

Silica

Silica concentration in the fluid provides the best-established geothermometer. The Si geothermometer is based on experimental data, and on thermodynamic calculations of the temperature dependence of Si concentrations in the fluid at equilibrium with quartz, chalcedony, cristobalite, and opal-A (Fournier and Rowe, 1966; Fournier, 1973; Von Damm *et al.*, 1991). The Si geothermometer applies to most natural fluids with pH $< \sim$8.7 in which the silica monomer is the dominant dissolved silica species. It is particularly useful at temperatures $>$150 °C, when quartz controls the silica concentration.

16.5 Summary and discussion

Most ridge flank fluid advection is channeled through basement outcrops and/or faults, as suggested by Fisher and Becker (2000). The volume fraction of recharge through sediment cover is as yet not well documented. It occurs mostly where basement is young, ≤5 Ma and up to $\sim$10 Ma, and the sediment cover is thin, <50–100 m thick, depending on lithology; fine-grained sediments – terrigeneous or hemipelagic, especially clay-rich, and calcareous oozes – seal basement with a thinner cover than silt size, siliceous, and pelagic sediments

(Schultz *et al.* 1992; Snelgrove and Forster, 1996; Giambalvo, *et al.*, 2000; Chapter 6). Accordingly, sediment contribution to ridge flank fluxes is through the spatially limited, by age and sedimentation, recharge and discharge sites, and by diffusion. The latter is driven by diagenesis that alters pore-fluid composition, thus gradients between the pore fluids and basement formation water, often having composition similar to bottom-seawater. The direction of exchange for various elements and isotopes that varies regionally, depending on sediment type and organic C content, is summarized in Table 16.1.

Quantifying fluxes is premature because of the very limited database. The suggestion of only ~1% of ridge flank fluid flux advection through sediments (Chapter 6) is based on end-member sediment distribution analysis and assumptions. If actual mixed-sediment distributions are used instead of just end members, as in Chapter 6, at least 5–10% and probably up to one-third of the fluid flux advects through sediments. (For example, in the eastern equatorial Pacific, the calcareous sediments contain $\geq$20% siliceous–mostly diatomaceous-ooze). Thus a maximum fluid flux of 95–90% and probably only ~70%, respectively, may be channeled through basement outcrops and/or faults. Even if only 5–10% advects through the sediments, based on an assumed axial heat flux of 7×10^{12} W and ridge flank fluid flux at 5–15 °C of 3.7–11×10^{15} kg yr^{-1} (Elderfield and Schultz, 1996), it provides a fluid flux of ~2.5–5×10^{14} kg yr^{-1}, respectively, approximately equal to 1% of the river flux of 3.7–4.2×10^{16} kg yr^{-1} (Edmond *et al.*, 1979). Because of the very large advective fluid fluxes involved, together with the diffusional fluxes across the basal sediment/oceanic basement interface discussed above, the contributions from ridge flank sediments may have an important impact on fluxes into the ocean or basement in comparison with river fluxes, even with just small concentration reductions or increases or changes in isotope ratios. When quantified, these two ridge flank exchange processes may help to resolve some of the global budget imbalances, in particular of Li, Sr, and oxygen isotope ratios; and alkalis, Mg, and Si, concentrations.

Despite the poor knowledge of low- to moderate-temperature fluxes from the ridge flank, based on the few documented fluid flux rates through sediments at young ridge flanks (Table 16.2), we present flux calculations for Si, Sr, and ^{87}Sr/^{86}Sr, for the first 5 million years, during which sediment cover is thin, assuming two boundary conditions: at a very low flow rate of 5 mm yr^{-1}, and at an average flow rate of 5 cm yr^{-1}. The concentrations used in the calculations are shown in Fig. 16.7. The 400 μmol kg^{-1} Si in Fig. 16.7a, is the most typical concentration value observed in pelagic sediment pore fluids, and the 1,000 μmol kg^{-1} Si is the value observed in diatom-rich sediments and in the Galapagos Mounds.

The ridge flank silica flux was calculated from:

$$Si = v\phi\Sigma A\Delta Si \qquad (16.13)$$

where v = fluid velocity in meters per year, ϕ = sediment porosity (~0.6), ΣA = cumulative area of ridge flank ($\Sigma A = C_o t$, where C_o is the creation rate = 3.45 km^2 yr^{-1}), and ΔSi is the silica anomaly. Figure 16.7a indicates that at 5 mm yr^{-1} advection rate, the silica flux through the sediment is insignificant, <1% of river flux. However, at 5 cm yr^{-1} fluid advection is ~10% of the river flux (Dixon, *et al.* 2002).

Table 16.1 *Direction of elemental and isotopic fluxes across the basal sediments/oceanic basement interface*

Element or isotope	Pore fluid concentration relative to seawater	Direction of flux
Li and other alkalis	Depleted at low temperatures and enriched at higher temperatures, threshold temperatures are as yet poorly defined	At low temperatures into sediment and at higher temperatures into basement
B	Mostly higher, except where illite or serpentine actively form	Mostly into basement
Ca	Depending on lithology, higher, lower, or no change. Highest values observed where volcanic ash is present	Depending on concentration
Mg	Lower; except in high organic C clay-rich sediment, ammonium produced expels Mg from exchange sites. Such sediment is rather rare on ridge flanks	Into basement
Sr	Mostly higher	Into basement
Ba	Higher in sediments with ≥ 0.5 wt. % organic C	Into basement or none
Fe, Mn	Higher in sediments with ≥ 0.5 wt. % organic C containing terrigenous matter	Into basement or none
Si	Higher	Into basement
SO_4	Lower when organic C content is ≥ 0.5 wt. %, otherwise no change	Into sediment or none
^6Li	Mostly ^6Li depleted except in organic C and clay-rich sediment with ammonium occupying exchange sites	^7Li into basement
^{10}B	Mostly enriched; depleted where illite or serpentine actively form	into basement; at sites of illite or serpentine formation ^{11}B into basement
$^{87}Sr/^{86}Sr$	Mostly less radiogenic	Into basement because of the higher than seawater Sr concentration

Table 16.2 *Published fluxes through ridge flank sediments*

Site	Location	Fluid flux (cm a^{-1})	Heat flow anomaly
Alarcon Ridge[a]	Gulf of California	0.2 ~ 1	5 MW km^{-1} of ridge
Baby Bare[b]	48° N Juan de Fuca Ridge	>300	2 ~ 3 MW
Western flank[c]	48° N Juan de Fuca Ridge		3 MW km^{-1} of ridge
ODP 677, 678[d]	Costa Rica Rift Flank	0.1 ~ 0.2	
Galapagos Mounds[e]	Galapagos Rift	1 ~ 30	
W Flank EPR 20° S[f]	Southern EPR	1 ~ 5	
Mariana Mounds[g]	Mariana spreading center	>10^5	

[a] Fisher *et al.* (2001).
[b] Mottl *et al.* (1998).
[c] Davis *et al.* (1999).
[d] Mottl. (1989).
[e] Maris *et al.* (1984).
[f] Bender *et al.* (1985/86).
[g] Wheat *et al.* (1995).

The average pore-fluid ^{87}Sr/^{86}Sr ratios used in our calculations for sediments overlying young basement, of ≤5 M.y. is 0.7070, derived from mixing of modern seawater Sr having ^{87}Sr/^{86}Sr = 0.70916 with the basement and sediment volcanic ash value of 0.7025. The very low ΔSr concentration value used in our calculation (10 μ mol kg^{-1}) is appropriate for pore fluids overlying young oceanic basement, with insignificant carbonate diagenesis that buffers pore fluid ^{87}Sr/^{86}Sr.

Recent mass-balance calculations (Davis *et al.*, 2003) indicate that hydrothermal circulation together with the benthic flux do not balance the oceanic Sr isotope budget; a hydrothermal input of ~6 × 10^9 mol yr^{-1} of ^{87}Sr/^{86}Sr at 0.7037 is missing. Could ridge flank circulation provide the missing flux? Our calculation for Sr, based on ΔSr = 10 μmol kg^{-1} and a ^{87}Sr/^{86}Sr ratio of 0.7070, shows in Fig. 16.7b that at 5 cm yr^{-1} fluid advection for 5 M.y. basement age, the Sr flux is more significant than the axial hydrothermal flux, as well as of the benthic flux. The "reduced" ^{87}Sr/^{86}Sr flux, derived from the mass-balance equation (16.14) of the weighted Sr inputs into the ocean that maintains the steady-state oceanic Sr isotope ratio of 0.70916, is shown in Fig. 16.7c. Interestingly, at the average

Fig. 16.7 Examples of global flux calculations for Si, Sr, and ^{87}Sr/^{86}Sr through ridge flank sediments overlying oceanic basement of ≤5 M.y. Advection rates used are 5 mm yr^{-1} and 5 cm yr^{-1}: (*a*) for Si fluxes with ΔSi concentrations of 400 and 1,000 μmol light gray dashed line near the top is the riverine Si flux; (*b*) for ΔSr concentrations of 10 and 20 μmol – riverine, the benthic, and axial hydrothermal Sr fluxes are also shown; (*c*) calculated ^{87}Sr/^{86}Sr fluxes at 5 mm yr^{-1} and 5 cm yr^{-1} fluid advection rates and reduced ^{87}Sr/^{86}Sr anomaly are shown. At fluid advection rate of 5 cm yr^{-1} and ΔSr = 20 μmol, Sr fluxes are significant, are approaching the reduced Sr isotopes anomaly.

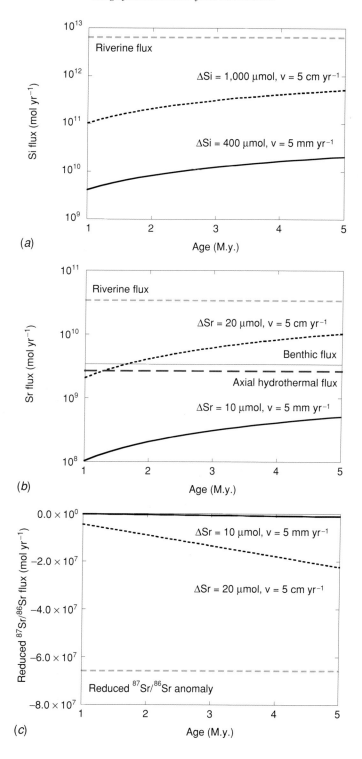

(a)

(b)

(c)

rate of fluid advection of 5 cm yr^{-1}, the calculated ridge flank contribution of lower than seawater and riverine ratios is significant, it approaches the "reduced" ^{87}Sr/^{86}Sr anomaly.

$$(f_{rivers}(0.7116 - 0.70916) + f_{axial}(0.7037 - 0.70916)$$
$$+ f_{benthic}(0.7084 - 0.70916) + f_{flank}(0.7070 - 0.70916) = 0 \qquad (16.14)$$

The above calculations indicate the importance of acquiring considerable more chemical and isotopic data on ridge flank circulation through sediments for better constraining oceanic mass balances.

16.6 Concluding remarks and recommendations

At present, although it is clear that marine sediments may be altered by seawater recharge and discharge through them, by the circulation of basement fluids, and by the effects of warm basement temperatures, we are not able to evaluate the impact of these processes on sediment diagenesis or on the chemistry of seawater. The known sites of basement and through sediment fluid flow which have been sampled are too few to cover the range of basement ages and sediment compositions and thicknesses over which fluid flow may occur. The chief concern is to identify the range of conditions over which fluids may circulate such that they may be altered chemically by sediment interaction. This could be achieved by obtaining high-resolution pore-fluid chemical and isotopic data, and simultaneously thermal and physical property analyses along transects perpendicular to active ridge crest at fast- to slow-spreading rates and a range of sediment cover, as well as along strike transects for variations in lithologies. For this it is important to select locations and sites where signals can be obtained and interpreted. Although more data are required to determine the importance of sediment interactions in terms of global chemical fluxes, the modeling of pore-fluid chemical concentrations are of great importance in determining the movement of pore fluids in response to basement fluid flow. The sensitivity obtained by using geochemical means, as compared to geophysical methods, to calculate rates of pore-fluid advection allows useful information to be gained about pore-fluid movements with respect to basement fluid under- or overpressures, and therefore about the nature of the underlying circulation systems. The integration of these data with the direct sampling of warm springs should place obvious sites of focused fluid upwelling into a more regional context for basement fluid movement.

Based on the above analyses and syntheses our principal observations and conclusions are:

- Recharge and discharge of fluid on ridge flanks is at least 5–10% through the sediment cover and probably up to a third advects through sediments. Thus, a maximum of 95–90% and probably just 70%, respectively, of fluid discharge is focused through basement outcrops and/or fault systems, as suggested by Fisher and Becker (2000).
- Recharge and discharge through sediments, occur where ocean crust is young, mostly <5 Ma and up to ~10 Ma, and the sediment cover is thin, mostly <50–100 m thick, depending on sediment lithology. The fluid flow is non-uniform, guided by porosity and permeability distribution.

- Based on the available limited database it is impossible to quantify the solute fluxes associated with the above two ridge flank sediment – related geochemical fluid flow regimes. The large fluid fluxes involved and the chemical gradients observed, however, suggest an important contribution to some global geochemical cycles.
- The above flow regimes are manifested by higher or lower than expected heat flow in localized areas or regionally.
- Diffusive communication between basal sediment pore fluids and the near-seawater formation fluid in permeable upper oceanic basement impacts basal sediment diagenesis and solute fluxes into and out of the sediment and formation fluids. The thickness of the basal layer affected depends on the residence time of the fluid flow regime. As yet an unquantified fraction of the fluxes from sediments into the upper basement fluid flow regimes will be discharged into the ocean via channeled flow.
- Three types of fluid–solid processes may imprint the chemical and isotopic signatures of ridge flank pore fluids and effect fluxes: (1) reactions between pore fluid and sediment, (2) reactions between advecting fluid and sediment, and (3) reactions between formation fluid and oceanic basement. Therefore, to decipher the origins, relative ratios, and approximate reaction temperatures at the fluid sources, more than one, at least two, and, if possible, several of the above geochemical proxies and geothermometers must be utilized simultaneously. At low to moderate temperatures and high rates of fluid flow, $>1–2$ cm yr^{-1}, the geochemical signatures of the fluids are subtle, therefore neither the fluid origin(s) nor the approximate reaction temperature(s) may be easily identified.
- Upper oceanic basement permeable layers, supporting high water/rock ratio lateral fluid flow systems, are widespread and regionally extensive in the Pacific in oceanic basement of up to 65–70 Ma. As yet no such system has been documented in the Atlantic.

References

Abbott, D. H., Menke, W. and Morin, R. 1983. Constraints upon water advection in sediments of the Mariana Trough. *J. Geophys. Res.* **88**: 1,075–1,093.

Anderson, R. N. and Hobart, M. A. 1976. The relation between heat flow, sediment thickness and age in Eastern Pacific. *J. Geophys. Res.* **81**: 2,968–2,989.

Baker, P. A. and Bloomer, S. H. 1988. The origin of celestite in deep-sea carbonates. *Geochim. Cosmochim. Acta* **52**: 335–339.

Baker, P. A., Gieskes, J. M., and Elderfield, H. 1982. Diagenesis of carbonates in deep-sea sediments – evidence from Sr/Ca ratios and interstitial dissolved Sr^{2+} data. *J. Sed. Petrol.* **52**: 71–82.

Baker, P. A., Stout, P. M., Kastner, M., and Elderfield, H. 1991. Large-scale lateral advection of seawater through oceanic crust in the central equatorial Pacific. *Earth Planet. Sci. Lett.* **105**: 522–533.

Barckhausen, U., Ranero, C. R., Von Huene, R., Cande, S. C., and Roeser, H. A. 2001. Revised tectonic boundaries in the Cocos plate off Costa Rica: implications for the segmentation of the convergent margin and for plate tectonic models. *J. Geophys. Res.* **106**: 19,207–19,220.

Becker, K., Langseth, M. G., and Hyndman, R. D. 1984. Temperature measurements in Hole 395A, Leg 78B. In *Initial Reports of The Deep Sea Drilling Project*, Vol. 78B, eds. R. Hyndman, M. Salisbury, *et al.* Washington, DC: US Govt. Printing Office, pp. 689–698.

Bender, M. L. 1983. Pore water chemistry of the Mounds Hydrothermal Field, Galapagos Spreading Center: results from *Glomar Challenger* piston coring. *J. Geophys. Res.* **88**: 1,049–1,056.

Bender, M. L., Hudson, A., Graham, D. W., Barnes, R. O., Leinen, M., and Kahn, D. 1985–1986. Diagenesis and convection reflected in pore water chemistry on the western flank of the East Pacific Rise, 20 degrees south. *Earth Planet. Sci. Lett.* **76**: 71–83.

Berner, R. A. 1978. Sulfate reduction and the rate of deposition of marine sediments. *Earth Planet. Sci. Lett.* **37**: 492–498.

1980. *Early Diagenesis*. Princeton, NJ: Princeton University Press.

Boetius, A., Ravenschlag, K., Schubert, C. J., Rickert, D., Widdel, F., Gieseke, A., Amann, R., Jørgensen, B. B., White, U., and Pfannkuche, O. 2000. A marine microbial consortium apparently mediating anaerobic oxidation of methane. *Nature* **407**: 623–626.

Borowski, W. S., Paull, C. K., and Ussler, W. 1999. Global and local variations of interstitial sulfate gradients in deep-water, continental margin sediments: sensitivity to underlying methane and gas hydrate. *Mar. Geol.* **159**: 131–154.

Boudreau, B. P. 1997. *Diagenetic Models and Their Implementation*. Berlin: Springer-Verläg.

Boudreau, B. P. and Westrich, J. T. 1984. The dependence of bacterial sulfate reduction on sulfate concentration in marine sediments. *Geochim. Cosmochim. Acta* **48**: 2,503–2,516.

Chan, L.-H. and Kastner, M. 2000. Lithium isotopic compositions of pore fluids and sediments in the Costa Rica subduction zone: implications for fluid processes and sediment contribution to the arc volcanoes. *Earth Planet. Sci. Lett.* **183**: 275–290.

Claypool, G. E. and Kaplan, I. R. 1974. The origin and distribution of methane in marine sediments. In *Natural Gases in Marine Sediments*, ed. I. R. Kaplan. New York: Plenum, pp. 99–139.

Craig, H. 1969. Abyssal carbon and radiocarbon in the Pacific. *J. Geophys. Res.* **74**: 5,491–5,506.

Dadey, K. 1991. The Mariana Mounds hydrothermal area: processes and products, Ph.D. thesis, University of Rhode Island.

Davis, E. E., Chapman, D. S., Forster, C. B., and Villinger, H. 1989. Heat flow variations correlated with buried basement topography on the Juan de Fuca Ridge Flank. *Nature* **342**: 533–537.

Davis, E. E., Chapman, D. S., Mottl, M. J., Bentkowski, W. J., Dadey, K., Forster, C., Harris, R., Nagihara, S., Rohr, K., Wheat, G., and Whiticar, M. 1992. FlankFlux: an experiment to study the nature of hydrothermal circulation in young oceanic crust. *Can. J. Earth Sci.* **29**: 925–952.

Davis, E. E., Fisher, A. T., and Firth, J. V., eds. 1997. Hydrothermal circulation in the oceanic crust: eastern flank of the Juan de Fuca Ridge. *Proceedings of the Ocean Drilling Program, Part A, Initial Reports*, Vol. 168. College Station, TX: Ocean Drilling Program. pp. 1–470.

Davis, E. E., Chapman, D. S., Wang, K., Villinger, H., Fisher, A. T., Robinson, S. W., Grigel, J., Prinbow, D., Stein, J., and Becker, K. 1999. Regional heat flow variations across the sedimented Juan de Fuca Ridge eastern flank: constraints on lithospheric cooling and lateral hydrothermal heat transport. *J. Geophys. Res.* **104**: 17,675–17,688.

Davis, A. C., Bickle, M. J., and Teagle, D. A. H. 2003. Imbalance in the oceanic strontium budget. *Earth Planet. Sci. Lett.,* **211**: 173–187.

D'Hondt, S., Rutherford S., and Spivack, A. J. 2002. Metabolic activity of sub-surface life in deep-sea sediments. *Science* **295**: 2,067–2,070.

Dixon, J. E., Leist, L. Langmuir, C., and Schilling, J.G. 2002. Recycled hydrated lithosphere observed in plume-influenced mid-ocean-ridge basalt. *Nature* **420**: 385–389.

Dunnebier, F. and Blackinton, G. 1980. A man-made hot spring on the ocean floor. *Nature* **284**: 338–340.

Edmond, J. M., Measures, C. I., McDuff, R. E., Chan, L. H., Coller, R., *et al.* 1979. Ridge crest hydrothermal activity and the balance of the major and minor elements in the ocean: the Galapagos data. *Earth Planet. Sci. Lett.* **46**: 1–18.

Elderfield, H. and Schultz, A. 1996. Mid-ocean hydrothermal fluxes and the chemical composition of the ocean. *Ann. Rev. Earth Planet. Sci.* **24**: 191–224.

Elderfield, H., Wheat, C. G., Mottl, M. J., Monnin, C., and Spiro, B. 1999. Fluid and geochemical transport through oceanic crust: a transect across the flank of the Juan de Fuca Ridge. *Earth Planet. Sci. Lett.* **172**: 151–165.

Fang, W. W., Langseth, M. G., and Schultheiss, P. J. 1993. Analysis and application of *in situ* pore pressure measurements in marine sediments. *J. Geophys. Res.* **98**: 7,921–7,938.

Fisher, A. T., and Becker, K. 2000. Channelized fluid flow in oceanic crust reconciles heat-flow and permeability data. *Nature* **403**: 71–74.

Fisher, A. T., Giambalvo, E., Sclater, J., Kastner, M., Ransom, B., Weinstein, Y., and Lonsdale, P. 2001. Heat flow, sediment and pore fluid chemistry, and hydrothermal circulation on the east flank of the Alarcon Ridge, Gulf of California. *Earth Planet. Sci. Lett.* **188**: 521–534.

Fouillac, C. and Michard, G. 1981. Sodium/lithium ratio in water applied to geothermometry of geothermal reservoirs. *Geothermics* **10**: 55–70.

Fournier, R. O. 1973. Silica in thermal waters: laboratory and field investigation. In *Proceedings of the International Symposium on Hydrogeochemistry and Biogeochemistry*, Japan 1970. Washington, DC: J. W. Clarke, pp. 122–139.

  1979. A revised equation for the Na/K thermometer. *Trans. Geotherm. Resour. Counc.* **3**: 221–224.

Fournier, R. O. and Rowe, J. J. 1966. Estimation of underground temperatures from the silica content of water, hot springs and steam wells. *Am. J. Sci.* **264**: 685–697.

Giambalvo, E., Fisher, A. T., Martin, J. T., Darty, L., and Lowell, R. P. 2000. Origin of elevated sediment permeability in a hydrothermal seepage zone, eastern flank of the Juan de Fuca Ridge, and implications for transport of fluid and heat. *J. Geophys. Res.* **105**: 913–928.

Giambalvo, E., Steefel, C. I., Fisher, A. T., Rosenberg, N. D., and Wheat, C. G. 2002. Effect of fluid–sediment reaction on hydrothermal fluxes of major elements, eastern flank of the Juan de Fuca Ridge. *Geochim. Cosmochim. Acta* **66**: 1,739–1,757.

Gieskes, J. M. and Lawrence, J. R. 1981. Alteration of volcanic matter in deep sea sediments: evidence from the chemical composition of interstitial waters from deep sea drilling cores. *Geochim. Cosmochim. Acta* **45**: 1,687–1,703.

Goloway, F. and Bender, M. L. 1982. Diagenetic models of interstitial nitrate profiles in deep-sea suboxic sediments. *Limnol. Oceanogr.* **27**: 624–638.

Hobart, M. A., Langseth, M. G., and Anderson, R. N. 1985. A geothermal and geophysical survey on the south flank of the Costa Rica Rift: Sites 504 and 505. In

Initial Reports of the Deep Sea Drilling Project, Vol. 83, eds. R. N. Anderson, J. Horrores, K. Backer, *et al.* Washington, DC: US Govt. Printing Office, 379–404.

Hunter, F. M. and Elderfield, H. 2001. Isotope geochemistry of low temperature hydrothermal fluids from upwelling sites on the Juan de Fuca Ridge. *Eos, Trans. Am. Geophys. Union* **82** (47): 638.

James, R. H. and Palmer, M. R. 2000. Marine geochemical cycles of the alkali elements and boron: the role of sediments. *Geochim. Cosmochim. Acta* **64**: 3,111–3,122.

Kastner, M., 1981. Authigenic silicates in deep-sea sediments: formation and diagenesis. In *The Sea, Vol. 7, The Oceanic Lithosphere*, ed. C. Emiliani. New York: Wiley, pp. 915–980.

Kastner, M. and Stonecipher, S. A. 1978. Zeolites in pelagic sediments of the Atlantic, Pacific, and Indian oceans. In eds. L. B. Sand and F. A. Mumpton, *Natural Zeolites: Occurrences, Properties, Use.* Oxford: Pergamon Press, pp. 199–220.

Kastner, M., Gieskes, J. M., and Hu, J.-Y. 1986. Carbonate recrystallization in basal sediments: evidence for convective fluid flow on a ridge flank. *Nature* **321**: 158–161.

Kastner, M., Morris, J., Chan, L.-H., Saether, O., Lückge, A., and Silver, E. 2000. Three distinct fluid systems at the Costa Rica subduction zone: chemistry, hydrology, and fluxes. *Extended Abstracts of the Goldschmidt Conference.* Cambridge: Cambridge Press, pp. 572–573.

Kharaka, Y. K. and Mariner, R. H. 1989. Chemical geothermometers and their applications to formation waters from sedimentary basins. In *Thermal History of Sedimentary Basins: Methods and Case Histories*, eds. D. Naesser and T. H. McCulloch. New York: Springer-Verlag. pp. 99–117.

Kvenvolden, K. A. 1993. Gas hydrates: geological perspective and global change. *Rev. Geophys.* **31**: 173–187.

Langseth, M. G. and Silver, E. A. 1996. The Nicoya convergent margin: A region of exceptionally low heat flow. *Geophys. Res. Lett.* **23**: 891–894.

Langseth, M. G., Cann, J. R., Natland, J. H., and Hobart, M. A. 1983. Geothermal phenomena at the Costa Rica Rift: background and objectives for drilling at Deep Sea Drilling Project Sites 501, 504 and 505. In *Initial Reports of The Deep Sea Drilling Program*, **69**; eds. J. R. Cann, M. G. Langseth, J. Horrolez, *et al.* Washington, DC: US Govt. Printing Office, 5–30.

Langseth, M. G., Mottl, M. J., Hobart, M. A., and Fisher, A. 1988. The distribution of geothermal and geochemical gradients near Site 501/504: implication for hydrothermal circulation in the oceanic crust. In *Proceedings of the Ocean Drilling Program, Initial Reports*, Vol. 111. eds. K. Becker, H. Sakai, *et al.* College Station, TX: Ocean Drilling Program, pp. 23–32.

Langseth, M. G., Becker, K., Von Herzen, R. P., and Schulthesis, P. 1992. Heat and fluid flux through sediment on western flank of the Mid-Atlantic Ridge: a hydrogrological study of North Pond. *Geophys. Res. Lett.* **19**: 517–520.

Lawrence, J. R. and Gieskes, J. M. 1981. Constraints on water transport and alteration in the oceanic crust from the isotopic composition of pore water. *J. Geophys. Res.* **86**: 7,924–7,934.

Lawrence, J. R., Gieskes, J. M., and Anderson, T. F., 1976. Oxygen isotope material mass balance calculations. In *Initial Reports of the Deep Sea Drilling Program*, eds. C. D. Hollister, C. Craddock, *et al.* Washington, DC: US Govt. Printing Office, pp. 507–512.

Lawrence, J. R., Drever, J. L., Anderson, T. F., and Brueckner, H. K. 1979. Importance of volcanic alteration in sediments of site 323: chemistry, O^{18}/O^{16} Sr^{87}/Sr^{86}. *Geochim. Cosmochim. Acta* **43**: 573–588.

Leinen, M., Rea, D. K., *et al.*, eds. 1986. *Initial Reports of the Deep Sea Drilling Project*, Vol. 92. Washington, DC: US Govt. Printing Office.

Lonsdale, P. 1977. Deep-tow observations at the Mounds abyssal hydrothermal field. *Earth Planet. Sci. Lett.* **36**: 92–110.

Magenheim, A. J., Spivack, A. J., Volpe, C., and Ransom, B. 1994. Precise determination of stable chlorine isotopic ratios in low-concentration natural samples. *Geochim. Cosmochim. Acta* **58**: 3,117–3,123.

Maris, C. R. P., Bender, M. L., Froelich, P. N., Barnes, R., and Luedtke, N. A. 1984. Chemical evidence for advection of hydrothermal solutions in the sediments of the Galapagos Mounds Hydrothermal Field. *Geochim. Cosmochim. Acta* **48**: 2,331–2,346.

Martin, J. A. and Kastner, M. 1996. Chemical and isotopic evidence for sources of fluids in a mud volcano field seaward of the Barbados accretionary wedge. *J. Geophys. Res.* **101**: 20,325–20,345.

McDuff, R. E. and Ellis, R. A. 1979. Determining diffusion coefficients in marine sediments: a laboratory study of resistivity techniques. *Am. J. Sci.* **279**: 666–675.

Monod, J. 1949. The growth of bacterial cultures. *Ann. Rev. Microbiol.* **3**: 371–394.

Mottl, M. J. 1989. Hydrothermal convection, reaction and diffusion in sediments on the Costa Rica Ridge flank: pore-water evidence from ODP sites 677 and 678. In *Proceedings of the Ocean Drilling Program, Scientific Reports*, Vols. 111, eds. K. Becker, H. Sakai, *et al.* College Station, TX: Ocean Drilling Program, pp. 195–213.

Mottl, M. J., Lawrence, J. R., and Keigwin, L. D. 1983. Elemental and stable-isotope composition of pore waters and carbonate sediments from Deep-Sea Drilling Project sites 501/504 and 505. In *Initial Reports of the Deep Sea Drilling Project*, Vol. 69, eds. J. R. Cann, M. G. Langseth, J. Honnorez, and R. P. Von Herzen. Washington, DC: US Govt. Printing Office, pp. 461–473.

Mottl, M. J., Wheat, C. G., Baker, E., Becker, N., Davis, E., Feely, R., Grehan, A., Kadko, D., Lilley, M., Massoth, G., Moyer, C., and Sansore, F. 1998. Warm spring discovered on 3.5 Ma-old oceanic crost, eastern flank of the Jum de Fuca Ridge. *Geology* **26**: 51–54.

Peacock, S. M. 2001. Are the lower planes of double seismic zones caused by serpentine dehydration in subducting oceanic mantle? *Geology* **29**: 299–302.

Perry, E. and Hower, J. 1970. Burial diagenesis in Gulf Coast pelitic sediments. *Clays and Clay Minerals* **18**: 165–177.

Ransom, B., Spivak, A. J., and Kastner, M. 1995. Stable Cl isotopes in subduction-zone pore waters: implications for fluid–rock reactions and the recycling of chlorine. *Geology* **23**: 715–718.

Richter, F. M. 1993. Fluid flow in deep-sea carbonates: estimates based on porewater Sr. *Earth Planet. Sci. Lett.* **119**: 133–141.

1996. Models for the coupled Sr–sulfate budget in deep-sea carbonates. *Earth Planet. Sci. Lett.* **141**: 199–211.

Richter, F. M. and DePaolo, D. J. 1987. Numerical models for diagenesis and the Neogene Sr isotopic evolution of seawater from DSDP Site 590B. *Earth Planet. Sci. Lett.* **83**: 27–38.

Rudnicki, M. D., Elderfield, H., and Mottl, M. J. 2001. Pore fluid advection and reaction in sediments of the eastern flank, Juan de Fuca Ridge, 48° N. *Earth Planet. Sci. Lett.* **187**: 173–189.

Schrag, D. P., Adkins, J. F., McIntyre, K., Alexander, J. I., Hodell, D. A., Charles, C. D., and McManus, J. F. 2002. The oxygen isotopic composition of seawater during the last glaciation maximum. *Quat. Sci. Rev.* **21**: 331–342.

Schultz, A. and Elderfield, H. 1997. Controls on the physics and chemistry of seafloor hydrothermal circulation. *Phil. Trans. Roy. Soc. Lond. A* **355**: 387–425.

Schultz, A., Delaney, J., and McDuff, R. E. 1992. On the partitioning of heat flux between diffuse and point source seafloor venting. *J. Geophys. Res.* **97**: 12,299–12,314.

Sclater, J. G., Crowe, J., and Anderson, R. N. 1976. On the reliability of oceanic heat flow averages. *J. Geophys. Res.* **81**: 2,997–3006.

Seyfried, W. E., Janecky, D. R., and Mottl, M. J. 1984. Alteration of the oceanic crust: implications for the geochemical cycle of lithium and boron. *Geochim. Cosmochim. Acta* **48**: 557–569.

Seyfried, W. E., Chen, X., and Chan, L.-H. 1998. Trace element mobility and lithium isotope exchange during hydrothermal alteration of the seafloor weathered basalt: an experimental study at 350 °C, 500 bars. *Geochim. Cosmochim. Acta* **62**: 940–960.

Silver, E. A., Kastner, M., Fisher, A. T., McIntosh, K. D., and Saffer, D. M. 2000. Fluid flow paths in the crust of the Middle American Trench, Costa Rica Margin. *Geology* **28**: 679–682.

Snelgrove, S. H. and Forster, C. B. 1996. Impact of seafloor sediment permeability and thickness on off-axis hydrothermal circulation: Juan de Fuca Ridge eastern flank. *J. Geophys. Res.* **101**: 2,915–2,925.

Spivack, A. J., Palmer, M. R., and Edmond, J. M. 1987. The sedimentary cycle of the boron isotopes. *Geochim. Cosmochim. Acta* **57**: 1,033–1,047.

Spivack, A. J., Kastner, M., and Ransom, B. 2002. Elemental and isotopic chloride geochemistry and fluid flow in Nankai Trough. *Geophys. Res. Lett.* **29**, 6–6-4.

Srodon, J. 1999. Nature of mixed layer clays and mechanisms of their formation and alteration. *Ann. Rev. Earth Planet. Sci.* **27**: 19–53.

Stein, C. A. and Stein, S. 1992. A model for the global variation in oceanic depth and heat flow with lithospheric age. *Nature* **359**: 123–129.

  1994. Constraints on hydrothermal heat flux through the oceanic lithosphere from global heat flow. *J. Geophys. Res.* **99**: 3,081–3,095.

Stout, P. M. 1985. Interstitial water chemistry and diagenesis of biogenic sediments from the eastern equatorial Pacific. In *Initial Reports of the Deep Sea Drilling Project*, Vol. 85, eds. L. Mayer, F. Theyer, *et al.* Washington, DC: US Government Printing Office, pp. 805–820.

Toth, D. J. and Lerman, A. 1977. Organic matter reactivity and sedimentation rates in the ocean. *Am. J. Sci.* **277**: 465–485.

Tromp, T. K., Van Cappellen, P., and Key, R. M. 1995. A global model for the early diagenesis of organic carbon and organic phosphorus in marine sediments. *Geochim. Cosmochim. Acta* **59**: 1,259–1,284.

Truesdell, A. H. 1976. Geochemical techniques in exploration, summary of section III. In *Proceedings of the Second United Nations Symposium on the Development and Use of Geothermal Resources*, San Francisco, Vol. 1. Washington, DC: US Govt. Printing Office. pp. 53–79.

Von Damm, K. L., Bischoff, J. L., and Rosenbauer, R. J. 1991. Quartz solubility in hydrothermal seawater: an experimental study and equation describing quartz solubility for up to 0.5 M NaCl solutions. *Am. J. Sci.* **291**: 977–1,007.

Wheat, C. G. and McDuff, R. E. 1994. Hydrothermal flow through the Mariana Mounds: dissolution of amorphous silica and degradation of organic matter on a mid-ocean ridge flank. *Geochim. Cosmochim. Acta* **58**: 2,461–2,475.

  1995. Mapping the fluid flow of the Mariana Mounds ridge flank hydrothermal system: pore water chemical tracers. *J. Geophys. Res.* **100**: 8,115–8,131.

Wheat, C. G. and Mottl, M. J. 1994. Hydrothermal circulation, Juan de Fuca Ridge eastern flank: factors controlling basement water composition, *J. Geophys. Res.* **99**: 3,067–3,080.

Wheat, C. G., Elderfield, H., Mottl, M. J., and Monnin, C. 2000a. Chemical composition of basement fluids within an oceanic ridge flank: implications for along-strike and across-strike hydrothermal circulation. *J. Geophys. Res.* **105**: 13,437–13,447.

Wheat, C. G., Jannasch, H., and Kastner, M. 2000b. Seawater transport and reaction in upper oceanic basement: chemical data from continuous monitoring of sealed boreholes. *Eos, Trans. Am. Geophys. Union* **81**: F457.

Wheat, C. G., Mottl, M. J., and Rudnicki, M. D. 2002. Trace element and REE composition of a low-temperature ridge flank hydrothermal spring. *Geochim. Cosmochim. Acta* **66**: 3,693–3,705.

Williams, D. L., Von Herzen, R. P., Sclater, J. G., and Anderson, R. N. 1974. The Galapagos Spreading Center: lithospheric cooling and hydrothermal circulation. *Geophys. J. Roy. Astron. Soc.* **38**: 587–608.

You, C.-F. and Gieskes, J. M. 2001. Hydrothermal alteration of hemi-pelagic sediments: experimental evaluation of geochemical processes in shallow subduction zones. *Appl. Geochem.* **16**: 1,055–1,066.

You, C.-F., Castillo, P. R., Gieskes, J. M., Chan, L.-H., and Spivack, A. J. 1997. Trace element behavior in hydrothermal experiments: implications for fluid processes at shallow depth in subduction zones. *Earth Planet. Sci. Lett.* **140**: 41–52.

17

Microbial reactions in marine sediments

Jon P. Telling, Edward R. C. Hornibrook, and R. John Parkes

17.1 Introduction

Marine sediments cover approximately 70% of the Earth's surface and contain substantial amounts of organic matter in sedimentary sequences up to 10 km in thickness (Fowler, 1990). Until relatively recently, it was thought that sediments at depths greater than $\sim$10 m below the seafloor (mbsf), were biologically inactive (Morita and Zobell, 1955). However, it is now known that bacteria are present and active to at least 800 mbsf, and almost certainly at greater depths (Parkes *et al.*, 1994; Parkes *et al.*, 2000; Wellsbury *et al.*, 2002). Moreover, it has been estimated that bacteria within marine sediments constitute about 76% of global bacterial biomass (Whitman *et al.*, 1998). Because of their abundance and activity at great depths in sediments, it is likely that bacteria influence many sedimentary processes that previously were attributed to geosphere reactions.

The organic carbon cycle can be divided into two parts: the biological cycle and geological cycle (Fig. 17.1). Collectively, reservoirs within the biological cycle contain $\sim$3 $\times$ 10^{12} tonnes carbon (t C), of which all but a tiny proportion originates from the synthesis of organic material from CO_2 (g) or HCO_3^- (aq) by the process of photosynthesis (Tissot and Welte, 1984).

In marine environments, planktonic microbial communities carry out the majority of primary production in surface waters; however, most organic matter sedimenting from the photic zone is recycled by a "microbial filter" before it reaches the seafloor. This filter consists of microorganisms in the water column that utilize organic detritus for heterotrophic growth (Hedges and Keil, 1995). The amount of organic matter that settles to the seafloor is largely a function of the mean net primary production in the photic zone and the depth of the water column (Suess, 1980). On average, less than 0.5% of original primary production is deposited in marine sediments (Hedges and Keil, 1995) and preservation typically is positively correlated with sedimentation rate, varying from 5 to 20% in coastal sediments and from 0.5 to 3.5% in deep-sea sediments (Henrichs and Reeburgh, 1987). Organic material also can be derived from re-deposition of marine sediments after erosion, such as by turbidity currents along continental margins (Meyers *et al.*, 1996).

Hydrogeology of the Oceanic Lithosphere, eds. E. E. Davis and H. Elderfield. Published by Cambridge University Press.
© Cambridge University Press 2004.

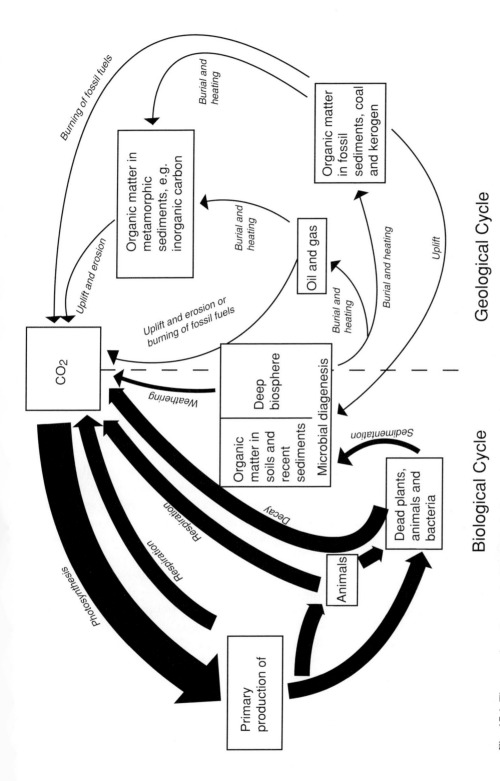

Fig. 17.1 The organic carbon cycle. The biological and geological organic carbon cycles are linked through the burial of organic matter in sediments and the uplift and erosion of rocks. Thicker lines indicate more rapid fluxes (adapted from Tissot and Welte, 1984).

The composition of organic matter reaching the seafloor is very different from that of the original material formed in the photic zone. Wakeham *et al.* (1997) showed that organic matter sinking through a 1,000 m water column retains ~1% of its original content of carbohydrates, amino acids and lipids, consisting mainly of uncharacterized refractory compounds by the time it reaches the seafloor. However, a proportion of organic matter in marine sediments can be intrinsically labile material that has been stabilized through sorption onto mineral lattices. Once desorbed, this material can be readily mineralized by bacteria (Keil *et al.*, 1994).

Leakage of organic carbon from the biological to the geological cycle eventually leads to the formation of natural gas, oil, and coal, or metamorphic forms of carbon. The amount of material transferred per year is small, but extremely significant when integrated over geological time. Sediments and rocks contain several orders of magnitude more organic carbon that the entire biological cycle (6.4×10^{15} versus 3×10^{12} t C; Tissot and Welte, 1984), however, turnover of organic carbon within the geological cycle is extremely slow, normally occurring on a time scale of millions of years. Re-oxidation to CO_2 can be enhanced by uplift and weathering (including microbial metabolism) of sedimentary rocks or more recently, by the combustion of fossil fuels upon re-entry into the biosphere.

The aim of this paper is to review the microbiological processes that drive degradation (diagenesis) of organic matter in marine sediments, and thus influence the transfer of organic carbon between the biological and geological cycles.

17.2 Bacteria as environmental catalysts

Marine sediments can be considered an "extreme" environment for most eukaryotic organisms. Approximately 85% of the Earth's seafloor is at a temperature of less than 5 °C and nearly 60% is consistently at ~2 °C (Russell, 1990). Conversely, in geologically active regions, and within very deep sediments possessing an average thermal gradient of 30 °C km^{-1}, temperatures can exceed 100 °C. In addition, the majority of the seafloor and underlying sediments are under hundreds of atmospheres of hydrostatic pressure (Saunders and Fofonoff, 1976) and brine solutions flowing through sediments can have ionic strengths several times that of seawater (Karl, 1995).

In contrast to eukaryotes, prokaryotes can thrive under "extreme" conditions. For example, the highest known temperature for growth of eukaryotes, such as algae and fungi, is ~60 °C (Madigan *et al.*, 2000), whereas prokaryotes have been isolated which are capable of growth under elevated pressures and at temperatures of up to 113 °C (Blochl *et al.*, 1997). Some prokaryotes are active at temperatures down to −20 °C (Rivkina *et al.*, 2000), in extremely acid (<pH 2) or alkali solutions (>pH 11), at high pressures (>1,000 atm), and in highly concentrated saline solutions (>30% NaCl; Madigan *et al.*, 2000).

17.2.1 Metabolic diversity

Eukaryotes typically are limited to gaining energy through either photosynthesis or the oxidation of organic compounds with oxygen. In contrast, prokaryotes can use a diverse range of energy sources, including metabolisms based purely on the oxidation of inorganic substrates and many that require strictly anaerobic conditions.

The amount of energy that can be obtained from a particular chemical reaction ultimately is dependent upon the Gibbs free energy (ΔG^0) of the reaction:

$$\Delta G^0 = -nF\Delta E'_0 \, (\text{kJ mol}^{-1}) \tag{17.1}$$

where ΔG^0 is the free energy change of the reaction at standard conditions, n is the number of electrons transferred by the reaction, F is Faraday's constant (96,485 coulomb mol^{-1}), and $\Delta E'_0$ is the redox potential of the compounds (i.e. the capacity of different compounds to gain or donate electrons relative to a hydrogen standard; Stumm and Morgan, 1996). Thauer *et al.* (1977) discuss in detail the theory and calculation of ΔG^0 for a variety of microbial processes.

It is appropriate to consider various biogeochemical reactions in terms of redox potentials because all known life is based on redox reactions (i.e. the transfer of electrons or hydrogen atoms). The redox potentials of important biogeochemical reactions are shown in Fig. 17.2. Strong reductants are situated at the top of the diagram and strong oxidants are placed at the bottom. Net reactions are exergonic (energy yielding) when electron donors that have a more negative E'_0 are coupled to electron acceptors that possess a more positive E'_0. Such coupled reactions may provide sufficient energy to support bacterial metabolism.

Bacterial metabolisms can be divided into those that are based upon (i) an organic energy substrate, such as acetate or glucose (organotrophs), and (ii) an inorganic energy substrate, such as Fe^{2+} or H_2S (lithotrophs). Energy released from these reactions must be conserved in order to drive endergonic (energy consuming) biochemical reactions, or else it will be lost rapidly to the surrounding environment as heat. In living cells, energy ultimately is transferred and stored in a variety of compounds possessing high-energy phosphate bonds, the most important of which is adenosine triphosphate (ATP). Other high-energy compounds can also be used, in particular acetyl-CoA within fermenting organisms. As a general rule, processes which release small amounts of energy (<7 kJ mol^{-1}) insufficient for the formation of ATP, are ones that cannot sustain life (Thauer *et al.*, 1977). However, some novel anaerobic bacteria, which form syntrophic relationships to achieve a subtle balance between concentrations of metabolic reactants and products, can function at close to zero ΔG^0, and thus operate extremely efficient metabolic systems (Jackson and McInerny, 2002).

Substrates are oxidized in cells either by fermentation or respiration. Fermentation is an internally balanced reaction, in which energy is released and ATP is synthesized by enzymatic reactions directly connected to the metabolism of the substrate. This process is known as substrate-level phosphorylation, an example of which is the fermentation of ethanol:

$$CH_3CH_2OH + H_2O \rightarrow CH_3COO^- + 2H_2 + H^+ \tag{17.2}$$

$$\text{ethanol} \qquad\qquad \text{acetate} \qquad \text{hydrogen}$$

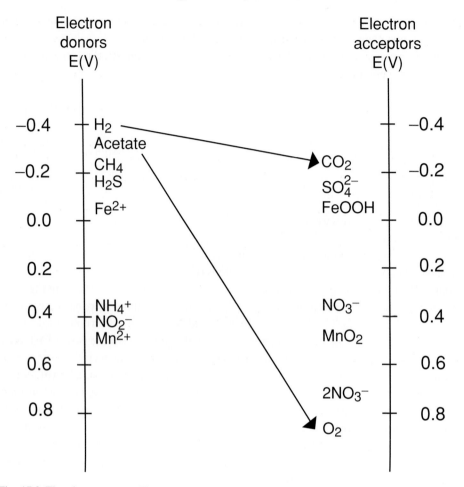

Fig. 17.2 The electron tower. The strongest reductants are at the top and the strongest oxidants at the bottom of the diagram. Coupled reactions between electron donors having more negative E'_o with electron acceptors possessing more positive E'_o are exergonic and may provide sufficient energy to support bacterial metabolism. In the example cited in the text, oxidation of acetate with oxygen has a steep negative gradient, and hence a high free-energy yield. The oxidation of hydrogen with carbon dioxide (methanogenesis or acetogenesis) has a shallower gradient and consequently, a lower free-energy yield.

Only a small amount of the potential energy of an organic compound is released during fermentation and thus products formed can be metabolized further by other organisms (Madigan *et al.*, 2000).

Respiration involves oxidation of an energy source by an external electron acceptor, such as O_2 or SO_4^{2-}. For example, the oxidation of glucose:

$$C_6H_{12}O_6 + 6O_2 \rightarrow 6CO_2 + 6H_2O \qquad (17.3)$$
glucose

The energy released is used to synthesize ATP via membrane-mediated processes not connected directly to the metabolism of specific substrates. This process is known as oxidative phosphorylation, and involves using a proton gradient established across the cell membrane to drive the synthesis of ATP or other activities (e.g. substrate uptake, locomotion via a flagellum).

In the natural environment, conditions deviate from those used to establish ΔG^0 and the actual free-energy change for a process will depend upon the exact pressure, temperature, and activities of reactants and products (Stumm and Morgan, 1996). A good example is the fermentation of ethanol to acetate (17.2) for which the free-energy change at standard conditions is endergonic ($\Delta G^0 = +10$ kJ mol^{-1}). However, under sufficiently low H_2 partial pressures, the reaction becomes exergonic. In the natural environment, the change to a negative ΔG^0 is achieved by the close proximity of anaerobic respiratory bacteria, which use H_2 as an energy substrate. The diffusional transfer of H_2 from one organism to another is termed "interspecies hydrogen transfer" (Conrad *et al.*, 1986) and represents a syntrophic relationship between different types of anaerobic bacteria.

17.3 Total bacterial biomass and activity estimates

Bacterial biomass and activities are generally highest in the top centimeter of marine sediments, decreasing rapidly with depth (Novitsky, 1983; Parkes *et al.*, 1994; Pfannkuche *et al.*, 1999; Dixon and Turley, 2000; Fig. 17.3). The decrease in bacterial biomass and activity with depth results from labile carbon being preferentially utilized in shallow layers, which leaves only the more refractory organic material to be buried in deeper sediments. Such a process has been successfully approximated in theoretical models (the so-called "multi-G models") by the inclusion of several different groups of organic carbon, each possessing a different susceptibility to degradation and each being degraded exponentially with time and depth (Westrich and Berner, 1984).

The total rate of organic matter mineralization in marine sediments can be estimated by measuring oxygen fluxes in the benthic zone. Under steady-state conditions, oxygen flux into sediment is equivalent to the rate of aerobic respiration plus underlying anaerobic processes because oxygen is used directly or indirectly to re-oxidize products of anaerobic respiration that diffuse upwards from the underlying anoxic zone (e.g. NH_4^+, Mn^{2+}, Fe^{2+}, H_2S, CH_4; Table 17.1 and Fig. 17.4; Jørgensen, 1982; Canfield *et al.*, 1993a,b).

Oxygen fluxes can be determined either by measuring oxygen concentration gradients using microelectrodes (e.g. Revsbech *et al.*, 1980; Gundersen and Jørgensen, 1990) or fiber-optic optodes (Klimant *et al.*, 1995), or alternatively by measuring time-dependent concentration changes of dissolved species in closed benthic chambers, preferably *in situ* to avoid sampling disturbance associated with sediment collection (e.g. Jahnke *et al.*, 1997). Fluxes determined by both methods are generally in good agreement at deep open water sites where the abundance of organic matter in sediments is low; however, in continental margin sediments, oxygen fluxes are generally larger than values calculated from pore-water gradients because of bioirrigation by macrofauna in the sub-oxic zone (Jahnke *et al.*,

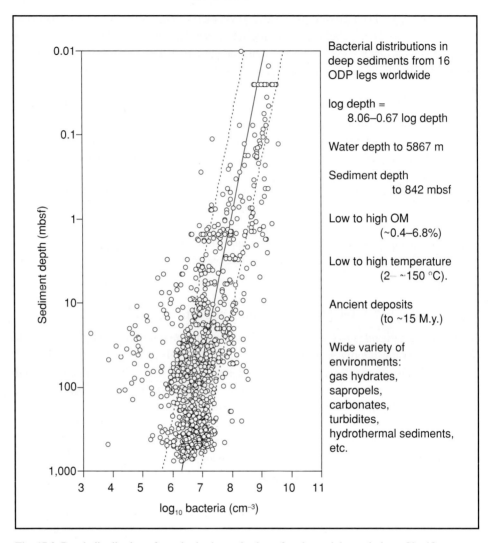

Fig. 17.3 Depth distribution of non-hydrothermal sub-surface bacterial populations. Significant populations are found at depths >800 m and are likely to be present at much greater depth (from Parkes *et al.*, 2000).

1990; Glud *et al.*, 1994; Sayles and Martin, 1995). Oxygen fluxes also will underestimate total respiration rates in sediments if (i) reduced inorganic products become trapped in the underlying sediment as a result of the formation of solid phases, such as pyrite (FeS_2) or gas hydrates: or (ii) reduced inorganic products diffuse into the water column (e.g. at euxinic sites possessing anoxic bottom waters).

Finally, rates of anaerobic respiration in sediments can be measured directly through the use of radiotracer compounds (e.g. sulfate reduction and methanogenesis; Wellsbury

Table 17.1 *Organic carbon oxidation reactions and the subsequent oxidation of reduced species (adapted from Canfield* et al., *1993a)*

Organic mineralization	Re-oxidation of inorganics[a]	Net reaction
O_2 respiration $CH_2O + O_2 \Rightarrow H_2O + CO_2$		$CH_2O + O_2 \Rightarrow H_2O + CO_2$
Dentrification $4/5H^+ + CH_2O + 4/5NO_3^- \Rightarrow$ $CO_2 + 2/5N_2 + 7/5H_2O$		$4/5H^+ + CH_2O + 4/5NO_3^- \Rightarrow$ $CO_2 + 2/5N_2 + 7/5H_2O$
Mn(IV) reduction $4H^+ + CH_2O + 2MnO_2 \Rightarrow$ $2Mn^{2+} + 3H_2O + CO_2$	$2H_2O + 2Mn^{2+} + O_2 \Rightarrow$ $2MnO_2 + 4H^+$	$CH_2O + O_2 \Rightarrow H_2O + CO_2$
Fe(III) reduction $8H^+ + CH_2O + 4FeOOH \Rightarrow$ $4Fe^{2+} + CO_2 + 7H_2O$	$6H_2O + O_2 + 4Fe^{2+} \Rightarrow$ $4FeOOH + 8H^+$	$CH_2O + O_2 \Rightarrow H_2O + CO_2$
Sulfate reduction $H^+ + CH_2O + 1/2SO_4^{2-} \Rightarrow$ $CO_2 + 1/2H_2S + H_2O$	$O_2 + 1/2H_2S \Rightarrow 1/2SO_4^{2-} + H^+$	$CH_2O + O_2 \Rightarrow H_2O + CO_2$
Methanogenesis (from acetate) $2CH_2O \Rightarrow CO_2 + CH_4$	$CH_4 + O_2 \Rightarrow H_2O + CO_2$ or	$CH_2O + O_2 \Rightarrow H_2O + CO_2$
	Anaerobic methane oxidation: $CH_4 + SO_4^{2-} + 2H^+ = H_2S +$ $CO_2 + 2H_2O$ and	
	$2O_2 + H_2S = SO_4^{2-} + H^+$	$CH_2O + O_2 \Rightarrow H_2O + CO_2$

[a] The reduction reactions combined with the re-oxidation reactions have the net stoichiometry of O_2 respiration, with the exception of denitrification where the reduction product N_2 is not involved further in sedimentary redox processes. The amount of nitrate used in denitrification can be considered as approximately equal to the amount of oxygen used to oxidize ammonia formed during mineralization of organic matter via: $NH_4^+ + 2O_2 \rightarrow NO_3^- + H_2O + 2H^+$. Anaerobic methane oxidation is included because of its importance in carbon remineralization, but other anaerobic re-oxidation pathways are not (e.g. oxidation of Fe^{2+} by NO_3^-).

et al., 1996), inhibitors (Trimmer *et al.*, 1998) or other procedures (e.g. isotope pairing; Nielsen, 1992), or indirectly via measurements of NH_3 or CO_2 production (Canfield *et al.*, 1993a,b).

17.4 Mineralization of organic matter

The main pathways for mineralization of organic matter in marine sediments are shown in Fig. 17.4.

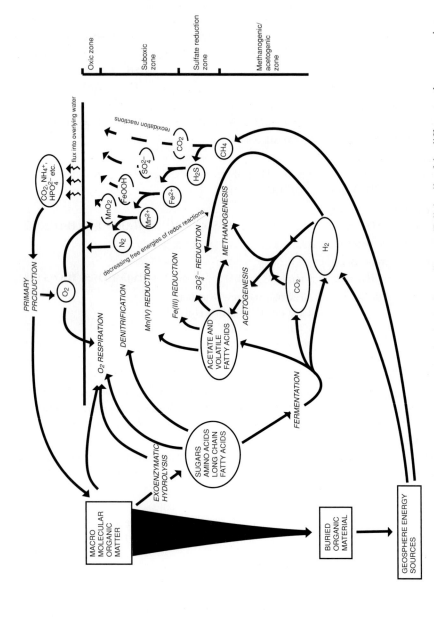

Fig. 17.4 Pathways of organic matter mineralization in marine sediments. Full details of the different pathways are given in the text.

17.4.1 Aerobic degradation

In the vast majority of seafloor environments, aerobic eukaryotes and prokaryotes are the first organisms to begin degradation of organic matter. The depth to which oxygen penetrates sediments (the oxic zone) can vary from nil in euxinic basins to several meters in organic-poor, open water sites (Jørgensen, 1982; Canfield, 1989). Because prokaryotes can assimilate only small organic molecules (<600 daltons; Weiss *et al.*, 1991), they must first break macromolecules into their composite monomers by the process of depolymerization, which they achieve by direct contact through enzymes associated with their cell membranes, or by excreting exoenzymes into their surroundings. The latter approach involves an initial investment of energy and carbon, for which the microorganisms are later compensated via energy gained from the oxidation of the labile, shorter-chained, organic molecules (Vetter *et al.*, 1998). Depolymerization generally is the rate-limiting step in the process of mineralization and it is controlled mainly by the lability of organic matter.

Intense aerobic respiration rapidly removes oxygen from sediment pore water, enabling anaerobic degradation to proceed, and potentially restricting aerobic mineralization to near-surface sediments were oxygen can be re-supplied from the overlying water column (Jørgensen, 1982; Revsbech *et al.*, 1980). Measurements of oxygen flux and oxidation of upward-diffusing reduced inorganic compounds have been used to estimate that aerobic respiration dominates organic mineralization in organic-poor, deep-sea sediments and is of either equal or lesser importance relative to anaerobic processes in ocean margin and coastal sediments (Jørgensen, 1982; Canfield, 1989; Wellsbury *et al.*, 1996).

17.4.2 Anaerobic degradation

Anaerobic degradation of organic matter typically drives sequential changes in pore-water chemistry which reflect the following sequence of microbial processes: denitrification–nitrification, Mn(IV)-oxide reduction, Fe(III)-oxide reduction, sulfate reduction, and methanogenesis (Figs. 17.4 and 17.5; Froelich *et al.*, 1979). This sequence of metabolic pathways coincides with decreasing free energies of the redox reactions (Fig. 17.2) and re-oxidation of reduced products, as well as the growth rates of the different groups of microorganisms. Hence bacteria possessing the most energy-yielding metabolism will dominate until their supply of electron acceptor becomes a limiting factor.

Complex interactions in anaerobic microbial ecosystems

Anaerobic bacteria (with the exception of denitrifiers which are facultatively aerobic bacteria) can only utilize a relatively narrow range of substrates compared to aerobic microorganisms. As a consequence, anaerobic bacteria are largely dependent upon fermenting bacteria to produce low molecular weight organic compounds, such as acetate, formate, propionate, lactate, various alcohols, H_2 and CO_2, from complex biopolymers. Further metabolism of fatty acids and alcohols by heterotrophic acetogenic bacteria yields additional H_2 and acetate (Fig. 17.4), making these compounds key intermediates in the anaerobic chain of decay.

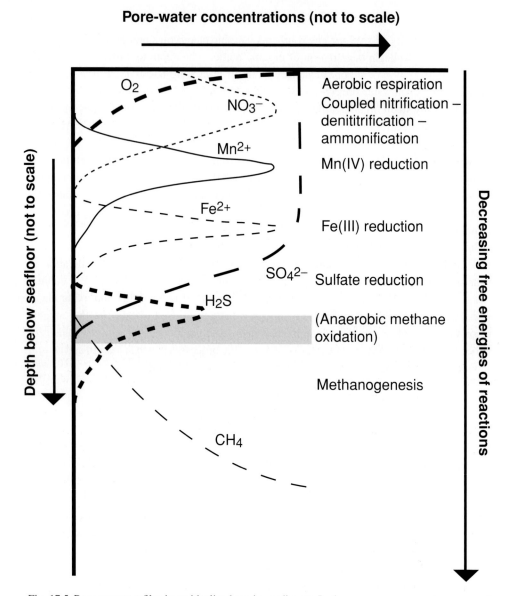

Fig. 17.5 Pore-water profiles in an idealized marine sediment. In the sequence, oxygen is consumed first, followed by nitrate, Mn(IV) oxides, Fe(III) oxides, sulfate, and finally CO_2, creating a zonation of reactants and products. The sequence is determined by the free-energy yield of the various redox reactions. Aerobic ammonia oxidation (nitrification) and anaerobic methane oxidation (shaded zone) are also shown. Depth is not to scale.

The importance of hydrogen and acetate in anaerobic biogeochemical cycling is reflected in their rapid turnover rates and low concentrations in the natural environment. Typically H_2 is present in nanomole quantities and acetate at micromole levels compared to millimole abundances for most electron acceptors (Lovley and Goodwin, 1988; Novelli *et al.*, 1988; McMahon and Chapelle, 1991; Hordijk *et al.*, 1994; Chapelle *et al.*, 1995). Therefore, in contrast to oxic environments where complete oxidation of macromolecular organic matter to CO_2 can be achieved by one type of microorganism, anaerobic environments require a complex community of physiologically different bacteria for complete mineralization of organic matter (Fig. 17.4).

H_2-uptake kinetics

Different groups of anaerobic respiratory bacteria (e.g. Fe(III)-oxide reducers, sulfate reducers, methanogens, acetogens) compete for fermentation products, in particular H_2 (e.g. Hoehler *et al.*, 1998). Bacteria possessing more energy efficient metabolisms tend to have faster H_2-uptake kinetics, which results in lower minimum H_2-threshold concentrations (i.e. the lowest H_2 concentration at which the bacteria can still metabolize). Consequently, under H_2-limited conditions these bacteria can maintain lower ambient concentrations of H_2 and effectively exclude other bacteria that have higher H_2 thresholds (Cord-Ruwisch *et al.*, 1988; Conrad and Wetter, 1990; Seitz *et al.*, 1990). Hence sulfate-reducing bacteria can maintain the pore-water concentration of H_2 at a level that is low enough to eliminate competition from methanogens (Lovley *et al.*, 1982; Lovley and Klug, 1983; Conrad *et al.*, 1986; Lovley and Goodwin, 1988). Similarly methanogens also can outcompete acetogens for available H_2 (Cord-Ruwisch *et al.*, 1988; Conrad and Wetter, 1990; Seitz *et al.*, 1990).

This theory is only applicable to bacterial growth within optimum physiological temperature ranges because outside these ranges, H_2 thresholds increase dramatically. For example, in a mixed culture study conducted at low H_2 partial pressures, a methanogen outcompeted an acetogen at temperatures greater than 10 °C because of its lower minimum H_2 threshold. However, at temperatures lower than 10 °C, the methanogen was outside its optimum physiological conditions, causing its H_2 minimum threshold to rise, allowing the acetogen to dominate (Kotsyurbento *et al.*, 2001). Under non-limiting concentrations of H_2, acetogens can also dominate because of their higher maximum rate of H_2 uptake (V_{max}) (Lovley and Goodwin, 1988; Kotsyurbenko *et al.*, 2001).

Another common example of a sequence of bacterial activity appearing contrary to expectations based upon relative free-energy changes is the occurrence of methanogenesis in open ocean sediments when high concentrations of sulfate are present (D'Hondt *et al.*, 2002). The reasons for this are still not fully understood, but may indicate the presence of methanogen specific substrates (Oremland *et al.*, 1982; although this suggestion is not always supported by radiotracer studies, e.g. Parkes *et al.*, 1990), that methanogens are better adapted physiologically to the particular environmental conditions (perhaps in close syntrophy with fermenters), or that methanogens are able to survive on lower energy yields than sulfate-reducers (Hoehler *et al.*, 2001). Even in sediments where no methane is evident in porewaters, radiotracer experiments indicate that methanogenesis can occur alongside sulfate reduction (Parkes *et al.*, 1990). The absence of CH_4 under such conditions probably reflects efficient anaerobic oxidation of CH_4 (Parkes *et al.*, 2000).

17.4.3 Denitrification/nitrate reduction

Denitrification is microbial reduction of NO_3^- to N_2, and proceeds in three stages: conversion of NO_3^- to NO_2^-, then to N_2O, and finally to N_2 (Seitzinger, 1998). Certain bacteria also are able to reduce NO_2^- a stage further to NH_4^+ in a process termed ammonification (Jørgensen and Sørensen, 1985; Sørensen, 1987). Denitrifying bacteria can use a diverse variety of organic substrates, ranging in complexity from monomers to humic compounds (Sørensen, 1987; Coates *et al.*, 2002).

Most studies of denitrification suggest that it accounts for only a minor portion (3–11%) of total mineralization of organic matter in continental margin sediments (Sørensen, 1979; Jørgensen, 1983; Canfield 1993; Middleburgh *et al.*, 1996). One exception is the report by Devol (1991) that denitrification accounted for up to 30% of benthic respiration in Washington coast sediments. However, studies by Luther *et al.* (1997) have demonstrated that reduced nitrogen compounds, such as ammonia, can be chemically oxidized directly to N_2 by MnO_2, bypassing bacterial denitrification and hence organic matter oxidation:

$$2NH_3 + 3MnO_2 + 6H^+ \rightarrow 3Mn^{2+} + N_2 + 6H_2O \tag{17.4}$$

This process could account for up to 90% of N_2 production in continental margin sediments, resulting in substantially lower estimates of the contribution of denitrification to organic matter degradation.

17.4.4 Mn(IV) and Fe(III) reduction

The importance of dissimilatory metal (Mn(IV) and Fe(III) oxides) reduction in organic mineralization has only relatively recently been established because of difficulties in separating purely chemical reactions from biologically mediated processes (Lovley, 1991). Fe(III) and Mn(IV) oxides are readily reduced abiotically by a wide range of compounds. In marine sediments, H_2S is of particular importance because of the significance of anaerobic sulfate reduction coupled to the formation of pyrite (FeS_2). Other known reductants include organic compounds, such as organic acids and phenolic acids, and quinones in humic complexes (Nealson and Saffarani, 1994).

Despite the difficulties of separating abiotic from biologically mediated reactions, studies involving pure cultures and natural populations indicate that Fe(III)-reducing microorganisms are responsible for a significant portion of Fe(III) reduction in sedimentary environments (Lovley and Philips, 1988; Lovley, 1991). In some cases, it has been demonstrated that quinones (present naturally in the environment, e.g. humic compounds or possibly compounds synthesized by bacteria) can act as electron shuttles between solid Fe(III) oxides and bacteria, which greatly enhance rates of Fe(III) reduction (Lovley *et al.*, 1998; Nevin and Lovley, 2002). Indeed in some circumstances, the humic compounds may serve as important electron acceptors with humic quinone groups having a redox potential intermediate between that of sulfate and Fe(III) oxides (Lovley *et al.*, 1996; Osterberg and Shirshova, 1997; Coates *et al.*, 2002; Cervantes *et al.*, 2000, 2002).

Because of methodological difficulties, only a few detailed studies have quantified the importance of bacterial dissimilatory metal reduction in the diagenesis of marine sediments (Aller, 1990; Canfield 1993a,b; Thamdrup and Canfield, 1996). These studies have demonstrated that in relatively shallow continental margin sediments, possessing high Fe and Mn concentrations, Fe(III) and Mn(IV) reduction individually can account for a large percentage of overall organic carbon oxidation in the upper 10 cm of sediment: 50–80% in near-coastal sediments between Denmark and Norway (Canfield *et al.*, 1993a,b), $\sim 80\%$ in near-shore Atlantic sediments (Aller, 1990), and up to 100% in the upper centimeter of Mn-enrich, hydrothermal sediments in near-shore Pacific sediments (Aller, 1990). The importance of metal reduction at these sites results from high metal concentrations combined with rapid recycling of Mn and Fe in the sub-oxic and oxic zones due to intense bioturbation (Canfield *et al.*, 1993a,b). In open ocean sediments, metal reduction is thought to be of minor importance because of lower metal concentrations and a lack of bioturbation (Canfield *et al.*, 1993a,b).

17.4.5 Sulfate reduction

Sulfate reduction is the most important anaerobic process in marine sediments because of the high concentration of the electron acceptor sulfate in seawater (~ 29 mM). Rates of sulfate reduction in marine sediments vary over six orders of magnitude, from a few picomole per cubic centimeter per day in very deep sediments to thousands of nonomole per cubic centimeter per day in near-surface coastal sediments (Canfield and Teske, 1996; Rudnicki *et al.*, 2001). The single most important control on rates of sulfate reduction is the lability of organic matter (Westrich and Berner, 1984). However, in marine environments sulfate concentrations below ~ 3 mM will begin to limit the activity of sulfate-reducing bacteria (Capone and Kiene, 1988). Rates of sulfate reduction in organic-poor, deep-sea sediments generally are low because the majority of labile carbon is utilized by aerobic processes before sulfate reduction can begin (Jørgensen, 1982). Consequently, sulfate is still often present in pore waters hundreds of meters below the seabed. In contrast, within continental margin sediments, all electron acceptors more energetically favorable than sulfate typically are fully consumed between a few centimeters and a few meters depth (Jørgensen, 1983; Canfield, 1989), which permits sulfate reduction to become the dominant process for anaerobic degradation of organic carbon. Because the concentration of sulfate in seawater is ~ 100 times greater than that of oxygen, the sulfate reduction zone correspondingly is two or three orders of magnitude thicker than the oxic zone (Jørgensen, 1983).

The main substrates for sulfate reduction are acetate, lactate, H_2, propionate, butyrate, valerate, and amino acids (Parkes *et al.*, 1989). Although acetate is the main substrate, accounting for up to 100% of sulfate reduction, an additional 17 different substrates are also used. Typically, the higher the rate of sulfate reduction in a sediment, the greater the diversity of substrate utilization (Parkes *et al.*, 1989). Sulfate reduction can account for up to 75% of all organic matter degradation in organic-rich, shallow water sediments (Wellsbury *et al.*, 1996), although this high percentage of anaerobic respiration usually decreases with increasing water depth (Jørgensen, 1982). Finally, a significant proportion of sulfate reduction is fuelled by anaerobic oxidation of methane rather than substrates derived from organic matter in sediments (Kasten and Jørgensen, 2000).

17.4.6 Anaerobic methane oxidation

The absence of methane within the sulfate reduction zone indicates that sub-surface methane oxidation by sulfate reduction is the primary means of CH_4 destruction within continental margin sediments (Fig. 17.5; Reeburgh, 1980; Devol and Ahmed 1981; Iversen and Blackburn, 1981; Devol, 1983; Iversen and Jørgensen, 1985; Niewohner *et al.*, 1998). It has been estimated that anaerobic methane oxidation globally accounts for 5–20% of total sulfate reduction at continental margins (Kasten and Jørgensen, 2000). The net equation for this process is:

$$CH_4 + SO_4^{2-} \rightarrow HCO_3^- + HS^- + H_2O \qquad (17.5)$$

Although the bacteria associated with this process have not yet been isolated (Niewohner *et al.*, 1998; Kasten and Jørgensen, 2000), a combination of genetic, biomarker, and rate process studies using radioactive and stable isotopic tracers indicate that the process of anaerobic methane oxidation requires a coupling of sulfate reduction and methanogenesis, involving both bacteria (*Desulfosarcina/Desulfococcus*) and methanogenic *Archaea* (closely related to *Methanosacrinales*), with *Archaea* directly oxidizing CH_4 (Devol and Ahmed, 1981; Devol, 1983; Alperin and Reeburgh, 1985; Hoehler *et al.*, 1994; Hansen *et al.*, 1998; Hinrichs *et al.*, 1999; Boetius *et al.*, 2000; Pancost *et al.*, 2000, 2001; Orphan *et al.*, 2001a,b).

17.4.7 Methanogensis and autotrophic acetogenesis

After depletion of sulfate from pore water, the dominant remaining electron acceptor is CO_2, with the possible exceptions of recalcitrant Fe(III) and Mn(IV)/Mn(III) minerals (Bottrell *et al.*, 2000) and humic compounds (Lovley *et al.*, 1998; Cervantes *et al.*, 2000, 2002). Carbon dioxide can be used as an electron acceptor by both methanogens and acetogens according to the net reactions:

$$4H_2 + H^+ + HCO_3^- \rightarrow CH_4 + 3H_2O \qquad (17.6)$$

$$4H_2 + H^+ + 2HCO_3^- \rightarrow CH_3COO^- + 4H_2O \qquad (17.7)$$

The terminal product of H_2 and CO_2 in sediments is CH_4 via methanogenesis because any acetate produced by (17.7) can be used subsequently as a substrate for methanogenesis (Fig. 17.4). The competition for common substrates, such as H_2, influences the flow of carbon and electrons, and subsequent rate of reaction, as well as the isotopic composition of the CH_4 (Hornibrook *et al.*, 1997, 2000; Avery *et al.*, 1999). Under typical substrate-limited conditions in deep sediments, it would be expected that methanogens should dominate because of their higher free-energy yields (Fig. 17.2) and consequently, lower minimum H_2 thresholds (Kotsyurbenko *et al.*, 2001).

Acetate formation and turnover in deeply buried sediments has been documented in several studies (Chapelle and Bradley, 1996; Parkes *et al.*, 2000), which suggests that microbial rather than abiotic processes dominate the formation of organic acids during low-temperature diagenesis. Other studies have combined acetate turnover rates with rates of methanogenesis (e.g. Wellsbury *et al.*, 2000, 2002). Site 1109 of ODP Leg 180 in the Woodlark extensional Basin, offshore from Papua New Guinea was drilled to over 800 mbsf. Site 1109 is situated at an intermediate location between continental margin sediments and typical open ocean sediments, with a water depth of 2211 m. Sediments have a low organic carbon content ($\sim$0.4% C) and an average thermal gradient (33 °C km^{-1}). A massive dolerite layer at the base of the sediment profile between 773 and 802 mbsf effectively seals the bottom of the sequence, preventing vertical diffusion of solutes from deeper layers from having any effect on bacterial activity (Wellsbury *et al.*, 2002); a feature common to all previous microbiological investigations of deep sediments (Parkes *et al.*, 2000). The presence of the dolerite layer also meant that for the first time a complete sediment column was sampled. The depth profiles of bacterial populations and activities at Site 1109 are shown in Fig. 17.6.

Similar to previous studies of sub-seafloor sediments (see references in Parkes *et al.*, 2000), maximum bacterial populations and activities were in the upper 20 m of the profile, with much lower rates (up to ×10,000) in deeper layers (Fig. 17.6). Highest rates of sulfate reduction were in the top 5 m of sediment, and generally close to zero below 87 m. A small peak in activity at the sulfate–methane interface at 87 m was indicative of anaerobic methane oxidation. Below 100 mbsf, methane started to accumulate, coincident with similar amounts of active methanogenesis from both acetate and $H_2 : CO_2$. Despite active rates of acetate methanogenesis, acetate concentrations remained relatively constant ($<$10 μm), demonstrating that deep formation of acetate was occurring, presumably by fermenters and/or heterotrophic or autotrophic acetogens, which were present in the sediment, even in the deeper layers (Fig. 17.6*b*). These results show that acetate is an important substrate for methanogenesis in deep sediments, in contrast to previous δ^{13}C studies which suggest that $H_2 : CO_2$ is the major substrate for methanogenesis in marine environments (Whiticar, 1999). Surprisingly, depth integration of data showed that the majority of bacterial cells and activity (with the exception of sulfate reduction) occurred in the sub-surface ($>$20 mbsf; Table 17.2).

At oceanic margin sites, flux calculations based on porewater profiles indicate that the majority of sulfate reduction below 1.5 m depth is used to oxidize methane produced in

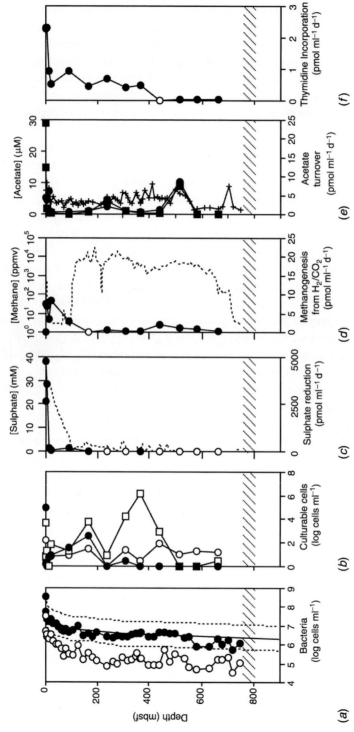

Fig. 17.6 Depth profiles of bacterial populations and activities in sediments at Site 1109, ODP Leg 180, Woodlark Basin. (*a*) Total bacterial populations (solid circles) and dividing and divided cells (open circles). (*b*) Culturable cells from MPN enrichments; heterotrophic acetogens (open circles), autotrophic acetogens (closed circles) and fermentative heterotrophs (open squares). (*c*) Sulfate reduction rates (solid circles) and pore-water sulfate (dashed line). (*d*) Methanogenesis from H_2 : CO_2 (solid circles) and *in situ* methane (dashed line). (*e*) Acetate metabolism to CO_2 (solid circles) and CH_4 (solid squares), and pore-water acetate (+). (*f*) Thymidine incorporation into bacterial DNA – a measure of bacterial growth (solid circles). Hollow symbols denote zero values and /// represents the dolerite layer (from Wellsbury *et al.*, 2002).

Table 17.2 *Relative percentages of depth-integrated bacterial activities occurring in three depth zones of deep marine sediments in the Woodlark Basin Site 1109 (from Wellsbury et al., 2002)*

	Sulfate reduction	Methanogenesis from $CO_2 : H_2$	Methanogenesis from acetate	Acetate oxidation to CO_2	Thymidine incorporation[a]	Bacterial populations
Upper 20 m	12	12	5	7	11	21
20–500 m	34	76	57	55	87	68
500 m-bottom	0.02	12	38	38	2	11

[a]Thymidine incorporation is a measure of bacterial growth.

underlying sediments (D'Hondt *et al.*, 2002). However, under specific conditions some of the methane can be incorporated into solid gas hydrates. Gas hydrates are composed of water molecules in the form of a rigid lattice of cages that contain molecules of natural gas, consisting mainly of methane (Kvenvolden, 1988). The formation of gas hydrates is controlled by a combination of low temperature, high pressure, and a suitable supply and composition of gas. In marine sediments, these conditions are met at 2000 mbsf in polar region continental shelves and the outer continental margins of deep oceanic regions, where cold bottom water is present and there is sufficient CH_4 present (Kvenvolden, 1988). The amount of carbon in hydrates, often largely composed of biogenic methane (Kvenvolden, 1988), is estimated to exceed the total carbon considerably in all known oil, gas, and coal deposits (Kvenvolden, 1988; Dickens *et al.*, 1996), and thus, may be of future importance as an energy resource. However, because of their potential instability, gas hydrates are a major concern for climate change and the atmospheric budget of greenhouse gases (Suess *et al.*, 1999; Dickens *et al.*, 2001).

Recent microbiological studies have shown that hydrate-bearing sediments contain a unique deep bacterial habitat, in which deep acetate formation below the gas hydrate zone provides substrates for methanogens (Wellsbury *et al.*, 2000; Fig. 17.7). Maximum rates of methane production from acetate are similar to those of $H_2 : CO_2$ methanogensis, but because of increased acetate concentrations with depth (up to 15 mM), acetate methanogenesis dominates in deeper sediment layers (Wellsbury *et al.*, 2000). Significant bacterial populations and activities have been found in other deep sediment gas-hydrate deposits (Cragg *et al.*, 1996; Marchesi *et al.*, 2001; Reed *et al.*, 2002) and given the global significance of gas hydrates as an organic carbon reservoir, they may represent a globally important deep biosphere habitat.

17.4.8 Substrates in the deep biosphere

Research by Wellsbury *et al.* (1997) has demonstrated that the elevated sub-surface bacterial activities in Blake Ridge sediments, Atlantic Ocean (Fig. 17.7), and lower but significant

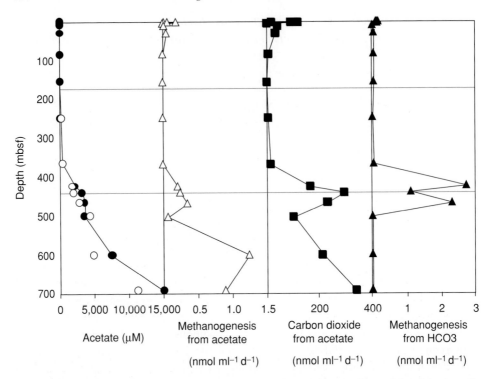

Fig. 17.7 Depth profiles of pore-water acetate concentrations and rates of bacterial activity in sediments from Blake Ridge, ODP Leg 164, Site 995. Dotted lines indicate the upper and lower boundaries of the gas hydrate stability field. Bacterial activities are greatly elevated at the base of the hydrate zone (modified from Wellsbury *et al.*, 1997).

activities in Woodlark Basin sediments (Fig. 17.6), can be explained, in part, by thermal gradients in the sub-surface. Organic material, which is refractory at low surface temperatures, can become significantly more labile upon heating with burial, leading to the formation of acetate and other acids (Wellsbury *et al.*, 1997; Fig. 17.8). In other regions, sub-surface flow can be important in sustaining bacterial activity deep in sediments.

In very deep sediments and those with steep thermal gradients, temperatures eventually become too high to support even bacterial life. Thermogenic formation of oil in sediment is believed to occur abiotically within the so-called "oil window" between 100 and 150 °C (Quigley and Mackenzie, 1988). However, bacteria can grow up to 113 °C (Blochl *et al.*, 1997) and possibly higher temperatures (Stetter, in press), and, consequently, they may play a role during the early stages of oil formation. Consistent with this suggestion, is the presence of high-temperature bacteria in hot oil reservoirs (Stetter *et al.*, 1993). Bacteria are known to be able to oxidize anaerobically a variety of short-chain hydrocarbons (alkanes, alkenes, and alkylbenzenes – see Heider *et al.*, 1998; Wilkes, 2000), and the number of hydrocarbons identified that can be bacterially oxidized is continually increasing. A complex bacterial community that can oxidize hexadecane

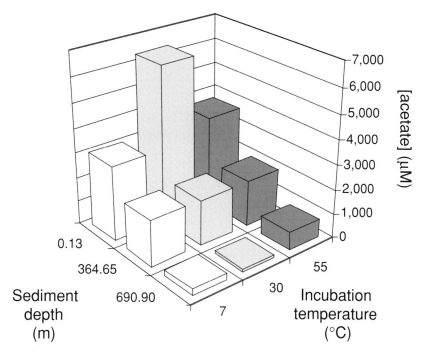

Fig. 17.8 Experimental generation of acetate from deep-sea sediment samples subjected to low-temperature heating. Sediment was collected from Blake Ridge, ODP Leg 164, Site 995. Concentrations of acetate produced in these experiments matched *in situ* concentrations, supporting the hypothesis that burial heating is stimulating acetate generation (from Wellsbury *et al.*, 1997).

($C_{16}H_{34}$) under strictly anaerobic conditions has recently been isolated (Zengler *et al.*, 1999). The end product of the degradation is CH_4, which reinforces previous suggestions that a proportion of CH_4 in petroleum reservoirs may be of biogenic origin (see Parkes, 1999). The presence and activity of bacteria in deep and ancient marine deposits is consistent with recent results from other deep environments (Parkes and Wellsbury, 2003).

17.5 The re-oxidation of reduced products from anaerobic respiration

The majority of reduced products from anaerobic respiration do not remain in sediment, but instead diffuse upwards to be eventually oxidized, directly or indirectly, by oxygen (Fig. 17.4). With the exception of anaerobic methane oxidation (and possibly anaerobic pyrite oxidation; Bottrell *et al.*, 2000), the majority of re-oxidation reactions in marine sediments are carried out in the shallow ($\sim$0–1 mbsf) oxic and sub-oxic zones, where bioturbation (advection of solid particles by macrofauna) and bioirrigation (solute exchange between bottom seawater and tubes in which macrofauna pump water) are active

(Aller, 1988; Glud *et al.*, 1994). Within this zone, exists a complex network of closely coupled oxidation and reduction cycles that are described below.

17.5.1 Nitrogen

The oxidation of reduced nitrogen compounds (NH_3, NO_2^-) to NO_3^- (nitrification) is an important process because it is the dominant source of nitrate for denitrification (Middleburg *et al.*, 1996) and a source of NO_3^- flux to seawater (Jørgensen, 1983). Microbial nitrification takes place in a series of steps via different groups of obligately aerobic bacteria in the family Nitrobacteraceae (Seitzinger, 1998). However, as mentioned previously, abiological NH_3 oxidation to N_2 by MnO_2 and reduction of NO_3^- to N_2 by Mn(II) may "short-circuit" the biologically catalysed nitrification–denitrification couple (Luther *et al.*, 1997). Under some circumstances ammonia can be anaerobically oxidized by nitrate with the production of N_2 (van der Graaf *et al.*, 1995; Strous *et al.*, 1999).

17.5.2 Manganese and iron

Rapid bacterial or abiotic oxidation of Mn(II) and Fe(II) by oxygen (Hanert, 1981; Juniper and Tebo, 1995; Straub *et al.*, 1996; Luther *et al.*, 1997) in bioturbated sediments regenerates reactive Mn(IV) and Fe(III) oxides, which are then available to serve as electron acceptors for organic matter mineralization. This process can result in continuous recycling of Mn and Fe in marine sediments, greatly increasing their relative importance in organic matter respiration and, in turn, diminishing the importance of aerobic respiration and sulfate reduction in sediments (Canfield *et al.*, 1993a; Thamdrup and Canfield, 1996). In addition, Fe(II) can chemically reduce MnO_2, such that chemical rather than dissimilatory reduction of Mn(IV) is thought to be important in some sediments (Canfield *et al.*, 1993a). These abiotic and dissimilatory interactions are illustrated in Fig. 17.9.

Iron redox chemistry also is very important in the cycling of other biologically important elements, in particular, the nutrient phosphorus because iron oxides and oxyhydroxides have a high adsorption capacity for anions at pH levels typical of seawater and marine sediment pore waters (Goldberg and Sposito, 1984). Consequently, the reduction and oxidation of Fe in marine environments is thought to drive the sedimentary phosphate cycle (Krom and Berner, 1980; Slomp *et al.*, 1996a,b). However, bacteria are also directly involved in the phosphate cycle, for example, via the oxidation of phosphite coupled to sulfate reduction (Schink and Friedrich, 2000; Schink *et al.*, 2002).

17.5.3 Sulfur

Redox reactions involving sulfur in sediments are of particular importance because of the significance of sulfate reduction and H_2S formation during diagenesis. Moreover, sulfur

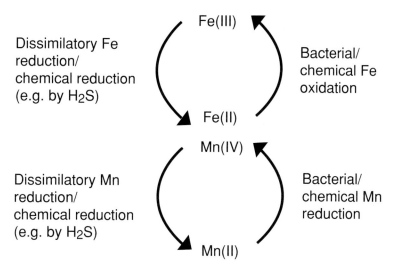

Fig. 17.9 Interactions of dissimilatory Fe/Mn reduction and abiotic reaction of Fe(II) with Mn(IV).

possesses a wide range of oxidation states (-2 to $+6$) under common environmental conditions. A schematic representation of the sulfur redox cycle is shown in Fig. 17.10.

Typically, $\sim$90% of H_2S produced by sulfate reduction is reoxidized, some by O_2, but a large proportion under anaerobic conditions by Fe(III) and Mn(IV) oxides (Jørgensen and Bak, 1991). Although Mn(IV) oxides can oxidize H_2S directly to SO_4^{2-}, the main products of anaerobic oxidation of H_2S tend to be elemental sulfur, thiosulfate, and polysulfides (Schippers and Jørgensen, 2001). These intermediate sulfur compounds may then be either (i) reduced back to H_2S by sulfate/sulfur reducing bacteria, (ii) chemically or bacterially oxidized via $O_2/MnO_2/NO_3^-$, (iii) bacterially disproportionated (Bak and Pfennig, 1987; Thamdrup *et al.*, 1993), or (iv) precipitated as FeS or FeS_2 by reaction with Fe^{2+} or iron oxides and oxyhydroxides.

Bacterial disproportionation is a form of inorganic fermentation in which a sulfur compound of intermediate oxidation state, such as elemental sulfur or thiosulphate, is partly oxidized to sulfate, and partly reduced to sulfide, with no net change in overall oxidation state (Eqs. (17.8) and (17.9)).

$$S_2O_3^{2-} + H_2O \rightarrow SO_4^{2-} + HS^- \tag{17.8}$$
$$4S° + 4H_2O \rightarrow SO_4^{2-} + 3H_2S + 2H^+ \tag{17.9}$$

The presence of Fe(III) and Mn(IV) oxides greatly enhances the free-energy yields of these reactions because they act as sulfide buffers (Thamdrup *et al.*, 1993; Lovley and Philips, 1994; Finster *et al.*, 1998). Furthermore, bacteria capable of lithoautotrophic growth via the complete oxidation of thiosulfate to sulfate using MnO_2 as an electron acceptor, but not FeOOH, have been enriched (Madrid, 2000; reported in Taylor *et al.*, 2001). Radiotracer experiments have established that the disproportionation of sulfur compounds rather than

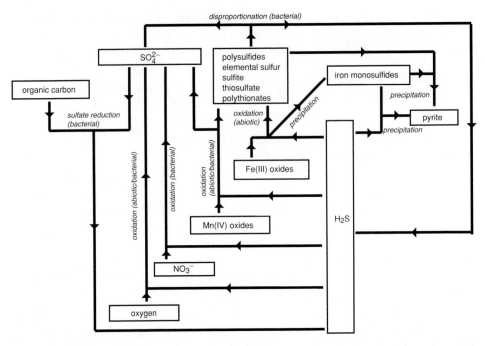

Fig. 17.10 Schematic representation of the sedimentary sulfur cycle. Note that interactions with dissolved metals (e.g. Mn^{2+}, Fe^{2+}) and oxidation reactions of monosulfides and pyrite are not shown for clarity.

net oxidation or reduction can dominate in bioturbated marine sediments, with thiosulfate as the main intermediate product (Jørgensen and Bak, 1991). This "thiosulfate shunt" is therefore potentially important for regulating electron flow in sediments (Jørgensen and Bak, 1991).

Although the majority of H_2S is reoxidized in marine sediments, up to 10% reacts with reactive Fe species, precipitating as iron sulfides, generally amorphous FeS (Berner, 1970; Eq. (17.10)). The amorphous FeS is converted over time to pyrite (FeS_2) by reaction with elemental sulfur or polysulfides (Eq. (17.11); Fig. 17.10; Berner, 1970). It has also been proposed that in certain environments, such as salt marshes, pyrite may precipitate directly from solution (Drobner *et al.*, 1990; Rickard and Luther, 1997; Eq. (17.12)).

$$H_2S + Fe^{2+} \rightarrow FeS + 2H^+ \tag{17.10}$$

$$FeS + S \rightarrow FeS_2 \tag{17.11}$$

$$FeS + H_2S \rightarrow FeS_2 + H_2 \tag{17.12}$$

Recently, it has been determined that even solid iron sulfides can be oxidized within anoxic marine sediments. Both FeS and FeS_2 can be oxidized abiotically by MnO_2 at near neutral pH, but not by nitrate. Evidence exists that Fe(III) oxides also can oxidize pyrite in deep sediments on geological time scales (Bottrell *et al.*, 2000). In the case of MnO_2, the actual oxidant is believed to be Fe(III) cyclically reduced by the pyrite surface and reoxidized by

MnO$_2$ (Schippers and Jørgensen, 2001). Oxidation of FeS can be coupled to the reduction of nitrate by the bacterium *Thiobacillus denitrificans* (Garcia-Gil and Golterman, 1993) and by Fe(II)-oxidizing nitrate-reducing bacteria (Straub *et al.*, 1996; Benz *et al.*, 1998).

17.6 Influence of advective flow

17.6.1 Divergent tectonic regions

It has been estimated that 25–30% of the Earth's total heat flux is transferred from the lithosphere to the hydrosphere by the circulation of seawater through oceanic spreading centers (Stein and Stein, 1994; Lowell, 1991). The most dramatic manifestation of this process are the high-temperature (250–400 °C) hydrothermal vents (Karl, 1995); however, off-axial diffuse low-temperature discharge is thought to be quantitatively more important, accounting for 70–80% of total heat loss at oceanic ridges (Stein and Stein, 1994). Off-axial discharge provides electron donors (labile dissolved organic matter) and electron acceptors (primarily seawater-derived sulfate) into deep sediments, which tend to be depleted of these compounds. A good example is Site 1027 of ODP Leg 168 along the eastern flank of the Juan de Fuca Ridge (Fig. 17.11) where seawater sulfate is absent from sediments 200 mbsf because of sulfate reduction; however, a fresh influx of sulfate enters sediments below 500 mbsf driven by thermal advection. The increase in sulfate abundance appears to stimulate bacterial activity as shown by an increase in bacterial cells and a marked decrease in CH$_4$ concentrations, which is indicative of anaerobic methane oxidation (Mather and Parkes, 2000).

17.6.2 Convergent tectonic regions

Low-temperature venting of fluid at convergent plate margins is an important global process which involves squeezing of fluids from thick sequences of organic-rich sediments trapped in accretionary wedges along subduction trenches. High concentrations of CH$_4$ and high rates of flow result in seabed cold seeps being able to support dense microbial communities and very high rates of metabolic activity. Estimates of fluid flux at subduction zones suggest that the volume of water in the world's oceans is recycled every 500 Ma, compared to 5–10 Ma along mid-ocean ridges (von Huene *et al.*, 1998; Wallman *et al.*, 1997; Suess *et al.*, 1998). Pressure differences drive advective flow, transporting CH$_4$ along faults and fractures, which provides abundant CH$_4$ for oxidation supporting the high metabolic rates and dense microbial communities at or near the sediment surface (Suess *et al.*, 1999; Boetius *et al.*, 2000; Tryon and Brown, 2001), and possibly at depth in sediments (Cragg *et al.*, 1996). Rates of CH$_4$ oxidation in these advective systems can be several orders of magnitude greater than within diffusive systems, such as Blake Ridge.

17.7 Summary

Research over the last two decades has demonstrated the importance of anaerobic processes in the mineralization of organic matter and the presence of active bacteria in the

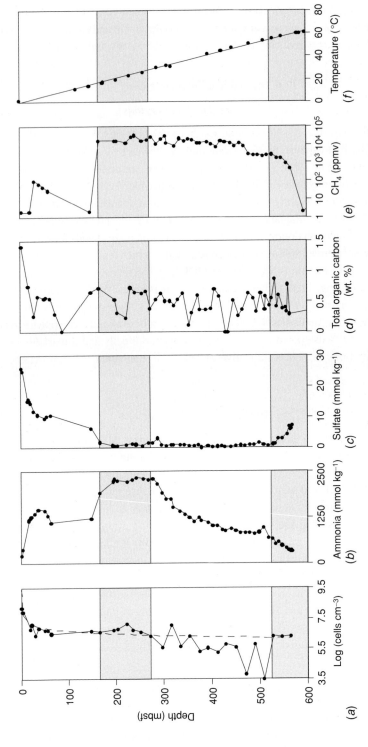

Fig. 17.11 Geochemical and bacterial depth profiles to ~565 mbsf at ODP Leg 168, Eastern Flank of the Juan de Fuca Ridge, Site 1027. (a) Total bacterial count. (b) Ammonia. (c) Sulfate. (d) Total organic carbon. (e) Methane. (f) Temperature. Horizontal lines and shading mark the regions of elevated bacterial populations (from Mather and Parkes, 2000).

deep sub-surface of marine environments. Anaerobic communities consist of complex syntrophic assemblages of fermentive and respiratory bacteria, which enable energy efficient metabolism of a diverse range of organic and inorganic compounds. Re-oxidation of the reduced products of anaerobic microbial activity, even under anaerobic conditions, is an important component of biogeochemical cycling in marine sediments, and new microbial pathways continue to be discovered.

Although bacterial communities in deep sediments generally metabolize organic matter at rates several orders of magnitude lower than surface communities, these rates can be extremely important when integrated across the enormous volume of sediment they inhabit and over geological time scales. Significant bacterial activity in deep sediments can be sustained and even stimulated by heating during burial because of re-activation of recalcitrant organic matter. In addition, in geologically active regions, thermal- or pressure-driven advective systems can greatly enhance rates of bacterial activity in deep sediments. Hence, microbial reactions in marine sediments play an important role in enhancing the flux of material between the biological and geological carbon cycles.

References

Aller, R. C. 1988. Benthic fauna and biogeochemical processes in marine sediments: the role of burrow structures. In *Nitrogen Cycling in Coastal Marine Environments*, eds. T. H. Blackburn and J. Sorensen. Chichester: Wiley, pp. 301–338.
  1990. Bioturbation and manganese cycling in hemipelagic sediments. *Phil. Trans. Roy. Soc. Lond. A* **331**: 51–68.
Alperin, M. J. and Reeburgh, W. S. 1985. Inhibition experiments on anaerobic methane oxidation. *Appl. Environ. Microbiol.* **50**: 940–945.
Avery, G. B., Jr., Shannon, R. D., White, J. R., Martens, C. S., and Alperin, M. J. 1999. Effect of seasonal changes in the pathways of methanogenesis on the $\delta^{13}C$ values of pore water methane in a Michigan peatland. *Global Biogeochem. Cycles* **13**: 475–484.
Bak, F. and Pfennig, N. 1987. Chemolithotrophic growth of *Desulfovibrio sulfdismutans* sp. nov. by disproportionation of inorganic sulfur compounds. *Arch. Microbiol.* **147**: 184–189.
Benz, M., Brune, A., and Schink, B. 1998. Anaerobic and aerobic oxidation of ferrous iron at neutral pH by chemoheterotrophic nitrate-reducing bacteria. *Arch. Microbiol.* **169**: 159–165
Berner, R. A. 1970. Sedimentary pyrite formation. *Am. J. Sci.* **268**: 1–23.
Blochl, E., Rachel, R., Burggraf, S., Hafenbradl, D., Jannasch, H. W., and Stetter, K. O. 1997. *Pyrolobus fumarii*, gen. and sp. nov., represents a novel group of Archaea, extending the upper limit for life to 113 degrees C. *Extermophiles* **1**: 14–21.
Boetius, A., Ravensclag, K., Schubert, C. J., Rickert, D., Widdel, F., Gieseke, A., Amann, R., Jorgensen, B. B., Witte, U., and Pfannkuche, O. 2000. A marine microbial consortium apparently mediating anaerobic oxidation of methane. *Nature* **407**: 623–626.
Bottrell, S. H., Parkes, R. J., Cragg, B. A., and Raiswell, R. 2000. Isotopic evidence for anoxic pyrite oxidation and stimulation of bacterial sulphate reduction in marine sediments. *J. Geol. Soc. Lond.* **157**: 711–714.

Canfield, D. E. 1989. Reactive iron in marine sediments. *Geochim. Cosmochim. Acta* **53**: 619–632.

Canfield, D. E. and Teske, A. 1996. Late Proterozoic rise in atmospheric oxygen concentration inferred from phylogenetic and sulphur-isotope studies. *Nature* **382**: 127–132.

Canfield, D. E., Thamdrup, B., and Hanse, J. W. 1993a. The anaerobic degradation of organic matter in Danish coastal sediments: iron reduction, manganese reduction, and sulfate reduction. *Geochim. Cosmochim. Acta* **57**: 3,867–3,883.

Canfield, D. E., Jorgensen, B. B., Fossing, H., Glud, R., Gundersen, J., Ramsing, N. B., Thamdrup, B., Hansen, J. W., Nielsen, L. P., and Hall, P. O. J. 1993b. Pathways of organic carbon oxidation in three continental margin sediments. *Mar. Geol.* **113**: 27–40.

Capone, D. G. and Kiene, R. P. 1988. Comparison of microbial dynamics in marine and freshwater sediments: contrasts in anaerobic carbon catabolism. *Limnol. Oceanogr.* **33**: 725–749.

Cervantes, F. J., van der Velde, S., Lettinga, G., and Field, J. A. 2000. Competition between methanogenesis and quinone respiration for ecologically important substrates in anaerobic consortia. *FEMS Microbiol. Ecol.* **34**: 161–171.

Cervantes, F. J., de Bok, A. M., Duong-Dac, T., Stams, A. J. M., Lettinga, G., and Field, J. A. 2002. Reduction of humic substances by halorespiring, sulphate-reducing and methanogenic microorganisms. *Environ. Microbiol.* **4**: 51–57.

Chapelle, F. H. and Bradley, P. M. 1996. Microbial acetogenesis as a source of organic acids in ancient Atlantic Coastal Plain sediments. *Geology* **24**: 925–928.

Chapelle, F. H., McMahon, P. B., Dubrovsky, N. M., Fujii, R. F., Oaksford, E. T., and Vroblesky, D. A. 1995. Deducing the distribution of terminal electron-accepting processes in hydrologically diverse groundwater systems. *Water Resources Res.* **31**: 359–371.

Coates, J. D., Cole, K. A., Chakraborty, R., O'Connor, S. M., and Achenbach, L. A. 2002. Diversity and ubiquity of bacteria capable of utilizing humic substances as electron donors for anaerobic respiration. *Appl. Environ. Microbiol.* **68**: 2,445–2,452.

Conrad, R. and Wetter, B. 1990. Influence of temperature on energetics of hydrogen metabolism in homoacetogenic, methanogenic, and other anaerobic bacteria. *Arch. Microbiol.* **155**: 94–98.

Conrad, R., Schink, B., and Phelps, T. J. 1986. Thermodynamics of H_2-consuming and H_2-producing metabolic reactions in diverse methanogenic environments under in situ conditions. *FEMS Microbio. Ecol.* **38**: 353–360.

Cord-Ruwisch, R., Seitz, H. J., and Conrad, R. 1988. The capacity of hydrogenotrophic anaerobic bacteria to compete for traces of hydrogen depends on the redox potential of the terminal electron acceptor. *Arch. Microbiol.* **149**: 350–357.

Cragg, B. A., Parkes, R. J., Fry, J. C., Weightman, A. J., Rochelle, P. A., and Maxwell, J. R. 1996. Bacterial populations and processes in sediments containing gas hydrates (ODP Leg 146: Cascadia margin). *Earth Planet. Sci. Lett.* **139**: 497–507.

D'Hondt, S., Rutherford, S., and Spivack, A. J. 2002. Metabolic activity of subsurface life in deep-sea sediments. *Science* **295**: 2,067–2,070.

Devol, A. H. 1983. Methane oxidation rates in the anaerobic sediments of Saanich Inlet. *Limnol. Oceanogr.* **28**: 738–742.

Devol, A. H. and Ahmed, S. I. 1981. Are high rates of sulfate reduction associated with anaerobic oxidation of methane? *Nature* **291**: 407–408.

Dickens, G. 2001. On the fate of past gas: what happens to methane released from a bacterially mediated gas hydrate capacitor? *Geochem. Geophys. Geosys.* **2**: 2000GC000131.

Dickens, G. R., Paull, C. K., Wallace, P., and ODP Leg 164 Scientific Party 1996. Direct measurement of in situ methane quantities in a large gas-hydrate reservoir. *Nature* **385**: 426–428.

Dixon, J. L. and Turley, C. M. 2000. The effect of water depth on bacterial numbers, thymidine incorporation rates and C : N ratios in northeast Atlantic surficial sediments. *Hydrobiol.* **440**: 217–225.

Drobner, E., Huber, H., Wachtershauser, G., Rose, D., and Stetter, K. 1990. Pyrite formation linked with hydrogen evolution under anaerobic conditions. *Nature* **346**: 742–744.

Finster, K., Liesack, W., and Thamdrup, B. 1998. Elemental sulfur and thiosulfate disproportionation by *Desulfocapsa sulfoexigens* sp nov, a new anaerobic bacterium isolated from marine surface sediment. *Appl. Environ. Microbiol.* **64**: 119–125.

Fowler, C. M. R. 1990. *The Solid Earth, An Introduction to Global Geophysics.* Cambridge: Cambridge University Press.

Froelich, P. N., Klinkhammer, G. P., Bender, M. L., Luedtke, N. A., Heath, G. R., Cullen, D., Dauphin, P., Hammond, D., Hartman, B., and Maynard, V. 1979. Early oxidation of organic matter in pelagic sediments of the eastern equatorial Atlantic: suboxic diagenesis. *Geochim. Cosmochim. Acta* **43**: 1,075–1,090.

Garcia-Gil, L. J. and Golterman, H. L. 1993. Kinetics of FeS-mediated denitrification in sediments from the Camargue (Rhône delta, southern France). *FEMS Microbiol. Ecol.* **13**: 85–92.

Glud, R. N., Gundersen, J. K., Jorgensen, B. B., Revsbech, N. P., and Schultz, H. D. 1994. Diffusive and total oxygen uptake of deep-sea sediments in the eastern South Atlantic Ocean: *in situ* and laboratory measurements. *Deep-Sea Res.* **41**: 1,767–1,788.

Goldberg, E. D. and Sposito, G. 1984. A chemical model of phosphate adsorption by soils. I. Reference oxide minerals. *Soil Sci. Amer. J.* **48**: 772–778.

Gundersen, J. K. and Jørgensen, B. B. 1990. Microstructure of the diffusive boundary layer and the oxygen uptake of the sea floor. *Nature* **345**: 604–607.

Hanert, H. H. 1981. The Genus *Gallionella*. In *The Prokaryotes*, eds. M. P. Star, H. Stolp, H. H.Truper, A. Balows, and H. G. Schelger. Berlin: Springer-Verlag p. 509.

Hansen, L. B., Finster, K., Fossing, H., and Iversen, N. 1998. Anaerobic methane oxidation in sulfate depleted sediments: effects of sulfate and molybdate additions. *Aquat. Microb. Ecol.* **14**: 195–204.

Hedges, J. I. and Keil, R. G. 1995. Marine chemistry discussion paper. Sedimentary organic matter preservation: an assessment and speculative synthesis. *Mar. Chem.* **4**: 81–115.

Heider, J., Spormann, A. M., Beller, H. R., and Widdell, F. 1999. Anaerobic bacterial metabolism of hydrocarbons. *FEMS Microbiol. Rev.* **22**: 459–473.

Henrichs, S. M. and Reeburgh, W. S. 1987. Anaerobic mineralization of marine sediment organic matter: rates and the role of anaerobic processes in the oceanic carbon economy. *Geomicrobiol. J.* **5**: 191–237.

Hinrichs, K.-U., Hayes, J. M., Sylva, S. P., Brewer, P. G., and DeLong, E. F. 1999. Methane-consuming archaebacteria in marine sediments. *Nature* **398**: 802–805.

Hoehler, T. M., Alperin, M. J., Albert, D. B., and Martens, C. S. 1994. Field and laboratory studies of methane oxidation in an anoxic marine sediment – evidence for a methanogen–sulfate reducer consortium. *Global Biogeochem. Cycles* **8**: 451–463.

1998. Thermodynamic control on hydrogen concentrations in anoxic sediments. *Geochim. Cosmochim. Acta* **62**: 1,745–1,756.

2001. Apparent minimum free energy requirements for methanogenic Archaea and sulfate-reducing bacteria in an anoxic marine sediment. *FEMS Microbiol. Ecol.* **38**: 33–41.

Hordijk, C. A., Kamminga, H., and Cappenberg, T. E. 1994. Kinetic studies of acetate in freshwater sediments: use of stable isotope tracers. *Geochim. Cosmochim. Acta* **58**: 683–694.

Hornibrook, E. R. C., Longstaffe, F. J., and Fyfe, W. S. 1997. Spatial distribution of microbial methane production pathways in temperate zone wetland soils: stable carbon and hydrogen isotope evidence. *Geochim. Cosmochim. Acta* **61**: 745–753.

2000. Evolution of stable carbon isotope compositions for methane and carbon dioxide in freshwater wetlands and other anaerobic environments. *Geochim. Cosmochim. Acta* **64**: 1,013–1,027.

Iversen, N. and Blackburn, T. H. 1981. Seasonal rates of methane oxidation in anoxic marine sediments. *Appl. Environ. Microbiol.* **41**: 1,295–1,300.

Iversen, N. and Jørgensen, B. B. 1985. Anaerobic methane oxidation rates at the sulfate–methane transition in marine sediments from Kattegat and Skagerrak (Denmark). *Limnol. Oceanogr.* **30**: 944–955.

Jackson, B. E. and McInerney, M. J. 2002. Anaerobic microbial metabolism can proceed close to thermodynamic limits. *Nature* **415**: 454–456.

Jahnke, R. A., Reimers, C. E., and Craven, D. B. 1990. Intensification of recycling of organic matter at the seafloor near oceanic margins. *Nature* **348**: 50–54.

Jahnke, R. A., Craven, D. B., McCorkle, D. C., and Reimers, C. E. 1997. $CaCO_3$ dissolution in California continental sediments: the influence of organic matter remineralization. *Geochim. Cosmochim. Acta* **61**: 3,587–3,604.

Jørgensen, B. B. 1982. Mineralization of organic matter in the sea-bed: the role of sulphate reduction. *Nature* **296**: 643–645.

1983. Processes at the sediment–water interface. In *The Major Biogeochemical Cycles and their Interactions*, eds. B. Bolin and R. B. Cook. New York: Wiley, pp. 477–509.

Jørgensen, B. B. and Bak, F. 1991. Pathways and microbiology of thiosulphate transformations and sulphate reduction in a marine sediment (Kattegat, Denmark). *Appl. Environ. Microbiol.* **57**: 847–856.

Jørgensen, B. B. and Sørensen, J. 1985. Seasonal cycles of O_2, NO_3^- and SO_4^{2-} reduction in estuarine sediments: the significance of an NO_3^- reduction maximum in spring. *Mar. Ecol. Prog. Ser.* **24**: 65–74.

Juniper, S. K. and Tebo, B. M. 1995. Microbe–metal interactions and mineral deposition at hydrothermal vents, In *The Microbiology of Deep-Sea Hydrothermal Vents*, ed. D. M. Karl. Boca Raton, FL: CRC Press, pp. 219–253.

Karl, D. M. 1995. The ecology of free-living bacteria at deep-sea hydrothermal vents, In *The Microbiology of Deep-Sea Hydrothermal Vents*, ed. D. M. Karl. Boca Raton, FL: CRC Press, pp. 35–124.

Kasten, S. and Jørgensen, B. B. 2000. Sulfate reduction in marine sediments. In *Marine Geochemistry*, eds. H. D. Schultz and M. Zabel. New York: Springer-Verlag, pp. 263–281.

Keil, R. G., Montlucon, D. B., Prahl, F. G., and Hedges, J. I. 1994. Sorptive preservation of labile organic matter in marine sediments. *Nature* **370**: 549–552.

Klimant, I., Meyer, V., and Kuhl, M. 1995. Fibre-optic oxygen microsensors, a new tool in aquatic biology. *Limnol. Oceanogr.* **40**: 1,159–1,165.

Kotsyurbenko, O. R., Glagolev, M. V., Nozhevnikova, A. N., and Conrad, R. 2001. Competition between homoacetogenic bacteria and methanogenic Archaea for hydrogen at low temperature. *FEMS Microbiol. Ecol.* **38**: 153–159.

Krom, M. D. and Berner, R. A. 1980. Adsorption of phosphate in anoxic marine sediments. *Limnol. Oceanogr.* **25**: 797–806.

Kvenvolden, K. M. 1988. Methane hydrates – a major reservoir of carbon in the shallow geosphere. *Chem. Geol.* **71**: 41–51.

Lovley, D. R. 1991. Dissimilatory Fe(III) and Mn(IV) reduction. *Microbiol. Rev.* **55**: 259–287.

Lovley, D. R. and Goodwin, S. 1988. Hydrogen concentrations as an indicator of the predominant terminal electron-accepting reactions in aquatic sediments. *Geochim. Cosmochim. Acta* **52**: 2,993–3,003.

Lovley, D. R. and Klug, M. J. 1983. Sulfate reducers can outcompete methanogens at freshwater sulfate concentrations. *Appl. Environ. Microbiol.* **45**: 187–192.

Lovley, D. R. and Philips, E. J. P. 1988. Novel mode of microbial energy metabolism: organic carbon oxidation coupled to dissimilatory reduction of iron and manganese. *Appl. Environ. Microbiol.* **54**: 1,472–1,480.

  1994. Novel processes for anaerobic sulfate production from elemental sulfur by sulfate-reducing bacteria. *Appl. Environ. Microbiol.* **60**: 2,394–2,399.

Lovley, D. R., Dwyer, D. F., and Klug, M. J. 1982. Kinetic analysis of competition between sulfate reducers and methanogens for hydrogen in sediments. *Appl. Environ. Microbiol.* **43**: 1,373–1,379.

Lovley, D. R., Coated, J. D., Blunt-Harris, E. L., Phillips, E. L., and Woodward, J. C. 1996. Humic substances as electron acceptors for microbial respiration. *Nature* **382**: 445–448.

Lovley, D. R., Fraga, J. L., Blunt-Harris, E. L., Hayes, L. A., Philips, E. J. P., and Coates, J. D. 1998. Humic substances as a mediator for microbially catalysed metal reduction. *Acta Hydrochim. Hydrobiol.* **26**: 152–157.

Lowell, R. P. 1991. Modeling continental and submarine hydrothermal systems. *Rev. Geophys.* **29**: 457–476.

Luther III, G. W., Sundby, B., Lewis, B. L., Brendel, P. J., and Silverberg, N. 1997. Interactions of manganese with the nitrogen cycle: alternative pathways to dinitrogen. *Geochim. Cosmochim. Acta* **19**: 4,043–4,052.

Madigan, M. T., Martinko, J. M., and Parker, J. 2000. *Biology of Microorganisms.* London: Prentice Hall.

Marchesi, J. R., Weightman, A. J., Cragg, B. A., Parkes, R. J., and Fry, J. C. 2001. Methanogen and bacterial diversity and distribution in deep gas hydrate sediments from the Cascadia Margin as revealed by 16S rRNA molecular analysis. *FEMS Microbiol. Ecol.* **34**: 221–228.

Mather, I. D. and Parkes, R. J. 2000. Bacterial profiles in sediments of the eastern flank of the Juan de Fuca Ridge, Sites 1026 and 1027. In *Proceeding of the Ocean Drilling Program, Scientific Results*, Vol. 168, eds. A. Fisher, E. E. Davis, and C. Escutia. College Station TX: Ocean Drilling Program, pp. 161–165.

McMahon, P. B. and Chapelle, F. H. 1991. Microbial production of organic acids in aquitard sediments and its role in aquifer geochemistry. *Nature* **349**: 233–235.

Meyers, P. A., Silliman, J. E., and Shaw, T. J. 1996. Effects of turbidity flows on organic matter accumulation, sulfate reduction, and methane generation in deep-sea sediments on the Iberia Abyssal Plain. *Org. Geochem.* **25**: 69–78.

Middleburg, J. J., Soetart, K., Herman, P. J. M., and Heip, C. H. R. 1996. Denitrification in marine sediments: a model study. *Global Biogeochem. Cycles* **10**: 661–673.

Morita, R. Y. and Zobell, C. E. 1955. Occurrence of bacteria in pelagic sediments collected during the Mid-Pacific Expedition. *Deep-Sea Res.* **3**: 66–73.

Nealson, K. H. and Saffarini, D. 1994. Iron and manganese in anaerobic respiration: environmental significance, physiology, and regulation. *Annu. Rev. Microbiol.* **48**: 311–343.

Nevin, K. P. and Lovley, D. R. 2002. Mechanisms for Fe(III) oxide reduction in sedimentary environments. *Geomicrobiol. J.* **19**: 141–159.

Nielsen, L. P. 1992. Denitrification in sediment determined from nitrogen isotope pairing. *FEMS Microbiol. Ecol.* **86**: 357–362.

Niewohner, C., Hensen, C., Kasten, S., Zabel, M., and Schulz, H. D. 1998. Deep sulfate reduction completely mediated by anaerobic methane oxidation in sediments of the upwelling area off Namibia. *Geochim. Cosmochim. Acta* **62**: 455–464.

Novelli, P. C., Michelson, A. R., Scranton, M. I., Banta, G. T., Hobbie, J. E., and Howarth, R. W. 1988. Hydrogen and acetate in two sulfate-reducing sediments: Buzzards Bay and Town Cove, Mass. *Geochim. Cosmochim. Acta* **52**: 2,477–2,486.

Novitsky, J. A. 1983. Microbial activity at the sediment–water interface in Halifax Harbor, Canada. *Appl. Environ. Microbiol.* **45**: 1,761–1,766.

Oremland, R. S., Marsh, L. M., and Polcin, S. 1982. Methane production and simultaneous sulphate reduction in anoxic, salt marsh sediments. *Nature* **401**: 217–218.

Orphan, V. J., Hinrichs, K.-U., Ussler, W., Paull, C. K., Taylor, L. T., Sylva, S. P., Hayes, J. M., and DeLong, E. F. 2001a. Comparative analysis of methane-oxidizing Archaea and sulfate-reducing bacteria in anoxic marine sediments. *Appl. Environ. Microbiol.* **67**: 1,922–1,934.

Orphan, V. J., House, C. H., Hinrichs, K.-U., McKeegan, K. D., and DeLong, E. F. 2001b. Methane-consuming archaea revealed by directly coupled isotopic and phylogenetic analysis. *Science* **293**: 484–487.

Osterberg, R. and Shirshova, L. 1997. Oscillating, nonequilibrium redox properties of humic acids. *Geochim. Cosmochim. Acta* **61**: 4,599–4,604.

Pancost, R. D., Sinninghe Damste, J. S. S., De Lint, S., Van der Maarel, M. J. E. C., Gottshal, J. C., and Science Party 2000. Biomarker evidence for widespread anaerobic methane oxidation in Mediterranean sediments by a consortium of methanogenic Archaea and Bacteria. *Appl. Environ. Microbiol.* **66**: 1,126–1,132.

Pancost, R. D., Hopmans, E. C., and Sinninghe Damste, J. S. S. 2001. Archaeal lipids in Mediterranean cold seeps: molecular proxies for anaerobic methane oxidation. *Geochim. Cosmochim. Acta* **65**: 1,611–1,627.

Parkes, R. J. 1999. Cracking anaerobic bacteria. *Nature* **401**: 217–218.

Parkes, R. J. and Wellsbury, P. 2003. Deep biospheres. In *Microbial Diversity and Bioprospecting*, ed. A. T. Bull. Washington, DC: ASM Press, pp. 120–129.

Parkes, R. J., Gibson, G. R., Mueller-Harvey, I., Buckingham, W. J., and Hernert, R. A. 1989. Determination of the substrates for sulphate-reducing bacteria within marine and estuarine sediments with different rates of sulphate reduction. *J. Gen. Microbiol.* **135**: 175–187.

Parkes, R. J., Cragg, B. A., Fry, J. C., Herbert, R. A., and Wimpenny, J. W. T. 1990. Bacterial biomass and activity in deep sediment layers from the Peru margin. *Phil. Trans. Roy. Soc. Lond. A* **331**: 139–153.

Parkes, R. J., Cragg. B. A., Bale, S. J., Getliff, J. M., Goodman, K., Rochelle, P. A., Fry, J. C., Weightman, A. J., and Harvey, S. M. 1994. A deep bacterial biosphere in Pacific Ocean sediments. *Nature* **371**: 410–413.

Parkes, R. J., Cragg, B. A., and Wellsbury, P. 2000. Recent studies on bacterial populations and processes in subseafloor sediments: a review. *Hydrogeol. J.* **8**: 11–28.

Pfannkuche, O., Boetius, A., Lochte, K., Lungreen, U., and Thiel, H. 1999. Responses of deep-sea benthos to sedimentation patterns in the North-East Atlantic in 1992. *Deep-Sea Res.* **46**: 573–596.

Quigley, T. M. and Mackenzie, A. S. 1988. The temperatures of oil and gas formation in the sub-surface. *Nature* **333**: 549–552.

Reeburgh, W. S. 1980. Anaerobic methane oxidation: rate depth distributions in Skan Bay sediments. *Earth Planet. Sci. Lett.* **47**: 345–352.

Reed, D. W., Fujita, Y., Delwiche, M. E., Blackwelder, D. B., Sheridan, P. P., Uchida, T., and Colwell, F. S. 2002. Microbial communities from methane hydrate-bearing deep marine sediments in a forearc basin. *Appl. Environ. Microbiol.* **68**: 3,759–3,770.

Revsbech, N. P., Jorgensen, B. B., and Blackburn, T. H. 1980. Oxygen in the sea bottom measured with a microelectrode. *Science* **207**: 1,355–1,356.

Rickard, D. and Luther III, G. W. 1997. Kinetics of pyrite formation by the H_2S oxidation of iron(II) monosulphide in aqueous solutions between 25 and 125 °C: the rate equation. *Geochim. Cosmochim. Acta* **61**: 115–134.

Rivkina, E. M., Friedmann, E. I., McKay, C. P., and Gilichinsky, D. A. 2000. Metabolic activity of permafrost bacteria below the freezing point. *Appl. Environ. Microbiol.* **66**: 3,230–3,233.

Rudnicki, M. D., Elderfield, H., and Spiro, B. 2001. Fractionation of sulfur isotopes during bacterial sulfate reduction in deep ocean sediments at elevated temperatures. *Geochim. Cosmochim. Acta* **65**: 777–789.

Russell, N. J. 1990. Cold adaption of microorganisms. *Phil. Trans. Roy. Soc. Lond. B* **326**: 595–611.

Saunders, P. M. and Fofonoff, N. P. 1976. Conversion of pressure to depth in the ocean. *Deep-Sea Res.* **23**: 109–111.

Sayles, F. L. and Martin, W. R. 1995. In situ tracer studies of solute transport across the sediment–water interface at the Bermuda Time Series site. *Deep-Sea Res.* **42**: 31–52.

Schink, B. and Friedrich, M. 2000. Bacterial metabolism – phosphite oxidation by sulphate reduction. *Nature* **406**: 37.

Schink, B., Thiemann, V., Laue, H., and Friedrich, M. W. 2002. *Desulfotignum phosphitoxidans* sp nov., a new marine sulfate reducer that oxidises phosphite to phosphate. *Arch. Microbiol.* **177**: 381–391.

Schippers, A. and Jørgensen, B. B. 2001. Oxidation of pyrite and iron sulfide by manganese dioxide in marine sediments. *Geochim Cosmochim. Acta* **65**: 915–922.

Seitz, H. J., Schink, B., Pfennig, N., and Conrad, R. 1990. Energetics of syntrophic ethanol oxidation in defined chemostat co-cultures. 1. Energy requirement for H_2 production and H_2 oxidation. *Arch. Microbiol.* **155**: 82–88.

Seitzinger, S. P. 1998. Denitrification in freshwater and coastal marine environments: ecological and geochemical significance. *Limnol. Oceanogr.* **33**: 702–724.

Slomp, C. P., Epping, E. H. G., Helder, W., and Van Raaphorst, W. 1996b. A key role for iron-bound phosphorus in authigenic apatite formation in North Atlantic continental platform sediments. *J. Mar. Res.* **54**: 1,179–1,205.

Slomp, C. P., Van der Gaast, S. J., and Van Raaphorst, W. 1996a. Phosphorus binding by poorly crystalline iron oxides in North Sea sediments. *Mar. Geochem.* **52**: 55–73.

Sørensen, J. 1987. Nitrate reduction in marine sediment: pathways and interactions with iron and sulfur cycling. *Geomicrobiol. J.* **5**: 401–421.

Stein, C. A. and Stein, S. 1994. Constraints on hydrothermal heat flux through the oceanic lithosphere from global heat flow. *J. Geophys. Res.* **99**: 3,081–3,095.

Stetter, K. O. in press. Volcanoes, hydrothermal venting and the origin of life. In
    Volcanoes and the Environment, eds. J. Marti and G. G. J. Ernst. Cambridge:
    Cambridge University Press.
Stetter, K. O., Huber, R., Blochl, E., Kurr, M., Eden, R. O., Fielder, M., Cash, H., and
    Vance, I. 1993. Hyperthermophilic Archaea are thriving in deep North-Sea and
    Alaskan oil reservoirs. *Nature* **365**: 743–745.
Straub, K. L., Benz, M., Schink, B., and Widdel, F. 1996. Anaerobic, nitrate-dependent
    microbial oxidation of ferrous iron. *Appl. Environ. Microbiol.* **62**: 1,458–1,460.
Strous, M., Fuerst, J. A., Kramer, E. H. M., Logemann, S., Muyzer, G., van de
    Pas-Schoonen, K. T., Webb, R., Kuenen, J. G., and Jetten, M. S. M. 1999. Missing
    lithotroph identifies as new planctomycete. *Nature* **400**: 446–449.
Stumm, W. and Morgan, J. J. 1996. *Aquatic Chemistry*. New York: Wiley.
Suess, E. 1980. Particulate organic carbon flux in the oceans – surface productivity and
    oxygen utilization. *Nature* **288**: 260–263.
Suess, E., Bohrmann, G., von Huene, R., Linke, P., Wallmann, K., Lammers, S., and
    Sahling, H. 1998. Fluid venting in the eastern Aleutian subduction zone. *J. Geophys.
    Res.* **103**: 2,597–2,614.
Suess, E., Torres, M. E., Bohrmann, G., Collier, R. W., Greinert, J., Linke, P., Rehder,
    G., Trehu, A., Wallmann, K., Winckler, G., and Zuleger, E. 1999. Gas hydrate
    destabilization: enhanced dewatering, benthic material turnover and large methane
    plumes at the Cascadia convergent margin. *Earth Planet. Sci. Lett.* **170**:
    1–15.
Taylor, G. T., Iabichella, M., Tung-Yuan, H., Scranton, M. I., Thunell, R. C.,
    Muller-Karger, F., and Varela, R. 2001. Chemoautotrophy in the redox transition
    zone of the Cariaco Basin: a significant midwater source of organic carbon
    production. *Limnol. Oceanogr.* **46**: 148–163.
Thamdrup, B. and Canfield, D. E. 1996. Pathways of carbon oxidation in continental
    margin sediments off central Chile. *Limnol. Oceanogr.* **41**: 1,629–1,650.
Thamdrup, F., Finster, K., Hansen, J. W., and Bak, F. 1993. Bacterial disproportionation of
    elemental sulfur coupled to chemical reduction of iron or manganese. *Appl. Environ.
    Microbiol.* **59**: 101–108.
Thauer, R. K., Jungermann, K., and Decker, K. 1977. Energy conservation in
    chemotrophic anaerobic bacteria. *Bacter. Rev.* **41**: 100–180.
Tissot, B. P. and Welte, D. H. 1984. *Petroleum Formation and Occurrence*. Heidelberg:
    Springer Verlag.
Trimmer, M, Nedwell, D. B., Sivyer, D. B., and Malcolm, S. J. 1998. Nitrogen fluxes
    through the lower estuary of the river Great Ouse, England: the role of the bottom
    sediments. *Mar. Ecol. Progr. Ser.* **163**: 109–124.
Tryon, M. D. and Brown, K. M. 2001. Complex flow patterns through Hydrate Ridge and
    their impact on seep biota. *Geophys. Res. Lett.* **28**: 2,863–2,866.
van der Graaf, A. A., Mulder, A., de Bruijn, P., Jetten, M. S. M., Robertson, L. A., and
    Kuenen, J. G. 1995. Anaerobic oxidation of ammonium is a biologically mediated
    process. *Appl. Environ. Microbiol.* **61**: 1,246–1,251.
Vetter, Y. A., Deming, J. W., Jumars, P. A., and Kriegerbrockett, B. B. 1998. A predictive
    model of bacterial foraging by means of freely released extracellular enzymes.
    Microbiol. Ecol. **36**: 75–92.
von Huene, R., Klaeschen, D., Gutscher, M., and Fruehn, J. 1998. Mass and
    fluid flux during accretion at the Alaskan margin. *Geol. Soc. Am. Bull.* **110**:
    468–482.

Wakeham, S. G., Lee, C., Hedges, J. I., Hernes, P. J., and Peterson, M. L. 1997. Molecular indicators of diagenetic status in marine organic matter. *Geochim. Cosmochim. Acta* **61**: 5,363–5,369.

Wallman, K., Linke, P., Suess, E., Bohrmann, G., Sahling, H., Schluter, M., Dahlman, M., Lammers, S., Greinhert, J., and von Mirbach, N. 1997. Quantifying fluid flow, solute mixing and biogeochemical turnover at cold vents of the eastern Aleutian subduction zone. *Geochim. Cosmochim. Acta* **61**: 5,209–5,219.

Weiss, M. S., Abele, U., Weckesser, J., Welte, W., and Schultz, G. E. 1991. Molecular architecture and electrostatic properties of a bacterial porin. *Science* **254**: 1,627–1,630.

Wellsbury, P., Herbert, R. A., and Parkes, R. J. 1996. Bacterial activity and production in near-surface estuarine and freshwater sediments. *FEMS Microbiol. Ecol.* **19**: 203–214.

Wellsbury, P., Goodman, K., Barth, T., Cragg, B. A., Barnes, S. P., and Parkes, R. J. 1997. Deep marine biosphere fuelled by increasing organic matter availability during burial and heating. *Nature* **388**: 573–576.

Wellsbury, P., Goodman, K., Cragg, B. A., and Parkes, R. J. 2000. The geomicrobiology of deep marine sediments from Blake Ridge containing methane hydrate (Sites 994, 995 and 997). In *Proceedings of the Ocean Drilling Program, Scientific Results*, Vol. 164, eds. C. K. Paull, R. Matsunoto, P. J. Wallace and W. P. Dillon. College Station, TX: Ocean Drilling Program, pp. 379–391.

Wellsbury, P., Mather, I., and Parkes, R. J. 2002. Geomicrobiology of deep, low organic carbon sediments in the Woodlark Basin, Pacific Ocean. *FEMS Microbiol. Ecol.* **42**: 59–70.

Westrich, J. T. and Berner, R. A. 1984. The role of sedimentary organic matter in bacterial sulfate reduction: the G model tested. *Limnol. Oceanogr.* **29**: 236–249.

Whiticar, M. J. 1999. Carbon and hydrogen isotope systematics of bacterial formation and oxidation of methane. *Chem. Geol.* **161**: 291–314.

Whitman, W. B., Coleman, D. C., and Wiebe, W. J. 1998. Prokaryotes: the unseen majority. *Proc. Natl. Acad. Sci. USA* **95**: 6,578–6,583.

Wilkes, H., Boreham, C., Harms, G., Zengler, K., and Rabus, R. 2000. Anaerobic degradation and carbon isotopic fractionation of alkylbenzenes in crude oil by sulphate-reducing bacteria. *Org. Geochem.* **31**: 101–115.

Zengler, K., Richnow, H. H., Rossello-Mora, R., Michaelis, W., and Widdel, F. 1999. Methane formation from long-chain alkanes by anaerobic microorganisms. *Nature* **401**: 266–269.

18

Microbial mediation of oceanic crust alteration

Hubert Staudigel and Harald Furnes

18.1 Introduction

During the last decade it has become increasingly obvious that ocean crust hydrothermal systems offer a wide range of microbial habitats. Microbial activity was found in and near mid-ocean ridge hydrothermal vents (e.g. Deming and Baross, 1993; Jannasch, 1995; Juniper and Tebo, 1995) and inside the oceanic crust (e.g. Thorseth *et al.*, 1995a; Furnes *et al.*, 1996; Fisk *et al.*, 1998; Furnes and Staudigel, 1999). Microbial activity in the oceanic crust has far-reaching consequences by influencing (a) chemical fluxes between seawater and basalt and (b) the composition of the oceanic crust before it is recycled into the mantle. This establishes a connection between the biosphere, lithosphere, and hydrosphere. This connection of microbial processes with the massive global fluxes associated with the oceanic crust generation–subduction cycle could be a cornerstone for a growing new paradigm in Earth sciences. This paradigm gives biology a much larger role in our view of the Earth as a complex biological, chemical, and physical system with important interactions between all these disciplines. In this new paradigm, for example, the biosphere has an impact on mantle geochemistry, and biological evolution of the planet has a fundamental impact on global solid Earth geochemical fluxes. Biological activity in the oceanic crust may be viewed as the "rock-bottom" of the global food chain where new biomass is produced by using the energy and nutrients from the oceanic crust. This bottom of the food chain is linked to upper trophic levels through hydrothermal circulation whereby deep biomass is ejected through hydrothermal venting and recycled back into the crust through the return flow of seawater into an ocean crustal biosphere (OCB).

Evidence for life in the oceanic crust comes from a wide range of data types, from forensic studies in ancient crust, *in situ* observations, and from experimental investigations. The evidence is compelling, but there are many open questions and uncertainties and a balanced review of this topic has to consider the evidence and its limitations in the context of the key open questions. For this reason, we will review a range of different types of data, but we will also evaluate in some detail the limitations of these data, and discuss open problems. Key aspects of life in the oceanic crust include observations on the presence of

Hydrogeology of the Oceanic Lithosphere, eds. E. E. Davis and H. Elderfield. Published by Cambridge University Press.
© Cambridge University Press 2004.

life in the oceanic crust, its effects of microbial activity on water–rock interaction and its potential to influence geochemical cycles. We will also discuss evidence for how long life could exist in the oceanic crust after its formation at mid-ocean ridges. This longevity of life is likely linked to the presence of hydrothermal circulation in the oceanic crust and it determines the size of the active deep biosphere. We will show in this review that the evidence for microbial mediation of ocean floor alteration is strong, and that there are some surprising relationships between microbial activity and geochemical interaction between seawater and the oceanic crust. However we will also show that there are important open problems and uncertainties that need to be addressed, a shortcoming that is typical for an emerging field of research.

18.2 Life in the ocean crustal biosphere (OCB)

Our current understanding of life in the OCB comes from a range of studies, from experimental to forensic evidence and to *in situ* observations of active systems. Most of the microbiological work so far is based on sampling hydrothermal fluids, either near and around black smokers, or from "megaplumes," massive eruptions of low-temperature fluids with large quantities of microbial material (e.g. Jannasch, 1995). These studies showed that the oceanic crust is a source of a wide range of microbial materials, much of it adapted to higher temperatures and a very large fraction of life from the kingdom of Archea. This work shows that these life forms are associated with hydrothermal systems, but they say very little about their actual adaptations in the oceanic crust, and their interaction with this environment. In an extreme view, even the origin of this life is not very clear, because the microbial material coming out of the crust could have been introduced from seawater, and any differences in populations may just reflect an enrichment of the ejected populations relative to the ambient seawater, without any actual adaptation to the oceanic crust.

The most critical proof for life in the oceanic crust must come from within the OCB and the discovery of apparently microbial alteration of volcanic glass in the oceanic crust (Thorseth *et al.*, 1995a; Furnes *et al.*, 1996). This is about the closest we have come to such a proof. The process of microbial glass alteration is still relatively poorly understood. Thorseth *et al.* (1992) argued that tubular or spongy alteration of glass may be caused by colonizing microbes that etch their way into glass through the use of local changes in pH from organic acids that may be produced during their metabolism. Welch and Ullmann (1996) showed that organic acids are particularly effective in dissolving silicates when compared to inorganic (single protenated) acids like HCl. The relationships between microbial activity and glass corrosion have been demonstrated in a series of experiments by Staudigel *et al.* (1995) and Thorseth *et al.* (1995a) and it was shown that microbes appear to dissolve the glass including the formation of hydration rinds and the development of biofilms. An inadvertent enhancement of glass dissolution may be beneficial to some microbes because there are a number of nutrients and trace metals in basaltic glass. Oxidation of reduced iron may be one of the major forms of energy these microbes may obtain from the glass. For this reason recent studies focused on iron-oxidizing microbes. Thorseth *et al.* (2001)

on observing typical "twisted stalk" shaped microbes on glass from the Knipovich Ridge suspected the activity of the bacterium *Gallionella*. Emerson and Moyers (2002) observed obligate microaerophilic Fe oxidizers in hydrothermal vents from Loihi Seamount. None of these studies, however, used materials from inside the oceanic crust, nor have they shown any direct relationships between microbes and glass dissolution.

18.3 Studying the OCB by drilling

18.3.1 Drilling-related problems

All data from the interior of the OCB are critically dependent on problems related to drilling, in particular the very low recovery rates in ocean crustal drilling and the drilling-induced microbial and chemical contamination of a drill-hole. Recovery rates in most ocean drill-holes into volcanic oceanic crust are less than 30%, which rarely allows the recovery of a cylindrical core but rather individual, often rounded fragments. During rotary drilling on the ocean floor, the formation breaks typically along fractures, between breccia fragments, or along flow contacts, grinding up any biofilms or other delicate microbial textures that could be found in the spaces between pillow lavas or within fractures. This typically limits the recovered materials to the least altered, most solid rock fragments, while grinding up and washing from the hole the less altered and soft biological materials. During drilling, a borehole is continuously flushed with surface seawater to lubricate and cool the drill bit and to remove drill cuttings. This water penetrates the formation according to its *in situ* permeability and all surfaces of drilled rock fragments are coated with surface seawater, contaminating all but the interiors of the recovered coherent rock fragments. However, physical stresses imposed on the formation and the core from drilling cause vibrations that are likely to allow drilling fluids to enter the rock much deeper that one could expect from the (unstrained) permeability structure. This makes it very likely that most of the potential host environments for microbes are contaminated with surface seawater microbes, in particular in unsealed crust and in crust that is recovered only with low recovery rates.

Smith *et al.* (2000) evaluated the potential for such contamination with two types of quantitative tests, the injection of a perfluorocarbon tracer (PFT) with the drilling fluids and another involving the delivery of fluorescent microbeads to drilled core when it enters the core liner. Both tests were successful in the delivery of the tracers (Smith *et al.*, 2000), but either technique has a few significant shortcomings. The PFT experiments during Leg 185 were generally positive for drilling fluid contamination suggesting that the drilling process invariably injects a few nanoliters of drilling fluid into non-fractured rock, equivalent to the addition of a few bacteria per gram of core material (Smith *et al.*, 2000). Smith *et al.* (2000) suggested that this is a rather minor amount of contamination, given the fact that the PFT test is overly conservative and tends to lead to false positives, due to potential for volatile PFT transfer from the outer contaminated surfaces to the sample interior during sample preparation. Thus, if a sample is shown to be devoid of PFTs it is actually quite likely that these samples are uncontaminated. However, all the samples with the lowest contamination studied by Smith *et al.* (2000) came from the interiors of unfractured core materials, portions of the crust that are rather unlikely hosts for any microbial activity. All cracks and outer surfaces of individual recovered rock fragments, are extensively coated with PFTs, and therefore with surface seawater. Therefore, the most promising environments for microbial research in the oceanic crust remain compromised for many microbial studies.

The microbead test resulted in several truly negative results in thin sections, and Smith *et al.* (2000) cautioned that even the thin sections with positive results may actually be free of drilling-related contamination because beads may have been introduced during the thin sectioning process. This finding shows that the samples were *not* contaminated during entry and residence in the core liner, but it does not show that they were not contaminated *during drilling*, which offers the most likely opportunity for contamination with surface seawater. Thus, it is quite possible that drilling fluids actually penetrated the core during drilling, while the microbeads remained near the surface of the (then water-saturated, and less vibrating) core. Absence of beads in these core materials is not sufficient to rule out contamination in the core once it is introduced into the core liner.

As a result, much progress has been made in out understanding of contamination and its quantification, but the most promising part of the oceanic crust for microbial habitats is likely not to be accessible by drilling without contamination with surface microbes. For this reason, there are still major hurdles in the way of *in situ* study of the microbiology of the ocean crust by drilling and experimental investigations remain an important tool as an analog system for studies of the microbial alteration of the oceanic crust.

18.3.2 Traces of microbial activity in the oceanic crust

It has been known for a long time that microbial activity leaves visible traces on altering glass surfaces (Mellor, 1922; Krumbein *et al.*, 1991), first seen on basaltic glass by Ross and Fisher (1986) and subsequently on sub-glacial hyaloclastites by Thorseth *et al.* (1992). The last made the first convincing case for a biotic origin of these features by suggesting that colonizing microbes may cause localized changes in pH at their contact area through the development of metabolic byproducts. This causes characteristic pitting, sponge textures, and microtubes. Subsequently, biopitting on basaltic glass in submarine lava flows in the oceanic crust at the Costa Rica Ridge was observed by Thorseth *et al.* (1995a), Furnes *et al.* (1996), and Giovannoni *et al.* (1996). Subsequently, Fisk *et al.* (1998) pointed out that those features could be found in a number of ocean crustal settings. Furnes and Staudigel (1999) determined the relative abundance of biotic and abiotic glass alteration and showed that most microbial alteration is found in the upper 300 m of the oceanic crust, decreasing to near-absence at about 500 m. Fisk (1999) reported such features in crust of 170 Ma, showing that the entire oceanic crust may carry these features, at all ages. It is curious that microtubes have not yet been identified in pillow lavas directly exposed to seawater on the ocean floor, neither have there been any microbes identified that are responsible for these features. This may suggest that the microbes causing these features may be limited to the interior of the

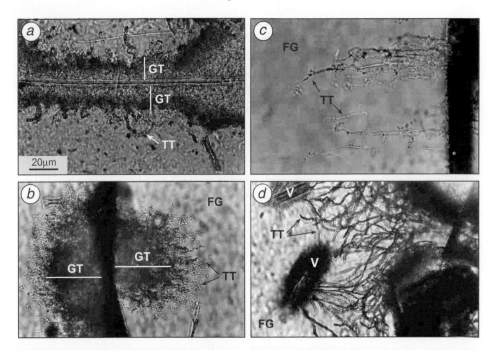

Fig. 18.1 Photomicrographs showing bioalteration textures. (*a*) and (*b*) show typical granular texture
(GT), with tubular texture (TT) extending into the fresh glass (FG). (*c*) and (*d*) show typical tubular
textures. (*c*) Shows tubes rooted in a fracture, whereas (*d*) shows that tubes may also be rooted in
variolites (V). (*a*) From ODP sample 146–896A-9R-1, 17–21: (*b*) from DSDP sample 46–396B-17R-
3, 62–70; (*c*) from DSDP sample 70–504B-35–1, 106–113; (*d*) from DSDP sample 46–396B-20R-4,
112–122. The scale (shown in (*a*)) is similar for all pictures.

oceanic crust, with a somewhat more shielded environment than the one reflected in dredge
samples.

   Furnes and Staudigel (1999) and Furnes *et al.* (2001) distinguish two major types of
microbial glass alteration, which is very distinct from abiotic alteration. Most abiotic
glass alteration features are apparently controlled by diffusion – controlled hydration in
an isotropic medium, with concentric, surface parallel alteration fronts progressing from
exterior surfaces of fresh glass toward the fresh interior. Hydration along cracks typically
extends symmetrically from the crack into the glass on both sides of the crack. Thin,
wedge-shaped intersections of cracks typically produce alteration rinds that are increasingly rounded off as they migrate inward from a pointy corner. This behavior continuously
decreases the total dissolving glass surface area with time until the glass alters entirely.
There are two major types of microbial alteration, a tubular type (TT; in Fig. 18.1) and
granular type (GT; Furnes and Staudigel, 1999). Both of them cannot be reconciled with
diffusion-controlled hydration or congruent dissolution of an isotropic glass medium. The
tubular type of biotic alteration includes irregular channels several micrometers thick and

up to 100 μm in length. The tubes are rooted typically in fractures or in vesicles, and they are often closely connected to varioles (Fig. 18.1*c,d*). The granular type (GT; Fig. 18.1) is formed by agglomerations of 0.2–0.6-μm sub-spherical grains in irregular, spongy linings along cracks. Such granular alteration may display half-spherical structures protruding into the fresh glass from crack surfaces (Fig. 18.1*b*), closely resembling a microbial culture on an agar plate. These features cluster at crack intersections and they become very rare at the terminations of thinning cracks. Most importantly, tubes and granular alteration display no apparent symmetry with respect to opposing sides of cracks. These asymmetries, and the microbe size of these features offer strong morphological evidence for a microbial origin for these features.

There are two additional lines of evidence for a microbial origin of tubular or granular alteration features: the use of molecular probes and microbeam chemical mapping. Thorseth *et al.* (1995a), Giovannoni *et al.* (1996), and Torsvik *et al.* (1998) used various types of fluorescent dyes that bind specifically to nucleic acids to stain thin sections and found DNA, in particular, in the tips of microtubes. They demonstrated that this DNA actually belongs to the two kingdoms of Bacteria and Archea, the latter a group of microbes that is particularly abundant in hydrothermal systems. This observation proves habitation of these tubes by microbes, but it does not strictly show that they were the ones that produced them. Furthermore, there is always a potential for these microbes to be introduced during drilling, even though this might be unlikely, because fluorescent staining is found within the inner part of the alteration zone, furthest away from the contaminating solution.

Microbeam geochemical mapping allows us to study the very fine-scale distribution of elements in the vicinity of microbial alteration features. In Fig. 18.2, we have shown a series of element maps of a region of granular alteration in a thin section. The SEM image is shown in Fig. 18.2*a*, displaying granular clusters near a crack that is about 2 μm wide. The zones of granular alteration are well correlated with zones of significant depletion in sodium, calcium, iron, and magnesium, and very slight depletion in sulfur and phosphorus. Granular alteration features show pronounced enrichment of potassium and slight enrichment of aluminium. The crack itself is enriched in iron and magnesium, probably indicating the presence of iron oxides and/or clays. Nitrogen and carbon also seem to be enriched in the crack relative to the basalt glass, with one particularly bright spot that also appears to be enriched in carbon. Carbon shows more patchy enrichments along the crack and some bright spots in the granular alteration zone. It is important to note that the carbon enrichments do not correlate with any enrichment in Ca, which eliminates calcium carbonate as a carbon-bearing phase, suggesting that the carbon is probably organic. Phosphorus is slightly depleted in the granular alteration zone, but it shows abundant bright spots. In particular the enrichments in carbon, nitrogen, and phosphorus are very indicative of a microbial involvement in the granular alteration but it does not prove that these cavities are actually made by these microbes.

Each one of the three lines of evidence for microbial alteration by itself is insufficient to prove a biotic origin of these features. In combination, however, all these lines of evidence make a microbial origin of these tubes much more likely than an abiotic origin.

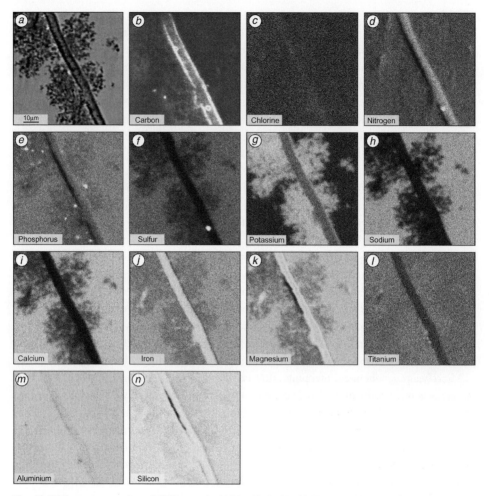

Fig. 18.2 Element maps from DSDP sample 418A, 49–2, 41–45. The SEM image (*a*) shows granular texture adjacent to a fracture. Within the granular area the carbon map (*b*) shows generally weak C enrichment, with strong enrichment in two spots. Within the fracture, however, there is a general enrichment and several bright spots. The depletion in Ca (i) shows that the carbon in (*b*) is not bound in calcite. Further, the map of chlorine (*c*) shows no trace of that element. This eliminates the possibility of glue-contamination along the fracture (the glue has a marked Cl concentration). The carbon is therefore taken to represent organic matter. Nitrogen (*d*) is also present within the fracture. Phosphorus is slightly depleted within the granular area, but contains several bright spots (about 1–3 μm in diameter). Sulfur (*f*) is depleted within the granular area, but contains one bright spot at the same location as high carbon. Potassium (*g*) is strongly enriched within the granular area, whereas sodium (*h*) is strongly depleted. Iron and magnesium, (*j*) and (*k*), respectively, are slightly depleted within the granular area, but enriched within the fracture. For titanium, aluminum and silicon, (*l*), (*m*), (*n*), respectively, there is only minor difference in the concentration level of the fresh glass and the granular area.

The development of high surface-area features like granular or tubular alteration cannot be expected from diffusion and congruent dissolution processes acting on a glassy isotropic medium. Increases in surface area influence the dissolution–precipitation equilibrium by favoring dissolution, and thus increasing the ionic strength of the solution. Staudigel *et al.* (2004) made estimates of the sizes of tubular and granular alteration to determine the development of dissolving surface areas between abiotic and biotic alteration. They measured the tube thicknesses and granule sizes, as well as their abundances in samples from a series of DSDP and ODP sites. There are approximately 10^{10} tubes per square meter of crack surface and the tubes are on average 1.5 μm thick and 53 μm long. This results in a 2.4-fold increase in the surface area of cracks due to microtube formation. For the granular texture they estimated an average granule diameter of 0.4 μm and an average granular layer of 30 μm thickness. This results in a 200-fold increase in surface area. Thus, microbial alteration substantially increases the surface available for dissolution, accelerating microbially mediated dissolution relative to abiotic dissolution.

Given that biotic and abiotic alteration can offer very distinct textures, their relative abundance can be determined microscopically. Furnes and Staudigel (1999) and Furnes *et al.* (2001) made such a determination for several ocean drill-holes and plotted them versus depth (Fig. 18.3). We plotted all data from these studies, but there are two sites that stand out for their deep penetration into basement and the geological detail known about them. Site 504B is the deepest reference section (1,840 m) so far drilled into the oceanic crust. It has been studied in great detail for alteration processes and for its geological characteristics (Alt *et al.*, 1993) even though it is characterized by relatively low recovery rates (22%). This recovery was sufficient to recover abundant volcanic glass with bioalteration textures and well logging helped in deciphering the geological characteristics (Alt *et al.*, 1993). Site 418A is substantially less deep (555 m) but it is similarly well-studied geologically, chemically, and mineralogically. This site is unique in its very high recovery rates (>70%; Robinson *et al.*, 1979). The remaining sites studied from the Lau Basin and the North Atlantic display less penetration into the crust and generally very low recovery rates.

The extent of bioalteration at Sites 504B and 417/418 displays very consistent results that define overall the upper bound of the data distribution at any depth (Fig. 18.3). Furnes and Staudigel (1999) and Furnes *et al.* (2001) show about 75% bioalteration in the top 200 m decreases to less than 10% at 500 m. The remaining sites consistently show more scatter, even though they display the same upper bound of their data distribution. Furnes *et al.* (2001) pointed out that the percentage of bioalteration can be a function of recovery rate, whereby the thinnest cracks show the smallest amount of bioalteration and they are favored in samples from holes with low recovery rates. Furthermore, they suggested that the volumetric estimates of bioalteration always are a minimum estimate because ambiguous alteration features were always counted as abiotic, to make the estimate most conservative. Biotic alteration features are very delicate and likely to be destroyed during subsequent hydration of glass that is associated with swelling. For these reasons, it is not too surprising that some values are deflected to low estimates of bioalteration. Furnes *et al.* (2001) suggested that this is a likely reason for the variation of percent biotic alteration in the Lau Basin and the

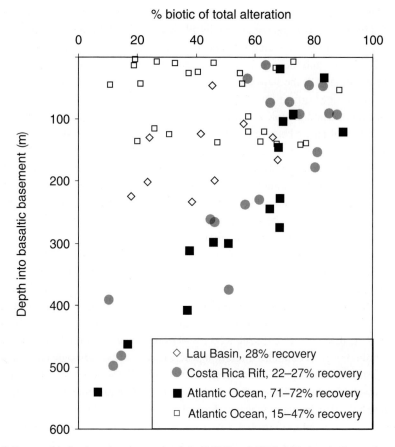

Fig. 18.3 Percent biotic alteration versus depth in DSDP and ODP drill sites (redrawn from Furnes *et al.*, 2001). Data from high-recovery sites (417/418) in the Atlantic and Site 504B form a well-defined trend that suggests 50–90% bioalteration in the upper 300 m of the oceanic crust, decreasing to <20% at 500 m depth. Low-recovery sites from the Lau Basin and the North Atlantic Ocean have a wider range in percent bioalteration, but the upper end of their data distribution is identical to the trend displayed by high-recovery rate sites.

low recovery in North Atlantic sites. However, it is interesting to note that Site 504B shows a rather systematic variation, despite the fact that recovery rates are rather low.

Bioalteration at Sites 504B and 417/418 is well correlated with permeability and abundance of alteration minerals indicating relatively oxygenated waters (e.g. celadonite; Furnes and Staudigel, 1999; Furnes *et al.*, 2001). At Site 504B the decrease in biotic alteration is also inversely correlated with temperature. Maximum bioalteration is found at temperatures up to 80 °C and it almost entirely disappears at temperatures above 120 °C. Correlation of high degrees of bioalteration with access to oxygen and its limitation to temperatures less than 120 °C further support the biological origin of these features.

The results of Furnes and Staudigel (1999) offer a depth constraint on (at least one type of) microbial activity in the oceanic crust: microbial activity dominates glass alteration in the upper 300 m and it is present down to about 500 m. This gives a minimum depth for the OCB.

18.3.3 Size of the ocean crustal biosphere: duration of bioalteration

After evaluating constraints on the depth of the OCB, it is important to determine the constraints on the total area of the OCB in the oceans. This area is potentially vast, given the fact that bioalteration was found in all drill-holes that recovered fresh volcanic glass. Drilling into oceanic crust involved a great number of ocean crust types and ages up to the oldest oceanic crust drilled at the 170 Ma Site 801C (Fisk, 1999). This offers the possibility that the OCB literally includes all of the ocean basins, more than 60% of the Earth's total surface area. So far, however, there is no irrevocable evidence for ongoing microbial activity in any age crust, and it is possible that all of these features are produced very early in the geological history of the oceanic crust and all life in the OCB ends a very short time after the oceanic crust is produced.

This scenario is, however, unlikely. Basaltic glass is unstable in the presence of water and it will continue to hydrate until it is effectively sealed off from water circulation. Textural evidence suggests that bioalteration lasts as long as abiotic alteration. If this were not the case, there would be many sites with glass hydration without any bioalteration textures. Therefore, the presence of bioalteration in glass surfaces anywhere in the oceanic crust proves that bioalteration generally outlasts hydration of glass in the oceanic crust. This allows us to reduce the question of the duration of bioalteration to the duration of glass hydration or hydrothermal circulation.

Hydration of glass is likely to continue for as long as hydrothermal alteration can be found in the oceanic crust. There are two ways to determine the duration of hydrothermal alteration in the oceanic crust: through direct isotopic dating of secondary phases in the oceanic crust (Hart and Staudigel, 1978; Booij *et al.*, 1995) or by constraining the actual convective heat loss in the oceanic crust (Stein and Stein, 1994; Stein *et al.*, 1995). The latter can be done by independently constraining the actual total heat flow and measuring the convective heat flow through *in situ* heat flow probes in the oceans. The differences between these estimates constrain the total convective heat flow at a particular place. To first order, the oceanic crust loses one-third of its total hydrothermal heat before it reaches 1 Ma and two-thirds before 10 Ma, and it is commonly assumed that hydrothermal heat loss continues to an age of about 50 Ma (Stein and Stein, 1994; Stein *et al.*, 1995), even though these estimates are quite crude. In particular, the termination of hydrothermal alteration cannot be realistically constrained by heat flow data alone, because the assumption of zero convective heat loss is an important component of cooling models for old plates, which makes the determination of zero heat loss into a circular argument. In any case, it is probably realistic to assume that convective heat loss in crust older than about 50 Ma is very small when compared to the convective heat loss in a near-axis regime.

Overall, we believe it is most realistic to assume that biotic alteration lasts at least to about 50 Ma but it may last as long as 170 Ma, as shown by the presence of bioalteration features at ODP Site 801C in the oldest oceanic crust found in the ocean basins (Fisk, 1999).

18.4 Ocean crust bioalteration in experimental studies

It became very obvious in the previous section that drilling technology imposes substantial limitations for the *in situ* study of the OCB. In particular drilling-related contamination plays an important role, but also the complexity of the alteration history and alteration conditions through time make it very difficult to correlate particular biological activities with particular conditions of alteration. Also we are not able to determine whether bioalteration is any different from abiotic alteration of glass and to determine what the differences are. In this section we will review experimental results that help us understand how biological processes might influence alteration of the oceanic crust. This includes the simulation of biotic alteration textures, and the effects of microbial alteration on the chemistry of glass alteration, in closed and open system experiments.

18.4.1 Simulation of alteration textures

Laboratory experiments allow for the simulation of microbial corrosion effects in the context of abiotic controls. Two studies succeeded in simulating such effects (Staudigel *et al.*, 1995; Thorseth *et al.*, 1995b), but both had some important limitations. Thorseth *et al.* (1995b) inoculated basaltic glass with microbes from submarine hyaloclastites from Surtsey volcano and attributed the development of very distinct glass corrosion features to microbial activity. However, all of their successful experiments were run with a 1% glucose solution, which is not very realistic for ocean conditions. Staudigel *et al.* (1995) succeeded in simulating biocorrosion with natural seawater without any additional nutrients and or bacterial inoculations other than those typical of shallow seawater microbial populations. This experiment showed very clearly the effects of biocorrosion, but these experiments were carried out using a (non-radioactive) borosilicate nuclear waste glass analog, and thus they are only indirectly applicable to basalt glass alteration. Taken together, the experiments of Staudigel *et al.* (1995) and Thorseth *et al.* (1995b) show rather convincing evidence that microbes are likely to have an effect on basaltic glass alteration, even without stimulation from high nutrient contents added to solutions.

The experiments of Thorseth *et al.* (1995b) included the formation of hydration rinds of about 1 μm and Staudigel *et al.* (1995) reported on the formation of biofilms on the order of 10 μm within about a year. Hydration rinds and biofilms are very important to the rates and the character of water–glass interaction. Hydration rinds create a diffusion barrier that becomes increasingly effective with time, isolating the residual glass, and slowing chemical transfer. Biofilms are important for several reasons. They may create a localized chemical environment that allows microbes to thrive that could otherwise not grow in the surrounding hydrous environment and they may locally change the pH such that glass can dissolve particularly effectively. Biofilms also contain much extracellular materials with high chemical affinity for a series of elements that can be taken out of solution, and biofilms serve as a source and filter for particles, capturing, for example, particles from downwelling seawater or releasing some biofilm material as it grows and sheds material.

18.4.2 Closed system water–rock interaction with abiotic controls

The previous sections made the case for an association of microbial activity with the alteration of volcanic glass, but it did not become clear whether these processes are any different for biotic or abiotic conditions. This can be shown only in closed system experiments allowing for rigorous sterile control experiments. Staudigel *et al.* (1998) carried out such seawater–basaltic glass dissolution experiments to explore how microbes influence glass dissolution and authigenic mineral formation. They constructed a flow-through apparatus, where a reactor vessel filled with basaltic glass sand was continuously flushed with seawater from a 50-l seawater reservoir in a closed system, with exception of the approximately weekly sampling of ~5–10 ml of solution for chemical monitoring (Si, particles and pH). They carried out two biotic experiments inoculated with the natural microbe population from the surface seawater, and one sterile experiment after autoclaving. All of these experiments were run for periods close to a year, whereby the sterile control did not yield any culturable microbes after 450 days. This experimental setup has some resemblances to an ocean crust alteration model system, with circulating seawater that may remove elements by dissolution and solution transport, and ocean crust materials that are highly reactive, with a porous structure that allows secondary minerals to form in open spaces and finally reduce porosity. At the end of the experiment, they separated the fine-grained, newly formed sediment from the reactor vessel to constrain the element fixation in the latter, and they studied solution chemistry during the experiment.

The main results were:

1. The biotic experiments produced almost an order of magnitude more reaction products than the sterile control. This shows that the biotic system is very effective at sequestering elements from solution and basalt inside the reactor. It also shows that a biotic system is more likely to fill up pore space with secondary minerals than an abiotic system. This suggests that biotic systems seal themselves more rapidly than abiotic systems.
2. The compositions of biotic and abiotic reaction products are very different. The abiotic experiment sequestered mostly Mg from seawater (in palygorskite, a Mg-rich layer silicate), but released almost all other elements to solution, while the biotic experiment fixed most of the basaltic elements released from dissolution, and it sequestered substantial amounts of Ca from seawater.
3. The solutions in the sterile experiment become continuously enriched with Si, until they reach solution–precipitation equilibrium after ~400 days. The silica in biotic experiments remained at near-zero concentrations for experiments with daylight and were only slightly elevated in experiments without access to light. Diatoms were abundant in the experiments under light conditions, but

the Si-utilizing phase in the dark experiment remains uncertain. The sterile experiment apparently removes more silica from the reactor than the biotic experiment.

4. In biotic experiments, glass takes up four times as much seawater Sr than abiotic experiments and they release about 20 times as much basaltic Sr from glass than the abiotic ones. This shows that microbial processes enhance the chemical exchange of Sr between seawater and basalt. This is consistent with the observed increase in surface area from microbial alteration in the oceanic crust, but the specific processes are not understood.

Staudigel *et al.*'s (1998) experimental model has many imperfections that limit its applicability to true ocean crust exposures. In particular, they used nutrient-poor surface seawater, with microbial populations that are quite different from the deeper ocean where most active mid-ocean ridge hydrothermal systems are found. Furthermore, some experiments were carried out in daylight, supporting the formation of diatoms that are not likely to form in true mid-ocean ridge hydrothermal systems. Also, while glass is one of the major phases to participate in seawater–ocean crust interaction, it is also quite clear that many other phases are present that may contribute to the alteration budget. Despite these imperfections, many important lessons may be learned from these experiments. As might be expected, biologically active hydrothermal systems appear to be more effective at dissolving ocean crust materials than biotic systems. Somewhat unexpectedly, however, biotic systems appear also to be more effective at keeping the products of alteration inside the crust, and actually appear to reduce the actual fluxes of elements from the crust into the oceans. A notable exception may be Sr, which shows dramatically enhanced chemical exchange between glass and solution: this will yield an increase in basaltic Sr flux into the oceans, even if the concentration of Sr in hydrothermal solutions does not differ for the composition of seawater. Another important side effect of the intense secondary mineral formation is the decrease in pore space and permeability, which may ultimately contribute to effective sealing of the oceanic crust.

18.4.3 Open system studies

The abovementioned experiments have important shortcomings as analog systems, in particular because the oceanic crust is an open chemical system that has very large water/rock ratios. Such large water/rock ratio systems are important for the study of the chemical behavior of elements that are in short supply in seawater relative to their inventories in basalt. Large water/rock ratios also support life by giving microbes more access to dilute nutrients or trace elements needed. Most experiments, in particular the ones with sterile controls, are typically run at very low water/rock ratios. Open system experiments have not had any abiotic controls (Staudigel *et al.*, 2004), and therefore any chemical changes cannot be uniquely attributed to biological effects versus inorganic phase equilibria. Staudigel *et al.* (2004) "simulated" a hydrothermal system by running approximately $10 \, \text{kg day}^{-1}$ of sand-filtered seawater over a series of reaction columns filled with basaltic glass. Some experiments were kept entirely at room temperature ($20 \, ^{\circ}\text{C}$) and others had a range of temperatures up to boiling, in line with columns at lower temperatures. They did a preliminary microbial characterization of some experiments and studied, in particular, the trace element characteristics of the reactor sediments. Staudigel *et al.* (2003) also analyzed the chemical composition of the fine-grained suspended matter in the incoming seawater.

Key findings from these experiments include:

1. Microbial colonization was found in all experiments, including a range of heterotrophs, chemolithotrophs, and photoautotrophs that were not identified in detail. No culturable microbes were found in the first few months of the experiment at 60 °C, but after four months this experiment developed very active populations of thermophilic heterotrophs, sulfur oxidizers, nitrite oxidizers, and ammonia oxidizers. All experiments started with the same inoculum, but these populations had to adjust to different temperature conditions. The time needed for adjustment was relatively short compared to geological time scales.

2. At all experimental conditions used, reactor sediments were enriched in Cs, Rb, Th, U, Sr, and Pb relative to basalt, while all other elements analyzed displayed relatively minor effects (Ba, Nb, REE, Hf, Zr, Sc, V, Cr, Co, Zn, Ni, Cu).

3. U and Pb showed dramatic fractionation with respect to each other and a strong dependence on temperature. U is enriched in reactor sediments, in particular in reactors at temperatures >50 °C; while Pb is most enriched in low-temperature reactors. The fractionation of Pb appears to be mostly linked to particles, while U is likely to be deposited as a function of its redox-dependent differences in solubilities. Seawater contains highly soluble (oxidized) $UO_2(CO_3)_2{}^{2-}$, which is likely to be reduced during water–rock reaction with the (highly reducing) basalt glass. Reduced U precipitates as uraninite (UO_2) or coffinite ($USiO_4$; Langmuir, 1978). Redox reactions of U may be linked to the activity of *Desulfovibrio vulcaris* (Lovley, 1991, 1993a,b).

4. In particular, the Th and Pb abundances in reactor sediments, but also their REE patterns, highlight the importance of particulate matter in these experimental systems. These observations are consistent with bulk ocean crust data that also demand some particle-based addition to the oceanic crust (Staudigel *et al.*, 1996). This shows that particulate transport is very important for hydrothermal systems.

18.5 Discussion: biological mediation of chemical processes in the OCB

Forensic investigations of microbial glass alteration in drill-holes and a range of other observations helped us understand some key controls of bioalteration in the oceanic crust and point to some important consequences of bioalteration on the chemical mass fluxes between seawater and the oceanic crust.

It is quite obvious that glass is rapidly colonized after its emplacement on the ocean floor (Thorseth *et al.*, 2001). Colonizing microbes excrete a series of organic acids that may be more effective at dissolving glass, more so than single protenated inorganic, completely dissociated acids at equivalent pH, as suggested by feldspar dissolution experiments by Welch and Ullmann (1996). This process results in pitting of the glass surface (e.g. Staudigel *et al.*, 1995; Thorseth *et al.*, 1995b) and surface roughing that increases the surface area of glass available for dissolution (Staudigel *et al.*, 2004). Microtube formation more than doubles the surface area, while mature granular alteration increases the surface area by more than two orders of magnitude. These dissolution processes create open spaces through congruent dissolution of glass that is made available for microbial utilization or for incorporation in authigenic mineral phases. Ferrous iron is one of the key nutrients in glass (>10 wt. %) and its oxidation may serve as an electron donor. The potential use of ferrous iron by microbes is inferred by the possible identification of Fe-oxidizing *Gallionella* (Thorseth *et al.*, 2001) or other obligate microaerophilic Fe oxidizers identified in hydrothermal waters from Loihi Seamount (Emerson and Moyers, 2002). The abundance of iron oxides on the seafloor near hydrothermal vents shows that iron oxidation is an important process in hydrothermal systems.

Immediately following initial colonization, exterior surfaces of glass are rapidly coated by biofilms (e.g. Staudigel *et al.*, 1995), which are made of extracellular polysaccharides, authigenic minerals, and microbes. These biofilms have not been studied in detail but they may have a profound effect on the dissolution process. Biofilms are very complex environments that control the local chemical environment, harboring a large variety of bacteria that may influence glass dissolution and they may contribute or remove particles from hydrothermal solutions. Microbial communities within biofilms are likely to be complex, but the nutrient availability in hydrothermal systems may favor sulfur- and sulfate-reducing bacteria (SRB), fermentative bacteria, sulfur-oxidizing bacteria, and iron-oxidizing bacteria. All of these bacteria may thrive in those biofilms, even when the bulk solution conditions may not be appropriate (Marshall, 1992; Davey and O'Toole, 2000). Despite the general recognition of the importance of the processes of colonization of surfaces and the formation of biofilms they remain an area of much future research (Cooksey and Wigglesworth-Cooksey, 1995; Dang and Lovell, 2000).

Suspended particles in the water column are among the most telling prospecting characteristics for hydrothermal plumes (e.g. Baker *et al.*, 1995), but their role within hydrothermal systems has not been systematically investigated. Hydrothermal plumes may include a range of solid (or colloidal) products that are related to hydrothermal activity or come from the water column. Their chemical composition may be inherited from biotic or abiotic water–rock interaction inside the hydrothermal system, or from the exchange (and co-precipitation) with seawater after venting. It is generally accepted that such materials play an important role in the deposition of hydrothermal sediments, or the chemistry of the oceans, and in deep-sea hydrothermal vent ecosystems. However, it is quite obvious that some fraction of these particles actually returns into hydrothermal systems, in particular, in major recharge aquifers with relatively rapid fluid motion. REE abundance patterns and Nd isotope composition of bulk ocean crust show that seawater particles play an important role, at least in the upper few hundred meters of the oceanic crust (Staudigel *et al.*, 1996). The experiments of Staudigel *et al.* (2003) showed that particles played an important role for the transport and deposition of relatively insoluble elements like Pb, Th, and the REE. One largely unexplored possibility is that seawater recharge and returning hydrothermal waters may introduce organic materials into the oceanic crust that could offer an organic carbon source for life in the oceanic crust providing nutrients for heterotrophic microbial communities.

The formation of biofilms and the growth of microbes in the oceanic crust offer great potential for fixation of elements that do not have any appropriate inorganic mineral hosts. This process will be particularly efficient for elements that play an important role in the metabolism of microbes (P, N, K, Fe), but it also extends to a wide range of micronutrients or bioreactive elements that may actually be toxic. Current experimental studies flag as candidates, in particular, Th, Pb, U, Rb, and Cs, but so far there is very little work done in this field. This process creates organic materials that will ultimately break down during diagenesis, freeing up their chemical inventory ultimately to be removed by pore waters or to be taken up by authigenic mineral phases like carbonates, clays, or zeolites. In particular, the insoluble elements will stay in place and be incorporated into these minerals. This type of process is likely to be a major control and substantially complicates our understanding of authigenic mineral formation, which is typically seen as a simple abiotic dissolution–precipitation process.

18.6 Concluding remarks

This review has offered a series of insights into how microbial activity might influence the chemical transfer between the oceanic crust and seawater and how it might depend on or influence the permeability of the oceanic crust. This shows that physical, chemical, and biological processes are interrelated and they all control the massive chemical and heat transfers that occur during water–rock interaction between seawater and the oceanic crust. The involvement of biology in these systems has been under-appreciated in the past, and it is quite clear that it does play an important role, at least within the upper 500 m of the oceanic crust, possibly covering one-half of the planet's surface. These discoveries offer a cornerstone for a new paradigm in the Earth sciences that increasingly sees the Earth as a truly integrated biological, chemical, and physical system. In particular the recycling of microbially altered ocean crust into the Earth's mantle and the utilization of mantle-derived nutrients show very clearly that there are much stronger connections between the biosphere and the lithosphere than previously thought. These connections challenge our largely disciplinary thinking about the Earth, and they demonstrate the need for multi-disciplinary study of the Earth. However, very important aspects of these new-found connections remain largely unknown, such as there is almost nothing known about the actual microbes involved, and how they actually dissolve volcanic rock, and/ or influence the chemical exchange. Studying these problems *in situ* on the ocean floor has proven to be extremely difficult, leaving this entire field open for much more exciting research.

Acknowledgments

We acknowledge the help of many collaborators in our previous work on microbially mediated glass alteration, including in particular R. Chastain, P. Keizer, B. Tebo, I. Thorseth, T. Torsvik, O. Tumyr, and A. Yayanos. This paper benefited greatly from reviews by

E. Davis, H. Elderfield, M. Fisk, and M. Mottl. Funding was provided though the National Science Foundation (OCE 0218945), and IGPP – Lawrence Livermore National Laboratory grant.

References

Alt, J. C., Kinoshita, H., Stokking, L. B. *et al.*, eds. 1993. *Leg 148 Preliminary Report: Hole 504B. Proceedings of the Ocean Drilling Program, Initial Reports*, Vol. 148. Washington, DC: Ocean Drilling Program.

Baker, E. T., German, C. R., and Elderfield, H. 1995. Hydrothermal plumes over spreadingcenter axes: global distribution and geological inferences. In *Seafloor Hydrothermal Systems: Physical, Chemical, and Geological Interactions.* Geophysical Monograph 91, eds. S. Humphris, R. Zierenberg, L. S. Mullineaux, and R. Thomson. Washington, DC: American Geophysical Union, pp. 47–71.

Booij, E., Gallahan, W. E., and Staudigel, H. 1995. Duration of low temperature alteration in the Troodos Ophiolite. *Chem. Geol.* **126**: 155–167.

Cooksey, K. E. and Wigglesworth-Cooksey, B. 1995. Adhesion of bacteria and diatoms to surfaces in the sea: a review. *Aquat. Microbiol. Ecol.* **9**: 87–96.

Dang, H. and Lovell, C. R. 2000. Bacterial primary colonization and early succession on surfaces in marine waters as determined by amplified rRNA gene restriction analysis and sequence analysis of 16S rRNA genes. *Appl. Environ. Microbiol.* **66**: 467–475.

Davey, M. E. and O'Toole, G. A. 2000. Microbial biofilms: from ecology to molecular genetics. *Microbiol. Molec. Biol. Rev.* **64**: 847–867.

Deming, J. W. and Baross, J. A. 1993. Deep-sea smokers: windows to a subsurface biosphere? *Geochim. Cosmochim. Acta* **57**: 3,219–3,230.

Emerson, D. and Moyers, C. L. 2002. Neutrophilic bacteria are abundant at the Loihi Seamount hydrothreaml vents and play a major role in Fe oxide deposition. *Appl. Environ. Microbiol.* **68**: 3,085–3,093.

Fisk, M. R. 1999. New shipboard laboratory may answer questions about deep biosphere. *Eos, Trans. Am. Geophys. Union* **80**: 580.

Fisk, M. R., Giovannoni, S. J., and Thorseth, I. H. 1998. The extent of microbial life in the volcanic crust of the ocean basins. *Science* **281**: 978–979.

Furnes, H. and Staudigel, H. 1999. Biological mediation in ocean crust alteration: how deep is the deep biosphere? *Earth Planet. Sci. Lett.* **166**: 97–103.

Furnes, H., Thorseth, I. H., Tumyr, O., Torsvik, T., and Fisk, M. R. 1996. Microbial activity in the alteration of glass pillow lavas from Hole 896A. In *Proceedings of the Ocean Drilling Program, Science Results*, eds. J. C. Alt, H. Kinoshita, L. B. Stokking, and P. J. Michael. College Station, TX: Ocean Drilling Program, pp. 191–206.

Furnes, H. , Staudigel, H., Thorseth, I. H., Torsvik, T., Muehlenbachs, K., and Tumyr, O. 2001. Bioalteration of basaltic glass in the oceanic crust, *Geochem. Geophys. Geosys.* **2**: ido: 2000GC000150.

Giovannoni, S. J., Fisk, M. R., Mullins, T. D., and Furnes, H. 1996. Genetic evidence for endolithic microbial life colonizing basaltic glass/seawater interfaces. In *Proceedings of the Ocean Drilling Program, Science Results*, Vol. 148, eds. J. C. Alt, H. Kinoshita, L. B. Stokking, and P. J. Michael. College Station, TX: Ocean Drilling Program, pp. 207–214.

Hart, S. R. and Staudigel, H. 1978. Oceanic crust: age of hydrothermal alteration. *Geophys. Res. Lett.* **5**: 1,009–1,012.

Jannasch, H. 1995. Microbial interactions with hydrothermal fluids. In *Seafloor Hydrothermal Systems: Physical, Chemical, and Geological Interactions*, Geophysical Monograph 91, eds. S. Humphris, R. Zierenberg, L. S. Mullineaux, and R. E. Thomson. Washington, DC: American Geophysical Union, pp. 273–296.

Juniper, S. K. and Tebo, B. M. 1995. Microbe–metal interactions and mineral deposition at hydrothermal vents. In *The Microbiology of Deep-Sea Hydrothermal Vents*, ed. D. M. Karl. Boca Raton, FL: CRC Press, pp. 219–253.

Krumbein, W. E., Urzi, C. E. C. A., and Gehrman, C. 1991. Biocorrosion and biodeterioration of antique and medieval Glass. *Geomicrob. J.* **9**: 139–160.

Langmuir, D. 1978. Uranium solution–mineral equilibria at low temperatures with applications to sedimentary ore deposits. *Geochim. Cosmochim. Acta* **42**: 547–569.

Lovley, D. R., Phillips, E. J. P., Gorby, Y. A., and Landa, E. R. 1991. Microbial reduction of uranium. *Nature* **350**: 413–416.

Lovley, D. R., Roden, E. E., Phillips, E. J. P., and Woodward, J. C. 1993a. Enzymatic iron and uranium reduction by sulfate-reducing bacteria. *Mar. Geol.* **113**: 41–53.

Lovley, D. R., Widman, P. K., Woodward, J. C., and Phillips, E. J. P. 1993b. Reduction of uranium by cytochrome c3 of Desulfovibrio vulgaris. *Appl. Environ. Microbiol.* **59**: 3,572–3,576.

Marshall, K. C. 1992. Biofilms: an overview of bacterial adhesion, activity, and control at surfaces. *ASM News* **58**: 202–207.

Mellor, E. 1922. Les lichen vitricole et la déterioration dex vitraux d'église, Thèse de dochert thesis, Sorbonne, Paris.

Robinson, P. T., Flower, M. F., Staudigel, H., and Swanson, D. A. 1979. Lithology and eruptive stratigraphy of Cretaceous oceanic crust, western Atlantic Ocean. In *Initial Reports of the Deep Sea Drilling Project*, Vols. 51–53, eds. T. Donnelly and J. Francheteau. Washington, DC: US Government Printing Office, pp. 1,535–1,556.

Ross, K. A. and Fisher, R. V. 1986. Biogenic grooving on glass shards. *Geology* **14**: 571–573.

Smith, D. C., Spivack, A. J., Fisk, M. R., Haveman, S. A., and Staudigel, H. 2000. Tracer-based estimates of drilling-induced microbial contamination of deep sea crust. *Geomicrobiol. J.* **17**: 207–219.

Staudigel, H., Chastain, R. A., Yayanos, A., and Bourcier, W. 1995. Biologically mediated dissolution of glass. *Chem. Geol.* **126**: 147–154.

Staudigel, H., Plank, T., White, W. M., and Schmincke, H.-U. 1996. Geochemical fluxes during seafloor alteration of the upper oceanic crust: DSDP Sites 417 and 418. In *SUBCON: Subduction From Top to Bottom*, Vol. 96, eds. G. E. Bebout and S. H. Kirby. Washington, DC: American Geophysical Union, pp. 19–38.

Staudigel, H., Yayanos, A., Chastain, R., Davies, G., Verdurmen, E. A. T., Schiffman, P., Bourcier, R., and De Baar, H. 1998. Biologically mediated dissolution of volcanic glass in seawater. *Earth Planet. Sci. Lett.* **164**: 233–244.

Staudigel, H., Furnes, H., Kelley, K., Plank, T., Muehlenbachs, K., Tebo, B., and Yayanos, A. 2004. Deep subsurface biosphere at mid-Ocean ridges. In *The Subseafloor Biosphere at Mid-Ocean Ridges*, Geophysical Monograph, Vol. 144, eds. W. Wilcock, E. Delong, D. Kelley, J. Baross, and S. Cary. Washington, DC: American Geophysical Union.

Stein, C. A. and Stein, S. 1994. Constraints on hydrothermal heat flux through the oceanic lithosphere from global heat flow. *J. Geophys. Res.* **99**: 3,081–3,096.

Stein, C. A., Stein, S., and Pelayo, A. 1995. Heat flow and hydrothermal circulation. In *Seafloor Hydrothermal Processes*. Geophysical Monograph, Vol. 91, eds. S. E.

Humphris, R. A. Zierenberg, L. S. Mullineaux, and R. E. Thomson. Washington, DC: American Geophysical Union, pp. 425–445.

Thorseth, I. H., Furnes, H., and Heldal, M. 1992. The importance of microbial activity in the alteration zone of natural basaltic glass. *Geochim. Cosmochim. Acta* **56**: 845–850.

Thorseth, I. H., Torsvik, T., Furnes, H., and Muehlenbachs, K. 1995a. Microbes play an important role in the alteration of oceanic crust. *Chem. Geol.* **126**: 137–146.

Thorseth, I. H., Furnes, H., and Tumyr, O. 1995b. Textural and chemical effects of bacterial activity on basaltic glass: an experimental approach. *Chem. Geol.* **119**: 139–160.

Thorseth, I. H., Torsvik, T., Torsvik, V., Daae, F. L., Pedersen, R. B., and Keldysh-98 Scientific Party 2001. Diversity of life in ocean floor basalt. *Earth Planet. Sci. Lett.* **194**: 31–37.

Torsvik, T., Furnes, H., Muehlenbachs, K., Thorseth, I. H., and Tumyr, O. 1998. Evidence for microbial activity at the glass–alteration interface in oceanic basalts. *Earth Planet. Sci. Lett.* **162**: 165–176.

Welch, S. A. and Ullman, W. J. 1996. Feldspar dissolution in acidic and organic solutions: compositional and PH dependence of dissolution rate. *Geochim. Cosmochim. Acta* **60**: 2,939–2,948.

Part V

Geochemical fluxes

19

Geochemical fluxes through mid-ocean ridge flanks

C. Geoffrey Wheat and Michael J. Mottl

19.1 Introduction

One of the fundamental missions for oceanic geochemists is to elucidate the geochemical cycle of the elements in the oceans. Historically, scientists have studied processes that center on inputs to the oceans from riverine and atmospheric sources and outputs resulting from deposition of material to the seafloor. Nearly 40 years ago Mackenzie and Garrels (1966) summed these inputs and outputs. They concluded that for a steady-state ocean additional sinks are required to balance geochemical cycles for many of the major and minor elements in seawater. This steady-state criterion is justified by the long residence time in the oceans for many elements relative to the oceanic mixing or turnover rate. To balance geochemical budgets they invoked a process called "reverse weathering" in which secondary minerals precipitate within the crust. This hypothesis was somewhat unsatisfying because of the lack of evidence for such reactions. Thirteen years later Edmond et al. (1979) measured the chemical composition of spring water that vented from a mid-ocean ridge (MOR) hydrothermal system. They used these data along with estimates of oceanic convective heat loss to calculate global geochemical budgets for high-temperature hydrothermal systems. They estimated that hydrothermal fluxes (a term used in this chapter without regard to area, i.e. meaning total flow of a species in a regional or global ocean-crustal sense) are comparable to those from rivers and concluded that these fluxes could solve many of the elemental imbalances in the oceans. In essence, reaction of seawater with basalt could be the process described as "reverse weathering" by Mackenzie and Garrels (1966). Although Edmond et al. (1979) overestimated hydrothermal fluxes because they used an excessively high figure for total heat loss (Sclater et al., 1980; Mottl and Wheat, 1994; Stein and Stein, 1994; Elderfield and Schultz, 1996), their pivotal paper highlighted the possibility that flow of seawater through oceanic crust and the accompanying chemical exchange between seawater and basalt could result in fluxes that are important for global geochemical cycles.

More than two decades later we are still striving to estimate geochemical fluxes from oceanic hydrothermal systems. Although the study of hot springs on MORs has garnered

Hydrogeology of the Oceanic Lithosphere, eds. E. E. Davis and H. Elderfield. Published by Cambridge University Press.
© Cambridge University Press 2004.

most of the emphasis and funding, most of the seawater flux and convective heat transport through oceanic crust occurs on ridge flanks (Sclater *et al.*, 1980; Stein and Stein, 1994; Schultz and Elderfield, 1997). Because heat is removed by flow of seawater through crust up to several tens of millions of years old and possibly older (e.g. see Chapter 13), we can infer that ridge flank hydrothermal systems underlie more than one-third of the seafloor. Convection within the crust, moreover, is not limited to those hydrothermal systems that vent to the oceans. Seawater flows through much of the rest of the crust as well, but these waters are trapped by an overlying blanket of impermeable sediment and so do not affect geochemical mass balances. The huge area and great variety of seafloor underlain by seawater convection requires us to focus on understanding processes and representative features if we are to constrain global geochemical fluxes from these systems. Two general approaches have been taken: (1) examination of secondary minerals and alteration halos to constrain the extent of crustal alteration in various units of the crust, and (2) determination of the composition of formation water in basaltic basement within a geologic and physical context and comparison of these fluids with seawater.

The two approaches are complementary. Crustal alteration is a cumulative process that occurs as the crust ages. Altered sections of crust, typically sampled by drilling, provide an integrated view of all the alteration that has affected a given crustal section from the time of its formation. Chemical fluxes can be estimated simply from the difference in composition between an altered section of sufficient age and its fresh equivalent, combined with an estimate of altered crustal thickness and the rate of crustal production and subduction. To understand the full range of alteration processes, however, and the role each type plays in overall chemical fluxes, more detailed studies of crustal sections over a range of ages are required. In some cases it is even possible to estimate the seawater flux involved in a given type of alteration, from chemical evidence regarding water/rock ratios that prevailed during alteration.

By contrast, the composition of seawater-derived formation waters in basement integrates the effects of crustal alteration over smaller distances and shorter times. Estimating global fluxes from data on solution composition requires that waters be collected from numerous settings in crust of a range of ages, representing different temperatures and redox conditions. The mass rate of seawater flux through each of these settings must then be estimated to derive elemental fluxes, which can then be integrated for the whole range of conditions over the entire seafloor. While it is difficult to determine the seawater flux directly, we do know the total convective heat loss from the oceanic crust and its distribution with crustal age (e.g. Sclater *et al.*, 1980; Stein and Stein, 1994). By partitioning this heat loss over a range of exit temperatures (temperature in uppermost basement) and crustal ages, and relating these temperatures to the composition of formation waters, we can estimate both the mass flux and the associated chemical fluxes.

At this time we do not know the magnitude of chemical fluxes that result from seawater circulation through and reaction with the crust on MOR flanks. In the following sections we present known constraints and our best estimate for some chemical fluxes. We discuss the partition of convective heat loss into distinct regions, i.e. ridges and flanks. We then present an overview of the methods used and results from studies of crustal alteration and formation water composition. Results from each of these sections are then merged to place constraints on geochemical fluxes from hydrothermal systems on ridge flanks. Finally we estimate the flux of phosphate as an example. The results are considered in the context of global ocean chemical budgets in Chapter 21.

19.2 Convective heat loss

Seawater flow within the crust is ultimately driven by heat from the mantle. Along the MOR axis this heat ascends advectively from the mantle in the form of basaltic magma. On MOR flanks this heat ascends conductively from the cooling lithospheric mantle. Adding heat to seawater lowers its density and makes it buoyant, causing it to flow at a rate that is related directly to the permeability structure of the crust via Darcy's Law. The heat transferred through the seafloor by hydrothermal circulation can be estimated from the difference between total heat flow predicted from models of lithospheric cooling and the conductive heat flow through the seafloor measured by conventional methods. Relying heavily on the data of Sclater *et al.* (1980) and Stein and Stein (1994), Mottl (2003) recently estimated this amount to be 9.9 ± 2 TW. The partitioning of this heat between ridge axis and flanks depends in part on the definition of where the axis ends and the flanks begin. Mottl and Wheat (1994) present a rationale for partitioning this heat and use Mg to illustrate the importance of hydrothermal processes in regulating geochemical cycles. Mottl (2003) has refined these arguments. Here we present a synopsis of his findings, but his paper should be consulted for a more detailed discussion.

The portion of the total convective power output that is available for fueling axial hydrothermal systems is limited by the amount of new oceanic crust produced each year. Parsons (1981) estimated an areal production rate of 3.3 ± 0.2 km^2 yr^{-1} that includes inputs from back-arc spreading. For an average crustal thickness of 6.5 ± 0.8 km (0.7 km for flows, 1.2 km for dikes, and 4.6 km for gabbros; see Chapter 3) and a density of 2.8 g cm^{-3}, this production rate yields $6.0 \pm 0.8 \times 10^{16}$ g of new oceanic crust per year. Not all of the heat associated with this new crust drives seawater circulation through the crust. The upper 0.7 km of extrusive flows cool by direct contact with ocean bottom water and so are not counted in the following arguments. Heat associated with formation of the remainder of the crust includes the latent heat (transition from liquid to solid; 676 J g^{-1}, Fukuyama, 1985) and the heat liberated upon cooling from a magmatic temperature (1175 ± 25 °C) to a hydrothermal temperature (350 ± 30 °C; e.g. Von Damm, 1995). For a basaltic heat capacity of 1.2 ± 0.15 J g^{-1} °C^{-1} the maximum heat available to drive high-temperature MOR hydrothermal systems is 2.8 ± 0.3 TW, for crystallizing and cooling the entire crust to 350 °C, all the way to the Moho. This may be reasonable for slow-spreading axes that have deep crustal faults. For intermediate- to fast-spreading ridges a lower limit is 1.5 ± 0.18 TW, a number derived by assuming the latent heat for the entire section of crust is removed, but only the dikes cool to 350 °C, thus allowing a narrow magma chamber to persist in the gabbroic section. Accounting for the global distribution of spreading rates

(Baker *et al.*, 1996), a reasonable value for hydrothermal heat loss on axis is 1.8 ± 0.3 TW. Therefore the available heat for non-axis hydrothermal systems is 8.1 ± 2 TW (9.9 ± 2 TW minus 1.8 ± 0.3 TW).

Here we define the ridge flanks as that portion of the oceanic crust that starts outside the zone influenced by thermal input from magmatic intrusions and ends at subduction zones. Considering that gabbros from Hess Deep, formed at the fast-spreading East Pacific Rise, lack evidence for hydrothermal circulation at temperatures lower than 250 °C (Gillis, 1995; Lecuyer and Reynard, 1996; 180 °C for the MARK gabbros on the Mid-Atlantic Ridge, Gillis *et al.*, 1993) and that these gabbros were cooled well within 1.3 Ma of formation (Gillis, 1995), we suggest that the axial heat influence is exhausted by 1 Ma, 10–80 km off-axis for the global range of half-spreading rates of 10–80 mm yr^{-1}. Thus the cumulative heat loss up to 1 Ma is 2.8 ± 0.3 TW, which crystallizes the entire crust and cools it to 350 °C, plus 0.15 TW to cool the dikes from 350 to the 2 °C of bottom seawater and an additional 0.22 TW to cool the gabbros from 350 to 215 °C, which is the average of 180 and 250 °C mentioned above. The resulting total of 3.2 ± 0.4 TW is equal to the acknowledged overestimate of Stein and Stein (1994) and is at the upper end of the range estimated by Mottl (2003), who noted that if axial heat loss is maximally efficient then the near-axial crust from 0.1 to 1 Ma in age may experience reheating rather than cooling. If this crust really is still cooling, then the additional 0.37 TW would result in hydrothermal systems that have different chemistries than high-temperature MOR systems, and this heat loss would be distributed over tens of kilometers from the ridge axis.

The remaining 6.7 ± 2 TW or 68% of the convective heat loss is distributed on the ridge flanks (Table 19.1), in crust as old as 80 Ma (Embley *et al.*, 1983; Chapter 10). The major unknown is the distribution of this heat loss with temperature and the resulting seawater flux. Is most of this heat lost from basement at 5, 20, or as high as 64 °C, which is the source temperature of the only ridge flank spring sampled to date (Mottl *et al.*, 1998; Wheat and Mottl, 2000; Wheat *et al.*, 2002)? How does the variation in temperature over this range affect crustal alteration and the composition of formation waters in basaltic basement?

19.3 Crustal alteration

It is clear from global heat flux estimates that seawater circulates through oceanic crust on a massive scale. Compilations of permeability measurements in ocean crustal boreholes (Fisher, 1998; Becker and Fisher, 2000) imply that most circulation is probably concentrated in the upper few hundred meters of basaltic crust (Fig. 19.1). Most older measurements in the uppermost part of the crust are lower than 10^{-12} m^2, yet this unit is clearly cooled and altered (Fig. 19.1). This and other observations point to the importance of highly permeable zones such as faults which are under-represented by drilling but contribute to a much higher effective formation permeability (e.g. Davis and Becker, 2002; Becker and Davis, 2003; Chapters 7 and 8). Few zones of extreme permeability have been observed, but

Table 19.1 *Cumulative oceanic advective heat loss with age*[a]

Age (Ma)		Heat Loss (TW)
0.1	(axial)	1.8
0.2	(near-axial)	2.0
1		3.2
2		3.4
3		3.8
4		4.1
5		4.5
6		4.8
7		5.1
8		5.4
9		5.6
20		7.5
35		9.0
52		9.8
65		9.9

[a]Modified from Mottl (2003) and Stein and Stein (1994).

their contribution to a high formation-scale permeability is strongly suggested on the basis of observations and simulations of regional heat flow patterns (Fisher and Becker, 1995; Davis *et al.*, 1997a), lateral gradients in basement temperature and pressure (Davis and Becker, 2002; Chapter 8), and an emerging scale dependence of permeability (Becker and Davis, 2003; Chapter 7).

Below the extrusive section borehole permeabilities decrease to about 10^{-16}–10^{-18} m^2 (although the formation-scale permeability may be larger here also; see Chapter 8). The upper extrusive igneous section should therefore host most of the seawater flow on ridge flanks, become the most altered part of the crust, and generate the most significant ocean-scale chemical fluxes. We note below that this generalization is consistent with results from low-temperature rock alteration studies, but that deeper sections can be highly altered at high-temperature at the ridge axis. Chemical fluxes from these high-temperature MOR hydrothermal systems may have an effect on global budgets for some elements (Elderfield and Schultz, 1996), but quantifying these fluxes is not in the scope of this chapter.

Mottl (2003) estimated that typical oceanic crust consists of about 0.7 km of extrusive flows, 1.2 km of sheeted dikes, and 4.6 km of gabbros. Each of these sections is exposed to seawater circulation at some point and thus is subject to alteration via inorganic and/or microbially mediated reactions. The result is a continuum of alteration that records the complex thermal and redox history of the circulating seawater. At one extreme, portions of the gabbroic section have experienced water–rock reaction at temperatures as high as

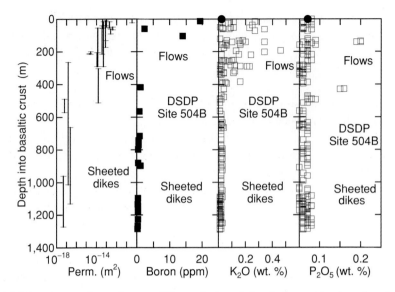

Fig. 19.1 Measurements of crustal permeability and selected chemical concentrations from bulk-rock analyses of DSDP and ODP cores. Permeability data illustrate a difference of several orders of magnitude between the upper flows and the lower sheeted dikes (Fisher, 1998; Becker and Fisher, 2000). Chemical data highlight changes in the bulk composition of basalt resulting from low-temperature hydrothermal alteration of the upper permeable basaltic flows. Seawater is a source of boron, potassium, and phosphate during the formation and deposition of secondary minerals in basaltic crust (modified from Alt *et al.*, 1996b).

600–800 °C with highly altered and reduced seawater (e.g. Gillis, 1995). In contrast, the upper few hundred meters of oceanic crust comprising pillow and sheet flows have oxidized halos that form by reaction with relatively unaltered, oxidized seawater at temperatures close to those of bottom seawater (e.g. Alt *et al.*, 1996a,b; Chapter 15).

Reactions characterizing these different thermal and chemical regimes have been partitioned in recent reviews between those that occur within a few kilometers of the ridge axis (Alt and Bach, 2003; Gillis, 2003) and those that occur in crust greater than about 1 Ma in age (Staudigel *et al.*, 1996; Chapter 15). The net effect of seawater–rock alteration on chemical exchange and oceanic budgets must include both high- and low-temperature processes, with the possibility that low-temperature exchange can reverse fluxes that occurred at high temperature at the ridge axis. Sulfate, for example, is removed from circulating seawater by anhydrite precipitation as seawater warms to about 150 °C within the crust (Seyfried and Bischoff, 1979). Anhydrite dissolves when the hydrothermal system cools below 150 °C, sometimes within the boundary of the axial valley (e.g. Mottl *et al.*, 1994; Alt, 1995a). Crustal alteration products and reactions have recently been thoroughly reviewed elsewhere (Alt, 1995b; Staudigel *et al.*, 1996; Alt and Bach, 2003; Gillis, 2003; Chapter 15), but for completeness, we present a brief summary of these works in the context of estimating chemical fluxes from ridge flanks.

Only a handful of deep-sea drilling sites have penetrated more than 100 m into basaltic crust. Much of the available data is from Deep Sea Drilling Project (DSDP) Sites 396 (Bohlke *et al.*, 1981), 417, 418 (Thompson, 1983; Staudigel *et al.*, 1996), and 504; and Ocean Drilling Program (ODP) Sites 896 (Alt *et al.*, 1986 and 1996a,b; Teagle *et al.*, 1996) and 735 (Bach *et al.*, 2001). Studies of basalt weathering from dredged samples (e.g. Hart, 1973) and ophiolites (e.g. Nehlig *et al.*, 1994; Gillis and Banerjee, 2000) have also contributed to our understanding of crustal alteration.

As magma intrudes and extrudes, and crystallizes and cools at the ridge axis it experiences both cool and hot hydrothermal reactions. Cool reactions (10–50 °C) occur in the upper extrusive volcanic section (comprising pillows, breccias, and massive flows) where iron-rich secondary phases such as celadonite, Fe-oxides, and nontronite are present (Alt *et al.*, 1986; Teagle *et al.*, 1996). These phases form halos around mineral grains and veins that typically are millimeters thick but can be up to 5 cm thick (e.g. Bohlke *et al.*, 1981; Chapter 15). The water that deposits the vein minerals is rich in iron, so that the secondary materials have 40% more Fe than found in fresh basalt (Humphris *et al.*, 1980). This dissolved iron must have been reduced (Fe^{+2}) because the oxidized form (Fe^{+3}) is readily removed from solution (e.g. Feely *et al.*, 1990; Mottl and McConachy, 1990; Rudnicki and Elderfield, 1993). This water, which must have also been enriched in silica, probably originated by reaction of seawater with basalt at high temperature, that was then mixed with cooler, relatively unreacted seawater. Present analogs for this type of water are the hydrothermal fluids that vent from Loihi Seamount at 11–77 °C (Wheat *et al.*, 2000a). Spring waters from Loihi Seamount have Fe/Mn ratios as high as 58, about the ratio in basalt (Frey and Clague, 1983), and high Fe/H_2S ratios that would have to reflect molar ratios greater than one in the original high-temperature fluid. An alternative source for this iron could be overlying sediment, where microbial processes reduce iron oxides during the oxidation of organic matter. Dissolved silica could likewise be generated in sediment, by dissolution of diatoms. Dissolved iron and silica could then diffuse into basement where they would precipitate as secondary minerals (e.g. Wheat and McDuff, 1994, 1995). This scenario requires a sediment layer, however, which is typically not present at the ridge axis. A second more likely alternative is that the iron oxides originate from reduced-Fe-bearing minerals in basalt that have been oxidized by cold bottom seawater that recharges the hydrothermal system (Alt and Bach, 2003).

Deeper within the extrusives the dominant sheet silicate changes from smectite to mixed-layer chlorite-smectite at ~100–200 °C and to chlorite at >200 °C. The upper sheeted dike section is even warmer (300–400 °C) and contains sub-greenschist and greenschist facies minerals (Alt *et al.*, 1986; Alt and Bach, 2003). Reaction temperatures warm further to about 400 °C in the lower sheeted dikes where plagioclase and hornblende are present (Gillis, 1995; Vanko *et al.*, 1996). This region is considered the "root zone" for high-temperature black smoker hydrothermal fluids rich in Cu, Zn, and sulfide (e.g. Von Damm, 1995). Reaction temperatures are higher still (>700 °C) within the uppermost gabbroic section (Kelley and Delaney, 1987; Gillis *et al.*, 1993; Kelley *et al.*, 1993), where assemblages indicate seawater penetration at the axis. The extent of seawater penetration as indicated by the empirical water/rock ratio decreases by several orders of magnitude from the upper volcanics to the deeper gabbros (Chapter 15; Gillis, 2003), consistent with the likely permeability structure of the crust. Bulk-rock analyses indicate that MOR hydrothermal systems remove K, Li, Cu, Zn, and S from basalt while adding H_2O and CO_2 to basalt and lowering its $\delta^{18}O$ (Gillis, 1995; Alt and Bach, 2003). This picture is generally consistent with that inferred from ophiolite exposures (Chapter 9).

Alteration by seawater at low temperature on ridge flanks is ubiquitous within each of the fundamental crustal units, although more extensive alteration is confined to the uppermost several hundred meters of volcanics (Alt *et al.*, 1996a; Chapter 15; see also Chapter 9). Secondary minerals include carbonates, smectite, saponite, talc, zeolites, and pyrite (Alt *et al.*, 1996a, Bach *et al.*, 2001). Hydrothermal systems warm in response to an added sediment burden that diminishes direct contact between basalt and seawater. As the connectivity to bottom seawater decreases, exchange between reducing basaltic minerals and seawater can produce a reduced fluid as indicated by the formation of saponite in veins. Each of these minerals has been identified within the 1500-m-long gabbroic section at ODP Site 735, but they are much less abundant than in the upper volcanic section (Bach *et al.*, 2001). Bulk-rock analyses indicate that this low-temperature hydrothermal exchange on ridge flanks results in the removal of S, Mg, Si, and Ca from basalt; the addition of P, Fe, H_2O, K, Li, Rb, Cs, B, and U to the crust; and higher $\delta^{11}B$, $\delta^{18}O$, $^{87}Sr/^{86}Sr$, and Fe^{3+}/Fe^T (Alt, Chapter 15). Note that at high temperatures K is removed from basalt and added to hydrothermal fluids, whereas at low temperatures K is removed from seawater to form secondary minerals within the crust, consistent with experimental results (Seyfried, 1977; Seyfried and Bischoff, 1979). Similar responses to temperature have been documented for other elements (Li, Rb, Cu, Y, REE, and Mo; Wheat *et al.*, 2002).

The combination of on- and off-axis reactions within the volcanic section results in crust that typically contains 10–20% altered materials but can include zones that consist of 40–100% altered materials (Chapter 15). The combined effect of high-temperature exchange at the axis and low-temperature exchange on the flanks is the removal of acidity, Ba, Mn, Fe, Cu, Ni, Zn, Li, Si, and Ca from the crust; the addition of C, P, Mg, Na, K, B, Rb, H_2O, Cs, and U to the crust; higher $\delta^{11}B$, $\delta^{18}O$, δD, $^{87}Sr/^{86}Sr$, and Fe^{3+}/Fe^T in the crust; a lower crustal δ^6Li, and no change in S (e.g. Fig. 19.1; Hart, 1973; Bohlke *et al.*, 1981; Hart and Staudigel, 1982; Thompson, 1983; Chan *et al.*, 1992; Ishikawa and Nakamura, 1992; Spivack and Staudigel, 1994; Alt, 1995a,b; Alt *et al.*, 1996a; Teagle *et al.*, 1996; Bach *et al.*, 2001).

19.4 Composition of formation waters

Studies of secondary minerals and alteration halos within basaltic rocks provide an integrated view of crustal alteration at a given location. In essence, these rocks tell the evolutionary story as the crust ages and is buried beneath an ever-increasing sediment column. In contrast, the composition of formation waters characterizes the present state of alteration within the crust, both thermally and chemically. Studies of formation waters also represent a larger geographic area, unlike the pinpoint studies of basalt recovered by deep-sea drilling.

Ideally, one could react seawater with basalt in the laboratory to determine the composition of formation water. Such experiments have been conducted at temperatures below 100 °C, but rates of reaction are so slow that even experiments lasting several hundred days did not reach equilibrium (Seyfried, 1977; Seyfried and Bischoff, 1979). Because direct experimentation is not realistic at these low temperatures, in situ formation-water samples must be obtained. Four techniques have been employed for estimating the chemical composition of formation waters on ridge flanks: (1) sampling spring waters; (2) determining systematic variations in the chemical composition of sediment pore waters in areas where they are upwelling at several centimeters per year or more, or where they have been sampled adjacent to the basement interface; (3) sampling waters in boreholes that penetrate basaltic basement; and (4) using the ground rock interstitial normative determination (GRIND) technique developed by Wheat *et al.* (1994).

Sampling springs on the seafloor is an ideal way to collect uncontaminated formation waters, but this typically requires an underwater vehicle (ROV or submersible) and so is difficult and costly. Springs on ridge flanks produce a subtle signature both on the seafloor and in the water column, and hence they are difficult to locate. In fact, direct sampling has occurred at only one location, Baby Bare, a high-standing outcrop on an otherwise buried basement ridge 101 km east of the Juan de Fuca Ridge on 3.5 Ma crust. Four springs were found near the summit along a fault that strikes roughly parallel to the buried ridge and to the active spreading center to the west (Becker *et al.*, 2000). The springs support communities of Thysirid clams (Mottl *et al.*, 1998). Spring temperatures are 25.0 °C, although warmer (63–64 °C) formation waters lie only a few to at most 20 m below the seafloor along this fault (Wheat *et al.*, 2003a). These springs could be thoroughly analyzed because large (750 ml) samples devoid of sampling artifacts could be collected. To date 41 dissolved species and four stable isotopes have been determined (Fig. 19.2; Mottl *et al.*, 1998; Sansone *et al.*, 1998; Wheat and Mottl, 2000; Butterfield *et al.*, 2001; Monnin *et al.*, 2001; McManus *et al.*, 2002; Wheat *et al.*, 2002).

The second method for estimating the composition of formation water involves searching for systematic variations in chemical profiles of pore waters and extrapolating these profiles through the sediment section to the sediment–basement interface (e.g. Wheat and Mottl, 1994; Wheat and McDuff, 1995). This method is a relatively inexpensive way to estimate the concentration of many elements in formation water (Fig. 19.3). It does have shortcomings, however: (1) unless the entire sediment column can be penetrated, which usually requires a drill ship, only zones of focused and rapid pore-water upwelling can provide data for a reasonable extrapolation (Wheat and Mottl, 2000); (2) pore waters may be altered by sedimentary diagenetic processes as the waters pass through the sediment column (e.g. Maris *et al.*, 1984; Wheat and McDuff, 1994); (3) artifacts caused by sediment squeezing or centrifuging to separate pore waters may significantly alter their composition (de Lange *et al.*, 1992); and (4) this technique often does not provide a large enough sample for measurement of many trace elements and isotopes. Nevertheless, excellent agreement has

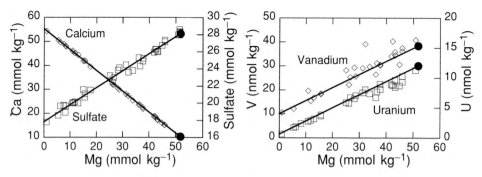

Fig. 19.2 Concentrations of Ca, sulfate, V, and U versus Mg in spring waters from Baby Bare seamount on 3.5 Ma crust of the eastern Juan de Fuca Ridge flank. The linear relationships indicate conservative mixing between two solutions, bottom seawater and basement formation water. Composition of formation water can be determined by extrapolating to 0 mmol kg^{-1} Mg. This extrapolation is reasonable because the lowest measured concentration of Mg was 0.98 mmol kg^{-1}. Filled circles are the concentration in seawater. These data indicate that at this site the oceanic crust is a net sink for sulfate, V, and U and a net source for Ca. Plots are modified from Wheat and Mottl (2000) and Wheat *et al.* (2002).

been obtained between the estimated composition of basement water using pore-water data from the upper few meters of the sediment column near springs and the Baby Bare spring waters for Cl, Mg, Ca, Sr, Na, Rb, sulfate, alkalinity, nitrate, ammonium, phosphate, and B, and good agreement for K and Li (Wheat and Mottl, 2000). This agreement is possible because coring operations targeted zones where pore water is seeping upward at volumetric fluxes of several centimeters per year or more. At these rates the advective flux greatly exceeds diagenetic fluxes for most elements. Only nine ridge flank hydrothermal sites have been examined using near-surface pore waters as a proxy for the composition of formation waters. Seven of these sites are summarized by Mottl and Wheat (1994). The two new sites are on 0–4 Ma crust on the eastern flank of East Pacific Rise (EPR) at 15° S (Mottl and Wheat, 2000) and on 20–25 Ma crust on the eastern flank of the EPR near the Middle American Trench at 9° N (Friedmann *et al.*, 2001).

Ridge flank hydrothermal sites have also been drilled as part of the DSDP and ODP (see Mottl and Wheat, 1994; Davis *et al.*, 1997b). Sediment pore waters sampled from drillcore have documented both upwelling and downwelling through the sediment at a few ridge flank sites (Mottl, 1989; Bender *et al.*, 1983; Davis *et al.*, 1997b). A significant benefit of drilling is that sediment from immediately above the basal contact can be recovered, squeezed, and the resultant pore waters analyzed. One can then use pore-water chemical gradients near the sediment–basement interface to extrapolate to the composition of the underlying formation water, regardless of the direction or lack of flow through the sediment section (e.g. Wheat *et al.*, 1996a).

In addition to providing an estimate for the composition of basement formation waters, sediment pore-water studies also provide important constraints on physical processes that guide seawater circulation through both the sediments and the igneous crust. For example,

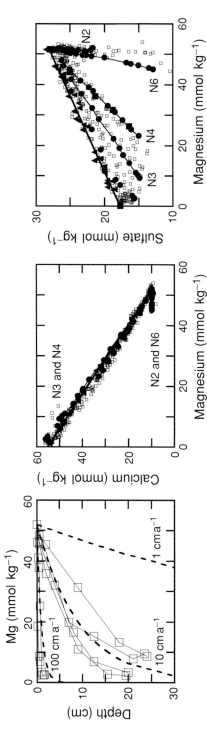

Fig. 19.3 Pore-water concentration–depth profiles and concentration cross-plots from sediments of the eastern flank of the Juan de Fuca Ridge illustrate how these data are used to determine seepage flux and constrain the composition of basement formation water. Fluxes (Darcy velocities) are estimated from concentration–depth profiles for Mg coupled with a simplified one-dimensional model that includes terms for time-dependent advection, diffusion, and reaction. The data are from push cores collected on Mama Bare and show a range of upwelling rates. The composition of the upwelling fluid can in some cases be determined from the asymptotic concentration; however, sub-surface mixing with bottom seawater can alter this asymptote (e.g. Wheat *et al.*, 1998). Concentration cross-plots provide a more robust means for estimating formation-water composition for those species that behave conservatively on mixing and thus that produce a single linear relationship that is identical to that for data from actual springs (triangles fit with solid line). Reactions in the sediment affecting Ca and especially sulfate produce deviations from a single linear relationship. With an increasing seepage velocity from cores N2 to N6 to N4 to N3, the advective flux increases relative to the reactive flux. When the reactive flux becomes negligible relative to the advective flux, the pore-water data match the spring data. On the eastern flank of the Juan de Fuca Ridge this happens at upwelling seepage velocities in excess of several centimeters per year. Plots are modified from Wheat and Mottl (2000) and Wheat *et al.* (2000b).

chemical data are useful as a measure of (1) the rate of fluid flow through the crust using both reactive (e.g. Maris and Bender, 1982; Maris *et al.*, 1984; Bender *et al.*, 1985/86; Wheat and McDuff, 1994) and conservative ions (e.g. Wheat and Mottl, 1994; Wheat and McDuff, 1995); and (2) regional transport processes, sources within the crust, and the potential distances over which flow may occur (e.g. Maris *et al.*, 1984; Mottl, 1989; Baker *et al.*, 1991; Wheat and Mottl, 1994; Elderfield *et al.*, 1999; Wheat *et al.*, 2000b, 2002). More details regarding reactions and flow through sediments can be found in Chapter 16.

The third method is to drill into basaltic basement and sample formation waters directly. Several DSDP and ODP holes have been drilled into basement, cased through the sediment section, and revisited days to years after drilling (see McDuff, 1984; Mottl and Gieskes, 1990; Gieskes and Magenheim, 1992; Magenheim *et al.*, 1992, 1995; and a review by Mottl and Gieskes, 1990). Most of the samples from these early attempts were collected without good knowledge of the hydrologic state of the holes, and were contaminated with a combination of bottom seawater and seawater that has reacted with rubble or drilling muds within the hole. Particularly problematic was that many holes were located in sediment ponds where basement is characteristically underpressured, and bottom seawater flows down open boreholes and into permeable horizons that have been penetrated. Such conditions make successful fluid sampling from open boreholes impossible.

Two approaches have been employed to overcome this difficulty. The simplest, which has provided the most uncontaminated formation waters collected to date, involves siting holes where natural super-hydrostatic conditions have been found typically to exist, i.e. in sediment-buried basement edifices (Chapter 8). Examples include holes drilled into local buried seamounts or intrusions in the sedimented Middle Valley and Escanaba Trough spreading centers during ODP Legs 139 and 169, and into buried basement ridges on the eastern flank of the Juan de Fuca Ridge during ODP Leg 168. At ODP Sites 1025 and 1026 on the Juan de Fuca Ridge flank, formation waters flowed up the open boreholes at rates of about $50,000 \, l \, h^{-1}$ shortly after they were drilled and before they were sealed with CORK instrumentation (Fisher *et al.*, 1997; Becker and Davis, 2003). Formation waters collected from these man-made "springs" (Wheat *et al.*, 2003c) vindicate the use of basal pore waters for estimating the composition of formation waters, show a change in concentration during the past four years in response to reaction with sediment at depth, and provide a measure of contamination that includes dissolved Fe, Cu, Co, Zn, and Pb even though the residence time for waters rising through the cased parts of the boreholes is short (Fig. 19.4). These waters are also loaded with Fe- and Si-rich particles that in part have a sedimentary component.

To allow direct sampling at both overpressured and underpressured sites, instrumented borehole seals (CORKs; Davis *et al.*, 1992) have been used to let the formation rebound to a pre-drilling state. Sampling of formation waters from CORKs has been accomplished using OsmoSamplers, which are continuous samplers that draw water using the osmotic pressure gradient created across a semi-permeable membrane by solutions of differing salinity (Theeuwes and Yum, 1976), into small-bore (0.8 or 1.2 mm ID) Teflon tubing (Jannasch *et al.*, 1994, 1997; Wheat *et al.*, 2000a). OsmoSamplers have been recovered from ODP Sites 1024 and 1027 (Wheat *et al.*, 2003b), both of which were underpressured

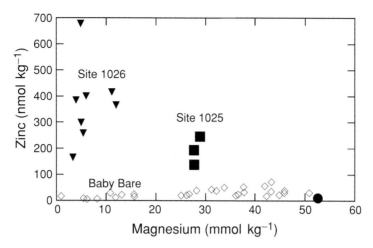

Fig. 19.4 Dissolved zinc versus magnesium in waters that vent from natural (Baby Bare) and man-made springs (ODP Holes 1025C and 1026B drilled into overpressured basement) on the eastern flank of the Juan de Fuca Ridge. The most Mg-depleted samples from the Baby Bare springs provide an estimate for the pristine formation water of about 15 nmol Zn kg^{-1} (Wheat *et al.*, 2002). Waters produced from ODP Holes 1025C and 1026B are contaminated with Zn, presumably from either the steel casing through which they have risen, or possibly from the basal sediments (Wheat *et al.*, 2003c).

and drew in seawater in the times after drilling and before sealing. Each sampler pumped about 8.4 ml per week for three years. Waters recovered from both sites show a drastic change in the major ion composition within the first 20–40 days after the borehole was sealed, followed by a more gradual change in composition. This gradual change ceased after 820 days at ODP Site 1024, but continued throughout the three-year deployment at ODP Site 1027. Waters within the borehole at ODP Site 1024 still contained remnants of surface seawater used to wash cuttings out of the hole, and were contaminated by reaction with drilling muds, the steel casing, and possibly the cement plug that was used in an attempt to isolate basaltic basement from the overlying sediment (Fig. 19.5). In contrast, waters from ODP Site 1027 are consistent with sediment pore-water results and with thermal data that indicate fluids flowed from a sill into the basaltic basement below (Davis and Becker, 2002). Trace element data show surprisingly little contamination given the presence of steel casing, Li-organic-rich grease at each joint, cement, and drilling muds.

In addition to the constraints provided on basement fluid composition, OsmoSampler records reveal the rate at which boreholes recover, and this provides insight into the rate at which fluid flows within the crust to carry away the initial transient drilling-induced contamination. Wheat *et al.* (2003b) have used the records from ODP Sites 1024 and 1027 (Fig. 19.5) to estimate natural rates of flow that are generally consistent with those estimated independently from thermal constraints (Chapter 8). These results illustrate that future borehole experiments, such as controlled tracer injections and microbial studies, can be conducted with modified CORKs that limit exposure to casing and overlying sediment and minimize the volume of the borehole.

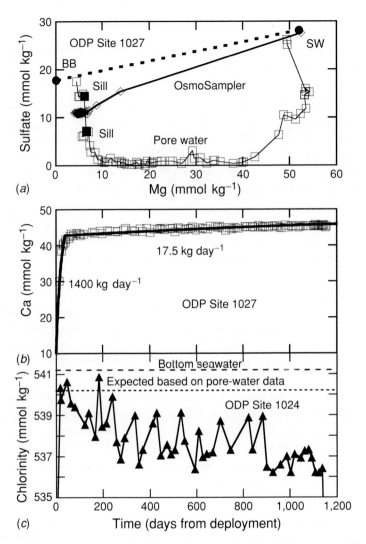

Fig. 19.5 Continuous water samplers, OsmoSamplers, were deployed in instrumented boreholes at ODP Holes 1024C and 1027C and recovered about three years later. (*a*) Sulfate versus Mg for ODP Hole 1027C (open diamonds connected by thick line) compared with consecutive pore-water samples from ODP Site 1027 (open squares connected by line; Davis *et al.*, 1997b). Filled squares indicate pore water from just above and just below the sill. The dashed line connects the composition of the Baby Bare springs (1 mmol Mg kg^{-1}) with that of bottom seawater (52 mmol Mg kg^{-1}). From these and similar data we conclude that water captured by the OsmoSampler is from the sill rather than from basaltic basement, consistent with thermal data (Becker and Davis, 2003; Wheat *et al.*, 2003b). (*b*) Change in water composition with time allows us to determine the water flux through the borehole. An initial flux of about 1400 kg day^{-1} is required to match the data from ODP Hole 1027C. After the initial 40 days a flux of 17.5 kg day^{-1} is required. If this flow is uniform through a surface area defined by a 10-m-thick sill with an effective porosity of 10%, then the seepage speed through the sill is about 8 m yr^{-1}. (*c*) Chlorinity of borehole water from ODP Hole 1024C is less than that of both bottom seawater and the value predicted from the gradient in pore water near the sediment–basement interface, indicating that surface seawater persists within the borehole even three years after drilling. There are additional sources of contamination, but these can be minimized in future deployments by modifications to the CORK. Plots are modified from Wheat *et al.* (2003b).

The fourth means for obtaining formation waters is the GRIND technique that was developed by Wheat *et al.* (1994). This method was employed because some highly altered sediment/rock samples yielded little, if any, pore water upon squeezing at the maximum pressure afforded by hydraulic presses aboard the *JOIDES Resolution.* The GRIND technique consists of fragmenting a freshly collected sample that is too hard to squeeze, grinding it with distilled water in a ball mill, squeezing the ground mixture, and analyzing the effluent. Chlorinity-adjusted concentrations of Mg, Ca, Na, K, Rb, and Sr in grind samples compared well with squeezed samples, but alkalinity, sulfate, dissolved silica, Ba, Mn, Li, and B did not because of ion exchange and precipitation reactions during sample manipulation. Although this technique has produced reliable results for some elements, it has not been tested on relatively unaltered basalt.

From all of these studies, it can be concluded that the net result of low-temperature alteration is the partial removal of Mg, K, Li, Rb, F, sulfate, alkalinity, nitrate, phosphate, V, U, Cu, Y, and REE (except for Ce) from circulating seawater; the addition of Ca, Cl, Sr, ammonium, boron, Si, Mn, Ba, Mo, Co, Ni, Cd, and Ce to circulating seawater; lower Na/Cl, $^{87}Sr/^{86}Sr$, and $\delta^{18}O$ in the altered seawater; and a remarkably young range of radiocarbon ages of the circulating seawater (1,000 to 10,000 years) along this transect of ridge flank sites (Elderfield *et al.*, 1999). These results are for the most part consistent with results from studies of basalt alteration noted above.

19.5 Geochemical fluxes from ridge flank hydrothermal systems

19.5.1 Constraints from basaltic alteration

Estimates of chemical exchange between oceanic basalt and bottom seawater have been made based on whole-rock analyses from DSDP Sites 417 and 418 (Hart and Staudigel, 1982; Thompson, 1983; Spivack and Staudigel, 1994; Staudigel *et al.*, 1996), DSDP Site 504 (Alt, 1995a; Alt *et al.*, 1996a), and ODP Site 735 (Bach *et al.*, 2001). The approach has been to compare the altered whole-rock composition with that of a presumed unaltered precursor. The average difference is calculated and multiplied by the depth interval that is altered under a particular set of conditions (e.g. low versus high temperature). In some cases the chemical difference over each depth interval is determined and the result integrated for the length of the sampled section. The extent of alteration is then adjusted to account for the yearly global production of new crust that is susceptible to the particular type of alteration. Some estimates include the entire crustal production to a given depth within the crust (e.g. Bloch and Bischoff, 1979). Others include only the crust from slow-spreading centers (Bach *et al.*, 2001). The end result is a global crustal chemical flux that can be compared with other globally significant fluxes such as riverine inputs (Table 19.2).

This analysis has several drawbacks. First, only a fraction of the core that is drilled is typically recovered. For example, only 30% of the volcanic section and transition zone in DSDP Hole 504B was recovered (Alt, 1995b). Core recovery was even worse in the dike section. One could assume that the characteristics of the recovered rocks are representative of the

Table 19.2 *Elemental fluxes based on studies of basalt alteration.*
Positive fluxes are into the ocean

	Flux (mol yr^{-1})	417/418[a] 500 m volcanics	504B[b] 600 m volcanics	735B[c] 1500 m gabbro	Factor of 8 higher	Riverine[d]
Si	10^9	−700	−740	8	−5900	6400
Mg	10^9	170	−340	20	−2700	5400
Ca	10^9	−1400	−6	−5	−11000	12000
Na	10^9	−20	−28	−1	−220	5700
K	10^9	−210	−60	−0.3	−1700	1300
P	10^9		−1	−0.1	−8	30
C	10^9	−2700	−120	−6	−22000	32000
H$_2$O	10^9	−5600		−27	−45000	
S	10^9		77		616	3200
Li	10^9		−2	−0.02	−16	14
Rb	10^6	−490	−55	−0.3	−3900	370
Cs	10^6	−6.5	−1	−0.01	−52	4.8
U	10^6	−9.7	−1	−0.01	−78	32

[a]Data from Hart and Staudigel (1982); Staudigel *et al.* (1996).
[b]Data from Alt *et al.* (1996a); Bach *et al.* (2001).
[c]Data from Bach *et al.* (2001).
[d]Data from Wheat and Mottl (2000); Bach *et al.* (2001); Wheat *et al.* (2002).

whole. This assumption is questionable given that highly altered and fractured intervals are difficult to recover with present drilling technologies and thus are likely to be undersampled. Second, one has to assume that the observed changes are globally applicable (see Chapter 20). DSDP Sites 417 and 418 illustrate the extent of heterogeneity in crustal compositions: they are greatly different even though they are only a few kilometers apart. Altered rocks from DSDP Site 417 have six times more K than rocks from DSDP Site 418, for example (Thompson *et al.*, 1983). Lastly, estimates exist from only a handful of deep-sea drilling sites. Even with complete recovery, recovered rock from these sites (less than one-trillionth of 1% of the area of the seafloor) cannot be considered globally representative.

Nevertheless, fluxes listed in Table 19.2 highlight some interesting differences and possible trends: (1) Alteration in the gabbroic section (ODP Hole 735B) that occurs from ridge flank hydrothermal processes has little effect on global geochemical budgets relative to riverine fluxes. This is not surprising given that the highly permeable upper volcanic section, which acts as a conduit for seawater circulation, is separated from the gabbros by an impermeable layer of lower dikes and sills that was altered at high temperature at the ridge axis. (2) Rocks from DSDP Hole 504B in 5.9 Ma crust are much less altered than those from DSDP Sites 418 and especially 417 on 110 Ma crust. This difference may be related to crustal age. A more likely explanation for DSDP Site 417, at least, is that the difference is related to geologic structure: this is located on a basement topographic high and may have been a long-lived conduit for venting of formation waters. Its setting is analogous to the present situation at Baby Bare, which hosts springs as noted above. (3) Mg is the only element for which the calculated direction of the flux at DSDP Sites 417 and 418 is different from that at DSDP Site 504. (4) Uptake of Rb, Cs, and U by the crust during ridge flank hydrothermal alteration produces globally significant geochemical fluxes. Similarly, but to a lesser extent, fluxes of Si, Mg, Ca, K, and Li from ridge flank hydrothermal systems account for 10–20% of riverine inputs. On the basis of these studies of rock alteration, low-temperature hydrothermal systems on ridge flanks do not contribute significantly to the global budget for Na, P, C, and S.

19.5.2 *Constraints from pore-, borehole, and spring water compositions*

Three general methods have been used to calculate ridge flank hydrothermal fluxes from the chemical composition of formation waters. The first is to multiply the concentrations of chemical elements in the venting water by a water flux, estimated from either the global advective heat loss and an average temperature of venting (Maris *et al.*, 1984), or as the flux required to remove the entire riverine input of Mg, given a Mg-depleted formation-water (Mottl and Wheat, 1994; Kadko *et al.*, 1995; Elderfield and Schultz, 1996; Sansone *et al.*, 1998; Wheat and Mottl, 2000; Wheat *et al.*, 2002). This method is ideal for "back-of-the-envelope" calculations, given the paucity of chemical data from sites with active seepage and/or venting. It indicates that warm (>45 °C; Mottl and Wheat, 1994) hydrothermal systems such as at Baby Bare play a minor role in chemical fluxes for most elements, but they could result in fluxes of Mg, Na, K, Li, Ca, Sr, S, B, Mn, Mo, and Cd that are greater than 10% of the riverine flux (Wheat and Mottl, 2000; Wheat *et al.*, 2002). A second method uses the observed variability in formation-water composition with temperature in basement (e.g. Mg, Mottl and Wheat, 1994; phosphate, Wheat *et al.*, 1996a). Although this method is the optimal one for calculating chemical fluxes, it requires that we know how the advective ocean/crustal exchange varies with temperature; we can only guess at this relationship at present. Estimates using this approach can only delimit the possible ranges in temperature over which geochemical fluxes may be significant to global budgets. The third method relies on estimates of other globally significant fluxes such as the riverine flux, and solves for the ridge flank flux based on the portion that is not accounted for (Palmer and Edmond, 1989; Chan *et al.*, 1992; Elderfield and Schultz, 1996; Butterfield *et al.*, 2001). This method relies on accurate estimates for the other chemical fluxes, some of which have been exaggerated, such as the high-temperature flux from venting along the MOR axis.

Fluxes from ridge flank hydrothermal processes are tabulated in the manuscripts cited above. We do not repeat any of these tabulations because some elements have only been analyzed in waters from Baby Bare, most of the data come from sites where the temperature in basement exceeds 45 °C, and the role of these warm ($\geq$45 °C) hydrothermal systems in controlling geochemical fluxes is probably minimal relative to fluxes from cooler (10–25 °C) hydrothermal systems.

19.5.3 The extent of a sediment contribution to ridge flank chemical fluxes

The composition of basement formation water is influenced both by reaction between seawater and basalt and by diffusion to and from the overlying sediment. This diffusion is driven by diagenetic reactions in the sediment that produce chemical gradients between sediment pore water and basement formation water (Chapter 16). Diffusive exchange with the sediment section removes sulfate from basement formation waters and supplies them with Fe, Mn, and other trace elements that are mobilized during redox reactions (Bender, 1983; Maris *et al.*, 1994; Wheat and McDuff, 1995; Elderfield *et al.*, 1999; Wheat *et al.*, 2000b, 2002).

This diffusive flux has been quantified for dissolved silica at the Mariana Mounds ridge flank site. Wheat (1990) found that the flux into basement across the sediment–basement interface is equivalent to the diffusive flux of dissolved silica to the oceans across the sediment–water interface. Some of this flux is removed in the form of vein filling in basement. The remaining dissolved silica is transported and ventilated by hydrothermal flow and ultimately contributes to the net hydrothermal flux of silica to the oceans, a net value that is about three times greater than the flux from dissolution alone near the sediment–water interface at this site. The oceanic flux of silica from dissolution near the sediment–basement interface, coupled with ridge flank hydrothermal processes, is conservatively about 25% of the silica flux from riverine sources (Wheat, 1990), indicating that contributions from the sediment column to formation waters can be a significant part of the geochemical flux delivered by hydrothermal processes.

A diffusive flux from the sediment section to basement has been observed for a variety of elements in several ridge flank settings. These observations have typically been used to constrain the path of seawater flow into basaltic basement. For example, dissolved Fe and Mn in pore waters in the Galapagos Mounds have a sediment source, indicating that seawater is recharged to formation waters in part via downwelling through the sediment section (Maris and Bender, 1982; Bender, 1983; Maris *et al.*, 1984). In contrast, gradients in porewater nitrate within the Mariana Mounds, coupled with a nitrate concentration in formation waters that is almost identical to bottom seawater, are consistent with 95% of the recharge entering basement through basalt exposed at the seafloor (Wheat and McDuff, 1995). Similar arguments have been used elsewhere to constrain regional patterns of seawater flow within oceanic basement (Sr, Baker *et al.*, 1991; SO_4, Elderfield *et al.*, 1999; SO_4, Ca, Li, Wheat *et al.*, 2000b; transition metals, Wheat *et al.*, 2002). However, the contribution of these sediment sources to hydrothermal fluxes has not been quantified because the particular environmental conditions studied so far are probably not typical globally.

An additional sediment component to ridge flank hydrothermal fluxes derives from reactions that occur as formation water upwells through the sediment. Diagenetic reactions can measurably alter the composition of upwelling formation water provided the rate of reaction is fast enough that the reactive flux is a measurable proportion of the advective flux. For the major ions, an upwelling Darcy seepage rate of at least 1 cm yr^{-1} is required for the advective flux to mask the reactive sediment flux in a sediment column that is <100 m thick (Wheat and Mottl, 2000). For more reactive species such as dissolved silica, upwelling seepage rates of tens of centimeters per year are required to mask diagenetic inputs (Wheat and McDuff, 1994; Wheat and Tribble, 1994). The net effect of these diagenetic reactions to fluxes is probably minimal given the distribution of fluid flow from ridge flank environments. At the Mariana Mounds, for example, 99% of the hydrothermal venting is at seepage rates >10 cm yr^{-1} (Wheat, 1990). At Baby Bare about 95% of the flow vents at rates >10 cm yr^{-1} (Wheat *et al.*, 1996b, 2003a). The conclusions drawn from these two studies are reinforced by constraints using the global database for sediment thickness, type, and permeability coupled with possible driving forces (Chapter 6), which indicate that the amount of hydrothermal venting through sediment is minimal. Thus, most of the global hydrothermal flow must vent directly through basement faults and basaltic outcrops.

19.5.4 Update for the estimation of chemical fluxes from ridge flank hydrothermal processes

Several methods for determining chemical fluxes from ridge flank hydrothermal processes are presented above. In principle, estimates based on basalt alteration should be the same as those from studies of formation waters. We indicated above that crustal alteration studies are best done by deep-sea drilling. In reality, however, such drilling has yet to penetrate more than half of the crustal thickness, and the sampling done by all drilling falls far short of adequately representing the highly heterogeneous section. The greatest depth achieved by Hole 504B reached only to the base of the sheeted dike section, 1.8 km below the top of the igneous crust, and it is likely at this and other sites that altered samples tend not to be recovered during the drilling process.

Studies of fluids provide a more regional integration of alteration but they cannot extract the historical information that is recorded by basalt alteration. Even in the well-sedimented environment of the eastern Juan de Fuca Ridge flank, basement formation waters dated using ^{14}C have been found to be only 1,000–10,000 years old; thus they represent only relatively recent conditions (Elderfield *et al.*, 1999). The advantage of using water to track alteration and chemical fluxes is that one does not have to rely on drilling alone. Basement–water compositions can be determined using sediment coring, heat flow, and submersible studies. This is not without its challenges, however, for locating zones of focused discharge is not an easy task given that the pattern of seawater circulation is much more complex than originally modeled (e.g. Fehn *et al.*, 1983).

Despite these qualifications, we feel the best approach for evaluating chemical fluxes from ridge flank hydrothermal processes is to use solution chemical data and recent global compilations of sediment thickness and convective heat loss. The best and most artifact-free samples are those collected directly from springs, yet only one site with springs, Baby Bare, has been sampled. Thus we must rely on samples from sediment pore waters collected from zones of focused up-flow and from near the sediment–basement interface. Here we use these global compilations of physical data and dissolved phosphate data to estimate the flux of phosphate into the crust as a result of ridge flank hydrothermal systems. Phosphate was chosen for three reasons. It is commonly used in rock alteration studies as a measure of the water/rock ratio, there is a reasonable data set for phosphate from DSDP and ODP cores, and earlier estimates are consistent with ridge flank hydrothermal systems playing an important role in the oceanic cycle of phosphate (Wheat *et al.*, 1996a).

We indicated above that one of the greatest uncertainties in estimating global exchange of seawater (see Fig. 10.3) and solutes between the ocean and crust arises from our limited knowledge of basement temperature. Currently, the best constraints on these temperatures are derived from the global data compilation of observed seafloor heat flux by Stein and Stein (1994) and the compilation of sediment thickness by David Divins at the National Geophysical Data Center (http://www.ngdc.noaa.gov/mgg/sedthick/sedthick.html). Temperatures are estimated from measured seafloor heat flux and local sediment thickness using a bulk thermal conductivity for the sediment section of 0.9 W m^{-1} °C^{-1}. The local heat flux deficit, defined by the difference between the local average seafloor heat flux and that predicted for the underlying lithosphere (using a lithospheric heat-flux vs. age relationship of 500 $A^{-0.5}$, where A is crustal age in Ma and heat flux is in mW m^{-2}; see Chapter 10), constrains the advective heat loss in a given area; this plus the local basement temperature estimate constrains the local water mass exchange between the crust and ocean and the variation of exchange as a function of temperature; and a global spatial integration of the exchange vs. temperature constrains the total (global ocean) mass and chemical exchange.

Estimates of basement temperatures calculated for 1 Myr bins on eastern and western flanks of spreading ridges in the Pacific and Atlantic Oceans between 20° S and 40° S are shown in Fig. 19.6. Values for the 0–4 Ma bins are adjusted to account for the proportion of heat loss from axial hydrothermal systems (Mottl, 2003). Calculated temperatures generally increase slowly but systematically with age from roughly 5 °C. This general pattern is the consequence of the competing effects of lithospheric aging, sedimentation, and waning advective heat loss. Younger crust has a high basal heat flow but less sediment to retain this heat, whereas older crust has a lower basal thermal input but thicker sediment. On average, virtually all of the convective heat loss from the oceanic crust comes from crust that is 5° to 15 °C. Exceptions occur on young, rapidly buried flanks such as on the Costa Rica Rift and Juan de Fuca Ridge where thick sediments cause much higher temperatures (e.g., Fig. 8.3), but the net contribution from such environments is minor. Our focus should thus be on reactions that take place in this relatively low temperature range.

With the generally low temperatures of Fig. 19.6 and the global partitioning of convective heat loss (Table 19.1), we calculate a ridge flank hydrothermal seawater flux by simply weighting results by area according to a partitioning between fast spreading systems

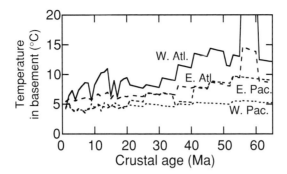

Fig. 19.6 Calculated temperature at the interface between sediment and upper basaltic basement as a function of age along transects east and west of active spreading ridges at 20° S to 40° S on the East Pacific Rise and Mid-Atlantic Ridge. Temperatures were calculated using global heat flow and sediment thickness compilations (see references in text), and a uniform thermal conductivity of $0.9\ \mathrm{W\,m^{-1}\,{}^{\circ}C^{-1}}$. Temperatures are typically less than 15 °C. Most of the best studied ridge flank hydrothermal systems are considerably warmer (e.g., the Juan de Fuca Ridge and Costa Rica Rift flanks). These warmer sites are ideal for studying hydrothermal processes but they are spatially rare and probably do not contribute significantly to global heat, fluid, and chemical fluxes. This figure is modified from Wheat *et al.*, 2003d.

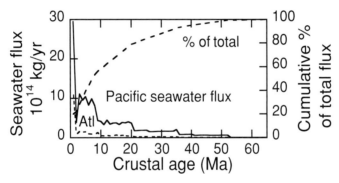

Fig. 19.7 Seawater flux from mid-ocean ridge flank hydrothermal systems of the Pacific and Atlantic Oceans plotted versus crustal age. The seawater mass flux is calculated following the scheme outlined in Fig. 10.3 using heat loss presented in Table 19.1 and basement temperatures in Fig. 19.6. Results are "integrated" by simply weighting results by area according to a partitioning between fast spreading systems (75%, Pacific Ocean) and slow spreading systems (25%, Atlantic Ocean) (Baker *et al.*, 1996). The estimated seawater flux from ridge flank hydrothermal systems equals $2.1 \times 10^{16}\ \mathrm{kg\,yr^{-1}}$ (~65% of global river flux, Mackenzie, 1992). The cumulative percentage shows that much of this seawater flux occurs through young crust.

(75%, Pacific Ocean) and slow spreading systems (25%, Atlantic Ocean) (Baker *et al.*, 1996) using the Atlantic and Pacific Oceans corridors above to represent "typical" oceanic crust (Fig. 19.7). The net flux of seawater through the crust thus estimated is $2.1 \times 10^{16}\ \mathrm{kg\,yr^{-1}}$, roughly 65% of the river flow (Mackenzie, 1992). Almost half of this flow is confined to crust that is less than 7 Ma and most (95%) of this flow occurs in crust less than 40 Ma in age.

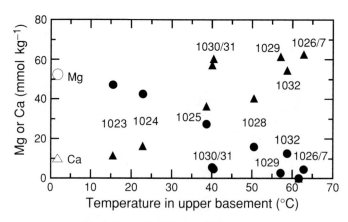

Fig. 19.8 Estimated concentrations of Mg and Ca in formation waters based on pore-water data versus the temperature in upper basaltic basement for sites drilled on the eastern flank of the Juan de Fuca Ridge during ODP Leg 168 (after Davis *et al.*, 1997b). This apparent composition–temperature trend may prove fruitful for estimating chemical fluxes from ridge flanks for these elements, although a more broadly based investigation is needed and underway.

This analysis provides a seawater flux for a given temperature. If chemical exchange is directly related to temperature in basement then one can determine chemical fluxes. The initial summary of work completed on ridge flanks indicates that for some chemical species there is a relationship between temperature in basement and formation-water composition (e.g. Fig. 19.8; Mottl and Wheat, 1994; Wheat and Mottl, 1994). Additional results using ODP data reinforce the assumption that temperature is the most significant parameter for chemical exchange for several elements (Wheat *et al.*, 1996a; Elderfield *et al.*, 1999). For some dissolved species such as sulfate, diffusive exchange with the overlying sediment section is the primary controlling factor (Elderfield *et al.*, 1999; Wheat *et al.*, 2000b). For sulfate this exchange is related to the rate of microbially mediated sulfate reduction in sediment and the residence time of water in basement. For such species chemical fluxes cannot be calculated from the seawater flux at a given temperature.

Using data from DSDP, ODP, and other ridge flank fluids, Wheat *et al.* (1996a) showed a distinct temperature dependence on the concentration of phosphate in basement water. We expand on this compilation by including additional sites and solve for the temperature dependence on the phosphate concentration (Fig. 19.9). This relationship is then used to calculate a phosphate concentration for a given temperature for a specified age domain. This concentration is then subtracted from the 2.7 μmol kg^{-1} in bottom seawater and multiplied by the seawater flux determined above to get an estimate of phosphate flux.

Similar to the heat budget, half of the flux of phosphate occurs through crust younger than 8 Ma and 95% occurs in crust younger than 42 Ma. The result is a phosphate flux from ridge-flank hydrothermal systems of 2.8×10^{10} mol y^{-1}, which is close to but higher than our earlier estimate of 6.5×10^9 mol yr^{-1} (Wheat *et al.*, 1996a). This flux is 130% of the dissolved inorganic riverine flux (Froelich *et al.*, 1982) and 26% of the total riverine flux of phosphorus (Fig. 19.10). Even if we use a sediment thickness that is one standard deviation

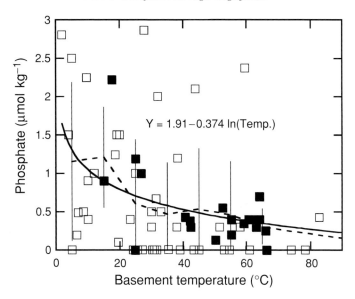

Fig. 19.9 Concentration of phosphate in basement formation waters plotted versus temperature estimated for the sediment–basement interface. Closed squares represent sites where studies of ridge flank hydrothermal fluxes were a priority. Open squares represent other DSDP and ODP sites. The solid line and equation are the best fit to all the data. The dashed line connects bin-averaged data using 10 °C bins. Vertical lines are one standard deviation about these bin averages. Bin-averaged data may be compared with the previous results of Wheat *et al.* (1996a).

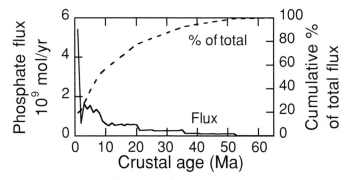

Fig. 19.10 Dissolved phosphate flux into the crust from ridge flank hydrothermal processes plotted versus crustal age. This flux is based on the phosphate–temperature relationship (Fig. 19.9) and seawater flux (Fig. 19.7). The net flux equals 130% of the dissolved inorganic riverine flux of phosphorus (Froelich *et al.*, 1982). The cumulative percentage demonstrates that much of the phosphate flux into the crust occurs in young crust (Wheat *et al.*, 2003d).

lower or higher than the average, the calculated phosphate flux into the crust is significant relative to the riverine dissolved inorganic flux

 This estimate of phosphorus transfer into the crust is eight times greater than that calculated from studies of rock alteration. Even using an extreme value for sediment thickness, the estimated flux exceeds that based on rock alteration by at least a factor of four. This difference may result from the incomplete recovery of altered rock during drilling, and especially the failure to recover zones that are faulted and highly altered. Pursuing this line of reasoning further, what if the other calculated fluxes based on bulk rock analyses were a factor of eight lower as well? In Table 19.2 we multiply the greatest flux for each element by eight, with the result that many of the calculated fluxes would be similar to riverine inputs.

This analysis for phosphate could also be applied to other elements that appear to show a trend with increasing temperature in basement, including Mg, Ca, K, and alkalinity (e.g. Fig. 19.8; Mottl and Wheat, 1994; Elderfield *et al.*, 1999). We presently are surveying these species in pore water from the DSDP and ODP data set to determine if a trend is apparent in all of the data. Other dissolved species, such as sulfate and silica, are so reactive in the sediment above basement that diffusive exchange between the sediment and basement predominantly controls the concentrations of these chemical species in formation waters, so similar estimates of fluxes are not possible at this time.

19.6 Concluding remarks and recommendations

We have presented two approaches for determining the extent of chemical exchange between the crust and oceans from low-temperature, ridge flank hydrothermal processes. One focuses on the composition of formation waters and its dependence on temperature in the upper igneous basement, coupled with the associated seawater flux through basement and its variation with temperature. We present a calculation for phosphate to illustrate the magnitude of hydrothermal fluxes from ridge flanks, and find the fluxes of several elements to be comparable to fluxes from rivers. This method for calculating fluxes is probably appropriate for a number of elements but the detailed analysis has yet to be completed.

From this analysis it is clear that additional work is required. First, water fluxes from crust of different ages must be constrained better. A systematic survey of a representative plate such as the Pacific needs to be conducted to define better the distribution and temperature of flow relative to crustal age, in the context of differing crustal topography, presence of basaltic outcrops and seamounts, and sedimentation rate and type. In the analysis above, the range for estimated temperatures for a particular age domain based on one standard deviation of the sediment thickness is about twice the average temperature. Applying one standard deviation to the lower bound of the sediment thickness for almost all of the age domains results in a temperature of about 2 °C and for phosphate results in a flux that is almost four times greater than the flux calculated using the average sediment thickness. What is an appropriate temperature range and magnitude of flow? Of particular interest is the role of low-temperature (2–10 °C) hydrothermal processes in the exchange of fluids between crust and oceans. Are these fluids important on a global scale? Second, the role of deep crustal faults needs to be examined. Waters from ODP Sites 1030 and 1031, for example, are more altered for a given temperature in uppermost basement than the other sites drilled on the Leg 168 transect (Davis *et al.*, 1997b). Another deep crustal fault has been located on 20–25 Ma crust east of the EPR (Friedmann *et al.*, 2001). These faults provide a potential pathway to vent highly altered fluids, the original temperature of which could be lowered by conduction before it reaches the sediment–basement interface, resulting in a chemical flux that is greater than projected for a specified basement temperature. Third, a systematic approach is required to determine which dissolved chemical species show a relationship with temperature in uppermost basement and to define those relationships with confidence. Fourth, springs must be located and sampled or producing boreholes must be established at sites where the upper crustal temperatures are below 50 °C, especially in the range of 10–25 °C, given that most of the ridge flank advective heat loss occurs at these temperatures on a global scale. Sampling such natural or man-made springs (below casing in the latter instance), and thus avoiding potential sampling artifacts resulting from the extraction of pore waters from sediment, is critical for determining fluxes of trace elements.

The alternative method focuses on the extent of crustal alteration determined from altered rock recovered by deep-sea drilling. This method provided the first holistic view that hydrothermal processes, especially those that occur at temperatures less than about 50 °C, provide a mechanism for transferring heat and chemicals between the oceans and basaltic crust. There are, however, only a handful of sites that have been drilled more than 100 m into basaltic basement. More sites are needed to get a better estimate of the chemical effects of crustal alteration. Recovery of basalt during these operations also needs to be improved in order to address the question of whether altered basalt is preferentially lost during drilling, as suggested by the eight-fold greater phosphate flux estimated from the formation-water method relative to the altered-rock method.

The ultimate outcome desired would be consistent estimates of chemical fluxes using the formation-water and altered-rock methods, and a model of these fluxes based on fluid flow on the scale of an oceanic plate that incorporates reaction kinetics.

Acknowledgments

This work was supported by the US National Science Foundation (OCE 9912367, OCE 9819454, and OCE 0002031 to CGW and OCE 98-19974 and 99-11933 to MJM). We thank E. Davis and M. Palmer for thoughtful reviews of this chapter. This is Global Undersea Research Unit contribution number 104 and contribution number 6365 from the School of Ocean and Earth Science and Technology of the University of Hawaii.

References

Alt, J. C. 1995a. Sulfur isotopic profile through the oceanic crust: sulfur mobility and seawater–crustal sulfur exchange during hydrothermal alteration. *Geology* **23**: 585–588.

　1995b. Subseafloor processes in mid-ocean ridge hydrothermal systems. In *Seafloor Hydrothermal Systems: Physical, Chemical, Biological, and Geological Interactions*, Geophysical Monograph 91. eds. S. Humphris, *et al.* Washington, DC: American Geophysical Union, pp. 85–114.

Alt, J. C. and Bach, W. 2003. Alteration of oceanic crust: subsurface rock–water interactions. In *Energy and Mass Transfer in Marine Hydrothermal Systems*, eds. P. E. Halbach, V. Tunnicliffe, and J. R. Hein. Berlin: Dahlem University Press, pp. 7–28.

Alt, J. C., Honnorez, J., Laverne, C., and Emmerman, R. 1986. Hydrothermal alteration of a 1 km section through the upper oceanic crust, DSDP Hole 504B: the mineralogy, chemistry, and evolution of seawater–basalt interactions. *J. Geophys. Res.* **91**: 10,309–10,335.

Alt, J. C., Teagle, D. A. H., Laverne, C., Vanko, D. A., Bach, W., Honnorez, J., Becker, K., Ayadi, M., and Pezard, P. A. 1996a. Ridge flank alteration of upper oceanic crust in the eastern Pacific: synthesis of results for volcanic rocks of Holes 504B and 896A. In *Proceedings of the Ocean Drilling Program, Scientific Results*, Vol. 148, eds. J. C. Alt, H. Kinoshita, L. B. Stokking, and P. J. Michael. College Station, TX: Ocean Drilling Program, pp. 435–450.

Alt, J. C., Laverne, D. C., Vanko, D. A., Tartarotti, P., Teagle, D. A. H., Bach, W., Zuleger, E., Erzinger, J., Honnorez, J., Pezard, P. A., Becker, K., Salisbury, M. H., and Wilkens, R. H. 1996b. Hydrothermal alteration of a section of upper oceanic crust in the eastern equatorial Pacific: a synthesis of results from Site 504 (DSDP Legs 69, 70, and 83, and ODP Legs 111, 137, 140, and 148). In *Proceedings of the Ocean Drilling Program, Scientific Results*, Vol. 148, eds. J. C. Alt, H. Kinoshita, L. B. Stokking, and P. J. Michael. College Station, TX: Ocean Drilling Program, pp. 417–434.

Bach, W., Alt, J. C., Niu, Y., Humphris, S. E., Erzinger, J., and Dick, H. J. B. 2001. The geochemical consequences of late-stage alteration of lower oceanic crust at the SW Indian Ridge: results from ODP Hole 735B (Leg 176). *Geochim. Cosmochim. Acta* **65**: 3,267–3,287.

Baker, E. T., Chen, Y. J., and Morgan, J. P. 1996. The relationship between near-axis hydrothermal cooling and the spreading rate of mid-ocean ridges. *Earth Planet. Sci. Lett.* **142**: 137–145.

Baker, P. A., Stout, P. M., Kastner, M., and Elderfield, H. 1991. Large-scale lateral advection of seawater through oceanic crust in the central equatorial Pacific. *Earth Planet. Sci. Lett.* **105**: 522–533.

Becker, K. and Davis, E. E. 2003. New evidence for age variation and scale effects of permeabilities of young oceanic crust from borehole thermal and pressure measurements. *Earth Planet. Sci. Lett.* **210**: 499–508.

Becker, K. and Fisher, A. T. 2000. Permeability of upper oceanic basement on the eastern flank of the Juan de Fuca Ridge determined with drill-string packer experiments. *J. Geophys. Res.* **105**: 897–912.

Becker, N. C., Wheat, C. G., Mottl, M. J., Karsten, J. L., and Davis, E. E. 2000. A geological and geophysical investigation of Baby Bare, locus of a ridge-flank hydrothermal system in the Cascadia Basin. *J. Geophys. Res.* **105**: 23,557–23,568.

Bender, M. L. 1983. Pore water chemistry of the mounds hydrothermal field, Galapagos Spreading Center: results from *Glomar Challenger* piston coring. *J. Geophys. Res.* **88**: 1,049–1,056.

Bender, M. L., Hudson, A., Graham, D. W., Barnes, R. O., Leinen, M., and Kahn, D. 1985/86. Diagenesis and convection reflected in pore water chemistry on the western flank of the East Pacific Rise, 20° S. *Earth Planet. Sci. Lett.* **76**: 71–83.

Bloch, S. and Bischoff, J. L. 1979. The effect of low-temperature alteration of basalt on the oceanic budget of potassium. *Geology* **7**: 193–196.

Bohlke, J. K., Honnorez, J., Honnorez-Guerstein, B.-M., Muehlenbachs, K., and Petersen, N. 1981. Heterogeneous alteration of the upper oceanic crust: correlation of rock chemistry, magnetic properties, and O isotope ratios with alteration patterns in basalts from Site 396B, DSDP. *J. Geophys. Res.* **86**: 7,935–7,950.

Butterfield, D. A., Nelson, B. K., Wheat, C. G., Mottl, M. J., and Roe, K. K. 2001. Evidence for basaltic Sr in midocean ridge-flank hydrothermal systems and implications for the global oceanic Sr isotope balance. *Geochim. Cosmochim. Acta* **65**: 4,141–4,153.

Chan, L. H., Edmond, J. M., Thompson, G., and Gillis, K. 1992. Lithium isotopic composition of submarine basalts: implications for the lithium cycle in the oceans. *Earth Planet. Sci. Lett.* **108**: 151–160.

Davis, E. E. and Becker, K. 2002. Observations of natural-state fluid pressures and temperatures in young oceanic crust and inferences regarding hydrothermal circulation. *Earth Planet. Sci. Lett.* **204**: 231–248.

Davis, E. E., Becker, K., Pettigrew, T., Carson, B., and MacDonald, R. 1992. CORK: a hydrologic seal and downhole observatory for deep-ocean boreholes. In *Proceedings of the Ocean Drilling Program, Initial Results*, Vol. 139, eds. E. E. Davis, M. J. Mottl, A. T. Fisher, *et al.* College Station, TX: Ocean Drilling Program, pp. 43–53.

Davis, E. E., Wang., K., He, J., Chapman, D. S., Villinger, H., and Rosenberger, A. 1997a. An unequivocal case for high Nusselt number hydrothermal convection in sediment-buried igneous oceanic crust. *Earth Planet. Sci. Lett.* **146**, 137–150.

Davis, E. E., Fisher, A. T., Firth, J. V., *et al.*, eds. 1997b. *Proceedings of the Ocean Drilling Program, Initial Reports*, Vol. 168. College Station, TX: Ocean Drilling Program.

de Lange G. J., Cranston, R. E., Hydes, D. H., and Boust, D. 1992. Extraction of pore water from marine sediments: a review of possible artifacts with pertinent examples from the North Atlantic. *Mar. Geol.* **109**: 53–76.

Edmond, J. M., Measures, C., McDuff, R. E., Chan, L. H., Collier, R., Grant, B., Gordon, L. I., and Corliss, J. B. 1979. Ridge crest hydrothermal activity and the balances of the major and minor elements in the ocean: the Galapagos data. *Earth Planet. Sci. Lett.* **46**: 1–18.

Elderfield, H. and Schultz, A. 1996. Mid-ocean ridge hydrothermal fluxes and the chemical composition of the ocean. *Ann. Rev. Earth Planet. Sci.* **24**: 191–224.

Elderfield, H, Wheat, C. G., Mottl, M. J., Monnin, C., and Spiro, B. 1999. Fluid and geochemical transport through oceanic crust: a transect across the eastern flank of the Juan de Fuca Ridge. *Earth Planet. Sci. Lett.* **172**: 151–169.

Embley, R. W., Hobart, M. A., Anderson, R. N., and Abbott, D. 1983. Anomalous heat flow in the Northwest Atlantic: a case for continued hydrothermal circulation in 80-M.Y. crust. *J. Geophys. Res.* **88**: 1,067–1,074.

Feely, R. A., Geiselman, T. L., Baker, E. T., Massoth, G. J., and Hammond, S. R. 1990. Distribution and composition of buoyant and non-buoyant hydrothermal plume particles from the ASHES vent at Axial Volcano, Juan de Fuca Ridge. *J. Geophys. Res.* **95**: 12,855–12,874.

Fehn, U., Green, K., von Herzen, R. P., and Cathles, L. 1983. Numerical models for the hydrothermal field at the Galapagos Spreading Center. *J. Geophys. Res.* **88**: 1,033–1,048.

Fisher, A. T. 1998. Permeability within basaltic oceanic crust. *Rev. Geophys.* **36**: 143–182.

Fisher, A. T. and Becker, K. 1995. Correlation between seafloor heat flow and basement relief: observational and numerical examples and implications for upper crustal permeability. *J. Geophys. Res.* **100**: 12,641–12,657.

Fisher, A. T., Becker, K., and Davis, E. E. 1997. The permeability of young oceanic crust east of the Juan de Fuca Ridge, as determined using borehole thermal measurements. *Geophys. Res. Lett.* **24**: 1,311–1,314.

Froelich P. N., Bender, M. L., Luedtke, N. A., Heath, G. R., and DeVries, T. 1982. The marine phosphorus cycle. *Am. J. Sci.* **282**: 474–511.

Frey, F. A. and Clague, D. A. 1983. Geochemistry of diverse basalt types from Loihi Seamount, Hawaii: petrogenetic implications. *Earth Planet. Sci. Lett.* **66**: 337–355.

Friedmann, P. K., Wheat, C. G., Underwood, M., Hoke, K., Fisher, A. T., Silver, E., Hutnak, M., Harris, R. N., and Stein, C. 2001. Evidence for hydrothermal circulation and alteration on 20–25 Ma crust about to be subducted in the Middle American Trench. *Eos, Trans. Am. Geophys. Union* **82**: F1147.

Fukuyama, H. 1985. Heat of fusion of basaltic magma. *Earth Planet. Sci. Lett.* **73**: 407–414.

Gieskes, J. M. and Magenheim, A. J. 1992. Borehole fluid chemistry of DSDP Holes 395A and 534A: results from Operation DIANAUT. *Geophys. Res. Lett.* **19**: 513–516.

Gillis, K. M. 1995. Controls on hydrothermal alteration in a section of fast-spreading oceanic crust. *Earth Planet. Sci. Lett.* **134**: 473–489.

2003. Subseafloor geology of hydrothermal root-zones at oceanic spreading centers. In *Energy and Mass Transfer in Marine Hydrothermal Systems*, eds. P. E. Halbach, V. Tunnicliffe, and J. R. Hein. Berlin: Dahlem University Press, pp. 53–70.

Gillis, K. M. and Banerjee, N. R. 2000. Hydrothermal alteration patterns in supra-subduction zone ophiolites. In *Ophiolites and Oceanic crust: New Insights from Field Studies and Ocean Drilling Program*, eds. Y. Dilek, E. Moores, D. Elthon, and A. Nicolas. Boulder, CO: Geological Society of America, pp. 349, 283–349, 297.

Gillis, K. M., Thompson, G., and Kelley, D. S. 1993. A view of the lower crustal component of hydrothermal systems at the Mid-Atlantic Ridge. *J. Geophys. Res.* **98**: 19,597–19,619.

Hart, S. R. 1973. A model for chemical exchange in the basalt–seawater system of oceanic layer II. *Can. J. Earth Sci.* **10**: 799–816.

Hart, S. R. and Staudigel, H. 1982. The control of alkalies and uranium in seawater by oceanic crust alteration. *Earth Planet. Sci. Lett.* **58**: 202–212.

Humphris, S. E., Melson, W. G., and Thompson, R. N. 1980. Basalt weathering on the East Pacific Rise and Galapagos Spreading Center, DSDP Leg 54. In *Initial Reports of the Deep Sea Drilling Project*, Vol. 54, eds. B. R. Rosendahl, R. Hekinian *et al.* Washington, DC: US Govt. Printing Office, pp. 773–787.

Ishikawa, T. and Nakamura, E. 1992. Boron isotope geochemistry of the oceanic crust from DSDP/ODP Hole 504B. *Geochim. Cosmochim. Acta* **56**: 1,633–1,639.

Jannasch, H. W., Johnson, K. S., and Sakamoto, C. M. 1994. Submersible, osmotically pumped analyzers for continuous determination of nitrate in situ. *Anal. Chem.* **66**: 3,352–3,361.

Jannasch, H., Sakamoto, C., Wheat, C. G., and Johnson, K. 1997. Osmotically pumped chemical analyzers and samplers for long-term oceanographic monitoring, Extended Abstract. *International Workshop in Marine Analytical Chemistry for Monitoring and Oceanographic Research*, Brest, France, 17–19 November.

Kadko, D., Baross, J, and Alt, J. 1995. The magnitude and global implications of hydrothermal flux. In *Seafloor Hydrothermal Systems*, eds. S. E. Humphris, R. A.

Zierenberg, L. S. Mullineaux, and R. E. Thomson. Washington, DC: American Geophysical Union, pp. 446–466.

Kelley, D. S. and Delaney, J. R. 1987. Two-phase separation and fracturing in mid-ocean ridge gabbros at temperatures greater than 700 °C. *Earth Planet. Sci. Lett.* **83**: 53–66.

Kelley, D. S., Gillis, K. M., and Thompson, G. 1993. Fluid evolution in submarine magma–hydrothermal systems at the Mid-Atlantic Ridge. *J. Geophys. Res.* **98**: 19,579–19,596.

Lecuyer, C. and Reynard, B. 1996. High-temperature alteration of oceanic gabbros by seawater (Hess Deep, Ocean Drilling Program Leg 147): evidence from oxygen isotopes and elemental fluxes. *J. Geophys. Res.* **101**: 15,883–15,897.

Mackenzie, F. T. 1992. Chemical mass balance between rivers and oceans. In *Encyclopedia of Earth System Science*, Vol. 1. New York: Academic Press, pp. 431–445.

Mackenzie, F. T. and Garrles, R. T. 1966. Chemical mass balance between rivers and oceans. *Am. J. Sci.* **264**: 507–525.

Magenheim, A. J., Bayhurst, G., Alt, J. C., and Gieskes, J. M. 1992. ODP Leg 137, borehole fluid chemistry in Hole 504B. *J. Geophys. Res.* **19**: 521–524.

Magenheim, A. J., Spivack, A. J., Alt, J. C., Bayhurst, G., Chan, L.-H., Zuleger, E., and Gieskes, J. M. 1995. Borehole fluid chemistry in Hole 504B, Leg 137: formation water or in-situ reaction. In *Proceedings of the Ocean Drilling Program, Scientific Results*, Vol. 137/140, eds. J. Erzinger, K. Becker, H. J. B. Dick, and L. B. Stokking. College Station, TX: Ocean Drilling Program, pp. 141–152.

Maris, C. R. P. and Bender, M. L. 1982. Upwelling of hydrothermal solution through ridge flank sediments shown by porewater profiles. *Science* **216**: 623–626.

Maris, C. R. P., Bender, M. L., Froelich, P. N., Barnes, R. O., and Luedtke, N. A. 1984. Chemical evidence for advection of hydrothermal solutions in the sediments of the Galapagos Mounds hydrothermal field. *Geochim. Cosmochim. Acta* **48**: 2,331–2,346.

McDuff, R. E. 1984. The chemistry of interstitial waters from the upper oceanic crust, Site 395, Deep Sea Drilling Project Leg 78B. In *Initial Reports of the Deep Sea Drilling Project*, Vol. 78B, eds. R. D. Hyndman, M. H. Salisbury, *et al.* Washington, DC: US Govt. Printing Office, pp. 795–799.

McManus, J., Nägler, T. F., Siebert, C., Wheat, C. G., and Hammond, D. 2002. Oceanic molybdenum isotope fractionation: diagenesis and hydrothermal ridge-flank alteration. *Geochem. Geophys. Geosyst.* **3**(12): 1078, doi:10.1029/2002GC000356, 2002.

Monnin C., Wheat, C. G., Dupre, B., Elderfield, H., and Mottl, M. 2001. Barite stability in marine sediments: implication for the flux of hydrothermal barium from ridge flanks settings. *Geochem. Geophys. Geosyst.* **2**: doi: 2000GC000073.

Mottl, M. J. 1989. Hydrothermal convection, reaction, and diffusion in sediments on the Costa Rica Rift flank: pore-water evidence from ODP Sites 677 and 678. In *Proceedings of the Ocean Drilling Program, Scientific Results*, Vol. 111, eds. K. Becker, H. Sakai, *et al.* College Station, TX: Ocean Drilling Program, pp. 195–213.

   2003. Partitioning of energy and mass fluxes between mid-ocean ridge axes and flanks at high and low temperature. In *Energy and Mass Transfer in Marine Hydrothermal Systems*, eds. P. E. Halbach, V. Tunnicliffe, and J. R. Hein. Berlin: Dahlem University Press, pp. 271–286.

Mottl, M. J. and Gieskes, J. M. 1990 Chemistry of waters sampled from oceanic basement boreholes, 1979–1988. *J. Geophys. Res.* **95**: 9,327–9,342.

Mottl, M. J. and McConachy, T. F. 1990. Chemical processes in buoyant hydrothermal plumes on the East Pacific Rise near 21° N. *Geochim. Cosmochim. Acta* **59**: 1,911–1,927.

Mottl, M. J. and Wheat, C. G. 1994. Hydrothermal circulation through mid-ocean ridge flanks: fluxes of heat and magnesium. *Geochim. Cosmochim. Acta* **58**: 2,225–2,237.

2000. Hydrothermal fluxes on mid-ocean ridge flanks: EXCO II on the eastern flank of the East Pacific Rise near 14° S. *Eos, Trans. Am. Geophys. Union* **81**: F458.

Mottl, M. J., Wheat, C. G., and Boulegue, J. 1994. Timing of ore deposition and sill intrusion at Site 856: evidence from stratigraphy, alteration, and sediment pore-water composition. In *Proceedings of the Ocean Drilling Program, Scientific Results*, Vol. 139, eds. M. J. Mottl, E. E. Davis, A. T. Fisher, and J. F. Slack. College Station, TX: Ocean Drilling Program, pp. 679–693.

Mottl, M. J., Wheat, G., Baker, E., Becker, N., Davis, E., Feely, R., Grehan, A., Kadko, D., Lilley, M., Massoth, G., Moyer, C., and Sansone, F. 1998. Warm springs discovered on 3.5 Ma oceanic crust, eastern flank of the Juan de Fuca Ridge. *Geology* **26**: 51–54.

Nehlig, P., Juteau, T., Bendel, V., and Cotten, J. 1994. The root zones of oceanic hydrothermal systems: constraints from the Samail ophiolite (Oman). *J. Geophys. Res.* **99**: 4,703–4,713.

Palmer, M. R. and Edmond, J. M. 1989. The strontium isotope budget of the modern ocean. *Earth Planet. Sci. Lett.* **92**: 247–262.

Parsons, B. 1981. The rates of plate creation and consumption. *Geophys. J. Roy. Astron. Soc.* **67**: 437–448.

Rudnicki, M. D. and Elderfield, H. 1993. A chemical model of the buoyant and neutrally buoyant plume above the TAG vent field, 26 degrees N, Mid-Atlantic Ridge. *Geochim. Cosmochim. Acta* **57**: 2,939–2,957.

Sansone, F. J., Mottl, M. J., Olson, E. J., Wheat, C. G., and Lilley, M. D. 1998. CO_2-depleted fluids from mid-ocean ridge-flank hydrothermal springs. *Geochim. Cosmochim. Acta* **62**: 2,247–2,252.

Schultz, A. and Elderfield, H. 1997. Controls on the physics and chemistry of seafloor hydrothermal circulation. *Phil. Trans. Roy. Soc. Lond. A* **355**: 387–425.

Sclater, J. G., Jaupert, C., and Galson, D. 1980. The heat flow through oceanic and continental crust and the heat loss of the earth. *Rev. Geophys.* **18**: 269–311.

Seyfried, W. E. 1977. Seawater–basalt interaction from 25–300 °C and 1–500 bars: implications for the origin of submarine metal-bearing hydrothermal solutions and regulation of ocean chemistry. Ph.D. thesis, University of Southern California, 242 pp.

Seyfried, W. E. and Bischoff, J. L. 1979. Low temperature basalt alteration by seawater: an experimental study at 70 °C and 150 °C. *Geochim. Cosmochim. Acta* **43**: 1,937–1,947.

Spivack, A. J. and Staudigel, H. 1994. Low-temperature alteration of the upper oceanic crust and the alkalinity budget of seawater. *Chem. Geol.* **115**: 239–247.

Staudigel, H., Plank, T., White, B., and Schminche, H.-U. 1996. Geochemical fluxes during seafloor alteration of the basaltic upper oceanic crust: DSDP Sites 417 and 418. In *Subduction Top to Bottom*, eds. G. E. Bebout, D. W. Scholl, S. H. Kirby, and J. P. Platt. Washington, DC: American Geophysical Union, pp. 19–38.

Stein, C. A. and Stein, S. 1994. Constraints on hydrothermal heat flux through the oceanic lithosphere from global heat flow. *J. Geophys. Res.* **99**: 3,081–3,095.

Teagle, D. A. H., Alt, J. C., Bach, W., Halliday, A. N., and Eringer, J. 1996. Alteration of upper oceanic crust in a ridge-flank hydrothermal upflow zone: mineral, chemical, and isotopic constraints from Hole 896A. In *Proceedings of the Ocean Drilling Program, Scientific Results*, Vol. 148, eds. J. C. Alt, H. Kinoshita, L. B. Stokking, and P. J. Michael. College Station, TX: Ocean Drilling Program, pp. 119–150.

Theeuwes, F. and Yum, S. I. 1976. Principles of the design and operation of generic osmotic pumps for the delivery of semisolid or liquid drug formulations. *Ann. Biomed. Eng.* **4**: 343–353.

Thompson, G. 1983. Basalt-seawater interactions. In *Hydrothermal Processes at Seafloor Spreading Centers*, eds. P. A. Rona, K. Bostrom, L. Laubier, and K. L. Smith, Jr. New York: Plenum, pp. 225–278.

Vanko, D. A., Laverne, C., Tartarotti, P., and Alt, J. C. 1996. Chemistry and origin of secondary minerals from the deep sheeted dikes cored during Leg 148 (Hole 504B). In *Proceedings of the Ocean Drilling Program, Scientific Results*, Vol. 148, eds. J. C. Alt, H. Kinoshita, L. B. Stoking, and P. J. Michael. College Station, TX: Ocean Drilling Program, pp. 71–86.

Von Damm, K. L. 1995. Controls on the chemistry and temporal variability of seafloor hydrothermal fluids. In *Seafloor Hydrothermal Systems: Physical, Chemical, Biological, and Geological Interactions*, Geophysical Monograph 91, eds. S. Humphris, *et al.* Washington, DC: American Geophysical Union, pp. 222–247.

Wheat, C. G. 1990. Fluid circulation and diagenesis in an off-axis hydrothermal system: the Mariana Mounds. Ph.D. thesis, University of Washington, Seattle.

Wheat, C. G. and McDuff, R. E. 1994. Hydrothermal flow through the Mariana Mounds: dissolution of amorphous silica and degradation of organic matter on a mid-ocean ridge flank. *Geochim. Cosmochim. Acta* **58**: 2,461–2,475.

1995. Mapping the fluid flow of the Mariana Mounds ridge flank hydrothermal system: pore water chemical tracers. *J. Geophys. Res.* **100**: 8,115–8,131.

Wheat, C. G. and Mottl, M. J. 1994. Hydrothermal circulation, Juan de Fuca eastern flank: factors controlling basement water composition. *J. Geophys. Res.* **99**: 3,067–3,080.

2000. Composition of pore and spring waters from Baby Bare: global implications of geochemical fluxes from a ridge flank hydrothermal system. *Geochim. Cosmochim. Acta* **64**: 629–642.

Wheat, C. G. and Tribble, J. 1994. Diagenesis of amorphous silica in Middle Valley, Juan de Fuca Ridge. In *Proceedings of the Ocean Drilling Program, Scientific Results*, Vol. 139, eds. M. J. Mottl, E. E. Davis, A. T. Fisher, and J. F. Slack. College Station, TX: Ocean Drilling Program, pp. 341–349.

Wheat, C. G., Boulegue, J., and Mottl, M. J. 1994. A technique for obtaining pore water chemical composition from indurated and hydrothermally altered sediment and basalt: the Ground Rock Interstitial Normative Determination (GRIND). In *Proceedings of the Ocean Drilling Program, Scientific Results*, Vol. 139, eds. M. J. Mottl, E. E. Davis, A. T. Fisher, and J. F. Slack. College Station, TX: Ocean Drilling Program, pp. 429–437.

Wheat, C. G., Feely, R. A., and Mottl, M. J. 1996a. Phosphate removal by oceanic hydrothermal processes: an update of the phosphorus budget in the oceans. *Geochim. Cosmochim. Acta* **60**: 3,593–3,608.

Wheat, C. G., Mottl, M. J., and Davis, E. E. 1996b. Patterns of heat and chemical transfer from a basement outcrop on the eastern flank of the Juan de Fuca Ridge: scales from several m^2 to 0.4 km^2. *Eos, Trans. Am. Geophys. Union* **77**: S256.

Wheat, C. G., McManus, J., Dymond, J., Collier, R., and Whiticar, M. 1998. Hydrothermal fluid circulation through the sediment of Crater Lake, Oregon: pore water and heat flow constraints. *J. Geophys. Res.* **103**: 9,931–9,944.

Wheat, C. G., Jannasch, H. W., Plant, J. N., Moyer, C. L., Sansone, F. J., and McMurtry, G. M. 2000a. Continuous sampling of hydrothermal fluids from Loihi Seamount after the 1996 event. *J. Geophys. Res.* **105**: 19,353–19,368.

Wheat, C. G., Elderfield, H., Mottl, M. J., and Monnin, C. 2000b. Chemical composition of basement fluids within an oceanic ridge flank: implications for along-strike and across-strike hydrothermal circulation. *J. Geophys. Res.* **105**: 13,437–13,447.

Wheat, C. G., Mottl, M. J., and Rudniki, M. 2002. Trace element and REE composition of a low-temperature ridge flank hydrothermal spring. *Geochim. Cosmochim. Acta* **66**: 3,693–3,705.

Wheat, C. G., Mottl, M. J., Fisher, A. J., Kadko, D., Davis, E. E., and Baker, E. 2003a. Heat flow from a basaltic outcrop on a ridge flank. *Geochem. Geophys. Geosys.* submitted.

Wheat, C. G., Jannasch, H. W., Kastner, M., Plant, J. N., and DeCarlo, E. H. 2003b. Seawater transport and reaction in upper oceanic basaltic basement: chemical data from continuous monitoring of sealed boreholes in a mid-ocean ridge flank environment. *Earth Planet. Sci. Lett.* **216**: 549–564.

Wheat, C. G., Jannasch, H. W., Kastner, M., Plant, J. N., DeCarlo, E. H., and Lebon, G. 2003c. Venting formation fluids from deep sea boreholes in a ridge flank setting: ODP Sites 1025 and 1026. *Geochem. Geophys. Geosys.* submitted.

Wheat, C. G., McManus, J., Mottl, M. J., and Giambalvo, E. 2003d. Oceanic phosphorus imbalance: the magnitude of the ridge-flank hydrothermal sink. *Geophys. Res. Lett.* **30**(17), 1895, doi: 10.1029/2003GL017318.

20

Insight into the hydrogeology and alteration of oceanic lithosphere based on subduction zones and arc volcanism

Simon M. Peacock

20.1 Introduction

Many features of subduction zones are directly linked to the subduction of H_2O (e.g. see papers in Bebout *et al.*, 1996), which in turn is linked to the hydrogeology and alteration of the incoming oceanic lithosphere (Fig. 20.1). H_2O plays a critical role in subduction zones. In order for subduction to occur, the thrust boundary between the subducting and overriding plates must be relatively weak. H_2O released from subducting lithosphere increases fluid pressures along the plate interface, which reduces effective stresses and promotes slip. Inter- and intraplate subduction-zone earthquakes are linked to, and in some cases triggered by, H_2O released from subducted lithosphere. H_2O dramatically lowers the temperature at which rocks melt and H_2O flux-melting explains how magmas can be generated in subduction zones which are cool relative to other tectonic settings. Volcanic rocks erupted in oceanic and continental arcs are characterized by relatively high concentrations of H_2O and fluid-mobile elements derived from subducted lithosphere. In subduction zones, H_2O-rich fluids transfer material among the hydrosphere, crust, and mantle reservoirs.

Water is subducted both as pore H_2O in sediment and crustal porosity and as chemically bound H_2O in hydrous minerals in sediments, oceanic crust, and possibly oceanic mantle. In this paper, estimates of the pore H_2O and chemically bound H_2O fluxes into subduction zones based primarily on DSDP and ODP drill cores are presented. Perhaps 5–20% of the incoming chemically bound H_2O is incorporated into arc magmas. High-pressure metabasalts exposed in ancient subduction zones and detailed seismological observations suggest that oceanic crust and mantle entering subduction zones may be more hydrated than inferred from drill core observations.

Essentially all oceanic lithosphere is ultimately subducted and the hydrogeologic history of the incoming oceanic lithosphere controls the amount and distribution of subducted H_2O. During subduction, pore H_2O is expelled by porosity collapse at shallow depths (<5–10 km), whereas chemically bound H_2O is released at greater depths by metamorphic dehydration reactions. Permeable zones in the incoming oceanic lithosphere likely remain permeable during subduction, at least to shallow depths, and influence H_2O flow paths.

Hydrogeology of the Oceanic Lithosphere, eds. E. E. Davis and H. Elderfield. Published by Cambridge University Press.
© Cambridge University Press 2004.

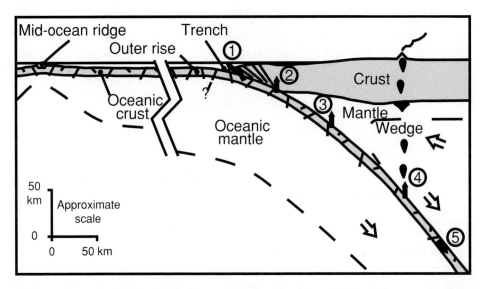

Fig. 20.1 Schematic cross-section through subduction zone showing important hydrogeologic processes. Faulting at the mid-ocean ridge, transforms (not shown), outer rise, and trench provide conduits for seawater to infiltrate and react with oceanic crust and mantle. Subduction of heterogeneously hydrated oceanic lithosphere releases H_2O as a result of porosity collapse and metamorphic dehydration reactions. Released H_2O (black arrows) may (1) reach the seafloor via updip fluid flow, (2) react with fore-arc crust, (3) react with fore-arc mantle, (4) trigger arc magmatism, or (5) be subducted past the volcanic front. Dashed line represents lithosphere–asthenosphere boundary.

Recent work on subduction-zone earthquakes, described below, suggests that faults in the incoming oceanic lithosphere are reactivated during subduction. Hydrogeologic studies of the oceanic lithosphere, therefore, provide insight into the hydrogeology of subduction zones. Conversely, studies of modern and ancient subduction zones can provide insight into hydrogeology of the incoming oceanic lithosphere.

20.2 H_2O entering subduction zones

Calculations based largely on DSDP and ODP drill cores suggest that $\sim 10^{12}$ kg of pore H_2O and $\sim 10^{12}$ kg of chemically bound H_2O is subducted globally each year (Tables 20.1 and 20.2). Subducted sediment dominates the pore H_2O budget whereas hydrous minerals in subducted oceanic crust dominate the chemically bound H_2O budget. Roughly 0.1–0.4×10^{12} kg of H_2O is returned to the crust each year by arc magmatism (Table 20.3), which is an order of magnitude less than estimates of the pore H_2O and chemically bound H_2O subducted each year.

The volume and composition of sediments subducted in different subduction zones varies considerably due to variable input, convergence rates, and degree of frontal offscraping (e.g. von Huene and Scholl, 1991). More than 1 km of sediment is subducted along parts

Table 20.1 *Estimated pore H₂O fluxes into subduction zones*

Flux (10^{12} kg yr^{-1})	Comments	Reference
Sediments		
0.9	Seismic reflection studies and depth–porosity relations	von Huene and Scholl (1991)
0.9	DSDP, ODP drill cores	Rea and Ruff (1996)
Uppermost oceanic crust		
0.07	5% porosity in uppermost 500 m	Wallmann (2001)
0.07 to 0.3	5–10% porosity in uppermost 0.5–1 km	Possible bounds, this study

Table 20.2 *Estimated chemically bound H₂O fluxes into subduction zones*

Flux (10^{12} kg yr^{-1})	Comments	Reference
Sediments		
0.09–0.17	3–6 wt. % H_2O in sediment, 500 m subducted	Ito *et al.* (1983)
0.07	5 wt. % H_2O in sediment, 200 m subducted past prism	Peacock (1990)
0.03–0.14	2–4 wt. % H_2O in sediment and range in subduction flux	Bebout (1996)
0.095	Global average (7 wt. % H_2O) based on DSDP, ODP cores	Plank and Langmuir (1998)
Oceanic crust		
0.6–1.2	Average 1–2 wt. % H_2O	Ito *et al.* (1983)
0.8	Average 2 wt. % H_2O in basalt, 1 wt. % H_2O in gabbro	Peacock (1990)
0.35	Upper 1 km of crust only, includes pore H_2O	Moore and Vrolijk (1992)
0.88–1.8	Average 1.5–3 wt. % H_2O	Bebout (1996)
1.8–2.3	5–6 wt. % H_2O in metabasalt, 1–2 wt. % H_2O in metagabbro	Schmidt and Poli (1998)
0.6–1.5	H_2O contents from DSDP, ODP drill cores	Wallmann (2001)
Oceanic mantle		
0.63	10% serpentine in uppermost 5 km	Schmidt and Poli (1998)
0.12–1.0	2.5–5% serpentinization of uppermost 5 to 20 km	Possible bounds, this study

Table 20.3 *Estimated chemically bound H$_2$O flux returned via arc magmas*

Flux (10^{12} kg yr^{-1})	Comments	Reference
0.01–0.2	Average 2 wt. % H$_2$O and wide range in magma flux	Ito *et al.* (1983)
0.14	Average 1 wt. % H$_2$O, 5 km^3 yr^{-1} arc magma flux	Peacock (1990)
0.1–0.4	1 to 3 wt. % H$_2$O, 2.5–5 km^3 yr^{-1} arc magma flux	Possible bounds, this study

of the Cascadia and southern Chile margins (von Huene and Scholl, 1991), whereas less than 100 m is subducted along the sediment-starved Tonga margin (Menard *et al.*, 1987). Von Huene and Scholl (1991) estimated that 0.9×10^{12} kg pore H$_2$O is subducted globally based on seismic reflection studies of trenches and accretionary prisms and porosity–depth relationships. Most of this pore H$_2$O is expelled at depths less than 5–10 km by sediment compaction and deformation (Moore and Vrolijk, 1992). In addition, much of the sedimentary section may be underplated at relatively shallow depth, and hence not returned to the mantle. Subducted sediment is dominantly terrigenous, with lesser amounts of calcareous and siliceous lithologies. Based on an extensive analysis of DSDP and ODP drill cores outboard of trenches, Plank and Langmuir (1998) determined that global subducting sediment (GLOSS) contains 7.29 ± 0.41 wt. % chemically bound H$_2$O. They estimated the chemically bound H$_2$O flux in sediments to be 0.095×10^{12} kg yr^{-1} (Plank and Langmuir, 1998), an order of magnitude less than the sediment pore H$_2$O flux.

Primary magmas erupted at mid-ocean ridges contain 0.09–0.52 wt. % H$_2$O based on chemical analysis of basaltic glasses (Fine and Stolper, 1986; Dixon *et al.*, 1988; Dixon and Stolper, 1995). High thermal gradients and faulting at mid-ocean ridges promote extensive hydrothermal circulation through the oceanic lithosphere. Additional hydrothermal circulation takes place on ridge flanks, along transform faults and fracture zones, and in trench–outer-rise regions. At elevated temperatures, infiltrating seawater reacts with the mafic oceanic crust to form hydrous minerals such as chlorite (13 wt. % H$_2$O), prehnite (4.4 wt. % H$_2$O), and actinolite (2.2 wt. % H$_2$O). The nature and distribution of hydrous minerals that form during hydrothermal circulation depend on the thermal history of the oceanic lithosphere and the extent to which seawater infiltrates the lithosphere (e.g. Staudigel *et al.*, 1981; Alt *et al.*, 1986, 1996; Chapter 15). Seafloor and borehole geophysical data (e.g. Becker *et al.*, 1989; Holmes and Johnson, 1993; Fisher, 1998), and studies of ophiolites (Chapter 9) show that the uppermost 0.5–1 km of the oceanic crust has high porosities of approximately 5–10%.

Hydrous minerals in the altered oceanic crust account for ~90% of the bound H$_2$O carried into subduction zones (Table 20.2; Ito *et al.*, 1983; Peacock, 1990; Bebout, 1996). Hydrothermal alteration of the oceanic crust is spatially heterogeneous and is localized along faults and other high-permeability horizons. Successive drilling legs at DSDP/ODP

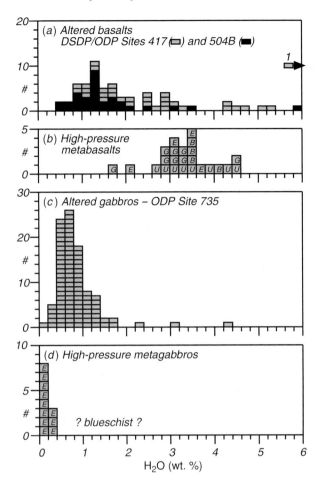

Fig. 20.2 H₂O contents of altered oceanic crust and high-pressure metamorphic equivalents. (*a*) Altered basalts, DSDP Site 417 (Alt and Honnorez, 1984) and DSDP/ODP Site 504 (Erzinger and Bach, 1996). (*b*) High-pressure metamorphosed pillow lavas, Corsican ophiolite, Corsica (Miller and Cartwright, 2000). (*c*) Altered gabbros, ODP Site 735B (Robinson *et al.*, 1992). (*d*) High-pressure metamorphosed gabbros, Monviso ophiolite, western Alps (Nadeau *et al.*, 1993). Metamorphic facies: B, blueschist facies; E, eclogite facies; G, greenschist facies; U, unmetamorphosed.

Hole 504B (5.9 Ma crust, Costa Rica rift) penetrated 275 m of sediment, 572 m of basaltic volcanic rocks, a 209-m transition zone, and 1050 m of a sheeted dike complex (Alt *et al.*, 1996; Alt, 2003). Basalts recovered from the volcanic section are 5–15% altered and typically contain 0.5–1.5 wt. % chemically bound H_2O. Typical chemically bound H_2O contents are 2–4 wt. % in the transition zone, and 1–3 wt. % in the sheeted dike complex (Alt *et al.*, 1996; Chapter 15). Chemical analyses of 32 representative samples from Hole 504B average 1.5 ± 1.0 wt. % H_2O (Fig. 20.2*a*) (Erzinger and Bach, 1996). Hydrothermal alteration may increase with increasing lithospheric age. Thirty-three basalts recovered from DSDP

Site 417 (109 Ma crust, North Atlantic) average 2.6 ± 1.5 wt. % H_2O (Fig. 20.2a; Alt and Honnorez, 1984). Deeper gabbroic sections of the oceanic crust appear to be less altered than the basalts and sheeted dike complex. Chemical analyses of 96 gabbroic samples from ODP Hole 735B (12 Ma crust, Southwest Indian Ridge) average 0.85 ± 0.55 wt. % H_2O (Fig. 20.2c; Robinson *et al.*, 1989), roughly double primary magma values.

No drill-holes have penetrated the oceanic mantle, so the extent to which the oceanic mantle is hydrated is not constrained by direct observations. At moderate- to fast-spreading ridges, faulting and hydrothermal circulation appear largely confined to crustal levels. At slow-spreading ridges, ultramafic rocks are tectonically emplaced at shallow depths, faulting extends into the mantle, and parts of the oceanic mantle may be hydrothermally altered (serpentinized; e.g. Buck *et al.*, 1998). The oxygen and hydrogen isotopic composition of some Alpine serpentinites indicate hydration occurred on the seafloor prior to subduction (e.g. Cartwright and Barnicoat, 1999). Wide-angle seismic studies yield P-wave velocities for the upper oceanic mantle ranging from 7.6 to 8.3 km s^{-1} and averaging 7.9–8.0 km s^{-1} (White *et al.*, 1992). Calculated P-wave velocities, for a confining pressure of 200 MPa and temperature appropriate for the lithospheric age, suggest the slowest P-wave speeds are likely harzburgite (ultramafic rocks rich in olivine and orthopyroxene) with up to 20% hydrous alteration (2.4 wt. % H_2O; Hacker *et al.*, 2003a).

Large outer-rise earthquakes seaward of subduction zones commonly rupture the oceanic mantle (e.g. Christensen and Ruff, 1988); if permeable, these faults may promote hydrothermal circulation and alteration of the oceanic mantle (Peacock, 2001; Ranero *et al.*, 2001). Because serpentine minerals contain 12.3 wt. % H_2O, even small amounts of serpentinization can contribute significantly to the total amount of H_2O entering a subduction zone. For example, if the uppermost 10 km of the oceanic mantle is, on average, 5% serpentinized, then hydrous minerals in the oceanic mantle would contribute 0.5×10^{12} kg yr^{-1} to the chemically bound H_2O flux entering subduction zones (Table 20.1).

20.3 H_2O release from subducted oceanic lithosphere

During subduction, pore H_2O is expelled by porosity collapse and chemically bound H_2O is liberated by metamorphic dehydration reactions. Low-density H_2O is expected to migrate upward and updip, primarily by focused flow along faults and high-permeability horizons (Peacock, 1990; Barnicoat and Cartwright, 1995). The fate of H_2O released from subducting lithosphere is not well constrained. Most of the H_2O released by porosity collapse and shallow dehydration reactions probably reaches the seafloor via focused and diffuse flow (Fig. 20.1). Low-salinity fluids observed seeping out of accretionary prisms (cold seeps) provide direct evidence for metamorphic dehydration fluids reaching the seafloor (e.g. Moore and Vrolijk, 1992).

At greater depth, H_2O released from subducting lithosphere may infiltrate and react with the overlying fore-arc crust and mantle (Fig. 20.1). Husen and Kissling (2001) interpret observed temporal changes in seismic velocities above the great Antofagasta subduction earthquake (Chile) as evidence for post-seismic fluid flow into the overlying continental crust. Active serpentine mud volcanoes observed in the Mariana fore-arc provide dramatic evidence for hydration of the shallow mantle wedge by H_2O fluids derived from the subducting lithosphere (Fryer, 1996; Fryer *et al.*, 1999). Similar inactive serpentine mud volcanoes are observed in the Izu-Bonin fore-arc (Fryer, 1992). The fore-arc mantle wedge in many subduction zones is characterized by relatively low seismic velocities and high Poisson's ratios; these observations are best explained by the presence of 10–20% serpentinite in the mantle wedge formed by reaction with H_2O released from the underlying subducting plate (Hyndman and Peacock, 2003). Locally, serpentinization of the fore-arc mantle may approach 50% (Kamiya and Kobayashi, 2000; Bostock *et al.*, 2002).

At depths of 80 to >150 km, water released from subducting lithosphere can trigger partial melting in the overlying mantle wedge (Fig. 20.1). Magmas erupted in subduction zones are relatively hydrous compared to magmas erupted at mid-ocean ridges and hotspots. The hydrous nature of subduction-zone magmas is indicated by the explosive nature of many arc volcanoes, the common presence of hornblende (2.5 wt. % H_2O) in arc lavas and plutons, and the ability of H_2O to lower melting temperatures such that magmatism can occur in the cool environment of a subduction zone (Gill, 1981; Tatsumi and Eggins, 1995). The H_2O content of primary (undifferentiated) subduction-zone magmas is difficult to measure directly. Many arc magmas have undergone magmatic fractionation and crustal assimilation which increases the H_2O content of a magma; in contrast, degassing at shallow depths prior to and during eruption decreases the H_2O content. Phenocrysts in basaltic arc lavas contain glass inclusions that exhibit a wide range in H_2O contents from 0.2 to 6 wt. % H_2O (e.g. Sisson and Grove, 1993; Sisson and Layne, 1993; Sobolev and Chaussidon, 1996; Roggensack *et al.*, 1997; Newman *et al.*, 2000). Tatsumi and Eggins (1995) concluded that primary subduction-zone magmas contain between 0 and ∼4 wt. % H_2O based on a review of natural samples and experiments. Sobolev and Chaussidon (1996) estimated that primitive island-arc basalts (tholeiites) contain 2–2.5 wt. % H_2O. Newman *et al.* (2000) proposed that primitive Mariana arc magmas contain 1–3 wt. % H_2O based on analyses of matrix glass and glass inclusions. Compared to mid-ocean ridge basalts (MORBs) and oceanic island basalts, arc basalts are enriched in a suite of minor and trace elements (e.g. K, B, Rb, Ba) that are inferred to be transferred from the subducting lithosphere to the mantle wedge by an H_2O-rich fluid (e.g. Gill, 1981; Hawkesworth *et al.*, 1993; Plank and Langmuir, 1993; Davidson, 1996; Elliot *et al.*, 1997). The concentration of fluid-mobile trace elements in arc lavas decreases with increasing distance from the trench suggesting declining H_2O input from the subducting lithosphere (Ryan *et al.*, 1996).

Many thermal–petrological models have been constructed to predict the location of metamorphic reactions in subducting lithosphere (Anderson *et al.*, 1976, 1978, 1980; Delany and Helgeson, 1978; Peacock, 1990, 1993, 1996; Poli and Schmidt, 1995; Schmidt and Poli, 1998; Iwamori, 1998; Peacock and Wang, 1999). Most dehydration reactions are temperature sensitive, although at pressures corresponding to depths of ∼100 km, H_2O is sufficiently compressible that some dehydration reactions become strongly pressure dependent. Important reactions that release chemically bound H_2O from subducting sediments include the transformation of opal to quartz (∼80 °C), the dehydration of clay mineral to form mica (100–180 °C), chlorite dehydration (400–600 °C), and mica dehydration/partial melting (750–1,000 °C; Ernst, 1990; Moore and Vrolijk, 1992; Johnson and Plank, 1999).

The progressive metamorphism of metabasalts involves complex, continuous reactions that occur over a range in pressure–temperature space (Spear, 1993). Phase equilibria experiments and field observations demonstrate that amphibole, lawsonite, phengite (mica), chlorite, talc, and zoisite are important hosts for H_2O in subducting oceanic crust (e.g. Peacock, 1990; Bebout, 1991; Pawley and Holloway, 1993; Poli and Schmidt, 1995). The most important reactions in subducting oceanic crust involve the transformation of hydrous metabasalt and metagabbro to largely anhydrous eclogite. In relatively warm subducting plates, like Nankai and Cascadia, subducting oceanic crust passes through the greenschist facies and transformation to eclogite is predicted to occur at ~50 km depth (Peacock and Wang, 1999). In cool subducting plates, like Honshu and Izu-Bonin, subducting oceanic crust passes through the blueschist facies and transformation to eclogite is not predicted to occur until depths >100 km (Peacock and Wang, 1999). Lawsonite (11.5 wt. % H_2O) is stable in cool subducting crust to depths >300 km (Pawley, 1994).

Important H_2O-releasing reactions in subducting oceanic mantle include brucite dehydration (500 °C), antigorite (serpentine) dehydration (600–700 °C), and chlorite dehydration (700–870 °C; Jenkins and Chernovsky, 1986; Ulmer and Trommsdorff, 1995; Wunder and Schreyer, 1997). In cool subducting plates, antigorite breaks down to form olivine + hydrous phase A + H_2O at depths of 150–250 km (Luth, 1995; Wunder and Schreyer, 1997; Bose and Navrotsky, 1998) and phase A (11.8 wt. % H_2O) is capable of transporting chemically bound H_2O to greater depth (Thompson, 1992; Luth, 1995).

20.4 Subduction-zone metabasalts and metagabbros

High-pressure metamorphic terranes provide information about the distribution of H_2O in ancient subduction zones and, in ideal cases, the hydration of the oceanic lithosphere prior to subduction. One of the more striking observations is that blueschist-facies metabasalts commonly contain more chemically bound H_2O than the 1–3 wt. % observed in altered basalts recovered in DSDP and ODP drill-holes (Figs. 20.2a, b; Bebout, 1996; Schmidt and Poli, 1998). For example, the Corsican ophiolite was subjected to high-pressure metamorphism during Alpine plate collision and pillow basalts from the ophiolite contain an average of 3.35 ± 0.67 wt. % chemically bound H_2O (Miller and Cartwright, 2000). Similarly, Bebout (1995) reported H_2O contents of 2.5–4 wt. % for lawsonite-blueschist facies metabasalts from Santa Catalina Island. Hacker et al. (2003a) used published mineral norms and compositions to calculate water contents for mid-ocean ridge basalts (MORBs) metamorphosed under a wide range of pressure–temperature conditions. Lawsonite-blueschist facies (pressure ~1 GPa, temperature ~300 °C) metabasalts contain 5.4 wt. % H_2O and epidote-blueschist facies (pressure ~1 GPa, temperature ~450 °C) metabasalts contain 3.1 wt. % H_2O (Hacker et al., 2003a). Schmidt and Poli's (1998) high estimate of the chemically bound H_2O in oceanic crust (Table 20.2) is based on the assumption that the basaltic layer is fully hydrated.

The higher H_2O contents of blueschist-facies MORBs compared to altered MORBs may be explained in several ways. (i) Additional hydration of oceanic basalts may take place in the trench–outer-rise region prior to subduction. (ii) During subduction, pore H_2O in the uppermost oceanic crust may hydrate surrounding basalt rather than escaping from the rock. (iii) During subduction, the basaltic layer may gain H_2O from dehydration reactions located deeper in the plate or downdip. (iv) Blueschist-facies basalts may gain water from underthrust units during and after transfer from the subducting plate to the overriding plate. Distinct geochemical patterns of ocean floor hydrothermal veins are recognizable in eclogite-facies metabasalts from the Zermatt–Saas Fee region in the Alps (Widmer *et al.*, 2000). Metamorphosed basalts and gabbros in high-pressure (subducted) ophiolites exhibit oxygen and hydrogen isotope patterns that reflect seafloor alteration prior to subduction (Barnicoat and Cartwright, 1995; Miller and Cartwright, 2000; Putlitz *et al.*, 2000). The range in observed stable isotope values and lack of homogeneity rule out the possibility of pervasive fluid infiltration during high-pressure metamorphism (Barnicoat and Cartwright, 1995; Putlitz *et al.*, 2000). This suggests that the relatively high H_2O contents observed in subduction-zone metabasalts may have been obtained prior to subduction and that we should be cautious in assuming that altered basalts recovered in DSDP and ODP drill cores well outboard of trenches represent a complete picture of the hydration state of oceanic crust prior to subduction.

In contrast to metabasalts, high-pressure metagabbros are commonly incompletely hydrated (Hacker, 1996), consistent with drill core observations of decreasing hydrothermal alteration with depth in the oceanic crust. The incomplete transformation of gabbro to blueschist- and eclogite-facies mineral assemblages at $T < 550\ °C$ is probably due to the limited availability of H_2O (Hacker, 1996; Barnicoat and Cartwright, 1997). To my knowledge, no H_2O contents for blueschist- or greenschist-facies metagabbros have been reported in the literature. Eclogite-facies gabbros from Monviso (western Alps) contain 0.04–0.27 wt. % H_2O (Fig. 20.2*d*; Nadeau *et al.*, 1993), but these rocks may have lost H_2O during peak prograde metamorphism.

20.5 Subduction-zone earthquakes and wave guides

Most of the world's earthquakes, including the deepest, occur in subduction zones. In extensional, transform, and intraplate settings, earthquakes are restricted to the upper 20–30 km of the Earth's crust. In contrast, earthquakes routinely occur at depths greater than 40 km in almost all subduction zones and reach depths of 660 km in western Pacific subduction zones. As depth increases, increasing lithostatic pressure and temperature inhibit normal brittle failure. Three major hypotheses have been proposed to explain brittle behavior at depths greater than 40 km in subduction zones: dehydration embrittlement (Raleigh and Paterson, 1965; Meade and Jeanloz, 1991; Kirby *et al.*, 1996), transformational faulting (Kirby, 1987; Green and Burnley, 1989), and melt shear instabilities (Kanamori *et al.*, 1998). While the origin of deep (300–660 km) earthquakes remains uncertain, a number of observations suggest that intermediate-depth earthquakes (40–300 km) are triggered by dehydration embrittlement linked to metamorphic dehydration reactions in the subducting oceanic lithosphere (Kirby *et al.*, 1996; Peacock and Wang, 1999; Hacker *et al.*, 2003b).

In the dehydration embrittlement model, intermediate-depth earthquakes are triggered by the metamorphic dehydration of hydrous minerals that form along faults at mid-ocean ridges and in the trench–outer-rise region prior to subduction (Kirby *et al.*, 1996). Most metamorphic dehydration reactions are temperature sensitive and intraslab earthquakes occur at greater depths in cooler subducting plates (Kirby *et al.*, 1996). Thermal models constructed for specific subduction zones show very good agreement between the predicted location of slab dehydration reactions and intermediate-depth earthquake hypocenters (Peacock and Wang, 1999; Peacock, 2001; Hacker *et al.*, 2003b). In the Tonga subduction zone, intraslab earthquakes down to 450 km depth occur along faults with the same asymmetry as outer-rise faults (Jiao *et al.*, 2000), consistent with idea that dehydration embrittlement reactivates pre-existing faults (Kirby *et al.*, 1996). If the dehydration embrittlement hypothesis is correct, then the distribution of intermediate-depth earthquakes within the subducting plate provides insight into the distribution of hydrous minerals and the extent of suboceanic hydrothermal alteration.

Intermediate-depth earthquakes occur within the subducting plate (Fig. 20.3) rather than along the plate boundary (Hasegawa *et al.*, 1978; Kirby *et al.*, 1996). In most subduction zones, Wadati–Benioff seismicity defines a relatively narrow zone approximately 10 km thick. In principle, the width of the Wadati–Benioff zone could be used to determine whether intraslab earthquakes are confined to the subducting oceanic crust or extend deeper into the subducting oceanic mantle. In practice, formal hypocenter location uncertainties are ~5 km and the complex three-dimensional structure of subduction zones suggests actual uncertainties may be greater. Statistical analysis of intraslab earthquake hypocenters in the Nankai (SW Japan) and Costa Rica subduction zones have statistical distributions consistent with an origin within the subducting, 7–8 km thick, oceanic crust (Hacker *et al.*, 2003b).

Beneath Tohoku (NE Japan), intermediate-depth earthquakes define a double seismic zone (Fig. 20.3). The upper seismic zone is located within the uppermost 10 km of the subducting plate, just below the plate interface as defined independently from P–S and S–P converted phases; the lower seismic zone is located within the subducting plate some 30–40 km deeper (Hasegawa *et al.*, 1978, 1994; Matsuzawa *et al.*, 1986; Igarashi *et al.*, 2001). Double seismic zones are observed in a number of subduction zones, including Alaska, the Aleutians, northern Chile, northeast Japan (Honshu), Kamchatka, the Kuriles, New Britain, and Tonga (see references in Seno and Yamanaka, 1996; Peacock, 2001). Lower-zone earthquakes clearly occur within the subducting oceanic mantle. Calculated pressure–temperature conditions of lower-zone earthquakes coincide with serpentine dehydration reactions consistent with a dehydration embrittlement mechanism (Seno and Yamanaka, 1996; Peacock, 2001; Seno *et al.*, 2001). If this mechanism is correct, it suggests that the upper oceanic mantle may be locally hydrated to depths of several tens of kilometers. Peacock (2001) proposed that submarine faulting outboard of the trench promotes the infiltration of seawater deep into the lithosphere and the formation of hydrous minerals (serpentinization) in the oceanic mantle. Phipps Morgan (2001) proposed that mantle serpentinization along deep faults is an important driving force for plate bending at trenches.

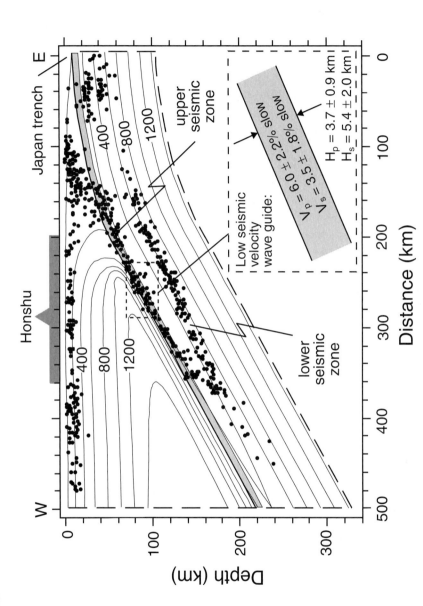

Fig. 20.3 Cross-section through NE Japan (Honshu ~39° N) subduction zone showing earthquake hypocenters (Hasegawa *et al.*, 1994), calculated thermal structure (Peacock and Wang, 1999; Peacock, 2001), and location of low-seismic-velocity wave guide (Abers, 2000). Earthquakes between 50 and 200 km depth define a double seismic zone. Low-seismic-velocity wave guide overlaps upper seismic zone. Contour interval = 200 °C. H_p and H_s = wave-guide thickness based on P-wave and S-wave dispersion, respectively (Abers, 2000). V_p and V_s = P-wave and S-wave velocities, respectively, of wave guide relative to surrounding medium (Abers, 2000).

Seismological observations can also be used to estimate the H_2O content of subducted oceanic crust. Seismic waves traveling updip or horizontally along-strike of the subducting plate commonly exhibit dispersion with high-frequency energy being retarded with respect to low-frequency energy (Hori *et al.*, 1985; Iidaki and Mizoue, 1991; Abers and Sarker, 1996). This dispersion is best explained by the presence of a low seismic-velocity wave guide located at the top of the subducting plate corresponding to the subducting oceanic crust (Fig. 20.3; e.g. Hori, 1990; Abers, 2000). In five northern Pacific subduction zones (Alaska, central Aleutians, Honshu, Mariana, and northern Kuriles), P-wave velocities within the 1–7-km-thick wave guide are 5–8% slower than adjacent mantle at depths of 100–150 km (Abers, 2000). The low velocities may represent a mixture of eclogite and untransformed gabbro or, more likely, partially hydrated blueschist-facies metagabbro (Abers, 2000) containing 1.5–3.5 wt. % chemically bound H_2O (Abers *et al.*, 2003). Beneath Nicaragua, P-wave velocities are $14.5 \pm 2.2\%$ slower, indicating the subducting Cocos crust is saturated with >5 wt. % H_2O (Abers *et al.*, 2003). As Abers *et al.* (2003) point out, the unusually hydrated nature of the subducting Cocos plate is consistent with the very high slab fluid "signal" seen in the geochemistry of the Nicaraguan volcanic arc (Noll *et al.*, 1996; Patino *et al.*, 2000) and the extremely low heat flow values measured off of the Nicoya peninsula indicating vigorous hydrothermal circulation in the oceanic crust (Langseth and Silver, 1996).

20.6 Conclusions

Many important subduction-zone processes, from earthquakes to arc magmatism, are linked to the subduction of H_2O. Approximately 10^{12} kg of pore H_2O and $\sim 10^{12}$ kg of chemically bound H_2O is subducted globally each year; approximately 5–20% of this H_2O (0.1–0.4×10^{12} kg H_2O yr^{-1}) is subsequently incorporated into arc magmas. Subducted sediment dominates the pore H_2O input to subduction zones, whereas hydrous minerals in the altered oceanic crust dominate the chemically bound H_2O input. DSDP and ODP drill cores suggest average H_2O contents of 1–3 wt. % for altered oceanic basalts and 0.5–1 wt. % for altered oceanic gabbros. Geological and geophysical observations, however, suggest oceanic crust and mantle entering subduction zones may be more hydrated than inferred from DSDP and ODP drill core observations. In ancient subduction zones, blueschist-facies metabasalts contain ~ 3–5 wt. % chemically bound H_2O. At depths of 100–150 km, low-seismic-velocity wave guides observed in several subduction zones indicate subducting oceanic crust contains 1.5–3.5 wt. % H_2O. Intermediate-depth earthquakes, which occur at 40–300 km, likely result from dehydration embrittlement and suggest that subducting oceanic crust and mantle are partially hydrated. While additional hydration may occur within subduction zones, lithospheric hydration at mid-ocean ridges, along transform faults and fracture zones, and particularly in the trench–outer-rise region may be more extensive than previously thought. Hydration of upper mantle rocks (i.e. serpentinization) may contribute significantly to the total H_2O transported by the lithosphere into subduction zones.

Acknowledgments

I thank E. Davis and A. Barnicoat for their constructive reviews and the National Science Foundation (EAR-0215565) for supporting this research.

References

Abers, G. A. 2000. Hydrated subducted crust at 100–250 km depth. *Earth Planet. Sci. Lett.* **176**: 323–330.

Abers, G. A. and Sarker, G. 1996. Dispersion of regional body waves at 100- to 150-km depth beneath Alaska: in situ constraints on metamorphism of subducted crust. *Geophys. Res. Lett.* **23**: 1,171–1,174.

Abers, G. A., Plank, T., and Hacker, B. R. 2003. The wet Nicaraguan slab. *Geophys. Res. Lett.* **30**: doi: 10.1029/2002GL015649.

Alt, J. C. 2003. Alteration of the upper oceanic crust: mineralogy, chemistry, and processes. In *Hydrogeology of the Oceanic Lithosphere*, eds. E. Davis and H. Elderfield. Cambridge: Cambridge University Press, Chapter 15, this volume.

Alt, J. C. and Honnorez, J. 1984. Alteration of the upper oceanic crust, DSDP site 417: mineralogy and chemistry. *Contrib. Mineral. Petrol.* **87**: 149–169.

Alt, J. C., Honnorez, J., Laverne, C., and Emmermann, R. 1986. Hydrothermal alteration of a 1 km section through the upper oceanic crust, Deep Sea Drilling Project Hole 504B: Mineralogy, chemistry, and evolution of seawater–basalt interactions. *J. Geophys. Res.* **91**: (10), 309–10, 335.

Alt, J. C., Laverne, C., Vanko, D. A., Tartarotti, P., Teagle, D. A. H., Bach, W., Zuleger, E., Erzinger, J., Honnorez, J., Pezard, P., Becker, K., Salisbury, M. H., and Wilkens, R. H. 1996. Hydrothermal alteration of a section of upper oceanic crust in the eastern equatorial Pacific: a synthesis of results from Site 504 (DSDP Legs 69, 70, and 83, and ODP Legs 111, 137, 140, and 148). In *Proceedings of the Ocean Drilling Program, Scientific Results*, Vol. 148, eds. J. C. Alt, H. Kinoshita, *et al.* College Station, TX: Ocean Drilling Program, pp. 417–434.

Anderson, R. N., Uyeda, S., and Miyashiro, A. 1976. Geophysical and geochemical constraints at converging plate boundaries – Part I: Dehydration in the downgoing slab. *Geophys. J. Int., RAS*, **44**: 333–357.

Anderson, R. N., DeLong, S. E., and Schwarz, W. M. 1978. Thermal model for subduction with dehydration in the downgoing slab. *J. Geol.* **86**: 731–739.

1980. Dehydration, asthenospheric convection and seismicity in subduction zones. *J. Geol.* **88**: 445–451.

Barnicoat, A. C. and Cartwright, I. 1995. Focused fluid flow during subduction: oxygen isotope data from high-pressure ophiolites of the western Alps. *Earth Planet. Sci. Lett.* **132**: 53–61.

1997. The gabbro–eclogite transformation: an oxygen isotope and petrographic study of west Alpine ophiolites. *J. Metam. Geol.* **15**: 93–104.

Bebout, G. E. 1991. Field-based evidence for devolatilization in subduction zones: implications for arc magmatism. *Science* **251**: 413–416.

1995. The impact of subduction-zone metamorphism on mantle–ocean chemical cycling. *Chem. Geol.* **126**: 191–218.

1996. Volatile transfer and recycling at convergent margins: mass-balance and insights from high-P/T metamorphic rocks. In *Subduction: Top to Bottom*, Geophysical Monograph 96, eds. G. E. Bebout, D. W. Scholl, *et al.* Washington, DC: American Geophysical Union, pp. 179–193.

Bebout, G. E., Scholl, D., W., Kirby, S. H., and Platt, J. P. (eds.) 1996. *Subduction: Top to Bottom*, Geophysical Monograph 96. Washington, DC: American Geophysical Union, pp. 384.

Becker, K., Sakai, H., Adamson, A. C., Alexandrovich, J., Alt, J. C., Anderson, R. N., Bideau, D., Gable, R., Herzig, P. M., Houghton, S., Ishizuka, H., Kawahata, H., Kinoshita, H., Langseth, M. G., Lovell, M. A., Malpas, J., Masuda, H., Merrill, R. B., Morin, R. H., Mottl, M. J., Pariso, J. E., Pezard, P., Phillips, J., Sparks, J., and Uhlig, S. 1989. Drilling deep into young oceanic crust, hole 504B, Costa Rica rift. *Rev. Geophys.* **27**: 79–102.

Bose, K. and Navrotsky, A. 1998. Thermochemistry and phase equilibria of hydrous phases in the system $MgO–SiO_2–H_2O$: implications for volatile transport to the mantle. *J. Geophys. Res.* **103**: 9,713–9,719.

Bostock, M. G., Hyndman, R. D., Rondenay, S., and Peacock, S. M. 2002. An inverted continental Moho and serpentinizatin of the forearc mantle. *Nature* **417**: 536–538.

Buck, W. R., Delaney, P. T., Karson, J. A., and Lagabrielle, Y. (eds.) 1998. *Faulting and Magmatism at Mid-Ocean Ridges*, Geophysical Monograph 106. Washington, DC: American Geophysical Union.

Cartwright, I. and Barnicoat, A. C. 1999. Stable isotope geochemistry of Alpine ophiolites: a window into ocean-floor hydrothermal alteration and constraints on fluid–rock interaction during high-pressure metamorphism. *Int. J. Earth Sci.* **88**: 219–235.

Christensen, D. H. and Ruff, L. J. 1988. Seismic coupling and outer rise earthquakes. *J. Geophys. Res.* **93**: 13,421–13,444.

Davidson, J. P. 1996. Deciphering mantle and crustal signatures in subduction zone magmatism. In *Subduction: Top to Bottom*, Geophysical Monograph 96, eds. G. E. Bebout, D. W. Scholl, *et al.* Washington, DC: American Geophysical Union, pp. 251–262.

Delany, J. M. and Helgeson, H. C. 1978. Calculation of the thermodynamic consequences of dehydration in subducting oceanic crust to 100 kb and >800 °C. *Am. J. Sci.* **278**: 638–686.

Dixon, J. E. and Stolper, E. M. 1995. An experimental study of water and carbon dioxide solubilities in mid-ocean ridge basaltic liquids. Part II: Applications to degassing. *J. Petrol.* **36**: 1,633–1,646.

Dixon, J. E., Stolper, E., and Delaney, J. R. 1988. Infrared spectroscopic measurements of CO_2 and H_2O in Juan de Fuca ridge basaltic glasses. *Earth Planet. Sci. Lett.* **90**: 87–104.

Elliott, T., Plank, T., Zindler, A., White, W., and Bourdon, B. 1997. Element transport from slab to volcanic front at the Mariana arc. *J. Geophys. Res.* **102**: 14,991–15,019.

Ernst, W. G. 1990. Thermobarometric and fluid expulsion history of subduction zones. *J. Geophys. Res.* **95**: 9,047–9,053.

Erzinger, J. and Bach, W. 1996. Downhole variation of molecular nitrogen in DSDP/ODP Hole 504B: preliminary results. In *Proceedings of the Ocean Drilling Program, Scientific Results*, Vol. 148, eds. J. C. Alt, H. Kinoshita, *et al.* College Station, TX: Ocean Drilling Program, pp. 3–8.

Fine, G. and Stolper, E. 1986. Dissolved carbon-dioxide in basaltic glasses: concentrations and speciation. *Earth Planet. Sci. Lett.* **76**: 263–278.

Fisher, A. T. 1998. Permeability within the basaltic oceanic crust. *Rev. Geophys.* **36**: 143–182.

Fryer, P. 1992. A synthesis of Leg 125 drilling of serpentine seamounts on the Mariana and Izu Bonin forearcs. In *Proceedings of the Ocean Drilling Program, Scientific*

Results, Vol. 125, eds. P. Fryer, J. A. Pearce, *et al.* College Station, TX: Ocean Drilling Program, pp. 593–614.

1996. Evolution of the Mariana convergent plate margin system. *Rev. Geophys.* **34**: 89–125.

Fryer, P., Wheat, C. G., and Mottl, M. J. 1999. Mariana blueschist mud volcanism: implications for conditions within the subduction zone. *Geology* **27**: 103–106.

Gill, J. 1981. *Orogenic Andesites and Plate Tectonics*. New York: Springer-Verlag, 390 pp.

Green, H. W. I. and Burnley, P. C. 1989. A new self-organizing mechanism for deep-focus earthquakes. *Nature* **341**: 733–737.

Hacker, B. R. 1996. Eclogite formation and the rheology, buoyancy, seismicity, and H_2O content of oceanic crust. In *Subduction: Top to Bottom*, Geophysical Monograph 96: eds. G. E. Bebout, D. W. Scholl *et al.* Washington, DC: American Geophysical Union, pp. 337–346.

Hacker, B. R., Abers, G. A., and Peacock, S. M. 2003a. Subduction Factory 1. Theoretical mineralogy, densities, seismic wave speeds, and H_2O contents. *J. Geophys. Res.* **108**: doi: 10.1029/2001JB001127.

Hacker, B. R., Peacock, S. M., Abers, G. A., and Holloway, S. D. 2003b. Subduction Factory 2. Are intermediate-depth earthquakes in subducting slabs linked to metamorphic dehydration reactions? *J. Geophys. Res.* **108**: doi: 10.1029/2001JB001129.

Hasegawa, A., Umino, N., and Takagi, A. 1978. Double-planed deep seismic zone and upper mantle structure in the northeastern Japan arc. *Geophys. J. Int., RAS*, **54**: 281–296.

Hasegawa, A., Horiuchi, S., and Umino, N. 1994. Seismic structure of the northeastern Japan convergent plate margin: a synthesis. *J. Geophys. Res.* **99**: 22,295–22,311.

Hawkesworth, C. J., Gallagher, K., Hergt, J. M., and McDermott, F. 1993. Mantle and slab contributions in arc magmas. *Ann. Rev. Earth Planet. Sci.* **21**: 175–204.

Holmes, M. L. and Johnson, H. P. 1993. Upper crustal densities derived from sea floor gravity measurements: northern Juan de Fuca ridge. *Geophys. Res. Lett.* **20**: 1,871–1,874.

Hori, S. 1990. Seismic waves guided by untransformed oceanic crust subducting into the mantle: the case of the Kanto district, central Japan. *Tectonophysics* **176**: 355–376.

Hori, S., Inoue, H., Fukao, Y., and Ukawa, M. 1985. Seismic detection of the untransformed "basaltic" oceanic crust subducting into the mantle. *Geophys. J. Int., RAS* **83**: 169–197.

Husen, S. and Kissling, E. 2001. Postseismic fluid flow after the large subduction earthquake of Antofagasta, Chile. *Geology* **29**: 847–850.

Hyndman, R. D. and Peacock, S. M. 2003. Serpentinization of the forearc mantle. *Earth Planet. Sci. Lett.* **212**: 417–432.

Igarashi, T., Matsuzawa, T., Umino, N., and Hasegawa, A. 2001. Spatial distribution of focal mechanimsms for interplate and intraplate earthquakes associated with the subducting Pacific plate beneath the northeastern Japan arc: a triple-planed deep seismic zone. *J. Geophys. Res.* **106**: 2,177–2,191.

Iidaki, T. and Mizoue, M. 1991. P-wave velocity structure inside the subducting Pacific Plate beneath the Japan region. *Phys. Earth Planet. Int.* **66**: 203–213.

Ito, E., Harris, D. M., and Anderson, A. T. J. 1983. Alteration of oceanic crust and geologic cycling of chlorine and water. *Geochim. Cosmochim. Acta* **47**: 1,613–1,624.

Iwamori, H. 1998. Transportation of H_2O and melting in subduction zones. *Earth Planet. Sci. Lett.* **160**: 65–80.

Jenkins, D. M. and Chernovsky, J. V. 1986. Phase equilibria and crystallochemical properties of Mg-chlorite. *Am. Mineral.* **71**: 924–936.

Jiao, W., Silver, P. G., Fei, Y., and Prewitt, C. T. 2000. Do intermediate- and deep-focus earthquakes occur on preexisting weak zones? An examination of the Tonga subduction zone. *J. Geophys. Res.* **105**: 28,125–28,138.

Johnson, M. C. and Plank, T. 1999. Dehydration and melting experiments constrain the fate of subducted sediments. *Geochem. Geophys. Geosys.* **1**: doi: 1999GC000014.

Kamiya, S. and Kobayashi, Y. 2000. Seismological evidence for the existence of serpentinized wedge mantle. *Geophys. Res. Lett.* **27**: 819–822.

Kanamori, H., Anderson, D. L., and Heaton, T. H. 1998. Frictional melting during the rupture of the 1994 Bolivian earthquake. *Science* **279**: 839–842.

Kirby, S. H. 1987. Localized polymorphic phase transitions in high-pressure faults and applications to the physical mechanism of deep earthquakes. *J. Geophys. Res.* **92**: 12,789–13,800.

Kirby, S. H., Engdahl, E. R., and Denlinger, R. 1996. Intraslab earthquakes and arc volcanism: dual physical expressions of crustal and uppermost mantle metamorphism in subducting slabs. In *Subduction: Top to Bottom*, Geophysical Monograph 96, eds. G. E. Bebout, D. W. Scholl, *et al*. Washington, DC: American Geophysical Union, pp. 195–214.

Langseth, M. G. and Silver, E. A. 1996. The Nicoya convergent margin – a region of exceptionally low heat flow. *Geophys. Res. Lett.* **23**: 891–894.

Luth, R. W. 1995. Is phase A relevant to the Earth's mantle? *Geochim. Cosmochim. Acta* **59**: 679–682.

Matsuzawa, T., Umino, N., Hasegawa, A., and Takagi, A. 1986. Normal fault type events in the upper plane of the double-planed deep seismic zone beneath the northeastern Japan Arc. *J. Phys. Earth* **34**: 85–94.

Meade, C. and Jeanloz, R. 1991. Deep-focus earthquakes and recycling of water into the Earth's mantle. *Science* **252**: 68–72.

Menard, H. W., Natland, J., Jordan, T. H., and Orcutt, J. A., eds. 1987. *Initial Reports of the Deep Ocean Drilling Project*, Vol. 91. Washington, DC: US Govt. Printing Office.

Miller, J. A. and Cartwright, I. 2000. Distinguishing between seafloor alteration and fluid flow during subduction using stable isotope geochemistry: examples from Tethyan ophiolites in the Western Alps. *J. Metam. Geol.* **18**: 467–482.

Moore, J. C. and Vrolijk, P. 1992. Fluids in accretionary prisms. *Rev. Geophys.* **30**: 113–135.

Nadeau, S., Philippot, P., and Pineau, F. 1993. Fluid inclusion and mineral isotopic compositions (H–C–O) in eclogitic rocks as tracers of local fluid migration during high-pressure metamorphism. *Earth Planet. Sci. Lett.* **114**: 431–448.

Newman, S., Stolper, E., and Stern, R. 2000. H_2O and CO_2 in magmas from the Mariana arc and back-arc systems. *Geochem. Geophys. Geosyst.* **1**: doi: 1999GC000027.

Noll, P. D., Jr., Newsom, H. E., and Ryan, J. G. 1996. The role of hydrothermal fluids in the production of subduction zone magmas: evidence from siderophile and chalcophile trace elements and boron. *Geochim. Cosmochim. Acta* **60**: 587–611.

Patino, L. C., Carr, M. J., and Feigenson, M. D. 2000. Local and regional variations in Central American arc lavas controlled by variations in subducted sediment input. *Contrib. Mineral. Petrol.* **138**: 265–283.

Pawley, A. R. 1994. The pressure and temperature stability limits of lawsonite: implications for H_2O recycling in subduction zones. *Contrib. Mineral. Petrol.* **118**: 99–108.

Pawley, A. R. and Holloway, J. R. 1993. Water sources for subduction zone volcanism: new experimental constraints. *Science* **260**: 664–667.

Peacock, S. M. 1990. Fluid processes in subduction zones. *Science* **248**: 329–337.

1993. The importance of the blueschist → eclogite dehydration reactions in subducting oceanic crust. *Geol. Soc. Am. Bull.* **105**: 684–694.

1996. Thermal and petrologic structure of subduction zones. In *Subduction: Top to Bottom*, Geophysical Monograph 96, eds. G. E. Bebout, D. W. Scholl, *et al.* Washington, DC: American Geophysical Union, pp. 119–133.

2001. Are the lower planes of double seismic zones caused by serpentine dehydration in subducting oceanic mantle? *Geology* **29**: 299–302.

Peacock, S. M. and Wang, K. 1999. Seismic consequences of warm versus cool subduction zone metamorphism: examples from northeast and southwest Japan. *Science* **286**: 937–939.

Phipps Morgan, J. 2001. The role of serpentinization and deserpentinization in bending and unbending the subducting slab (abs.). *Eos, Trans. Am. Geophys. Union* **82**: F1154.

Plank, T. A. and Langmuir, C. H. 1993. Tracing trace elements from sediment input to volcanic output at subduction zones. *Nature* **362**: 739–742.

1998. The chemical composition of subducting sediment and its consequences for the crust and mantle. *Chem. Geol.* **145**: 325–394.

Poli, S. and Schmidt, M. W. 1995. H_2O transport and release in subduction zones: experimental constraints on basaltic and andesitic systems. *J. Geophys. Res.* **100**: 22,299–22,314.

Putlitz, B., Matthews, A., and Valley, J. W. 2000. Oxygen and hydrogen isotope study of high-pressure metagabbros and metabasalts (Cyclades, Greece): implications for the subduction of oceanic crust. *Contrib. Mineral. Petrol.* **138**: 114–126.

Raleigh, C. B. and Paterson, M. S. 1965. Experimental deformation of serpentinite and its tectonic implications. *J. Geophys. Res.* **70**: 3,965–3,985.

Ranero, C. R., Phipps Morgan, J., McIntosh, K. D., and Reichert, C. 2001. Flexural faulting and mantle serpentinization at the Middle America Trench. *Eos, Trans. Am. Geophys. Union* **82**: F1154.

Rea, D. K. and Ruff, L. J. 1996. Composition and mass flux of sediment entering the world's subduction zones: implications for global sediment budgets, great earthquakes, and volcanism. *Earth Planet. Sci. Lett.* **140**: 1–12.

Robinson, P. T., Von Herzen, R. P., *et al.*, eds. 1989. *Proceedings of the Ocean Drilling Program, Initial Reports*, Vol. 118. College Station, TX: Ocean Drilling Program.

Roggensack, K., Hervig, R. L., McKnight, S. B., and Williams, S. N. 1997. Explosive basaltic volcanism from Cerro Negro volcano: influence of volatiles on eruptive style. *Science* **277**: 1,639–1,642.

Ryan, J., Morris, J., Bebout, G. E., and Leeman, B. 1996. Describing chemical fluxes in subduction zones: insights from "depth-profiling" studies of arc and forearc rocks. In *Subduction: Top to Bottom*, Geophysical Monograph 96, eds. G. E. Bebout, D. W. Scholl, *et al.* Washington, DC: American Geophysical Union, pp. 263–268.

Schmidt, M. W. and Poli, S. 1998. Experimentally based water budgets for dehydrating slabs and consequences for arc magma generation. *Earth Planet. Sci. Lett.* **163**: 361–379.

Seno, Y. and Yamanaka, Y. 1996. Double seismic zones, compressional deep outer-rise events, and superplumes. In *Subduction: Top to Bottom*, Geophysical Monograph 96, eds. G. Bebout, D. W. Scholl, *et al.* Washington, DC: American Geophysical Union, pp. 347–355.

Seno, T., Zhao, D., Kobayashi, Y., and Nakamura, M. 2001. Dehydration of serpentinized slab mantle: seismic evidence from southwest Japan. *Earth Planets Space* **53**: 861–871.

Sisson, T. W. and Grove, T. L. 1993. Temperatures and H_2O contents of low-MgO high-alumina basalts. *Contrib. Mineral. Petrol.* **113**: 167–184.

Sisson, T. W. and Layne, G. D. 1993. H_2O in basalt and basaltic andesite glass inclusions from 4 subduction-related volanoes. *Earth Planet. Sci. Lett.* **117**: 619–635.

Sobolev, A. V. and Chaussidon, M. 1996. H_2O concentrations in primary melts from supra-subduction zones and mid-ocean ridges: implications for H_2O storage and recycling in the mantle. *Earth Planet. Sci. Lett.* **137**: 45–55.

Spear, F. S. 1993. *Metamorphic Phase Equilibria and Pressure–Temperature–Time Paths*, Mineralogical Society of America Monograph 1. Washington, DC: Mineralogical Society of America.

Staudigel, H., Hart, S. R., and Richardson, S. H. 1981. Alteration of the oceanic crust: processes and timing. *Earth Planet. Sci. Lett.* **52**: 311–327.

Tatsumi, Y. and Eggins, S. (eds.) 1995. *Subduction Zone Magmatism*, Frontiers in Earth Sciences. Ann Arbor, MI: Blackwell Science, Inc.

Thompson, A. B. 1992. Water in the earth's upper mantle. *Nature* **358**: 295–302.

Ulmer, P. and Trommsdorff, V. 1995. Serpentine stability to mantle depths and subduction-related magmatism. *Science* **268**: 858–861.

von Huene, R. and Scholl, D. W. 1991. Observations at convergent margins concerning sediment subduction, subduction erosion, and the growth of continental crust. *Rev. Geophys.* **29**: 279–316.

Wallmann, K. 2001. The geological water cycle and the evolution of marine $\delta^{18}O$ values. *Geochim. Cosmochim. Acta* **65**: 2,469–2,485.

White, R. S., McKenzie, D., and O'Nions, R. K. 1992. Oceanic crustal thickness from seismic measurements and rare earth element inversions. *J. Geophys. Res.* **97**: 19,683–19,715.

Widmer, T., Ganguin, J., and Thompson, A. B. 2000. Ocean floor hydrothermal veins in eclogite facies rocks of the Zermatt–Saas Zone, Switzerland. *Schw. Min. Petrogr. Mitt.* **80**: 63–73.

Wunder, B. and Schreyer, W. 1997. Antigorite: high-pressure stability in the system $MgO–SiO_2–H_2O$ (MSH). *Lithos* **41**: 213–227.

21

Hydrothermal fluxes in a global context

Mike Bickle and Harry Elderfield

21.1 Introduction

The chemical composition of the oceans, and its chemical evolution, is controlled by the interplay of a variety of processes, hydrologic, tectonic, geochemical, biological, and sedimentological, many of which have a profound influence on the Earth's climate. The concept that contrasting geochemical cycles somehow interact to control ocean composition within narrow limits has been a long-standing idea. For example, it formed the basis for Sillén's (1967) equilibrium ocean model involving the titration of acids (volatiles) from the mantle with bases from weathering of continental crust. A more recent model, but with long-standing roots, concerns long-term controls on Cenozoic climate based on the carbonate–silicate cycle and assumes that atmospheric levels of CO_2 are maintained by inputs from the solid Earth which are balanced by removal by weathering of silicate rocks and precipitation of carbonate minerals in the ocean (Walker et al., 1981; Berner et al., 1983).

The general idea that ocean composition is controlled by a balance of fluxes derived from alteration of the oceanic crust and the continental crust was strongly underpinned by the discovery of the presence of high-temperature hydrothermal vents on ocean ridges and calculations that global chemical fluxes from the vents are of the same magnitude of global river fluxes (Edmond et al., 1979). A good example of this concerns the "Mg problem" posed by Drever (1974), a few years before the first discovery of submarine hot springs at Galapagos. Drever estimated the magnitude of the input fluxes of Mg to the oceans by rivers and found that it was impossible to balance inputs with known outputs. It is now recognized that removal of Mg from seawater by reaction with oceanic crust provides the major oceanic Mg sink.

The purpose of this chapter is to consider hydrothermal fluxes (a term used here without regard to area, meaning total flow of a species) in a global context so as to estimate their global importance to ocean composition. This is a complex issue because hydrothermal fluxes operate on a range of both space and time scales; methods of evaluating hydrothermal fluxes must also consider such scales (Fig. 21.1). Because hydrothermal fluxes of chemicals

Hydrogeology of the Oceanic Lithosphere, eds. E. E. Davis and H. Elderfield. Published by Cambridge University Press.
© Cambridge University Press 2004.

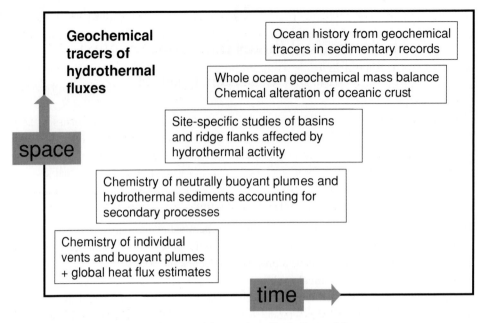

Fig. 21.1 Space and time scales of hydrothermal fluxes.

and isotopes are, of course, associated with fluxes of heat and of water, we consider these first. Next, we consider modern expressions of hydrothermal flux on axis (ridge crest) and off axis (ridge flank). Following this we consider integrated fluid and chemical fluxes using evidence of altered oceanic crust, in the modern oceans and from ophiolites.

A book and three recent reviews of this subject are available. Humphris *et al.* (1995) is an overview of axial hydrothermal processes. (Elderfield and Schultz, 1996) focused on the relationship between geophysical and geochemical estimates of hydrothermal fluxes. German and Von Damm (2003) have considered the geochemistry of modern hydrothermal activity at ridge crest hydrothermal systems, and Mott (2003) has considered the partition of hydrothermal energy and mass fluxes between mid-ocean ridges and flanks. Chapter 19 is specifically concerned with fluxes through ridge flanks: fluxes through sediments are discussed in Chapter 16.

21.2 Methods for calculation of chemical fluxes

Chemical elements are added to or removed from seawater as a consequence of reaction of seawater with oceanic crust at elevated temperature and fluid flow driven by thermal buoyancy.

There are several methods by which fluid chemistry has been converted to global estimates of hydrothermal fluxes of chemical constituents. One method is simply to combine fluid chemistry with estimates of fluid flux calculated, as discussed in Section 21.3, using estimates of the heat fluxes. An alternative method is based on δ^3He/heat data for individual vent sites (Jenkins *et al.*, 1978) together with the estimate of the flux of global primordial He which is extracted by hydrothermal activity from the upper mantle. The difficulty with this method is that significant variability has now been observed in δ^3He/heat ratios for individual vent sites (see Elderfield and Schultz, 1996). Chemical fluxes resulting from fluid flow through sediments may be estimated using advection–diffusion–reaction transport models (Chapter 16).

Two other methods take a global approach. One is based on geochemical mass balance for the oceans, where the hydrothermal input is calculated as that necessary to balance the known river input. The other is to calculate element fluxes from the oceanic crust by the difference in composition between fresh oceanic basalt and old altered oceanic crust. Significant discrepancies exist between the estimates made by all of these methods and between estimates made using different elements in the oceanic mass-balance calculations.

21.3 Thermal constraints on water fluxes

21.3.1 Heat flow methods

There are three principal hydrological regimes in the oceans (COSOD II 1987; Preface): (i) flow at ridge axes, driven by buoyancy generated by highly localized magmatic heat; (ii) flow in off-axis settings driven by deep-seated heat from the cooling lithosphere; and (iii) flow at continental margins driven by topographically or compositionally derived "head" and by consolidation caused by gravitational or tectonic stress. The last may be significant but is beyond the scope of this book. Flow at ridge axes is expressed by the well-known hot springs on seafloor. These are short-lived systems in the Lagrangian reference frame of the lithospheric plate (Introduction), but, nevertheless, of great significance to chemical budgets. Flow in off-axis settings, when considered over the lifetime of the lithosphere or over the full areal extent of oceanic crust in a Eulerian reference frame, makes the greatest contribution to the total volumetric flow of seawater in the crust.

One approach to estimating hydrothermal fluxes from data such as these was carried out by Stein and Stein (1994) who used heat flow data from the literature to estimate average heat flow values for crust in different age ranges (Fig. 21.2; see Chapter 10). This method allowed estimation of the difference between predicted and observed values (the so-called "missing heat"). The global heat flow anomaly so calculated and attributable to hydrothermal flux is about 9 TW of a total oceanic heat flux of 32 TW. This topic, including a more extensive summary of papers on this subject, is discussed in detail in Chapter 10 where the view is expressed that "the cumulative value of approximately 10 TW is unlikely to change significantly by the addition of new data, unless detailed surveys alter our view of the processes that cause heat flow deficits." This heat flux can be partitioned between crust of different ages. For example, it implies that about 20% of the hydrothermal flux is associated with oceanic crust younger than 1 Ma, 50% with crust younger than 4 Ma (Chapter 10), and the remainder with crust between 1 and 65 Ma (Fig. 21.2). The age of 65 Ma, where predicted and observed values converge in this treatment, has been termed the "sealing age" of the crust and has been considered to represent the age when secondary mineral formation

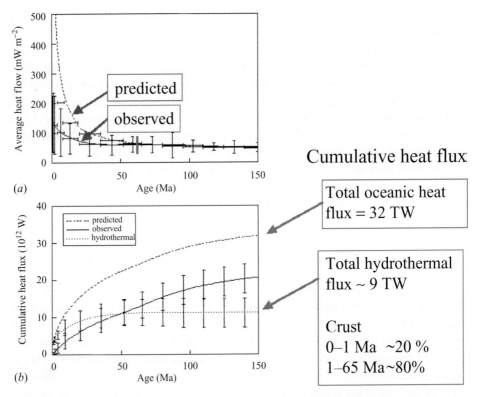

Fig. 21.2 Comparison of averaged measured heat flux versus age and values predicted of conductive cooling.

prevents further hydrothermal circulation or sediment accumulation restricts hydrothermal exchange between the crust and the ocean.

The method outlined above has significant uncertainties. Measured conductive heat flow requires insertion of a heat flow probe into sediment. This means that the method cannot be applied in many circumstances on young oceanic crust, which is often unsedimented. Therefore, except in exceptional circumstances, axial or near-axial heat flow cannot be estimated by this method. The scatter in conductive heat flow values (Chapter 10) results in a large uncertainty on the hydrothermal heat flux estimates. In order to estimate hydrothermal chemical fluxes we need to know the mass fluxes of seawater associated with the heat transport. Hydrothermally advected heat flux (f) is related to the volumetric flux of seawater (q) by the water temperature anomaly (ΔT) and the volumetric heat capacity of seawater (ρc)

$$q = \frac{f}{\rho c \Delta T}$$

Heat capacity varies with temperature, pressure, and composition. Specific heat c becomes very large close to the two-phase boundary and especially near the critical point ($\sim405\ °C$ at 30 MPa) with the range for c at seafloor pressures at which most black smoker vents have been found is about 4.1–7.5 × 10³ J m⁻³ K⁻¹. The hydrothermal water temperature anomaly (Δ*T*) is also poorly constrained in young crust with a corresponding uncertainty in the calculated hydrothermal fluxes.

Because of the scatter in heat flow values and uncertainty in hydrothermal water temperatures, the method proposed by Stein and Stein places only the broadest constraints on chemical fluxes. For these reasons it is necessary to rely on alternative methods of calculating chemical fluxes. At ridge crests it is possible to place constraints on the high-temperature hydrothermal fluxes by consideration of the heat available and the efficiency of the hydrothermal systems. No such constraint is available on the flank fluxes and these can only be constrained by oceanic mass-balance and the amount of alteration seen in old oceanic crust.

21.3.2 High-temperature fluxes

The maximum available heat to drive high-temperature hydrothermal circulation at mid-ocean ridges is the magmatic heat flux. This has two components: the latent heat of crystallization of newly formed cooling ocean crust (estimated at ∼500 J g⁻¹ by Davis *et al.*, 2003) and the heat of cooling from magmatic temperatures (∼1,200 °C) to the hydrothermal temperature structure, estimated to average 816 J g⁻¹ by Davis *et al.* (2003). Hydrothermal systems probably operate at about 50% efficiency from the modeling of hydrothermal heat loss by Morton and Sleep (1985) and Sleep (1991), constrained by the depth to the top of the magma chamber.

Given a 7.1 km thick average oceanic crust, a global spreading rate of 3.45 km² yr⁻¹ and 50% efficiency, the heat flux carried by high-temperature flow at the axis would be ∼1.3 TW. Were all this flow carried by black smoker fluid heated to 400 °C at 350 bar, then the total water flux would be ∼2.3 × 10¹³ kg yr⁻¹ (Davis *et al.*, 2003). This estimate is less than the combined estimates by Kadko *et al.* (1994) of 1.8 × 10¹³ kg yr⁻¹ for loss of latent heat only and by Mottl (2003) of 4.9 × 10¹³ kg yr⁻¹ for the latent heat plus heat of cooling to 350 °C (see Chapter 19), partly because of choice of thermal parameters but mainly because of the more conservative assumption of the hydrothermal efficiency of only 50%. A complication is that part of the heat flux may be carried by a larger water flux heated to only intermediate temperatures although such fluids have not been sampled at ocean ridges (they might be hard to find). It should be noted that the lower temperature so-called "diffuse flow" at high-temperature vent fields, which may carry as much as one order of magnitude greater heat out of the system than high-temperature flow (Rona and Trivett, 1992; Schultz *et al.*, 1992), is a consequence of mixing high-temperature hydrothermal fluid with seawater (e.g. James and Elderfield, 1996; Teagle *et al.*, 1998a). This has no effect on high-temperature fluxes although it has a profound effect on the chemistry of the vent fluids. Therefore, axial diffuse flow systems can be considered, along with hydrothermal plumes, purely in terms of their modification of primary hydrothermal input of chemicals associated with axial venting. This is beyond the scope of this chapter.

The estimates of axial high-temperature hydrothermal fluxes quoted above are upper limits calculated on the assumption that all the ocean-ridge magma is delivered to a

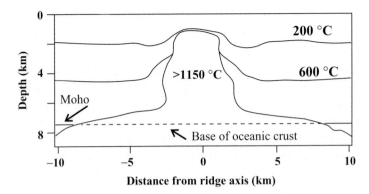

Fig. 21.3 Thermal structure beneath East Pacific Rise at 9° 30′ N redrawn after Dunn and Toomey (2000) from three-dimensional seismic tomography with temperature calculated assuming both anharmonic and anelastic effects. Note that the steep gradients in temperature away from the ridge axis imply efficient removal of heat by hydrothermal circulation throughout the crust.

high-level magma chamber where all the latent heat is available to drive the axial circulation. However there is little consensus on how the interdependent igneous and hydrothermal processes at ridge axes operate. Observations of magma chamber formation throughout the lower crust in the Oman ophiolite (Kelemen *et al.*, 1997), and interpretation of seafloor compliance (Crawford *et al.*, 1999) and seismic data from the EPR (Dunn and Toomey, 2000), indicate magma chambers at two or more depths in the lower oceanic crust, and thus are inconsistent with this simple model. If a significant fraction of the magma is intruded within the lower crust, this would substantially reduce the heat available to drive high-temperature axial hydrothermal circulation and the lower crust would take correspondingly longer to cool (Chen, 2001). Surprisingly, despite the evidence for distributed magma chambers in the lower crust, seismic observations indicate that the lower crust does cool rapidly. Magma chambers are restricted to within 1 or 2 km of the ridge axis and a discernable Moho develops within 3 km of the ridge axis (Fig. 21.3). The hydrothermal structure which might cause such cooling is not understood, although lateral extraction of heat is likely to be inefficient.

Direct constraints on the recharge flux to the high-temperature hydrothermal systems might resolve this controversy, but these are limited to one site, DSDP/ODP Hole 504B. Here, Teagle *et al.* (2003) used $^{87}Sr/^{86}Sr$ ratios in anhydrite, precipitated from the recharge fluid, and altered rock to model the recharge, assuming that fluid–solid exchange followed a linear–kinetic rate law. The estimate of the time-integrated recharge of $1.7 \pm 0.2 \times 10^6$ kg m^{-2} (1σ error) is equivalent to a global mass flow rate of only 6×10^{12} kg yr^{-1} for the global average spreading rate of 3.45 km^2 yr^{-1}. This is only ∼25% of the estimated maximum possible high-temperature flux of 2.3×10^{13} kg yr^{-1} calculated above. However, it is premature to generalize from only one estimate of the recharge flux in oceanic crust, which is likely to be heterogeneous.

21.3.3 Flank fluxes

The heat transferred through the seafloor by hydrothermal circulation has been estimated from the difference between total heat flow predicted from models of lithospheric cooling and the conductive heat flow through the seafloor measured by heat flow probes. The partition of heat and fluid fluxes between axial and flank flow is very difficult to estimate. This has been discussed by many authors (e.g. Kadko *et al.*, 1994; Stein and Stein, 1994; Elderfield and Schultz, 1996; Mottl, 2003; Chapter 19). Using a mean total hydrothermal flux of 8.8 ± 2 TW, Elderfield and Schultz (1996) proposed a partition of 2.6 ± 0.5 TW as axial (defined as 0–1 Ma crust) and 6.2 ± 1.2 TW off axis. Alternatively, if we assume that hydrothermal systems operate at about 50% efficiency (Section 21.3.1), the resultant axial heat flux would be $\sim$1.3 TW. Wheat and Mottl (Chapter 19) assume that this value is 1.8 ± 0.3 TW (they assume 100% efficiency) and arrive at a consequent flank heat flux of 8.1 ± 2 TW (their estimate of total heat flux of 9.9 ± 2 TW minus 1.8 ± 0.3 TW). Therefore, we might conclude that the off-axis heat flux is about 7 TW, with an uncertainty of about ± 2 TW.

The second major source of uncertainty in estimating fluid fluxes at flanks is fluid temperature. Harris and Chapman (Chapter 10) consider that the principal reason the same robustness in estimation of cumulative hydrothermal heat flux cannot be ascribed to estimation of volumetric or mass flow is that the effective discharge temperature for ridge flank hydrothermal circulation systems is poorly defined. For example, Elderfield and Schultz (1996) give the example that the fluid flow would be very large, $\sim$500 $\times$ 10^{13} kg yr^{-1}, if the fluid temperature was 5 °C. Temperatures up to about 65 °C have been recorded for flanks (Mottl *et al.*, 1998; Elderfield *et al.*, 1999; Chapter 8), but these values are only typical of young and highly sedimented flanks which probably make a minor contribution to total flank fluxes. Wheat and Mottl (Chapter 19) estimate that, on average, virtually all of the convective heat loss from the oceanic crust comes from crust that is 10–25 °C.

21.4 Controls on hydrothermal fluid chemistry

21.4.1 High-temperature ridge crest hydrothermal fluids

High-temperature hydrothermal fluids have been collected and analysed for more than two decades. Fluid chemistries vary between sites and temporally, but nevertheless show common overall features (Table 21.1). German and Von Damm (2003) recognized four factors that control that composition of vent fluids: (i) water–rock reaction, (ii) phase separation, (iii) magmatic degassing, and (iv) biological influence. The effect of biological processes operative at high temperatures are poorly known and will not be discussed. Magmatic degassing contributes only a minor part of the volatile flux but has been recognized by high δ^3He/heat ratios and high concentrations of CO_2, presumably associated with recent magma supply within the oceanic crust.

Water–rock reaction significantly modifies the chemical composition of seawater entering the reaction zone, converting it from an alkaline oxidizing solution under pH control by

Table 21.1 *Comparison of primary axial high-temperature hydrothermal chemical fluxes and river chemical fluxes*[a]

Element	$C_{\text{hydrothermal fluid}}$ $(\text{mol kg}^{-1})^b$	C_{seawater} $(\text{mol kg}^{-1})^b$	$Q_{\text{hydrothermal}}$ mol yr^{-1}	Q_{rivers} $(10^{10}\ \text{mol yr}^{-1})$
Li	$411 - 1322\ \mu$	$26\ \mu$	$1.2 - 3.9 \times 10^{10}$	1.4
K	$17 - 32.9\ m$	$9.8\ m$	$2.3 - 6.9 \times 10^{11}$	190
Rb	$10 - 33\ \mu$	$1.3\ \mu$	$2.6 - 9.5 \times 10^{8}$	0.037
Cs	$100 - 202\ n$	$2.0\ n$	$2.9 - 6.0 \times 10^{6}$	0.00048
Be	$10 - 38.5\ n$	0	$3.0 - 12 \times 10^{5}$	0.0037
Mg	0	$53\ m$	-1.6×10^{12}	530
Ca	$10.5 - 55\ m$	$10.2\ m$	$9.0 - 1300 \times 10^{9}$	1200
Sr	$87\ \mu$	$87\ \mu$	0	2.2
Ba	$>8 - >42.6\ \mu$	$0.14\ \mu$	$>2.4 - 13 \times 10^{8}$	1.0
SO_4	$0 - 0.6\ m$	$28\ m$	-8.4×10^{11}	370
Alk	$-0.1 - -1.0\ m$	$2.3\ m$	$-7.2 - 9.9 \times 10^{10}$	3000
Si	$14.3 - 22.0\ m$	$0.05\ m$	$4.3 - 6.6 \times 10^{11}$	640
P	$0.5\ \mu$	$2\ \mu$	-4.5×10^{7}	3.3
B	$451 - 565\ \mu$	$416\ \mu$	$1.1 - 4.5 \times 10^{9}$	5.4
Al	$4 - 20\ \mu$	$0.02\ \mu$	$1.2 - 6.0 \times 10^{8}$	6.0
Mn	$360 - 1140\ \mu$	0	$1.1 - 3.4 \times 10^{10}$	0.49
Fe	$750 - 6470\ \mu$	0	$2.3 - 19 \times 10^{10}$	2.3
Co	$22 - 227\ n$	$0.03\ n$	$6.6 - 68 \times 10^{5}$	0.011
Cu	$9.7 - 44\ \mu$	$0.007\ \mu$	$3.0 - 13 \times 10^{8}$	0.50
Zn	$40 - 106\ \mu$	$0.01\ \mu$	$1.2 - 3.2 \times 10^{9}$	1.4
Ag	$26 - 38\ n$	$0.02\ n$	$7.8 - 11 \times 10^{5}$	0.0088
Pb	$9 - 359\ n$	$0.01\ n$	$2.7 - 110 \times 10^{5}$	0.015
As	$30 - 452\ n$	$27\ n$	$0.9 - 140 \times 10^{5}$	0.072
Se	$1 - 72\ n$	$2.5\ n$	$3.0 - 220 \times 10^{4}$	0.0079
CO_2	$5.7 - 16.7\ m$	$2.3\ m$	$1.0 - 12 \times 10^{11}$	
CH_4	$25 - 100\ \mu$	$0\ \mu$	$0.67 - 2.4 \times 10^{10}$	
H_2	$0.05 - 1\ m$	$0\ m$	$0.3 - 1.5 \times 10^{10}$	
H_2S	$2.9 - 12.2\ m$	$0\ m$	$0.85 - 9.6 \times 10^{11}$	

[a] Data from Elderfield and Schultz (1996) based on Table 21.1 of Kadko *et al.* (1994) using water flux of $3(\pm 1.5) \times 10^{13}\ \text{kg yr}^{-1}$.
[b] $m = 10^{-3}$, $\mu = 10^{-6}$, $n = 10^{-9}$.

carbonate buffering to an acidic, reducing, metal-rich NaCl solution. Seawater is modified by a series of fluid–rock reactions at progressively higher temperatures as it passes through the hydrothermal systems (Bowers and Taylor, 1985). Observations on samples of oceanic crust and ophiolites indicate that the lowest temperature reactions form clay minerals (smectites) in the upper crust, followed by alteration of mafic minerals to mixed-layer chlorite-smectites and then chlorite and plagioclase to albite (e.g. Chapter 15). These reactions quantitatively remove Mg from seawater and elevate Ca at the expense of Na in the hydrothermal fluids. The fate of seawater sulfate is important but uncertain. Sulfate may be precipitated as anhydrite as seawater is heated to $> 120\,°C$ (Bowers and Taylor, 1985) or reduced to sulfide by reaction with FeO-bearing silicate minerals (Seyfried and Ding, 1995). Anhydrite is observed in DSP/ODP Hole 504B (Teagle *et al.*, 1998b), but the degree of oxidation of altered oceanic crust rocks and in ophiolites (Alt *et al.*, 1996) implies significant sulfate reduction.

Hydrothermal fluid is heated rapidly to temperatures of $\sim 400\,°C$ in narrow reaction zones close to the underlying convecting magma chambers (Cann and Strens, 1982). The chemistry of the vent fluids is consistent with equilibrium with epidote–chlorite–quartz assemblages expected at such temperatures (Bernt and Seyfried, 1993), an observation consistent with the occurrence of bands of epidosites (epidote–quartz and chlorite–quartz rocks) in ophiolites (Richardson *et al.*, 1987), although such assemblages have rarely been found in oceanic crust (Banerjee *et al.*, 2000).

As a consequence of the water–rock reactions, chloride becomes the only major anion present in the hydrothermal fluid. Its behavior is near-conservative during water–rock reaction (hydration and dehydration reactions are thought to be minor), but it is affected by phase separation which is a common process in hydrothermal systems. Chloride concentrations are crucial in determining fluid chemistry because most cations are present as chloro-complexes.

Estimates of hydrothermal fluxes for a range of elements as compared with fluxes of elements to the oceans from the world's rivers are given in Table 21.1 and Fig. 21.4. Taken at face value, they show that the fluxes of chemical elements from high-temperature hydrothermal venting are within about one order of magnitude of the input of elements to oceans from rivers.

Apart from the uncertainties in this type of global calculation, a caveat must be placed upon these estimates which relates to the modification of these primary fluxes by processes at the seafloor or within the water column by mixing with seawater. Seawater interacts with high-temperature hydrothermal systems in shallow sub-surface diffuse flow systems and through entrainment in buoyant hydrothermal plumes. The mixing process results in chemical reaction as fluids are cooled. The consequences on seawater composition are beyond the scope of this chapter but are reviewed elsewhere (German and Von Damm, 2003). Such consequences are significant, including the precipitation of ore minerals as well as hydrothermal plume particles that scavenge metals from seawater to produce metal-rich hydrothermal sediments.

21.4.2 Hydrothermal fluxes at ridge flanks

This topic has been discussed thoroughly in Chapters 16 and 19. The prevailing view is that chemical exchange at flanks is controlled primarily by basement temperature (e.g. Mottl and Wheat, 1994; Wheat and Mottl, 1994; Wheat *et al.*, 1996; Elderfield *et al.*, 1999).

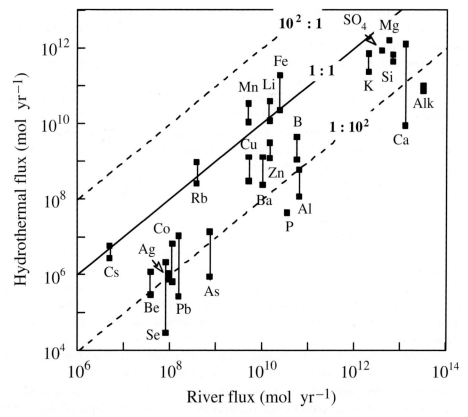

Fig. 21.4 Estimates of hydrothermal fluxes for a range of elements as compared with fluxes of elements to the oceans from the world's rivers (Elderfield and Schultz, 1996).

Additionally, where fluid flow occurs through sediments, diffusional exchange through and reaction with sediments is an important factor.

21.5 Oceanic mass balance

Mass-balance inputs to the oceans should also put constraints on hydrothermal fluxes given that the riverine input is known. However, the large and poorly constrained flank fluxes complicate mass balance for most tracers (Mg, Li, Ge/Si). Palmer and Edmond (1989) first showed that oceanic mass balance of Sr-isotopic compositions apparently require high-temperature hydrothermal fluxes six times higher than the magmatic heat available, given that Sr is not particularly mobile in lower temperature fluids. However, Butterfield *et al.* (2001) disputed this and estimated flank Sr fluxes from analysis of intermediate-temperature flanks fluids and calculated that they could supply the difference. The problem with this estimate is that it depends on the very poorly known flank fluxes and water chemistry. Davis *et al.* (2003) showed that available samples of the upper crust are nowhere near sufficiently altered to supply the flux inferred by Butterfield *et al.* (2001). Additionally, Sr-isotopic alteration of the sampled oceanic sections is compatible with likely high-temperature hydrothermal fluxes calculated from the magmatic heat constraints. An alternative possibility is that imbalance in the oceanic Sr-isotope budgets arises because the riverine input varies on time scales short compared with the residence time of Sr in the oceans.

Ophiolites have also provided complete sections through altered oceanic crust, although their formation in back-arc settings and degree and style of alteration preclude their use as direct analogs of most oceanic crust. Bickle and Teagle (1992) noted that $^{87}Sr/^{86}Sr$ ratios in the Troodos ophiolits are elevated to ~0.705 over the ~2.5 km upper crustal section and that the epidosite zones of high-temperature concentrated flow had identical values, and they inferred that this alteration was coincident with high-temperature circulation and that the vent fluids would have had corresponding Sr-isotopic ratios. Bickle and Teagle concluded from simple modelling of Sr-isotopic tracer transport in Troodos that recharge fluxes responsible for this alteration were at least 2.9×10^7 kg m^{-2}, a factor of three higher than the flux sustainable by magmatic heat available from average thickness oceanic crust. The modeling predicts that the hydrothermal fluxes of $<5 \times 10^6$ kg m^{-2} calculated for normal oceanic crust should result in vent fluids with $^{87}Sr/^{86}Sr$ ratios only marginally elevated above fresh oceanic basalts, as is observed. The Oman ophiolite also has elevated $^{87}Sr/^{86}Sr$ ratios and Davis *et al.* (2003) speculated that the arc crust may have been substantially thicker than normal oceanic crust. The potential differences between hydrothermal systems in back-arc environments and normal oceanic crust are reflected in higher $^{87}Sr/^{86}Sr$ ratios for high-temperature vent fluids from back-arc environments.

Despite the poor knowledge of low–moderate temperature fluxes from ridge flanks, Kastner and Rudnicki (Chapter 16) and Wheat and Mottl (Chapter 19) have presented results of interesting calculations of chemical fluxes on flanks. Kastner and Rudnicki (Chapter 16) have presented results of the calculation of fluxes for Si, Sr, and $^{87}Sr/^{86}Sr$ through sediments to seawater for the first 5 million years of crustal evolution, during which sediment cover is thin, where they assumed two boundary conditions: one at a very low spatially averaged fluid flux of 5 mm yr^{-1}, and the second at an average flux of 50 mm yr^{-1}. At the lower rate, the silica flux through the sediment is insignificant, <1% of river flux, whereas at the higher rate, the total silica flux becomes ~10% of the river flux. When applied to Sr and $^{87}Sr/^{86}Sr$, fluid advection at 50 mm yr^{-1} causes the Sr flux to be more significant than both the axial hydrothermal flux and the benthic flux of Sr from porewater diffusion. Kastner and Rudnicki speculate that the low $^{87}Sr/^{86}Sr$ is derived from volcanogenic sediments. However, the same mass-balance limitations raised by Davis *et al.* (2003) must apply and it seems very unlikely that ridge-derived volcanogenic sediments are sufficiently voluminous to supply significant Sr.

Wheat and Mottl (Chapter 19) argue that the largest uncertainty in estimating ocean–crustal exchange lies with the uncertainties of the distribution of advective heat loss with basement temperature. They have constrained this using global sediment thickness and heat flux data sets, and a compilation of convective heat loss from the oceanic crust with age. Calculated temperatures range from ~20 to 30 °C over 0–65 Ma, the crucial age range for global budgets. If, as suggested above (Section 21.4.2), chemical exchange is directly related to basement temperature, then it is possible to estimate chemical fluxes. Wheat and Mottl suggest a flux of phosphate to seawater of significance relative to river input, but almost an order of magnitude greater than estimates made from studies of rock alteration.

These above calculations and discussion provide a measure of current uncertainties in estimating the global importance of the hydrology of the oceanic lithosphere to oceanic mass balance of chemical elements. One type of uncertainty is in estimating heat and fluid fluxes both on and off axis. What is the thermal efficiency of hydrothermal systems? How accurate are conventional methods of estimating heat flux on flanks? What is the temperature of the basement as a function of crustal age? How important is flow through sediments compared with focused discharge? A second class of uncertainty is in the global scale of chemical fluxes: can we reconcile differences in estimates of fluxes from studies of crustal alteration and from fluid chemistry? Perhaps the problems in mass balance are because estimates of river flux are incorrect.

Acknowledgments

We thank NERC for supporting our research.

References

Alt, J. C., Laverne, C., Vanko, D. A., *et al.* 1996. Hydrothermal alteration of a section of oceanic crust in the eastern equatorial pacific: a synthesis of results from site 504 (DSDP legs 69, 70, and 83, and ODP legs 111, 137, 140 and 148. In *Proceedings of the Ocean Drilling Program, Scientific Results*, Vol. 148, eds. J. C. Alt, H. Kinoshita, L. B. Stokking, and P. J. Michael. College Station, TX: Ocean Drilling Program, pp. 417–434.

Banerjee, N. R., Gillis, K. M., and Muehlenbachs, K. 2000. Discovery of epidosites in a modern oceanic setting, the Tonga forearc. *Geology* **28**(2): 151–154.

Berner, R. A., Lasaga, A. C., and Garrels, R. H. 1983. The carbonate–silicate geochemical cycle and its effect on atmospheric carbon dioxide over the past 100 million years. *Am. J. Sci.* **283**: 641–683.

Bernt, M. E. and Seyfried, W. E. 1993. Calcium and sodium exchange during hydrothermal alteration of calcic plagioclase at 400 °C and 400 bars. *Geochim. Cosmochim. Acta* **57**: 4,445–4,451.

Bickle, M. J. and Teagle, D. A. H. 1992. Strontium alteration in the Troodos ophiolite: implications for fluid fluxes and geochemical transport in mid-ocean ridge hydrothermal systems. *Earth Planet. Sci. Lett.* **113**: 219–237.

Bowers, T. S. and Taylor, H. P. 1985. An integrated chemical and stable-isotope model of the origin of mid-ocean ridge hot spring systems. *J. Geophys. Res.* **90**(B14): 12,583–12,606.

Butterfield, D. A., Nelson, B. K., Wheat, C. G., Mottl. M. J., and Roe, K. K. 2001. Evidence for basaltic Sr in midocean ridge-flank hydrothermal systems and implications for the global oceanic Sr isotope balance. *Geochim. Cosmochim. Acta* **65**: 4,141–4,153.

Cann, J. R. and Strens, M. R. 1982. Black smokers fuelled by freezing magma. *Nature* **298**: 147–149.

Chen, Y. J. 2001. Thermal effects of gobbro accretion from a deeper second melt lens at the fast spreading East Pacific Rise. *J. Geophys. Res.* **106** (B5): 8,581–8,588.

COSOD II 1987. *Report of the Second Conference on Scientific Ocean Drilling*, 6–8 July, Strasbourg. 142 pp.

Crawford, W. C., Webb, S. C., and Hildebrand, J. A. 1999. Constraints on melt in the lower crust and Moho at the East Pacific Rise, 9° 48′ N, using seafloor compliance measurements. *J. Geophys. Res.* **104**(B2): 2,923–2,939.

Davis, A., Bickle, M. J., and Teagle, D. A. H. 2003. Imbalance in the oceanic strontium budget. *Earth Planet. Sci. Lett.* **211**: 173–187.

Drever, J. I. 1974. The magnesium problem. In *The Sea*, Vol. 5, ed. E. D. Goldberg. New York: Wiley-Interscience, pp. 337–358.

Dunn, R. A. and Toomey, D. R. 2000. Three-dimensional seismic structure and physical properties of the crust and shallow mantle beneath the East Pacific Rise at 9° 30′ N. *J. Geophys. Res.* **105**(B10): 23,537–23,555.

Edmond, J. M., Measures, C., McDuff, R. E., Chan, L. H., Collier, R., Grant, B., Gordon, L. I., and Corliss, J. B. 1979. Ridge crest hydrothermal activity and the balances of the major and minor elements in the ocean: The Galapagos data. *Earth Planet. Sci. Lett.* **46**: 1–18.

Elderfield, H. and Schultz, A. 1996. Mid-ocean ridge hydrothermal fluxes and the chemical composition of the ocean. *Ann. Rev. Earth Planet. Sci.* **24**: 191–224.

Elderfield, H., Wheat, C. G., *et al.* 1999. Fluid and geochemical transport through oceanic crust: a transect across the eastern flank of the Juan de Fuca Ridge. *Earth Planet. Sci. Lett.* **172**: 151–165.

German, C. R. and Von Damm, K. 2003. Hydrothermal processes. In *Oceans and Marine Geochemistry*, Treatise on Geochemistry, Vol. 6, ed. H. Elderfield. Oxford: Elsevier-Pergamon, pp. 181–222.

Humphris, S. E., Zierenberg, R. A., Mullineaux, L. S., and Thomson, R. E. (eds.) 1995. *Seafloor Hydrothermal Systems*, Geophysical Monograph 91. Washington, DC: American Geophysical Union, 466 pp.

James, R. H. and Elderfield, H. 1996. Chemistry of ore-forming fluids and mineral formation rates in an active hydrothermal sulfide deposit on the mid-Atlantic Ridge. *Geology* **24**: 1,147–1,150.

Jenkins, W. J., Edmond, J. M., and Corliss, J. B. 1978. Excess 3He and 4He in Galapagos submarine hydrothermal waters. *Nature* **272**: 156–158.

Kadko, D., Baker, E., Alt, J., and Baross, J. 1994. *Global Impact of Submarine Hydrothermal Processes*, Final Report of RIDGE/VENT Workshop. 55pp.

Kelemen, P. B., Koga, K., and Shimizu, N. 1997. Geochemistry of gabbro sills in the crust/mantle transition zone of the Oman ophiolite: implications for the origin of the lower oceanic crust. *Earth Planet. Sci. Lett.* **146**: 475–488.

Morton, J. L. and Sleep, N. H. 1985. A mid-ocean ridge thermal model: constraints on the volume of axial hydrothermal heat flux. *J. Geophys. Res.* **90**(B13): 11,345–11,353.

Mottl, M. J. 2003. Partitioning of energy and mass fluxes between mid-ocean ridge axes and flanks at high and low temperature. In *Energy and Mass Transfer in Marine Hydrothermal Systems*, eds. P. E. Halbach, V. Tunnicliffe, and J. R. Hein. Berlin: Dahlem University Press, pp. 271–286.

Mottl, M. J. and Wheat, C. G. 1994. Hydrothermal circulation through mid-ocean ridge flanks: fluxes of heat and magnesium. *Geochim. Cosmoch. Acta* **58**(10): 2,225–2,237.

Mottl, M. J., Wheat, G., Baker, E., Becker, N., Davis, E., Feely, R., Grehan, A., Kadko, D., Lilley, M., Maesoth, G., Moyer, C., and Sansore, F. 1998. Warm springs discovered on 3.5 Ma oceanic crust, eastern flank of the Juan de Fuca Ridge. *Geology* **26**: 51–54.

Palmer, M. and Edmond, J. M. 1989. The strontium isotope budget of the modern ocean. *Earth Planet. Sci. Lett.* **92**: 11–26.

Richardson, C. J., Cann, J. R., Richards, H. G., and Cowan, J. G. 1987. Metal-depleted root zones of the Troodos ore-forming hydrothermal systems, Cyprus. *Earth Planet. Sci. Lett.* **84**: 243–253.

Rona, P. A. and Trivett, D. A. 1992. Discrete and diffuse heat transfer at ASHES vent field, Axial Volcano, Juan de Fuca Ridge. *Earth Planet. Sci. Lett.* **109**: 57–71.

Schultz, A., Delany, J. R., and McDuff, R. E. 1992. On the partitioning of heat flux between diffuse and point source venting. *J. Geophys. Res.* **97**: 12,299–12,314.

Seyfried, W. E. and Ding, K. 1995. Phase equilibria in seafloor hydrothermal systems: a review of the role of redox, temperature, pH and dissolved Cl on the chemistry of hot spring fluids at mid-ocean ridges. In *Seafloor Hydrothermal Systems*, Geophysical Monograph. 91, eds. S. E. Humphris, R. A. Zeirenberg, L. S. Mullineaux, and R. E. Thomson. Washington, DC: American Geophysical Union, pp. 248–272.

Sillén L. G. 1967. How have sea water and air got their present compositions? *Chem. Britain* **3**: 291–297.

Sleep, N. H. 1991. Hydrothermal circulation, anhydrite precipitation, and thermal structure at ridge axes. *J. Geophys. Res.* **96**(B2): 2,375–2,387.

Stein, C. A. and Stein, S. 1994. Constraints on hydrothermal heat flux through the oceanic lithosphere from global heat flow. *J. Geophys. Res.* **99**: 3,081–3,096.

Teagle, D. A. H., Alt, J. C., Chiba, H., Humphics, S. E., and Halliday, A. N. 1998a. Strontium and oxygen isotopic constraints on fluid mixing, alteration and mineralization in the TAG hydrothermal deposit. *Chem. Geol.* **149**: 1–24.

Teagle, D. A. H., Alt, J. C., and Halliday, A. N. 1998b. Tracing the chemical evolution of fluids during hydrothermal recharge: constraints from anhydrite recovered in ODP Hole 504B. *Earth Planet. Sci. Lett.* **155**: 167–182.

Teagle, D. A. H., Bickle, M. J., and Alt, J. C. 2003. Recharge flux to ocean-ridge black smoker systems: a geochemical estimate from ODP Hole 504B. *Earth Planet. Sci. Lett.* **210**(1–2): 81–89.

Walker, J. C. G., Hays, P. B., and Kasting, J. F. 1981. A negative feedback mechanism for the long-term stabilisation of Earth's surface temperature. *J. Geophys. Res.* **86**: 9,776–9,782.

Wheat, C. G. and Mottl, M. J. 1994. Hydrothermal circulation, Juan de Fuca Ridge eastern flank: factors controlling basement water composition. *J. Geophys. Res.* **99**(B2): 3,067–3,080.

Wheat, C. G., Feely, R. A., and Mottl, M. J. 1996. Phosphate removal by oceanic hydrothermal processes: an update of the phosphorus budget in the oceans. *Geochim. Cosmochim. Acta* **60**: 3,593–3,608.

Index